W. Schwabhäuser
W. Szmielew A. Tarski

Metamathematische Methoden in der Geometrie

Mit 167 Abbildungen

Teil I: Ein axiomatischer Aufbau der euklidischen Geometrie
von W. Schwabhäuser, W. Szmielew und A. Tarski

Teil II: Metamathematische Betrachtungen
von W. Schwabhäuser

ISHI PRESS INTERNATIONAL

Metamathematische Methoden in der Geometrie

Teil I: Ein axiomatischer Aufbau der euklidischen Geometrie
by Wolfram Schwabhäuser, Wanda Szmielew and Alfred Tarski

Teil II: Metamathematische Betrachtungen (Hochschultext)
by Wolfram Schwabhäuser

or

Metamathematical Methods in Geometry
Part I: An axiomatic structure of Euclidean geometry
Part II: Metamathematical Views (university text)

by Wolfram Schwabhäuser, Wanda Szmielew and Alfred Tarski

Published in 1983 in Berlin Heidelberg, Germany

This Printing in November, 2011
by Ishi Press in New York and Tokyo

with a new introduction by Michael Beeson

ISBN 4-87187-707-8
978-4-87187-707-7

Ishi Press International
1664 Davidson Avenue, Suite 1B
Bronx NY 10453-7877
USA

1-917-507-7226

Printed in the United States of America

Foreword

Michael Beeson

December 23, 2011

The book in your hands is an important contribution to Euclidean geometry. In order to put that contribution into perspective, we begin by reviewing its ancestors. Its primordial ancestor, of course, is Euclid's *Elements*. We know nothing about Euclid except what he wrote: not his date or place of birth or death, not how much of his book represented his own work, not even where he lived. Possibly Euclid lived and worked in Alexandria, which was the intellectual center of the Greek civilization at the time, around 300 BCE. His textbook comprised thirteen Books (a "book" was in those days more like a "chapter" today, but it occupied one papyrus scroll). This book is famous not only for codifying what was then known of geometry, but also for introducing the modern style of mathematics: it contains axioms and proofs of numbered propositions, each of which is supposed to be rigorously derived from the axioms and the previous propositions.

Over the centuries (beginning already with Geminus and Proclus[1]), there has been criticism of some of Euclid's arguments. In particular, Euclid's Postulate 5 was the center of attention. That postulate says that if K and L are two lines, traversed by a third line M making interior angles (totalling) less than two right angles on one side of M, then K and L must meet on that side of M. In other words, K and L are not parallel (lines are called "parallel" if they do not meet). It was felt that this postulate was not so "self-evident" as the other postulates, and great efforts were expended by several mathematicians in attempts to prove it from the other postulates, or from some new postulate that might be more self-evident. Along the way some errors in Euclid were discovered. For example, Euclid implicitly assumed that certain lines must intersect. He actually needed "Pasch's axiom", which asserts that if a line meets one side of a triangle, then it meets one of the other two sides or a vertex.

Several of these efforts foundered, as mistakes crept into the reasoning,

[1]Proclus, *A commentary on the first book of Euclid's Elements*, translated with introduction and notes by Glenn R. Morrow, Princeton University Press (1992).

for example by implicitly assuming some geometrical fact that is actually equivalent to Postulate 5 itself. These difficulties led to a greater focus on ways to set out the reasoning in such a way as to assist in verifying its correctness. In Italy and Germany, formal logic was born from these efforts: geometry was the mother of logic. Eventually it was realized that the fifth postulate *cannot* be proved from the others, because there are "models" of the other axioms in which the fifth postulate fails. This realization, that "non-Euclidean" geometry is possible, was an eye-opener, as people had previously thought that geometry was fixed by the nature of the universe, or (as Kant said) by the nature of the human mind.

Logic grew up, left its mother, and developed further in the hands of Peano (who introduced many of the modern symbols) and Frege. It married set theory, whose fathers were Cantor and Dedekind[2] , and together they gave birth to the "axiomatic method", according to which each branch of mathematics should be developed from appropriate axioms, by logical deductions from the axioms, *without regard to what the axioms were about.* Non-Euclidean geometry had shown that the same axioms could be "about" different things; the axiomatic method would allow mathematics to deal with several (or even many) examples simultaneously. This method was championed by David Hilbert, and in his 1899 book, *Foundations of Geometry*, he applied the method to geometry. He famously said that if geometry were done correctly, one could speak of "tables, chairs, and beer mugs" instead of "points, lines, and planes", and all the deductions should still be correct. For example, instead of assuming that there is exactly one line through two given points, you would assume there is exactly one chair through two given tables. That makes no sense as a statement about real tables and chairs, but if we *assume* it, and continue calling points, lines, and planes by these new names, all the reasoning should still be valid. That is, nothing about the nature of actual points, lines, and planes is allowed to be used, except what is captured in the axioms.

But the marriage of set theory and logic gave birth, not only to this beautiful and talented child (the axiomatic method), but also to a monster: the "paradoxes". For example, the well-known "Russell's paradox" about the set of sets which are not members of themselves. It appeared that self-evident postulates and correct logic could lead to a contradiction! Revulsion at these misbegotten creatures led to an even more careful examination of the principles of reasoning. The blame was laid on a too-liberal use of set theory; a restraining order was placed on set theory and mathe-

[2]Two fathers? Well, in Cantor's case the mother was Fourier analysis, since Cantor started by studying the sets on which Fourier series converge. In Dedekind's case the mother was number theory, since he started by studying "ideal numbers", needed to fix the failure of unique factorization of algebraic integers.

matics was henceforth to live with this restraining order. The new regime was called "first order logic".

Hilbert, however, had used set theory freely in his *Foundations of Geometry.* He needed some "continuity axioms" to guarantee that a line could not be extended to a larger line. He used the axiom of Archimedes to guarantee that it could not be made *longer.* That axiom mentions the concept of "natural number", which goes beyond geometry. Then he also needed to guarantee that a line cannot be enlarged by inserting new points, i.e. filling "gaps". This problem had been solved by Dedekind using Dedekind cuts, which are defined as sets of rational numbers. For example, the point on L at distance $\sqrt{2}$ from a point called 0 exists because we can define the set of points whose distance x from 0 satisfies $x^2 < 2$. In modern terminology, the axiom that Dedekind cuts can be filled is called "second order", because it involves sets of points. Hilbert did not take this for a completeness axiom, but stated more directly that a line cannot be extended; that axiom requires mentioning not only sets of points, but sets of unspecified "new points" in a hypothetical extension.

By the 1920s, it was realized that Hilbert's axiom system was not first-order, because of Archimedes's axiom and the completeness axiom, as well as because of treating segments, rays, and angles as sets. It was therefore necessary to go back over the axiomatization of geometry and remove this flaw, providing a first-order axiomatization of geometry. This task was undertaken by Alfred Tarski, in his 1926-27 lectures at the University of Warsaw. Readers new to this subject must understand that there are now *three* different kinds of Euclidean geometry to think about:

- Full Euclidean geometry, with a second-order completeness axiom, mentioning arbitrary sets of points. This theory has only one model, the Euclidean plane $\mathbb{R}^2$.

- First order Euclidean geometry, or Tarski geometry, in which Archimedes's axiom is not used, and the completeness axiom is replaced by a first-order schema, saying in essence that all first-order definable cuts can be filled. One might, like Tarski, have only variables for points, or one might have variables for lines, circles, etc. as well.

- Ruler and compass geometry, which is like first-order Euclidean geometry except that the completeness schema is replaced by "line-circle continuity", which says that a line that contains a point inside a circle (nearer to its center than some point on the circle) must meet the circle.

Once it was realized that the first order theories would have more than one model, and only the second order theories would uniquely characterize the plane, the question immediately arose whether one can characterize the models of such theories as planes F^2 coordinatized by certain kinds of fields F. Such a theorem is called a "representation theorem" for a geometry. Starting with his 1926-27 lectures, Tarski began working towards such a theorem, and by 1930 he had achieved some significant results: a first-order axiomatization of Euclidean geometry, and the representation theorem that its models are planes over real-closed fields. A real-closed field is a field in which "gaps" defined as places where a polynomial changes sign are filled. (Technically, polynomials of odd degree have roots, and positive elements have square roots.) Tarski showed that his first order geometry is equivalent (using coordinates) to the first order theory of real closed fields. Similarly, the models of ruler and compass geometry are planes F^2 over some Euclidean field F. (A Euclidean field is an ordered field in which positive elements have square roots.)

Tarski proved an amazing theorem about the theory of real-closed fields, known as "elimination of quantifiers", which meant that every formula of the theory is equivalent to one without quantifiers. It follows that there is a "decision method" for Euclidean geometry; in principle the truth or falsity of any given formula must (a) follow from the axioms and (b) be determinable in a mechanical way.

By contrast, there is no such decision procedure for the theory of Euclidean fields, and hence, also not for ruler and compass geometry, as was proved by Ziegler in 1982.[3] The result is not mentioned in this book, but would certainly have been if the publication dates had not so nearly overlapped. Ziegler's theorem applies to any finitely axiomatizable extension of field theory, so it gives another proof of Tarski's theorem that the theory of real closed fields is not finitely axiomatizable.

Tarski's results first were published in a RAND Corporation technical report in 1948, but without any details about geometry; the focus was on the decision procedure for algebra (i.e. real-closed fields). What happened between 1930 and 1948 will be discussed below. For now, we fast-forward to 1955. In 1955-56 Tarski lectured on geometry at the University of California at Berkeley, and enlisted his students to help with the problems of simplifying the axiom system, studying the relationships between the axioms, and checking that the axioms were really sufficient to develop Euclidean geometry. There were many details to check. In 1959 a *very* brief description of the theory and statement of the 1930 results was published.[4] In 1959–60, Wanda Szmielew and Tarski began the project of (as

[3]Ziegler, Martin, Einige Unentscheidbare Körpertheorien, *Enseignement Math.* **28** 269–280 (1982).

[4]Alfred Tarski, What is elementary geometry?, in *The axiomatic method, with special*

it was described much later by Steven Givant) "preparing a treatise on the foundations of geometry developed within the framework of contemporary mathematical logic." The first part of this work was to be a systematic development of Euclidean geometry based on Tarski's axioms. Drafts of this part were written, but the project was never completed along these lines. Szmielew's well-known book *Foundations of Geometry*, with Karl Borsuk as co-author, was published in 1960 and, judging by its content, must have been completed before she worked with Tarski, as the axiomatic system is quite different, closer to Hilbert's (and not first order). Szmielew did the writing, although she no doubt discussed matters with Tarski. When Wolfram Schwabhäuser came to work with Tarski in Berkeley in 1965, he was assigned to teach the course "Foundations of geometry" in 1965-66 that Szmielew had taught at Berkeley during her visit five years earlier; she sent him (or Tarski gave him) a copy of the manuscript to assist him in his lectures. When he returned to West Germany, he also taught the subject there. He probably had access not only to Szmielew's manuscript, but also to the 1965 Ph. D. thesis of Tarski's student H. N. Gupta, whose amazing proofs were never published separately, although he did publish a two-page summary of his results.[5] In 1967, Szmielew made another visit to Berkeley, and wrote another manuscript, as described in Schwabhäuser's introduction, page 6 of this book. She and Tarski were planning to write a book together based on this manuscript, and indeed a project report to NSF in 1970 says their manuscript is "now being prepared for publication", but that book was still unwritten in 1976, when Szmielew died.

Schwabhäuser started writing this book in 1974 (according to his wife). As originally planned, it contains two parts. The first part is a systematic development of Euclidean geometry from (a variant of) Tarski's axioms. The second part presents metamathematical results about the theory. Schwabhäuser is the sole author of the second part. The first part (specifically sections 1-8, 10, and the first part of 9) is Szmielew's first manuscript, with changes that Schwabhäuser himself calls "inessential" in his introduction, which is why she is named as a joint author, and the axiom system itself is due to Tarski, which is partly why he is named as a joint author. The joint authorship was agreed to by Tarski in 1978, two years after the death of Szmielew, and represented the conclusion of

---

reference to geometry and physics. Proceedings of an International Symposium held at the Univ. of Calif., Berkeley, Dec. 26, 1957–Jan. 4, 1958, edited by Henkin, Suppes, and Tarski. Studies in Logic and the Foundations of Mathematics, North-Holland, Amsteram (1959). Available as a 2007 reprint, Brouwer Press, ISBN 1-443-72812-8

[5]Gupta, Haragauri N., An axiomatization of finite-dimensional Cartesian spaces over arbitrary ordered fields. *Bulletin de l'Académia Polonaise des Sceinces, série des sciences mathématiques, astronomiques, et physiques 13:551-552 (1965).*

a substantial dispute between Szmielew (represented by Tarski after her death) and Schwabhäuser about his intent to publish material based on her manuscript.[6]

Now that Tarski was a co-author, Schwabhäuser asked Tarski for suggestions for material to be included. Tarski and Steven Givant replied in a long letter. That letter, though not originally intended for publication, was polished up for publication 21 years later by its original co-author Steven Givant.[7] Very likely most readers of this book will also want to study that letter, which in published form is 39 pages long, containing a detailed list of the various forms of the axioms that have been considered and a history of their evolution. Tarski made no direct contribution to the book, i.e., no part of the text was written by Tarski.

What changes in Hilbert's system were required to achieve Tarski's goal of a first-order axiom system for geometry? Hilbert's planar foundations had two primitive objects (points and lines) and four primitive relations (incidence, betweenness, congruence of segments and congruence of angles). Veblen (in 1904) and Pieri (in 1908) used only one primitive object (point) and two primitive relations betweenness and equidistance. Tarski followed Veblen and Pieri's approach.[8] But the reduction to two primitives is not the main issue. Nor is the the fact that Hilbert treated segments and circles as sets of points an important issue; that is easily fixable. The main problem is the axiom of continuity. Hilbert had two axioms that are not first order. Tarski's system replaced Hilbert's continuity axiom with a schema that says certain Dedekind cuts can be filled, namely those cuts that can be defined in the first-order language, i.e. by reference to geometric concepts only, as opposed to allowing arbitrary definitions of sets or even sets with no definitions at all. In retrospect at least, this seems fairly straightforward. But Tarski, Szmielew, Gupta, and Schwabhäuser placed other demands on the axioms: They should be as

[6]Szmielew's 1967 manuscript apparently contained her proof of the representation theorem. Givant showed this manuscript to Schwabhäuser in 1975, but he wrote to Tarski that he had never read it, so the work in the last part of Part I was independently done. Tarski's correspondence can be found in the Tarski Nachlass in the Bancroft Library at the University of California, Berkeley.

[7]Alfred Tarski and Steven Givant, Tarski's system of geometry, *Bulletin of Symbolic Logic* 5 (2), 1999.

[8]In *The Legacy of Mario Pieri in Geometry and Arithmetic* (Birkhauser, 2007), authors Elene Marchisotto and James Smith translate some of Pieri's work and discuss his important contributions. On page 357 one discovers that the Ph. D. advisor of Smith was Tarski's student H. N. Gupta, and that Tarski came to his thesis defense, and asked a single question: where did this field begin? Smith naively answered, with Hilbert, and it was Tarski who pointed him to Pieri. Section 3.10 of that book contains relevant historical notes.

simple as possible, and as few as possible, and they should be mutually independent. Achieving these goals meant that one had to investigate alternative forms of several axioms, check that simpler forms could prove more complex forms (or could not), and consider whether certain axioms might be superfluous or were really necessary. Except for the continuity schema, the axioms should have a particular simple form, which I here call "Euclidean form". This kind of formula asserts that if some points are given bearing certain relations, then some other points exist bearing certain relations to each other and the original points. All of Euclid's theorems are of this form (as has to be checked one proposition at a time). Moreover, one should also pay attention to the axiomatization of "ruler-and-compass" geometry, in which the full continuity schema is replaced by "line-circle" and "circle-circle" continuity. The line-circle continuity axiom asserts that if a line L has a point inside circle C, then L meets C. The circle-circle continuity axiom is similar, but with "line L" replaced by "circle K." As an example of the issues that arise: is there a way to construct the perpendicular to line L from point P without using any continuity axiom at all? The usual way involves constructing the intersection points of two circles. This problem was beautifully solved by H. N. Gupta, and the solution is in the book you are holding. Here is another issue: the axiom of Pasch, mentioned above, has an "or" (disjunction) in the conclusion, so it isn't in Euclidean form; it says that if a line meets side AC of a triangle ABC, then one of these alternatives holds: either it meets AB, or it meets BC, or it meets B, or it contains A and C both. This disjunction is ugly; disjunctions do not occur in Euclid. One therefore considers two disjunction-free versions, known as "inner Pasch" and "outer Pasch". Are they equivalent? Do they imply Pasch's axiom? Which one is simpler in the sense of developing the rest of the theory? Schwabhäuser chose "inner Pasch". A footnote in the Tarski-Givant letter indicates that Tarski favored "outer Pasch"; at any rate the relations between the two are studied in this book. Inner and outer Pasch are illustrated in Fig. 1.

Another issue is the treatment of lines, rays, and angles. Hilbert treated segments and rays as sets of points, but Tarski's theories have only variables for points. Lines and rays are treated implicitly, as pairs of points. Hilbert defined an angle to be a pair of rays, so it is not even a set of points, but a set of sets of points. But then, although angles are defined, Hilbert takes the relation of congruence of angles to be primitive. On the other hand, Tarski treated angles indirectly, so theorems about angles become theorems about triples of non-collinear points. For example, the side-angle-side principle of triangle congruence (SAS) has to be expressed using points only; this becomes the "five-segment axiom" in

Figure 1: Inner Pasch (left) and Outer Pasch (right). Line pb meets triangle acq in one side. The open circles show the points asserted to exist on the other side.

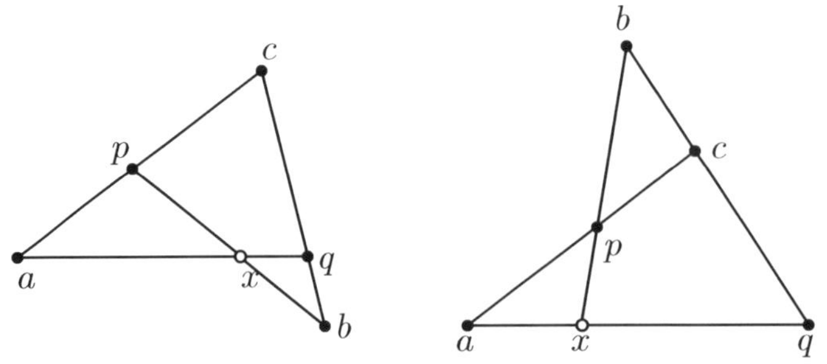

Tarski's theories; see Fig. 2. In that figure, the congruent triangles are dbc and DBC; the congruence of angle dbc and angle DBC is expressed by the congruence of triangles ABD and abd (which is expressed by the three congruences of their sides, which in turn is expressed using the equidistance relation on points). It may take some time for geometers accustomed to thinking about angles as sets (or even as pictures) to learn to think about angles as being given by three points, but after this change of viewpoint, one often feels that the first-order versions state more succinctly "what is really going on."

Figure 2: Tarski's 5-segment axiom. cd is determined. That is, if $ab = AB$, $bc = BC$, $ad = AD$, and $bd = BD$, then the fifth segment $cd = CD$ as well.

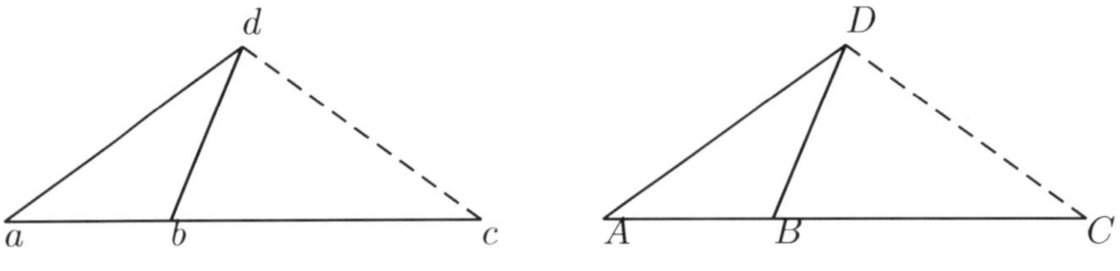

Part I of this book, then, contains the details of these axiomatic developments; geometry is built up step by step, using first only a few axioms, then introducing more axioms as they are needed. In a sense this geometry is "pre-Euclidean", supplying the infrastructure on which Euclid is (or should be) based. Throughout this subject there is the danger of making a mistake by accidentally assuming something that is "intuitively obvious" but has not been rigorously proved. Therefore it is necessary to construct and exhibit detailed proofs. Schwabhäuser was, by accounts of those who knew him, a very careful man; and in modern times much of Part I of this book has been computer-checked for correctness.[9] In that computer-checking, it seems that only one trivial and easily fixable error was found.

Part II of the book is devoted to metamathematical results about formal theories of geometry. This part of the book does not share the joint authorship of Part I; it is Schwabhäuser's alone; yet Tarski's presence is felt, because many of the theorems either are Tarski's, or are generalizations to other geometrical theories of theorems Tarski proved for his "elementary geometry" (that corresponds to real-closed fields). Schwabhäuser considers questions of decidability, completeness, finite axiomatizability (or not), and definability; and discusses model-theoretic proofs as well as proofs by quantifier elimination. By 1983, there was considerable discussion in print of quantifier elimination, so there was no need to publish the details that had been missing in 1959; instead Schwabhäuser extends the results to hyperbolic geometry, absolute (sometimes called neutral) geometry, as well as affine and projective geometry. To study these different geometries, one needs to study the process of coordinatizing them, and check that coordinatization can be done within the confines of certain specified first-order axiom systems. Hyperbolic geometry is coordinatized in a different way than Euclidean geometry; and each of these geometries offers some complications. The struggle to complete Part II was Schwabhäuser's alone, so we know less about the steps along the way than we do about Part I.

Every reader naturally has some curiosity about the authors. What kind of people were Tarski, Szmielew, and Schwabhäuser, and what else happened in their lives besides the work in this book? Tarski was a very colorful character, and his life has been chronicled in the biography by the Fefermans[10], where one can find numerous anecdotes and details about Tarski's life. We therefore give only the barest outline here. (Quotes about Tarski's life below are from this biography.)

Tarski was born in 1901 in Warsaw, as Alfred Tajtelbaum (or Teitelbaum in the Germanic spelling). His mother was "brilliant, well educated, and willful", as well as an heiress. His father was a businessman from a family of businessmen. Polish was spoken at home, not Yiddish, although the family was Jewish. Poland was the center of the world in

[9] Frédérique Guilhot, Formalisation en coq et visualisation d'un cours de géométrie pour le lycée, *Revue des Sciences et Technologies de l'Information, Technique et Science Informatiques, Langages applicatifs*, **24**, 1113–1138, 2005.

[10] Feferman, Anita, and Feferman, Solomon, *Alfred Tarski: Life and Logic*, Cambridge University Press (2004).

logic in the post-World War I period. Tarski's teachers included Sierpiński and Kuratowski (big names in set theory and topology) and in logic, Jan Łukasiewicz and Stanislaw Leśniewski, whose work established the "Polish school" in axiomatic logic. Tarski's first two papers were about problems posed by Leśniewski; the first was published under the name Tajtelbaum, the second under "Tajtelbaum-Tarski", and from 1924 on, he was officially Tarski. There are three amusing stories in his biography about how that particular name was chosen; but the point was, "Tarski" did not sound Jewish. To make the point certain, Tarski converted to Catholicism in 1924. This conversion and name change made possible a professional life in Polish academia, which would not have been open to a Jew. His biographers speculate that more was involved: it was fashionable in certain circles to be "more Polish than Jewish." Tarski was a professed atheist, but "if you were going to be Polish then you had to say you were Catholic."

He received his Ph. D. degree in 1924, from the University of Warsaw, under Leśniewski. That same year he and Banach published their famous "Banach-Tarski paradox", about cutting a sphere into pieces that can be re-assembled into a larger sphere. He was appointed docent at Warsaw University, which gave him the right to lecture, some recognition, but not much money. So he continued to live in his parents' home; and once when he asked his father for money, Ignacy Teitelbaum replied, "Money? You need money? Well, why don't you go see your old man Tarski?" He lived with his parents until 1929, when he married Maria Witkowska. She was "small, dark, pretty, socially adept, and above all warm, understanding, and loyal.... She was never in competition with Alfred and had no need for the limelight; in short, she was the perfect mate for him."

As mentioned above, already in 1927 Tarski had done significant work on axiomatic geometry. How did he get from set theory to geometry? Through logic. To understand the intellectual setting of Tarski's work on geometry, we must review the development of logic 1900-1927. Russell and Whitehead's work *Principia Mathematica* was a masterpiece of axiomatic formalism; remember that first-order logic did not exist at the time, and Russell and Whitehead went straight to a theory of first-order, second-order, third-order, etc. classes called "types". Intuitively, we have some "objects" and then sets of those objects, sets of those sets, and so on to sets of sets of sets of sets, level after level. This was seen as the way to avoid the paradoxes, for the "set of sets that are not members of themselves" cannot occur at any level; no set can be a member of itself since it would have to have two different levels. Zermelo worked out an axiom system based on this idea in 1904-1908, but it seems to have not made an impact in Poland. Russell and Whitehead's theory was more influential among logicians. It was very complicated, and the natural way to proceed was to isolate the first-order part and study that more deeply. That is the background to the famous theorems of Löwenheim and Skolem, which in the period from 1915 into the 1920s began the subject of first-order logic.

Thus in 1925 and 1926, first-order logic was in its infancy; but it was clear that Hilbert's famous 1899 book *Foundations of Geometry*, which at the time had been at the cutting edge of rigor, was no longer "rigorous enough". It made too much use of set theory; it could speak of arbitrary sets of points. In 1927–29, Tarski conducted the "exercise sessions" for Łukasiewicz's logic seminar, which apparently involved some lecturing and independent research. He took up the subject of quantifier elimination.

For readers who are not logicians: a "quantifier" is either "there exists" or "for every". If one has to decide whether a statement involving such phrases is true or false, in general there will be infinitely many cases to consider, so it will be hard to give a rule. But if the vocabulary is sufficiently restricted it may be possible to "eliminate quantifiers" and show that every statement is equivalent to one without quantifiers. As an example, "there exists a (real) solution of $ax^2 + bx + c = 0$" is equivalent to $b^2 - 4ac \geq 0$, as readers who know the quadratic formula will recognize. This equivalence eliminates the "there exists" in the first statement. We don't need to check infinitely many possible values of x.

Several examples of situations in which quantifiers could be eliminated were already known at this early stage; Skolem had given a couple, and the American logician C. H. Langford had already obtained the result about discrete linear orderings, which every logic student now learns. Tarski found some new results, and since he was running an "exercise session", he suggested to his student Mojżes Presburger that he find a quantifier elimination procedure for the theory of natural numbers with addition. That theory is known today as "Presburger arithmetic", testifying to Presburger's successful solution of the problem in the spring of 1928. It served as his master's thesis in 1930.[11]

In these early years, Tarski must have begun to formulate a first-order version of Euclidean geometry, and realized that it would, or should, be equivalent in some sense to a first-order algebraic theory. In geometry one can erect two perpendicular lines as coordinate axes, and give geometric definitions of addition and multiplication, as was done already by Descartes in his *La Geometrie*, which was at the cutting edge of rigor in

[11] In a just world this would have been the beginning of a long and brilliant career. Presburger's paper was only f pages; and its significance may not have been fully recognized at the time; so he only got a master's degree, not a Ph. D., and left the university to work in the insurance industry. But there are fates worse than that: Presburger was a Jew, and died in a Nazi death camp in 1943.

its time. And, in the subject known as "analytic geometry", one can discuss points, lines, and circles in terms of algebraic formulae. Once these theories were formulated, even roughly, the question whether they admit elimination of quantifiers would have occurred to Tarski. One needs something like the quadratic-formula example above, but *much* more general: given any system of algebraic equations and inequalities (for example $ax^2 + bx + c = 0$ is a very simple case) needs a way to find conditions on the coefficients (analogous to $b^2 - 4ac \geq 0$, which involves the coefficients a, b, and c, but not x) that tell you whether the system has a solution or not. There can be many variables x, not just one, and many equations and inequalities. Of course this is complicated, but special cases were already known (Sturm's theorem, for example); nevertheless it took Tarski until 1930 to work it out. However, *nothing* was published about it until 1948, and that publication was only a statement of results, with no details about the actual method.

The reason, of course, was that a lot of other pressing things happened in the period 1930-1948. Tarski went to Vienna, met Carnap, Gödel, and Quine. He lost a competition for a professorship in Lvov; then in 1937 he was a candidate for a professorship at the University of Poznan, but (in spite of his name change and conversion) was not hired because he was a Jew. On his way back from a conference in Holland in 1938, he stopped in Berlin, and heard Hitler "giving his violent speech after his return from Munich", and then stayed up until three or four o'clock discussing logic (not politics) with "the three logicians who had not yet fled the Nazi terror."

In August, 1939, Tarski was on the ocean liner *Pilsudski* bound for New York, not to flee the Nazis, but to attend a conference in Cambridge, Massachusetts called the *Unity of Science.* He thought he was going to stay a month, and came with just one suitcase, no winter clothes, and no wife. On September 1, the Germans bombed Warsaw and invaded. At the conference, Tarski met the American logicians, including Church and Curry, and Church's students Kleene and Rosser. (He had met Quine in Vienna.) These people and many others wrote letters resulting in Tarski being granted a permanent U.S. visa, but he still had no job. As his biographers put it, "jobs were scarce and competition was fierce. To make matters worse, there was an influx of brilliant intellectual refugees from Europe also desperate for positions; and finally, ... mathematical logic was not a mainstream field at any university. To tide him over until he found something, funds were cobbled together at Harvard to appoint him as a research associate." Not only did he have job trouble, his family was still in Poland, in the middle of the war. He tried to get his family to Copenhagen and from there to the U.S. But the Germans invaded Denmark, blocking this escape route. He later tried to get them out through Sweden, with the aid of a philosopher whose father was on the Swedish Supreme Court, but this failed too, since submarine warfare had intensified and it was now impossible to travel from Sweden to the U.S. Meantime, Russell and Carnap were also at Harvard that year, and had a seminar with Tarski and Quine. In January, 1942, Tarski became a visiting fellow at the Institute for Advanced Studies, where he became friends with Gödel and his wife. Finally in the fall of 1942, Tarski moved to California to accept a new job at the University of California in Berkeley, where he worked for the rest of his life. Not, however, without a nerve-wracking encounter with his draft board in Princeton. Since he was now a permanent resident, he could be drafted, and it took "a flurry of letters" to get him an occupational deferment. He remained separated from his family until after the war.

Tarski thus had a number of concerns more pressing than publishing his work on geometry, but he *had* already in 1939 prepared a manuscript for publication. The Fefermans (in Interlude IV of their biography) have assembled a number of details about the various stages in its publication. The 1939 monograph, entitled *The Completeness of Elementary Algebra and Geometry*, was to be published in Paris. But the German invasion "disrupted the publication process." Tarski did have two sets of page proofs that remained. In 1948, J. C. C. McKinsey, a logician who had befriended Tarski at Harvard, was working at RAND Corporation. He "may have suggested that the procedure could be programmed for computer calculations of the optimization of strategies in the theory of games." As a result of this suggestion he wound up revising the 1939 manuscript under Tarski's supervision, and it was published as a RAND technical report in 1948, under the new title, *A Decision Method for Elementary Algebra and Geometry*, which was reprinted more publicly by UC Berkeley in 1951, and again in Volume I of Tarski's *Collected Works.* In 1967, a "lightly edited" version of the original 1939 page proofs was published in France, 27 years late. But better late than never!

These presentations of Tarski's work focused on the algebraic side of *Elementary Algebra and Geometry.* The metamathematical results are about polynomials and inequalities, not about points, lines, and circles. All one needs about the geometry is that it should support coordinatization, so that one can prove in an appropriate sense that it is "equivalent" to an algebraic theory. What went unpublished (until this book appeared in 1983) was Tarski's work on the axiomatization(s) of geometry itself; his improvements on Hilbert, so to speak. Of course he had *some* axiomatic theory of geometry in his 1926-27 lectures at the University of Warsaw, in the "exercise section". Thanks to the Tarski-Givant letter (*op. cit.*) we even know what this axiom system was, and the story of how Szmielew and Gupta and Schwabhäuser improved and studied the theory, resulting in this volume, has been told at the beginning of this foreword.

The second author of this book is Wanda Szmielew. Wanda was born April 5, 1918, in Warsaw, and started her university studies in 1935. She met and worked with Tarski and Lindenbaum, and formed a personal relationship with Tarski, as we know because she and her husband went hiking in the mountains with Tarski (who kept a journal of his hikes and hiking companions). During the war, she worked as a surveyor, but also found time to do the mathematical work that eventually became her dissertation. After the war, she worked at the Łódź Institute of Technology and at Łódź University. She earned her M. A. at the University of Warsaw in 1947 and became senior assistant to the Chair of Mathematics.[12]

She came to Berkeley in 1949 under Tarski's sponsorship. She had already proved the decidability of the theory of Abelian groups, and the plan was to write this up as a Ph. D. dissertation. This plan was successful. Since the theorem had already been proved, though, there may well have been time for Tarski and Szmielew to discuss geometry, but no details are known about their mathematical discussions that year. For details of their personal lives, the reader may consult the Tarski biography *op. cit.* Wanda returned to Poland in 1950, Ph. D. in hand, and became assistant professor at the University of Warsaw, rising to docent in 1954 and to associate professor in 1957.

Nine years after her first visit, she spent another year in Berkeley (1959–60). Her book *Foundations of Geometry*, co-authored by Karol Borsuk, was published by North-Holland in 1960, so it must have been close to final form by the time her visit to Berkeley began; but as discussed above, it does not follow Tarski's theories. On this visit, she and Tarski definitely discussed geometry: they worked together on the manuscript already discussed on the first page of this foreword. Tarski had worked with students on problems in the axiomatization of geometry since 1955, so he must have hoped that Wanda would help him bring the axiomatization and results about it to final form. But as discussed above, that didn't happen until Schwabhäuser wrote this book. Wanda did have a Ph. D. student, Zenon Piesyk, who wrote a dissertation in 1965 (in Polish) entitled *Axiomatic System of Alfred Tarski.* Ten years later, Piesyk published a book (also in Polish): *Wykłady z geometrii elementarnej* (Lectures on elementary geometry). It seems one would have to go to Warsaw to see

[12] These details are from *Wanda Szmielew, 1918–1976*, an obituary in *Studia Logica* **36** 4 (1977). The original Polish obituary is referenced in that English translation.

either of these works.

We turn now to Wolfram Schwabhäuser, who was born on May 20, 1931, in Riesa, a small town on the River Elbe about forty kilometers from Dresden. He remained in Riesa, attending high school while World War II raged, more or less all around him. In 1950, he began his university studies in mathematics at Humboldt University in Berlin, earning his degree in 1956, and then his Ph. D. in 1960. From 1956 on, he held a position as Scientific Assistant in the Institute for Mathematical Logic at Humboldt University under Prof. Schröter, and with his Ph. D. in 1960 he was promoted to "Oberassistent". Schwabhauser was invited to speak at both the 1960 Congress at Stanford and the 1963 Model Theory Symposium at Berkeley, but in both cases he could not obtain a visa from the U. S. Department of State. In 1964, he attended the International Congress for Logic, Methodology, and Philosophy of Science in Jerusalem, where he presented a paper with the same title as this book. There, he met Tarski, and they discussed the possibility of Schwabhäuser coming to Berkeley the following year. After the conference, he did not return to East Germany. According to his wife, this decision was not at all spontaneous, but was carefully planned; however, he dared not tell anyone, even his parents; he had given no sign of his plan. For example, he had even ordered the coal for the winter, knowing he would not be there to use it. The need for secrecy led him to abandon all his personal items, including books, diaries, and letters. After the Congress, he took a position as assistant to Prof. Hans Hermes, at the Institute for Foundational Studies of the University of Münster, West Germany. In July 1965, he earned the post-Ph. D. credential known as "habilitation" at Münster. Then in the academic year 1965–66, the conversation with Tarski in Jerusalem came to fruition: Schwabhäuser visited Berkeley, where he worked with Tarski on the foundations of geometry.

The year of Schwabhäuser's visit to Berkeley was also the year of Gupta's Ph. D. thesis, so Schwabhäuser had company. During that year, Schwabhäuser taught two courses: one on mathematical logic, and one on the foundations of geometry; and he also received support from Tarski's NSF grant "Foundations of Geometry." Schwabhäuser returned to West Germany in 1966, and became docent at the University of Bonn, where he became Professor in 1969. In 1967, he married Inge-Marie Scholl, who had studied mathematics and physics at the Free University of Berlin (in West Berlin). In his final career move, he became professor of Informatik (Computer Science) in Stuttgart in 1973, where he served the rest of his life.

Schwabhäuser's hobbies were playing the piano and the flute, photography, and hiking. His wife, Inge-Marie, wrote that "Our greatest pleasure was the birth of our son Thomas in 1976."

Schwabhäuser's book was published in 1983; and according to Inge-Marie, he worked on it intensely for ten years. So his 1978 letter to Tarski asking for material for inclusion came in the middle of the effort. Two years after the publication of his book, he became ill, in spring 1985, and his student, research assistant, and friend Uwe Schöning filled in for him as a teacher. He died December 27 of that year, of cancer. His obituary[13] mentions that he is survived by his nine-year-old son.

Schöning described Schwabhäuser as "a nice, calm, friendly, serious person, a gentleman of science, very well respected in the faculty." Schöning also has some memories of the book project: "Writing this book was a tough work and effort which lasted many years. The epsilon which he tolerated for having mistakes in the book was very very small, if not to say, zero. According to this small epsilon, the time to finish this work was accordingly long. His secretary, Ms. Sonnenschein, worked with an IBM ball-head typewriter. She had to use six such heads to supply all the needed special symbols and Greek letters." Mathematicians of a certain age will fondly recall those IBM Selectric type balls; you who grew up with TeX will never know what hard work it used to be to write a book.

This book is a testament to the enduring attractions of points, lines, and circles, and to the dedication and perseverance of its authors, who kept at their work as long as the spark of life flowed in their blood, and who, when that spark was gone, passed the torch to others. Though their progress was impeded by world wars and by the Cold War, and though they, like the rest of us, had jobs and families (or sometimes, had no jobs, and had families halfway around the world, or a country to which they could not return), they still managed to think about points, lines, circles, and axioms, and check details.

Now that the book is no longer out of print, it is also in digital form, which makes it safe to say: nine-tenths of the readers of this book have not yet been born. Five hundred years from now, when technology has developed beyond our wildest imagination, today's pdf (portable document format) files will still be readable and a small minority of people will still be interested in points, lines, circles, and axioms. They will have no difficulty finding this book, and the fundamental results it contains will still be studied.

[13] This obituary, by Walter Knödel, was published in the Stuttgart university magazine *Uni-kurier* in February, 1986.

Wolfram Schwabhäuser, date unknown

Wanda Szmielew in 1938

Alfred Tarski in 1971 at the Tarski Symposium

Hochschultext

Vorwort

Das vorliegende Buch besteht aus zwei Teilen.

Teil I enthält einen axiomatischen Aufbau der euklidischen Geometrie auf Grund eines Axiomensystems von Tarski, das in einem gewissen Sinne (auch für die absolute Geometrie) gleichwertig ist mit dem Hilbertschen Axiomensystem, aber formalisiert ist in einer Sprache, die für die Betrachtungen in Teil II besonders geeignet ist. Mehrere solche Axiomensysteme wurden schon vor langer Zeit von Tarski veröffentlicht. Hier wird nun die Durchführung eines Aufbaus der Geometrie auf Grund eines solchen Axiomensystems - unter Benutzung von Resultaten von H. N. Gupta - allgemein zugänglich gemacht. Die vorliegende Darstellung wurde vom zuerst genannten Autor allein geschrieben, aber sie beruht zum Teil auf unveröffentlichten Resultaten von Alfred Tarski und Wanda Szmielew; daher gebührt ihnen ein Teil der Autorschaft. Mehr über Entstehung und Inhalt von Teil I sowie über die Geschichte der Tarskischen Axiomensysteme wird in der Einleitung (Abschnitt I.0) gesagt.

Teil II enthält metamathematische Untersuchungen und Ergebnisse über verschiedene Geometrien, was vielfach auf eine Anwendung von Methoden und Sätzen der mathematischen Logik auf Geometrien hinausläuft (vgl. die Einführung am Anfang des zweiten Teils [II.1.1]). In den ersten beiden Abschnitten werden zunächst die wichtigsten Hilfsmittel aus beiden Gebieten - Logik und Geometrie - zusammengestellt. Die weiteren Abschnitte sind dann der eigentlichen "Metamathematik der Geometrie" gewidmet. Es wurde versucht, über den Entwicklungsstand dieses von Tarski inspirierten Gebiets einen möglichst vollständigen Überblick zu geben, wenn auch an manchen Stellen nur durch Angabe von Literaturzitaten. Andererseits wurden große Teile ausführlich (mit allen Beweisen) dargestellt, so daß das Buch als Grundlage für Vorlesungen und

Seminare geeignet sein dürfte. Auch der zweite Teil enthält eine Reihe von neuen Resultaten, die in der bisherigen Literatur gar nicht oder nur unter stärkeren Voraussetzungen behandelt wurden. Zur Gliederung und zum Inhalt von Teil II vergleiche man das Inhaltsverzeichnis und die Überblicke am Anfang der einzelnen Abschnitte (bei II.1 als Nr. II.1.2).

Ausführliche Register und zahlreiche Verweise innerhalb des Textes sollen das Nachschlagen erleichtern. Für die Verweise innerhalb des Textes wurden die Unterabschnittsnummern verwendet, die auch in den Kopfleisten der einzelnen Seiten innen angegeben sind. Bei Verweisen innerhalb desselben Teils wurde dabei die Nummer dieses Teils (I. bzw. II.) weggelassen.

Viele Kollegen, Freunde und Mitarbeiter haben zur Entstehung dieses Buches beigetragen. Die Herren W. Benz, E. Engeler, H. N. Gupta, C. F. Moppert, G. H. Müller, A. Prestel und L. W. Szczerba haben eine frühere (Manuskript-)Fassung angesehen und Anregungen, hilfreiche Kommentare oder Verbesserungsvorschläge gegeben. A. Tarski schickte mir umfangreiche Notizen mit wertvollen Ergänzungen und Ratschlägen, die an verschiedenen Stellen in die endgültige Fassung eingearbeitet oder berücksichtigt wurden; bei der Abfassung dieser Notizen wurde er von St. Givant unterstützt. L. W. Szczerba und A. Prestel gaben ihre freundliche Zustimmung zur Aufnahme von anderweitig noch nicht veröffentlichten Resultaten (vgl. die entsprechenden Hinweise auf S. 393, 400, 414 bzw. 445, 447). Anregende Gespräche über hier behandelte Probleme hatte ich mit vielen Kollegen, die ich nicht alle aufzählen kann. Korrekturhinweise zu der früheren Fassung erhielt ich von den Herren G. Getto, H. Gilg, A. Prestel, H. Seeland und W. Tischhauser. Herr Getto legte außerdem ein ausführliches Register für die frühere Fassung an, auf das ich mich für die jetzige Fassung weitgehend stützen konnte. Die Schreibarbeiten für die frühere Fassung wurden von Frau I. Geisselhart (Teil I) und Frau H. Sonnenschein (Teil II) ausgeführt. Die äußerst mühevollen Schreibmaschinenarbeiten (mit vielen Kugelköpfen) für die gesamte Reinschrift wurden ebenfalls von Frau Sonnenschein übernommen. Die Zeichnungen wurden von Herrn K. Adler angefertigt.

Allen genannten und nicht genannten Personen, die mir geholfen haben, sage ich meinen herzlichen Dank. Außerdem danke ich dem Deutschen Akademischen Austauschdienst (DAAD) für die Finanzierung des auf S. 414 erwähnten Gastaufenthalts sowie dem Springer-Verlag für die stets gute und verständnisvolle Zusammenarbeit.

Stuttgart, im Juni 1983 Wolfram Schwabhäuser

Inhaltsverzeichnis

Genauere Angaben über den Inhalt von Teil II sind enthalten in den Überblicken am Anfang der einzelnen Abschnitte (bei Abschnitt II.1 als Nr. II.1.2).

Teil I: Ein axiomatischer Aufbau der euklidischen Geometrie

von Wolfram Schwabhäuser, Wanda Szmielew und Alfred Tarski

0. Einleitung

(von W. Schwabhäuser)

0.1 Zum Hilbertschen Axiomensystem. Ein axiomatischer Aufbau der Geometrie ist schon in den "Elementen" von Euklid (um -300) konzipiert, aber erst gegen Ende des vorigen Jahrhunderts wurden verschiedene Wege zu einem axiomatischen Aufbau in der heute üblichen Strenge für diese "euklidische Geometrie" angegeben. Am bekanntesten davon wurde der Aufbau auf Grund des Axiomensystems in Hilberts "Grundlagen der Geometrie" (erste Fassung 1899, bald danach entstand die heute übliche Fassung, die seit vielen Auflagen unverändert beibehalten wurde, s. etwa Hilbert 1977). Darin werden (für die räumliche Geometrie) drei Sorten von geometrischen Objekten unterschieden, nämlich Punkte, Geraden und Ebenen. Die Grundbegriffe dieses Axiomensystems sind die Inzidenz zwischen Punkten und Geraden, die Inzidenz zwischen Punkten und Ebenen, die Zwischenbeziehung, die Streckenkongruenz und die Winkelkongruenz. Die letzten beiden lassen sich als eine vierstellige bzw. als eine sechsstellige Relation zwischen Punkten (die die Strecken bzw. Winkel bestimmen) auffassen; die Grundbegriffe sind damit Relationen zwischen den betrachteten geometrischen Objekten.

0.2 Besonderheiten des Tarskischen Axiomensystems. Von derselben Art ist das Axiomensystem von Tarski, das für den hier dargestellten Aufbau zugrundegelegt wird. Im Unterschied zum Hilbertschen Axiomensystem treten darin jedoch nur Punkte als die einzigen geometrischen Objekte auf, und als Grundbegriffe werden nur zwei Relationen zwischen Punkten verwendet, und zwar eine Streckenkongruenz und eine Zwischenbeziehung (mit geringen Abweichungen gegenüber den entsprechenden Hilbertschen Begriffen, s. 1.1). Dieses Vorgehen hat verschiedene Vorteile.

Für die späteren metamathematischen Betrachtungen in Teil II ist eine Formalisierung der geometrischen Sätze unerläßlich. Ein Vorteil des Tarskischen Ansatzes besteht darin, daß diese Formalisierung besonders einfach wird; fast der ganze Aufbau der Geometrie läßt sich formalisieren in einer Sprache L(D,B) im Rahmen der (einsortigen) Prädikatenlogik der ersten Stufe (mit Identität) mit Variablen für Punkte und zwei Relationszeichen (Prädikatensymbolen) D und B für die Streckenkongruenz und die Zwischenbeziehung. (Konsequenterweise werden Punkte daher mit kleinen Buchstaben bezeichnet, während Hilbert Großbuchstaben verwendet.)

Schon vom rein geometrischen Standpunkt aus kann man es als Vorteil ansehen, daß Geraden und Ebenen ebenso als Punktmengen behandelt werden, wie andere Punktmengen, die in elementargeometrischen Überlegungen auftreten, z.B. Halbgeraden ("Strahlen"), Halbebenen, Strecken, Winkel, Kreise, Kegelschnitte (trotzdem lassen sich Sätze über Punktmengen dieser speziellen Art im oben genannten Rahmen formalisieren, vgl. Anmerkung 6.26(i)). Das ist zunächst eine Vereinheitlichung, es führt aber auch dazu, daß dieselben geometrischen Objekte und Grundbegriffe für Geometrien verschiedener Dimension verwendet werden können. (Im Hilbertschen Aufbau der ebenen Geometrie fallen dagegen Ebenen als Objekte und die zweite Inzidenzrelation als Grundbegriff weg; eine analoge Behandlung von Geometrien höherer Dimension könnte mit zusätzlichen Objekten und Grundbegriffen durchgeführt werden.) Vom metamathematischen Standpunkt aus bedeutet das, daß dieselbe formalisierte Sprache L(D,B) für Geometrien verschiedener Dimension verwendet wird. Erst dadurch wird es möglich, Sätze zu formulieren, die für jede Dimension größer-gleich 2 gelten und zusammen die "dimensionsfreie Geometrie" bilden (tatsächlich gehören nahezu alle Sätze dieses Aufbaus zur dimensionsfreien Geometrie, s. auch 0.4, 1.7.4).

Die bisher genannten Vorteile des Tarskischen Ansatzes bezogen sich nur auf die Wahl der Objekte und Grundbegriffe bzw. auf die Wahl der formalisierten Sprache für die Geometrie. Darüberhinaus hat das Axiomensystem den ästhetischen Vorzug, aus besonders wenigen und einfachen Axiomen zu bestehen. Die Einfachheit zeigt sich vor allem darin, daß - im Unterschied zu den sonst verwendeten Systemen - keinerlei Hilfsdefinitionen für zusätzliche geometrische Begriffe benötigt werden, vielmehr ist jedes Axiom in einfacher Weise allein mit Hilfe der Grundbegriffe D und B formuliert. (Natürlich werden zusätzliche Begriffe beim späteren Aufbau der Geometrie eingeführt.)

Es läßt sich wohl nicht vermeiden, daß dafür einige Beweise länger oder komplizierter ausfallen als im Aufbau auf Grund des umfangreicheren Hilbertschen Axiomensystems. An einigen wichtigen Stellen wird auf Unterschiede gegenüber dem Hilbertschen Aufbau hingewiesen.

0.3 Zur Geschichte des Tarskischen Systems (nach Notizen von Alfred Tarski). Tarski stellte ein Axiomensystem der genannten Art schon vor mehr als 50 Jahren auf und verwendete es in seiner Vorlesung an der Universität Warschau im akademischen Jahr 1926-27. Diese erste Version ist angegeben in Tarski 1940 (S. 33f.) und 1974. Eine zweite, vereinfachte Version erschien in Tarski 1948 und 1951 (S. 55f.). Weitere Vereinfachungen (durch Weglassen und Ändern von Axiomen) wurden 1956-57 in gemeinsamen Untersuchungen von Eva Kallin, Scott Taylor und Tarski erzielt. Dadurch entstand die dritte veröffentlichte Version des Axiomensystems, die in Tarski 1959 erschien und von Tarski in einer Vorlesung über Grundlagen der Geometrie an der University of California, Berkeley 1956-57 verwendet wurde. In allen Fällen wurde das Axiomensystem für die ebene euklidische Geometrie formuliert, aber darauf hingewiesen, daß man durch Abänderung der beiden Dimensionsaxiome (hier A8 und A9, s. 1.2, 1.7.5) zu einer beliebigen Dimension $n \geq 2$ übergehen kann. Zwei Axiome der dritten Version wurden später als entbehrlich nachgewiesen in Gupta 1965.

Das hier in Teil I dieses Buches verwendete Axiomensystem ist eine weitere Version, die sich an einigen Stellen von der (vereinfachten) dritten Version unterscheidet (insbes. A7, vgl. 1.7.5, s.a. 5.1). Das Resultat, daß auch dieses Axiomensystem (bis auf eine geringfügige Änderung von A7 und A10, s. 1.7.5) für den Aufbau der euklidischen Geometrie verwendet werden kann (bis zur Charakterisierung der Modelle im Darstellungssatz, hier 16.15) stammt von H.N. Gupta und ist ebenfalls enthalten in seiner Dissertation Gupta 1965. Diese Arbeit hat jedoch nicht das Ziel, einen direkten Aufbau auf Grund dieses Axiomensystems durchzuführen. Vielmehr wird (neben anderen Untersuchungen) der Aufbau nur so weit ausgeführt, bis die Gleichwertigkeit mit der dritten Version aus Tarskis Vorlesung gezeigt werden kann; außerdem wird aber eine schwächere euklidische Geometrie (mit Modellen über beliebigen angeordneten Körpern) axiomatisch aufgebaut, und aus dem Darstellungssatz für diese ergibt sich unser Darstellungssatz als Spezialfall.

Ebenfalls 1965 arbeitete Wanda Szmielew als Visiting Research Worker in Berkeley in einem von der U.S. National Science Foundation unterstützten Forschungsprojekt und schrieb dabei in Zusammenarbeit mit Tarski ein Manuskript über den ersten Teil eines direkten Aufbaus der ebenen Geometrie auf Grund der Axiome, die auch hier (als A1 bis A9) verwendet werden. Dieses Manuskript enthält eine Reihe eigener, neuer Ideen der Verfasserin. Natürlich wurde vieles übernommen aus Gupta 1965, andererseits wird dort (S. 4) darauf hingewiesen, daß Anregungen von Wanda Szmielew zu Verbesserungen der Arbeit führten.

Wanda Szmielew benutzte dieses Manuskript für Vorlesungen in Warschau. Außerdem stellte sie es mir im Herbst 1965 dankenswerterweise zur Verfügung als Grundlage für eine Vorlesung "Foundations of Geometry", die ich im akademischen Jahr 1965-66 an der University of California, Berkeley, hielt und in der im ersten Teil der Aufbau auf Grund des Tarskischen Axiomensystems behandelt wurde. Der Inhalt des Manuskripts ist im wesentlichen derselbe wie hier in den Abschnitten 1 bis 8, 10 und am Anfang von Abschnitt 9 (bis Satz 9.6), also etwa das, was man braucht, um mit den vom Hilbertschen Aufbau her geläufigen Methoden weiterzukommen. Anläßlich dieser und weiterer Vorlesungen über das gleiche Thema in Bonn nahm ich naturgemäß Änderungen gegenüber dem Manuskript vor, die in der vorliegenden Darstellung ihren Niederschlag finden. Die Änderungen beziehen sich vor allem auf die Formulierung und Anordnung der Sätze, was an einigen Stellen zu Vereinfachungen und zur Verringerung der Anzahl der Hilfssätze führt. Dagegen wurden die geometrischen Ideen für die Beweise im wesentlichen unverändert aus dem Manuskript übernommen.

Bei einem weiteren Forschungsaufenthalt in Berkeley im Jahre 1967 schrieb Wanda Szmielew ein neues Manuskript, in dem der Aufbau der ebenen euklidischen Geometrie auf Grund des hier verwendeten Axiomensystems (bis auf eine unwesentliche Änderung) bis zum Darstellungssatz vollständig durchgeführt ist. Dieses Manuskript ist hier nicht berücksichtigt; ich erfuhr davon erst nach Abschluß der vorliegenden Darstellung von Teil I.

Beide Manuskripte wurden bisher nicht veröffentlicht. Sie waren ursprünglich gedacht als Vorarbeit für ein gemeinsames Buch von Szmielew und Tarski über Grundlagen der Geometrie, in dem ein systematischer Aufbau der euklidischen Geometrie für beliebige Dimension und metamathematische Fragen, auch zu anderen Geometrien, behandelt werden sollten (ähnlich wie in diesem Buch). Die Durchführung dieses Plans verzögerte sich und konnte wegen des vorzeitigen Todes von Wanda Szmielew im Jahre 1976 nicht mehr verwirklicht werden.

Mit dem vorliegenden Teil I wird unseres Wissens die erste in der Literatur allgemein zugängliche Darstellung eines systematischen Aufbaus auf Grund irgendeiner Version des Tarskischen Axiomensystems gegeben. Sie wurde vom zuerst genannten Verfasser (Schwabhäuser) geschrieben. Weil anderweitig nicht veröffentlichte Ergebnisse von Tarski und Szmielew aus dem genannten Manuskript wesentlich verwendet werden, schlug er vor, den Teil I unter den Namen aller drei Autoren zu veröffentlichen. Diesem Vorschlag stimmte Tarski 1978, auch im Namen von Wanda Szmielew, zu und gab wertvolle Ergänzungen, die an verschiedenen Stellen eingearbeitet wurden.

<u>0.4 Zum Inhalt von Teil I</u>. Über die vom Hilbertschen Aufbau verschiedene Hälfte (bis Satz 9.6 und Abschnitt 10), die aus dem Szmielewschen Manuskript entstand, wurde schon berichtet. Im weiteren Aufbau lehnte sich der Verfasser enger an Hilbert an, als es sonst in der Tarski-Schule üblich ist.

Insbesondere wurde auch die Winkelkongruenz behandelt, was zum Beweis des Darstellungssatzes gar nicht nötig ist (vgl. Gupta 1965). Erst dadurch ergeben sich aber die Hilbertschen Axiome als Sätze in unserem Aufbau. Da man umgekehrt die Tarskischen Axiome auch leicht als Sätze auf Grund des Hilbertschen Axiomensystems erhalten kann, ergibt sich auf diesem Wege die Gleichwertigkeit beider Axiomensysteme und auch die Gleichwertigkeit der Gruppen I. bis III. von Hilbert (für die Ebene) mit den Axiomen A1 bis A9 des betrachteten Tarskischen Systems (unter Zugrundelegung geeigneter Definitionen für den jeweiligen Übergang zu den anderen Grundbegriffen). Damit überträgt sich auch der Darstellungssatz von Pejas (vgl. II.2.14(iv)) auf die "absolute Geometrie" auf Grund der Axiome A1 bis A9.

Im Unterschied zu Hilbert erfolgt nahezu der ganze Aufbau dimensionsfrei (vgl. 0.2), d.h. hier ohne Verwendung des "oberen Dimensionsaxioms" A9 (s. 1.2). Wie üblich wird das Parallelenaxiom (hier in der Form A10, s. 1.2, 1.7.5) möglichst spät verwendet. Vorher (bis einschließlich Abschnitt 11) wird also eine "dimensionsfreie absolute Geometrie" (auf Grund der Axiome A1 bis A8) aufgebaut.

Außerdem wurde die Behandlung von Unterräumen beliebiger endlicher Dimension innerhalb der dimensionsfreien absoluten Geometrie hinzugefügt (s. Abschnitte 9, 11, 16). Eine solche Behandlung (ohne Parallelenaxiom!) findet sich schon in Gupta 1965 und wird dort (S. 4) Wanda Szmielew zugeschrieben. Während dort aber der Begriff des n-dimensionalen Simplex als wesentliches Hilfsmittel verwendet wird, habe ich mich - motiviert durch "Hilberts Grundlagen" - auf die Betrachtung von Seiteneinteilungen und Halbräumen gestützt. Der Darstellungssatz 16.15 wird dann für beliebige endliche Dimension $n\geq 2$ bewiesen.

Die Einführung eines angeordneten Körpers erfolgte im Sinne der "affinen Streckenrechnung" aus Hilbert 1977, §24. Das ist zwar aufwendiger als die Streckenrechnung unter Benutzung des Kongruenzbegriffs (s. Hilbert 1977, §15), aber dafür steht die Konstruktion auch für die in Teil II betrachteten affinen Geometrien zur Verfügung.

0.5 Bezeichnungen und Abkürzungen. Die Darstellung des axiomatischen Aufbaus geschieht in der (mathematischen) Umgangssprache. Zum Aufschreiben von Sätzen werden zwar die üblichen logischen Zeichen ($\neg$, $\wedge$, $\vee$, $\rightarrow$, $\leftrightarrow$, $\forall$, $\exists$, $=$ für "nicht", "und", "oder", "wenn - so", "genau dann wenn [gdw]", "für jedes", "es gibt ein", "gleich") benutzt, um die im zweiten Teil benötigte Formalisierung von Sätzen (vgl. II.1.3f.) vorzubereiten. Innerhalb des ersten Teils können diese Zeichen als umgangssprachliche Abkürzungen aufgefaßt werden. Der Leser, der darüber hinaus an formalisierten Beweisen (etwa mittels Schlußregeln aus der Prädikatenlogik) interessiert ist, kann eine solche Formalisierung selbst vornehmen.

Beim Aufschreiben von Sätzen (auch Axiomen) schließen wir uns der üblichen Konvention an, daß Allquantoren ($\forall$) mit den zugehörigen Variablen weggelassen werden können, wenn sie am Anfang stehen und sich auf den ganzen Satz beziehen; so wird z.B.

$ab\equiv cc\rightarrow a=b$ (Axiom A3, s. 1.2)

verwendet als gleichwertig mit

$\forall a\forall b\forall c(ab\equiv cc\rightarrow a=b)$.

Es werden auch die folgenden Bezeichnungen und Abkürzungen aus der Logik (im selben Sinne wie oben) verwendet:

$\iota x\alpha(x)$ für dasjenige x mit der Eigenschaft $\alpha(x)$,

$\bigwedge_{i=1}^{n}\alpha_i$, $\bigvee_{i=1}^{n}\alpha_i$, $\bigwedge_{1\leq i<j\leq n}\alpha_{i,j}$, ... für "allgemeine Konjunktionen"

(Und-Verbindungen) und "allgemeine Disjunktionen" (Oder-Verbindungen) mit Gliedern α_i, $\alpha_{i,j}$, ..., die den angegebenen Bedingungen genügen,

$a \neq b$ für $\neg a = b$ (a ist von b verschieden),

$\neq(a_1 \ldots a_n)$ für $\bigwedge_{1 \leq i < j \leq n} a_i \neq a_j$ ($a_1, \ldots, a_n$ sind paarweise verschieden),

$\overset{\geq k}{\exists} x\alpha(x)$ für $\exists x_1 \ldots \exists x_k[\neq(x_1 \ldots x_k) \wedge \bigwedge_{\kappa=1}^{k} \alpha(x_\kappa)]$ (es gibt wenigstens k Dinge [Objekte der betrachteten Art] mit der Eigenschaft $\alpha(x)$),

$\overset{\leq k}{\exists} x\alpha(x)$ für $\neg \overset{\geq k+1}{\exists} x\alpha(x)$ (es gibt höchstens k Dinge mit der Eigenschaft $\alpha(x)$),

$\overset{=k}{\exists} x\alpha(x)$ für $\overset{\geq k}{\exists} x\alpha(x) \wedge \overset{\leq k}{\exists} x\alpha(x)$ (es gibt genau k Dinge x mit der Eigenschaft $\alpha(x)$),

$\overset{\geq k}{\exists}$ für $\exists x_1 \ldots x_k \neq(x_1 \ldots x_k)$ (es gibt wenigstens k Objekte [der betrachteten Art, z.B. Punkte]).

Doppelpunkte in der Verbindung $:\leftrightarrow$ und := kennzeichnen (explizite) Definitionen.

Sonstige Abkürzungen:
□ : Ende des Beweises.
o.B.d.A.: ohne Beschränkung der Allgemeinheit.

1. Das Tarskische Axiomensystem, kartesische Räume

1.1 Vorbemerkungen. Die geometrischen Objekte, auf die sich unser Axiomensystem bezieht, heißen *Punkte*, sie mögen eine nicht-leere Menge bilden und seien durch kleine lateinische Buchstaben bezeichnet. Grundbegriffe des Axiomensystems sind eine vierstellige Relation D, die *Streckenkongruenz* (equidistance relation), und eine dreistellige Relation B, die *Zwischenbeziehung* (betweenness), auf der Menge der Punkte.

D$abcd$ wird gelesen als "ab ist kongruent zu cd ", wir schreiben dafür auch $ab \equiv cd$. Das soll anschaulich bedeuten, daß der Abstand der Punkte a und b genau so groß ist wie der Abstand der Punkte c und d, mit anderen Worten, daß die Strecken ab und cd gleiche Längen haben. Babc wird gelesen als "b liegt zwischen a und c "; das soll anschaulich bedeuten, daß b auf der Verbindungsstrecke der Punkte a und c liegt. Im Gegensatz zu den Hilbertschen Grundbegriffen ist im Falle $ab \equiv cd$ auch zugelassen, daß $a=b$ und (damit) $c=d$ ist, und im Falle Babc auch zugelassen, daß b mit einem der Punkte a, c (evtl. mit beiden) zusammenfällt.

1.2 Das Axiomensystem. Das hier zugrundegelegte Tarskische Axiomensystem besteht aus den Axiomen A1 bis A11, die im folgenden angegeben und erläutert werden.

A1: $ab \equiv ba$.

A2: $ab \equiv pq \wedge ab \equiv rs \rightarrow pq \equiv rs$.

A3 (Identitätsaxiom für die Streckenkongruenz):
$ab \equiv cc \rightarrow a=b$.

Das sind wohl unmittelbar einleuchtende Forderungen.

A4 (Axiom der Streckenabtragung, Abb. 1):

$\exists x(\mathrm{B}qax \wedge ax \equiv bc)$.

(Zu jedem [zu beliebigem] q, a, b, c gibt es ein x mit der genannten Eigenschaft, vgl. die Konvention über Allquantoren in 0.5.)

Das ist im wesentlichen eine kürzere Formulierung des auch bei Hilbert 1977 auftretenden Streckenabtragungsaxioms (III.1) "Auf einer beliebigen von a ausgehenden Halbgeraden gibt es stets einen Punkt x, so daß $ax \equiv bc$ ist". Die genannte Halbgerade wird hier festgelegt durch einen Punkt q auf der entgegengesetzten Halbgeraden; jedoch ist auch der Fall $q=a$ zugelassen, in dem keine Halbgerade festgelegt wird. Die Eindeutigkeit des Punktes x im Falle $q \neq a$ wird nicht gefordert, läßt sich aber (wie bei Hilbert) später beweisen.

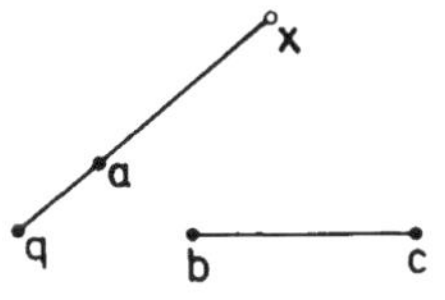

Abb. 1

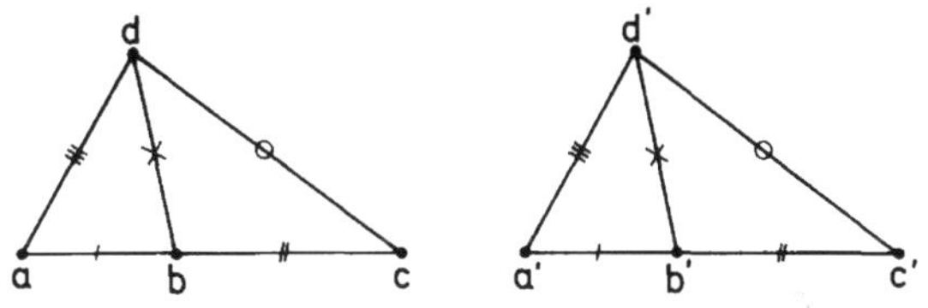

Abb. 2

A5 (Fünf-Strecken-Axiom, Abb. 2):

$a \neq b \wedge \mathrm{B}abc \wedge \mathrm{B}a'b'c' \wedge ab \equiv a'b' \wedge bc \equiv b'c' \wedge ad \equiv a'd' \wedge bd \equiv b'd' \rightarrow cd \equiv c'd'$.

In Abbildungen werden zueinander kongruente Strecken häufig (wie hier) durch gleiche Markierungen gekennzeichnet. Die beiden Teilfiguren in Abb. 2 können natürlich eine beliebige Lage zueinander im Raum haben.

Axiom A5 ist das einzige, das es erlaubt, aus der Kongruenz gewisser Strecken auf die Kongruenz von Strecken mit einer anderen Richtung zu schließen. Etwas Ähnliches wird im Hilbertschen System nur erreicht durch das Axiom III.5 (einen Teil des ersten Kongruenzsatzes), in dem die Winkelkongruenz eine wesentliche Rolle spielt. Da durch die Kongruenz der "linken Teildreiecke" in Abb. 2 die Kongruenz der Winkel $\sphericalangle cbd$ und $\sphericalangle c'b'd'$ gesichert wird, kann man auch A5 als einen (anderen) Teil des ersten Kongruenzsatzes (hier Satz 11.49) auffassen.

A6 (Identitätsaxiom für die Zwischenbeziehung):

$\mathrm{B}aba \rightarrow a=b$.

A7 (Axiom von Pasch, innere Form, Abb. 3):

$$Bapc \wedge Bbqc \rightarrow \exists x(Bpxb \wedge Bqxa).$$

Die klassische Form des Axioms von Pasch (II.4 bei Hilbert 1977) besagt, daß eine mit einem Dreieck (hier acq) in einer Ebene liegende Gerade, die eine Seite des Dreiecks (hier ac) innen (hier in p) schneidet und durch keine Ecke geht, noch eine weitere Seite innen schneidet. Schneidet sie also insbesondere die Seite cq außen (in ihrer Verlängerung) in einem Punkt b, so muß sie die Seite aq innen schneiden. Von dieser Feststellung wird in A7 (abgesehen vom möglichen Zusammenfallen von

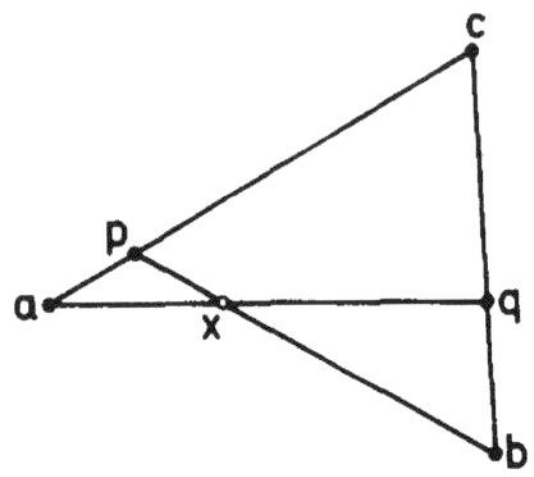

Abb. 3

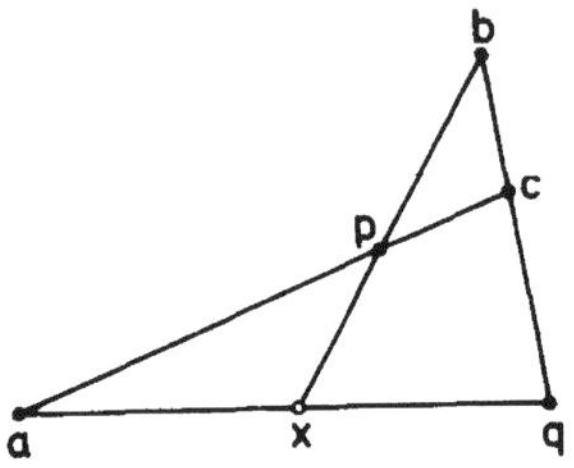

Abb. 4

Punkten oder Geraden) der Spezialfall $Bbqc$ ausgedrückt, in dem der Schnittpunkt x i n n e r h a l b der Strecke pb liegt. Daneben kann man den Spezialfall $Bqcb$ (b auf der entgegengesetzten Verlängerung der Seite cq, Abb. 4) betrachten, in dem der Schnittpunkt x a u ß e r h a l b der Strecke pb liegt; dadurch entsteht die "äußere Form"

AP: $Bapc \wedge Bqcb \rightarrow \exists x(Baxq \wedge Bbpx)$.

Diese wird im vorliegenden Aufbau erst ziemlich spät (Satz 9.6) bewiesen (vgl. auch 1.7.5). Außer dieser inhaltlichen Bedeutung ist in der Formulierung von A7 zunächst auch die Reihenfolge der Punkte zu beachten, da die "Symmetrie der Zwischenbeziehung" (Satz 3.2) erst später bewiesen wird.

A8 (unteres Dimensionsaxiom):

$$\exists a \exists b \exists c(\neg Babc \wedge \neg Bbca \wedge \neg Bcab).$$

A8 besagt, daß es Punkte a, b, c gibt, die nicht "kollinear sind", d.h. nicht "auf einer Geraden liegen" (was durch das Zutreffen einer der genannten Zwischenbeziehungen ausgedrückt würde). Das bedeutet, daß die Dimension des ganzen Raumes wenigstens 2 ist.

A9 (oberes Dimensionsaxiom, Abb. 5):
$p \neq q \wedge ap \equiv aq \wedge bp \equiv bq \wedge cp \equiv cq \rightarrow Babc \vee Bbca \vee Bcab$.

A9 besagt, daß in der Menge derjenigen Punkte, die von zwei (verschiedenen) Punkten p und q denselben Abstand haben, je drei Elemente a, b, c auf einer Geraden liegen (wie vorher) oder zusammenfallen; damit ist diese Menge selbst eine Gerade ("das Mittelot zu p und q ") oder einelementig. Für Dimensionen größer 2 ist diese Forderung verletzt (für Dimension 3 ist die genannte Menge z.B. eine Ebene). Somit wird durch A9 ausgedrückt, daß die Dimension des ganzen Raumes höchstens 2 ist.

Da der Aufbau der Geometrie dimensionsfrei durchgeführt wird (vgl. 0.4), wird A9 erst ganz zum Schluß verwendet. Dabei werden dann auch untere und obere Dimensionsaxiome für eine beliebige endliche Dimension $n \geq 2$ betrachtet, s.a. 1.7.5.

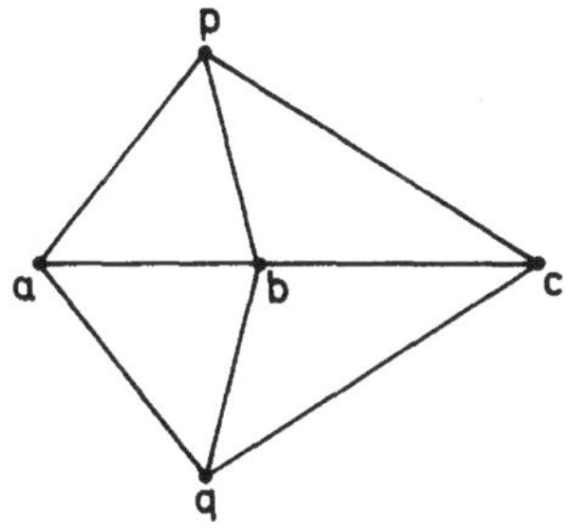

Abb. 5

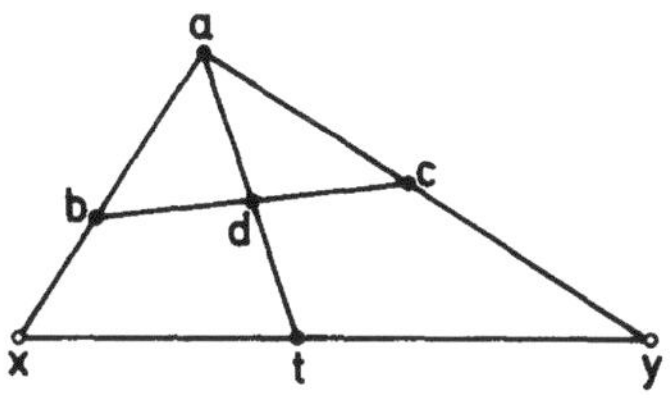

Abb. 6

A10 (Euklidisches Axiom, Abb. 6):
$Badt \wedge Bbdc \wedge a \neq d \rightarrow \exists x \exists y (Babx \wedge Bacy \wedge Bxty)$.

A10 besagt im wesentlichen: Durch einen Punkt t im Inneren eines Winkels $\sphericalangle bac$ gibt es stets eine Gerade, die beide Schenkel dieses Winkels trifft. Diese Forderung ist bekannt als äquivalent zum Parallelenaxiom (IV bei Hilbert 1977, hier als Satz 12.11), s.a. 1.7.5. In der hier angegebenen Tarskischen Formulierung läßt sie sich besonders einfach mit Hilfe des Grundbegriffs B (ohne vorherige Einführung irgendwelcher Hilfsbegriffe) ausdrücken.

A11 (Stetigkeitsaxiom, Abb. 7):
$\forall X \forall Y \{\exists a \forall x \forall y [x \in X \wedge y \in Y \rightarrow Baxy] \rightarrow \exists b \forall x \forall y [x \in X \wedge y \in Y \rightarrow Bxby]\}$.
(Zu beliebigen Punktmengen X, Y, deren Elemente im Vergleich zu einem beliebigen Punkt a in der angegebenen Weise angeordnet sind, gibt es stets einen Punkt b, der beide Mengen "trennt", d.h.

zwischen beliebigen Punkten $x \in X$ und $y \in Y$ liegt.)

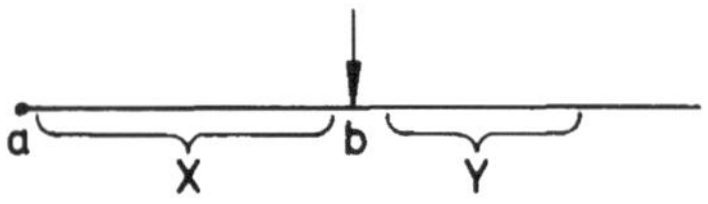

Abb. 7

In dem Fall, daß eine der Mengen X, Y leer oder X gleich der Einermenge $\{a\}$ ist, gelten Voraussetzung und Behauptung trivialerweise (z.B. leistet $b=a$ das Verlangte), und es spielt keine Rolle, daß dann die andere Menge nicht unbedingt in einer Geraden enthalten sein muß. In allen anderen Fällen folgt aus der Voraussetzung (und früheren Axiomen), daß es eine Gerade durch den genannten Punkt a gibt, in der die Punktmengen X und Y enthalten sind (Abb. 7).

A11 ist damit analog (und in einem gewissen Sinne äquivalent) zum bekannten Axiom vom Dedekindschen Schnitt für reelle Zahlen. Durch den Verzicht auf die Voraussetzung, daß die "Unterklasse" X und die "Oberklasse" Y eine erschöpfende Einteilung einer Geraden (oder einer von a ausgehenden Halbgeraden) bilden, wird die Formulierung (mittels B) erleichtert und der (weniger interessante) Fall zugelassen, daß es mehrere Punkte b mit der genannten Eigenschaft ("Trennpunkte" oder "Schnittelemente") gibt.

A11 ist in unserem System das einzige Axiom, das nicht im Rahmen der Prädikatenlogik der ersten Stufe formuliert ist (da Variablen für Mengen von Punkten verwendet werden) und sich auch nicht in diesem Rahmen formulieren läßt (vgl. II.1.56).

In vielen Anwendungen von A11 werden aber nur Punktmengen X, Y benutzt, die sich im genannten Rahmen definieren lassen, d.h. in der Form $x \in X \leftrightarrow \alpha(x)$, $y \in Y \leftrightarrow \beta(y)$, wobei $\alpha(x)$, $\beta(y)$ im genannten Rahmen Eigenschaften von Punkten beschreiben, die im Zusammenhang von A11 sinnvoll sind. Das läßt sich präzisieren zu

(*) $\alpha(x)$ und $\beta(y)$ sind Formeln im Rahmen der Prädikatenlogik der ersten Stufe mit Identität mit den Relationszeichen D und B (kurz: Formeln von L(D,B)) derart, daß die Variablen a, b, y nicht frei in $\alpha(x)$ und die Variablen a, b, x nicht frei in $\beta(y)$ vorkommen (vgl. II.1.8).

Die Möglichkeit solcher Anwendungen, d.h. die Gültigkeit der Forderung von A11 für solche Punktmengen X, Y wird ausgedrückt in der Abschwächung

A11' (Schema der elementaren Stetigkeitsaxiome):

$$\exists a \forall x \forall y[\alpha(x) \wedge \beta(y) \rightarrow \mathrm{B}axy] \rightarrow \exists b \forall x \forall y[\alpha(x) \wedge \beta(y) \rightarrow \mathrm{B}xby]$$

für beliebige Formeln $\alpha(x)$, $\beta(y)$ mit (*).

Jedes Paar von solchen Formeln liefert ein Axiom, d.h., A11' besteht aus unendlich vielen Axiomen, die sich aber alle im genannten Rahmen (in der ersten Stufe) formulieren lassen. Hierbei ist es wesentlich, daß in den Definitionen der obigen Mengen X, Y "Parameter" auftreten dürfen, d.h., daß die Formeln $\alpha(x)$, $\beta(y)$ außer x bzw. y noch weitere freie Variablen enthalten dürfen (andernfalls wäre A11' aus den übrigen Axiomen beweisbar und damit völlig überflüssig, vgl. II.4.37ff.).

Als weitere Abschwächung (eines Spezialfalls) von A11' betrachten wir noch das sog. Kreisaxiom ("circle axiom").

CA (Kreisaxiom, Abb. 8):
$\mathrm{B}cqp \wedge \mathrm{B}cpr \wedge ca \equiv cq \wedge cb \equiv cr \rightarrow \exists x[cx \equiv cp \wedge \mathrm{B}axb]$.

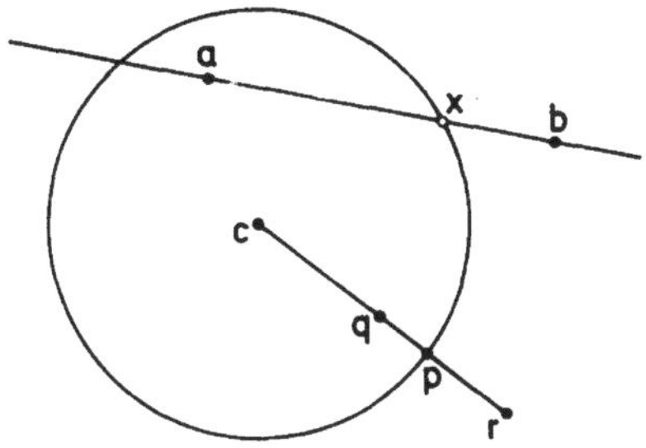

Abb. 8

CA erweist sich als gleichwertig zu dem oft als Axiom benutzten Satz: "Eine Gerade, die einen Punkt im Inneren eines Kreises enthält (und mit ihm in einer Ebene liegt), schneidet diesen Kreis (d.h., sie hat wenigstens einen Punkt mit ihm gemein)." Hier wird etwas Ähnliches (was sich mit unseren Grundbegriffen leichter formulieren läßt) ausgesagt für den Kreis mit dem Mittelpunkt c und dem Radius cp in einer Ebene durch die Punkte a, b, c. Die Voraussetzung von CA besagt nämlich, daß a ein Punkt im Inneren oder auf diesem Kreis und b ein Punkt im Äußeren oder auf diesem Kreis ist; die Behauptung besagt, daß es einen Punkt x auf diesem Kreis gibt, der zwischen a und b liegt. (Hierbei spielt es keine Rolle, ob p, q, r in der genannten Ebene liegen. Für Dimension ≥ 3 kann man statt des Kreises auch die "Hyperkugel" mit dem Mittelpunkt c und dem Radius cp verwenden, d.h. die Menge aller Punkte x mit $cx \equiv cp$.)

<u>1.3 Modelle, kartesische Räume</u>. Vor Beginn des axiomatischen Aufbaus sei noch etwas über Modelle unserer Axiome gesagt. Auf Grund der Wahl der Grundbegriffe (bzw. der zugehörigen formalisierten Sprache) kommen ganz

allgemein als Modelle passende Relationalstrukturen in Frage, d.h. Strukturen der Form $\mathfrak{A}=\langle A,D,B\rangle$, wobei A eine nicht-leere Menge ist (interpretiert als die Menge der Punkte) und D und B eine vierstellige bzw. eine dreistellige Relation auf A (interpretiert als Streckenkongruenz bzw. Zwischenbeziehung). Als spezielle solche Strukturen führen wir *kartesische Räume* ein, um überhaupt erst einmal gewisse Modelle unserer Axiome zur Verfügung zu haben.

Sei $\mathfrak{F}=\langle F,+,\cdot,\leq\rangle$ ein angeordneter Körper. Mit F^n ($n\in\mathbb{N}$) bezeichnen wir wie üblich die Menge der n-tupel von Elementen von F und mit x_ν die ν-te Komponente des n-tupels $x\in F^n$ (d.h. $x=\langle x_1,\ldots,x_n\rangle$). Für n-tupel verwenden wir die üblichen Bezeichnungen aus der Vektorrechnung; insbesondere bedeute $x-y$ oder auch $\overrightarrow{yx}$ das n-tupel mit $(x-y)_\nu=x_\nu-y_\nu$ und λx für $\lambda\in F$ das n-tupel mit $(\lambda x)_\nu=\lambda x_\nu$ $(1\leq\nu\leq n)$.

1.4 Definition. Unter dem *n-dimensionalen kartesischen Raum* über dem angeordneten Körper $\mathfrak{F}$ verstehen wir (mit den obigen Bezeichnungen) die Relationalstruktur

(1) $\mathfrak{L}_n(\mathfrak{F})=\langle F^n,\mathrm{D}^n_{\mathfrak{F}},\mathrm{B}^n_{\mathfrak{F}}\rangle$,

wobei die Relationen $\mathrm{D}^n_{\mathfrak{F}}$, $\mathrm{B}^n_{\mathfrak{F}}$ festgelegt sind durch

(2) $\mathrm{D}^n_{\mathfrak{F}}xyuv$ gdw $\sum_{\nu=1}^{n}(x_\nu-y_\nu)^2=\sum_{\nu=1}^{n}(u_\nu-v_\nu)^2$,

(3) $\mathrm{B}^n_{\mathfrak{F}}xyz$ gdw es gibt ein $\lambda\in F$ mit $0\leq\lambda\leq 1$ und $y-x=\lambda(z-x)$.

1.5 Anmerkung. Wir rechnen hier (durchweg in diesem Buch) auch die Zahl 0 zur Menge $\mathbb{N}$ der natürlichen Zahlen. Für $n=0$ gibt es genau ein n-tupel (das durch 0 Komponenten festgelegt wird und in den obigen Definitionen wie der Nullvektor behandelt wird), also besteht $\mathfrak{L}_0(\mathfrak{F})$ aus nur einem Punkt (und die Relationen $\mathrm{D}^0_{\mathfrak{F}}$, $\mathrm{B}^0_{\mathfrak{F}}$ treffen darauf zu).

Man kann nun die Gültigkeit unserer Axiome in den kartesischen Räumen $\mathfrak{L}_n(\mathfrak{F})$ nachprüfen und erhält folgendes Ergebnis.

1.6 Satz. *Vor.:* $\mathfrak{F}$ *ist ein angeordneter Körper,* $n\in\mathbb{N}$ *beliebig.*

Beh.: (i) A1 *bis* A3, A5 *bis* A7 *und* A10 *gelten in* $\mathfrak{L}_n(\mathfrak{F})$.

(ii) A4 *(Streckenabtragung) gilt in* $\mathfrak{L}_n(\mathfrak{F})$ *gdw* $n=0$ *oder* $n=1$ *oder* $\mathfrak{F}$ *ist pythagoreisch (d.h. zu jedem* $a,b\in F$ *gibt es ein* $c\in F$ *mit* $a^2+b^2=c^2$*).*

(iii) A8 *gilt in* $\mathfrak{L}_n(\mathfrak{F})$ *gdw* $n \geq 2$.

(iv) A9 *gilt in* $\mathfrak{L}_n(\mathfrak{F})$ *gdw* $n \leq 2$.

(v) A11 *gilt in* $\mathfrak{L}_n(\mathfrak{F})$ *gdw* $n=0$ *oder* $\mathfrak{F}$ *ist isomorph zum angeordneten Körper* $\mathbb{R}$ *der reellen Zahlen.*

(vi) *Wenn* $n \geq 2$, *so gelten die Axiome aus* A11' *in* $\mathfrak{L}_n(\mathfrak{F})$ *gdw* $\mathfrak{F}$ *reell-abgeschlossen (vgl.* II.3.16*) ist. (Für* $n=1$ *vgl.* II.3.30*).*

(vii) CA *gilt in* $\mathfrak{L}_n(\mathfrak{F})$ *gdw* $n=0$ *oder* $n=1$ *oder* $\mathfrak{F}$ *ist euklidisch (d.h., zu jedem positiven Element* a *von* F *gibt es in* F *eine "Quadratwurzel" [d.h. ein Element* c *mit* $c^2=a$*]).*

Zum Beweis: (i) bis (iv) und (vii) ergeben sich ohne Schwierigkeiten mit Methoden, die aus der analytischen Geometrie geläufig sind, und können dem Leser überlassen bleiben. Zum Beweis von (v) kann man zeigen, daß die Gültigkeit von A11 in $\mathfrak{L}_n(\mathfrak{F})$ gleichwertig ist mit der Gültigkeit des Axioms vom Dedekindschen Schnitt in $\mathfrak{F}$. (vi) ist etwas schwieriger (genauer s. II.3.28f.): Ähnlich wie in (v) (aber mit genauerer Untersuchung der auftretenden Formeln) kann man zeigen, daß die Gültigkeit der Axiome aus A11' gleichwertig ist mit der Gültigkeit der Axiome eines Schemas DS' der "elementaren Axiome vom Dedekindschen Schnitt" (das sich zum Axiom vom Dedekindschen Schnitt verhält wie A11' zu A11) in $\mathfrak{F}$. Die Behauptung ergibt sich dann aus dem Resultat aus Tarski 1948/51, daß DS' zusammen mit den Axiomen für angeordnete Körper gerade ein Axiomensystem für die reell-abgeschlossenen Körper bildet. □

Der im nächsten Abschnitt beginnende axiomatische Aufbau der Geometrie wird zu dem Ergebnis führen, daß sogar jedes endlich-dimensionale Modell der Axiome A1 bis A8 und A10 isomorph ist zu einem kartesischen Raum über einem gewissen angeordneten Körper $\mathfrak{F}$. Daraus ergibt sich der Darstellungssatz (16.15), in dem die Modelle von A1 bis A10, von den betrachteten Erweiterungssystemen sowie von entsprechenden Systemen mit anderen Dimensionsaxiomen bis auf Isomorphie charakterisiert werden als diejenigen kartesischen Räume, die nach Satz 1.6 in Frage kommen.

1.7 Bemerkungen zum Axiomensystem und zu einzelnen Axiomen (nach Notizen von Alfred Tarski). Das Tarskische Axiomensystem (in jeder der in 0.3 genannten Versionen) unterscheidet sich in einer Reihe von Merkmalen von den meisten, wenn nicht von allen, vorher aus der Literatur bekannten Axiomensystemen für die euklidische Geometrie. Von diesen früheren Systemen sind wohl die beiden Systeme von Pieri 1908 und von Veblen 1914 dem jetzigen von der Konzeption her am ähnlichsten.

1.7.1. Ein wesentliches Merkmal des Tarskischen Aufbaus ist die strenge Unterscheidung zwischen der vollen Geometrie (hier mit Benutzung von A11) und ihrem elementaren Teil. Als "elementar" wird dabei, anschaulich gesagt, der Teil bezeichnet, der sich ohne Benutzung mengentheoretischer Begriffe aufbauen läßt. Die Präzisierung läuft darauf hinaus, daß die volle Geometrie in (der Prädikatenlogik) der zweiten Stufe und ihr elementarer Teil (wie auch andere elementare Geometrien) in der ersten Stufe formalisiert wird (vgl. II.1.3 u. 43 bis 46) - für die Dimension n führt das zu den Geometrien P_n^2 bzw. P_n^* in II.2.1.

Das Axiomensystem für die volle Geometrie P_n^2 ist, ebenso wie das Axiomensystem von Hilbert und andere, kategorisch, d.h., je zwei Modelle sind zueinander isomorph. Dagegen ist es ein besonderes Merkmal des Tarskischen Systems für die elementare Geometrie P_n^*, daß dieses vollständig ist. Das läuft darauf hinaus, daß ein im "Standardmodell" $\mathcal{L}_n(\mathbb{R})$ (oder in einem beliebigen, nicht notwendig dazu isomorphen Modell) gültiger Satz schon aus den Axiomen (A1 bis A10 und Schema A11') folgt, wenn er sich nur elementar (d.h. in der ersten Stufe) formulieren läßt. Dieses Ergebnis und andere Charakterisierungen des Begriffs Vollständigkeit werden in Teil II ausführlich behandelt, s. insbesondere II.3.15ff., 32, II.1.39. Außerdem ist die elementare Geometrie P_n^* entscheidbar (s. II.3.4., 31).

1.7.2. Ein weiteres Merkmal des Tarskischen Systems ist die formale Einfachheit der Axiome. Wie schon gesagt, ist jedes Axiom in einfacher Weise allein mit Hilfe der Grundbegriffe D und B formuliert. In allen anderen aus der Literatur bekannten Axiomensystemen werden dagegen zumindest einige und manchmal die meisten Axiome nicht allein mit Hilfe der Grundbegriffe formuliert, sondern unter Verwendung zusätzlicher Begriffe (z.B. Halbgerade, Halbebene, Gerade), die vorher definiert werden.

Natürlich ist der Begriff der formalen Einfachheit nur vage und verschiedener Deutungen fähig. Für Geometrien mit endlichen Axiomensystemen wie hier die volle euklidische Geometrie kann man als besonders augenfälliges Maß für die Einfachheit die Länge des ganzen Systems verwenden, d.h. die Summe der Längen der einzelnen Axiome (unter der Länge werde dabei die Anzahl des Auftretens von Grundzeichen mit Ausnahme von Klammern im Sinne der Formalisierung gemäß II.1.3f. verstanden).

In diesem Sinne ist das hier betrachtete System besonders kurz. Bei einem Vergleich mit anderen Systemen wären natürlich die dort auftretenden zusätzlichen Begriffe mit Hilfe der Grundbegriffe auszudrücken.

Besonders groß ist dann der Unterschied gegenüber dem System von Pieri 1908. Darin wird als einziger Grundbegriff eine dreistellige Relation zwischen Punkten verwendet, die hier P genannt wird (mit unseren Bezeichnungen: $Pabc \leftrightarrow ab \equiv ac$, s. auch II.4.12ff.). Das ganze Axiomensystem für die volle dreidimensionale Geometrie besteht aus 24 Axiomen. Für die meisten Axiome wird neben der "normalen Formulierung" die durch Rückführung auf den Grundbegriff P entstehende Form tatsächlich angegeben. Dabei zeigt sich, daß die Länge des Axioms XXI (eine Version des Axioms von Pasch) nicht viel kleiner ist als die des ganzen hier verwendeten Tarskischen Axiomensystems (mit entsprechender Abänderung der Dimensionsaxiome für die Dimension 3). Bei Hinzunahme von nur einem der kürzeren Axiome von Pieri, z.B. XI, ergibt sich schon eine größere Länge als die des ganzen Tarskischen Systems.

Man könnte vermuten, daß die Kürze des Tarskischen Axiomensystems erreicht wurde durch die Verwendung neuer Axiome, die besonders stark sind und sich kurz und einfach formulieren lassen. Das ist jedoch nicht der Fall. Tatsächlich wurde jedes der Tarskischen Axiome oder eine einfache Variante davon schon als Axiom in einem früheren System in der Literatur benutzt; entsprechende Hinweise werden in 1.7.5 bei den Bemerkungen zu einzelnen Axiomen gegeben.

Nach unserem Eindruck haben zwei Faktoren zur Kürze des Axiomensystems beigetragen. Der erste ist die Auswahl der Zwischenbeziehung und der Streckenkongruenz als Grundbegriffe. Beide Begriffe haben eine klare und einfache geometrische Bedeutung; der erste ist typisch für den affinen und der zweite für den metrischen Aspekt der Geometrie. Die beiden Begriffe haben zusammen eine große Ausdrucksfähigkeit in dem Sinne, daß sich mit ihnen die meisten grundlegenden Gesetze und Definitionen, die beim Aufbau der Geometrie auftreten, auf natürliche und kurze Weise formulieren lassen. Hier ist zu erwähnen, daß auch im Axiomensystem von Veblen 1914 als einzige Grundbegriffe eine Zwischenbeziehung und eine Streckenkongruenz verwendet werden, allerdings im Hilbertschen Sinne, d.h., die Zwischenbeziehung trifft nur zu auf paarweise verschiedene Punkte und die Streckenkongruenz nur auf Strecken mit verschiedenen Endpunkten.

Zur Beschreibung des zweiten Faktors beachte man, daß bei der Formulierung elementargeometrischer Gesetze traditionellerweise gewisse Grenzfälle, auch Trivial- oder Ausartungsfälle genannt, durch einschränkende Voraussetzungen ausgeschlossen werden. So wird z.B. bei der Formulierung des Axioms von Pasch üblicherweise vorausgesetzt, daß die Dreiecksecken (hier a, c, q in A7) nicht kollinear sind, und als Axiom der Streckenabtragung wird üblicherweise A4 mit den Voraussetzungen $q \neq a$ und $b \neq c$ verwendet. Hier wurden solche einschränkenden Voraussetzungen stets weggelassen, soweit sich dadurch an der Gültigkeit des betreffenden Satzes nichts änderte. Dadurch wurden die betreffenden Axiome nicht nur verkürzt, sondern auch verschärft, so daß sich einfachere Gesetze beweisen lassen, die sonst als zusätzliche Axiome benötigt würden oder einen komplizierteren Beweis erforderten. In den genannten Beispielen etwa führt dieses Vorgehen dazu, daß sich die späteren Sätze 3.1, 3.2 und 3.5(1) besonders einfach beweisen lassen.

Das für den elementaren Teil P_2^* der Geometrie angegebene Axiomensystem (und analog das für P_n^*) ist notwendigerweise unendlich (vgl. II.3.33). Damit läßt sich die Summe der Längen aller Axiome in diesem Fall nicht mehr als Maß für die "Einfachheit" des Axiomensystems verwenden. Stattdessen könnte man den Begriff der Länge eines Axiomenschemas (hier A11') verwenden.

1.7.3. Ein ganz anderer und interessanterer Begriff der Einfachheit eines Axiomensystems, der für endliche ebenso wie für unendliche Systeme anwendbar ist, bezieht sich auf die Stellung der Quantoren $\forall$ und $\exists$ in den Axiomen. Damit zusammenhängende Fragen (für verschiedene Geometrien, Axiomensysteme und einzelne Axiome) werden in Abschnitt II.6 ausführlich behandelt. Es sei hier nur darauf hingewiesen, daß die Axiome A1 bis A10 sog. Allheits-Existenz-Aussagen ($\forall\exists$-Aussagen, vgl. II.6.2, 5) sind, d.h., es stehen (nach einer einfachen logischen Umformung) alle Allquantoren ($\forall$) mit den zugehörigen Variablen am Anfang des Axioms und dahinter alle Existenzquantoren ($\exists$). Die elementaren Stetigkeitsaxiome (in A11') sind nicht von dieser Form, jedoch wurde in Tarski 1959, S. 24, darauf hingewiesen, daß man für P_n^* auch ein Axiomensystem aus $\forall\exists$-Aussagen konstruieren kann. Ein (anderer) Beweis dafür (und für ein allgemeineres Resultat, nach Szczerba) wird in II.6 behandelt, s. insbesondere II.6.68, 85. Aus dieser Bauart der Axiome ergeben sich modelltheoretische Konsequenzen für die Klasse aller Modelle (vgl. II.6.10). Außerdem läßt sich der "Präfixtyp" $\forall\exists$ nicht verbessern, d.h., es gibt kein Axiomensystem für P_n^*, das nur aus Allheitsaussagen ($\forall$-Aussagen) oder nur aus Existenzaussagen ($\exists$-Aussagen) besteht (vgl. II.6.13).

1.7.4. Eine weitere Besonderheit des Tarskischen Axiomensystems ist es schließlich, daß man für jede endliche Dimension $n \geq 2$ zwei Dimensionsaxiome (ein unteres und ein oberes) verwenden kann, während die übrigen Axiome für alle Dimensionen dieselben sind. Damit ergibt sich die Frage, ob diese übrigen Axiome für die elementare Geometrie schon ausreichen zum Aufbau der dimensionsfreien Geometrie, welche aus den (elementaren) Sätzen besteht, die für jede endliche Dimension $n \geq 2$ gelten. Die Vermutung, daß das tatsächlich der Fall ist, wurde von Scott 1959 bewiesen für die Standardmodelle $\mathcal{L}_n(\mathbb{R})$ und ergibt sich damit für die vollständigen Geometrien P^*_n, deren dimensionsfreier Teil hier dementsprechend mit P^* bezeichnet wird. Es sei darauf hingewiesen, daß P^* auch Modelle unendlicher Dimension besitzt. Aus dem Resultat von Scott ergibt sich auch, daß man die Dimensionsaxiome weitgehend beliebig wählen kann; darüberhinaus wird gezeigt, daß sich an der Gültigkeit einer beliebigen Aussage α nichts ändert, wenn man nur die Dimension oberhalb einer gewissen Schranke abändert, dabei ist diese Schranke durch die Anzahl der verschiedenen Variablen in α bestimmt. Das Ergebnis von Scott wurde in Gupta 1965, S. 407, ausgedehnt auf die elementare euklidische Geometrie ohne Stetigkeitsschema (hier mit P bezeichnet). Eine weitere Ausdehnung auf die absolute Geometrie A wird hier in Teil II ausgeführt (s. II.3.70ff., insbes. II.3.74, 75), auf andere Ausdehnungen wird hingewiesen (II.3.83, 84).

Für die vollen Geometrien P^2_n ist die obige Frage dagegen negativ zu beantworten - jedenfalls, wenn man für den Begriff dimensionsfreie Geometrie weiterhin nur endliche Dimensionen berücksichtigt - ; hier läßt sich nämlich leicht eine Aussage formulieren, die nur in unendlichdimensionalen Modellen gilt. Andererseits könnte man auf diese Weise formulieren und als Axiom fordern, daß die Dimension endlich ist.

Zur Einbeziehung der Dimension 1 s. II.4.105f., 108f.

1.7.5 Zu einzelnen Axiomen. Zu A5: Die Idee, den Begriff der Winkelkongruenz aus dem Hilbertschen Axiomensystem zu eliminieren (vgl. Bem. zu A5 in 1.2), findet sich bereits bei Mollerup 1904, wo ein "Sechs-Strekken-Axiom" verwendet wird. Ein "Fünf-Strecken-Axiom" (von dem sich unser A5 nur unwesentlich unterscheidet) tritt schon auf in Veblen 1914.

Zu A7: Auf den Zusammenhang mit der klassischen Form des Axioms von Pasch wurde schon in 1.2 hingewiesen. Die dort genannte "äußere Form" AP wurde insbesondere verwendet in den früheren Versionen des Tarskischen Axiomensystems (s. 0.3). Damit folgte Tarski dem Vorgehen bei Schur 1909. Dort wiederum (S. 9) wird die Idee, statt der klassischen eine äußere Form (die im wesentlichen mit AP übereinstimmt) zu verwenden,

E. H. Moore zugeschrieben. Eine Abschwächung von AP wird als Axiom VIII in Veblen 1904 verwendet.

Die Frage, ob sich statt der äußeren auch eine "innere Form" verwenden läßt, wurde von Tarski aufgeworfen und in Gupta 1965 positiv beantwortet. Der Beweis der äußeren Form wird hier im wesentlichen aus Gupta 1965 übernommen und erfolgt verhältnismäßig spät (s. 9.5, 6). Die bei Gupta verwendete innere Form IP unterscheidet sich allerdings von unserem A7 dadurch, daß in den Zwischenbeziehungen die Punkte zum Teil in umgekehrter Reihenfolge angegeben sind. Beide Formen sind damit äquivalent unter Voraussetzung der "Symmetrie der Zwischenbeziehung" (hier Satz 3.2). Um den Darstellungssatz für unser Axiomensystem (16.15) aus dem entsprechenden Resultat von Gupta zu erhalten (vgl. 0.3), genügen also die hier angegebenen Beweise für die Sätze 3.1 und 3.2 (diese Bewiese sind sehr einfach und benutzen keine vorher bewiesenen Sätze).

Zu A8 und A9: Die folgenden beiden Formen eines "oberen Dimensionsaxioms für die ebene Geometrie" erweisen sich (auf Grund der früheren Axiome A1 bis A7) als zu A9 äquivalent.

$$A9^{(1)}: \quad \exists y\{[(\mathrm{B}axy \vee \mathrm{B}xya \vee \mathrm{B}yax) \wedge \mathrm{B}byc] \vee [(\mathrm{B}xyb \vee \mathrm{B}ybx \vee \mathrm{B}bxy) \wedge \mathrm{B}cya] \vee [(\mathrm{B}xyc \vee \mathrm{B}ycx \vee \mathrm{B}cxy) \wedge \mathrm{B}ayb]\}.$$

$$A9^{(2)}: \quad \exists y\{[(\mathrm{B}axy \vee \mathrm{B}xya \vee \mathrm{B}yax) \wedge \mathrm{B}byc] \vee [(\mathrm{B}xyb \vee \mathrm{B}ybx) \wedge \mathrm{B}cya] \vee [(\mathrm{B}xyc \vee \mathrm{B}ycx) \wedge \mathrm{B}ayb]\}.$$

Bei $A9^{(1)}$ beachte man, daß durch die drei Bestandteile in runden Klammern jeweils die Kollinearität der dort betrachteten Punkte ausgedrückt wird. In dem "wesentlichen" Fall, daß die Punkte a, b, c nicht kollinear sind, besagt $A9^{(1)}$ also: Jeder Punkt x liegt auf einer Geraden, die durch eine geeignete Ecke des Dreiecks abc und einen Punkt y der gegenüberliegenden Seite geht. Die verkürzte Form $A9^{(2)}$ besagt dasselbe; im zweiten und dritten Disjunktionsglied (Bestandteil in eckigen Klammern) ist lediglich der schon im ersten Disjunktionsglied erfaßte Fall weggelassen, daß x im Inneren des Dreiecks abc liegt. Durch beide Axiome wird also im wesentlichen Fall ausgedrückt, daß alle Punkte in einer Ebene liegen. Der (weniger interessierende) Fall, daß a, b, c kollinear sind, braucht aber nicht ausgeschlossen zu werden; man überlegt sich leicht, daß dann $A9^{(1)}$ und $A9^{(2)}$ jedenfalls gelten (sogar in Modellen beliebiger Dimension). Insgesamt wird durch diese Axiome also ausgedrückt, daß die

Dimension des ganzen Raumes höchstens 2 ist.

$A9^{(1)}$ wurde von Tarski in den ersten beiden Versionen seines Axiomensystems (s. 0.3) verwendet; diese Form wurde angeregt durch die Überlegungen in Enriques 1911, S. 85-86. $A9^{(1)}$ und $A9^{(2)}$ sind (im Unterschied zu A9) nur mit Hilfe der Zwischenbeziehung formuliert und damit auch geeignet für affine Geometrien (mit Anordnung, vgl. II.2.13(ii)). Tatsächlich wurde eine Variante von $A9^{(2)}$ (wiederum äquivalent auf Grund der Symmetrie der Zwischenbeziehung) für affine Geometrien verwendet in Szczerba u. Tarski 1965 (vgl. II.7.2).

$A9^{(2)}$ ist noch recht einfach im Vergleich zu anderen bekannten Aussagen, die nur mit Hilfe der Zwischenbeziehung ausgedrückt sind. Für höhere Dimension dagegen wird die Formulierung von Dimensionsaxiomen unter alleiniger Benutzung der Zwischenbeziehung recht kompliziert. Es sei daher hier nur verwiesen auf 11.70, wo solche Dimensionsaxiome unter Benutzung zusätzlicher Definitionen eingeführt werden, und zwar ein unteres und ein oberes Dimensionsaxiom Dim_n^- bzw. Dim_n^+ für die Dimension n, welches ausdrückt, daß die Dimension wenigstens bzw. höchstens n ist. Als Grundbegriff wird dafür nur die (hier mit der Zwischenbeziehung ausgedrückte) Kollinearität benötigt. Damit sind auch diese Axiome für verschiedene affine Geometrien geeignet. Für $n \geq 3$ werden diese Axiome auch im Sinne von 1.7.3 komplizierter; in Kordos 1969 wurde nämlich gezeigt, daß sich dann das untere Dimensionsaxiom nicht als Allheits-Existenz-Aussage mit dem einzigen Grundbegriff B formulieren läßt (s.a. II.6.43 bis 50).

Damit wird es nahegelegt, für beliebige Dimension n auch Dimensionsaxiome zu betrachten, in denen die Streckenkongruenz als Grundbegriff auftritt. Solche Axiome wurden verwendet in Gupta 1965; das untere Dimensionsaxiom für die Dimension n drückt dabei die Existenz eines n-dimensionalen kartesischen Koordinatensystems aus. Daraus entstehen durch eine technische Abänderung (möglichst geringe Benutzung der Rechtwinkel-Relation) die folgenden Dimensionsaxiome.

dim_n^- (unteres Dimensionsaxiom für die Dimension n):

$$\exists s \exists u_1' \exists u_1 \ldots \exists u_n [s \neq u_1' \wedge B u_1 s u_1' \wedge \bigwedge_{\nu=1}^{n} s u_\nu \equiv s u_1' \wedge \bigwedge_{1 \leq \mu < \nu \leq n} u_\mu u_\nu \equiv u_1' u_2] \qquad (n \geq 1).$$

(Es gibt ein n-dimensionales kartesisches Koordinatensystem [mit dem Ursprung s und den Einheitspunkten u_ν auf den Achsen].)

$\dim_n^+$ (oberes Dimensionsaxiom für die Dimension n): $\neg\dim_{n+1}^-$.

Das sind die (unter den Gesichtspunkten von 1.7.2 und 1.7.3) einfachsten uns bisher bekannten Dimensionsaxiome in den Grundbegriffen D und B: $\dim_n^-$ ist eine Existenzaussage mit genau $n+2$ paarweise verschiedenen Variablen, $\dim_n^+$ ist dementsprechend eine Allheitsaussage mit $n+3$ Variablen. Die Gleichwertigkeit von $\dim_n^-$ mit Dim_n^- (und damit auch die von $\dim_n^+$ mit Dim_n^+) auf Grund der früheren Axiome A1 bis A7 ergibt sich aus den Sätzen 16.2 bis 16.4.

Auf das Resultat von Scott, nach dem man die Dimensionsaxiome weitgehend beliebig wählen kann, wurde schon in 1.7.4 hingewiesen. Aus diesem Resultat ergibt sich insbesondere, daß jede zu $\dim_n^-$ äquivalente Aussage nicht weniger als $n+1$ paarweise verschiedene Punktvariablen enthalten kann, und zwar unabhängig vom Präfixtyp und unabhängig von der Wahl der geometrischen Grundbegriffe in einem sinnvollen Rahmen.

Zu A10: Bei Tarski 1959 und Gupta 1965 wird als Euklidisches Axiom eine Version verwendet, die sich von unserem A10 nur dadurch unterscheidet, daß in einer der Zwischenbeziehungen die Punkte in umgekehrter Reihenfolge angegeben sind. Dafür gilt das zu A7 Gesagte analog.

Die in 1.2 hinter A10 angegebene geometrische Feststellung tritt schon auf in Lorenz 1791, §44, S. 102. Sie wird dort zwar nicht zu den Axiomen gerechnet, sondern aus anschaulichen Vorstellungen gefolgert; aus ihr wird aber tatsächlich das Parallelenaxiom bewiesen. Zu weiteren zum Parallelenaxiom äquivalenten Forderungen sowie zur Geschichte des Parallelenaxioms s.a. Engel u. Stäckel 1895 und Baldus 1964.

Hier seien noch die folgenden zwei (auf Grund von A1 bis A8 bzw. A1 bis A8 und A11) zu A10 äquivalenten Forderungen erwähnt, in denen die Streckenkongruenz eine wesentliche Rolle spielt (während A10 nur mit Hilfe der Zwischenbeziehung formuliert ist).

A10$^{(1)}$: $\mathrm{B}abc \vee \mathrm{B}bca \vee \mathrm{B}cab \vee \exists x(ax\equiv bx \wedge ax\equiv cx)$.

A10$^{(2)}$: $\mathrm{B}adb \wedge da\equiv db \wedge \mathrm{B}aec \wedge ea\equiv ec \wedge \mathrm{B}def \wedge ed\equiv ef \rightarrow df\equiv bc$.

A10$^{(1)}$ besagt (für die Dimension 2) im wesentlichen: In einem nichtentarteten Dreieck schneiden sich zwei (gewählte) Mittelsenkrechten. Anders ausgedrückt: Ein nicht-entartetes Dreieck kann einem Kreis einbeschrieben werden. (Ohne Dimensionsvoraussetzung kann man eine entsprechende Verschärfung - d.h. mit der zusätzlichen Bedingung, daß x mit a, b, c komplanar ist - aus $A10^{(1)}$ folgern.) Aus dieser Forderung wird schon bei Bolyai 1832 das Parallelenaxiom bewiesen. $A10^{(2)}$ besagt, daß die Verbindungsstrecke der Mitten zweier Seiten eines Dreiecks halb so lang ist wie die dritte Seite. Das erweist sich als äquivalent zur Forderung, daß die Summe der Innenwinkel eines beliebigen Dreiecks gleich zwei Rechten ist. Ein Beweis des Parallelenaxioms daraus bei Legendre 1833 geht zurück auf Resultate von Saccheri 1733, s.a. II.6.28(iii).

Setzt man nicht nur die früheren Axiome A1 bis A9, sondern auch das Schema A11' der Stetigkeitsaxiome voraus, so wird die darauf beruhende absolute Geometrie A_2^* am Parallelenaxiom "gabelbar" (s. II.3.80). Damit läßt sich A10 sogar äquivalent ersetzen durch jede Aussage, die mit, aber nicht ohne A10 beweisbar ist.

Im Unterschied zu den anderen genannten Versionen ist $A10^{(2)}$ eine "Allheitsaussage", oder (mit anderen Worten) von dem besonders einfachen Präfixtyp $\forall$ im Sinne von II.6.2 (s.a. 1.7.3), d.h., bis auf die Allquantoren am Anfang (die hier nach 0.5 weggelassen sind) treten keine weiteren Quantoren auf. Damit erhebt sich die Frage, ob es auch eine zu A10 äquivalente Aussage dieses Präfixtyps gibt, die nur mit Hilfe der Zwischenbeziehung formuliert ist. In II.6.42 wird gezeigt, daß das nicht der Fall ist.

Zu A11: Auf den Zusammenhang von A11 mit dem Axiom vom Dedekindschen Schnitt (Dedekind 1872) für die reellen Zahlen und auf die für die Geometrie vorgenommenen Modifikationen wurde schon in 1.2 eingegangen. Die damit erreichte Version scheint einfacher zu sein als alle Formen von Stetigkeitsaxiomen, die sonst für die Geometrie in der Literatur benutzt werden.

An Stelle des Schemas A11' genügt ein entsprechendes Schema A11" in der Sprache L(B), d.h., daß die Streckenkongruenz D nicht als Grundbegriff in den Formeln $\alpha(x)$, $\beta(y)$ verwendet wird. Das ergibt sich aus dem Beweis von 1.6(vi) (s.a. II.6.86(iii)).

1.7.6 Unabhängigkeitsfragen. Ein Axiom α eines Axiomensystems Σ heißt bekanntlich *unabhängig* (in Σ), falls es nicht aus den übrigen Axiomen von Σ folgt. Der Beweis für eine solche Unabhängigkeit geschieht im allgemeinen durch Konstruktion eines Modells, in dem die übrigen Axiome, aber nicht α gelten. Ein Axiomensystem Σ heißt *unabhängig*, falls jedes seiner Axiome unabhängig (in Σ) ist. Im jeweils anderen Fall heißt α bzw. Σ *abhängig*.

Für die frühere Version des Tarskischen Axiomensystems in Tarski 1959 und Teile davon wurden Unabhängigkeitsfragen eingehend untersucht in Gupta 1965. Dabei blieb für nur zwei Axiome die Frage offen, ob sie unabhängig sind. Außerdem wurde dort eine weitere Version des Axiomensystems konstruiert, die unabhängig ist.

Für unser Axiomensystem ergibt sich aus Satz 1.6 ohne weiteres die Unabhängigkeit der Axiome A4, A8, A9, A11. Aus der Existenz eines Modells der hyperbolischen Geometrie (s. etwa II.2.4f.) ergibt sich die Unabhängigkeit von A10. Mit Hilfe der bei Gupta 1965 (S. 41ff.) konstruierten Modelle erhält man auch für unser Axiomensystem die Unabhängigkeit von A5 und von A6.

Die Unabhängigkeit der äußeren Form AP des Axioms von Pasch in einer geeigneten Version des Axiomensystems wurde gezeigt in Szczerba 1970 und gab Anlaß zur Einführung der sog. Pasch-freien Geometrie (s. auch II.8.2ff., II.8.7(ii), 8).

In einem anderen Sinne wird das Wort "unabhängig" bisweilen für Begriffe in einem System von Grundbegriffen (wie hier D und B) gebraucht. Die Abhängigkeit eines Begriffs ist dann dasselbe wie seine Definierbarkeit mit Hilfe der übrigen Begriffe des Systems im Sinne von II.4.2. Diesbezügliche Fragen werden ausführlich in Abschnitt II.4 behandelt. Im System der Grundbegriffe D und B insbesondere ist D unabhängig und B abhängig; die Abhängigkeit geht jedoch verloren bei Weglassen der Stetigkeitsaxiome (s. II.4.40, 24, 31).

2. Folgerungen aus A1 bis A5

Aus A1 und A2 erhält man leicht, daß die Streckenkongruenz (aufgefaßt als zweistellige Relation zwischen Paaren von Punkten) eine Äquivalenzrelation ist, d.h., daß die folgenden Sätze gelten:

2.1 Satz (*Reflexivität von* ≡). $ab \equiv ab$.

2.2 Satz (*Symmetrie von* ≡). $ab \equiv cd \rightarrow cd \equiv ab$.

2.3 Satz (*Transitivität von* ≡). $ab \equiv cd \wedge cd \equiv ef \rightarrow ab \equiv ef$.

Setzt man nämlich für beliebige Punkte a, b in A2 speziell $p=r=b$, $q=s=a$, so ist die Voraussetzung nach A1 erfüllt, und man erhält $ba \equiv ba$, also 2.1 (mit anderen Variablen). 2.2 ergibt sich durch Anwendung von A2 auf die beiden Kongruenzen $ab \equiv cd$ (Voraussetzung) und $ab \equiv ab$ (2.1). 2.3 ergibt sich dann mittels 2.2 aus A2. □

Aus A1 und 2.3 ergibt sich außerdem, daß die Streckenkongruenz von der Reihenfolge der Punkte in einem Paar unabhängig ist, d.h., daß folgende Sätze gelten:

2.4 Satz. $ab \equiv cd \rightarrow ba \equiv cd$.

2.5 Satz. $ab \equiv cd \rightarrow ab \equiv dc$.

Strecken definieren wir im wesentlichen nach Hilbert 1977.

2.6 Definition. Unter einer *Strecke* verstehen wir ein ungeordnetes Paar $\{a,b\}$ von Punkten, das wir auch mit ab oder ba bezeichnen; a und b nennen wir dann die *Endpunkte* der Strecke ab, und ab nennen wir die *Verbindungsstrecke* der Punkte a und b. Die Strecken der Form aa (mit zusammenfallenden Endpunkten) nennen wir *Nullstrecken* und die anderen *echte* oder *nicht-entartete Strecken*.

Die Sätze 2.1 bis 2.5 können wir dann kurz zusammenfassen in

2.7 Satz. *Die Streckenkongruenz ist eine Äquivalenzrelation zwischen Strecken.*

Umformungen von Streckenkongruenzen, die nur darauf beruhen (insbesondere auf den Sätzen 2.2, 2.4, 2.5), werden wir in weiteren Beweisen im allgemeinen nicht besonders erwähnen.

2.8 Satz. $aa \equiv bb$.
(Nullstrecken sind stets untereinander kongruent.)

Beweis: Sei x gemäß A4 (Streckenabtragung) ein Punkt mit $\mathrm{B}bax \wedge ax \equiv bb$. Nach A3 ist dann $x=a$, also $aa \equiv bb$. □

2.9 Anmerkung. Hier wird zum ersten Mal eine Beweismethode verwendet, die in diesem Aufbau der Geometrie besonders häufig vorkommt: Um zu zeigen, daß ein gegebener Punkt (hier a) eine gewisse Eigenschaft besitzt, wird zunächst auf Grund eines Existenzaxioms oder -satzes ein weiterer Punkt mit geeigneten Eigenschaften (hier x) eingeführt, von dem man nachträglich zeigen kann, daß er mit dem gegebenen Punkt zusammenfällt.

Zur besseren Übersicht bei Anwendungen des Fünf-Strecken-Axioms (A5) wollen wir folgende Abkürzung einführen.

2.10 Definition. Die Punkte a, b, c, d, a', b', c', d' bilden eine *äußere Fünf-Strecken-Konfiguration*, abgekürzt:

$$\mathrm{AFS}\begin{pmatrix} a & b & c & d \\ a' & b' & c' & d' \end{pmatrix} :\leftrightarrow \mathrm{B}abc \wedge \mathrm{B}a'b'c' \wedge ab \equiv a'b' \wedge bc \equiv b'c' \wedge ad \equiv a'd' \wedge bd \equiv b'd'.$$

Dann läßt sich das Fünf-Strecken-Axiom offenbar auch folgendermaßen formulieren:

$$\mathrm{A5:}\ \mathrm{AFS}\begin{pmatrix} a & b & c & d \\ a' & b' & c' & d' \end{pmatrix} \wedge a \neq b \rightarrow cd \equiv c'd'.$$

Anmerkung. Axiom A5 erlaubt, auf die Kongruenz der "äußeren Strecken" cd und $c'd'$ (vgl. Abb.2) zu schließen, während später (Satz 4.2) auf die Kongruenz der "inneren Strecken" bd und $b'd'$ geschlossen wird; daher die Bezeichnung.

<u>2.11 Satz</u> ("*Aneinanderlegen von Strecken*").

$$\mathrm{B}abc \wedge \mathrm{B}a'b'c' \wedge ab \equiv a'b' \wedge bc \equiv b'c' \rightarrow ac \equiv a'c'.$$

<u>Beweis</u>: Aus der Voraussetzung und den bisherigen Sätzen ergibt sich zunächst

$$\mathrm{AFS}\begin{pmatrix} a & b & c & a \\ a' & b' & c' & a' \end{pmatrix}.$$

Ist nun noch $a \neq b$, so ergibt sich die Behauptung sofort aus A5. Ist dagegen $a=b$, so ergibt sich aus den vorausgesetzten Kongruenzen mittels A3 auch $a'=b'$ und damit die Behauptung. □

<u>2.12 Satz</u> (*Eindeutigkeit der Streckenabtragung*).

$$q \neq a \rightarrow \overset{\leq 1}{\exists} x(\mathrm{B}qax \wedge ax \equiv bc).$$

<u>Beweis</u>: Seien x und x' Punkte mit der angegebenen Eigenschaft. Nach 2.11 ist dann $qx \equiv qx'$, nach 2.7 außerdem $ax \equiv ax'$ und damit

$$\mathrm{AFS}\begin{pmatrix} qaxx \\ qaxx' \end{pmatrix}.$$

Somit ist $xx \equiv xx'$, wegen A3 also $x = x'$. □

3. Einfache Sätze über die Zwischenbeziehung

3.1 Satz. *Es ist stets* $Babb$.

Beweis: Sei x gemäß A4 ein Punkt mit $Babx \wedge bx \equiv bb$. Nach A3 ist dann $b=x$, also $Babb$. □

3.2 Satz (*Symmetrie der Zwischenbeziehung*). $Babc \rightarrow Bcba$.

Beweis: Durch Anwendung des Axioms von Pasch (A7) auf $Babc$ (Vorauss.) und $Bbcc$ (3.1) erhalten wir einen Punkt x mit $Bbxb \wedge Bcxa$. Nach A6 ist dann $x=b$, also $Bcba$. □

Diese "Symmetrie" der dreistelligen Relation B kommt schon in der anschaulichen Bedeutung des Wortes "zwischen" zum Ausdruck. Nachdem sie einmal gezeigt ist, werden wir die Anwendung von 3.2 in weiteren Beweisen i.a. nicht besonders erwähnen. Als Beispiel sei nur als Folgerung aus 3.1 angeführt:

3.3 Satz. *Es ist stets* $Baab$.

3.4 Satz. $Babc \wedge Bbac \rightarrow a=b$.

Beweis: Durch Anwendung von A7 auf die Voraussetzung erhalten wir einen Punkt x mit $Bbxb \wedge Baxa$. Nach A6 ist dann $x=b$ und $x=a$, also $a=b$. □

3.5 Satz. $Babd \wedge Bbcd \rightarrow Babc \wedge Bacd$.

3.6 Satz. $Babc \wedge Bacd \rightarrow Bbcd \wedge Babd$.

3.7 Satz. $Babc \wedge Bbcd \wedge b \neq c \rightarrow Bacd \wedge Babd$.

<u>Beweise von 3.5 bis 3.7</u>: Mit (1) bzw. (2) sei derjenige Teil bezeichnet, der aus dem jeweiligen Satz durch Weglassen des zweiten bzw. ersten Konjunktionsglieds in der Behauptung entsteht.

Zu 3.5(1): Durch Anwendung von A7 auf die Voraussetzung erhalten wir einen Punkt x mit $\mathrm{B}xb \wedge \mathrm{B}cxa$. Nach A6 ist dann $x=b$, also $\mathrm{B}cba$ und damit $\mathrm{B}abc$.

Zu 3.6(1): Das ergibt sich mittels 3.2 aus 3.5(1) (mit anderen Variablen).

Zu 3.7(1): Sei x gemäß A4 ein Punkt mit $\mathrm{B}acx \wedge cx \equiv cd$. Nach 3.6(1) ist dann $\mathrm{B}bcx$. Auf Grund der Eindeutigkeit der Streckenabtragung (2.12) ergibt sich nun $x=d$ und damit die Behauptung.

Zu 3.5(2): Das ergibt sich im Falle $b=c$ aus der Voraussetzung, im Falle $b \neq c$ aus 3.5(1) und 3.7(1).

Zu 3.6(2), 3.7(2): Das ergibt sich mittels 3.2 aus 3.5(2) bzw. 3.7(1).

□

<u>Anmerkung</u>. Die vorausgesetzte Verschiedenheit der Punkte ist in 3.7 (und damit auch für die Eindeutigkeit der Streckenabtragung) nicht entbehrlich, wie sich etwa aus dem Beispiel $d=a \neq b=c$ ergibt.

Die Sätze 3.5 bis 3.7 erfassen alle Fälle, in denen man aus zwei der vier Zwischenbeziehungen $\mathrm{B}abc$, $\mathrm{B}abd$, $\mathrm{B}acd$, $\mathrm{B}bcd$ auf die anderen beiden schließen kann. (Zu anderen Fällen, in denen man nur auf Disjunktionen schließen kann, vgl. die Sätze 5.1 bis 5.3.) Gelten alle vier, so bedeutet das anschaulich, daß die Punkte a, b, c, d in dieser Reihenfolge auf einer Geraden angeordnet sind (Geraden werden erst in Abschnitt 6 eingeführt). Allgemeiner kann man die Anordnung von n Punkten a_1 bis a_n (auf einer Geraden) durch eine *verallgemeinerte Zwischenbeziehung* B_n folgendermaßen beschreiben:

<u>3.8 Definition</u>. $\mathrm{B}_n a_1 a_2 \ldots a_n :\leftrightarrow \bigwedge_{1 \leq i<j<k \leq n} \mathrm{B}a_i a_j a_k \qquad (n \geq 3).$

Aus den bisherigen Sätzen erhält man dann leicht die folgenden vier Sätze über B_n.

<u>3.9 Satz</u>. $\mathrm{B}_n a_1 a_2 \ldots a_n \rightarrow \mathrm{B}_n a_n a_{n-1} \ldots a_1$.

<u>3.10 Satz</u>. *Wenn* $1 \leq l_1 < l_2 < \ldots < l_s \leq n$, *so*

$$B_n a_1 \ldots a_n \rightarrow B_s a_{l_1} a_{l_2} \ldots a_{l_s}.$$

<u>3.11 Satz</u>. $B_n a_1 \ldots a_n \wedge B a_l p a_{l+1} \rightarrow B_{n+1} a_1 \ldots a_l p a_{l+1} \ldots a_n \qquad (1 \leq l < n).$

<u>3.12 Satz</u>. $B_n a_1 \ldots a_n \wedge B a_l a_n p$

$$\rightarrow B_{n-l+2} a_l \ldots a_n p \wedge [a_l \neq a_n \rightarrow B_{n+1} a_1 \ldots a_n p] \qquad (1 \leq l < n).$$

<u>3.13 Satz</u>. $\overset{\geq 2}{\exists}$. (*Es gibt wenigstens zwei Punkte.*)

<u>Beweis</u>: Seien *a*, *b*, *c* gemäß dem unteren Dimensionsaxiom (A8) gewählt. Dann sind *a*, *b*, *c* paarweise verschieden, da sonst nach 3.1 eine der in A8 genannten Zwischenbeziehungen gelten müßte. □

<u>Anmerkung</u>. Das untere Dimensionsaxiom A8 wird hier zum ersten Male verwendet. Es ist für diesen Satz nicht entbehrlich, da alle anderen Axiome auch im einpunktigen Raum $\mathfrak{L}_0(\mathbb{R})$ gelten. An Stelle von A8 genügt jedoch das (mit 3.13 gleichwertige) untere Dimensionsaxiom für die Dimension 1(s. 11.70, 9.46(2) und 1.7.5 [Bem. zu A8 und A9]).

<u>3.14 Satz</u>. $\exists c[Babc \wedge b \neq c]$.

<u>Beweis</u>: Seien *u*, *v* gemäß 3.13 verschiedene Punkte und *c* gemäß A4 ein Punkt mit $Babc \wedge bc \equiv uv$. Dann leistet *c* (wegen A3) das Verlangte. □

Ausgehend von zwei verschiedenen Punkten a_1, a_2 erhalten wir mittels 3.14 durch Induktion über *n*:

<u>3.15 Satz</u>. $a_1 \neq a_2 \rightarrow \exists a_3 \ldots \exists a_n [B_n a_1 \ldots a_n \wedge \neq(a_1 \ldots a_n)] \qquad (n \geq 3).$

Insbesondere gibt es also unendlich viele Punkte. Zum Beweis dafür wurde jedoch statt A8 nur der Satz 3.13 benötigt; über Modelle erhalten wir somit folgendes Resultat.

<u>3.16 Theorem</u>. *Wenn* $\mathfrak{M} = \langle M, D, B \rangle$ *Modell von* A1 *bis* A7 *und* $|M| \geq 2$ *ist, so ist* *M unendlich.*

Geometrien mit endlich vielen Punkten (außer der trivialen einpunktigen) werden also bereits durch unsere Axiome A1 bis A7 ausgeschlossen.

3.17 Satz. $Babc \wedge Ba'b'c \wedge Bapa'$

$\rightarrow \exists q[Bpqc \wedge Bbqb']$ (Abb. 9).

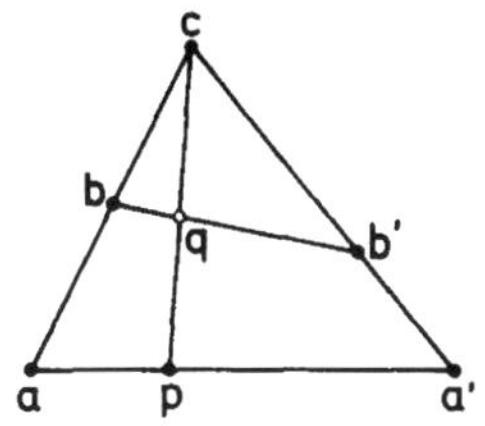

Abb.9

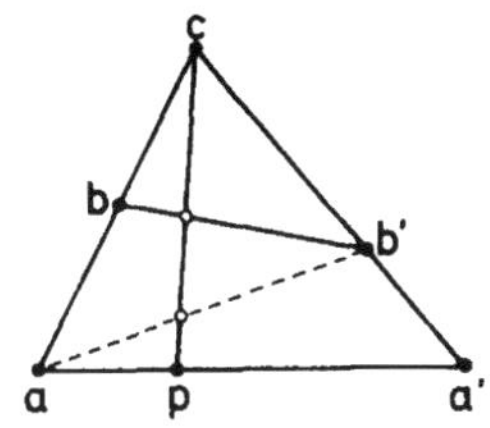

Abb.10

Beweis: Das erhält man leicht durch zweimalige Anwendung des Axioms von Pasch (A7) (unter Verwendung von 3.5, vgl. Abb. 10). □

4. Einfache Sätze über Kongruenz und Zwischenbeziehung

4.1 Definition. Die Punkte a, b, c, d, a', b', c', d' bilden eine *innere Fünf-Strecken-Konfiguration:*

$$\mathrm{IFS}\begin{pmatrix} a & b & c & d \\ a' & b' & c' & d' \end{pmatrix} :\leftrightarrow \mathrm{B}abc \wedge \mathrm{B}a'b'c' \wedge ac\equiv a'c' \wedge bc\equiv b'c' \wedge ad\equiv a'd' \wedge cd\equiv c'd'.$$

In den geforderten Kongruenzen ist hier also gegenüber der Definition 2.10 b mit c und b' mit c' vertauscht.

4.2 Satz. $\mathrm{IFS}\begin{pmatrix} a & b & c & d \\ a' & b' & c' & d' \end{pmatrix} \rightarrow bd\equiv b'd'.$

Beweis: Falls $a=c$ ist, ist nach A3 auch $a'=c'$, nach A6 also $b=c$ und $b'=c'$; damit ergibt sich die Behauptung in diesem Falle unmittelbar aus der Voraussetzung. Wir betrachten nun den Fall, daß $a\neq c$ ist. Sei e gemäß 3.14 ein Punkt mit $\mathrm{B}ace \wedge c\neq e$ und e' gemäß A4 ein Punkt mit

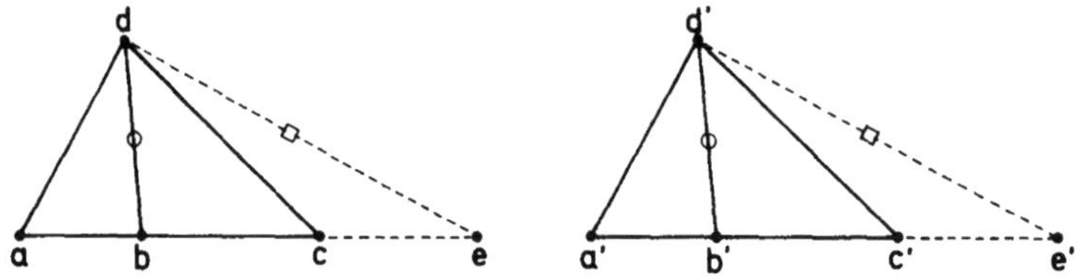

Abb. 11

$\mathrm{B}a'c'e' \wedge c'e'\equiv ce$ (Abb. 11). Dann ist

$$\mathrm{AFS}\begin{pmatrix} a & c & e & d \\ a' & c' & e' & d' \end{pmatrix},$$

nach A5 also $ed\equiv e'd'$.

Mit 3.6 erhalten wir weiter

$$\mathrm{AFS}\begin{pmatrix} e & c & b & d \\ e' & c' & b' & d' \end{pmatrix}.$$

Nochmalige Anwendung von A5 liefert dann die Behauptung $bd\equiv b'd'$. □

Analog zu 2.11 erhalten wir damit

<u>4.3 Satz</u> ("*Ineinanderlegen von Strecken*").

$$\mathrm{B}abc \wedge \mathrm{B}a'b'c' \wedge ac\equiv a'c' \wedge bc\equiv b'c' \to ab\equiv a'b'.$$

Zur bequemeren Beschreibung mehrerer Streckenkongruenzen führen wir noch die Kongruenz von "geometrischen Figuren" (aus endlich vielen Punkten) ein.

<u>4.4 Definition</u>. Die *n-tupel* $\langle a_1,\ldots,a_n\rangle$ und $\langle b_1,\ldots,b_n\rangle$ von Punkten sind zueinander *kongruent*:

$$(a_1\ldots a_n)\equiv(b_1\ldots b_n) :\leftrightarrow \bigwedge_{1\le i<j\le n} a_ia_j\equiv b_ib_j.$$

(Nach 2.7 und 2.8 läßt sich die Bedingung in der Konjunktion gleichwertig durch $1\le i,j\le n$ ersetzen.)

Diese Kongruenz ist offenbar eine Äquivalenzrelation zwischen n-tupeln von Punkten. Wir verwenden sie im folgenden vor allem für "*geordnete Dreiecke*" (Tripel von Punkten).

<u>4.5 Satz</u>. $\mathrm{B}abc \wedge ac\equiv a'c' \to \exists b'[\mathrm{B}a'b'c' \wedge (abc)\equiv(a'b'c')]$.

(*Ein Punkt zwischen a und c läßt sich in eine zu ac kongruente Strecke stets "kongruent übertragen".*)

<u>Beweis</u>: Um b' durch Streckenabtragung erhalten zu können, wählen wir zunächst nach 3.14 einen Punkt d' mit $\mathrm{B}c'a'd' \wedge a'\neq d'$ (Abb. 12). Gemäß A4 sei dann b' ein Punkt mit $\mathrm{B}d'a'b' \wedge a'b'\equiv ab$ und c'' ein Punkt mit

Abb. 12

$\mathrm{B}d'b'c'' \wedge b'c''\equiv bc$. Nach 3.6 ist dann $\mathrm{B}a'b'c'' \wedge \mathrm{B}d'a'c''$. Nach 2.11 ("Aneinanderlegen") ist auch $ac\equiv a'c''$ und nach 2.12 (Eindeutigkeit der Streckenabtragung) somit $c''=c'$. Also leistet b' das Verlangte. □

<u>4.6 Satz</u>. $\mathrm{B}abc \wedge (abc) \equiv (a'b'c') \rightarrow \mathrm{B}a'b'c'$.

<u>Beweis</u>: Sei b'' gemäß 4.5 ein Punkt mit $\mathrm{B}a'b''c' \wedge (abc) \equiv (a'b''c')$. Nach Voraussetzung ist dann $(a'b''c') \equiv (a'b'c')$ und damit

$$\mathrm{IFS}\begin{pmatrix} a'b''c'b'' \\ a'b''c'b' \end{pmatrix}.$$

Nach 4.2 ist nun $b''b'' \equiv b''b'$, nach A3 also $b''=b'$ und damit $\mathrm{B}a'b'c'$. □

Der Inhalt von 4.6 läßt sich offenbar auch so ausdrücken:

<u>4.7 Satz</u>. *Ist φ eine eindeutige Abbildung von einer (endlichen oder unendlichen) Menge N von Punkten in die Menge der Punkte, die "Längen" erhält (d.h. hier, daß jede Verbindungsstrecke von Punkten von N in eine dazu kongruente Strecke übergeführt wird), so bleibt unter φ auch die Zwischenbeziehung erhalten.*

<u>4.8 Definition</u>. Abbildungen der in 4.7 genannten Art werden auch *Isometrien* genannt; Isometrien des ganzen Raumes (d.h. der Menge aller Punkte) auf sich heißen *Bewegungen*.

Trivialerweise bleibt unter einer Isometrie auch die vierstellige Relation D der Streckenkongruenz erhalten. Damit haben wir

<u>4.9 Satz</u>. *Jede Bewegung ist ein Automorphismus des ganzen Raumes.*

Beispiele für Bewegungen sind die Punkt- und die Geradenspiegelungen, die später behandelt werden (Abschnitt 7 bzw. 10).

<u>4.10 Definition</u>. Die Punkte a, b, c sind *kollinear*:

$$\mathrm{Col}\, abc :\leftrightarrow \mathrm{B}abc \vee \mathrm{B}bca \vee \mathrm{B}cab.$$

Die Kollinearität von Punkten a, b, c bedeutet natürlich anschaulich, daß diese Punkte auf einer Geraden liegen. Wir werden jedoch umgekehrt Geraden erst später mit Hilfe des Begriffs der Kollinearität einführen (s. 6.14).

Aus 4.10 und 3.2 ergibt sich

<u>4.11 Satz</u>. $\mathrm{Col}\, abc \rightarrow \mathrm{Col}\, bca \wedge \mathrm{Col}\, cab \wedge \mathrm{Col}\, cba \wedge \mathrm{Col}\, bac \wedge \mathrm{Col}\, acb$.
(*Die Relation* Col *ist invariant gegenüber Permutationen.*)

Aus 3.3 ergibt sich sofort

4.12 Satz. Col aab .

Aus 4.6 erhält man durch Unterscheidung der Fälle in 4.10

4.13 Satz. Col $abc \wedge (abc) \equiv (a'b'c') \to$ Col $a'b'c'$.

4.14 Satz. Col $abc \wedge ab \equiv a'b' \to \exists c' (abc) \equiv (a'b'c')$.

Beweis: Falls Babc ist, sei c' gemäß A4 ein Punkt mit B$a'b'c' \wedge b'c' \equiv bc$; nach 2.11 ist dann auch $ac \equiv a'c'$, also leistet c' das Verlangte. Der Fall Bbac läßt sich durch Vertauschung von a mit b auf den ersten Fall zurückführen. Falls schließlich Bacb ist, ergibt sich die Behauptung nach 4.5. □

Aus jeder der beiden Bedingungen

$$\text{AFS}\begin{pmatrix} a & b & c & d \\ a' & b' & c' & d' \end{pmatrix}, \qquad \text{IFS}\begin{pmatrix} a & c & b & d \\ a' & c' & b' & d' \end{pmatrix}$$

ergibt sich nach 2.11 bzw. 4.3 die Kongruenz $(abc) \equiv (a'b'c')$. Damit liegt es nahe, den folgenden Oberbegriff einzuführen.

4.15 Definition. Die Punkte a, b, c, d, a', b', c', d' bilden eine *Fünf-Strecken-Konfiguration*:

$$\text{FS}\begin{pmatrix} a & b & c & d \\ a' & b' & c' & d' \end{pmatrix} :\leftrightarrow \text{Col}\ abc \wedge (abc) \equiv (a'b'c') \wedge ad \equiv a'd' \wedge bd \equiv b'd'.$$

4.16 Satz. $\text{FS}\begin{pmatrix} a & b & c & d \\ a' & b' & c' & d' \end{pmatrix} \wedge a \neq b \to cd \equiv c'd'$.

Beweis: In jedem der möglichen Fälle Babc, Bbca, Bbca gilt nach 4.6 die entsprechende Zwischenbeziehung auch für die unteren Punkte; somit ergibt sich die Behauptung nach den Sätzen A5 bzw. 4.2 (die im wesentlichen in 4.16 zusammengefaßt werden). □

4.17 Satz. $a \neq b \wedge$ Col $abc \wedge ap \equiv aq \wedge bp \equiv bq \to cp \equiv cq$ (Abb. 13).

Beweis: Aus der Voraussetzung erhalten wir

$$\text{FS}\begin{pmatrix} abcp \\ abcq \end{pmatrix},$$

woraus sich mittels 4.16 die Behauptung ergibt. □

Anmerkung. Solange das obere Dimensionsaxiom (A9) nicht vorausgesetzt

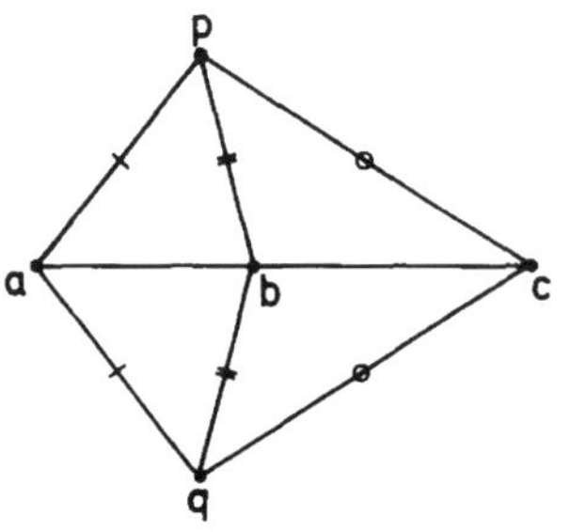

Abb. 13

wird, brauchen hierbei die Punkte a, b, c, p, q - anschaulich gesagt - nicht in einer Ebene zu liegen; die durch a und b bestimmte Gerade braucht also nicht das "Mittellot" zu p und q zu sein, sie ist vielmehr nur enthalten in der "mittelsenkrechten Hyperebene" zu p und q (falls $p \neq q$ vorausgesetzt wird). Für die kartesischen Räume $\mathcal{C}_n(\mathcal{F})$ läßt sich diese Anschauung natürlich mit bekannten Mitteln präzisieren. Für beliebige Modelle von A1 bis A8 ist das später ebenfalls möglich, wenn genügend Sätze zur Verfügung stehen (s. 11.65).

4.18 Satz. $a \neq b \wedge \mathrm{Col}\, abc \wedge ac \equiv ac' \wedge bc \equiv bc' \rightarrow c = c'$.

Dieser Satz besagt, daß ein mit verschiedenen Punkten a, b kollinearer Punkt durch seine Abstände von diesen Punkten eindeutig festgelegt ist. (Vgl. Abb. 14, der Satz besagt, daß die hypothetisch gestrichelten Linien mit der ausgezogenen Linie zusammenfallen.)

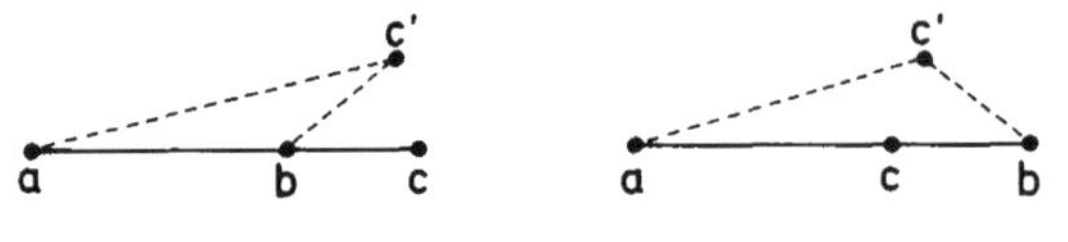

Abb. 14

Beweis von 4.18: Aus 4.17 mit $p=c$, $q=c'$ ergibt sich $cc \equiv cc'$, nach A3 also $c=c'$. □

Im speziellen Falle $\mathrm{B}acb$ ist die Voraussetzung $a \neq b$ natürlich entbehrlich; falls nämlich $a=b$ ist, fällt nach A6 der Punkt c und nach A3 auch c' mit diesen Punkten zusammen. Damit erhalten wir noch

4.19 Satz. $\mathrm{B}acb \wedge ac \equiv ac' \wedge bc \equiv bc' \rightarrow c = c'$.

5. Konnexität der Zwischenbeziehung und Streckenvergleich

In Ergänzung zu den "linearen Anordnungssätzen" 3.5 bis 3.7 wollen wir jetzt den folgenden Satz beweisen.

5.1 Satz. $a \neq b \wedge \mathrm{B}abc \wedge \mathrm{B}abd \rightarrow \mathrm{B}acd \vee \mathrm{B}adc$.

Anmerkung. In älteren Fassungen des Tarskischen Axiomensystems trat dieser Satz als zusätzliches Axiom auf. Der Beweis geht auf H.N. Gupta zurück (s. Gupta 1965, S. 35 ff., 47 f.).

Beweis von 5.1: Auf Grund des Axioms der Streckenabtragung wählen wir Punkte c', d', so daß

$$\mathrm{B}adc' \wedge dc' \equiv cd,\quad \mathrm{B}acd' \wedge cd' \equiv cd.$$

Dann genügt es offenbar zu zeigen, daß $c=c'$ oder $d=d'$ ist.
Ebenfalls durch Streckenabtragung erhalten wir Punkte b', b'', so daß

$$\mathrm{B}ac'b' \wedge c'b' \equiv cb,\quad \mathrm{B}ad'b'' \wedge d'b'' \equiv db.$$

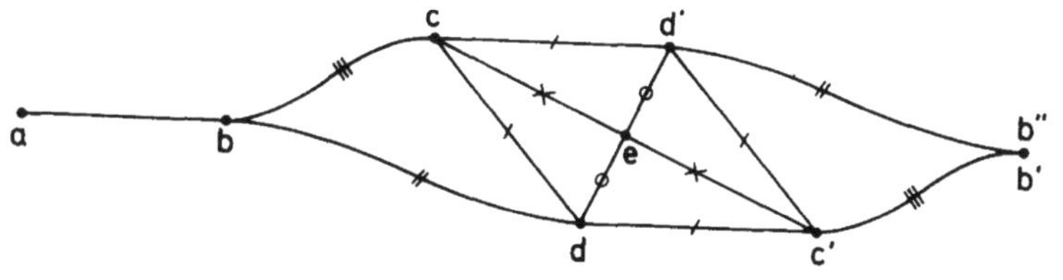

Abb. 15

Mittels 3.12 (mit $l=1$) erhalten wir die verallgemeinerten Zwischenbeziehungen

$$\mathrm{B}_5abcd'b'' \text{ und } \mathrm{B}_5abdc'b'$$

(vgl. Abb. 15, in der gleiche Markierungen kongruente Strecken andeuten).

Nach 2.11 (Aneinanderlegen kongruenter Strecken, die in Abb. 15 mit einfachen und doppelten Markierungen versehen sind) erhalten wir $bc'\equiv b''c$ und durch nochmalige Anwendung dieses Satzes (Anlegen der dreifach markierten Strecken) somit $bb'\equiv b''b$. Nach Satz 2.12 (Eindeutigkeit der Streckenabtragung auf dem von b ausgehenden, durch a bestimmten Strahl) ist also $b''=b'$.

Nun erhalten wir

$$\mathrm{AFS}\begin{pmatrix} b & c & d' & c' \\ b' & c' & d & c \end{pmatrix}.$$

Also ist auch noch

$$c'd'\equiv cd$$

(im Falle $b\neq c$ nach A5, im Trivialfall $b=c$ ist das unmittelbar zu sehen). Die bei der Wahl von c' und d' festgestellten Zwischenbeziehungen erlauben die Anwendung des Axioms von Pasch (A7); sei dementsprechend e ein Punkt mit

$$\mathrm{B}cec' \wedge \mathrm{B}ded'.$$

Dann erhalten wir

$$\mathrm{IFS}\begin{pmatrix} ded'c \\ ded'c' \end{pmatrix}, \quad \mathrm{IFS}\begin{pmatrix} cec'd \\ cec'd' \end{pmatrix},$$

also $ec\equiv ec'$, $ed\equiv ed'$.

Wir nehmen nun $c\neq c'$ an. Dann haben wir $d=d'$ nachzuweisen. Aus der Annahme ergibt sich zunächst $c\neq d'$, da sonst wegen der zu Anfang festgestellten Kongruenzen alle vier Punkte c, d', c', d zusammenfallen würden. Auf Grund des Axioms der Streckenabtragung wählen wir Punkte p, q, r, so daß (Abb. 16)

$$\mathrm{B}c'cp \wedge cp\equiv cd', \ \mathrm{B}d'cr \wedge cr\equiv ce, \ \mathrm{B}prq \wedge rq\equiv rp.$$

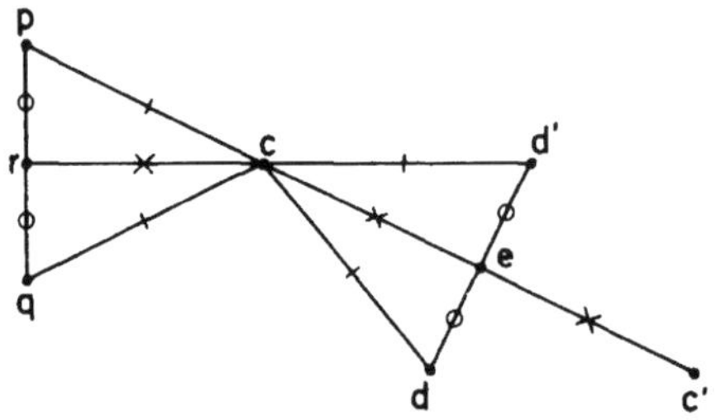

Abb. 16

Dann ist

$$\mathrm{AFS}\begin{pmatrix} d'crp \\ p\ ced' \end{pmatrix},$$

also $rp \equiv ed'$ und damit auch $rq \equiv ed$. Somit ist

$$\mathrm{AFS}\begin{pmatrix} d'edc \\ p\ rqc \end{pmatrix},$$

also $d'd \equiv pq$ (nach 2.11) und $cq \equiv cd$ (falls $d' \neq e$, nach A5, falls $d' = e$, weil dann auch $d' = d$ und $p = q$ ist). Damit ist auch $cp \equiv cq$.

Wir wollen nun den Satz 4.17 mehrfach auf die Punkte p und q anwenden. Da auch $rp \equiv rq$, $r \neq c$ (wegen $c \neq c'$) und Col rcd' ist, erhalten wir nach 4.17 $d'p \equiv d'q$. Aus $c \neq d'$, Col $cd'b$, Col $cd'b'$ ergibt sich ebenso $bp \equiv bq$ und $b'p \equiv b'q$. Wegen $c \neq d'$ und der zuerst festgestellten verallgemeinerten Zwischenbeziehung ist erst recht $b \neq b'$, wegen Col $bc'b'$ also $c'p \equiv c'q$. Wegen $c \neq c'$ und Col $c'cp$ ist schließlich $pp \equiv pq$, nach A3 also $p = q$. Wegen $pq \equiv d'd$ ist damit auch $d = d'$, wie noch zu zeigen war. □

Aus 5.1 ergibt sich mittels 3.6 sofort

<u>5.2 Satz</u>. $a \neq b \wedge \mathrm{B}abc \wedge \mathrm{B}abd \rightarrow \mathrm{B}bcd \vee \mathrm{B}bdc$.

<u>5.3 Satz</u>. $\mathrm{B}abd \wedge \mathrm{B}acd \rightarrow \mathrm{B}abc \vee \mathrm{B}acb$.

<u>Beweis</u>: Sei p gemäß 3.14 ein Punkt mit $\mathrm{B}dap \wedge p \neq a$. Nach 3.5 ist dann $\mathrm{B}pab \wedge \mathrm{B}pac$, woraus sich nach 5.2 die Behauptung ergibt. □

Aus 5.2 oder 5.3 kann man in ebenso einfacher Weise den Satz 5.1 zurückerhalten. Alle drei Sätze sind also in einem gewissen Sinne untereinander gleichwertig. Wir nennen die in ihnen ausgedrückte Eigenschaft die *Konnexität der Zwischenbeziehung*, da darauf die Konnexität (im üblichen Sinne) oder Linearität (Satz 5.10) der Kleiner-Gleich-Beziehung für Strecken beruht, die folgendermaßen eingeführt wird.

<u>5.4 Definition</u>. ab ist *kleiner-gleich* cd (cd *größer-gleich* ab):

$$ab \leq cd :\leftrightarrow \exists y[\mathrm{B}cyd \wedge ab \equiv cy],$$

$$cd \geq ab :\leftrightarrow ab \leq cd.$$

Als gleichwertige Charakterisierung erhält man

<u>5.5 Satz</u>. $ab \leq cd \leftrightarrow \exists x[\mathrm{B}abx \wedge ax \equiv cd]$.

Beweis: Ist y gemäß 5.4 bzw. x gemäß 5.5 gegeben, so sei x bzw. y nach 4.14 gewählt mit $(abx)\equiv(cyd)$.
Nach 4.6 ergibt sich dann das Gewünschte. □

Durch Rückgang auf 5.4 bzw. 5.5 kann man dann ohne Schwierigkeiten die folgenden Sätze beweisen:

5.6 Satz. $ab\leq cd \wedge ab\equiv a'b' \wedge cd\equiv c'd' \rightarrow a'b'\leq c'd'$.

5.7 Satz (*Reflexivität*). $ab\leq ab$.

5.8 Satz (*Transitivität*). $ab\leq cd \wedge cd\leq ef \rightarrow ab\leq ef$.

5.9 Satz. $ab\leq cd \wedge cd\leq ab \rightarrow ab\equiv cd$.

5.10 Satz (*Konnexität*). $ab\leq cd \vee cd\leq ab$.

5.11 Satz. $aa\leq cd$.
(*Nullstrecken sind minimal.*)

5.12 Satz. $\mathrm{Col}\, abc \rightarrow [\mathrm{B}abc \leftrightarrow ab\leq ac \wedge bc\leq ac]$.

5.13 Anmerkung. Auf Grund von 5.6 ist insbesondere $\leq$ - aufgefaßt als zweistellige Relation zwischen Punktepaaren - von der Reihenfolge der Punkte in einem Paar unabhängig, kann also als Relation zwischen Strecken aufgefaßt werden. Nach 5.6 kann man auch noch zu einer entsprechenden Kleiner-Gleich-Beziehung zwischen Äquivalenzklassen bezüglich der Streckenkongruenz übergehen; diese ist dann nach 5.9 antisymmetrisch, also eine lineare Ordnung zwischen Äquivalenzklassen kongruenter Strecken oder "Streckengrößen" (s. 13.4).

In gewohnter Weise kann man von $\leq$ zu einer Kleiner-Beziehung bzw. Grösser-Beziehung für Strecken übergehen mittels

5.14 Definition. $ab<cd :\leftrightarrow ab\leq cd \wedge \neg ab\equiv cd$.
$cd>ab :\leftrightarrow ab<cd$.

Dafür gelten ähnliche Sätze, die hier wohl nicht formuliert zu werden brauchen.

6. Halbgeraden und Geraden

Nachdem die Konnexität der Zwischenbeziehung zur Verfügung steht, können wir in naheliegender Weise Halbgeraden und Geraden einführen.

6.1 Definition. (i) Sind p, a, b paarweise verschiedene Punkte mit Bapb, so sagen wir dafür auch, daß a und b *auf entgegengesetzten Seiten* von p liegen.

(ii) Die Punkte a und b liegen *auf derselben Seite* des Punktes p oder sind *äquivalent* bezüglich p:

(1) $a \underset{p}{\approx} b :\leftrightarrow a \neq p \wedge b \neq p \wedge [\mathrm{B}pab \vee \mathrm{B}pba]$.

In Hinblick auf die spätere Definition 6.7 sagen wir dafür auch "*a und b bestimmen dieselbe von p ausgehende Halbgerade* oder *denselben von p ausgehenden Strahl*".

6.2 Satz. $a \neq p \wedge b \neq p \wedge c \neq p \wedge \mathrm{B}apc \rightarrow [\mathrm{B}bpc \leftrightarrow a \underset{p}{\approx} b]$.

Beweis: Die Richtung von rechts nach links ergibt sich sofort nach den Sätzen 3.6 und 3.7, die Umkehrung nach 5.2. □

Da man einen Punkt c gemäß der Voraussetzung von 6.2 nach 3.14 wählen kann, erhält man hieraus als weitere Charakterisierung der Relation $\underset{p}{\approx}$:

6.3 Satz. $a \underset{p}{\approx} b \leftrightarrow a \neq p \wedge b \neq p \wedge \exists c[c \neq p \wedge \mathrm{B}apc \wedge \mathrm{B}bpc]$.

Mittels 3.4 oder 6.2 erhält man leicht aus 6.1(1) die weitere Charakterisierung

6.4 Satz. $a \underset{p}{\approx} b \leftrightarrow \mathrm{Col}\ apb \wedge \neg \mathrm{B}apb$.

Aus einer der Charakterisierungen 6.1(1), 6.3 und aus 6.2 ergibt sich, daß die zweistellige Relation $\underset{p}{\cong}$ eine Äquivalenzrelation auf der Menge der von p verschiedenen Punkte ist, d.h., daß die folgenden Sätze gelten.

6.5 Satz (*Reflexivität*). $a \neq p \rightarrow a \underset{p}{\cong} a$.

6.6 Satz (*Symmetrie*). $a \underset{p}{\cong} b \rightarrow b \underset{p}{\cong} a$.

6.7 Satz (*Transitivität*). $a \underset{p}{\cong} b \wedge b \underset{p}{\cong} c \rightarrow a \underset{p}{\cong} c$.

6.8 Definition. Die Äquivalenzklasse (bezüglich $\underset{p}{\cong}$) eines beliebigen Punktes $a \neq p$, d.h. die Punktmenge

$\mathrm{H}(pa) := \{x \mid x \underset{p}{\cong} a\}$ (definiert für $a \neq p$)

nennen wir die *durch a bestimmte von p ausgehende Halbgerade* oder auch *die Halbgerade von p aus durch a* oder den entsprechend bestimmten *Strahl*. Eine Punktmenge K ist eine *Halbgerade* ("half-line") oder ein *Strahl*

$\mathrm{Hl}\, K :\leftrightarrow \exists p \exists a [a \neq p \wedge K = \mathrm{H}(pa)]$.

6.9 Definition. Von p ausgehende Halbgeraden $\mathrm{H}(pa)$, $\mathrm{H}(pb)$ heißen *zueinander entgegengesetzt* oder auch $\mathrm{H}(pa)$ *entgegengesetzt* zu b (mit $b \neq p$), falls $\mathrm{B}apb$ ist.

6.10 Anmerkung. Nach 6.2 ist diese Definition unabhängig von der Wahl der bestimmenden Punkte (Repräsentanten) auf $\mathrm{H}(pa)$ und (im ersten Falle) auf $\mathrm{H}(pb)$, und die zu einer Halbgeraden entgegengesetzte Halbgerade ist eindeutig bestimmt und von ihr verschieden.

Von den früheren Feststellungen der Möglichkeit und Eindeutigkeit der Streckenabtragung (A4 und 2.12) kann man jetzt natürlich zu der folgenden Formulierung übergehen, indem man statt q einen Punkt r auf der entgegengesetzten Seite von a verwendet (d.h. auf der Seite, nach der wirklich abgetragen wird).

6.11 Satz. $r \neq a \wedge b \neq c \rightarrow \overset{=1}{\exists} x [x \underset{a}{\cong} r \wedge ax \equiv bc]$.

6.12 Definition. Ist x dieser Punkt, so sagen wir auch, x entstehe durch *Abtragen der Strecke bc von a aus in Richtung nach r* bzw. *auf dem durch r bestimmten Strahl* (d.h. auf $\mathrm{H}(ar)$).

<u>6.13 Satz</u>. $a \underset{p}{\approx} b \rightarrow [pa \leq pb \leftrightarrow \mathrm{B}pab]$.

<u>Beweis</u>: Für die Richtung von links nach rechts sei y gemäß Definition von $\leq$ (5.4) gewählt; nach 6.11 ist dann $y=a$, also Bpab. Die Umkehrung ist trivial. □

<u>6.14 Definition</u>. Für verschiedene Punkte p, q verstehen wir unter der *durch p und q bestimmten Geraden* ("line") oder auch *der Geraden durch p und q* die Punktmenge

$$\mathrm{L}(pq) := \{x \mid \mathrm{Col}\, pqx\} \qquad \text{(definiert für } p \neq q).$$

Eine Punktmenge A ist eine *Gerade*:

$$\mathrm{Ln}\, A :\leftrightarrow \exists p \exists q [p \neq q \wedge A = \mathrm{L}(pq)].$$

Für $a \in A$ sagen wir dann auch "der Punkt *a liegt auf* der Geraden A " oder "... *inzidiert mit* der Geraden a " oder "die Gerade A *geht durch* den Punkt a "; für $a \notin A$ sagen wir auch "a ist (oder liegt) *außerhalb* (von) A ".

Eine Gerade L(pq) läßt sich dann folgendermaßen aus zwei entgegengesetzten Halbgeraden und dem Ausgangspunkt p zusammensetzen:

<u>6.15 Satz</u>. $p \neq q \wedge p \neq r \wedge \mathrm{B}qpr \rightarrow \mathrm{L}(pq) = \mathrm{H}(pq) \cup \{p\} \cup \mathrm{H}(pr)$.

<u>Beweis</u>: Col pqx ist nach Def. 4.10 (und 3.2) gleichbedeutend mit

$$(\mathrm{B}pqx \vee \mathrm{B}pxq) \vee \mathrm{B}xpq.$$

Für von p verschiedene Punkte x ist das nach 6.1(1) und 6.2 gleichbedeutend mit

$$x \underset{p}{\approx} q \vee x \underset{p}{\approx} r,$$

woraus man die Behauptung erhält. □

Nach 6.10 kann man also als zweiten Bestimmungspunkt von L(pq) statt q auch jeden anderen von p verschiedenen Punkt s von L(pq) (d.h. von H(pq) oder H(pr)) verwenden, d.h., es gilt

<u>6.16 Satz</u>. $p \neq q \wedge s \neq p \wedge s \in \mathrm{L}(pq) \rightarrow \mathrm{L}(pq) = \mathrm{L}(ps)$.

Auf Grund der Sätze 4.12, 4.11 über die Kollinearität gilt andererseits

<u>6.17 Satz</u>. $p \neq q \rightarrow p \in \mathrm{L}(pq) \wedge q \in \mathrm{L}(pq) \wedge \mathrm{L}(pq) = \mathrm{L}(qp)$.

<u>6.18 Satz</u>. $\mathrm{Ln}\, A \wedge a \neq b \wedge a \in A \wedge b \in A \rightarrow A = \mathrm{L}(ab)$.

Beweis: Nach 6.14 hat A die Darstellung $A=\mathrm{L}(pq)$ für geeignete Punkte p, q mit $p\neq q$. Nach Voraussetzung ist wenigstens einer der Punkte a, b von p verschieden; o.B.d.A. sei $b\neq p$. Nach 6.16 und 6.17 ist dann

$$A=\mathrm{L}(pq)=\mathrm{L}(pb)=\mathrm{L}(bp)=\mathrm{L}(ba)=\mathrm{L}(ab). \qquad \square$$

Nun erhalten wir den bei Hilbert 1977 in den Axiomen I.1 und I.2 enthaltenen Satz:

6.19 Satz. $a\neq b \rightarrow \overset{=1}{\exists} A[\mathrm{Ln}\,A \wedge a\in A \wedge b\in A]$.

(Zu zwei Punkten a, b gibt es stets genau eine Gerade, die durch a und b geht.)

6.20 Anmerkung. "Zwei" wird hier zur Bezeichnung der Anzahl gebraucht; die Feststellung, daß p, q zwei Punkte sind, bedeutet also insbesondere, daß $p\neq q$ ist.

Beweis von 6.19: Nach 6.17, 6.18 ist $\mathrm{L}(ab)$ die einzige Gerade der angegebenen Art. □

Als unmittelbare Folgerung aus der Eindeutigkeitsaussage in 6.19 ergibt sich

6.21 Satz. $\mathrm{Ln}\,A \wedge \mathrm{Ln}\,B \wedge A\neq B \rightarrow \overset{\leq 1}{\exists} x[x\in A \wedge x\in B]$.

(Zwei Geraden A, B haben höchstens einen gemeinsamen Punkt.)

6.22 Definition. Wenn der Punkt x in 6.21 existiert, so sagen wir auch, daß A und B *sich schneiden* in ("intersect at") x oder daß x der *Schnittpunkt* von A und B ist:

$$AB\ \mathrm{Is}\ x :\leftrightarrow \mathrm{Ln}\,A \wedge \mathrm{Ln}B \wedge A\neq B \wedge x\in A \wedge x\in B.$$

Die bei Definition 4.10 erwähnte anschauliche Bedeutung der Kollinearität läßt sich jetzt präzise fassen in

6.23 Satz. $\mathrm{Col}\,abc \leftrightarrow \exists A[\mathrm{Ln}A \wedge a\in A \wedge b\in A \wedge c\in A]$.

(Punkte a, b, c sind kollinear genau dann, wenn sie auf einer Geraden liegen.)

Beweis: Wegen 4.12 und 6.19 genügt es, den Fall zu betrachten, daß a, b, c paarweise verschieden sind. Ist dann $\mathrm{Col}\,abc$, so leistet $\mathrm{L}(ab)$ nach Def. 6.14 das für A Verlangte. Ist andererseits A eine solche Gerade, so ist nach 6.18 $A=\mathrm{L}(ab)$, wegen $c\in A$ und Def. 6.14 also $\mathrm{Col}\,abc$. □

Als einfachere Formulierung des unteren Dimensionsaxioms (A8) erhält man nach Def. 4.10 offenbar

6.24 Satz (*Unteres Dimensionsaxiom*). $\exists a \exists b \exists c \neg \mathrm{Col}\, abc$.
(*Es gibt nicht-kollineare Punkte a, b, c.*)

6.25 Satz. $\forall a \forall b [a \neq b \rightarrow \exists c \neg \mathrm{Col}\, abc]$.
(*Außerhalb jeder Geraden* [*hier* L(ab)] *gibt es noch einen Punkt.*)

Beweis: Sei $a \neq b$. Läge nun jeder Punkt c auf $A = \mathrm{L}(ab)$, so lägen auch beliebige Punkt x, y, z auf A, wären also nach 6.23 kollinear im Widerspruch zu 6.24. □

6.26 Anmerkungen. (i) Die Einführung der Geraden und Halbgeraden und die darüber bewiesenen Sätze dienen dazu, manche weiteren Beweise einfacher oder übersichtlicher zu machen. Andererseits läßt sich jedoch die Verwendung von Geraden im Aufbau der Geometrie (bis zu den Darstellungssätzen am Schluß von Teil I) prinzipiell vermeiden. Dazu kann man das, was über Geraden gesagt wird, durch bestimmende Punktepaare und die Relation der Kollinearität ausdrücken; an Stelle von 6.18 erhält man dann z.B. als Satz:

(1) $p \neq q \wedge a \neq b \wedge \mathrm{Col}\, pqa \wedge \mathrm{Col}\, pqb \rightarrow \forall x [\mathrm{Col}\, pqx \leftrightarrow \mathrm{Col}\, abx]$.

(Für ein allgemeines Verfahren zur Elimination von Geradenvariablen vgl. II.4.59.) Entsprechendes gilt für andere spezielle Arten von Punktmengen, die jeweils durch eine feste endliche Anzahl von Punkten festgelegt werden (Halbgeraden, Ebenen, Halbebenen, Kreise u. dgl.). Werden nur die Axiome A1 bis A10 und A11' (oder ein Teil davon) vorausgesetzt, so läßt sich der in diesem Teil angestrebte Aufbau im Rahmen der Prädikatenlogik der ersten Stufe mit Punktvariablen und den Grundbegriffen des Axiomensystems (d.h. in der Sprache L(D,B)) durchführen (vgl. 0.5). Das spielt für manche späteren metamathematischen Betrachtungen eine Rolle.

(ii) (andere Beweisanordnung). Man kann natürlich auch Geraden einführen und die wichtigsten Sätze darüber beweisen, ohne vorher Halbgeraden und die Relation $\underset{p}{\simeq}$ einzuführen. Das gleiche gilt für den Beweis entsprechender Sätze über die Kollinearität gemäß Anmerkung (i). In einem solchen Aufbau spielt der Satz (1) eine tragende Rolle. Daher sei hier noch ein direkter Beweis für diesen Satz (ohne Benutzung von Definitionen aus 6.) skizziert.

Für jede der beiden Richtungen in der Behauptung werden drei Kollinearitäten vorausgesetzt, von denen jede gemäß Definition 4.10 auf drei Fälle von Zwischenbeziehungen führt. Die sich ergebenden 27 Fälle könnte man natürlich einzeln betrachten. Um jedoch die Zahl der Fälle zu reduzieren, empfiehlt es sich, den Satz auf folgendes Lemma zurückzuführen (Übung).

(2) $p \neq q \wedge \mathrm{Col}\, pqa \wedge \mathrm{Col}\, pqb \rightarrow \mathrm{Col}\, pab \wedge \mathrm{Col}\, qab$.

Beweis (Skizze): Hier könnte man in entsprechender Weise 9 Fälle betrachten. Zur besseren Übersicht betrachten wir für zwei feste Punkte p, q die drei "Abschnitte" $\{x \mid \mathrm{B}pqx\}$, $\{x \mid \mathrm{B}qxp\}$, $\{x \mid \mathrm{B}xpq\}$, in die die Gerade $\mathrm{L}(pq)$ gemäß Definition 4.10 durch p, q "zerlegt" wird. Die genannten 9 Fälle lassen sich dann in zwei Gruppen zusammenfassen.

Fall 1: a und b liegen bezüglich p, q im selben Abschnitt (oder - ohne Benutzung von Mengen ausgedrückt - a und b stehen mit p, q in derselben der gemäß 4.10 möglichen Zwischenbeziehungen). Dann ergibt sich die Behauptung (geeignete Zwischenbeziehungen) aus den Sätzen 5.1 bis 5.3.

Fall 2: a und b liegen bezüglich p, q in verschiedenen Abschnitten. Dann ergibt sich die Behauptung aus den Sätzen 3.5 bis 3.7. □

Auch bei diesem Aufbau ist also die Konnexität der Zwischenbeziehung (Abschnitt 5) ein wichtiges Hilfsmittel.

Die Definition der Kollinearität können wir auf Grund von 6.23 folgendermaßen erweitern.

6.27 Definition. Die Punkte $a_1, \ldots, a_n$ sind *kollinear*

$$\mathrm{Col}\, a_1 \ldots a_n :\leftrightarrow \exists A [\mathrm{Ln}\, A \wedge \bigwedge_{\nu=1}^{n} a_\nu \in A].$$

7. Punktspiegelungen

7.1 Definition. m ist *Mittelpunkt* der Strecke ab oder auch a geht durch *Spiegelung* an m in b über:

$$\mathrm{M}amb :\leftrightarrow \mathrm{B}amb \wedge ma \equiv mb .$$

Mittels 3.2, 2.7 und A6 erhält man sofort die folgenden beiden Sätze.

7.2 Satz. $\mathrm{M}amb \rightarrow \mathrm{M}bma$.

7.3 Satz. $\mathrm{M}ama \leftrightarrow m=a$.

7.4 Satz. $\forall p \overset{=1}{\exists} p' \mathrm{M}pap'$.

Beweis: Falls $p \neq a$ ist, ergibt sich das aus der Möglichkeit (A4) und Eindeutigkeit (2.12) der Streckenabtragung; falls $p=a$ ist, leistet offenbar $p'=a$ das Verlangte und ist nach A3 auch der einzige solche Punkt. □

7.5 Definition. Der nach 7.4 eindeutig bestimmte Punkt p' heißt das *Spiegelbild von p am Punkt a*, abgekürzt:

$$\mathrm{S}_a p = \mathrm{S}_a(p) := \iota p' \mathrm{M}pap' .$$

Die dadurch erklärte Abbildung S_a (d.h. $\mathrm{S}_a := \{\langle p,p' \rangle \mid \mathrm{M}pap'\}$) heißt die *Spiegelung (Punktspiegelung) am Punkte a*. Für $\mathrm{M}pap'$ sagen wir deshalb auch, daß p und p' *spiegelbildlich bezüglich a* liegen.

Aus dem Gesagten ergeben sich unmittelbar die folgenden Sätze.

7.6 Satz. $\mathrm{S}_a p = p' \leftrightarrow \mathrm{M}pap'$.

7.7 Satz. $\mathrm{S}_a \mathrm{S}_a p = p$.

7.8 Satz. $\forall p' \overset{=1}{\exists} p\, S_a p = p'$.

7.9 Satz. $S_a p = S_a q \rightarrow p = q$.

7.10 Satz. $S_a p = p \leftrightarrow p = a$.

7.4 (bzw. 7.5) und 7.8 lassen sich zusammenfassen in

7.11 Satz. S_a *ist eine eineindeutige Abbildung des ganzen Raumes auf sich.*

Da es verschiedene Punkte gibt, ist diese Abbildung nach 7.10 nicht die identische Abbildung, aber ihre Verkettung mit sich selbst ist nach 7.7 die identische Abbildung. Das wird zusammengefaßt in

7.12 Satz. S_a *ist involutorisch.*

Wir wollen nun zeigen, daß S_a sogar eine Bewegung ist (4.8). Dazu benötigen wir noch, daß jede Strecke zu ihrer Bildstrecke kongruent ist, d.h.

7.13 Satz. $pq \equiv S_a(p) S_a(q)$.

Beweis: Sei $p' = S_a p$, $q' = S_a q$. Falls $p = a$ ist, ist nach 7.10 auch $p' = a$, nach Definition von q' andererseits $aq \equiv aq'$, also gilt die Behauptung. Für das folgende sei also $p \neq a$. Durch Streckenabtragung (A4) erhalten wir Punkte x, y, x', y' (Abb. 17) mit

$$Bp'px \wedge px \equiv qa, \qquad Bq'qy \wedge qy \equiv pa,$$

$$Bxp'x' \wedge p'x' \equiv qa, \qquad Byq'y' \wedge q'y' \equiv pa.$$

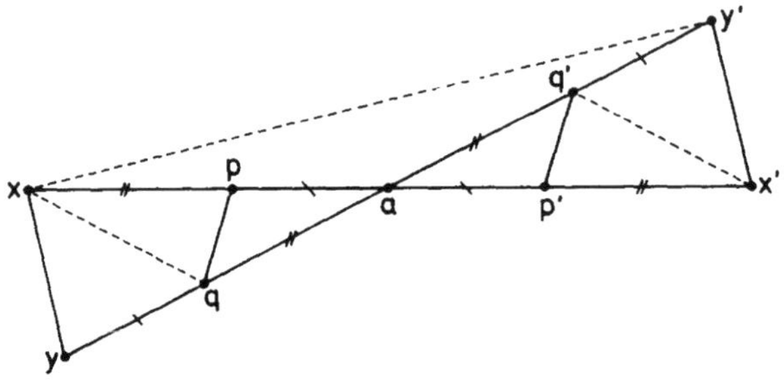

Abb. 17

Nach 3.12 (mit $l=1$) und 3.9 ergeben sich dann die verallgemeinerten Zwischenbeziehungen

$$B_5 xpap'x' \quad \text{und} \quad B_5 yqaq'y'.$$

Nach 2.11 (Aneinanderlegen von Strecken) ergeben sich die Streckenkongruenzen

$$ax \equiv ay \equiv ay' \equiv ax'.$$

Nun erhalten wir die äußere Fünf-Streckenkonfiguration

$$\mathrm{AFS}\begin{pmatrix} x & ax'y' \\ y'ay & x \end{pmatrix}.$$

Wegen $p \neq a$ ist erst recht $x \neq a$, nach A5 also $x'y' \equiv xy$. Unter Benutzung von 4.2 erhalten wir weiter

$$\mathrm{IFS}\begin{pmatrix} y & q & ax \\ y' & q' & ax' \end{pmatrix}, \quad \text{also} \quad qx \equiv q'x',$$

$$\mathrm{IFS}\begin{pmatrix} x & p & aq \\ x' & p' & aq' \end{pmatrix}, \quad \text{also} \quad pq \equiv p'q'. \qquad \square$$

Mit 4.8, 4.9 erhalten wir

<u>7.14 Satz</u>. *Jede Punktspiegelung* S_a *ist eine Bewegung und (damit) ein Automorphismus des ganzen Raumes.*

Im einzelnen wird die Eigenschaft, Automorphismus zu sein, ausgedrückt durch 7.11 und die folgenden beiden Sätze.

<u>7.15 Satz</u>. $Bpqr \leftrightarrow BS_a(p)S_a(q)S_a(r)$.

<u>7.16 Satz</u>. $pq \equiv rs \leftrightarrow S_a(p)S_a(q) \equiv S_a(r)S_a(s)$.

<u>7.17 Satz</u>. $Mpap' \wedge Mpbp' \rightarrow a=b$.
(Jede Strecke hat höchstens einen Mittelpunkt.)

<u>Anmerkung</u>. Die Existenz des Mittelpunktes zu jeder Strecke wird erst später bewiesen (Satz 8.22).

<u>Beweis von 7.17</u>. Es ist $pb \equiv p'b$, nach 7.13 (angewendet auf S_a) andererseits $p'b \equiv pS_a(b)$, also $pb \equiv pS_a(b)$. Durch Vertauschung von p mit p' ergibt sich ebenso $p'b \equiv p'S_a(b)$. Nach Voraussetzung ist auch $Bpbp'$, nach 4.19 also $b=S_a(b)$ und nach 7.10 somit $a=b$. $\square$

Nur eine andere Version von 7.17 ist

<u>7.18 Satz</u>. $S_a p = S_b p \rightarrow a = b$.

Die Verkettung (Hintereinanderausführung) von Punktspiegelungen ist eine nicht-kommutative Operation; für $a \neq b$ haben die beiden Abbildungen $S_a S_b$ und $S_b S_a$ sogar für jedes Argument p verschiedene Werte, wie sich aus dem folgenden Satz ergibt. (Mit Hilfsmitteln, die erst später eingeführt werden, läßt sich auch für unsere Geometrie der bekannte Satz beweisen, daß die genannten beiden Abbildungen "Translationen" in entgegengesetzten Richtungen längs $L(ab)$ um das "Doppelte" der Strecke ab sind.)

<u>7.19 Satz</u>. $S_a S_b p = S_b S_a p \leftrightarrow a = b$.

<u>Beweis</u>: Die Richtung von rechts nach links ist trivial. Gelte nun die linke Seite von 7.19, und sei $p' = S_a p$. Dann ist $Mpap'$ und nach Voraussetzung andererseits $S_a S_b(p) = S_b(p')$, d.h. $MS_b(p)aS_b(p')$. Aus der letzten Bedingung erhalten wir nach 7.14 ("S_b ist Automorphismus") und 7.7 auch $MpS_b(a)p'$. Wegen der Eindeutigkeit des Mittelpunkts der Strecke pp' (7.17) ist also $a = S_b(a)$ und nach 7.10 somit $a = b$. □

<u>7.20 Satz</u>. $\mathrm{Col}\, amb \wedge ma \equiv mb \rightarrow a = b \vee \mathrm{M}amb$.

<u>Beweis</u>: Entsprechend der Definition der Kollinearität unterscheiden wir folgende Fälle.

<u>Fall 1</u>: $\mathrm{B}amb$. Dann ist $\mathrm{M}amb$.

<u>Fall 2</u>: $\mathrm{B}abm$. Durch Anwendung von 4.3 (Ineinanderlegen von Strecken) auf die Zwischenbeziehungen $\mathrm{B}abm$ und $\mathrm{B}bbm$ (3.3) und die entsprechenden Kongruenzen ergibt sich $ab \equiv bb$ und daraus $a = b$ nach A3.

<u>Fall 3</u>: $\mathrm{B}bam$. Dieser Fall läßt sich durch Vertauschung von a mit b auf Fall 2 zurückführen. □

Wir beweisen noch zwei Lemmata über Mittelpunkte.

<u>7.21 Lemma</u> (Lemma vom [zentral-]symmetrischen Viereck).
$\neg \mathrm{Col}\, abc \wedge b \neq d \wedge ab \equiv cd \wedge bc \equiv da \wedge \mathrm{Col}\, apc \wedge \mathrm{Col}\, bpd \rightarrow \mathrm{M}apc \wedge \mathrm{M}bpd$.

(Sind in einem "ebenen Viereck" gegenüberliegende Seiten kongruent, so halbieren sich die Diagonalen, Abb. 18.)

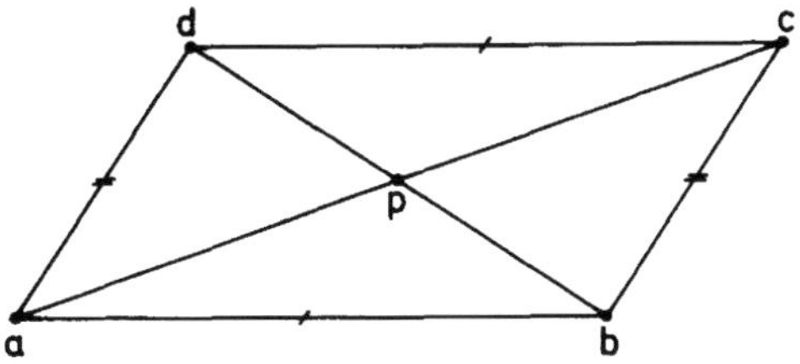

Abb. 18

Anmerkung. Die Existenz eines Punktes p gemäß Voraussetzung sichert, daß alle betrachteten Punkte in einer Ebene (in einem erst später präzisierten Sinne) liegen. In der euklidischen Geometrie (mit Verwendung von A10) sind Vierecke der vorausgesetzten Art natürlich dasselbe wie "Parallelogramme" (Vierecke mit parallelen gegenüberliegenden Seiten). In der absoluten Geometrie auf Grund der Axiome A1 bis A8, die hier nur vorausgesetzt werden, gilt das jedoch nicht allgemein; daher wird hier die Bezeichnung "symmetrisches Viereck" verwendet.

Beweis von 7.21: Durch Anwendung von 4.14 auf $\mathrm{Col}\,bdp \wedge bd{\equiv}db$ erhalten wir einen Punkt p' mit $(bdp){\equiv}(dbp')$, für den nach 4.13 auch $\mathrm{Col}\,bdp'$ ist. Gemäß Def. 4.15 ist

$$\mathrm{FS}\begin{pmatrix} bdp & a \\ dbp' & c \end{pmatrix} \quad \text{und} \quad \mathrm{FS}\begin{pmatrix} bdp & c \\ dbp' & a \end{pmatrix},$$

nach 4.16 also $pa{\equiv}p'c$ und $pc{\equiv}p'a$ und damit $(apc){\equiv}(cp'a)$. Wegen $\mathrm{Col}\,apc$ ist nach 4.13 auch $\mathrm{Col}\,acp'$. Somit ist sowohl p als auch p' gemeinsamer Punkt der Geraden $\mathrm{L}(ac)$ und $\mathrm{L}(bd)$, die wegen $\neg\mathrm{Col}\,abc$ voneinander verschieden sind. Nach 6.21 ist also $p=p'$. Nun ergeben sich beide Teile der Behauptung mittels 7.20. □

7.22 Lemma (Krippenlemma, bei Gupta 1965 als Theorem 3.45).

$$\mathrm{B}a_1ca_2 \wedge \mathrm{B}b_1cb_2 \wedge ca_1{\equiv}cb_1 \wedge ca_2{\equiv}cb_2$$
$$\wedge\, \mathrm{M}a_1m_1b_1 \wedge \mathrm{M}a_2m_2b_2 \rightarrow \mathrm{B}m_1cm_2.$$

7.23 Definition. Dafür, daß die Voraussetzung dieses Lemmas gilt, sagen wir auch, daß a_1, m_1, b_1, c, b_2, m_2, a_2 (in dieser Reihenfolge) eine *Krippenfigur* bilden (Abb. 19).

Beweis von 7.22: Nach 5.10 ist $ca_1{\le}ca_2 \vee ca_2{\le}ca_1$. Wegen der Symmetrie von Voraussetzung und Behauptung in bezug auf Vertauschung der Indizes

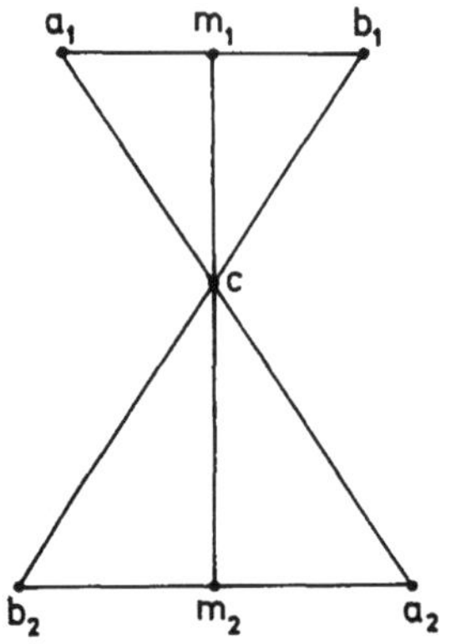

Abb. 19

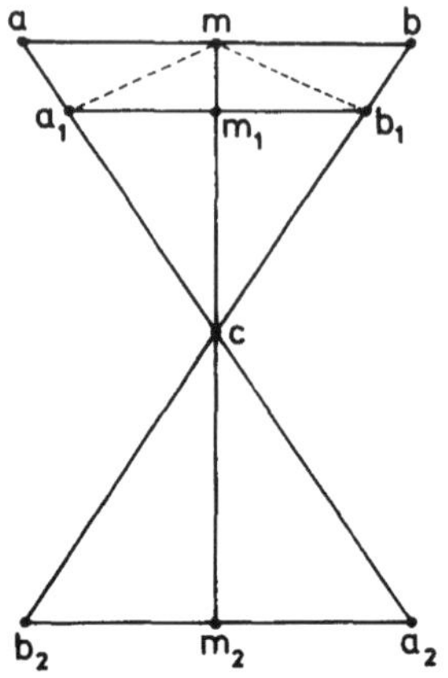

Abb. 20

können wir o. B. d. A. $ca_1 \leq ca_2$ annehmen. Falls $a_2 = c$ ist, ist dann erst recht $a_1 = c$, nach A3 auch $b_2 = b_1 = c$ und nach 7.3 somit $m_2 = m_1 = c$, woraus sich die Behauptung ergibt. Sei nun $a_2 \neq c$. Sei weiter $a = S_c a_2$, $b = S_c b_2$ und $m = S_c m_2$ (Abb. 20). Da S_c ein Automorphismus ist (7.14), ist dann auch $Mamb$. Nach 5.6 ist auch $ca_1 \leq ca$, woraus sich Bca_1a ergibt (im Falle $a_1 = c$ trivialerweise, sonst wegen $a_1 \underset{c}{\simeq} a$ nach 6.13). Ebenso ergibt sich Bcb_1b. Auf Grund von 3.17 wählen wir nun einen Punkt q mit $Bmqc \wedge Ba_1qb_1$, für den nach 3.6 auch $Bqcm_2$ gilt. Es genügt dann zu zeigen, daß $q = m_1$ ist. Nach Konstruktion ist

$$\mathrm{IFS}\begin{pmatrix} aa_1cm \\ bb_1cm \end{pmatrix},$$

nach 4.2 also $ma_1 \equiv mb_1$. Daraus ergibt sich $qa_1 \equiv qb_1$ (falls $m \neq c$, nach 4.17, falls $m = c$, weil dann nach A6 auch $q = m$ ist). Somit ist Ma_1qb_1, wegen der Eindeutigkeit des Mittelpunktes also $q = m_1$. □

7.24 Anmerkung. Es mag auf den ersten Blick störend wirken, daß im Beweis des Krippenlemmas die Fälle $ca_1 \leq ca_2$ und $ca_2 \leq ca_1$ unterschieden wurden. Diese Fallunterscheidung erscheint aber ganz natürlich, wenn man betrachtet, daß das Axiom von Pasch (A7, innere Form!) bisher das einzige Mittel ist, das es erlaubt, einen Punkt als Schnittpunkt zweier Geraden zu konstruieren, und daß dabei der zu konstruierende Punkt stets zwischen zwei schon vorher gegebene Punkte auf jeder der Geraden fällt. Es liegt also auf der Hand, daß man um einen als Schnittpunkt zu konstruierenden Punkt - wie hier q - erst einen "Rahmen" herumbaut, um vorher andere Punkte zu erhalten, zwischen die er fallen muß. Diese Methode wird auch noch an anderen Stellen angewendet, bevor der Satz von Pasch in allgemeinerer Form zur Verfügung steht. Der Beweis bei Gupta 1965 benutzt nicht die Konnexität 5.10; statt dessen wird der "Rahmen" nach beiden Seiten vergrößert (so ähnlich wie im Beweis von 7.13). Zur Rolle des Krippenlemmas s. Anmerkung 8.19(ii).

7.25 Lemma. $ca \equiv cb \rightarrow \exists x \mathrm{M}axb$.

Anmerkung. Die Existenz eines Mittelpunkts für die Strecke ab ergibt sich hier zunächst nur unter der Voraussetzung, daß schon ein Punkt c vorhanden ist, der von beiden Endpunkten denselben Abstand hat. Später wird die Existenz des Mittelpunkts von Strecken allgemein bewiesen (8.22). Der in 7.25 ausgedrückte Spezialfall wird aber schon vorher benötigt (insbesondere in 8.18).

Beweis von 7.25: Fall 1: $\mathrm{Col}\,abc$. Nach 7.20 ist dann $a=b$ oder $\mathrm{M}acb$. Dementsprechend leistet $x=a$ oder $x=c$ das Verlangte.

Fall 2: $\neg\mathrm{Col}\,abc$. Gemäß 3.14, A4 bzw. A7 wählen wir Punkte p, q und r, so daß $\mathrm{B}cap \wedge a \neq p$, $\mathrm{B}cbq \wedge bq \equiv ap$, $\mathrm{B}arq \wedge \mathrm{B}brp$ (Abb.21). Durch nochmalige Anwendung von A7 (etwa auf die Zwischenbeziehungen $\mathrm{B}cap$ und $\mathrm{B}brp$) erhal-

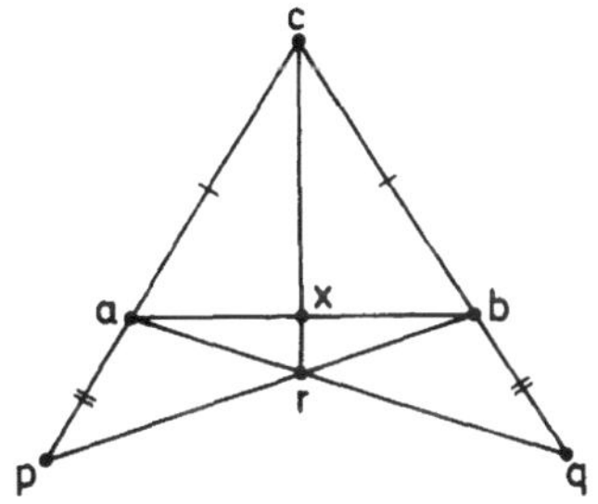

Abb. 21

ten wir einen Punkt x mit $\mathrm{B}axb \wedge \mathrm{B}rxc$. Wir wollen zeigen, daß x das Verlangte leistet, also auch $xa \equiv xb$ ist. Dazu genügt es zu zeigen, daß $ra \equiv rb$ ist; denn daraus ergibt sich das Gewünschte mittels 4.17 (und A6)(wie im vorigen Beweis).

Nach Konstruktion ist zunächst $c \neq a$ und

$$\mathrm{AFS}\begin{pmatrix} capb \\ cbqa \end{pmatrix},$$

nach A5 also $pb \equiv qa$. Da über r noch keine Streckenkongruenzen zur Verfügung stehen, wählen wir (vgl. Anmerkung 2.9) zunächst einen weiteren Punkt r' gemäß 4.5, so daß $\mathrm{B}ar'q \wedge (brp) \equiv (ar'q)$. Dann ist

$$\mathrm{IFS}\begin{pmatrix} br\ pa \\ ar'qb \end{pmatrix} \quad \text{und} \quad \mathrm{IFS}\begin{pmatrix} br\ pq \\ ar'qp \end{pmatrix},$$

nach 4.2 also $ra \equiv r'b$ und $rq \equiv r'p$ und damit $(arq) \equiv (br'p)$. Wegen $\mathrm{Col}\,arq$ und 4.13 ist somit auch $\mathrm{Col}\,br'p$. Damit ist sowohl r als auch r' ein Punkt der beiden Geraden $\mathrm{L}(aq)$ und $\mathrm{L}(bp)$, die nach Konstruktion (und Voraussetzung von Fall 2) voneinander verschieden sind. Nach 6.20 ist nun $r=r'$, also tatsächlich $ra \equiv rb$. □

8. Rechte Winkel

8.1 Definition. a, b, c bilden einen *rechten Winkel* (mit dem *Scheitel* b):

$$Rabc :\leftrightarrow ac \equiv aS_b(c).$$

Üblicherweise wird ein rechter Winkel definiert als ein solcher, der zu seinem Nebenwinkel kongruent ist. Das wird im wesentlichen auch durch die Definition (ohne Benutzung der Begriffe Winkel und Nebenwinkel) ausgedrückt (Abb. 22), es wird nur (ebenso wie bei der Zwischenbeziehung) zugelassen, daß a oder c mit dem Scheitel b zusammenfällt, daß also gar kein Winkel im üblichen Sinne (s. 11.1, 11.18) vorliegt. In Abbildungen werden wir rechte Winkel häufig (wie hier) durch einen Bogen mit eingesetztem Punkt markieren.

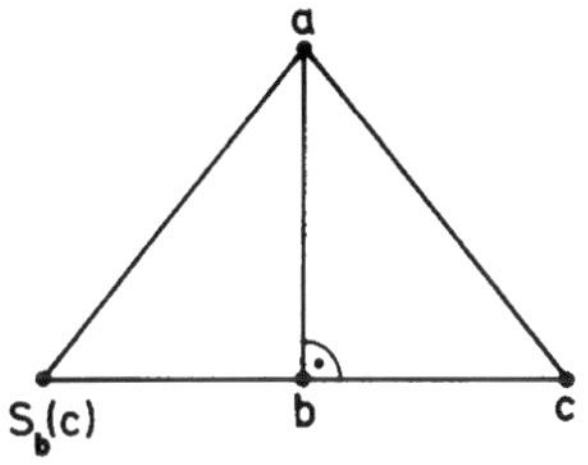

Abb. 22

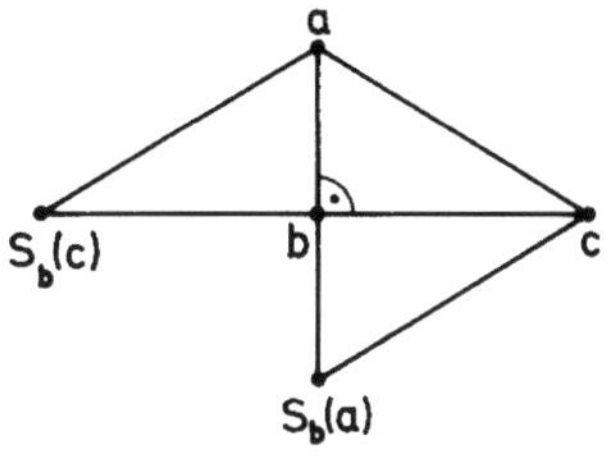

Abb. 23

8.2 Satz. $Rabc \rightarrow Rcba$.

Beweis: Nach 7.13 (angewandt auf S_b) ist $aS_b(c) \equiv S_b(a)c$. Daraus ergibt sich der Satz unmittelbar (Abb. 23). □

Die Anwendung dieser "Symmetrie" der Relation R werden wir in weiteren Beweisen i.a. nicht besonders erwähnen.

Die "Unabhängigkeit von der Wahl des bestimmenden Punktes auf einem Schenkel" wird ausgedrückt durch

<u>8.3 Satz</u>. $Rabc \wedge a \neq b \wedge \mathrm{Col}\, baa' \to Ra'bc$.

<u>Beweis</u>: Das ergibt sich sofort nach 4.17 (mit $p=q$, $q=S_b(c)$, vgl. Abbildungen 22 und 13).

Direkt aus der Definition 8.1 (und einfachsten Sätzen über Spiegelungen) erhalten wir die folgenden beiden Sätze.

<u>8.4 Satz</u>. $Rabc \to RabS_b(c)$.

<u>8.5 Satz</u>. $Rabb$.

<u>8.6 Satz</u>. $Rabc \wedge Ra'bc \wedge Baca' \to b=c$.

<u>Beweis</u>: Sei $c'=S_bc$ (Abb. 24). Nach Voraussetzung ist dann $ac \equiv ac'$ und $a'c \equiv a'c'$, nach 4.19 also $c=c'$ und nach 7.10 somit $b=c$.

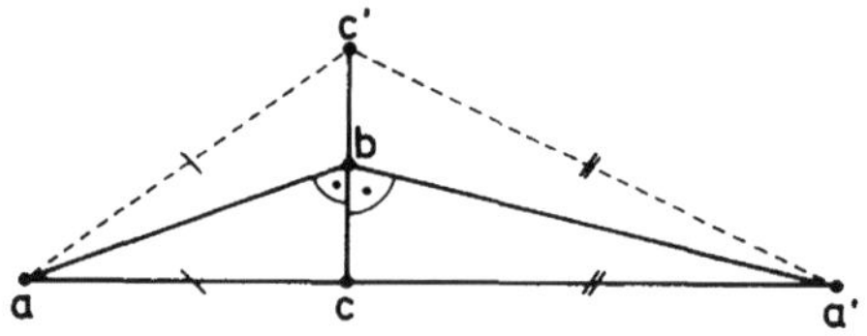

Abb. 24

<u>8.7 Satz</u>. $Rabc \wedge Racb \to b=c$.

<u>Beweis</u>: Sei wieder $c'=S_b(c)$ und $a'=S_c(a)$ (Abb. 24). Wir können noch $b \neq c$ annehmen (sonst ist nichts zu zeigen). Wegen $Racb$ ist dann nach 8.3 auch $Racc'$, also $ac' \equiv a'c'$. Nach Voraussetzung ist außerdem $ac' \equiv ac \equiv a'c$. Somit ist auch $a'c \equiv a'c'$, d.h. $Ra'bc$, nach 8.6 also doch $b=c$. □

<u>8.8 Satz</u>. $Raba \to a=b$.

<u>Beweis</u>: Da auch $Raab$ ist (8.5, 8.2), ergibt sich die Behauptung aus 8.7. □

<u>8.9 Satz</u>. $Rabc \wedge \mathrm{Col}\, abc \to a=b \vee c=b$.

<u>Beweis</u>: Wir nehmen noch $a \neq b$ an. Nach 8.3 ist dann $Rcbc$, nach 8.8 also $c=b$. □

<u>8.10 Satz</u>. $Rabc \wedge (abc) \equiv (a'b'c') \to Ra'b'c'$.

<u>Beweis</u>: Falls $b=c$ ist, ergibt sich die Behauptung aus $b'=c'$ und 8.5. Sei

nun $b \neq c$ (und damit $b' \neq c'$). Sei weiter $d = S_b(c)$, $d' = S_{b'}(c')$. Dann ergibt sich

$$\mathrm{AFS}\begin{pmatrix} c & b & d & a \\ c' & b' & d' & a' \end{pmatrix}$$

und damit $ad \equiv a'd'$. Nach Voraussetzung ist auch $a'c' \equiv ac \equiv ad$; also ist $a'c' \equiv a'd'$, d.h. $\mathrm{R}a'b'c'$. □

<u>8.11 Definition</u>. (i) Die Geraden A und A' *stehen* aufeinander *senkrecht* (oder *sind* zueinander *senkrecht* oder *orthogonal*) *im Punkte* x:

$$A \underset{x}{\perp} A' :\leftrightarrow \mathrm{Ln}\,A \wedge \mathrm{Ln}\,A' \wedge x \in A \wedge x \in A' \wedge \forall u \forall v [u \in A \wedge v \in A' \rightarrow \mathrm{R}uxv].$$

(ii) Die Geraden A und A' sind zueinander *senkrecht* (*orthogonal*) (ohne Angabe von x):

$$A \perp A' :\leftrightarrow \exists x A \underset{x}{\perp} A'.$$

(iii) (Abkürzungen bei der Beschreibung von Geraden durch Punkte) A bzw. ab ist *senkrecht* zu cd (in x):

$$A \underset{(x)}{\perp} cd :\leftrightarrow c \neq d \wedge A \underset{(x)}{\perp} \mathrm{L}(cd),$$

$$ab \underset{(x)}{\perp} cd :\leftrightarrow a \neq b \wedge c \neq d \wedge \mathrm{L}(ab) \underset{(x)}{\perp} \mathrm{L}(cd).$$

Die Angabe $\underset{(x)}{\perp}$ bedeutet hier wie im folgenden, daß die Feststellung sowohl für $\underset{x}{\perp}$ als auch für $\perp$ gilt. Die zweite Beziehung in (iii) läßt sich natürlich auch ohne Verwendung des Begriffs Gerade definieren (vgl. Anmerkung 6.26(i)).

Dann gilt offenbar

<u>8.12 Satz</u>. $A \underset{(x)}{\perp} A' \leftrightarrow A' \underset{(x)}{\perp} A$.

Auf Grund von 8.3 (und 8.2) ergibt sich als gleichwertige Charakterisierung an Stelle von 8.11(i):

<u>8.13 Satz</u>. $A \underset{x}{\perp} A' \leftrightarrow \mathrm{Ln}\,A \wedge \mathrm{Ln}\,A' \wedge x \in A \wedge x \in A'$
$\wedge \exists u \exists v [u \in A \wedge v \in A' \wedge u \neq x \wedge v \neq x \wedge \mathrm{R}uxv]$.

Ist $A \underset{x}{\perp} A'$, so enthalten A und A' also auf Grund von 8.9 nicht-kollineare Punkte und sind damit voneinander verschieden, also ist x der nach 6.21 eindeutig bestimmte Schnittpunkt von A und A'. Somit erhalten wir:

<u>8.14 Satz</u>. (i) $A \perp A' \rightarrow A \neq A'$.

(ii) $A \underset{x}{\perp} A' \leftrightarrow A \perp A' \wedge AA'\mathrm{Is}\, x$.

(iii) $\overset{\leq 1}{\exists} x A \underset{x}{\perp} A'$.

In den späteren Sätzen dieses Abschnitts haben wir es vor allem mit Orthogonalitätsbeziehungen der Form $ab \perp cx$, wobei $x \in \mathrm{L}(ab)$ ist, zu tun. Daher beweisen wir darüber zunächst zwei Hilfssätze.

8.15 Satz. $a \neq b \wedge \mathrm{Col}\, abx \rightarrow [ab \perp cx \leftrightarrow ab \underset{x}{\perp} cx]$.

Beweis: Die Richtung von links nach rechts ergibt sich aus 8.14(i), (ii); die Umkehrung ist trivial. □

8.16 Satz. $a \neq b \wedge \mathrm{Col}\, abx \wedge \mathrm{Col}\, abu \wedge u \neq x \rightarrow [ab \perp cx \leftrightarrow \neg\mathrm{Col}\, abc \wedge \mathrm{R}cxu]$.

Beweis: Gelte die Voraussetzung. Weiter sei zunächst $ab \perp cx$, nach 8.15 also $ab \underset{x}{\perp} cx$. Nach Definition 8.11(i), (iii) ist dann $\mathrm{R}cxu$, außerdem $c \neq x$, nach 8.9 also $\neg\mathrm{Col}\, uxc$ und wegen $\mathrm{L}(ab) = \mathrm{L}(ux)$ auch $\neg\mathrm{Col}\, abc$. Gelte nun umgekehrt die rechte Seite der $\leftrightarrow$-Beziehung. Wegen $\neg\mathrm{Col}\, abc$ ist zunächst $c \neq x$. Nach 8.13 (mit $v = c$) ist dann $\mathrm{L}(ab) \underset{x}{\perp} \mathrm{L}(cx)$ und damit $ab \perp cx$. □

8.17 Definition. Ist $ab \perp cx \wedge \mathrm{Col}\, abx$, so heißt die Gerade $\mathrm{L}(cx)$ eine *Senkrechte* im Punkt x auf der Geraden $\mathrm{L}(ab)$ oder ein (das, s. 8.18) *Lot* vom Punkt c auf die Gerade $\mathrm{L}(ab)$; x heißt dann der *Fußpunkt* dieses Lots. Dafür, daß man diese Gerade konstruiert (bei gegebenem x auf $\mathrm{L}(ab)$ bzw. c außerhalb $\mathrm{L}(ab)$, vgl. 8.21, 8.18), sagt man auch, daß man die Senkrechte *errichtet* bzw. das Lot *fällt*.

8.18 Satz (Lotsatz). $\neg\mathrm{Col}\, abc \rightarrow \overset{=1}{\exists} x [\mathrm{Col}\, abx \wedge ab \perp cx]$.
(Auf eine Gerade gibt es von einem Punkt außerhalb der Geraden stets genau ein Lot).

Beweis: 1. Eindeutigkeit: Sei $\mathrm{Col}\, abx_\nu \wedge ab \perp cx_\nu$ ($\nu = 1,2$). Nach 8.15, 12, 11(iii), (i) ist dann $\mathrm{R}cx_1x_2$ und $\mathrm{R}cx_2x_1$, nach 8.7 also $x_1 = x_2$.

2. Existenz (Konstruktion des Lotfällens, Abb. 25): Durch Streckenabtragung (A4) erhalten wir einen Punkt y mit $\mathrm{B}bay$ (statt dessen genügt $\mathrm{Col}\, aby$) und $ay \equiv ac$. Dann hat die Strecke cy nach 7.25 einen Mittelpunkt (da a von beiden Endpunkten denselben Abstand hat). Sei also $\mathrm{M}cpy$ und damit (nach 8.1) $\mathrm{R}apy$. Wieder mittels A4 (bzw. 7.5) erhalten wir Punkte

z, q, q', c', so daß $Bayz \wedge yz \equiv yp$, $Bpyq \wedge yq \equiv ya$, $q' = S_z(q)$, $Bq'yc' \wedge yc' \equiv yc$. Dann ist

$$\mathrm{AFS}\begin{pmatrix} ayzq \\ qypa \end{pmatrix},$$

also $zq \equiv pa$ und damit $(apy) \equiv (qzy)$. Nach 8.10 ist dann $Rqzy$ und damit $yq \equiv yq'$. Wegen $yc \equiv yc'$ hat die Strecke cc' nach 7.25 einen Mittelpunkt x, für den $Ryxc$ gilt. Nach Konstruktion bilden nun die Punkte q, z, q', y, c', x, c eine Krippenfigur (7.23); nach dem Krippenlemma (7.22) ist also $Bzyx$ und damit Col zyx. Außerdem ist $c \neq y \neq z$, woraus sich $y \neq x$ ergibt

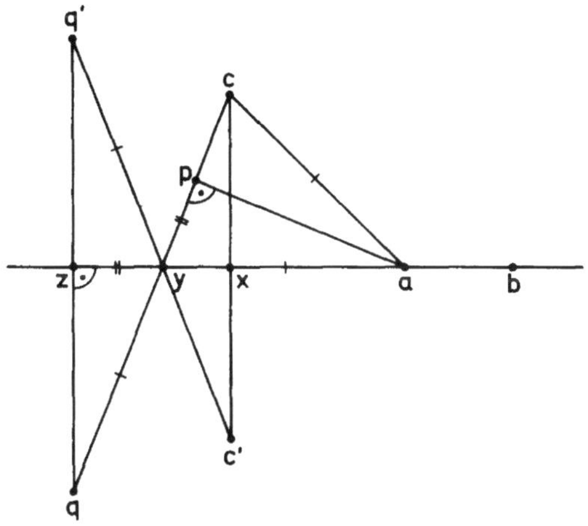

Abb. 25

(andernfalls lägen sämtliche Punkte der betrachteten Krippenfigur auf einer Geraden). Nach 8.13 ist also $yz \perp_x cx$, wegen $L(ab) = L(yz)$ also auch Col abx und $ab \perp cx$. □

<u>8.19 Anmerkungen.</u> (i) Üblicherweise konstruiert man das Lot von einem Punkt c auf eine Gerade A (in einer Ebene) durch Schlagen von Kreisen (schneidet ein Kreis um c mit genügend großen Radius die Gerade A in den beiden Punkten u und v und schneiden sich Kreise mit demselbem Radius um u und um v in dem von c verschiedenen Punkt c', so ist $L(cc')$ das gesuchte Lot). Diese Konstruktion läßt sich jedoch zum Beweis für die Existenz des Lotes aus unseren Axiomen (A1 bis A8) nicht verwenden; denn sie beruht auf dem Kreisaxiom (s. 1.2), das sich aus unseren Axiomen (sogar aus A1 bis A10) gar nicht beweisen läßt.

Aus demselben Grund wird diese Konstruktion auch im Hilbertschen Aufbau nicht zum Beweis des Lotsatzes verwendet. Statt dessen wird dort der Lotsatz mit Hilfe der Winkelantragung (III.4 in Hilbert 1977, S. 13) bewiesen (vgl. etwa Baldus 1964, Satz 23e). Auch diese zweite Methode ist hier nicht anwendbar, da im Gegensatz zum Hilbertschen Aufbau keine Axiome über die Winkelkongruenz zur Verfügung stehen. Zum Beweis entsprechender Sätze über die Winkelkongruenz werden vielmehr später (Abschnitt 11) schon die hier behandelten Sätze über rechte Winkel verwendet (vgl. den Beweis von 11.15).

Das hier über den Lotsatz Gesagte gilt ebenso für die Sätze 8.21 (Errichten von Senkrechten) und 8.22 (Existenz des Mittelpunkts von Strecken).

Der vorliegende Beweis von 8.18 läßt sich allerdings als sinngemäße Übertragung des Beweises aus Baldus 1964 ansehen. Dort wird nämlich nach Wahl beliebiger Punkte y und a auf der gegebenen Geraden (mittels Winkelantragung und Streckenabtragung) ein Punkt $c' \neq c$ konstruiert, so daß die Winkel $\sphericalangle ayc$ und $\sphericalangle ayc'$ zueinander kongruent sind (Def. 11.2) und $yc' \equiv yc$ ist. Das wird für einen speziellen Punkt y (mit anderen Hilfsmitteln) auch hier erreicht.

(ii) Im ursprünglichen Manuskript, das sich mit der ebenen Geometrie beschäftigt (s. 0.3), ist das Krippenlemma nicht enthalten. Statt dessen wird dort zum Beweis von 8.18 das obere Dimensionsaxiom (A9) verwendet: Mittels

$$\mathrm{AFS}\begin{pmatrix} q & yc & z \\ q' & yc' & z \end{pmatrix}$$

erhält man zunächst $zc \equiv zc'$ und dann nach A9 die gewünschte Kollinearität Col zyx; alles andere bleibt ungeändert.

Die Rolle des Krippenlemmas im hier dargestellten Aufbau besteht vor allem darin, diese Anwendung des oberen Dimensionsaxioms entbehrlich zu machen. Im Beweis von 8.18 wird durch das Krippenlemma - anschaulich gesagt - gesichert, daß der Punkt x mit den vorher betrachteten Punkten in einer Ebene liegt (was erst in 9.20 präzisiert wird).

<u>8.20 Lemma</u>. $\mathrm{R}abc \wedge \mathrm{MS}_a(c)p\mathrm{S}_b(c) \rightarrow \mathrm{R}bap \wedge [b \neq c \rightarrow a \neq p]$ (Abb. 26).

<u>Beweis</u>: Zur Abkürzung setzen wir $d = \mathrm{S}_b(c)$, $b' = \mathrm{S}_a(b)$, $c' = \mathrm{S}_a(c)$, $d' = \mathrm{S}_a(d)$, $p' = \mathrm{S}_a(p)$ (Abb. 27). Dann ist auch $\mathrm{R}b'bc$ (nach 8.3 bzw. trivialerweise im Falle $a=b$). Unter Benutzung des Satzes, daß S_a ein Automorphismus ist (7.14), erhält man nun $\mathrm{R}bb'c'$, also $bc' \equiv bd'$, und weiter

$$\mathrm{IFS}\begin{pmatrix} c'p & db \\ d'p'cb \end{pmatrix} .$$

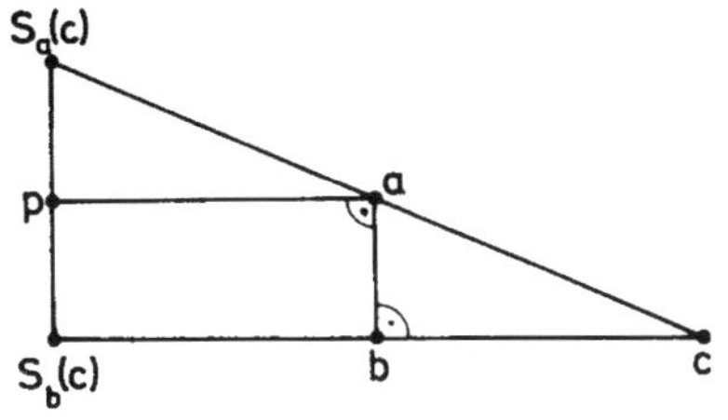

Abb. 26

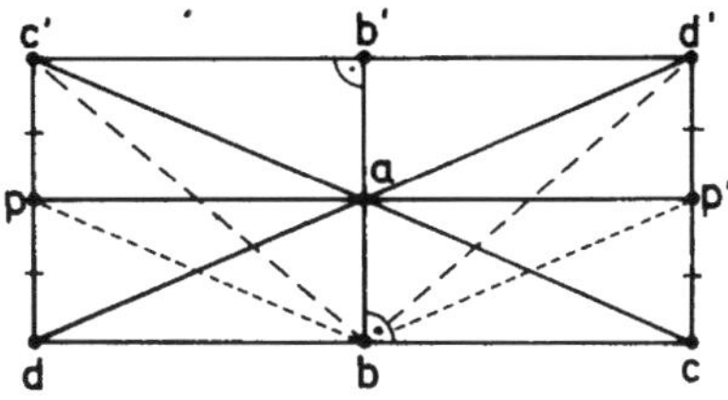

Abb. 27

Nach 4.2 ist also $bp \equiv bp'$, d.h. Rbap (Teil 1 der Behauptung). Wäre nun $a=p$, so wäre auch $c=S_a(c')=S_p(c')=d$ und nach 7.3 somit $b=c$ (Teil 2 der Behauptung). □

8.21 Satz. $a \neq b \rightarrow \exists p \exists t[ab \perp pa \wedge \mathrm{Col}\, abt \wedge \mathrm{B}ctp]$.
(*Auf einer Geraden* [*hier* L(ab)] *gibt es in einem Punkt* [*hier* a] *"in einer gegebenen Halbebene" eine Senkrechte,* Abb. 28.)

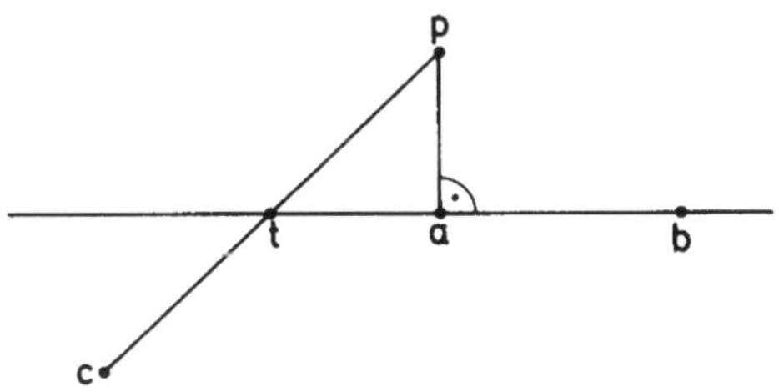

Abb. 28

Anmerkung. Falls c nicht auf L(ab) liegt, drücken die Forderungen über t hierbei aus, daß p in der "zu c entgegengesetzten Halbebene bezüglich L(ab)" liegt (d.h. derjenigen Halbebene mit der Begrenzung L(ab), die in der Ebene durch a, b und c enthalten ist und den Punkt c nicht enthält, Präzisierung in 9.14 f., s.a. 10.15). Falls c auf L(ab) liegt, wird durch diese Forderungen keine Halbebene festgelegt.

Beweis von 8.21: Fall 1: $\neg$ Col abc. Sei x gemäß 8.18 der Fußpunkt des Lots von c auf $L(ab)$. Dann ist $Raxc$ und damit $aS_x(c) \equiv ac \equiv aS_a(c)$ (Abb. 29). Nach 7.25 hat die Strecke $S_x(c)S_a(c)$ also einen Mittelpunkt p, für den

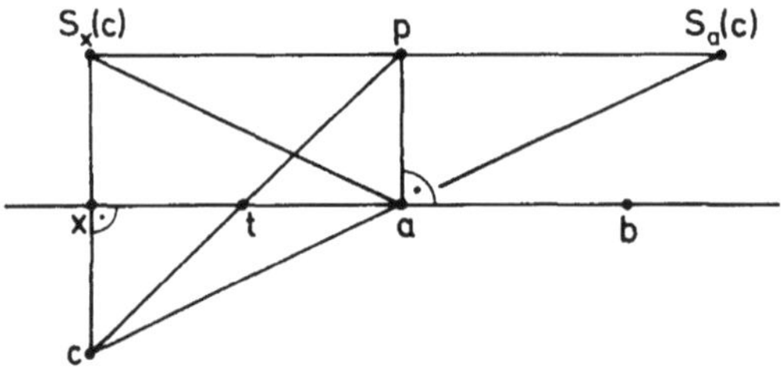

Abb. 29

nach 8.20 $Rxap$ und $a \neq p$ gilt. Sei t gemäß 3.17 ein Punkt mit $Bctp \wedge Bxta$. Dann ist insbesondere Col abt. Ist nun $x \neq a$, so ist $ab \perp pa$ nach 8.13. Ist dagegen $x=a$, so ist auch $t=a$, also $L(pa)=L(pt)=L(cx)$ und damit ebenfalls $ab \perp pa$ nach Wahl von x.

Fall 2: Col abc. Sei c' gemäß 6.25 ein Punkt mit $\neg$ Col abc'. Auf Grund von Fall 1 gibt es jedenfalls einen Punkt p mit $ab \perp pa$. Dieser leistet zusammen mit $t=c$ das Verlangte. □

8.22 Satz. $\forall a \forall b \overset{=1}{\exists} x Maxb$.

(*Jede Strecke hat genau einen Mittelpunkt.*)

Beweis: Die Eindeutigkeit des Mittelpunkts wurde schon in 7.17 bewiesen. Es bleibt die Existenz zu zeigen.

Fall 1: $a=b$. Dann leistet $x=a$ das Verlangte.

Fall 2: $a \neq b$ (Abb. 30). Durch zweimalige Anwendung von 8.21 (zuerst mit einem Punkt auf $L(ab)$) erhalten wir Punkte q, p, t mit $ab \perp qb$ und $ab \perp pa \wedge$ Col $abt \wedge Bqtp$. Wegen der Symmetrie des bisher Festgestellten in bezug auf Vertauschung von a mit b und p mit q können wir o. B. d. A. $ap \leq bq$ annehmen (vgl. Anm. 7.24). Sei nun r gemäß Def. 5.4 gewählt, so daß $Bbrq \wedge ap \equiv br$, und x gemäß A7 (angewandt auf $Bptq \wedge Bbrq$), so daß $Btxb \wedge Brxp$. Dann ist insbesondere Col abx, nach 8.9 aber $\neg$ Col $abp \wedge \neg$ Col abr. Es genügt nun zu zeigen, daß $bp \equiv ar$ ist; denn nach dem Lemma vom zentral-symmetrischen Viereck (7.21) ergibt sich daraus sofort $Maxb \wedge Mpxr$ (x leistet das Verlangte).

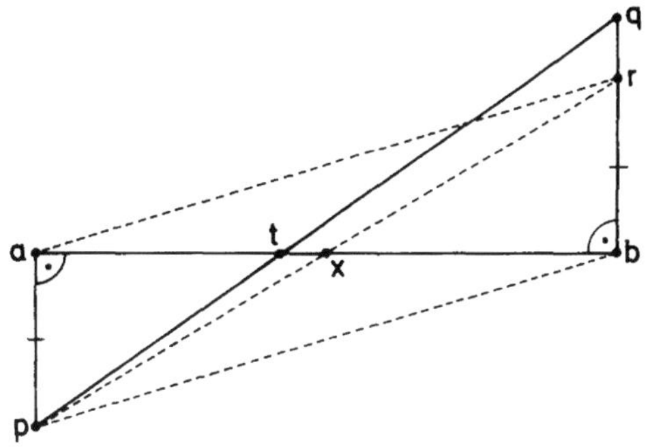

Abb. 30

Zunächst ist $x \neq a$ (sonst wäre Col par und damit $b=a$ wegen der Eindeutigkeit des Lotfußpunkts). Sei nun (Abb. 31) $p'=S_a p$, und seien r', m

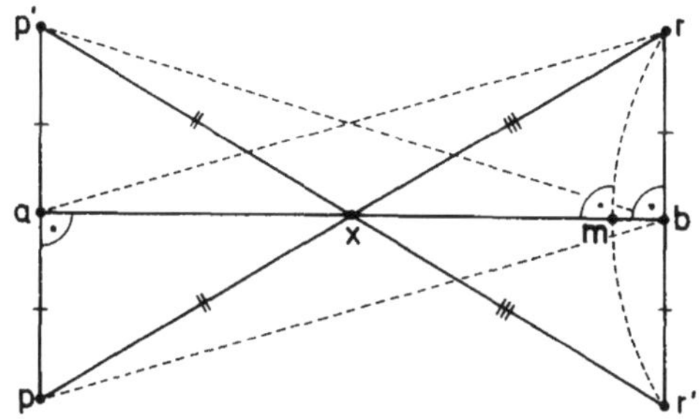

Abb. 31

gemäß A4 bzw. 7.25 Punkte mit $\mathrm{B}p'xr' \wedge xr' \equiv xr$ und $\mathrm{M}rmr'$, also auch $\mathrm{R}rmx$. Wegen $\mathrm{R}xap$ ist $xp \equiv xp'$ und (nach 8.9) $\neg$ Col xpp'. Nach dem Krippenlemma (7.22) ist $\mathrm{B}axm$, außerdem $m \neq x$ (wie im Beweis von 8.18). Somit ist $m \in \mathrm{L}(ax) = \mathrm{L}(ab)$ und $ab \perp rm$, wegen der Eindeutigkeit des Lotfußpunkts (8.18) also $m=b$ und damit $\mathrm{M}rbr'$. Nun ergibt sich

$$\mathrm{IFS}\begin{pmatrix} p' & a & p & r \\ r & b & r' & p' \end{pmatrix},$$

nach 4.2 also $ar \equiv bp'$. Wegen $\mathrm{R}bap$ ist andererseits $bp \equiv bp'$ und damit $bp \equiv ar$, wie zu zeigen. □

<u>8.23 Anmerkung</u>. Der erste Teil dieses Existenzbeweises (Fall 2) kann wieder (vgl. Anm. 8.19(i)) als Übertragung eines Spezialfalles der Hilbertschen Konstruktion (Hilbert 1977, Satz 26, S. 25f.) angesehen werden, wo statt der rechten Winkel beliebige zueinander kongruente Winkel

in a und b angetragen werden. Der zweite Teil des vorliegenden Beweises (der bei Hilbert nicht auftritt) wäre entbehrlich, wenn man schon den Satz zur Verfügung hätte, daß zwei rechtwinklige Dreiecke mit kongruenten entsprechenden Katheten auch kongruente Hypotenusen besitzen. Dieser Satz wird hier jedoch erst später als 10.12 bewiesen, und zwar mit Hilfe von 8.22. Auch die Kongruenz rechter Winkel (11.16), die im Hilbertschen Aufbau viel früher gezeigt werden kann, beruht im wesentlichen auf 10.12.

Durch Weglassen der ersten Schritte im vorangehenden Beweis erhält man unmittelbar das folgende Lemma, das später (in 9.4) benötigt wird.

8.24 Lemma. $pa \perp ab \wedge qb \perp ab \wedge \mathrm{Col}\, abt \wedge \mathrm{B}ptq \wedge \mathrm{B}brq \wedge ap \equiv br$
$\rightarrow \exists x[\mathrm{M}axb \wedge \mathrm{M}pxr]$ (Abb. 30).

9. Halbebenen und Ebenen, Unterräume

Hinweis. Für den ersten Teil von Abschnitt 10 (Sätze über Geradenspiegelungen bis 10.12) wird dieser Abschnitt nicht benötigt.

9.1 Definition. Die Punkte a, b liegen auf *entgegengesetzten Seiten* der Geraden A bzw. $\mathrm{L}(pq)$ oder auch *a liegt entgegengesetzt zu b* bezüglich A bzw. $\mathrm{L}(pq)$ (Abb. 32):

$$\mathrm{B}aAb :\leftrightarrow \mathrm{Ln}\,A \wedge a \notin A \wedge b \notin A \wedge \exists t[t \in A \wedge \mathrm{B}atb],$$

$$\mathrm{B}a{}^{p}_{q}b :\leftrightarrow p \neq q \wedge \mathrm{B}a\mathrm{L}(pq)b.$$

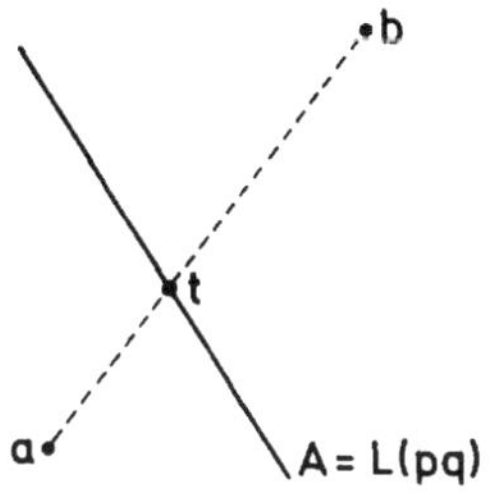

Abb. 32

Aus 3.2 erhalten wir unmittelbar die folgende "Symmetrie", die wir in weiteren Beweisen i.a. nicht besonders erwähnen werden.

9.2 Satz. $\mathrm{B}aAb \rightarrow \mathrm{B}bAa$.

Grundlegend für die Einführung von Halbebenen ist der Satz 9.5, für dessen Beweis zunächst zwei Lemmata behandelt werden.

9.3 Lemma. $BaAc \wedge m\in A \wedge Mamc \wedge r\in A \rightarrow \forall b[a \underset{r}{\simeq} b \rightarrow BbAc]$.
(*Wenn a und c auf entgegengesetzten Seiten der Geraden A liegen, und zwar spiegelbildlich bezüglich eines Punktes von A, und r auf A liegt, so liegt jeder Punkt b der Halbgeraden* H(*ra*) *entgegengesetzt zu c bezüglich A*, Abb. 33.)

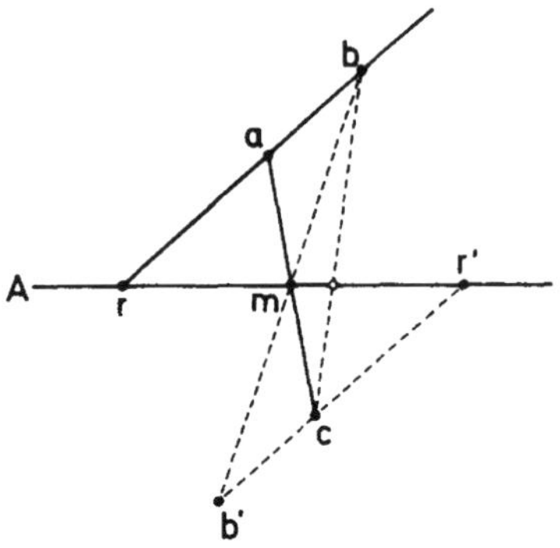

Abb. 33

Beweis: Sei $a \underset{r}{\simeq} b$. Nach Def. 6.1(ii) ist $Brba \vee Brab$.

Fall 1: $Brba$. Da auch $Bcma$ ist, erhalten wir nach A7 einen Punkt t mit $Bbtc \wedge Bmtr$. Dieser liegt insbesondere auf A und leistet somit das in der Definition von $BbAc$ Verlangte.

Fall 2: $Brab$ (Abb. 33). Sei $b' = S_m b$ und $r' = S_m r$. Nach 7.14 (S_m ist Automorphismus) ist $Br'cb'$. Da auch b' und b spiegelbildlich bezüglich m und damit entgegengesetzt zu A liegen, liefert Fall 1 die Beziehung $BcAb$. □

9.4 Lemma. $BaAc \wedge r\in A \wedge A \perp ar \wedge s\in A \wedge A \perp cs$

$\rightarrow [Mrms \rightarrow \forall u(u \underset{r}{\simeq} a \leftrightarrow S_m(u) \underset{s}{\simeq} c)]$

$\wedge \forall u \forall v[u \underset{r}{\simeq} a \wedge v \underset{s}{\simeq} c \rightarrow BuAv]$.

(*Wenn a, c auf entgegengesetzten Seiten von A liegen und r, s die Fußpunkte der Lote von a bzw. c auf A sind, so liegen die beiden Halbgeraden* H(*ra*) *und* H(*sc*) *"spiegelbildlich" bezüglich des Mittelpunkts von rs, und jeder Punkt der einen liegt bezüglich A entgegengesetzt zu jedem Punkt der anderen*, Abb. 34.)

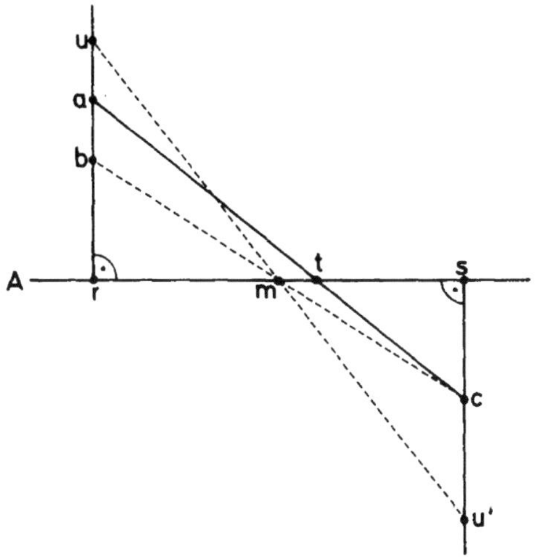

Abb. 34

Beweis: Sei t gemäß Def. von $BaAc$ (9.1) gewählt.

Fall 1: $r \neq s$. Wegen der Symmetrie von Voraussetzung und Behauptung in bezug auf Vertauschung von a mit c und r mit s können wir o. B. d. A. $sc \leq ra$ annehmen (Abb. 34). Gemäß Def. 5.4 und Lemma 8.24 wählen wir Punkte b, m mit $Brba \wedge sc \equiv rb$ und $Mrms \wedge Mbmc$. Nach 8.22 (7.17) ist m schon eindeutig festgelegt durch die Bedingung $Mrms$. Für beliebiges u ist nun $u \underset{r}{\approx} a$ gleichwertig mit $u \underset{r}{\approx} b$, und das wiederum - da S_m ein Automorphismus ist - mit $S_m(u) \underset{s}{\approx} c$. Somit gilt der erste Teil der Behauptung. Seien nun u, v beliebig mit $u \underset{r}{\approx} a$, $v \underset{s}{\approx} b$. Nach dem schon Bewiesenen ist dann $u' := S_m(u) \underset{s}{\approx} c$ und damit auch $u' \underset{s}{\approx} v$, außerdem $Bu'Au$, nach Lemma 9.3 also $BvAu$, wie zu zeigen.

Fall 2: $r=s$. Nach Voraussetzung ist $Rart \wedge Rcst$, wegen $Batc$ und 8.6 also $t=r=s$. Somit sind die Halbgeraden $H(ra)$ und $H(sc)$ zueinander entgegengesetzt (6.9). Daraus ergibt sich die Behauptung schon auf Grund von Sätzen über Halbgeraden (vgl. 6.10). □

9.5 Satz. $BaAc \wedge r \in A \rightarrow \forall b[a \underset{r}{\approx} b \rightarrow BbAc]$.

(Wenn a und c auf entgegengesetzten Seiten der Geraden A liegen und r auf A liegt, so liegt jeder Punkt b der Halbgeraden $H(ra)$ entgegengesetzt zu c bezüglich A.)

<u>Beweis</u> (nach Gupta 1965, Abb. 35): Seien x, y, z die Fußpunkte der Lote von a, b bzw. c auf A, m der Mittelpunkt von xz und

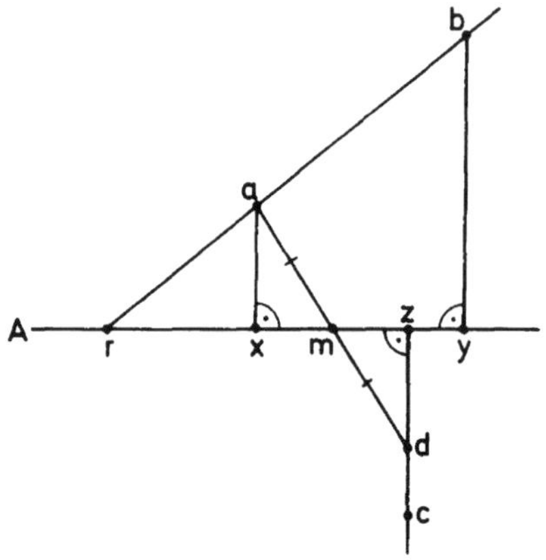

Abb. 35

$d = S_m(a)$. Nach 9.4 (Teile 1 und 2 der Beh.) ist dann $d \underset{z}{\equiv} c$ und $BaAd$. Nach 9.3 ist nun $BbAd$ und damit nach 9.4 (Teil 2) auch $BbAc$. □

<u>9.6 Satz</u>. (Satz von Pasch, äußere Form).

$$Bacp \wedge Bbqc \rightarrow \exists x[Baxb \wedge Bpqx] \qquad \text{(Abb. 36)}.$$

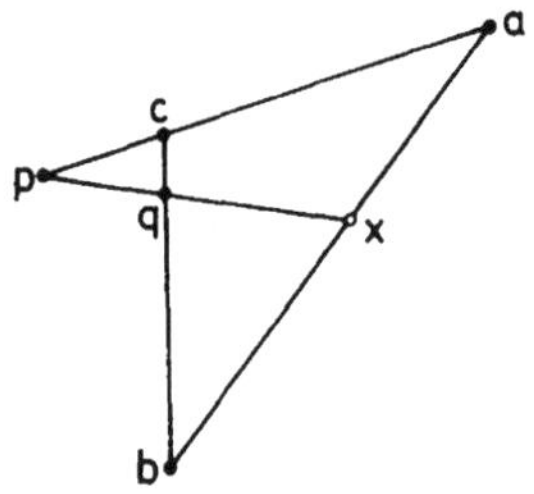

Abb. 36

<u>Beweis</u> (nach Gupta 1965, vgl. 0.3, 1.7.5): <u>Fall 1</u>: Col pqc.

<u>Fall 1a</u>: $Bpqc$. Nach 3.6 ist dann auch $Bpqa$, also leistet $x=a$ das Verlangte.

<u>Fall 1b</u>: $\neg Bpqc$. Nach 6.4 ist dann $p \underset{q}{\equiv} c$, wegen $Bcqb$ somit $Bpqb$, also leistet $x=b$ das Verlangte.

Fall 2: $\neg$ Col pqc. Dann ist $A:=L(pq)\neq L(cq)$.

Fall 2a: $b\in A$. Dann ist $b=q$ als Schnittpunkt der genannten Geraden, also leistet $x=b$ das Verlangte.

Fall 2b: $b\notin A$. Dann ist $BcAb \wedge c\underset{p}{\equiv}a$, nach 9.5 also $BaAb$. Dementsprechend sei x ein Punkt mit $Baxb \wedge x\in A$. Es bleibt $Bpqx$ zu zeigen. Gemäß A7 sei t ein Punkt mit $Bptx \wedge Bbtc$. Dann ist t Schnittpunkt der Geraden A und $L(cb)=L(cq)$, also $t=q$ und damit tatsächlich $Bpqx$. □

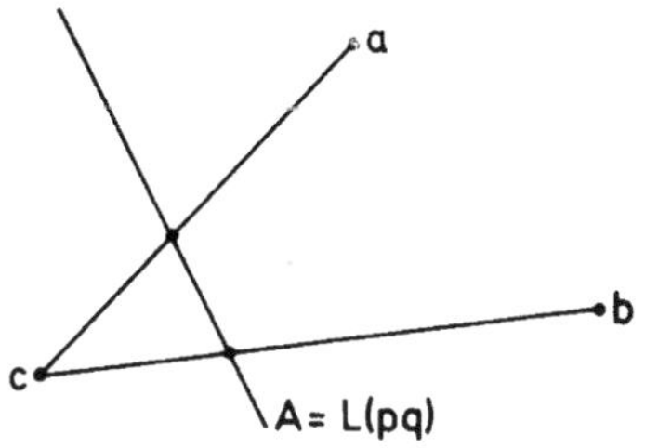

Abb. 37

9.7 Definition. Die Punkte a, b liegen auf *derselben Seite* der Geraden A bzw. $L(pq)$ oder sind *äquivalent* bezüglich pq (Abb. 37):

$$a\underset{A}{\equiv}b :\leftrightarrow \exists c[BaAc \wedge BbAc],$$

$$a\underset{pq}{\equiv}b :\leftrightarrow p\neq q \wedge a\underset{L(pq)}{\equiv}b.$$

In Hinblick auf die spätere Definition 9.14 sagen wir dafür auch a und b *liegen in derselben* oder *bestimmen dieselbe Halbebene* mit der Begrenzung A.

9.8 Satz. $BaAc \rightarrow (BbAc \leftrightarrow a\underset{A}{\equiv}b)$.

Beweis: Sei $BaAc$. Ist außerdem $BbAc$, so ist trivialerweise $a\underset{A}{\equiv}b$ nach Def. 9.7. Gelte nun $a\underset{A}{\equiv}b$, und seien d, x, y gemäß 9.7, 9.1 gewählt mit $BaAd \wedge BbAd$, $x,y\in A$, $Baxd$, $Bbyd$ (Abb. 38). Sei z ein Punkt gemäß A7 mit $Bxzb \wedge Byza$, und sei o. B. d. A. $z\notin A$ (falls $\neg$ Col abd, ist das nach Abschnitt 6 ohnehin der Fall, falls Col abd, d.h. $x=y$, leistet $z=a$ oder $z=b$ das Verlangte). Dann ist $a\underset{y}{\equiv}z$ und $z\underset{x}{\equiv}b$. Aus der Voraussetzung $BaAc$ ergibt sich durch zweimalige Anwendung von 9.5 somit $BzAc$ und $BbAc$. □

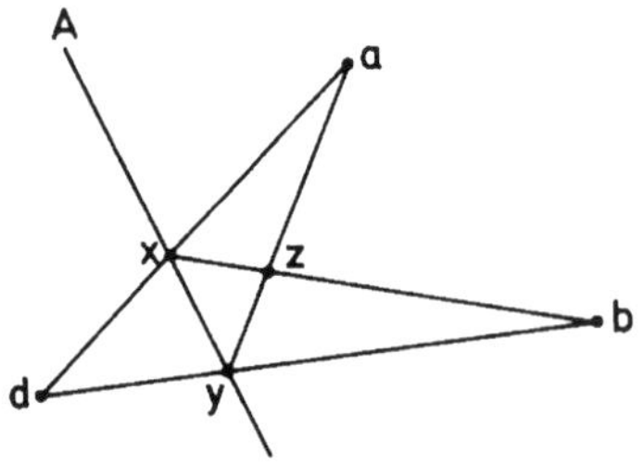

Abb. 38

9.9 Satz. $BaAb \to \neg a \underset{A}{\approx} b$.

Beweis: Aus $a \underset{A}{\approx} b$ würde sich nach 9.8 auch $BbAb$ ergeben, was nicht möglich ist (ein Punkt t gemäß 9.1 müßte sonst nach A6 mit b zusammenfallen im Widerspruch zu $b \notin A$). □

Mit 3.14 ergibt sich sofort

9.10 Lemma. $\mathrm{Ln}\,A \wedge a \notin A \to \exists c BaAc$.

Aus 9.7, 9.10, 9.8 ergibt sich, daß für eine beliebige Gerade A die zweistellige Relation $\underset{A}{\approx}$ eine Äquivalenzrelation auf der Menge der Punkte außerhalb A ist, d.h., daß die folgenden Sätze gelten.

9.11 Satz (*Reflexivität*). $\mathrm{Ln}\,A \wedge a \notin A \to a \underset{A}{\approx} a$.

9.12 Satz (*Symmetrie*). $a \underset{A}{\approx} b \to b \underset{A}{\approx} a$.

9.13 Satz (*Transitivität*). $a \underset{A}{\approx} b \wedge b \underset{A}{\approx} c \to a \underset{A}{\approx} c$.

9.14 Definition. Die Äquivalenzklassen bezüglich dieser Relation nennen wir die *von der Geraden A ausgehenden Halbebenen* oder die Halbebenen *mit der Begrenzung A* oder die *Seiten* von A. Die Äquivalenzklasse eines beliebigen Punktes a außerhalb A, d.h. die Punktmenge

$\mathrm{Hp}(A,a) := \{x \mid x \underset{A}{\approx} a\}$ (definiert für $\mathrm{Ln}\,A \wedge a \notin A$),

nennen wir die *von A ausgehende Halbebene ("half-plane") durch a* oder auch die *durch a bestimmte Seite von* oder *Halbebene mit der Begrenzung A*. Ist $A = \mathrm{L}(pq)$, so verwenden wir dafür auch die Bezeichnung

$\mathrm{Hp}(pq,a) := \mathrm{Hp}(\mathrm{L}(pq),a) = \{x \mid x \underset{pq}{\approx} a\}$ (definiert für $\neg\,\mathrm{Col}\,pqa$).

9.15 Definition. Von A ausgehende Halbebenen $\mathrm{Hp}(A,a)$, $\mathrm{Hp}(A,b)$ heißen *zueinander entgegengesetzt* oder auch $\mathrm{Hp}(A,a)$ *entgegengesetzt zu b* (mit

$b \notin A$), falls BaAb ist.

9.16 Anmerkung. Nach 9.8 ist diese Definition unabhängig von der Wahl der bestimmenden Punkte (Repräsentanten) auf Hp(A,a) und (im ersten Falle) auf Hp(A,b), und die zu einer Halbebene entgegengesetzte Halbebene ist nach 9.8, 9.10 eindeutig bestimmt und nach 9.9 von ihr verschieden.

9.17 Satz. $a \underset{A}{\simeq} b \wedge \mathrm{B}acb \to c \underset{A}{\simeq} a$.
(Halbebenen sind "konvexe" Punktmengen, d.h., sie enthalten mit Punkten a, b auch jeden Punkt c auf der Verbindungsstrecke).

Beweis: Seien d, x, y wie im Beweis von 9.8 gewählt und t gemäß 3.17 derart, daß $\mathrm{B}ctd \wedge \mathrm{B}xty$ ist (Abb. 39). Dann ist BcAd; denn t leistet das in der Definition (9.1) Verlangte, und es ist auch $c \notin A$ (andernfalls wäre BcAb entgegen 9.9). Zusammen mit BaAd liefert das die Behauptung. □

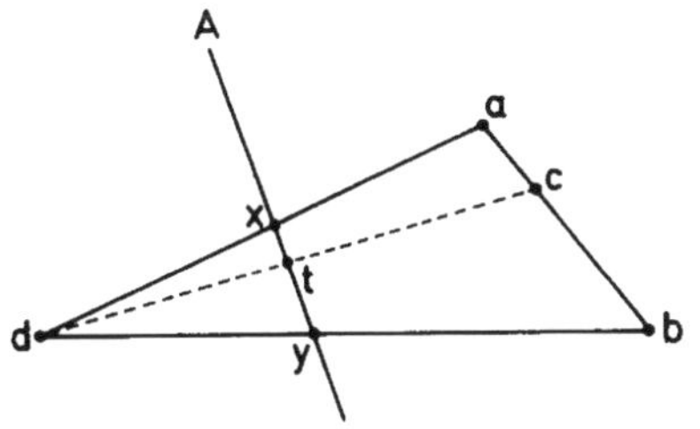

Abb. 39

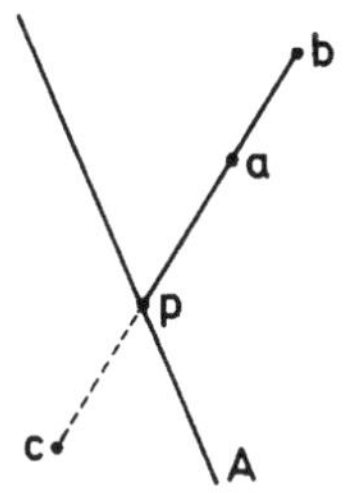

Abb. 40

Ein häufig benutzter Zusammenhang zwischen den Begriffen entgegengesetzt bzw. äquivalent bezüglich Geraden und bezüglich Punkten (6.1) wird in den folgenden beiden Sätzen angegeben.

9.18 Satz. $\mathrm{Ln}\, A \wedge p \in A \wedge \mathrm{Col}\, abp \to [\mathrm{B}aAb \leftrightarrow \mathrm{B}apb \wedge a \notin A \wedge b \notin A]$.

9.19 Satz. $\mathrm{Ln}\, A \wedge p \in A \wedge \mathrm{Col}\, abp \to [a \underset{A}{\simeq} b \leftrightarrow a \underset{p}{\simeq} b \wedge a \notin A]$.

Beweise: 9.18 ergibt sich leicht aus 9.1 und der Eindeutigkeit des Schnittpunkts von Geraden (Abb. 32); 9.19 aus 9.18 mittels 9.8, 6.2 unter Verwendung eines Punktes c auf der zu H(pa) entgegengesetzten Halbgeraden (Abb. 40). □

Ebenen setzen wir nun aus zueinander entgegengesetzten Halbebenen und der gemeinsamen Begrenzung zusammen.

9.20 Definition. Unter der *durch A und r bestimmten Ebene ("plane")* oder auch der *Ebene durch A und r* verstehen wir die Punktmenge

$\mathrm{Pl}(A,r):=\mathrm{Hp}(A,r)\cup A\cup \mathrm{Hp}(A,r')$ (definiert für $\mathrm{Ln}\,A \wedge r\notin A$),

wobei r' beliebig mit $\mathrm{Br}Ar'$ (nach 9.16 bzw. 9.8 ist diese Definition unabhängig von der Wahl von r'), d.h.

$\mathrm{Pl}(A,r):=\{x\,|\,x\underset{A}{\simeq}r \vee x\in A \vee \mathrm{Br}Ax\}$.

Ist $A=\mathrm{L}(pq)$, so verwenden wir dafür auch die Bezeichnung

$\mathrm{Pl}(pqr):=\mathrm{Pl}(\mathrm{L}(pq),r)$ (definiert für $\neg\mathrm{Col}\,pqr$).

Die Punktmenge E ist eine *Ebene*:

$\mathrm{Pl}\,E:\leftrightarrow \exists p\exists q\exists r[\neg\mathrm{Col}\,pqr \wedge E=\mathrm{Pl}(pqr)]$.

Für $a\in E$ sagen wir dann auch "*a liegt in* oder *inzidiert mit* der Ebene *E* " oder "*E geht durch* den Punkt *a* ", für $a\notin E$ auch "*a liegt außerhalb E* "; ist außerdem A eine Gerade, so sagen wir für $A\subseteq E$ auch "*A liegt in E* " oder "*E geht durch A* ".

Aus 9.15, 9.16 ergibt sich sofort

9.21 Satz. $\mathrm{Ln}\,A \wedge r\notin A \wedge s\in\mathrm{Pl}(A,r) \wedge s\notin A \rightarrow \mathrm{Pl}(A,r)=\mathrm{Pl}(A,s)$.

(*Zur Festlegung von* Pl(*A*,*r*) *kann man statt r jeden anderen Punkt dieser Ebene, der nicht auf A liegt, verwenden.*)

9.22 Lemma. $AA'\,\mathrm{Is}\,p \wedge r\in A' \wedge r\neq p \rightarrow A'\subseteq\mathrm{Pl}(A,r)$.

(*Schneiden sich die Geraden A, A' in p, so ist A' enthalten in der durch A und einen weiteren Punkt von A' bestimmten Ebene.*)

Beweis (Abb. 41, ohne die Punkte s, t): Nach 9.19 ist die Halbgerade $\mathrm{H}(pr)$ enthalten in der Halbebene $\mathrm{Hp}(A,r)$. Entsprechendes gilt für die entgegengesetzte Halbgerade und die entgegengesetzte Halbebene. Also gilt die Behauptung. □

9.23 Definition. Für sich schneidende Geraden A, A' definieren wir *die durch A und A' bestimmte Ebene*

$\mathrm{Pl}(A,A'):=\mathrm{Pl}(A,r)$,

wobei $r\in A'$ beliebig mit $r\notin A$. Nach 9.22, 9.21 ist diese Definition unabhängig von der Wahl von r.

9.24 Satz. $AA'\,\mathrm{Is}\,p \rightarrow A\subseteq\mathrm{Pl}(A,A') \wedge A'\subseteq\mathrm{Pl}(A,A') \wedge \mathrm{Pl}(A,A')=\mathrm{Pl}(A',A)$.

Beweis: Die ersten beiden Teile der Behauptung gelten nach 9.20, 9.22. Es genügt nun zu zeigen, daß $\mathrm{Pl}(A,A')\subseteq\mathrm{Pl}(A',A)$ ist (der Rest ergibt sich daraus durch Vertauschung von A mit A'). Sei r gemäß 9.23 gewählt

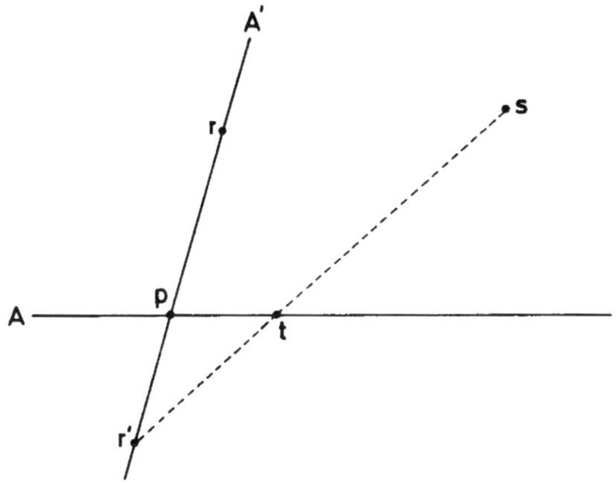

Abb. 41

und r' mit $Brpr' \wedge p \neq r'$, und sei $s \in \mathrm{Pl}(A,A')=\mathrm{Pl}(A,r)$ beliebig (Abb. 41). Dann liegt s auf A oder in einer der beiden Halbebenen $\mathrm{Hp}(A,r)$, $\mathrm{Hp}(A,r')$, o. B. d. A. etwa in $\mathrm{Hp}(A,r)$. Ist nun $s \in A'$, so ist trivialerweise $s \in \mathrm{Pl}(A',A)$. Andernfalls schneiden sich die Geraden A' und $\mathrm{L}(sr')$ in r'. Außerdem enthält die Gerade $\mathrm{L}(sr')$ einen Punkt t von A wegen $s \in A \vee Br'As$. Wegen $r' \notin A$ ist $t \neq r'$, nach 9.22 also $\mathrm{L}(sr') \subseteq \mathrm{Pl}(A',t)=\mathrm{Pl}(A',A)$ und damit $s \in \mathrm{Pl}(A',A)$. □

<u>9.25 Satz.</u> $\mathrm{Pl}\,E \wedge a \in E \wedge b \in E \wedge a \neq b \rightarrow \mathrm{L}(ab) \subseteq E \wedge \exists c\, E=\mathrm{Pl}(abc)$.
(*Eine Gerade durch zwei Punkte einer Ebene ist ganz in dieser Ebene enthalten, und die Ebene läßt sich durch diese Gerade und einen weiteren Punkt bestimmen.*)

<u>Beweis</u>: Nach 9.20 hat E die Darstellung $E=\mathrm{Pl}(A,r)$. Sei $A'=\mathrm{L}(ab)$.

<u>Fall 1</u>: $A'=A$. Dann ist die Behauptung trivial.

<u>Fall 2</u>: $A' \neq A$. Dann haben A und A' höchstens einen gemeinsamen Punkt, und wir können o. B. d. A. $b \notin A$ annehmen (Abb. 42). Sei c ein Punkt von A, der nicht auf A' liegt, und $A''=\mathrm{L}(cb)$. Dann schneidet A'' sowohl A (in c) als auch A' (in b). Somit ergibt sich

$E=\mathrm{Pl}(A,r)=\mathrm{Pl}(A,b)=\mathrm{Pl}(A,A'')$	(wegen $b \in E-A$, 9.21, 9.23)
$=\mathrm{Pl}(A'',A)$	(nach 9.24)
$=\mathrm{Pl}(A'',a)=\mathrm{Pl}(A'',A')$	(wegen $a \in E-A''$, 9.21, 9.23)
$=\mathrm{Pl}(A',A'')=\mathrm{Pl}(A',c)=\mathrm{Pl}(abc)$	(nach 9.24, 9.23).

Damit ist auch $A' \subseteq E$ (9.24). □

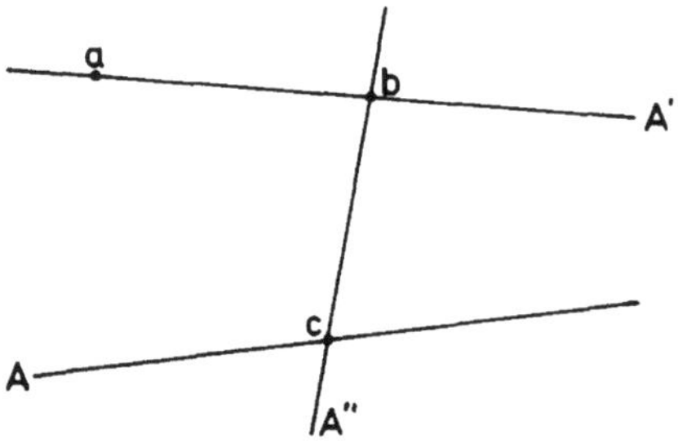

Abb. 42

9.26 Satz. $\neg \mathrm{Col}\, abc \wedge \mathrm{Pl}\, E \wedge a\in E \wedge b\in E \wedge c\in E \rightarrow E=\mathrm{Pl}(abc)$.

Beweis: Nach 9.25 ist E von der Form $E=\mathrm{Pl}(abr)=\mathrm{Pl}(A,r)$ mit $A:=\mathrm{L}(ab)$, wegen $c\in E-A$ und 9.21 also $E=\mathrm{Pl}(A,c)=\mathrm{Pl}(abc)$. □

Aus dem Bisherigen ergibt sich

9.27 Satz.

(1) $\neg \mathrm{Col}\, abc \rightarrow \mathrm{Pl}(abc)=\iota E[\mathrm{Pl}\, E \wedge a\in E \wedge b\in E \wedge c\in E]$.

(2) $\mathrm{Ln}\, A \wedge c\notin A \rightarrow \mathrm{Pl}(A,c)=\iota E[\mathrm{Pl}\, E \wedge A\subseteq E \wedge c\in E]$.

(3) $AA'\mathrm{Is}\, p \rightarrow \mathrm{Pl}(A,A')=\iota E[\mathrm{Pl}\, E \wedge A\subseteq E \wedge A'\subseteq E]$.

(Pl(abc) bzw. Pl(A,c) bzw. Pl(A,A') ist [unter den Voraussetzungen von 9.20 bzw. 9.23] diejenige Ebene, in der a, b, c bzw. A und c bzw. A und A' liegen.)

9.28 Folgerung. $\neg \mathrm{Col}\, abc \rightarrow \overset{=1}{\exists} E[\mathrm{Pl}\, E \wedge a\in E \wedge b\in E \wedge c\in E]$.
(Durch nicht-kollineare Punkte a, b, c geht stets genau eine Ebene.)

9.29 Folgerung. $\neg \mathrm{Col}\, abc \rightarrow \mathrm{Pl}(abc)=\mathrm{Pl}(bca)=\mathrm{Pl}(cab)$
$=\mathrm{Pl}(bac)=\mathrm{Pl}(acb)=\mathrm{Pl}(cba)$
(Unabhängigkeit von der Reihenfolge der Punkte in Def. 9.20).

9.30 Folgerung. $\mathrm{Ln}\, A \wedge \bigwedge_{\nu=1}^{2} [\mathrm{Pl}\, E_\nu \wedge A\subseteq E_\nu] \wedge E_1 \neq E_2 \rightarrow \forall x[x\in E_1 \wedge x\in E_2 \leftrightarrow x\in A]$.
(Verschiedene Ebenen mit einer gemeinsamen Geraden A haben als gemeinsame Punkte nur die Punkte dieser Geraden.)

9.31 Satz. $s \underset{pq}{\simeq} r \wedge s \underset{pr}{\simeq} q \rightarrow \mathrm{B}q^p_s r$ (Abb. 43).

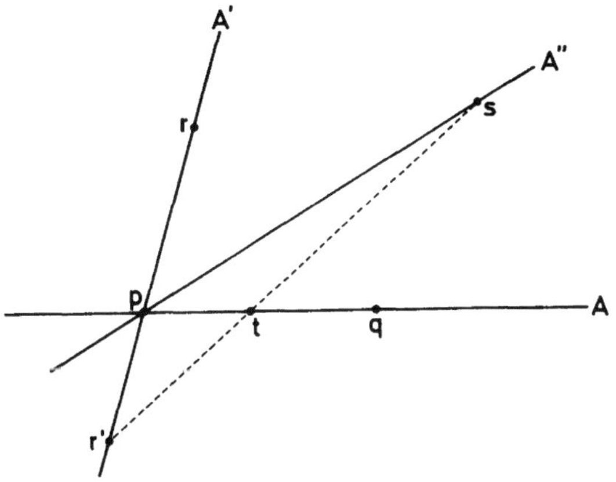

Abb. 43

Beweis: Sei $A=\mathrm{L}(pq)$, $A'=\mathrm{L}(pr)$, $A''=\mathrm{L}(ps)$, und seien r', t wie im Beweis von 9.24 gewählt. Dann ist $\mathrm{B}r'ts$, also insbesondere $t \underset{r'}{\simeq} s$. Durch mehrfache Anwendung von 9.19 ergibt sich somit $t \underset{A'}{\simeq} s \underset{A'}{\simeq} q$, $t \underset{p}{\simeq} q$ und $t \underset{A''}{\simeq} q$. Andererseits ist auch $t \underset{s}{\simeq} r'$, wegen $\mathrm{B}r'A''r$ und 9.5 also $\mathrm{B}tA''r$ und nach 9.8 somit $\mathrm{B}qA''r$. □

9.32 Definition. (i) Die Punkte $a_1, \ldots, a_n$ sind *komplanar*:

$$\mathrm{Cp}\, a_1 \ldots a_n :\leftrightarrow \exists E[\mathrm{Pl}\, E \wedge \bigwedge_{\nu=1}^{n} a_\nu \in E]$$

(d.h., $a_1, \ldots, a_n$ liegen in einer Ebene).

(ii) $A_1, \ldots, A_n$ sind *komplanare Geraden*:

$$\mathrm{Cp}\, A_1 \ldots A_n :\leftrightarrow \exists E[\mathrm{Pl}\, E \wedge \bigwedge_{\nu=1}^{n} (\mathrm{Ln}\, A_\nu \wedge A_\nu \subseteq E)].$$

9.33 Satz. $\mathrm{Cp}\, abcd \leftrightarrow \exists x\{[\mathrm{Col}\, abx \wedge \mathrm{Col}\, cdx]$
$\vee [\mathrm{Col}\, acx \wedge \mathrm{Col}\, bdx]$
$\vee [\mathrm{Col}\, adx \wedge \mathrm{Col}\, bcx]\}$.

(a, b, c, d sind komplanar genau dann, wenn sie sich in Paare zusammenfassen lassen, die auf Geraden A_1, A_2 mit einem gemeinsamen Punkt liegen.)

Beweis: Die Richtung von rechts nach links ist trivial; denn sind A_1, A_2 solche Geraden, so schneiden sie sich oder fallen zusammen, liegen also nach 9.24 in einer Ebene.

Gelte nun die linke Seite, und sei E eine Ebene, in der a, b, c, d liegen.

Fall 1: Die Punkte a, b, c, d liegen, bis auf höchstens einen, auf einer Geraden. O. B. d. A. sei $\mathrm{Col}\, abc$ (das übrige ergibt sich durch Vertauschung der Variablen). Sei A_1 eine Gerade durch a, b, c und A_2 eine Gerade durch c und d. Dann gilt die rechte Seite (hier der erste Fall der Disjunktion) mit $x=c$.

Fall 2: Fall 1 gilt nicht, d.h., a, b, c, d sind paarweise verschieden und je drei davon nicht kollinear. Sei $A=\mathrm{L}(ab)$, $A'=\mathrm{L}(ac)$ (Abb. 44). Nach 9.26, 9.29 ist $E=\mathrm{Pl}(abc)=\mathrm{Pl}(A,c)=\mathrm{Pl}(A',b)$ und außerdem $d\in E-(A\cup A')$. Gemäß Def. 9.20 verbleiben die folgenden Unterfälle.

Fall 2a: $\mathrm{B}cAd$. Nach 9.1 existiert dann ein Punkt $x\in A$ mit $\mathrm{B}cxd$, der das Verlangte leistet.

Fall 2b: $\mathrm{B}bA'd$. Analog zu 2a unter Vertauschung von b mit c.

Fall 2c: $d \underset{A}{\simeq} c \wedge d \underset{A'}{\simeq} b$. Nach 9.31 ist dann $\mathrm{B}b\,{}^{a}_{d}\,c$, nach 9.1 existiert also ein Punkt x mit $\mathrm{Col}\, adx \wedge \mathrm{B}bxc$, der das Verlangte leistet. □

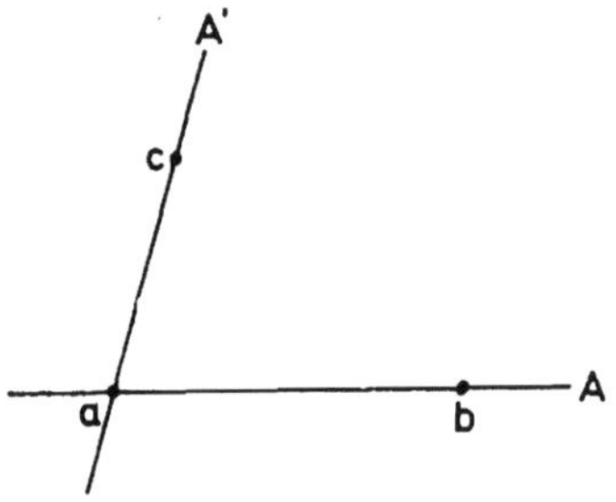

Abb. 44

9.34 Satz. *Wenn* $\neg\mathrm{Col}\, abc$, *so ist* $\mathrm{Pl}(abc)$ *die kleinste Menge* E *von Punkten mit den Eigenschaften*

(i) $a,b,c\in E$ und

(ii) *E ist abgeschlossen in bezug auf Bildung von Verbindungsgeraden (d.h.,* $u\in E \wedge v\in E \wedge u\neq v \rightarrow \mathrm{L}(uv)\subseteq E$*).*

Beweis: Nach 9.25, 9.27 hat Pl(*abc*) die genannten Eigenschaften. Sei nun *E* eine beliebige Punktmenge mit den Eigenschaften (i) und (ii) und $d \in$ Pl(*abc*), d.h. Cp *abcd*; dann ist zu zeigen, daß $d \in E$ ist. Nach 9.33 existiert ein Punkt *x* auf der Verbindungsgeraden von zweien der Punkte *a*, *b*, *c* derart, daß *d* auf der Verbindungsgeraden von *x* und dem dritten der Punkte *a*, *b*, *c* liegt. Wegen (i) und (ii) ist dann $d \in E$, wie zu zeigen. □

9.35 Anmerkungen. (i) Der Beweis von 9.34 liefert das stärkere Resultat, daß man jeden Punkt *d* von Pl(*abc*), von *a*, *b*, *c* ausgehend, nur durch *zweimaliges Bilden von Verbindungsgeraden* erhalten kann.

(ii) Auf Grund der Axiome A1 bis A9 läßt sich zeigen (s. 11.69), daß es genau eine Ebene gibt, und daß alle Punkte in dieser Ebene liegen. Die Modelle dieser Axiome sind also "zweidimensional".

9.36 Höhere Dimensionen. Das hier nur zugrunde gelegte System der Axiome A1 bis A8 hat dagegen auch Modelle höherer Dimension (s. 1.6). Für solche Modelle lassen sich die bisherigen Betrachtungen dieses Abschnitts leicht auf höhere Dimensionen übertragen und insbesondere (*n*+1)-*dimensionale affine Unterräume* einführen (induktiv über *n*), sofern nur der ganze Raum (d.h. das zugrunde gelegte Modell von A1 bis A8) keine kleinere Dimension hat. Die Übertragung wird hier nur skizziert; die Ausführung von Einzelheiten kann dem Leser überlassen bleiben. Zur Hervorhebung der Analogie zu dem Bisherigen (wo der Fall *n*=1 betrachtet wurde) werden die bisherigen Nummern von Definitionen und Sätzen mit dem angehängten Index *n* zusätzlich in Klammern angegeben.

Punkte (genauer: einelementige Punktmengen), Geraden und Ebenen werden auch als *nulldimensionale*, *eindimensionale* bzw. *zweidimensionale* (affine) Unterräume des ganzen Raumes bezeichnet. Für das Folgende setzen wir den Begriff des *n*-dimensionalen affinen Unterraumes als gegeben voraus; die Abkürzung $\mathrm{Sb}_n A$ soll bedeuten, daß die Punktmenge *A* ein (affiner) Unterraum ("subspace") der Dimension *n* ist. Außerdem setzen wir voraus, daß dafür die folgenden Sätze gelten.

(A) *Jeder n-dimensionale Unterraum ist abgeschlossen in bezug auf Bildung von Verbindungsgeraden.*

(B) *Zu n+1 Punkten* $p_0, \ldots, p_n$ *die nicht in einem höchstens (n-1)-dimensionalen Unterraum liegen, gibt es genau einen n-dimensionalen Unterraum A, in dem sie liegen; dieser werde bezeichnet mit* $\mathrm{Sb}_n(p_0 \ldots p_n)$.

Für $n \leq 2$ ist diese Induktionsannahme auf Grund der bisherigen Betrachtungen erfüllt. Von Interesse ist das Folgende nur unter der Zusatzvoraussetzung:

(C) *Außerhalb eines n-dimensionalen Unterraumes gibt es noch weitere Punkte.*

9.37 (9.1$_n$) Definition. Die Punkte a, b liegen auf *entgegengesetzten Seiten* des n-dimensionalen Unterraumes A:

$$\mathrm{B}^n aAb :\leftrightarrow \mathrm{Sb}_n A \wedge a \notin A \wedge b \notin A \wedge \exists t [t \in A \wedge \mathrm{B}atb].$$

Dann gilt offenbar

9.38 (9.2$_n$) Satz. $\mathrm{B}^n aAb \rightarrow \mathrm{B}^n bAa$.

9.39 (9.5$_n$) Satz. $\mathrm{B}^n aAc \wedge r \in A \rightarrow \forall b [a \underset{r}{\simeq} b \rightarrow \mathrm{B}^n bAc]$.

Beweis (Abb. 45): Sei t gemäß Def. von $\mathrm{B}^n aAc$ gewählt und L eine in A enthaltene Gerade durch r und t ($L = \mathrm{L}(rt)$, falls $r \neq t$, sonst $L = \mathrm{L}(ru)$, wobei u beliebig mit $u \neq r$, $u \in A$). Dann ist auch $\mathrm{B}aLc$, nach 9.5 also $\mathrm{B}bLc$ und damit auch $\mathrm{B}^n bAc$. □

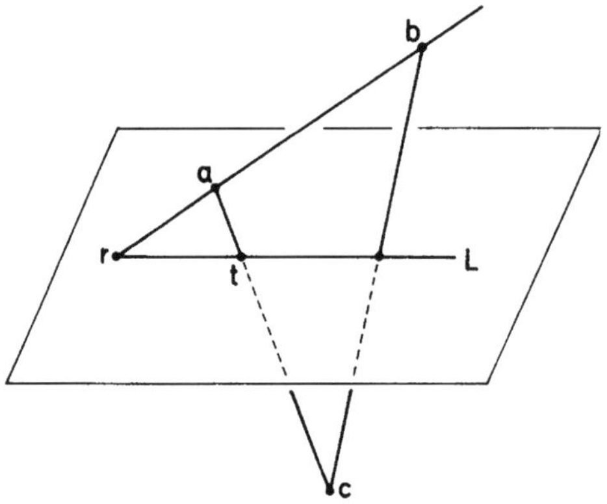

Abb. 45

9.40 (9.7$_n$) Definition. Die Punkte a, b liegen *auf derselben Seite* des n-dimensionalen Unterraums A:

$$a \overset{n}{\underset{A}{\simeq}} b :\leftrightarrow \exists c [\mathrm{B}^n aAc \wedge \mathrm{B}^n bAc].$$

9.41 (9.8_n) Satz. $B^n aAc \rightarrow (B^n bAc \leftrightarrow a \overset{n}{\underset{A}{\approx}} b)$.

Der Beweis verläuft analog zu dem von 9.8 (mit Hilfe von 9.5_n).

9.42 Anmerkung. Analog zu den früheren Sätzen erhält man auch Sätze 9.9_n bis 9.13_n, die wohl nicht mehr formuliert zu werden brauchen; insbesondere ist $\overset{n}{\underset{A}{\approx}}$ eine Äquivalenzrelation auf der Menge der Punkte außerhalb A, falls $Sb_n A$ ist.

9.43 (9.14_n) Definition. Der von A begrenzte $(n+1)$-*dimensionale Halbraum ("half-space") durch* a:

$Hs_{n+1}(A,a) := \{x \mid x \overset{n}{\underset{A}{\approx}} a\}$ (definiert für $Sb_n A \wedge a \notin A$).

9.44 (9.20_n) Definition. Der durch A und r bestimmte $(n+1)$-*dimensionale (affine) Unterraum*:

$Sb_{n+1}(A,r) := \{x \mid x \overset{n}{\underset{A}{\approx}} r \vee x \in A \vee B^n rAx\}$ (definiert für $Sb_n A \wedge r \notin A$)

$= Hs_{n+1}(A,r) \cup A \cup Hs_{n+1}(A,r')$,

wobei r' beliebig mit $B^n rAr'$. Ist $A = Sb_n(p_0 \ldots p_n)$, so verwenden wir dafür auch die Bezeichnung

$Sb_{n+1}(p_0 \ldots p_n r) := Sb_{n+1}(A,r)$ (definiert für Punkte $p_0, \ldots, p_n, r$, die nicht in einem höchstens n-dimensionalen Unterraum liegen).

Wir sagen dann auch, daß dieser Unterraum von $p_0, \ldots, p_n, r$ oder von A und r *aufgespannt* oder *erzeugt* wird (vgl. 9.51, 55f.).

Die Punktmenge E ist ein $(n+1)$-*dimensionaler Unterraum*

$Sb_{n+1}E :\leftrightarrow \exists r \exists A[Sb_n A \wedge r \notin A \wedge E = Sb_{n+1}(A,r)]$.

Eine induktive Charakterisierung der Relation Sb_n erhält man durch Hinzunahme der Anfangsbedingung

$Sb_0 A \leftrightarrow \exists a A = \{a\}$.

9.45 Anmerkungen. (i) Weiter erhält man ohne Schwierigkeiten Analoga 9.15_n bis 9.19_n und 9.21_n bis 9.31_n zu den früheren Sätzen und Definitionen; an die Stelle von p in 9.22 bis 9.24 und 9.31 tritt dabei jetzt ein $(n-1)$-dimensionaler Unterraum U, der der Durchschnitt der n-dimensionalen Unterräume A und A' ist; an die Stelle der Relation Col tritt die Beziehung, daß Punkte $p_0, \ldots, p_{n+1}$ in einem n-dimensionalen Unterraum liegen.

(ii) Damit ergeben sich insbesondere die als Induktionsvoraussetzung benutzten Sätze (A) und (B) aus 9.36 für $n+1$ an Stelle von n ("Induktionsbehauptung"). Folglich stehen die bisher bestrachteten Begriffe und Sätze für jedes n zur Verfügung (und sind nur unter der Zusatzvoraussetzung (C) aus 9.36 von Interesse).

(iii) Die in (i) zuletzt genannte Beziehung wird anschließend für sich betrachtet in Analogie zur Komplanarität von Punkten a_0, a_1, a_2, a_3 (9.32 f.)

9.46 (9.32$_{n-2}$) Definition. Die Punkte $a_0, \ldots, a_n$ sind *affin abhängig*:

(1) $\mathrm{Ah}_n a_0 \ldots a_n :\leftrightarrow \exists U [\bigvee_{\nu=0}^{n-1} \mathrm{Sb}_\nu U \wedge \bigwedge_{\nu=0}^{n} a_\nu \in U]$ $(n \geq 1)$

($a_0, \ldots, a_n$ liegen in einem höchstens $(n-1)$-dimensionalen Unterraum).

Punkte $a_0, \ldots, a_n$ mit $\neg \mathrm{Ah}_n a_0 \ldots a_n$ nennen wir dann *affin unabhängig* (das bedeutet, daß sie einen n-dimensionalen Unterraum A "aufspannen", d.h. gemäß 9.36(B) bestimmen).

Damit erhalten wir als Spezialfälle die Sätze:

(2) $\mathrm{Ah}_1 a_0 a_1 \leftrightarrow a_0 = a_1$.

(3) $\mathrm{Ah}_2 a_0 a_1 a_2 \leftrightarrow \mathrm{Col}\, a_0 a_1 a_2$.

(4) $\mathrm{Ah}_3 a_0 a_1 a_2 a_3 \leftrightarrow \mathrm{Cp}\, a_0 a_1 a_2 a_3$.

9.47 Anmerkungen. (i) In 9.46(1) wurde eine Disjunktion (und nicht nur das letzte Glied davon) verwendet, damit sich das Gewünschte auch dann ergibt, wenn die Dimension des ganzen Raumes kleiner als $n-1$ ist.

(ii) Eine gleichwertige Definition für die Relation Ah_n (s.a. 9.50(i)) ist enthalten in Gupta 1965, S. 174f. (dort als "linear dependence" bezeichnet). In der euklidischen Geometrie, in der man in gewohnter Weise Vektoren einführen kann, ist die durch $\mathrm{Ah}_n a_0 \ldots a_n$ ausgedrückte affine Abhängigkeit der $n+1$ Punkte $a_0, \ldots, a_n$ natürlich gleichbedeutend mit der linearen Abhängigkeit von n Vektoren, etwa $\overrightarrow{a_0 a_1}, \ldots, \overrightarrow{a_0 a_n}$. Wesentlich für unseren Aufbau ist es, daß wir Sätze über Ah_n und über die Dimension von Unterräumen für beliebige Modelle der absoluten Geometrie erhalten, für die i.a. kein so einfacher Zusammenhang mit Vektorräumen besteht. Ein solcher Aufbau wurde (anscheinend erstmalig) von Wanda Szmielew gegeben (mit etwas anderen Hilfsmitteln, vgl. 0.4).

9.48 (9.33$_n$) Satz.
$$\begin{aligned}\mathrm{Ah}_{n+2}\mathfrak{a}bcd \leftrightarrow \exists x\{&[\mathrm{Ah}_{n+1}\mathfrak{a}bx \wedge \mathrm{Col}\, cdx]\\ &\vee [\mathrm{Ah}_{n+1}\mathfrak{a}cx \wedge \mathrm{Col}\, bdx]\\ &\vee [\mathrm{Ah}_{n+1}\mathfrak{a}dx \wedge \mathrm{Col}\, bcx]\},\end{aligned}$$
wobei $\mathfrak{a}$ eine Abkürzung für die Variablenfolge $a_0 \ldots a_{n-1}$ ist.

Beweis: Die Richtung von rechts nach links ergibt sich wieder unmittelbar (ein höchstens n-dimensionaler Unterraum und eine Gerade durch einen Punkt x dieses Raumes sind zusammen in einem höchstens $(n+1)$-dimensionalen Unterraum enthalten).

Verwendet man von dieser Richtung nur die erste Zeile mit $x=c$, so erhält man das (offenbar auch für $n=0$ geltende) Lemma:

9.49 Lemma. $\mathrm{Ah}_{n+1}a_0\ldots a_{n+1} \rightarrow \mathrm{Ah}_{n+2}a_0\ldots a_{n+1}a_{n+2}$ $(n\geq 0)$.

Gelte nun die linke Seite von 9.48. Wir unterscheiden hier folgende Fälle.

Fall 1: $a_0,\ldots, a_{n-1}, b, c, d$ liegen, bis auf höchstens einen der Punkte b, c, d, in einem höchstens n-dimensionalen Unterraum.

Fall 2: Fall 1 gilt nicht. Dann ist $\neg\mathrm{Ah}_{n-1}\mathfrak{a}$ (da sonst nach 9.49 auch $\mathrm{Ah}_{n+1}\mathfrak{a}bc$ wäre), d.h., $a_0,\ldots, a_{n-1}$ spannen einen $(n-1)$-dimensionalen Unterraum U auf.

Der Rest des Beweises verläuft analog zu dem von 9.33 (unter Verwendung von 9.31_n). □

9.50 Anmerkungen. (i) 9.48 (9.33_n) liefert zusammen mit den vorhergehenden Sätzen 9.46(2), (3) eine induktive Charakterisierung der affinen Abhängigkeit, für die nur der Begriff der Kollinearität benötigt wird. (Statt 9.46(3) kann man auch 9.33_n mit $n=0$ verwenden.) Da die Relation Ah_n nach Definition invariant gegenüber Permutation der Argumente ist, gilt offenbar auch

(1) $\mathrm{Ah}_n a_0\ldots a_n \leftrightarrow \bigvee_{\mathrm{Perm}\,\langle i_0,\ldots,i_n\rangle} \exists x[\mathrm{Col}\, a_{i_0} a_{i_1} x \wedge \mathrm{Ah}_{n-1} a_{i_2}\ldots a_{i_n} x]$ $(n\geq 2)$,

wobei die Disjunktion über alle Permutationen $\langle i_0,\ldots,i_n\rangle$ der Indizes $0,\ldots, n$ (bzw. alle Permutationen der Variablen) erstreckt wird. (1) läßt sich als eine homogenere Form der Rekursionsbedingung auffassen und zu einer induktiven Charakterisierung der affinen Abhängigkeit von Punkten verwenden. Diese Charakterisierung wird bei Gupta 1965 als Definition zugrunde gelegt.

(ii) Aus 9.33_m (mit $m=n-2$) ergibt sich jedoch, daß man sich in (1) auf die Permutationen von drei beliebig herausgegriffenen Variablen beschränken kann. Von diesen sechs Permutationen führen - wegen der Symmetrie von Col - jeweils zwei zum selben Disjunktionsglied, so daß man für jedes n

mit drei Disjunktionsgliedern auskommt. Auch dieses Resultat wurde mir zuerst von H. N. Gupta mündlich mitgeteilt.

9.51 (9.34_{n-1}) Satz. Wenn $\neg \mathrm{Ah}_n a_0 \dots a_n$, *so ist* $\mathrm{Sb}_n(a_0 \dots a_n)$ *die kleinste Menge E von Punkten mit den Eigenschaften*

(i) $a_0, \dots, a_n \in E$ und

(ii) *E ist abgeschlossen in bezug auf Bildung von Verbindungsgeraden.*

Das ergibt sich analog zu 9.34. Aus dem Beweis kann man noch das folgende stärkere Resultat erhalten:

9.52 (9.35_{n-1}) Satz. *Jeder Punkt d von* $\mathrm{Sb}_n(a_0 \dots a_n)$ *läßt sich, von* $a_0, \dots, a_n$ *ausgehend, durch n-maliges Bilden von Verbindungsgeraden erhalten.*

Satz 9.51 gibt Anlaß zu den folgenden allgemeineren Definitionen (außerhalb des Rahmens der Prädikatenlogik der ersten Stufe).

9.53 Definition. Die Punktmenge *E* ist ein *(affiner) Unterraum*:

$\mathrm{Sb}\, U :\leftrightarrow U \neq \emptyset \wedge \forall u \forall v (u \in U \wedge v \in U \wedge u \neq v \rightarrow \mathrm{L}(uv) \subseteq U)$

(*U* ist nicht leer und abgeschlossen in bezug auf Bildung von Verbindungsgeraden).

Ist außerdem $\mathrm{Sb}_n U$, so nennen wir die Zahl *n* die *Dimension* von *U* und schreiben dafür $\dim U = n$.

9.54 Anmerkung. Es gibt auch unendlichdimensionale Modelle von A1 bis A8 und darin unendlichdimensionale Unterräume. Die Betrachtung der Unterräume von fester endlicher Dimension wurde hier vorangestellt, da sich diese (im Sinne von Anmerkung 6.26(i)) noch im Rahmen der Prädikatenlogik der ersten Stufe behandeln läßt.

9.55 Definition. Die *affine Hülle* der Punktmenge *A*:

$\mathrm{Cla}(A) := \bigcap \{U \mid \mathrm{Sb}\, U \wedge A \subseteq U\}$.

Damit wird Cla(*A*) charakterisiert als der "von *A erzeugte* Unterraum", d.h. der kleinste affine Unterraum, der *A* umfaßt; denn der Durchschnitt aller *A* umfassenden Unterräume ist offenbar selbst ein solcher Unterraum.

9.56 Satz.

(i) $\neg \mathrm{Ah}_n a_0 \dots a_n \rightarrow \mathrm{Cla}(\{a_0, \dots, a_n\}) = \mathrm{Sb}_n(a_0 \dots a_n)$.

(ii) $\forall a_0 \ldots \forall a_n \Big\{ \bigvee_{k=0}^{n} \bigvee_{1\le i_1<\ldots<i_k\le n} [\neg \mathrm{Ah}_k a_0 a_{i_1} \ldots a_{i_k}$

$\wedge \mathrm{Cla}(\{a_0,\ldots,a_n\}) = \mathrm{Sb}_k(a_0 a_{i_1} \ldots a_{i_k})] \Big\}.$

(iii) $x \in \mathrm{Cla}(A) \leftrightarrow \exists A'[A' \subseteq A \wedge A'$ endlich $\wedge\ x \in \mathrm{Cla}(A')]$.

Beweis: (i) ist nur eine andere Formulierung von 9.34_{n-1} und (ii) eine Folgerung daraus.

Zu (iii): Die Menge der Punkte x mit der auf der rechten Seite angegebenen Eigenschaft erweist sich auf Grund von (i) und (ii) als ein Unterraum (d.h. als abgeschlossen in bezug auf die Bildung von Verbindungsgeraden) und ist dann offenbar der kleinste Unterraum, der A umfaßt.

□

9.57 Definition. Das *Kompositum* der affinen Unterräume E_1 und E_2:

$E_1 \sqcup E_2 := \mathrm{Cla}(E_1 \cup E_2)$ (definiert für $\mathrm{Sb}\, E_1 \wedge \mathrm{Sb}\, E_2$)

(der kleinste affine Unterraum, der die Vereinigung $E_1 \cup E_2$ umfaßt).

Dieser Raum ist offenbar endlichdimensional, falls E_1 und E_2 es sind (s. 9.56(i), (ii)).

Für spätere Anwendungen beweisen wir zunächst noch einen speziellen Satz über Ebenen.

9.58 Satz. $\mathrm{Sb}_3 U \wedge \bigwedge_{\nu=1}^{2} [\mathrm{Pl}\, E_\nu \wedge E_\nu \subseteq U \wedge p \in E_\nu] \rightarrow \exists A[\mathrm{Ln}\, A \wedge p \in A \wedge A \subseteq E_1 \wedge A \subseteq E_2]$.

(*Sind Ebenen* E_1, E_2 *in einem dreidimensionalen Unterraum enthalten und haben einen gemeinsamen Punkt* p, *so haben sie eine Gerade durch* p *gemeinsam.*)

Beweis: Es genügt zu zeigen, daß E_1 und E_2 noch einen weiteren gemeinsamen Punkt $q \neq p$ besitzen; denn dann leistet $A = \mathrm{L}(pq)$ das Verlangte. Sei o. B. d. A. $E_1 \neq E_2$ (sonst ist alles klar). U läßt sich aufspannen durch E_1 und einen Punkt $s \in E_2 - E_1$ (Abb. 46). Sei $L = \mathrm{L}(ps)$. Dann ist $L \subseteq E_2$. E_2 läßt sich aufspannen durch L und einen weiteren Punkt $t \in E_2 - L$. Wegen $t \in U$ gilt einer der folgenden drei Fälle.

Fall 1: $\mathrm{B}^2 s E_1 t$. Nach Definition von B^2 existiert dann ein Punkt $q \in E_1$ mit $\mathrm{B} sqt$, der (wegen $q \notin L$, $q \in \mathrm{L}(st) \subseteq E_2$) das Verlangte leistet.

Fall 2: $s \overset{2}{\underset{E_1}{=}} t$. Dann läßt sich die Überlegung von Fall 1 anwenden auf einen Punkt s' mit $Bsps' \wedge s' \neq p$ an Stelle von s.

Fall 3: $t \in E_1$. Dann leistet $q=t$ das Verlangte. □

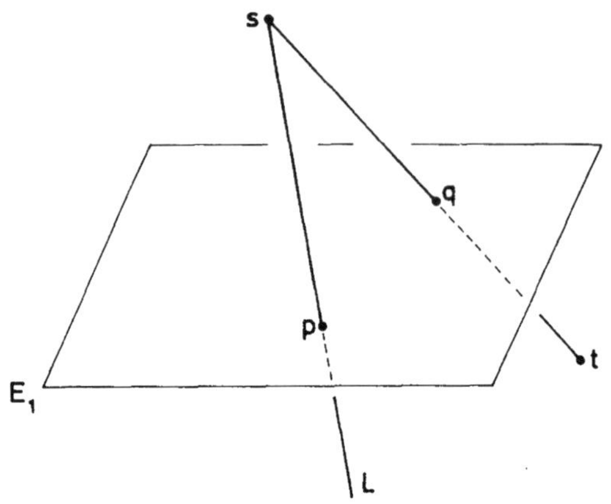

Abb. 46

9.59 Satz. $E_1 \sqcup E_2 = U \wedge E_1 \cap E_2 = D \neq \emptyset \wedge \mathrm{Sb}_n U \wedge \mathrm{Sb}_{n-m} E_1 \wedge \mathrm{Sb}_k E_2$
$\rightarrow \dim D = k-m$ $\quad (0 \leq m \leq n)$.

(*Sind* E_1, E_2 *affine Unterräume mit nicht-leerem Durchschnitt* [der offenbar wieder ein affiner Unterraum ist], *so ist die Summe ihrer Dimensionen gleich der Summe der Dimensionen ihres Kompositums und ihres Durchschnitts;* m.a.W. *ein* $(n-m)$-*dimensionaler Unterraum* E_1 *schneidet aus einem* k-*dimensionalen Unterraum* E_2 *mit* $\dim(E_1 \sqcup E_2) = n$ *einen* $(k-m)$-*dimensionalen Unterraum* D *aus, falls er überhaupt einen gemeinsamen Punkt mit ihm hat.*)

Beweis: Die Überlegung im Beweis von 9.58 (mit $E_1 \neq E_2$) funktioniert genau so, wenn U eine beliebige Dimension n hat und E_1 ein $(n-1)$-dimensionaler Unterraum von U ist. Damit haben wir 9.59 für den Spezialfall $m=1$, $k=2$. Eine Wiederholung der Überlegung (Hinzunahme weiterer Punkte ebenso wie t) liefert dasselbe für beliebiges $k \leq n$ (Induktion über k mit trivialem Anfang $k=1$). Somit gilt 9.59 zunächst für den Fall $m=1$ (n und k beliebig) (und trivialerweise für $m=0$). Daraus kann man das Ganze durch Induktion über m erhalten. Sei nämlich jetzt $\mathrm{Sb}_{n-(m+1)} E_1$ und alles andere wie in der Voraussetzung von 9.59, und gelte der Satz schon für die Zahl $m\geq 1$. Sei a ein Punkt von E_2-E_1 und $E_1':=\mathrm{Sb}_{n-m}(E_1,a)$. Dann ist auch $E_1'\sqcup E_2=U$, nach Induktionsvoraussetzung ist also $D':=E_1'\cap E_2$ von der Dimension $k-m$. Außerdem ist $a\in D'$. Nun ist

$$E_1\sqcup D'=E_1',$$

da E_1' beide Räume umfaßt und das Kompositium mindestens E_1 und a enthalten muß. Andererseits ist

$$\begin{aligned} E_1\cap D' &= E_1\cap E_1'\cap E_2 \\ &= E_1\cap E_2 \qquad \text{(wegen } E_1\subseteq E_1'\text{)} \\ &= D. \end{aligned}$$

Anwendung des schon bekannten Spezialfalls $m=1$ von 9.59 auf E_1' und die Unterräume E_1 und D' liefert, daß der Durchschnitt D die Dimension $(k-m)-1=k-(m+1)$ hat; das war gerade in der Induktionsbehauptung verlangt. □

10. Geradenspiegelungen

Nachdem die Existenz und Eindeutigkeit des Mittelpunkts einer Strecke gezeigt wurde (8.22), können wir definieren:

10.1 Definition. *Der Mittelpunkt* der Strecke ab:

$M(ab) := \iota x M axb$.

10.2 Satz. $\text{Ln}\, A \to \forall p \overset{=1}{\exists} p' [M(pp') \in A \wedge (A \perp pp' \vee p=p')]$.

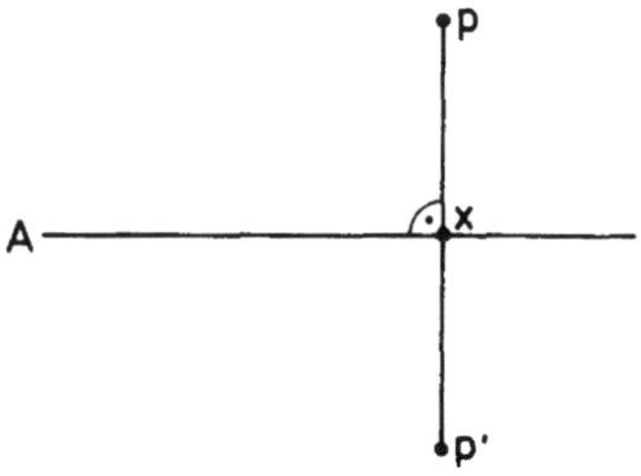

Abb. 47

Beweis: Zur Abkürzung setzen wir $x=M(pp')$ (Abb. 47). Dann gilt $p=p' \leftrightarrow p=x$ (nach 7.10) und wegen $\text{Col}\, pxp'$ auch $A\perp pp' \leftrightarrow A \perp px$.

Fall 1: $p \notin A$. Ist p' ein Punkt der genannten Art, so ist $p \neq x$ (wegen $x \in A$); auf Grund der Vorbemerkung ist also x der nach dem Lotsatz (8.18) eindeutig bestimmte Fußpunkt des Lots von p auf A und $p'=S_x p$. Diese beiden Punkte leisten aber auch das Verlangte.

Fall 2: $p \in A$. Ist p' ein Punkt der genannten Art, so muß $p=x$ sein; denn sonst wäre (da auch $x \in A$ ist) $L(px)=A$ und damit $A \perp A$ im Widerspruch zu 8.14(i). Damit ist auch $p'=S_x p=p$, d.h. p' ist eindeutig bestimmt. Andererseits leistet p das für p' Verlangte. □

10.3 Definition. Der durch 10.2 eindeutig bestimmte Punkt p' heißt das *Spiegelbild von p an der Geraden A*, abgekürzt:

$$S_A p = S_A(p) := \iota p'[M(pp') \in A \wedge (A \perp pp' \vee p = p')] \qquad \text{(definiert für Ln } A\text{)}.$$

Die dadurch erklärte Abbildung S_A heißt die *Spiegelung (Geradenspiegelung) an der Geraden A*, und A heißt die *Achse* dieser Spiegelung (sie erweist sich nach 10.8 als durch S_A eindeutig bestimmt). Für $S_A p = p'$ sagen wir auch, daß p und p' *spiegelbildlich bezüglich A* liegen. Zu Beschreibung der Spiegelung durch Punkte verwenden wir noch die Abkürzung

$$S_{ab} := \begin{cases} S_{L(ab)}, & \text{falls } a \neq b, \\ S_a, & \text{falls } a = b \quad \text{(Punktspiegelung, 7.5)}. \end{cases}$$

Die charakterisierende Eigenschaft in 10.3 geht bei Vertauschung von p mit p' in sich über. Daraus erhalten wir sofort die folgenden Sätze.

10.4 Satz. $S_A p = p' \leftrightarrow S_A p' = p$.

10.5 Satz. $\text{Ln } A \to S_A S_A p = p$.

10.6 Satz. $\text{Ln } A \to \forall p' \overset{=1}{\exists} p S_a p = p'$.

10.7 Satz. $S_A p = S_A q \to p = q$.

Aus der Fallunterscheidung im Beweis von 10.2 ergibt sich außerdem

10.8 Satz. $\text{Ln } A \to [S_A p = p \leftrightarrow p \in A]$.

Da es Punkte außerhalb jeder Geraden gibt (6.25), ist eine Geradenspiegelung also nicht die identische Abbildung. In Analogie zu 7.11, 7.12 erhalten wir also

10.9 Satz. *Jede Geradenspiegelung* S_A *ist eine eineindeutige Abbildung des ganzen Raumes auf sich und involutorisch.*

10.10 Satz. $\text{Ln } A \to pq \equiv S_A(p) S_A(q)$.

Beweis: Sei $p' = S_A p$, $q' = S_A q$, $x = M(pp')$, $y = M(qq')$, $z = M(xy)$, $r = S_z(p)$ und $r' = S_z(p')$ (Abb. 48; dabei brauchen q und q' nicht mit den übrigen Punkten in einer Ebene zu liegen, was durch Strichelung der Verbindung angedeutet wird). Dann ist $x, y, z \in A$, $p' = S_x p$ und $q' = S_y q$. Nun ist S_z ein

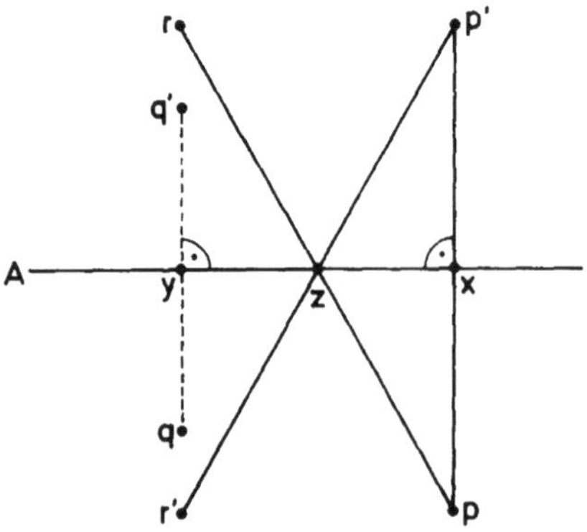

Abb. 48

Automorphismus (7.14), wegen $Mpxp'$ also auch $Mryr'$, d.h. $r'=S_y r$. Anwendung von 7.13 auf die Punktspiegelung S_y liefert also $qr \equiv q'r'$. Nach Konstruktion ist $A \perp px \vee p=x$; somit ist $Rzxp$ (nach 8.17, 8.11 bzw. 8.5), also

(1) $zp \equiv zp'$

und damit auch $zr \equiv zr'$. Analog ergibt sich

(2) $zq \equiv zq'$.

Somit erhalten wir

$$\mathrm{AFS}\begin{pmatrix} r & z & p & q \\ r' & z & p' & q' \end{pmatrix}.$$

<u>Fall 1</u>: $r \neq z$. Dann ergibt sich die Behauptung $pq \equiv p'q'$ aus A5.

<u>Fall 2</u>: $r=z$. Nach Wahl von r und (1) ist dann $p=z=p'$, und die Behauptung ergibt sich aus (2). □

Mit 10.9 und 10.10 haben wir (in Analogie zu 7.12, 7.14)

<u>10.11 Satz</u>. *Jede Geradenspiegelung S_A ist eine involutorische Bewegung und (damit) ein Automorphismus des ganzen Raumes.*

Wir können nun den folgenden, schon in Anmerkung 8.23 angekündigten Satz beweisen.

<u>10.12 Satz</u>. $Rabc \wedge Ra'b'c' \wedge ab \equiv a'b' \wedge bc \equiv b'c' \rightarrow ac \equiv a'c'$.

(Sind in rechtwinkligen Dreiecken entsprechende Katheten zueinander kongruent, so sind auch die Hypotenusen zueinander kongruent.)

Die Idee des Beweises besteht darin, eine Bewegung zu konstruieren, die die Punkte a', b', c' in a, b bzw. c überführt. (Aus der Isometrie-Eigenschaft ergibt sich dann die Behauptung.) Eine solche Bewegung läßt sich durch Hintereinanderausführung einiger Punkt- und Geradenspiegelungen erhalten.

Beweis von 10.12: Sei $x=\mathrm{M}(bb')$, $a_1=\mathrm{S}_x(a')$, $c_1=\mathrm{S}_x(c')$. Da auch $b=\mathrm{S}_x(b')$ und die Punktspiegelung S_x eine Bewegung ist, ist $(a'b'c')\equiv(a_1bc_1)$ und nach 8.10 somit $\mathrm{R}a_1bc_1$. Nach Voraussetzung ist auch $ab\equiv a_1b \wedge bc\equiv bc_1$, und es bleibt $ac\equiv a_1c_1$ zu zeigen. (Damit ist der Satz zurückgeführt auf den Spezialfall, daß die Scheitel der betrachteten rechten Winkel [jetzt $\sphericalangle abc$ und $\sphericalangle a_1bc_1$] zusammenfallen.)

Sei $y=\mathrm{M}(cc_1)$ (Abb. 49). Dann ist offenbar $\mathrm{R}byc$, woraus sich $c=\mathrm{S}_{by}(c_1)$ ergibt (s. 10.3, wobei jetzt auch $b=y$ zugelassen ist). Natürlich ist auch $b=\mathrm{S}_{by}(b)$ (nach 10.8 bzw. 7.10). Sei nun $a_2=\mathrm{S}_{by}(a_1)$. Da S_{by} als Geraden- oder Punktspiegelung eine Bewegung ist, ist $(a_1bc_1)\equiv(a_2bc)$ und nach 8.10 somit $\mathrm{R}a_2bc$. Nach den bisherigen Feststellungen ist auch $ab\equiv a_2b$, und es bleibt $ac\equiv a_2c$ zu zeigen. (Damit ist der zu beweisende Satz zurückgeführt auf den Spezialfall, daß die betrachteten rechtwinkligen Dreiecke [jetzt (abc) und (a_2bc)] eine gemeinsame Kathete haben.)

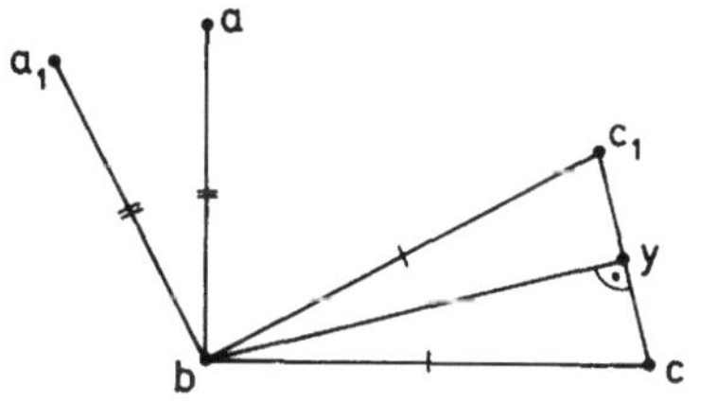

Abb. 49

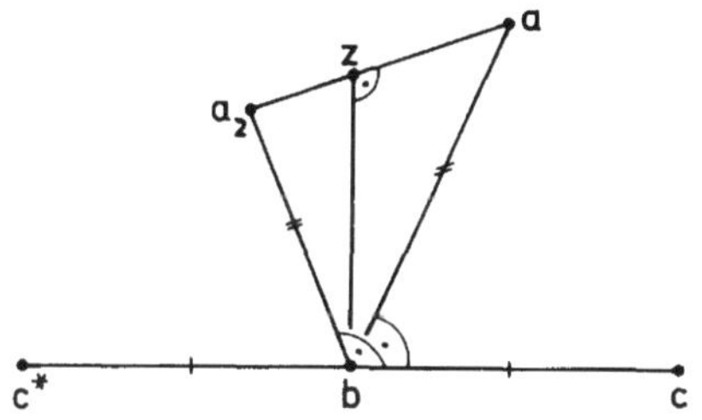

Abb. 50

Sei $z=\mathrm{M}(aa_2)$ (Abb. 50). Wie oben ist dann $\mathrm{R}bza$, $a=\mathrm{S}_{bz}(a_2)$ und $b=\mathrm{S}_{bz}(b)$. Sei $c^*=\mathrm{S}_b(c)$. Wegen der (im jetzt betrachteten Spezialfall) vorausgesetzten Rechtwinkel-Beziehungen ist dann $ac\equiv ac^*$ und $a_2c\equiv a_2c^*$. Wegen $\mathrm{Col}\, aza_2$ und 4.17 (im Falle $a=a_2$ trivialerweise) ist dann auch $zc\equiv zc^*$ und damit $\mathrm{R}zbc$, woraus sich (wie oben) $c^*=\mathrm{S}_{bz}(c)$ ergibt. Da S_{bz} eine Bewegung ist, ist $a_2c\equiv ac^*$, nach den oben festgestellten Kongruenzen also auch $ac\equiv a_2c$, wie zu zeigen. □

10.13 Anmerkungen. (i) Es hätte nahe gelegen, statt der letzten Spiegelung S_{bz} die Spiegelung an einer Ebene durch z, b und c zu betrachten (oder S_{bc}, falls $z=b$), um das Dreieck (a_2bc) direkt in (abc) überzuführen. Dazu hätten jedoch Spiegelungen an Ebenen erst eingeführt werden müssen, was durch die Verwendung von S_{bz} entbehrlich gemacht wird.

(ii) Die Hintereinanderausführung der betrachteten Spiegelungen S_x, S_{by} und S_{bz} führt $(a'b'c')$ in (abc^*) über, was zum Beweis genügt. Will man trotzdem, wie ursprünglich angedeutet, $(a'b'c')$ durch ein Produkt von (höchstens drei) Punkt- oder Geradenspiegelungen in (abc) überführen, so braucht man nur die Konstruktion des obigen Beweises auf die Dreiecke (abc^*) und $(a'b'c')$ anzuwenden.

Eine triviale Folgerung aus den Definitionen 10.3 und 9.1 ist

10.14 Satz. $S_A(p)=p' \wedge p \notin A \rightarrow \mathrm{B}pAp'$.

(Ein Punkt p außerhalb der Geraden A und sein Spiegelbild an A liegen auf entgegengesetzten Seiten von [d.h. Halbebenen mit der Begrenzung] A.)

Das in 8.21 behandelte Errichten von Senkrechten läßt sich unter Benutzung von Halbebenen folgendermaßen bequemer formulieren.

10.15 Satz. $\mathrm{Ln}\,A \wedge a \in A \wedge q \notin A \rightarrow \exists p[A \perp pa \wedge p \underset{A}{\approx} q]$.

Beweis: Sei o. B. d. A. $A=\mathrm{L}(ab)$, sei c gemäß 9.10 ein Punkt mit $\mathrm{B}qAc$, und seien p, t gemäß 8.21 gewählt. Dann ist $\mathrm{B}pAc$, nach 9.8 also $p \underset{A}{\approx} q$, d.h., p leistet das Verlangte. □

10.16 Satz (*Dreiecksantragung*).

$$\neg\mathrm{Col}\,abc \wedge \neg\mathrm{Col}\,a'b'p \wedge ab \equiv a'b' \rightarrow \overset{=1}{\exists} c'[(abc) \equiv (a'b'c') \wedge c' \underset{\widetilde{a'b'}}{\approx} p].$$

(Ein geordnetes Dreieck läßt sich mit einer Seite an eine Halbgerade H(a'q) stets so eindeutig antragen, daß die dritte Ecke in eine gegebene von L(a'q) begrenzte Halbebene fällt.)

Beweis: 1. Existenz (Abb. 51): Sei x gemäß 8.18 der Fußpunkt des Lots von c auf $\mathrm{L}(ab)$, d.h. $\mathrm{Col}\,abx \wedge ab \perp cx$, und sei x' gemäß 4.14 ein Punkt mit $(abx) \equiv (a'b'x')$. (Für die kollinearen Punkte a, b, x kann jede der drei möglichen Zwischenbeziehungen eintreten, die entsprechende gilt dann auch für a', b', x'.)

Sei $A=\mathrm{L}(a'b')$, und sei q gemäß 10.15 gewählt mit $a'b' \perp qx' \wedge q \underset{A}{\approx} p$ und c' gemäß 6.11 (Streckenabtragung) mit $c' \underset{x'}{\approx} q \wedge x'c' \equiv xc$.

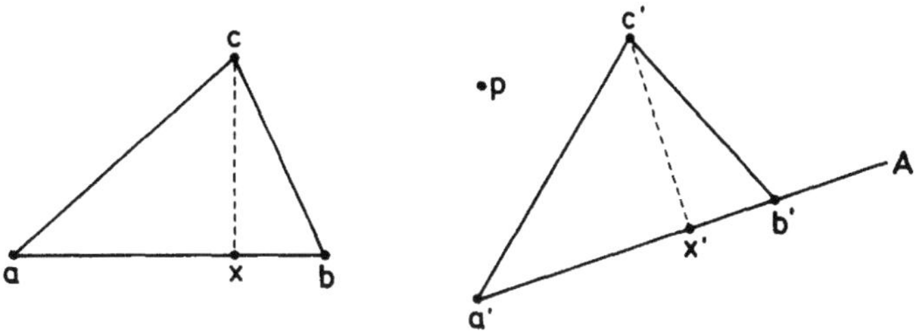

Abb. 51

Nach 9.19 ist dann auch $c'\underset{A}{\approx}q$ und damit $c'\underset{A}{\approx}p$. Aus den angegebenen Orthogonalitätsbeziehungen ergibt sich $\mathrm{R}axc \wedge \mathrm{R}a'x'c'$ sowie $\mathrm{R}bxc \wedge \mathrm{R}b'x'c'$. Nach 10.12 ergibt sich daraus und aus den zugehörigen Kongruenzen, daß auch $ac\equiv a'c'$ und $bc\equiv b'c'$ ist. Somit leistet c' das Verlangte.

2. Eindeutigkeit (Abb. 52): Seien c' und c'' Punkte der angegebenen Art, also $(a'b'c')\equiv(a'b'c'') \wedge c'\underset{A}{\approx}c''$. Sei $c^*=S_A(c'')$. Nach 10.14 ist $\mathrm{B}c''Ac^*$ und

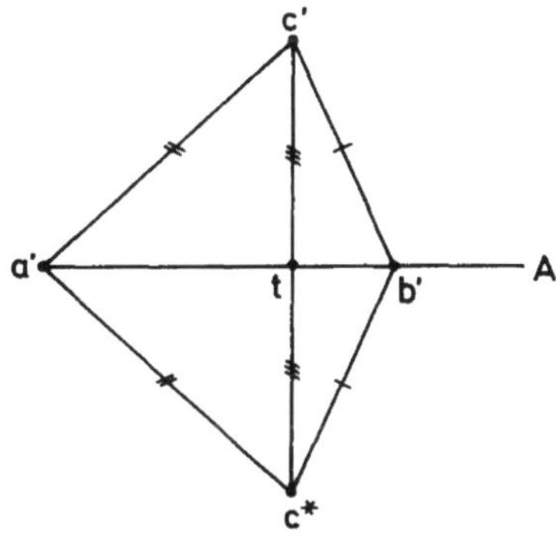

Abb. 52

nach 9.8 somit $\mathrm{B}c'Ac^*$. Dementsprechend sei t ein Punkt mit $\mathrm{Col}\,a'b't \wedge \mathrm{B}c'tc^*$. Da S_A eine Bewegung ist, ist auch $(a'b'c^*)\equiv(a'b'c'')\equiv(a'b'c')$. Mit a' und b' ist nach 4.17 auch t ein Punkt, der von c' und c^* denselben Abstand hat; somit ist $\mathrm{M}c'tc^*$. Aus den festgestellten Kongruenzen ergibt sich nun $\mathrm{R}a'tc'$ und $\mathrm{R}b'tc'$. Wegen $a'\neq b'$ ist aber wenigstens einer der beiden Punkte a', b' von t verschieden, also ist $A\underset{t}{\perp}c'c^*$ und damit $c^*=S_A(c')$. Somit haben c' und c'' unter S_A dasselbe Bild c^*, und nach 10.7 ist $c'=c''$. □

11. Kongruenz und Größenvergleich von Winkeln, Kongruenzsätze, Orthogonalität für Unterräume

11.1 Definition. Unter einem *Winkel* (*angle*) verstehen wir ein ungeordnetes Paar $\{K,L\}$ von Halbgeraden K, L mit einem gemeinsamen Ausgangspunkt. Ist K=H(ba) und L=H(bc), so wird der Winkel $\{K,L\}$ auch mit $\sphericalangle(K,L)$ oder mit $\sphericalangle abc$ bezeichnet; K und L heißen dann die *Schenkel* und b der *Scheitel* dieses Winkels. $\sphericalangle abc$ heißt ein

Nullwinkel, falls seine Schenkel zusammenfallen (d.h. $a \underset{b}{\approx} c$),

rechter Winkel, falls Rabc (s. 8.1 bis 8.3),

gestreckter Winkel, falls seine Schenkel zueinander entgegengesetzte Halbgeraden sind (d.h. Babc).

Nullwinkel und gestreckte Winkel fassen wir auch unter der Bezeichnung *entartete Winkel* zusammen. In Abbildungen werden wir die betrachteten Winkel häufig durch Bogen markieren.

Diese Definition der Winkel stimmt im wesentlichen mit der aus Hilbert 1977, Seite 13, überein; lediglich die entarteten Winkel sind dort ausgeschlossen.

Die *Kongruenz* von Winkeln, die bei Hilbert als Grundbegriff auftritt, haben wir hier zu definieren. In Übereinstimmung mit der Anschauung sollen Winkel $\sphericalangle abc$ und $\sphericalangle def$ zueinander kongruent oder gleich groß heissen, falls man durch Abtragen kongruenter Strecken auf entsprechenden Schenkeln zu Punkten a', c', d', f' derart gelangt, daß die Verbindungsstrecken $a'c'$ und $d'f'$ zueinander kongruent sind. Um eine Definition der Winkelkongruenz - zunächst als sechsstellige Relation zwischen Punkten - zu erhalten, die sich einfach mit den Grundbegriffen unseres Axiomensystems formulieren läßt, stellen wir diese Forderung zunächst für spezielle Punkte a', c', d', f', die entstehen, indem man die auf den Schenkeln gegebenen Strecken ba, bc, ed, ef um die entsprechenden Strecken auf den Schenkeln des anderen Winkels verlängert (Abb. 53).

Daß man auch zu anderen Punkten der genannten Art übergehen kann, wird dann in den folgenden beiden Sätzen ausgedrückt, die für Anwendungen bequemer sind.

<u>11.2 Definition</u>. *abc* ist *winkelkongruent* zu *def* (Abb. 53):

$$abc\underset{A}{\equiv}def :\leftrightarrow a\neq b \wedge c\neq b \wedge d\neq e \wedge f\neq e$$
$$\wedge \exists a'\exists c'\exists d'\exists f'[Bbaa' \wedge aa'\equiv ed \wedge Bbcc' \wedge cc'\equiv ef$$
$$\wedge Bedd' \wedge dd'\equiv ba \wedge Beff' \wedge ff'\equiv bc$$
$$\wedge a'c'\equiv d'f'].$$

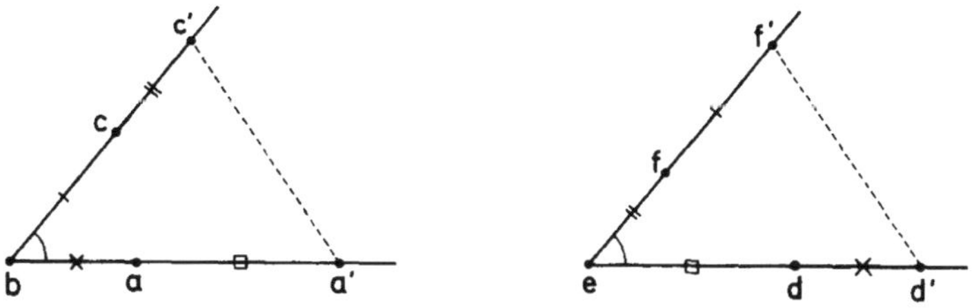

Abb. 53

Die Angabe A ("angles") lassen wir meist weg und schreiben statt dessen einfach $abc\equiv def$ (im Unterschied zu 4.4 ohne Klammern). In Hinblick auf 11.11 sagen wir dafür auch, daß die Winkel $\sphericalangle abc$ und $\sphericalangle def$ zueinander *kongruent* sind.

<u>11.3 Satz</u>. $abc\equiv def \leftrightarrow \exists a'\exists c'\exists d'\exists f'\,[a'\underset{b}{\simeq}a \wedge c'\underset{b}{\simeq}c \wedge d'\underset{e}{\simeq}d \wedge f'\underset{e}{\simeq}f$
$\wedge (a'bc')\equiv(d'ef')]$.

<u>11.4 Satz</u>. $abc\equiv def \leftrightarrow a\neq b \wedge c\neq b \wedge d\neq e \wedge f\neq e$
$\wedge \forall a'\forall c'\forall d'\forall f'\,[a'\underset{b}{\simeq}a \wedge c'\underset{b}{\simeq}c \wedge d'\underset{e}{\simeq}d \wedge f'\underset{e}{\simeq}f$
$\wedge ba'\equiv ed' \wedge bc'\equiv ef' \rightarrow a'c'\equiv d'f']$.

<u>Beweis von 11.3 und 11.4</u>: Die rechten Seiten der gdw-Beziehungen in 11.2, 11.3, 11.4 seien mit (2), (3) bzw. (4) bezeichnet. Dann ist die Gleichwertigkeit dieser drei Bedingungen zu zeigen. Wir setzen schon voraus, daß $a\neq b \wedge c\neq b$ und $d\neq e \wedge f\neq e$ ist (d.h. daß a, b, c und d, e, f je einen Winkel bestimmen) (das folgt ja aus jeder der Bedingungen (2), (3), (4)).

Auf Grund des Axioms der Streckenabtragung (A4) existieren Punkte a', c', d', e' derart, daß die ersten beiden Zeilen der Existenzforderung in (2) gelten (dadurch sind diese Punkte nach 2.12 auch eindeutig bestimmt). Außerdem gilt für diese Punkte nach 6.1(1) und 2.11 die Voraussetzung der in (4) ausgedrückten Implikation. Damit ergeben sich sofort die Implikationen (4) $\rightarrow$ (2) und (2) $\rightarrow$ (3).

(3) $\rightarrow$ (4): Seien a', c', d', f' Punkte mit der in (3) verlangten Eigenschaft. Seien zunächst a'' und d'' beliebige Punkte mit $a'' \underset{b}{\simeq} a \wedge d'' \underset{e}{\simeq} d \wedge ba'' \equiv ed''$ (Abb. 54). Nach 4.3 ("Ineinanderlegen von Strecken") ist dann (für jede der nach 5.10, 5.6, 6.13 möglichen beiden Zwischenbeziehungen) auch $a'a'' \equiv d'd''$ und damit $(ba'a'') \equiv (ed'd'')$. Nach dem Satz über die Fünf-Strecken-Konfiguration (4.16) ist also $a''c' \equiv d''f'$. Damit ist gezeigt, daß man auf den Schenkeln H(ba) und H(ed) zu beliebigen Punkten a'' und d'' der in (4) genannten Art übergehen kann. Eine Wiederholung der eben

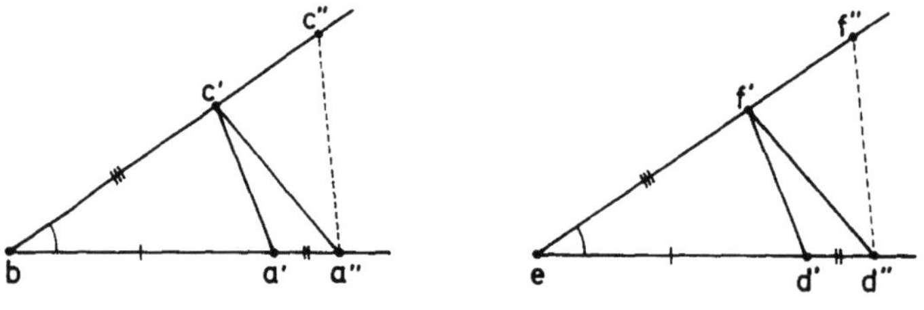

Abb. 54

angestellten Überlegung für die anderen Schenkel H(bc) und H(ef) liefert, daß man auch dort zu beliebigen Punkten c'' und f'' der genannten Art übergehen kann und dabei die gewünschte Kongruenz $a''c'' \equiv d''f''$ (in Abb. 54 gestrichelt) erhält. □

11.5 Anmerkung. Mit Hilfe von 11.3 bzw. 11.4 kann man nun viele weitere Sätze über die Winkelkongruenz zurückführen auf Sätze über die Streckenkongruenz (ggf., indem man zunächst kongruente Strecken auf entsprechenden Schenkeln abträgt). Das ist eine reine Routineabgelegenheit und kann dem Leser überlassen bleiben. Daher werden hier solche Sätze im allgemeinen nur formuliert und in manchen Fällen Hinweise auf zu verwendende frühere Sätze gegeben.

So ergibt sich, daß die Winkelkongruenz (aufgefaßt als zweistellige Relation) eine Äquivalenzrelation auf $\{\langle x,y,z\rangle \mid x \neq y \wedge z \neq y\}$ (d.h. auf der Menge der Punktetripel, die Winkel bestimmen) ist, m. a. W., daß die folgenden Sätze gelten.

11.6 Satz (*Reflexivität*). $a \neq b \wedge c \neq b \rightarrow abc \equiv abc$.

<u>11.7 Satz</u> (*Symmetrie*). $abc \equiv def \rightarrow def \equiv abc$.

<u>11.8 Satz</u> (*Transitivität*). $a_1b_1c_1 \equiv a_2b_2c_2 \wedge a_2b_2c_2 \equiv a_3b_3c_3 \rightarrow a_1b_1c_1 \equiv a_3b_3c_3$.

Durch Abtragen kongruenter Strecken auf beiden Schenkeln eines Winkels erhält man

<u>11.9 Satz</u>. $a \neq b \wedge c \neq b \rightarrow abc \equiv cba$.

Zusammen mit 11.8 besagt das, daß die Winkelkongruenz unabhängig von der Reihenfolge der Schenkel eines Winkels ist. Schließlich erhält man aus 11.3 oder 11.4 unmittelbar, daß die Winkelkongruenz auch unabhängig von der Wahl der Punkte auf den Schenkeln ist, d.h.

<u>11.10 Satz</u>. $abc \equiv def \wedge a' \underset{b}{\simeq} a \wedge c' \underset{b}{\simeq} c \wedge d' \underset{e}{\simeq} d \wedge f' \underset{e}{\simeq} f \rightarrow a'bc' \equiv d'ef'$.

Somit können wir 11.6 bis 11.10 zusammenfassen in

<u>11.11 Satz</u>. *Die Winkelkongruenz ist eine Äquivalenzrelation zwischen Winkeln.*

<u>11.12 Definition</u>. Ein Winkel W_2 heißt ein *Nebenwinkel* (oder *Supplementärwinkel*) eines Winkels W_1, falls W_1 und W_2 einen Schenkel gemeinsam haben und die verbleibenden Schenkel von W_1 und W_2 zueinander entgegengesetzte Halbgeraden sind (Abb. 55). W_2 heißt (der) *Scheitelwinkel* von W_1, falls seine Schenkel, die zu den Schenkeln von W_1 entgegengesetzten Halbgeraden sind (Abb. 56).

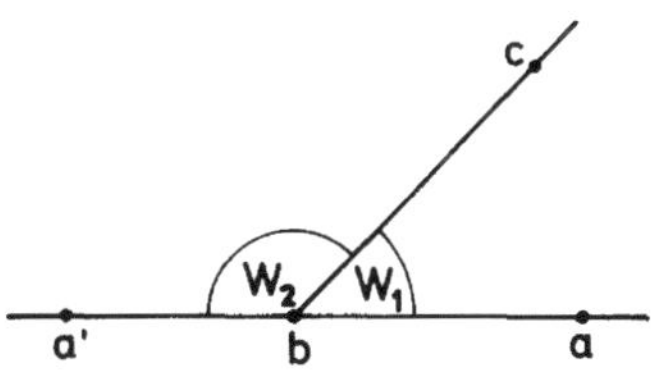

Abb. 55: Nebenwinkel

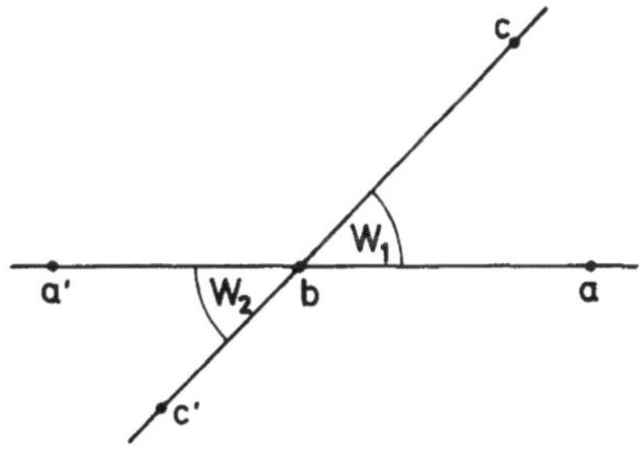

Abb. 56: Scheitelwinkel

Mit Hilfe der Überlegung im Beweis von 11.3 und 11.4 (zu (3) → (4)) kann man natürlich auch zu Punkten a'' und d'' auf der zu einem Schenkel entgegengesetzten Halbgeraden übergehen (unter Verwendung von 2.11 ["Aneinanderlegen von Strecken"] an Stelle von 4.3). Damit erhält man

11.13 Satz. $abc \equiv def \wedge \mathrm{B}aba' \wedge a' \neq b \wedge \mathrm{B}ded' \wedge d' \neq e \rightarrow a'bc \equiv d'ef$.
(Nebenwinkel von zueinander kongruenten Winkeln sind zueinander kongruent.)

Durch Anwendung von 11.13 auf 11.9 erhält man noch

11.14 Satz. $\mathrm{B}aba' \wedge \neq(aba') \wedge \mathrm{B}cbc' \wedge \neq(cbc') \rightarrow abc \equiv a'bc'$.
(Jeder Winkel ist kongruent zu seinem Scheitelwinkel.)

Die Möglichkeit und Eindeutigkeit der Winkelantragung, die bei Hilbert 1977 als Axiom III.4 gefordert wird, ergibt sich nun (gemäß 11.5) als leichte Folgerung aus dem Satz über die Dreiecksantragung (10.16).

11.15 Satz (*Winkelantragung*).

$$\neg \mathrm{Col}\, abc \wedge \neg \mathrm{Col}\, dep$$
$$\rightarrow \exists f[abc \equiv def \wedge f \underset{ed}{\simeq} p]$$
$$\wedge \forall f_1 \forall f_2 \left\{ \bigwedge_{\nu=1}^{2} [abc \equiv def_\nu \wedge f_\nu \underset{ed}{\simeq} p] \rightarrow f_1 \underset{e}{\simeq} f_2 \right\}.$$

(Ein nicht-entarteter Winkel $\sphericalangle abc$ läßt sich an einen gegebenen Schenkel H(ed) *in eine gegebene von* L(ed) *begrenzte Halbebene hinein eindeutig antragen.)*

Aus 10.12, 8.10 bzw. 8.1 ergeben sich gemäß 11.5 die folgenden drei Sätze.

11.16 Satz. $\mathrm{R}abc \wedge a \neq b \wedge c \neq b \wedge \mathrm{R}a'b'c' \wedge a' \neq b' \wedge c' \neq b' \rightarrow abc \equiv a'b'c'$.
(Rechte Winkel sind zueinander kongruent.)

11.17 Satz. $\mathrm{R}abc \wedge abc \equiv a'b'c' \rightarrow \mathrm{R}a'b'c'$.
(Jeder zu einem rechten Winkel kongruente Winkel ist ein Rechter.)

11.18 Satz. $\mathrm{B}cbd \wedge \neq(bcd) \wedge a \neq b \rightarrow [\mathrm{R}abc \leftrightarrow abc \equiv abd]$.
(Ein Winkel ist ein Rechter genau dann, wenn er zu einem seiner Nebenwinkel [und damit nach 11.14 oder 8.2 auch zu dem anderen] kongruent ist, Abb. 22.)

Damit erhalten wir als Spezialfall von 11.15 (neben der schon in 10.15 festgestellten Existenz) die Eindeutigkeit der Senkrechten in einer Halbebene, d.h.

<u>11.19 Satz</u>. $\mathrm{R}bap_1 \wedge \mathrm{R}bap_2 \wedge p_1 \underset{ab}{\simeq} p_2 \rightarrow p_1 \underset{a}{\simeq} p_2$.

Natürlich kann man von einer Halbebene zu der sie enthaltenden Ebene übergehen und erhält

<u>11.20 Satz</u>. $\mathrm{Ln}\,A \wedge \mathrm{Pl}\,E \wedge a \in A \wedge A \subseteq E \rightarrow \overset{=1}{\exists} B[\,\mathrm{Ln}\,B \wedge B \subseteq E \wedge A \underset{a}{\perp} B]$.
(*In einer Ebene E gibt es auf einer Geraden A in einem Punkt a genau eine Senkrechte B.*)

Aus 4.6 (die Zwischenbeziehung bleibt bei der Tripelkongruenz erhalten) erhält man noch

<u>11.21 Satz</u>.
(1) $a \underset{b}{\simeq} c \rightarrow [abc \equiv a'b'c' \leftrightarrow a' \underset{b'}{\simeq} c']$.
(2) $\mathrm{B}abc \wedge \neq(abc) \rightarrow [abc \equiv a'b'c' \leftrightarrow \mathrm{B}a'b'c' \wedge \neq(a'b'c')]$.
(*Ein Winkel* $\sphericalangle a'b'c'$ *ist kongruent zu einem Nullwinkel [gestreckten Winkel] genau dann, wenn er selbst ein Nullwinkel [gestreckter Winkel] ist.*)

<u>11.22 Satz</u>. ("*Aneinander- und Ineinanderlegen von Winkeln*").
$[(\mathrm{B}a\overset{b}{\underset{p}{}}c \wedge \mathrm{B}a'\overset{b'}{\underset{p'}{}}c') \vee (a \underset{bp}{\simeq} c \wedge a' \underset{b'p'}{\simeq} c')]$
$\wedge\, abp \equiv a'b'p' \wedge pbc \equiv p'b'c' \rightarrow abc \equiv a'b'c'$.

(*Durch Aneinanderlegen* [Abb. 57, 58] *bzw. Ineinanderlegen* [Abb. 59] *kongruenter Winkel erhält man wieder kongruente Winkel.* Gleiche Markierungen der Bogen in den Abbildungen bedeuten kongruente Winkel.)

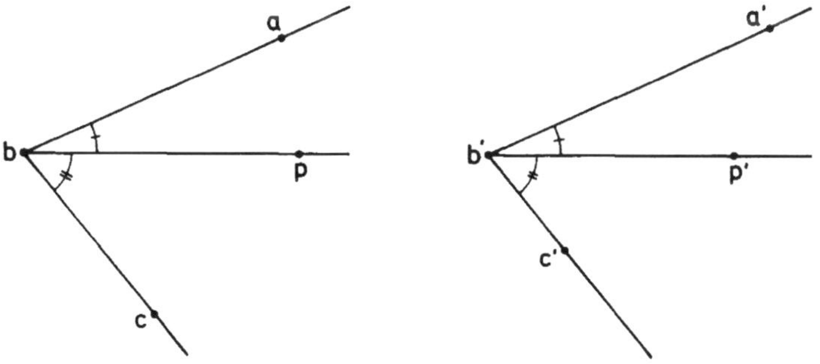

Abb. 57

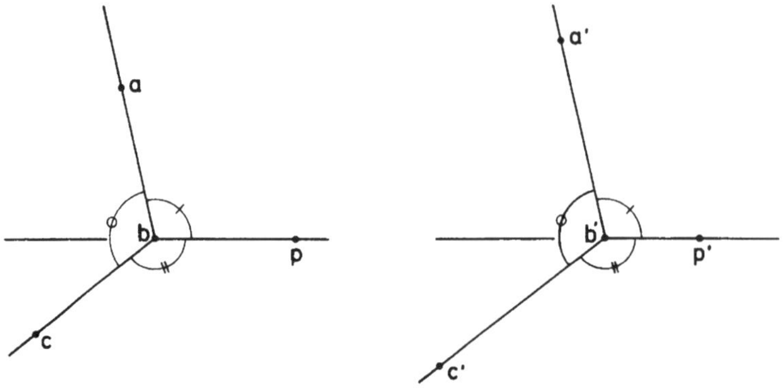

Abb. 58

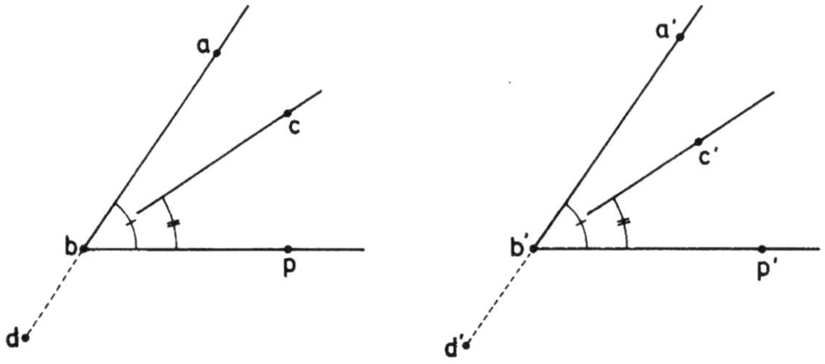

Abb. 59

Beweis: Fall 1: $\mathrm{B}a{}^{b}_{p}c \wedge \mathrm{B}a'{}^{b'}_{p'}c'$ (Aneinanderlegen). Sei d (gemäß 9.1) ein Punkt mit $\mathrm{Col}\,bpd \wedge \mathrm{B}adc$ und damit $a \neq d$. Durch Streckenabtragung erhalten wir Punkte a'', d'', c'' (Abb. 60) mit $a'' \underset{b}{\simeq} a' \wedge b'a'' \equiv ba$,

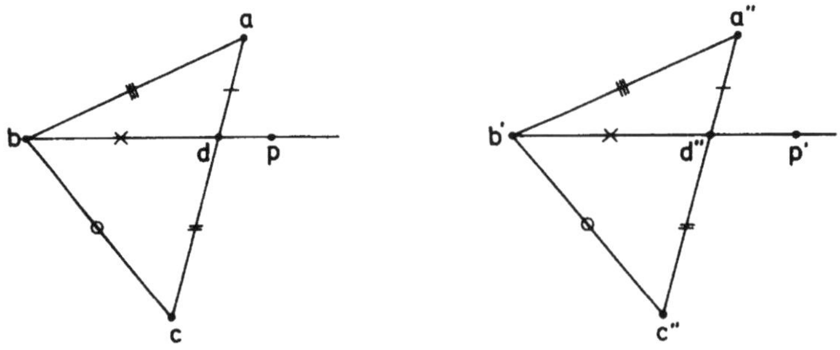

Abb. 60

Col $b'p'd'' \wedge [d''\underset{b'}{\simeq}p' \leftrightarrow d\underset{b}{\simeq}p] \wedge b'd''\equiv bd$, $Ba''d''c'' \wedge d''c''\equiv dc$. Nach 11.10 bzw. 11.13 gilt nun $d\neq b \rightarrow abd\equiv a'b'd''$, nach 11.4 (oder trivialerweise, falls $d=b$) ist also $ad\equiv a''d''$. Nach A5 ist also auch $bc\equiv b'c''$, somit $(abcd)\equiv(a''b'c''d'')$ und nach 11.3 insbesondere

(1) $abc\equiv a''b'c''$.

Aus der Konstruktion ergibt sich außerdem $pbc\equiv p'b'c''$ (falls $d=b$ nach 11.13, sonst wegen $dbc\equiv d''b'c''$ nach 11.10 bzw. 11.13). Nach Voraussetzung ist aber auch $pbc\equiv p'b'c'$, und c' und c'' liegen in derselben (in der zu a' entgegengesetzten) Halbebene mit der Begrenzung L$(b'p')$. Aus der Eindeutigkeit der Winkelantragung (11.15) ergibt sich somit $c'\underset{b'}{\simeq}c''$, aus (1) und 11.10 schließlich $abc\equiv a'b'c'$, wie zu zeigen.

Fall 2: $a\underset{bp}{\simeq}c \wedge a'\underset{b'p'}{\not\simeq}c'$ (Ineinanderlegen). Gemäß 3.14 seien d, d' Punkte mit B$abd \wedge d\neq b$, B$a'b'd' \wedge d'\neq b'$ (Abb. 59). Nach 11.13 läßt sich die Winkelkongruenz $abp\equiv a'b'p'$ gleichwertig ersetzen durch $dbp\equiv d'b'p'$. Nach Fall 1 ergibt sich dann $dbc\equiv d'b'c'$ und wieder nach 11.13 somit $abc\equiv a'b'c'$. □

Als Nächstes wollen wir Winkel der Größe nach vergleichen, d.h. eine Kleiner-Gleich-Beziehung einführen. Dazu erst eine Hilfsdefinition.

11.23 Definition. p (oder auch die Halbgerade H(bp)) *liegt im Winkel* $\sphericalangle abc$ (Abb. 61):

$$\mathrm{I}\,pabc :\leftrightarrow a\neq b \wedge c\neq b \wedge p\neq b \wedge \exists x\{\mathrm{B}axc \wedge [x=b \vee x\underset{b}{\simeq}p]\}.$$

(Der hier zugelassene Fall $x=b$ tritt für gestreckte Winkel $\sphericalangle abc$ ein.)

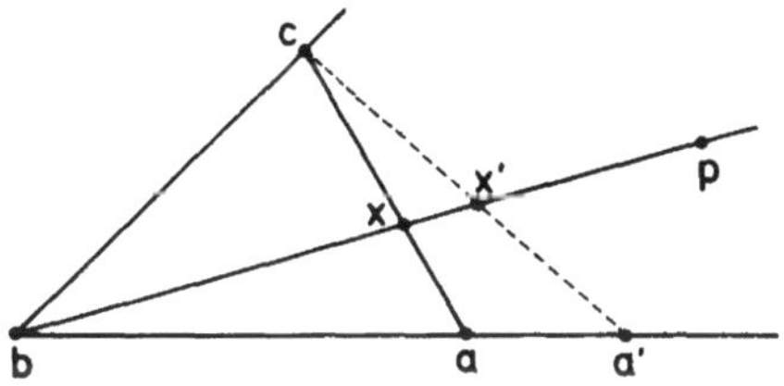

Abb. 61

Dann gilt offenbar

11.24 Satz. $\mathrm{I}\,pabc \rightarrow \mathrm{I}\,pcba$.

11.25 Satz. $\mathrm{I}\,pabc \wedge a'\underset{b}{\simeq}a \wedge c'\underset{b}{\simeq}c \wedge p'\underset{b}{\simeq}p \rightarrow \mathrm{I}\,p'a'bc'$.

Beweis: Im Falle Babc (abc gestreckt) ist alles klar. Sei also jetzt ¬Babc und x gemäß 11.23 gewählt. Mit Hilfe der inneren und äußeren Form des Satzes von Pasch (A7 bzw. 9.6) ergibt sich (Abb.61), daß man von a zu a' (und einem geeigneten x') übergehen kann, d.h., daß I $pa'bc$ gilt. Ebenso kann man von c zu c' übergehen; der Übergang von p zu p' ist nach Def. 11.23 trivial. □

Damit ist gezeigt, daß das Bestehen von I $pabc$ nur vom Winkel $\sphericalangle abc$ und der Halbgeraden H(bp) abhängt.

11.26 Anmerkung. Aus den Axiomen A1 bis A8 (und auch A9) folgt noch nicht, daß p selbst zwischen geeigneten Punkten auf den Schenkeln von $\sphericalangle abc$ liegen muß, wenn I $pabc$ gilt (vgl. die Bemerkungen zu A10 in 1.2, Gegenbeispiele liefert die hyperbolische Geometrie).

11.27 Definition. abc ist *kleiner-gleich* def (oder def *größer-gleich* abc):

$$abc \leq def :\leftrightarrow \exists p\,[\mathrm{I}\,pdef \wedge abc \equiv dep],$$

$$def \geq abc :\leftrightarrow abc \leq def.$$

In Hinblick auf 11.30, 11.37 sagen wir dafür auch, daß der Winkel $\sphericalangle abc$ kleiner-gleich dem Winkel $\sphericalangle def$ ist (analog für größer-gleich).

11.28 Lemma. $(abc) \equiv (a'b'c') \wedge \mathrm{Col}\, acd \rightarrow \exists d'\,(abcd) \equiv (a'b'c'd')$.

Beweis: Falls $a \neq c$ ist, sei d' gemäß 4.14 gewählt mit $(acd) \equiv (a'c'd')$, dieser Punkt leistet nach dem Satz über die Fünf-Strecken-Konfiguration (4.16) das Verlangte. Falls $a = c$ ist, erhält man einen Punkt d' der gesuchten Art durch Dreiecksantragung (10.16), falls $\neg\,\mathrm{Col}\, abd$, und wieder mittels 4.14, falls $\mathrm{Col}\, abd$. □

Mit Hilfe des Lemmas kann man die Situation von 11.23 auf einen zu $\sphericalangle abc$ oder einen zu $\sphericalangle abp$ kongruenten Winkel übertragen und erhält damit folgende weitere Charakterisierung der Kleiner-Gleich-Beziehung.

11.29 Satz. $abc \leq def \leftrightarrow \exists q\,[\mathrm{I}\,cabq \wedge abq \equiv def]$.

Durch Rückgang auf 11.27 bzw. 11.29 kann man dann ohne Schwierigkeiten die folgenden Sätze beweisen.

11.30 Satz. $abc \leq def \wedge abc \equiv a'b'c' \wedge def \equiv d'e'f' \rightarrow a'b'c' \leq d'e'f'$.

11.31 Satz.

(1) $a \underset{b}{\simeq} c \wedge d \neq e \wedge f \neq e \rightarrow abc \leq def$.

(2) $a \neq b \wedge c \neq b \wedge \mathrm{B}def \wedge \neq(def) \rightarrow abc \leq def$.
(Nullwinkel sind minimal, gestreckte Winkel sind maximal.)

11.32 Satz *(Reflexivität)*. $a \neq b \wedge c \neq b \rightarrow abc \leq abc$.

11.33 Satz *(Transitivität)*. $a_1b_1c_1 \leq a_2b_2c_2 \wedge a_2b_2c_2 \leq a_3b_3c_3 \rightarrow a_1b_1c_1 \leq a_3b_3c_3$.

11.34 Satz. $abc \leq def \wedge def \leq abc \rightarrow abc \equiv def$.

11.35 Satz *(Konnexität)*. $a \neq b \wedge c \neq b \wedge d \neq e \wedge f \neq e \rightarrow abc \leq def \vee def \leq abc$.

Beweis: Auf Grund von 11.31 können wir o. B. d. A. annehmen, daß $\sphericalangle abc$ und $\sphericalangle def$ nicht-entartete Winkel sind. Durch Winkelantragung erhalten wir dann einen Punkt c' mit $def \equiv abc' \wedge c' \underset{ba}{\simeq} c$. Dann ist $c' \in \mathrm{Pl}(bac) = \mathrm{Pl}(bca) = \mathrm{Pl}(\mathrm{L}(bc), a)$ (9.29, 9.20), also liegen c' und a in derselben oder in zueinander entgegengesetzten Halbebenen mit der Begrenzung $\mathrm{L}(bc)$, oder c' liegt auf $\mathrm{L}(bc)$.

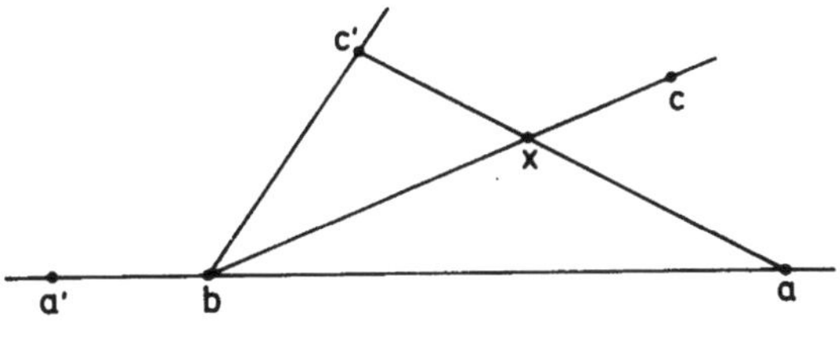

Abb. 62

Fall 1: $\mathrm{B}c'{}^{b}_{c}a \vee \mathrm{Col}\, bcc'$ (Abb. 62). Sei dann x ein Punkt mit $\mathrm{Col}\, bcx \wedge \mathrm{B}c'xa$ (gemäß 9.1 oder $x=c'$). Dann ist $x \underset{a}{\simeq} c'$, nach 9.19 also $x \underset{ba}{\simeq} c' \underset{ba}{\simeq} c$ und somit $x \underset{b}{\simeq} c$. Folglich ist $\mathrm{I}\, cabc'$, nach 11.29 also $abc \leq def$.

Fall 2: $c' \underset{bc}{\simeq} a$. Nach 9.31 ist dann $\mathrm{B}c{}^{b}_{c'}a$. Durch Vertauschung von c mit c' in Fall 1 ergibt sich dann $\mathrm{I}\, c'abc$ und nach 11.27 somit $def \leq abc$. □

Ist hierbei $\neg \mathrm{Col}\, bcc'$, so werden die eben betrachteten Fälle miteinander vertauscht, wenn man a durch einen Punkt a' mit $a' \neq b \wedge \mathrm{B}aba'$ (also insbesondere $\mathrm{B}a{}^{b}_{c}a'$) ersetzt, d.h. zu Nebenwinkeln übergeht. Daraus ergibt sich

11.36 Satz. $\mathrm{B}aba' \wedge \neq(aba') \wedge \mathrm{B}ded' \wedge \neq(ded') \rightarrow [abc \leq def \leftrightarrow d'ef \leq a'bc]$.
(Beim Übergang zu Nebenwinkeln wird die ≤-Beziehung umgekehrt.)

11.37 Anmerkung. Die in 11.30 vorausgesetzten Winkelkongruenzen gelten nach 11.11 insbesondere für Tripel, die denselben Winkel bestimmen. Somit kann $\leq$ als Relation zwischen Winkeln aufgefaßt werden. Nach 11.30 kann man auch noch zu einer entsprechenden Kleiner-Gleich-Beziehung zwischen Äquivalenzklassen bezüglich der Winkelkongruenz übergehen; diese ist dann nach 11.34 antisymmetrisch, also eine lineare Ordnung zwischen Äquivalenzklassen kongruenter Winkel oder "Winkelgrößen" (vgl. 13.4).

In gewohnter Weise kann man von $\leq$ zu einer Kleiner-Beziehung bzw. Grösser-Beziehung für Winkel übergehen mittels

11.38 Definition. $abc<def :\leftrightarrow abc\leq def \wedge \neg abc\equiv def$,
$def>abc :\leftrightarrow abc<def$.

Dafür erhält man offenbar ähnliche Sätze, deren Formulierung hier nicht ausgeführt wird.

11.39 Definition. *a, b, c* bilden einen *spitzen* bzw. *stumpfen Winkel*:

$\mathrm{Sp}\, abc :\leftrightarrow \exists a'\exists b'\exists c'[\mathrm{R}a'b'c' \wedge abc<a'b'c']$,

$\mathrm{St}\, abc :\leftrightarrow \exists a'\exists b'\exists c'[\mathrm{R}a'b'c' \wedge abc>a'b'c']$

(d.h. $\sphericalangle abc$ ist kleiner bzw. größer als ein Rechter).

11.40 Definition. Unter einem *Dreieck* verstehen wir eine Menge $\{a,b,c\}$ von paarweise verschiedenen Punkten *a, b, c*, die dann die *Ecken* dieses Dreiecks heißen; das Dreieck selbst bezeichnen wir auch kurz mit *abc*. Die Strecken *ab, bc, ca* heißen dann die *Seiten* dieses Dreiecks und die Winkel $\sphericalangle cab$, $\sphericalangle abc$ und $\sphericalangle bca$ die *Innenwinkel* dieses Dreiecks bei *a, b* bzw. *c*. Unter einem *Außenwinkel* des Dreiecks *abc* bei *a, b* bzw. *c* verstehen wir einen Nebenwinkel des Innenwinkels bei *a, b* bzw. *c*. Das Dreieck *abc* heißt *nicht-entartet*, falls seine Ecken nicht kollinear sind.

11.41 Satz (Satz vom Außenwinkel).

$\neg\, \mathrm{Col}\, abc \wedge \mathrm{B}bad \wedge d\neq a \rightarrow acb<cad \wedge abc<cad$.

(*Ein Außenwinkel in einem nicht-entarteten Dreieck ist stets grösser als jeder der beiden nicht-anliegenden Innenwinkel*, Abb. 63.)

11.42 Anmerkung. Mit Hilfe von A10, d.h. in der euklidischen Geometrie, ergibt sich (s. 12.23), daß ein Außenwinkel sogar stets "gleich der Summe der beiden nicht-anliegenden Innenwinkel" ist, d.h. durch Aneinanderlegen dieser Innenwinkel gemäß 11.22 erhalten werden kann. Aus den Axiomen A1 bis A8 (und auch A9) folgt das jedoch noch nicht.

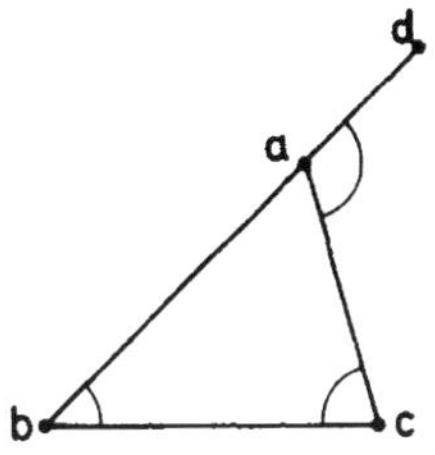

Abb. 63

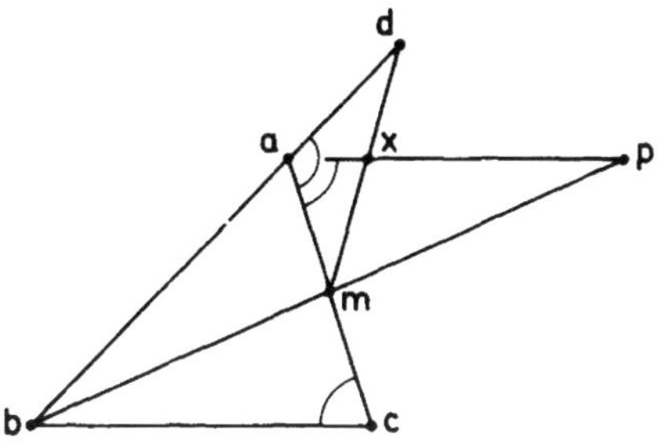

Abb. 64

Beweis von 11.41: Es genügt zu zeigen, daß $acb<cad$ ist (der Rest ergibt sich dann aus dem Bewiesenen und der Kongruenz [s. 11.14] der beiden Außenwinkel bei a). Sei m=M(ac) und $p=S_m b$ (Abb. 64). Durch Anwendung von 7.13 auf die Spiegelung S_m ergibt sich $(acb)\equiv(cap)$ und damit insbesondere $acb\equiv cap$. Nach dem Axiom von Pasch (A7) existiert ein Punkt x mit B$mxd\wedge$Baxp, also ist I$pcad$. Nach 11.27 ist also zunächst $acb\leq cad$. Nach Konstruktion ist andererseits $p\underset{ac}{\simeq}d$, aber $p\notin$L(ad) und damit $\neg p\underset{a}{\simeq}d$, wegen der Eindeutigkeit der Winkelantragung (11.15) also $\neg cap\equiv cad$ und somit auch $\neg acb\equiv cad$. □

11.43 Korollar. - Col $abc\wedge$ [Rbac v St bac] → Sp $abc\wedge$ Sp acb.

(Ist in einem nicht-entarteten Dreieck ein Innenwinkel ein Rechter oder stumpf, so sind die übrigen beiden Innenwinkel spitz.)

Beweis: Der (jeder der beiden) Außenwinkel bei a ist ebenfalls ein Rechter oder (nach 11.36) spitz; somit liefert 11.41 die Behauptung. □

11.44 Satz.

(1) ¬ Col abc → [$ab\equiv ac$ ↔ $acb\equiv abc$].

(2) ¬ Col abc → [$ab<ac$ ↔ $acb<abc$].

(Für nicht-entartete Dreiecke gilt: 1. Kongruenten Seiten liegen kongruente Winkel gegenüber und umgekehrt, 2. der größeren von zwei Seiten liegt der größere Winkel gegenüber und umgekehrt; Abb. 65)

Beweis: Zu (1)(→): Das ergibt sich sofort aus 11.3.

Zu (2)(→): Sei c' gemäß 5.4 ein Punkt mit B$ac'c\wedge ab\equiv ac'$ und damit $c'\neq c$. Dann ist I$c'abc$, und aus 11.41 (für das Dreieck $c'cb$), (1)(→) und

11.27 ergibt sich $acb < ac'b \equiv abc' \leq abc$.

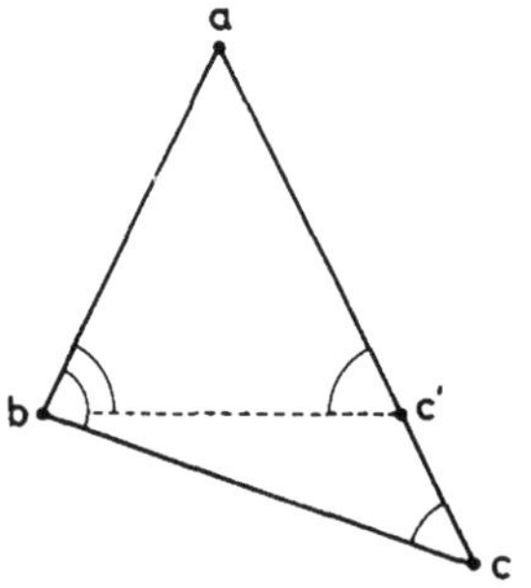

Abb. 65

Natürlich gilt (2)(→) auch mit > statt < (Vertauschung von *b* mit *c*). Da für zwei gegebene Strecken bzw. Winkel die Beziehungen <, ≡, > alle Möglichkeiten erschöpfen und sich gegenseitig ausschließen, gelten auch die Umkehrungen (1)(←) und (2)(←). □

<u>11.45 Definition</u>. Ein Dreieck *abc* mit $ab \equiv ac$ (wie in 11.44(1)) heißt ein *gleichschenkliges Dreieck* mit der *Spitze a*, mit den *Schenkeln ab* und *ac*, mit der *Grundlinie* oder *Basis bc* und mit den *Basiswinkeln* $\sphericalangle abc$ und $\sphericalangle acb$. Ein Dreieck *abc* mit $ab \equiv ac \equiv bc$ heißt *gleichseitig*. Ein Dreieck *abc* mit R*bac* heißt ein *rechtwinkliges Dreieck* mit den *Katheten ab* und *ac* und mit der *Hypotenuse bc*.

Aus 11.43, 11.44 erhält man sofort

<u>11.46 Korollar</u>. $\neg \mathrm{Col}\, abc \wedge [\mathrm{R}bac \vee \mathrm{St}\, bac] \rightarrow ab < bc \wedge ac < bc$.
(In einem Dreieck ist die einem rechten oder stumpfen Winkel gegenüberliegende Seite die größte Seite; insbesondere ist im rechtwinkligen Dreieck die Hypotenuse die größte Seite.)

<u>11.47 Satz</u>. $\mathrm{R}acb \wedge ch \underset{h}{\perp} ab \rightarrow \mathrm{B}ahb \wedge \neq(ahb)$.
(In einem [nicht-entarteten] rechtwinkligen Dreieck liegt der Höhenfußpunkt echt zwischen den Endpunkten der Hypotenuse, Abb. 66.)

<u>Beweis</u>: Aus 11.43 erhält man $\neq(ahb)$, aus 11.46 weiter $ah < ac < ab$ und $bh < bc < ab$ und aus 5.12 damit die Behauptung. □

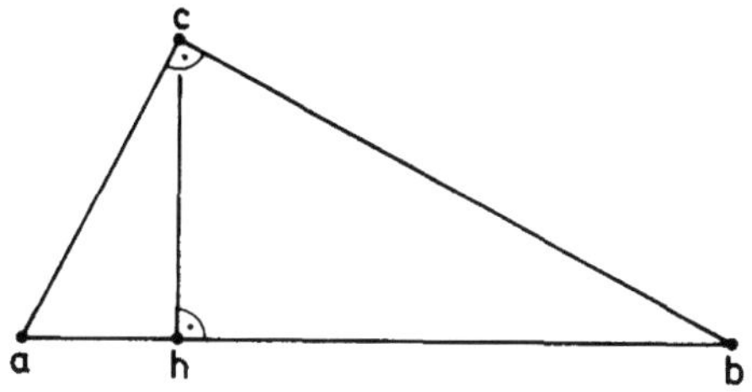

Abb. 66

11.48 Kongruenzsätze. Wir kommen nun zu den vier klassischen "Kongruenzsätzen für Dreiecke". Diese beziehen sich eigentlich auf geordnete Dreiecke (d.h. Tripel) oder auf Dreiecke abc und $a'b'c'$, für die zusätzlich eine eineindeutige Zuordnung zwischen den Ecken gegeben ist, die hier durch entsprechende Buchstaben ausgedrückt ist. Die Sätze besagen, daß die sechs "Bestimmungsstücke" (Seiten und Innenwinkel) des ersten Dreiecks zu den entsprechenden Bestimmungsstücken des zweiten Dreiecks kongruent sind, wenn das nur für gewisse drei dieser Bestimmungsstücke gilt. Es haben sich *Merkregeln* aus den Buchstaben S (für "Seiten") und W (für "Winkel") eingebürgert, die andeuten, für welche Bestimmungsstücke die Kongruenz in dem jeweiligen Satz schon vorausgesetzt wird. So bedeutet SWS "für zwei Seiten und den davon eingeschlossenen Winkel (dessen Schenkel diese Seiten enthalten)", WSW bedeutet "für zwei Winkel und die davon eingeschlossene Seite (die die Scheitel dieser Winkel verbindet)", WWS bedeutet "für zwei Winkel und eine nicht davon eingeschlossene Seite" usw.

11.49 Erster Kongruenzsatz (SWS).

$$abc \equiv a'b'c' \wedge ba \equiv b'a' \wedge bc \equiv b'c'$$
$$\rightarrow ac \equiv a'c' \wedge [a \neq c \rightarrow bac \equiv b'a'c' \wedge bca \equiv b'c'a'] \qquad \text{(Abb. 67).}$$

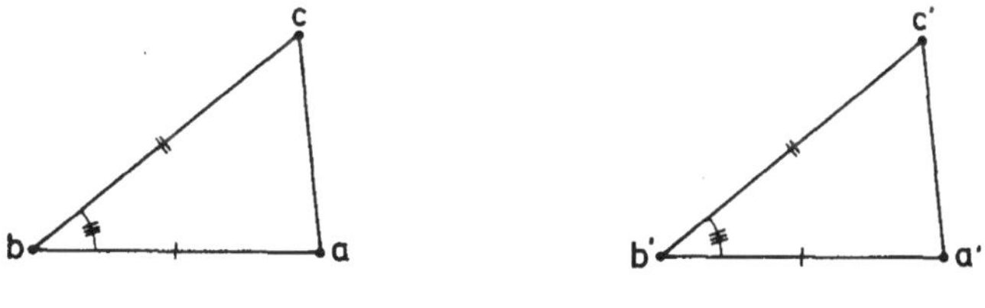

Abb. 67

Beweis: Aus der Voraussetzung ergibt sich mittels 11.4 sofort $ac \equiv a'c'$. Somit ist $(abc) \equiv (a'b'c')$, woraus sich nach 11.3 der Rest der Behauptung ergibt. □

11.50 Zweiter Kongruenzsatz.

(i) (WSW) $\neg \mathrm{Col}\, abc \wedge bac \equiv b'a'c' \wedge abc \equiv a'b'c' \wedge ab \equiv a'b'$
$\rightarrow ac \equiv a'c' \wedge bc \equiv b'c' \wedge acb \equiv a'c'b'$ (Abb. 68).

Abb. 68

(ii) (WWS) $\neg \mathrm{Col}\, abc \wedge bca \equiv b'c'a' \wedge abc \equiv a'b'c' \wedge ab \equiv a'b'$
$\rightarrow ac \equiv a'c' \wedge bc \equiv b'c' \wedge bac \equiv b'a'c'$ (Abb. 69).

Abb. 69

Beweis: Sei c'' der durch Streckenabtragung (6.11) entstehende Punkt mit $c'' \underset{b'}{\simeq} c' \wedge b'c'' \equiv bc$. Aus den in (i) und (ii) gemeinsamen Voraussetzungen ergibt sich mittels 11.4 (oder 11.49), daß $(abc) \equiv (a'b'c'')$ ist, d.h. (nach 11.3), daß in den Dreiecken abc und $a'b'c''$ alle entsprechenden Bestimmungsstücke zueinander kongruent sind. Somit genügt es, $c''=c'$ nachzuweisen.

Zu (i): Nach Vor. ist $b'a'c' \equiv bac \equiv b'a'c''$, nach Konstruktion außerdem $c'' \underset{a'b'}{\widetilde{\simeq}} c'$. Aus der Eindeutigkeit der Winkelantragung (11.15) ergibt sich $c'' \underset{a'}{\simeq} c'$; somit ist c'' der eindeutig bestimmte Schnittpunkt der Geraden $\mathrm{L}(b'c')$ und $\mathrm{L}(a'c')$, also $c''=c'$.

Zu (ii): Nach Vor. ist $b'c'a' \equiv bca \equiv b'c''a'$, nach Konstruktion außerdem $\mathrm{B}b'c'c'' \vee \mathrm{B}b'c''c'$. Wäre nun $c'' \neq c'$, so wäre einer der beiden Winkel $\sphericalangle b'c'a'$ und $\sphericalangle b'c''a'$ ein Außenwinkel und der andere ein nicht anliegender Innenwinkel im Dreieck $a'c'c''$, die beiden Winkel könnten also nach 11.41 nicht zueinander kongruent sein. □

Aus 11.3 ergibt sich sofort

11.51 Dritter Kongruenzsatz (SSS).

$\neq(abc) \wedge ab \equiv a'b' \wedge ac \equiv a'c' \wedge bc \equiv b'c'$
$\rightarrow bac \equiv b'a'c' \wedge abc \equiv a'b'c' \wedge bca \equiv b'c'a'$.

11.52 Vierter Kongruenzsatz (WSS).

$abc \equiv a'b'c' \wedge ac \equiv a'c' \wedge bc \equiv b'c' \wedge bc \leq ac$
$\rightarrow ba \equiv b'a' \wedge bac \equiv b'a'c' \wedge bca \equiv b'c'a'$.
(*Sind in einem Dreieck zwei Seiten und der der größeren gegenüberliegende Winkel kongruent zu den entsprechenden Bestimmungsstücken eines zweiten Dreiecks, so sind alle entsprechenden Bestimmungsstücke zueinander kongruent,* Abb. 70.)

Abb. 70

Beweis: Sei a'' der durch Streckenabtragung entstehende Punkt mit $a'' \simeq_b a' \wedge b'a'' \equiv ba$. Wie im Beweis von 11.50 ist dann $(abc) \equiv (a''b'c')$, und es genügt zu zeigen, daß $a''=a'$ ist. (Wegen $bc \leq ac$ ist $a \neq c$ und damit $\neq(abc)$, so daß sich aus der Tripelkongruenz nach 11.3 auch die gewünschten Winkelkongruenzen ergeben.) Wir nehmen $a'' \neq a'$ an. Es genügt dann zu zeigen, daß die Voraussetzung $bc \leq ac$ (die vorher nicht benutzt wird) nicht gelten kann. Nach Konstruktion und Annahme ist $c'a' \equiv ca \equiv c'a''$, $\neq(b'a'a'')$ und $\mathrm{B}b'a'a'' \vee \mathrm{B}b'a''a'$.

Fall 1: $\mathrm{B}b'a'a''$.

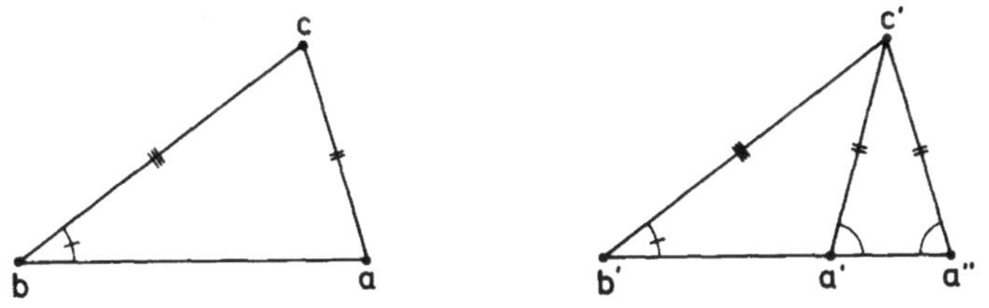

Abb. 71

Fall 1a: $\neg\,\mathrm{Col}\,abc$ (Abb. 71). Dann ist $\sphericalangle a''a'c'$ ein Außenwinkel im Dreieck $a'b'c'$, nach 11.41 also $a'b'c' < a''a'c'$. Nach 11.44(1) ist $a''a'c' \equiv a'a''c'$. Somit ist $a''b'c' \equiv a'b'c' < a'a''c' \equiv b'a''c'$, nach 11.44(2) also $a''c' < b'c'$

und damit auch $ac<bc$, wie zu zeigen.

Fall 1b: Col abc. Dann liegt auch c' auf der Geraden durch b', a', a''. Nach 7.20 ist dann M$a'c'a''$, insbesondere also B$a'c'a''$, nach 3.5 folglich B$b'a'c'$ und damit $a'c'\leq b'c'$, wegen $b'\neq a'$ (nach 6.11) also sogar $a'c'<b'c'$ und damit auch $ac<bc$, wie zu zeigen.

Fall 2: B$b'a''a'$. Hier erhält man das Gewünschte, indem man in der Behandlung von Fall 1 a' mit a'' vertauscht. □

Für spätere Anwendungen formulieren wir noch das folgende Lemma.

11.53 Lemma. R$adc \wedge c\neq d \wedge \neq(abd) \wedge$ B$dab \rightarrow dbc<dac \wedge ac<bc$ (Abb. 72).

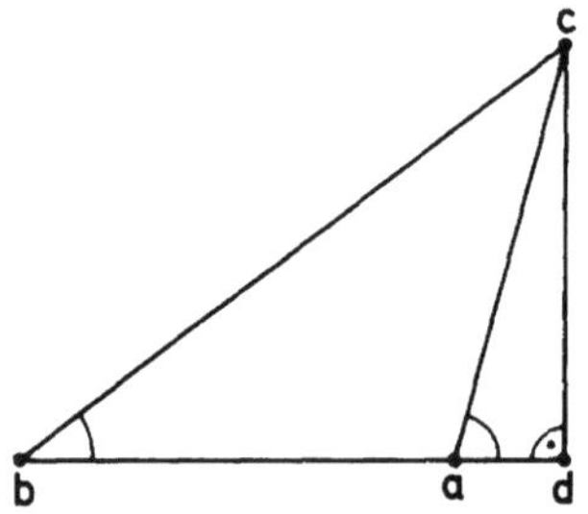

Abb. 72

Beweis: Setzt man $a'=S_d(a)$, so erhält man das Gewünschte sofort (unter Weglassung von Strichen) aus der Betrachtung von Fall 1a des vorigen Beweises. □

11.54 Anmerkungen. (i) Ein Teil des ersten Kongruenzsatzes wird bei Hilbert 1977 als Axiom verwendet. Nach seinem Beweis stehen alle Axiome der Hilbertschen Axiomgruppen I. bis III. in unserem Aufbau zur Verfügung (vgl. 0.4); das Entsprechende gilt auch für die räumliche Version der Hilbertschen Axiome bei Verwendung eines zugehörigen Dimensionsaxioms (s. 1.7.5, Bem. zu A8 und A9).

(ii) Bisweilen wird nur 11.50(i) als zweiter Kongruenzsatz bezeichnet.

(iii) Aus den vorangehenden Überlegungen erhält man noch, daß die Voraussetzung $bc\leq ac$ im vierten Kongruenzsatz nicht entbehrlich ist. Man kann sogar von einem beliebigen Dreieck $a'b'c'$ mit $a'c'<b'c'$ und ¬R$b'a'c'$ ausgehen. Sei dann $a''=S_{d'}(a')$, wobei d' der Fußpunkt des Lots von c' auf die Gerade L$(b'a')$ ist bzw. $d'=c'$, falls Col $a'b'c'$. Dann

erhält man (Übung): 1. Die Dreiecke $a'b'c'$ und $a''b'c'$ erfüllen die übrigen Voraussetzungen von 11.52, aber nicht die Behauptung. 2. Es gilt jedoch für jedes Dreieck abc, das (zusammen mit $a'b'c'$) den übrigen Voraussetzungen von 11.52 genügt, daß seine Bestimmungsstücke kongruent sind zu den entsprechenden Bestimmungsstücken des Dreiecks $a'b'c'$ o d e r zu denen des Dreiecks $a''b'c'$ (mittels 11.53 ergibt sich nämlich, daß a' und a'' die einzigen Punkte auf $L(b'a')$ sind, die von c' den gewünschten Abstand haben).

11.55 Orthogonalität für Unterräume. Der Begriff der Orthogonalität wurde in 8.11 für Geraden eingeführt. Im Folgenden wird eine Ausdehnung dieses Begriffs auf Unterräume behandelt. Mit den bereitgestellten Mitteln kann man die wichtigsten Resultate über die Orthogonalität von endlichdimensionalen affinen Unterräumen erhalten. Wir beschränken uns jedoch auf die Orthogonalität zwischen Geraden und Unterräumen (nur das wird zum Beweis des Darstellungssatzes 16.15 gebraucht). Das Ganze ist - analog zu 9.36ff. - natürlich nur von Interesse für den Fall, daß das Axiom A9 verletzt ist, d.h. nicht alle Punkte in einer Ebene liegen. Bei Verwendung von A9 wird der Rest dieses Abschnitts nicht gebraucht.

11.56 Definition. Unter einem *Winkel zwischen Halbebenen H, K mit gemeinsamer Begrenzung A* verstehen wir einen beliebigen Winkel $\sphericalangle bac$, dessen Scheitel a auf A liegt und dessen Schenkel die auf A im Punkte a in die Halbebenen hinein errichteten (nach 11.19 eindeutig bestimmten) Senkrechten sind (Abb. 73).

Der folgende Satz zeigt, daß dieser Winkel bis auf Kongruenz eindeutig bestimmt (also von der Wahl von a unabhängig) ist.

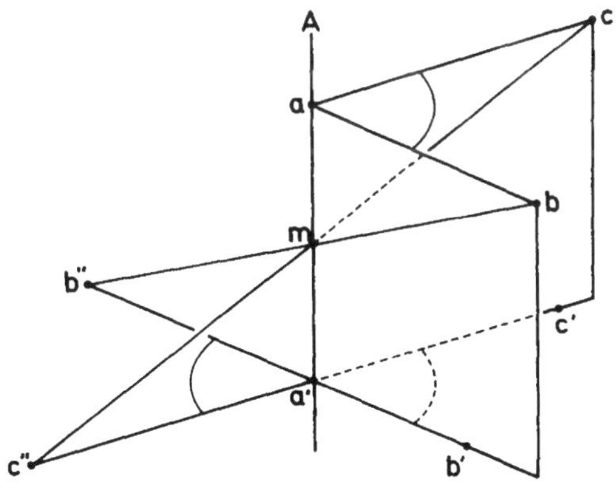

Abb. 73

<u>11.57 Satz.</u> $b \underset{aa'}{\simeq} b' \wedge c \underset{aa'}{\simeq} c' \wedge \mathrm{R}baa' \wedge \mathrm{R}caa' \wedge \mathrm{R}b'a'a \wedge \mathrm{R}c'a'a \rightarrow bac \equiv b'a'c'$.

<u>Beweis</u>: Sei $A=\mathrm{L}(aa')$, sei m der Mittelpunkt von aa' und $b''=\mathrm{S}_m(b)$, $c''=\mathrm{S}_m(c)$ (Abb. 73). Da S_m eine Bewegung ist, ist zunächst $bac \equiv b''a'c''$. Außerdem ist $\mathrm{R}b''a'a$ und $b'' \in \mathrm{Pl}(A,b)$, wegen der Eindeutigkeit der Senkrechten in einer Ebene (11.20) also $\mathrm{Col}\, b''a'b'$ und wegen $\mathrm{B}bAb''$ nach 9.8, 9.18 sogar $\mathrm{B}b''a'b'$. Ebenso ergibt sich $\mathrm{B}c''a'c'$. Somit ist auch $b''a'c'' \equiv b'a'c'$ (Kongruenz von Scheitelwinkeln, 11.14), also gilt die Behauptung. □

Für das Folgende brauchen wir den Spezialfall, daß der genannte Winkel ein Rechter ist. Dann ist es sinnvoll, Def. 11.56 von den obigen Halbebenen auf die sie enthaltenden Ebenen zu übertragen.

<u>11.58 Definition</u>. Die Ebenen ("planes") E, F bilden (miteinander) einen *rechten Winkel*:

$$\mathrm{Rp}\, EF :\leftrightarrow \mathrm{Pl}\, E \wedge \mathrm{Pl}\, F \wedge \exists a \exists A \exists B \exists C [\mathrm{Ln}\, A \wedge \mathrm{Ln}\, B \wedge \mathrm{Ln}\, C \wedge E \cap F = A \wedge B \subseteq E \wedge C \subseteq F \wedge A \underset{a}{\perp} B \wedge A \underset{a}{\perp} C \wedge B \underset{a}{\perp} C].$$

<u>11.59 Definition</u>. Die Gerade A *steht senkrecht auf* (oder ist *orthogonal zu* oder ist ein *Lot auf*) dem affinen Unterraum U (im Punkte x):

$$A \underset{x}{\hat{\perp}} U :\leftrightarrow \mathrm{Ln}\, A \wedge \mathrm{Sb}\, U \wedge A \cap U = \{x\} \wedge \forall B [\mathrm{Ln}\, B \wedge x \in B \wedge B \subseteq U \rightarrow A \underset{x}{\perp} B]$$

(d.h., A schneidet U in x und steht senkrecht auf jeder in U enthaltenen Geraden durch den Punkt x).

$$A \hat{\perp} U :\leftrightarrow \exists x A \underset{x}{\hat{\perp}} U.$$

$$ab \underset{(x)}{\hat{\perp}} U :\leftrightarrow a \neq b \wedge \mathrm{L}(ab) \underset{(x)}{\hat{\perp}} U.$$

Die Betrachtung beliebiger Unterräume überschreitet natürlich den Rahmen der Prädikatenlogik der ersten Stufe; es läßt sich aber für jede natürliche Dimension n noch der folgende Spezialfall in diesem Rahmen behandeln (vgl. Anm. 6.26(i), 9.54):

$$A \overset{n}{\underset{(x)}{\perp}} U :\leftrightarrow \mathrm{Sb}_n U \wedge A \underset{(x)}{\hat{\perp}} U.$$

$$ab \overset{n}{\underset{(x)}{\perp}} U :\leftrightarrow \mathrm{Sb}_n U \wedge ab \underset{(x)}{\hat{\perp}} U.$$

<u>11.60 Satz.</u> $U=\mathrm{Sb}_n(xx_1 \ldots x_n) \wedge \bigwedge_{\nu=1}^{n} A \underset{x}{\perp} \mathrm{L}(xx_\nu) \rightarrow A \underset{x}{\hat{\perp}} U \qquad (n \geq 1)$.

(Eine Gerade steht schon dann senkrecht auf einem n-dimensionalen Unterraum, wenn sie auf n Geraden senkrecht steht, die diesen Raum "aufspannen".)

Beweis: Sei p ein von x verschiedener Punkt von A und $q=S_x p$. Es genügt zu zeigen, daß für jeden Punkt a von U die Beziehung Rpxa gilt, d.h. (nach 8.1 und 8.2) $ap\equiv aq$. Nach Voraussetzung gilt das Gewünschte schon für $a=x$, x_1, ..., x_n. Da man jeden Punkt von U, von diesen Punkten ausgehend, durch (n-maliges) Bilden von Verbindungsgeraden erhalten kann (9.52), genügt es zu zeigen, daß die gewünschte Eigenschaft beim Bilden von Verbindungsgeraden erhalten bleibt, d.h.

$$a\neq b \wedge \mathrm{Col}\,abc \wedge ap\equiv aq \wedge bp\equiv bq \rightarrow cp\equiv cq.$$

Das wurde aber bereits als Satz 4.17 gezeigt. □

11.61 Lemma. $\mathrm{Pl}\,E \wedge \mathrm{Pl}\,F \wedge \mathrm{Ln}\,A \wedge E\cap F=A \wedge \mathrm{Ln}\,C \wedge C\subseteq F \wedge C\perp A$
$\rightarrow [\mathrm{Rp}\,EF \leftrightarrow C \hat{\perp} E]$.
(Schneiden sich die Ebenen E, F in der Geraden A und ist C eine Senkrechte auf A innerhalb F, so bilden E und F einen rechten Winkel genau dann, wenn C auf E senkrecht steht.)

Beweis: Die Richtung von rechts nach links ergibt sich unmittelbar aus den Definitionen (und 11.20), die Umkehrung aus 11.57 und 11.60. □

11.62 Satz (Lotsatz für Unterräume).
$\mathrm{Sb}_n U \wedge p\notin U \rightarrow \overset{=1}{\exists} L[\,\mathrm{Ln}L \wedge p\in L \wedge L \hat{\perp} U]$.
(Auf einen endlichdimensionalen Unterraum U gibt es von einem Punkt außerhalb U stets genau ein Lot.)

11.63 Satz. $\mathrm{Sb}_n U \wedge p\notin U \wedge q\in U \rightarrow \overset{=1}{\exists} L[\mathrm{Ln}\,L \wedge L \underset{q}{\hat{\perp}} U \wedge L\subseteq \mathrm{Sb}_{n+1}(U,p)]$.
(Auf einem n-dimensionalen Unterraum U gibt es in einem Punkt q dieses Raumes genau eine Senkrechte, die in einem gegebenen (n+1)-dimensionalen Raum U' mit U⊆U' enthalten ist.)

Beweis von 11.62 und 11.63:

1. Eindeutigkeit in 11.62: Sind L_1, L_2 verschiedene solche Lote mit den Fußpunkten x_1 bzw. x_2, d.h. $L_\nu \perp_{x_\nu} U$ ($\nu=1,2$), so sind sie insbesondere verschiedene Lote von p auf die Verbindungsgerade $L(x_1 x_2)$ im Widerspruch zum Lotsatz 8.18.

2. Eindeutigkeit in 11.63: Seien L_1, L_2 verschiedene solche Senkrechten. Dann ist die Ebene $E:=\mathrm{Pl}(L_1,L_2)$ auch noch im $(n+1)$-dimensionalen Raum $\mathrm{Sb}_{n+1}(U,p)$ enthalten, schneidet also U nach 9.59 in einer Geraden A.

Dann sind L_1, L_2 insbesondere verschiedene Senkrechten innerhalb der Ebene E auf der Geraden A im Punkt q im Widerspruch zur Eindeutigkeitsbehauptung in 11.20.

<u>Existenz</u>: Wir beweisen die Existenzbehauptungen beider Sätze (kurz 11.62_n und 11.63_n) gemeinsam durch Induktion über n. Für $n=1$ gelten sie nach 8.18 und 11.20. Wir setzen sie nun schon für n voraus.

Zu 11.62_{n+1} (Abb. 74): Sei U ein $(n+1)$-dimensionaler Unterraum, etwa $U=\mathrm{Sb}_{n+1}(U',a)$, und $p\notin U$. Sei A das (nach 11.62_n existierende) Lot von p auf den n-dimensionalen Unterraum U' mit dem Fußpunkt r, und sei C die (nach 11.63_n existierende) Senkrechte innerhalb U auf U' im Punkt r.

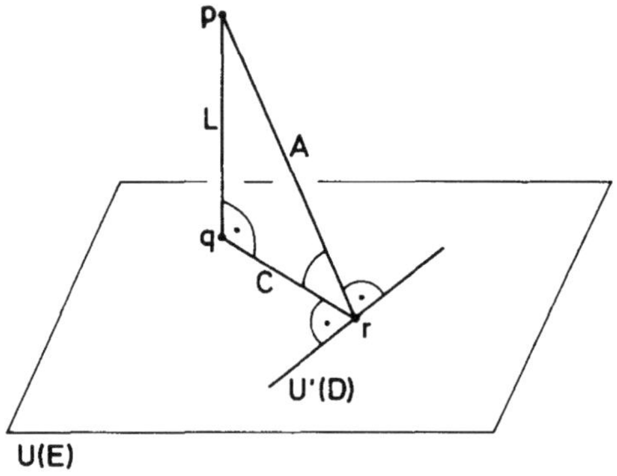

Abb. 74

Schließlich sei L (gemäß 8.18) das Lot von p auf C mit dem Fußpunkt q. Um zu zeigen, daß L das Verlangte ($L \mathbin{\hat{\perp}} U$) leistet, sei B eine beliebige in U enthaltene Gerade durch q. Dann ist $L \perp B$ zu zeigen. Für $B=C$ gilt das nach Konstruktion. Sei also jetzt $B\neq C$ und $E=\mathrm{Pl}(B,C)$. Diese in U enthaltene Ebene schneidet U' nach Satz 9.59 in einer Geraden D mit $r\in D$. Nach Wahl von A und C steht D senkrecht auf diesen beiden Geraden und damit auf der Ebene $F=\mathrm{Pl}(A,C)$. Nach dem Lemma 11.61 ergibt sich nun, erstens, daß die Ebenen E und F einen rechten Winkel bilden, und zweitens, daß L senkrecht auf E steht. Damit ist insbesondere $L \perp B$, wie zu zeigen.

Zu 11.63_{n+1} (Abb. 75): Sei wieder U ein $(n+1)$-dimensionaler Unterraum, $p\notin U$ und $q\in U$. Sei A das (nach 11.62_{n+1} existierende) Lot von p auf U mit dem Fußpunkt r.

<u>Fall 1</u>: $r=q$. Dann leistet A das für L Verlangte.

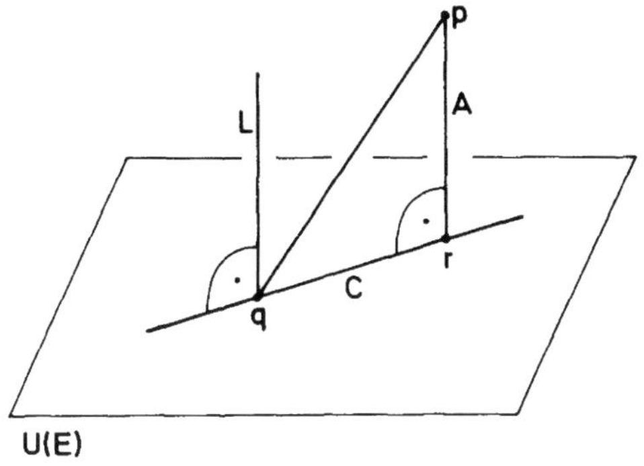

Abb. 75

<u>Fall 2</u>: $r \neq q$. Sei dann C=L(qr), F=Pl(pqr) und L (gemäß 11.20) die Senkrechte innerhalb F auf C im Punkt q. Ist nun wieder B eine beliebige in U enthaltene und von C verschiedene Gerade durch q, so ist A insbesondere senkrecht auf E:=Pl(B,C), nach 11.61 also RpEF und $L \hat{\perp} E$ und damit $L \perp B$. Wie in 11.62_{n+1} ist also $L \hat{\perp} U$, wie zu zeigen. □

<u>11.64 Anmerkung</u>. 11.62 und 11.63 gelten nicht allgemein für unendlichdimensionale Unterräume U; es gibt vielmehr Modelle von A1 bis A8 mit echten unendlichdimensionalen Unterräumen, auf denen es keine senkrechte Gerade gibt. Ein Beispiel, auf das mich H. König (Saarbrücken) hinwies, läßt sich folgendermaßen konstruieren.

Euklidische Räume seien wie in der analytischen Geometrie üblich eingeführt, d.h. festgelegt durch einen Vektorraum über einen angeordneten Körper und ein *inneres Produkt* (das ist eine symmetrische, positiv-definite Bilinearform). Sei $\mathfrak{U}$ ein euklidischer Raum über $\mathbb{R}$ mit einer abzählbaren Orthonormalbasis $B=\{b_n \mid n \in \mathbb{N}\}$ (ein solcher läßt sich aus einem Vektorraum mit der Basis B durch geeignete Definition des inneren Produkts erhalten). Sei $\mathfrak{V}$ die (nach bekannten Sätzen existierende und bis auf Isomorphie eindeutig festgelegte) metrische Vervollständigung von $\mathfrak{U}$, d.h. ein euklidischer Raum über $\mathbb{R}$ mit den folgenden beiden Eigenschaften.

(i) Jede Cauchy-Folge von Vektoren aus $\mathfrak{V}$ hat in $\mathfrak{V}$ einen Grenzwert (Vollständigkeit).

(ii) Jeder Vektor aus $\mathfrak{V}$ läßt sich schon als Grenzwert einer Cauchy-Folge von Vektoren aus $\mathfrak{U}$ darstellen (Minimalität der Vervollständigung).

Nach Wahl von $\mathfrak{U}$ gilt natürlich

(iii) Jeder Vektor aus $\mathfrak{U}$ läßt sich (eindeutig) als Linearkombination von endlich vielen Basisvektoren b_n darstellen.

Damit ergibt sich

(iv) $\mathfrak{U}$ ist nicht vollständig und somit ein echter Unterraum von $\mathfrak{V}$.

Es liegt nämlich z.B. der Grenzwert der Cauchy-Folge $(a_n)_n$ mit

(1) $a_n = \sum_{\nu=0}^{n} \frac{1}{2^\nu} b_\nu$

nicht in $\mathfrak{U}$. Allgemeiner erhält man:

(v) Jeder Vektor aus läßt sich eindeutig als "verallgemeinerte Linearkombination von Basisvektoren" darstellen, d.h. als Grenzwert einer unendlichen Reihe der Form

(2) $a = \sum_{\nu=0}^{\infty} c_\nu \cdot b_\nu$, wobei stets $c_\nu \in \mathbb{R}$ (nämlich $c_\nu = a \cdot b_\nu$).

Damit ergibt sich schließlich

(vi) Kein vom Nullvektor verschiedener Vektor a von $\mathfrak{V}$ steht senkrecht auf allen Vektoren von $\mathfrak{U}$.

Wenn man in einem euklidischen Raum eine Streckenkongruenz und eine Zwischenbeziehung in der üblichen Weise einführt, so erhält man ein Modell von A1 bis A8 und A10, sofern nur die Dimension ≥ 2 und der angeordnete Grundkörper $\mathfrak{f}$ pythagoreisch ist. Für $\mathfrak{f} = \mathbb{R}$ gilt in diesem Modell auch das Stetigkeitsaxiom A11. Aus dem konstruierten Raum $\mathfrak{V}$ erhalten wir also insbesondere ein Modell von A1 bis A8, A10 und A11 mit einem (dem durch $\mathfrak{U}$ bestimmten) Unterraum U der oben angegebenen Art.

<u>11.65 Definition</u>. Sind p und q verschiedene Punkte eines Unterraumes U^*, so verstehen wir unter dem *Mittellotraum* oder der *mittelsenkrechten Hyperebene* zu p und q in U^* die Menge der Punkte von U^*, die von p und q dieselben Abstände haben, Bezeichnung:

$\mathrm{Ml}_{U^*}(pq) := \{a \mid a \in U^* \wedge pa \equiv qa\}$ (definiert für $\mathrm{Sb}\, U^* \wedge p \neq q \wedge p \in U^* \wedge q \in U^*$).

Der Index U^* und die Angabe "in U^*" werden auch weggelassen, wenn U^* die Menge aller Punkte ist. Die genannte Punktmenge heißt auch das *Mit-*

tellot bzw. die *Mittelebene* zu p und q (in U^*), falls U^* die Dimension 2 bzw. 3 hat. Die Spezialfälle, daß U^* eine endliche Dimension hat oder die Menge aller Punkte ist, lassen sich hierbei (und in den folgenden Sätzen) wieder im Rahmen der Prädikatenlogik der ersten Stufe behandeln.

11.66 Satz.

$$\mathrm{Sb}U^* \wedge p \neq q \wedge p \in U^* \wedge q \in U^* \wedge U = \mathrm{Ml}_{U^*}(pq) \wedge x = \mathrm{M}(pq)$$

$$\rightarrow [a \in U \leftrightarrow a \in U^* \wedge \mathrm{R}pxa] \qquad (1)$$

$$\wedge\, pq \mathrel{\underset{x}{\hat{\perp}}} U \wedge U \subseteq U^* \qquad (2)$$

$$\wedge\, U^* = U \cup \mathrm{L}(pq) \qquad (3)$$

$$\wedge\, [\mathrm{Sb}_{n+1}U^* \rightarrow \mathrm{Sb}_n U] \qquad (4)$$

$$\wedge\, [\mathrm{Sb}_{n+1}U^* \rightarrow (U' = U \leftrightarrow pq \mathrel{\underset{x}{\hat{\perp}}} U' \wedge U' \subseteq U^* \wedge \mathrm{Sb}_n U') \qquad (5)$$

$$\wedge\, [U' = U \leftrightarrow pq \mathrel{\underset{x}{\hat{\perp}}} U' \wedge U^* = U' \cup \mathrm{L}(pq)]. \qquad (6)$$

(*Ist U der Mittellotraum zu p und q in U^* und x der Mittelpunkt von pq, so gilt Folgendes:* (1) *U besteht aus den Punkten a, die mit p und x einen rechten Winkel bilden.* (2) *U ist ein Unterraum von U^*, der in x zu $L(pq)$ orthogonal ist* (daher der Name). (3) *U^* ist das Kompositum von U und $L(pq)$.* (4) *Wenn U^* $(n+1)$-dimensional ist, so ist U n-dimensional* $(n \geq 0)$. (5) *Im Falle* $\mathrm{Sb}_{n+1}U^*$ *ist der Mittellotraum eindeutig bestimmt durch die Bedingungen* (2) *und* (4). (6) *Der Mittellotraum ist* [*ohne Dimensionsvoraussetzung*] *eindeutig bestimmt durch die Bedingungen* (2) *und* (3).)

Beweis: Nach Definitionen der Rechtwinkelrelation ist $\mathrm{R}pxa$ gleichwertig mit $pa \equiv qa$, also gilt (1).

Zu (2): Nach 4.17 ist U abgeschlossen in bezug auf Bildung von Verbindungsgeraden, d.h. ein Unterraum. Alles andere ergibt sich aus (1).

Zu (3): Die Inklusion $\supseteq$ ist trivial. Sei nun $a \in U^*$ beliebig und o. B. d. A. $a \notin \mathrm{L}(pq)$. Sei $E = \mathrm{Pl}(pqa)$ und B die Senkrechte (gemäß 11.20) auf $\mathrm{L}(pq)$ in x innerhalb E. Wegen (1) ist dann $B \subseteq U$, nach 9.27 also $a \in E = \mathrm{Pl}(B,p) \subseteq U \cup \mathrm{L}(pq)$. Damit gilt auch die umgekehrte Inklusion.

(4) ergibt sich nach dem Dimensionssatz 9.59 aus (2) (insbesondere $\mathrm{L}(pq) \cap U = \{x\}$) und (3).

Zu (5) und (6): Die Richtungen ($\to$) sind schon in (2) bis (4) enthalten. Genüge nun U' den jeweils rechts genannten Bedingungen. Wegen (1) ist dann $U'\subseteq U$. In (5) ist wegen der Gleichheit der Dimensionen also $U'=U$. Zur umgekehrten Inklusion in (6) sei $a\in U$, d.h. $a\in U^* \wedge \mathrm{R}pxa$. Aus der Charakterisierung des Kompositums U^* als affine Hülle und 9.56(iii), (ii) ergibt sich $a\in \mathrm{Sb}_{n+1}(xpa_1\ldots a_n)$ für geeignetes $n\in\mathbb{N}$ und geeignete Punkte $a_1,\ldots, a_n\in U'$. Durch Anwendung von (5) auf die Unterräume $U^*_0=\mathrm{Sb}_{n+1}(xpa_1\ldots a_n)$, $U'_0=\mathrm{Sb}_n(xa_1\ldots a_n)$ und $U_0=\mathrm{Ml}_{U^*_0}(pq)$ erhält man $a\in U'_0$ und wegen $U'_0\subseteq U'$ erst recht $a\in U'$. □

Wählt man zu einem gegebenen Punkt x auf einer Geraden A weitere Punkte p, q auf A mit $\mathrm{M}pxq$, so erhält man aus 11.66(5), (1) die folgenden Korollare.

11.67 Korollar.

$$\mathrm{Sb}_{n+1}U^* \wedge \mathrm{Ln}\,A \wedge x\in A \to \exists^{=1} U[\mathrm{Sb}_n U \wedge U\subseteq U^* \wedge A \underset{x}{\hat{\perp}} U] \qquad (n\geq 0).$$

(In einem (n+1)-dimensionalen (Unter-)Raum U gibt es zu einer Geraden A und einem Punkt x auf A genau einen n-dimensionalen Unterraum U, der in x auf A senkrecht steht.)*

11.68 Korollar.

$$A \underset{x}{\hat{\perp}} U \wedge \mathrm{Sb}_n U \wedge \mathrm{Sb}_{n+1}U^* \wedge A\subseteq U^* \wedge U\subseteq U^* \to U=\bigcup\{B \mid \mathrm{Ln}\,B \wedge A \underset{x}{\perp} B \wedge B\subseteq U^*\} \qquad (n\geq 1).$$

(Wenn A auf dem n-dimensionalen Unterraum U senkrecht steht und beide enthalten sind in dem (n+1)-dimensionalen Unterraum U, so ist U die Vereinigung der Menge aller Geraden, die im Schnittpunkt x auf A senkrecht stehen und innerhalb U* liegen.)*

11.69 Satz. $\mathrm{A9} \leftrightarrow \forall a\forall b\forall c\forall d\; \mathrm{Cp}\,abcd$.

(Das obere Dimensionsaxiom A9 *gilt genau dann, wenn alle Punkte in einer Ebene liegen.)*

Beweis: Sei $p\neq q$. Der Rest der Voraussetzung von A9 bedeutet nach 11.66, daß die Punkte a, b, c in einem Mittellotraum U geeigneter Dimension zu p und q liegen. Wenn alle Punkte in einer Ebene liegen, kann U höchstens die Dimension 1 haben, d.h., die genannten Punkte sind kollinear. Andernfalls gibt es einen wenigstens zweidimensionalen Mittellotraum U zu p und q und darin nicht-kollineare Punkte a, b, c, die ein Gegenbeispiel zu A9 liefern. □

An Stelle von A8 und A9 führen wir nun noch Dimensionsaxiome zur Festlegung einer beliebigen endlichen Dimension n ein.

11.70 Definition.

(1) $\mathrm{Dim}_n^-: \leftrightarrow \exists a_0 \dots \exists a_n \neg \mathrm{Ah}_n a_0 \dots a_n$ $(n \geq 1)$.

(*Unteres Dimensionsaxiom für die Dimension* n: Die Dimension [des ganzen Raumes, d.h. der Menge aller Punkte] ist wenigstens n.)

(2) $\mathrm{Dim}_n^+: \leftrightarrow \neg \mathrm{Dim}_{n+1}^-$ $(n \geq 1)$.

(*Oberes Dimensionsaxiom für die Dimension* n: Die Dimension ist höchstens n.)

(3) $\mathrm{Dim}_n: \leftrightarrow \mathrm{Dim}_n^- \wedge \mathrm{Dim}_n^+$ $(n \geq 1)$.

(*Dimensionsaxiom für die Dimension* n: Die Dimension ist genau n.)

Hierbei werde für Ah_n die induktive Charakterisierung gemäß 9.50(1), 9.46(2) zugrundegelegt (um eine möglichst einfache Charakterisierung mit Hilfe der als Grundbegriff verwendeten Zwischenbeziehung bzw. der Kollinearität zu erhalten).

11.71 Satz. *Schon aus* A1 *bis* A7 *ergibt sich*

(1) $\mathrm{Dim}_2^- \leftrightarrow$ A8.

(2) $\mathrm{Dim}_n^- \rightarrow$ A8 $(n \geq 2)$.

Alles, was bisher aus A1 *bis* A8 *bewiesen wurde, läßt sich also auch aus* A1 *bis* A7 *und* Dim_n^- *beweisen* $(n \geq 2)$.

Beweis: (1) gilt nach Definition. Aus der induktiven Charakterisierung von Ah_n kann man Lemma 9.49 ohne vorherige Benutzung von A8 erhalten (nämlich mittels Satz 3.1 und der Definition 4.10 der Kollinearität) und damit auch (2). □

Aus 11.69 ergibt sich noch

11.72 Satz. (i) $\mathrm{Dim}_2^+ \leftrightarrow$ A9.

(ii) A8 *und* A9 *zusammen sind also gleichwertig mit dem obigen Dimensionsaxiom* Dim_2.

11.73 Definition. Ein beliebiger Unterraum U bildet zusammen mit der Streckenkongruenz und der Zwischenbeziehung - eingeschränkt auf U - eine Relationalstruktur $\mathfrak{U} = \langle U, D|_U, B|_U \rangle$, die wir die *durch* U *bestimmte Relationalstruktur* nennen wollen.

11.74 Satz. *Für ein beliebiges Modell* $\mathfrak{A}$ *von* A1 *bis* A8 *gilt: Ist* U *ein* n*-dimensionaler Unterraum, so ist die dadurch bestimmte Relationalstruktur* $\mathfrak{U}$ *ein Modell von* A1 *bis* A7 *und* Dim_n*; jedes der Zusatzaxiome* A10, A11, CA, A11' (*als Schema*), *das in* $\mathfrak{A}$ *gilt, gilt dabei auch in* $\mathfrak{U}$. *Ins-*

besondere ist also jede Ebene ein Modell von A1 *bis* A9.

Beweis: Die Gültigkeit von Dim_n in $\mathfrak{U}$ ist klar (wegen $\mathrm{Sb}_n U$). Es bleibt also zu zeigen, daß die Gültigkeit der genannten Axiome sich von $\mathfrak{A}$ auf $\mathfrak{U}$ überträgt. Für die Axiome A1 bis A3, A5, A6 sieht man das unmittelbar (da diese Axiome nur mit Allquantoren bzw. als offene Formeln geschrieben sind, ergibt es sich schon daraus, daß $\mathfrak{U}$ eine Unterstruktur von $\mathfrak{A}$ ist). In A4, A7, A10, A11, CA wird zu gegebenen Objekten (Punkten bzw. Punktmengen) die Existenz weiterer Punkte (x, $\dot{x}$ und y bzw. b) gefordert. Zu gegebenen Objekten in $\mathfrak{U}$ existieren diese weiteren Punkte nach Voraussetzung zunächst in $\mathfrak{A}$. Sie liegen aber in den nicht-trivialen Fällen jeweils auf einer Geraden L (oder lassen sich so wählen), wobei L durch zwei Punkte von $\mathfrak{U}$ geht und damit ganz in $\mathfrak{U}$ enthalten ist. Damit existieren Punkte der gewünschten Art schon in $\mathfrak{U}$. Auf diese Weise erhält man auch die Übertragung für das Schema A11', wenn man noch folgendes beachtet: Eine Punktmenge, die sich in $\mathfrak{U}$ durch eine Formel $\alpha(x)$ (mit Parametern) beschreiben läßt, läßt sich auch in $\mathfrak{A}$ durch eine entsprechende Formel $\alpha'(x)$ beschreiben, und zwar, wenn nötig, unter Benutzung zusätzlicher Parameter $a_0,\ldots,a_n$ zur Beschreibung von U in $\mathfrak{A}$ ($U=\mathrm{Sb}_n(a_0\ldots a_n)$). □

Für kartesische Räume erhält man leicht die folgende Ergänzung zu Satz 1.6 (Übung).

11.75 Satz.

(i) Dim_n^- gilt in $\mathfrak{L}_m(\mathfrak{F})$ gdw $m \geq n$.

(ii) Dim_n^+ gilt in $\mathfrak{L}_m(\mathfrak{F})$ gdw $m \leq n$.

(iii) Dim_n gilt in $\mathfrak{L}_m(\mathfrak{F})$ gdw $m = n$.

12. Parallelität (im euklidischen Sinne)

12.1 Anmerkung. Bisher haben wir nur die Axiome A1 bis A8 verwendet. Die bisher bewiesenen Sätze gehören damit zur "dimensionsfreien absoluten Geometrie" (vgl. 0.4).

Von jetzt ab verwenden wir außer A1 bis A8 auch das "euklidische Axiom" A10, das sich als gleichwertig mit dem klassischen Parallelenaxiom erweist, und treiben damit (ebenfalls dimensionsfreie) *euklidische Geometrie*. Gelegentlich noch auftretende Sätze, die sich ohne Verwendung von A10 beweisen lassen, werden von jetzt ab durch Anhängen der Bezeichnung "(a)" gekennzeichnet, um auf ihre Zugehörigkeit zur absoluten Geometrie hinzuweisen.

Die hier als grundlegender Begriff verwendete Parallelität von Geraden kann man als Parallelität "im euklidischen Sinne" bezeichnen im Unterschied zu der anders definierten hyperbolischen Parallelität, die in der hyperbolischen Geometrie verwendet wird (s. II.2.3(ii)).

12.2 Definition. Die Gerade A ist zur Geraden B *parallel im engeren Sinne* oder *echt parallel:*

$$A \overset{\vee}{\parallel} B :\leftrightarrow \mathrm{Cp}\,AB \wedge \neg\exists x[x\in A \wedge x\in B]$$

(d.h., A und B sind komplanare Geraden und haben keinen gemeinsamen Punkt).

$$ab \overset{\vee}{\parallel} cd :\leftrightarrow a\neq b \wedge c\neq d \wedge \mathrm{L}(ab) \overset{\vee}{\parallel} \mathrm{L}(cd).$$

Bei der Parallelität im weiteren Sinne ist noch das Zusammenfallen beider Geraden zugelassen.

12.3 Definition. Die Gerade A ist zur Geraden B *parallel (im weiteren Sinne)*:

$A \| B :\leftrightarrow A \overset{\vee}{\|} B \vee [\mathrm{Ln}\, A \wedge \mathrm{Ln}\, B \wedge A=B]$,
$ab \| cd :\leftrightarrow a \neq b \wedge c \neq d \wedge \mathrm{L}(ab) \| \mathrm{L}(cd)$,
$A \| cd :\leftrightarrow c \neq d \wedge A \| \mathrm{L}(cd)$.

Aus den Definitionen erhält man sofort:

12.4 Satz (a) (*Reflexivität*). $\mathrm{Ln}\, A \to A \| A$.

12.5 Satz (a) (*Symmetrie*).
(1) $A \overset{\vee}{\|} B \to B \overset{\vee}{\|} A$.
(2) $A \| B \to B \| A$.

12.6 Satz (a) $A \overset{\vee}{\|} B \to \forall b \forall b' [b \in B \wedge b' \in B \to b \underset{A}{\approx} b']$.
(*Ist* $A \overset{\vee}{\|} B$, *so liegen alle Punkte von* B *auf derselben Seite von* A.)

Beweis: Sei $b, b' \in B$. Die A und B enthaltende Ebene E ist nach 9.27 eindeutig bestimmt als $E = \mathrm{Pl}(A,b)$. Der Punkt b' von E kann nicht auf A und nicht auf der zu b entgegengesetzten Seite von A liegen, da sonst A und B einen gemeinsamen Punkt hätten. Nach 9.20 bleibt also nur die Möglichkeit $b \underset{A}{\approx} b'$. □

12.7 Folgerung (a). $ab \overset{\vee}{\|} cd \leftrightarrow c \underset{ab}{\approx} d \wedge \neg \exists x [\mathrm{Col}\, abx \wedge \mathrm{Col}\, cdx]$.

12.8 Anmerkung. Will man nur mit Punkten (ohne Punktmengen) arbeiten und neue Begriffe möglichst unmittelbar auf die Grundbegriffe D und B zurückführen (vgl. Anm. 6.26(i)), so ist 12.7 eine besser geeignete Charakterisierung der Parallelität im engeren Sinne.

12.9 Satz (a). $\mathrm{Cp}\, ABC \wedge A \perp C \wedge B \perp C \to A \| B$.
(*Geraden, die eine gemeinsame Senkrechte haben und mit dieser in einer Ebene liegen, sind parallel.*)

Beweis: Sei E eine Ebene, in der A, B, C liegen. Wenn A und B überhaupt einen gemeinsamen Punkt x haben, muß $A=B$ sein wegen der Eindeutigkeit des Lots (8.18) (falls $x \notin C$) bzw. der Eindeutigkeit der Senkrechten innerhalb E (11.20) (falls $x \in C$). □

12.10 Satz (a). $\mathrm{Ln}\, A \to \exists B [A \| B \wedge a \in B]$.
(*Zu einer Geraden* A *gibt es durch einen beliebigen Punkt* a *wenigstens eine Parallele.*)

Beweis: Ist $a\in A$, so ist offenbar $B=A$ die einzige Gerade der genannten Art. Ist $a\notin A$, so sei C das Lot von a auf A und B die Senkrechte auf C im Punkte a innerhalb der Ebene $\mathrm{Pl}(A,a)$; diese leistet nach 12.9 das Verlangte.

Das bei Hilbert 1977 verwendete Parallelenaxiom können wir nun als Satz mittels A10 beweisen.

12.11 Satz (*Parallelenaxiom*).

$$\mathrm{Ln}\,A \wedge a\notin A \to \overset{\leq 1}{\exists} B[A\| B \wedge a\in B].$$

(*Zu einer Geraden A gibt es durch einen beliebigen Punkt außerhalb A höchstens eine Parallele.*)

Beweis (Abb. 76): Wir nehmen an, daß B und C voneinander verschiedene Parallelen zu A durch a sind, und haben zu zeigen, daß dann A10 nicht gilt. Sei t ein beliebiger Punkt von A. Auf Grund der Annahme sind A, B,

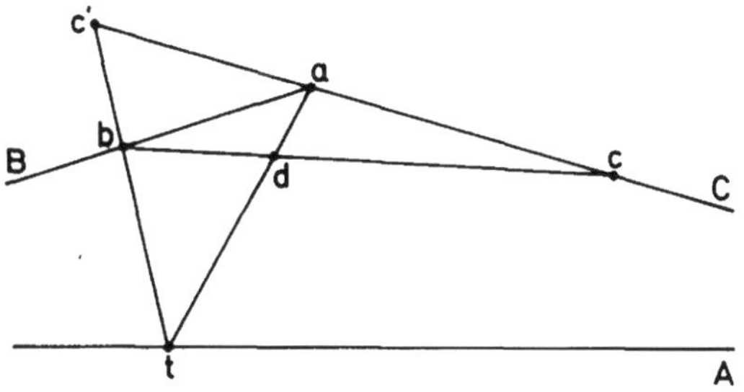

Abb. 76

C enthalten in der Ebene $E=\mathrm{Pl}(A,a)=\mathrm{Pl}(B,t)$. Sei c' ein Punkt von C, der auf der zu t entgegengesetzten Seite von B liegt (ein solcher existiert, da B und C sich in a schneiden). Seien dann b, c Punkte mit $b\in B \wedge \mathrm{B}c'bt$ und $\mathrm{B}c'ac \wedge c\neq a$, und sei schließlich d gemäß A7 ein Punkt mit $\mathrm{B}bdc \wedge \mathrm{B}adt$. Nach Konstruktion ist auch $b\neq a$, somit $a\notin \mathrm{L}(bc)$ und insbesondere $a\neq d$. Damit ist die Voraussetzung von A10 erfüllt, nach A10 müßte also $\mathrm{B}abx \wedge \mathrm{B}acy \wedge \mathrm{B}xty$ gelten für gewisse Punkte x, y. Als Punkte von B bzw. C müßten diese Punkte nach 12.6 beide mit a auf derselben Seite von A liegen. Andererseits müßten sie wegen $\mathrm{B}xty$ auf entgegengesetzten Seiten von A oder auf A liegen, was nicht möglich ist. □

12.12 Anmerkung. Auf Grund von A1 bis A8 ist 12.11 sogar gleichwertig mit A10, d.h., man kann auch umgekehrt A10 als Satz aus 12.11 zurückerhalten. Sei nämlich die Voraussetzung von A10 erfüllt. Ist $\mathrm{Col}\,abc$, so liegen alle gegebenen Punkte auf einer Geraden, und für jede der für diese Punkte möglichen Zwischenbeziehungen erhält man Punkte x, y der gewünschten Art (sogar ohne Benutzung von 12.11). Sei nun $\neg \mathrm{Col}\, abc$, und sei A die Parallele zu $\mathrm{L}(bc)$ durch den Punkt t. Dann ist $\mathrm{L}(ab)$ verschieden von der eindeutig bestimmten Parallelen $\mathrm{L}(bc)$ zu A durch c, also schneidet $\mathrm{L}(ab)$ die Gerade A in einem gewissen Punkt x. Ebenso schneiden sich $\mathrm{L}(ac)$ und A in einem gewissen Punkt y. Für die Schnittpunkte erhält man leicht die geforderten Zwischenbeziehungen.

12.10 und 12.11 lassen sich zusammenfassen in

<u>12.13 Satz.</u> $\mathrm{Ln}\, A \to \overset{=1}{\exists} B[A \| B \wedge a \in B]$.
(Zu einer beliebigen Geraden A gibt es durch einen beliebigen Punkt a stets genau eine Parallele.)

<u>12.14 Satz (a).</u> $\mathrm{Pl}\, E \wedge A \subseteq E \wedge a \in E \wedge A \| B \wedge a \in B \to B \subseteq E$.
(Parallelenbildung gemäß 12.11 *führt aus einer Ebene nicht heraus, d.h., mit A und a liegt auch die [eine] Parallele zu A durch a in der Ebene E.)*

<u>Beweis</u>: Für $A=B$ ist das klar. Sonst liegen A und B nach Definition jedenfalls in einer Ebene; diese enthält A als Teilmenge und a als Element, ist also nach 9.27 gleich E. □

Mit Hilfe des Parallelenaxioms ergibt sich nun, daß die Parallelitätsrelation auch transitiv, also eine Äquivalenzrelation auf der Menge der Geraden ist.

<u>12.15 Satz</u> *(Transivität)*. $A \| B \wedge B \| C \to A \| C$.

<u>Beweis</u>: <u>Fall 1</u>: A, B, C liegen in einer Ebene. Wäre dann nicht $A \| C$, so müßten sich A und C in einem Punkt x schneiden, damit wären sie aber verschiedene Parallelen zu B durch den Punkt x im Widerspruch zur Eindeutigkeit der Parallelen (12.11).

<u>Fall 2</u>: A, B, C liegen nicht in einer Ebene (Abb. 77). Dann ist insbesondere $A \neq B$ und damit $A \overset{\vee}{\|} B$ Sei F die eindeutig bestimmte Ebene, in der A und B liegen, und sei c ein nicht in F liegender Punkt von C. Sei $E_A = \mathrm{Pl}(A,c)$ und $E_B = \mathrm{Pl}(B,c)$. Diese beiden Ebenen liegen im dreidimensionalen Unterraum $\mathrm{Sb}_3(F,c)$, schneiden sich also nach 9.58 in einer Geraden C'. Dann hat C' keinen Punkt mit der Ebene F gemein; denn sonst

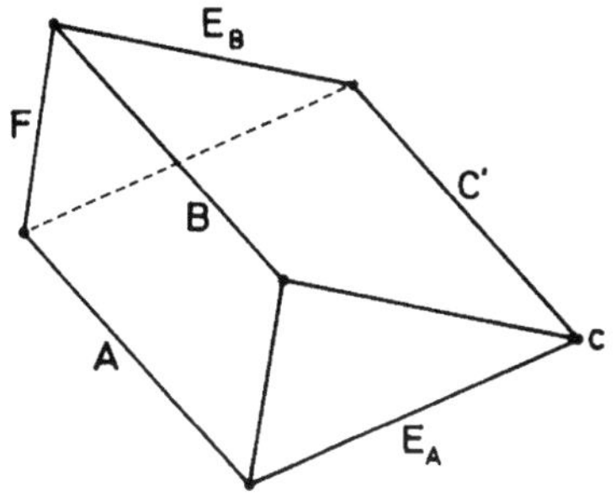

Abb. 77

wäre ein solcher Punkt x allen drei betrachteten Ebenen gemein, er läge also nach 9.30 auf A (als gemeinsamer Punkt der sich dort schneidenden Ebenen) und ebenso auf B im Widerspruch zu $A \overset{\vee}{\|} B$. Insbesondere hat also die Gerade C' mit A und ebenso mit B keinen Punkt gemein. Da sie mit jeder dieser Geraden in einer Ebene liegt, ist $A \| C'$ und $B \| C'$. Wegen der Eindeutigkeit der Parallelen zu B durch c ist nun $C' = C$ und damit $A \| C$. □

Nur eine andere Version von 12.15 (Fall 1) ist der folgende Satz über die Existenz von Schnittpunkten, der später häufig benutzt wird.

<u>12.16 Satz</u>. $A \| B \wedge \mathrm{Cp}\, ABC \wedge CA \;\mathrm{Is}\; a \rightarrow \exists b\; CB \;\mathrm{Is}\; b$.
(Eine Gerade, die A schneidet, schneidet auch jede in derselben Ebene liegende Parallele zu A.)

<u>12.17 Satz (a)</u>. $\mathrm{M}apc \wedge \mathrm{M}bpd \wedge a \neq b \rightarrow ab \| cd$ (Abb. 78).

<u>Beweis</u>: <u>Fall 1</u>: $\mathrm{Col}\, abp$. Dann ist offenbar $\mathrm{L}(ab) = \mathrm{L}(cd)$ und damit $ab \| cd$.

<u>Fall 2</u>: $\neg \mathrm{Col}\, abp$. Sei L das Lot von p auf $\mathrm{L}(ab)$ mit dem Fußpunkt e. Da die Punktspiegelung S_p ein Automorphismus ist, ist L auch das Lot von p auf $\mathrm{L}(cd)$ mit dem Fußpunkt $f = \mathrm{S}_p(e)$. Damit haben $\mathrm{L}(ab)$ und $\mathrm{L}(cd)$ die

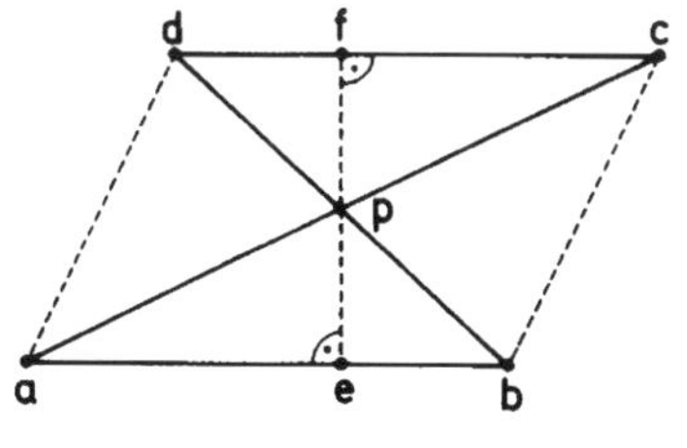

Abb. 78

gemeinsame Senkrechte *L*. Außerdem liegen alle betrachteten Punkte und Geraden in der Ebene Pl(*abp*). Somit ergibt sich die Behauptung nach 12.9. □

Aus dem Lemma vom zentralsymmetrischen Viereck (7.21) und 12.17 erhält man ohne weiteres

12.18 Satz (a). $ab \equiv cd \wedge bc \equiv da \wedge \neg \mathrm{Col}\, abc \wedge b \neq d \wedge \mathrm{Col}\, apc \wedge \mathrm{Col}\, bpd$

$\rightarrow ab \parallel cd \wedge bc \parallel da \wedge \mathrm{B}b{}_{c}^{a}d \wedge \mathrm{B}a{}_{d}^{b}c.$

(*Jedes zentralsymmetrische Viereck* [*gemäß* 7.21] *ist ein Parallelogramm* [*mit den bekannten Lagebeziehungen für die gegenüberliegenden Ecken*], Abb. 78 oder 18).

12.19 Satz. $\neg \mathrm{Col}\, abc \wedge ab \parallel cd \wedge bc \parallel da$

$\rightarrow ab \equiv cd \wedge bc \equiv da \wedge \mathrm{B}b{}_{c}^{a}d \wedge \mathrm{B}a{}_{d}^{b}c.$

(*Jedes Parallelogramm ist ein zentralsymmetrisches Viereck.*)

Beweis: Sei *p* der Mittelpunkt von *ac* und $d' = \mathrm{S}_p b$. Nach 7.13 ist dann $ab \equiv cd'$ und $bc \equiv d'a$. Damit läßt sich 12.17 mit *d'* an Stelle von *d* anwenden. Aus dem Resultat und der Eindeutigkeit der Parallelen ergibt sich L(*cd*)=L(*cd'*) und L(*ad*)=L(*ad'*), aus der Eindeutigkeit des Schnittpunkts von Geraden also *d'*=*d* und damit die Behauptung. □

Ohne Schwierigkeiten erhält man auch die folgenden "klassischen" Sätze.

12.20 Satz. $ab \parallel cd \wedge ab \equiv cd \wedge \mathrm{B}b{}_{c}^{a}d \rightarrow bc \parallel da \wedge bc \equiv da \wedge \mathrm{B}a{}_{d}^{b}c.$

(*Jedes nicht-entartete Viereck mit einem Paar von zueinander kongruenten und parallelen Gegenseiten ist* [*bei geeigneter Reihenfolge der Ecken*] *ein Parallelogramm und zentralsymmetrisch.*)

12.21 Satz. $\mathrm{B}b{}_{c}^{a}d \rightarrow [ab \parallel cd \leftrightarrow bac \equiv dca].$

(*Zwei Geraden sind zueinander parallel genau dann, wenn sie mit einer weiteren Geraden kongruente "Wechselwinkel" bilden*, Abb. 79.)

12.22 Satz. $a \underset{p}{\simeq} c \wedge b \underset{pa}{\simeq} d \rightarrow [ab \parallel cd \leftrightarrow bap \equiv dcp].$

(Analog für "*Stufenwinkel*" oder "*Gegenwinkel*", Abb. 80.)

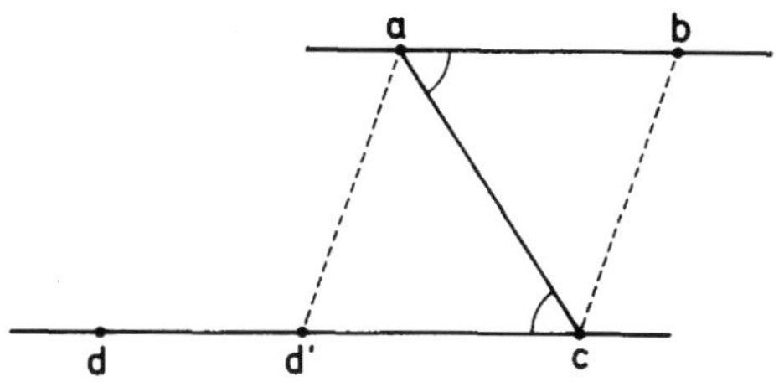

Abb. 79

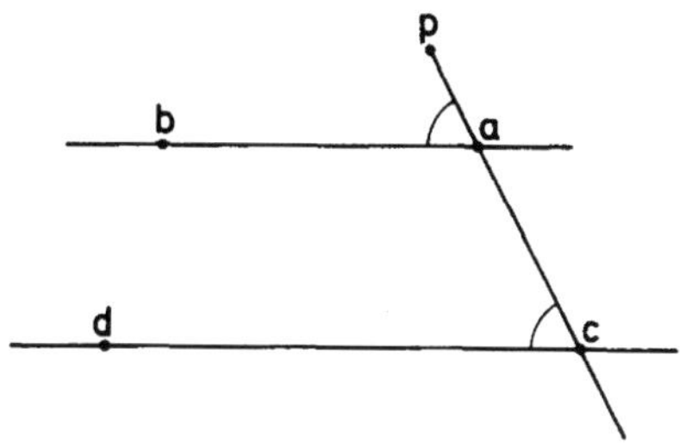

Abb. 80

<u>12.23 Satz.</u> $\neg \mathrm{Col}\, abc \rightarrow \exists b' \exists c' [\mathrm{B} b \overset{a}{\underset{c}{}} b' \wedge \mathrm{B} c \overset{a}{\underset{b}{}} c' \wedge \mathrm{B} b'ac' \wedge abc \equiv bac' \wedge acb \equiv cab']$.
(Die "Summe" der Innenwinkel eines Dreiecks ist ein gestreckter Winkel [genauer: durch Aneinanderlegen der Innenwinkel erhält man einen gestreckten Winkel], Abb. 81.)

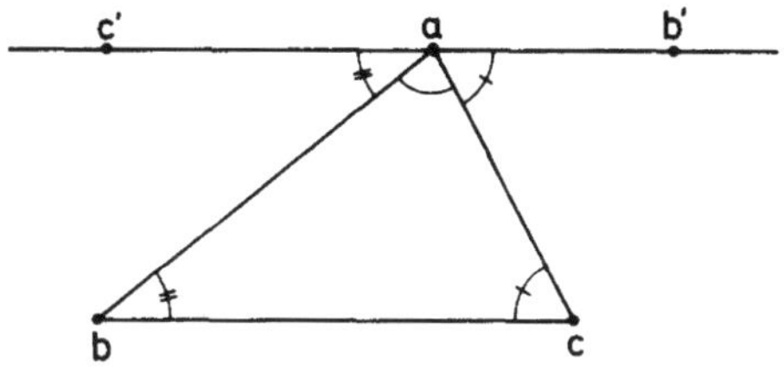

Abb. 81

Der Begriff der Parallelität läßt sich auf Unterräume beliebiger Dimension übertragen. Wir beschränken uns hier auf die später einmal verwendete Parallelität von Ebenen und die dafür benötigten Sätze.

<u>12.24 Definition.</u> E_1, E_2 sind (zueinander) *parallele Ebenen*:

$$E_1 \overset{2}{\parallel} E_2 :\leftrightarrow \exists a_1 \exists A_1 \exists B_1 \exists a_2 \exists A_2 \exists B_2 [\bigwedge_{\nu=1}^{2} (\mathrm{Pl}\, E_\nu \wedge A_\nu B_\nu \;\mathrm{Is}\; a_\nu \wedge A_\nu \subseteq E_\nu \wedge B_\nu \subseteq E_\nu) \wedge A_1 \parallel A_2 \wedge B_1 \parallel B_2]$$

(d.h., es gibt sich schneidende Geraden A_1, B_1 in E_1 mit entsprechenden Parallelen, die sich ebenfalls schneiden und in E_2 liegen).

<u>12.25 Satz.</u> $E_1 \overset{2}{\parallel} E_2 \leftrightarrow \mathrm{Pl}\, E_1 \wedge \mathrm{Pl}\, E_2 \wedge \{E_1 = E_2 \vee \exists U [\mathrm{Sb}_3 U \wedge E_1 \subseteq U \wedge E_2 \subseteq U \wedge E_1 \cap E_2 = \emptyset]\}$.
(Ebenen E_1, E_2 sind parallel genau dann, wenn sie gleich sind oder in einem dreidimensionalen Unterraum enthalten sind und keinen gemeinsamen Punkt besitzen.)

<u>12.26 Satz.</u> $E_1 \overset{2}{\parallel} E_2 \leftrightarrow \mathrm{Pl}\, E_1 \wedge \mathrm{Pl}\, E_2 \wedge \forall F \forall A_1 \forall p [\mathrm{Pl}\, F \wedge \mathrm{Ln}\, A_1 \wedge E_1 \cap F = A_1 \wedge p \in E_2 \wedge p \in F \rightarrow \exists A_2 (A_1 \parallel A_2 \wedge E_2 \cap F = A_2)]$.

(Ebenen E_1, E_2 sind parallel genau dann, wenn jede Ebene F, die E_1 in einer Geraden A_1 schneidet und einen Punkt p von E_2 enthält, die Ebene E_2 in einer zu A_1 parallelen Geraden A_2 schneidet.)

Beweis von 12.25 und 12.26: Die rechten Seiten von 12.24, 12.25 und 12.26 bezeichnen wir der Reihe nach mit (i), (ii), (iii). Es ist die Gleichwertigkeit dieser Bedingungen zu zeigen.

(i)→(ii): Seien a_ν, A_ν, B_ν (ν=1,2) wie in (i).

Fall 1: $a_2 \in E_1$. Nach 12.14 sind dann auch A_2 und B_2 in E_1 enthalten, also ist $E_2 = \mathrm{Pl}(A_2, B_2) = E_1$.

Falls 2: $a_2 \notin E_1$. Dann ist zunächst $U = \mathrm{Sb}_3(E_1, a_2)$ ein dreidimensionaler Unterraum, in dem auch A_2 und B_2 und damit E_2 enthalten sind. Wir nehmen nun an, daß E_1 und E_2 doch einen gemeinsamen Punkt besitzen. Dann besitzen sie nach 9.58 eine gemeinsame Gerade, sei C eine solche. Da die Parallelität von Geraden eine Äquivalenzrelation ist, ist C zu wenigstens einer der Geraden A_1, B_1 nicht parallel, o. B. d. A. zu A_1. Wegen $A_1 \| A_2$ ist sie dann auch nicht parallel zu A_2. Da C mit A_ν aber in der Ebene E_ν liegt, schneidet sie A_ν, und es ist

(1) $E_\nu = \mathrm{Pl}(C, A_\nu)$ (ν=1,2).

Außerdem ist A_2 die Parallele zu A_1 durch einen Punkt von E_1 (nämlich von C), nach 12.14 ist also $A_2 \subseteq E_1$ und wegen (1) somit $E_1 = E_2$ im Widerspruch zur Voraussetzung von Fall 2.

(ii)→(iii): Sei F eine beliebige Ebene, die E_1 in einer Geraden A_1 schneidet und einen Punkt p von E_2 enthält.

Fall 1: $E_1 = E_2$. Dann leistet $A_2 \doteq A_1$ das Verlangte.

Fall 2: $E_1 \neq E_2$. Mit A_1 und p liegt auch F in U, und nach 9.58 schneiden sich F und E_2 jedenfalls in einer Geraden A_2. Diese kann jedoch mit A_1 keinen gemeinsamen Punkt haben, da sonst auch E_1 und E_2 diesen Punkt gemein hätten. Also ist $A_1 \overset{\vee}{\|} A_2$, d.h. A_2 leistet das Verlangte.

(iii)→(i): Für $E_1 = E_2$ ist die Behauptung (i) trivial. Sei nun $E_1 \neq E_2$. Seien $p \in E_2 - E_1$ und sich schneidende Geraden A_1, B_1 in E_1 beliebig gewählt. Anwendung von (iii) auf A_1 und auf B_1 liefert dann entsprechende Parallelen A_2 und B_2 in E_2, die zueinander nicht parallel sein können, also sich schneiden. □

Über das Senkrechtstehen von Geraden auf Unterräumen beweisen wir noch

12.27 Satz.

$$\mathrm{Sb}_n U \wedge A \mathbin{\hat{\perp}} U \wedge A \parallel B \wedge \exists U^*[\mathrm{Sb}_{n+1} U^* \wedge U \subseteq U^* \wedge B \subseteq U^*] \rightarrow B \mathbin{\hat{\perp}} U \qquad (n \geq 1).$$

Beweis: Sei o. B. d. A. $A \neq B$. Die Ebene F durch A und B schneidet U nach 9.59 in einer Geraden C. B steht nach 12.21 senkrecht auf C in einem gewissen Punkt q innerhalb F und ist damit die wie in 11.63_n konstruierte Senkrechte L in q auf U innerhalb des A und U enthaltenden $(n+1)$-dimensionalen Raumes U^* (Abb. 75) □

13. Die Sätze von Pappus-Pascal und von Desargues

13.1 Satz (a). $\neg \mathrm{Col}\,abc \wedge \mathrm{M}bpc \wedge \mathrm{M}cqa \wedge \mathrm{M}arb$
$\rightarrow \exists A[\,\mathrm{Ln}\,A \wedge A \underset{r}{\perp} ab \wedge A \perp pq\,].$

(*Das Mittellot zu einer Seite eines Dreiecks steht senkrecht auf der Verbindungsgeraden der Mitten der anderen beiden Seiten*, Abb. 82.)

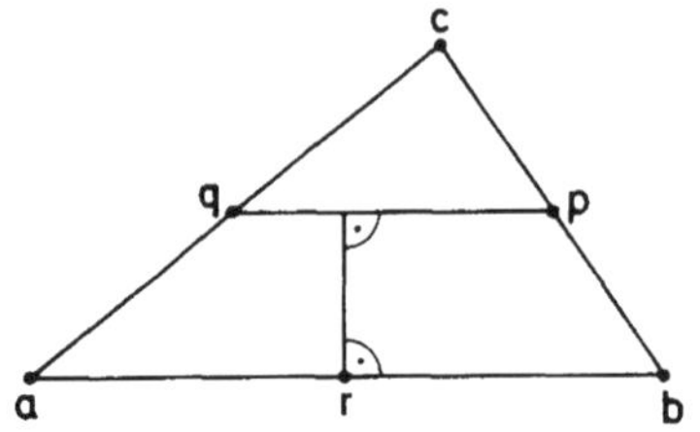

Abb. 82

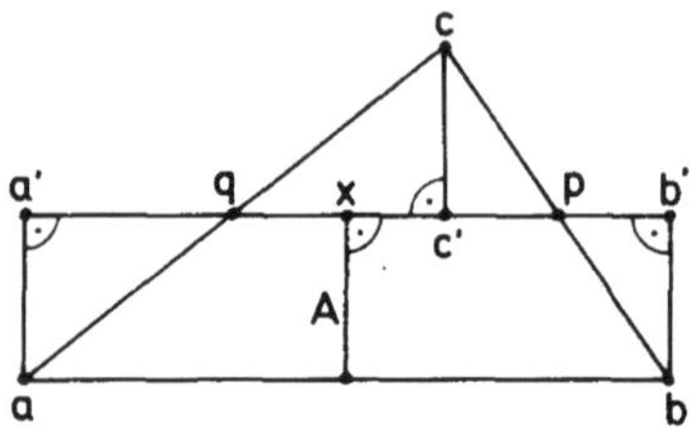

Abb. 83

Beweis: Sei c' der Fußpunkt des Lots von c auf $\mathrm{L}(pq)$ und $a'=\mathrm{S}_q c'$, $b'=\mathrm{S}_p c'$ (Abb. 83). Da die betrachteten Punktspiegelungen Bewegungen sind (7.14), die c in a bzw. b überführen, sind auch a' und b' die Fußpunkte der Lote von a bzw. b auf $\mathrm{L}(pq)$, und es ist $aa' \equiv cc' \equiv bb'$. Sei nun x der Mittelpunkt der Strecke $a'b'$ und A die Senkrechte auf $\mathrm{L}(pq)$ in x in der Ebene $\mathrm{Pl}(abc)$ (in der alle hier betrachteten Punkte liegen). Die Geradenspiegelungen S_A führt nun a' in b' über und damit (als Bewegung nach 10.11) auch den Punkt a in einen Punkt b'' mit den Eigenschaften $\mathrm{R}a'b'b''$, $aa' \equiv bb''$ und $b'' \underset{pq}{\approx} a$. Durch diese Eigenschaften ist aber gerade der Punkt b eindeutig festgelegt, also liegen a und b spiegelbildlich bezüglich A. Damit steht A senkrecht auf $\mathrm{L}(ab)$ und geht durch den Mittelpunkt r von ab. A leistet somit das Verlangte. □

13.2 Satz (a) (Hjelmslev).

$\mathrm{B}c_b^a d \wedge \mathrm{R}bca \wedge \mathrm{R}bda \wedge \mathrm{Col}\, cde \wedge ae \perp cd$

$\rightarrow bac \equiv dae \wedge bad \equiv cae \wedge \mathrm{B}ced.$

(*In einer Figur wie in Abb.* 84 - *mit rechten Winkeln wie markiert - bestehen die angegebenen Winkelkongruenzen und Zwischenbeziehungen.*)

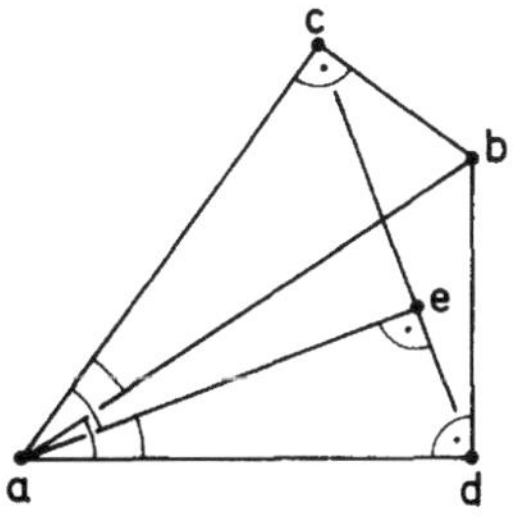

Abb. 84

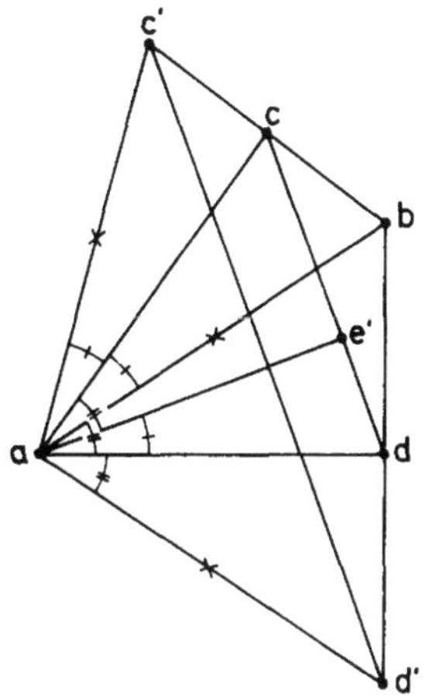

Abb. 85

Beweis: Mit Hilfe der Sätze über die Winkelantragung (11.15) und über das Aneinander- und Ineinanderlegen von Winkeln (11.22) erhält man zunächst einen Punkt e' mit der in der Behauptung für e verlangten Eigenschaft (Übung!). Sei nun $c'=S_c b$ und $d'=S_d b$ (Abb. 85). Dann ist

(1) $ac' \equiv ab \equiv ad'$

sowie $bac \equiv c'ac$ und $bad \equiv d'ad$. Durch nochmalige Anwendung von 11.22 ergibt sich

(2) $c'ae \equiv e'ad'$.

Durch die Geradenspiegelung an $L(ae')$ geht c' wegen (1) und (2) in d' über. Damit ist $A=L(ae')$ das Mittellot auf der Seite $c'd'$ des Dreiecks $c'd'b$, das nach 13.1 auch auf $L(cd)$ senkrecht steht. Wegen der Eindeutigkeit des Lots von a auf $L(cd)$ ist also $e'=e$, wie zu zeigen. □

13.3 Anmerkungen. (i) 13.2 ist ein Spezialfall eines in Hjelmslev 1907 für die absolute Geometrie (auf anderem Wege) bewiesenen Satzes. In Hilbert 1977 (S. 55) ist ein Beweis von 13.2 für die euklidische Geometrie ausgeführt.

(ii) In Anlehnung an Hilbert 1977 führen wir nun "Größen" von Strecken und Winkeln und eine Projektionsoperation ein, um den Satz von Pappus-Pascal zu beweisen. Für diese Größen wird nicht verlangt, daß es reelle

Zahlen sind (was man für beliebige Modelle von A1 bis A8 oder A10 auch nicht erwarten kann), sondern es wird (in diesem Zusammenhang) nur gebraucht, daß Strecken bzw. Winkel dieselbe Größe haben genau dann, wenn sie zueinander kongruent sind. Als Größen definieren wir daher einfach die Äquivalenzklassen nach der Streckenkongruenz bzw. der Winkelkongruenz (vgl. 2.1 bis 2.3 und 11.6 bis 11.8). Da jede Äquivalenzklasse durch zwei bzw. drei Punkte festgelegt werden kann, läßt sich auch diese Überlegung im Rahmen der Prädikatenlogik der ersten Stufe (wenn auch etwas umständlicher) behandeln (vgl. Anm. 6.26(i)).

13.4 Definition. A ist eine *Streckengröße* (*Größe einer Strecke*):

$$\text{Sg}\,A :\leftrightarrow \exists a\exists b[A=\{\langle x,y\rangle \mid xy\equiv ab\}].$$

α ist eine *Winkelgröße* (*Größe eines Winkels*):

$$\text{Wg}\,\alpha :\leftrightarrow \exists a\exists b\exists c[a\neq b \wedge c\neq b \wedge \alpha=\{\langle x,y,z\rangle \mid xyz\equiv abc\}].$$

α ist *Größe eines spitzen Winkels*:

$$\text{Wgs}\,\alpha :\leftrightarrow \exists a\exists b\exists c[\,\text{Sp}\,abc \wedge \alpha=\{\langle x,y,z\rangle \mid xyz\equiv abc\}].$$

Die $\langle a,b\rangle$ enthaltende Äquivalenzklasse (Streckengröße) nennen wir auch die *Größe der Strecke ab*; die $\langle a,b,c\rangle$ enthaltende Äquivalenzklasse (Winkelgröße) nennen wir auch die *Größe des Winkels* $\sphericalangle abc$.

13.5 Definition. Die (zweistellige) Hilbertsche *Projektionsoperation* ordnet jeder Größe α eines spitzen Winkels und jeder Streckengröße C wieder eine Streckengröße zu, die kurz mit αC bezeichnet wird und festgelegt ist als Größe der Kathete eines rechtwinkligen Dreiecks, die mit der Hypotenuse der Größe C einen Winkel der Größe α einschließt (Abb.86), d.h.

(1) $B=\alpha C \leftrightarrow \text{Sg}\,B \wedge \text{Sg}\,C \wedge \text{Wgs}\,\alpha$

$$\wedge \exists a\exists b\exists c[\text{R}bca \wedge \langle b,a,c\rangle\in\alpha \wedge \langle a,b\rangle\in C \wedge \langle a,c\rangle\in B].$$

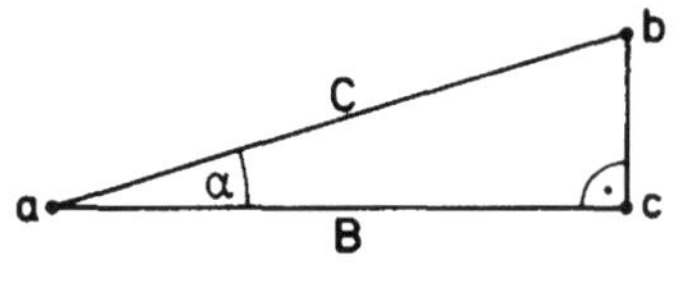

Abb. 86

Nach dem zweiten Kongruenzsatz (11.50) ist ein rechtwinkliges Dreieck mit den genannten "Bestimmungsgrößen" C und α bzw. B und α bis auf Kongruenz eindeutig bestimmt. Somit wird durch die Bedingung in 13.5(1) die Größe αC eindeutig festgelegt, und es gilt außerdem

13.6 Lemma (a) (*Kürzungsregel*). $\text{Wgs}\,\alpha \wedge \alpha C=\alpha D \rightarrow C=D$.

Das wichtigste Hilfsmittel für das Folgende ist

<u>13.7 Lemma (a)</u>. $\alpha\beta C=\beta\alpha C$.

<u>Beweis</u>: Falls α oder β Größe eines Nullwinkels ist, ist die Behauptung trivial. Sonst (Abb. 84) seien a, b gewählt mit $\langle a,b\rangle \in C$. Durch Antragen von Winkeln der Größen α bzw. β an $\mathrm{H}(ab)$ in zueinander entgegengesetzte Halbebenen und Fällen der Lote von b aus auf die anderen Schenkel erhält man Punkte c, d mit $\langle a,c\rangle \in \alpha C$, $\langle a,d\rangle \in \beta C$. Sei nun e der Fußpunkt des Lots von a auf $\mathrm{L}(cd)$. Nach 13.2 ist dann auch $\langle d,a,e\rangle \in \alpha$ und $\langle c,a,e\rangle \in \beta$, d.h. $\langle a,e\rangle \in \alpha\beta C$ und $\langle a,e\rangle \in \beta\alpha C$ und damit $\alpha\beta C=\beta\alpha C$. □

<u>13.8 Lemma (a)</u> (Abb. 87).

$$\mathrm{Col}\, opq \wedge \mathrm{Col}\, ouv \wedge u \neq o \wedge v \neq o \wedge \mathrm{R}puo \wedge \mathrm{R}qvo \rightarrow [p \underset{o}{\simeq} q \leftrightarrow u \underset{o}{\simeq} v].$$

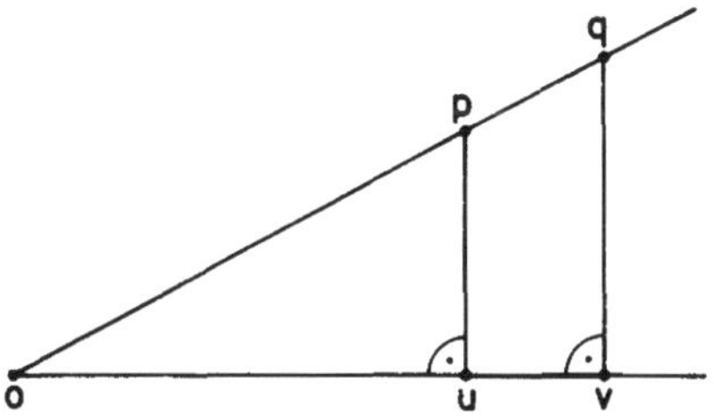

Abb. 87

Das erhält man leicht aus bekannten Sätzen, z.B. 7.22 □

<u>13.9 Definition</u>. Die Geraden $\mathrm{L}(pr)$ und $\mathrm{L}(qs)$ *haben eine gemeinsame Senkrechte durch* den Punkt o:

$$pr \underset{o}{\perp\!\!\!\perp} qs :\leftrightarrow \exists A[\,\mathrm{Ln}\, A \wedge o \in A \wedge A \perp pr \wedge A \perp qs].$$

Mit Hilfe dieser Relation können wir den Satz von Pappus-Pascal zunächst in einer Version für die absolute Geometrie formulieren (Beweis s. hinter Anm. 13.12).

<u>13.10 Satz (a)</u> (Pappus-Pascal).

$$\neg \mathrm{Col}\, oaa' \wedge \mathrm{Col}\, oabc \wedge b \neq o \wedge c \neq o \wedge \mathrm{Col}\, oa'b'c' \wedge b' \neq o \wedge c' \neq o \wedge bc' \underset{o}{\perp\!\!\!\perp} cb' \wedge ca' \underset{o}{\perp\!\!\!\perp} ac' \rightarrow ab' \underset{o}{\perp\!\!\!\perp} ba'.$$

(*Liegen die von o verschiedenen Ecken eines Sechsecks abwechselnd auf zwei Geraden, die sich in o schneiden* [Abb. 88], *und gibt es für zwei Paare von gegenüberliegenden Seiten [kurz "Gegenseitenpaare"] jeweils eine gemeinsame Senkrechte durch o, so auch für das dritte Gegenseitenpaar.*)

Unter der einer Seite gegenüberliegenden Seite verstehen wir hierbei natürlich die dritte danach in der Reihenfolge, in der die Seiten durchlaufen werden; analog für Ecken. Gegenüberliegende Ecken sind hier also mit demselben Buchstaben (einmal mit und einmal ohne Strich) bezeichnet.

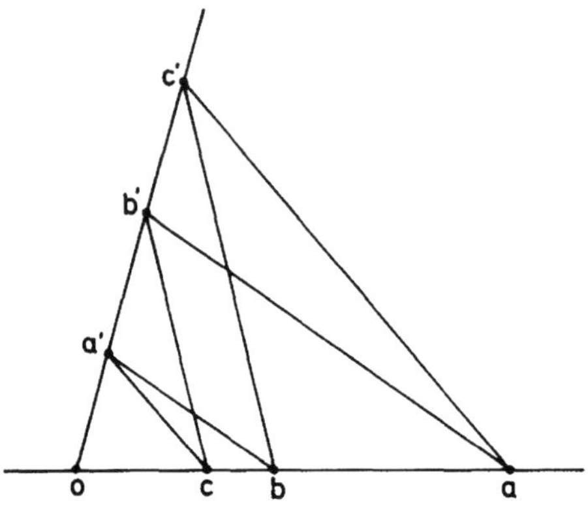

Abb. 88

Alle hier genannten Geraden liegen nach Voraussetzung in einer Ebene. Nach § 12 ist in der euklidischen Geometrie das Zutreffen der Relation $\underset{o}{\perp\!\!\!\perp}$ auf solche Geraden (bzw. die sie bestimmenden Punktepaare) gleichwertig mit der Parallelität dieser Geraden und damit Satz 13.10 gleichwertig mit

13.11 Satz (Pappus-Pascal).

$\neg \mathrm{Col}\, oaa' \wedge \mathrm{Col}\, oabc \wedge b \neq o \wedge c \neq o \wedge \mathrm{Col}\, oa'b'c' \wedge b' \neq o \wedge c' \neq o$
$\wedge\, bc' \parallel cb' \wedge ca' \parallel ac' \wedge ab' \parallel ba'$.

(*Wenn in einem Sechseck wie in* 13.10 *zwei Gegenseitenpaare aus parallelen Geraden bestehen, so auch das dritte Gegenseitenpaar.*)

13.12 Anmerkungen. (i) Satz 13.11 wird manchmal auch nur nach Pappus benannt und ist ein Spezialfall des folgenden Satzes von Pascal für die projektive Geometrie: *Liegen die Ecken eines Sechsecks auf einem Kegelschnitt, so liegen die drei Punkte, in denen sich jeweils gegenüberliegende Seiten schneiden, auf einer Geraden.* Von einer euklidischen Ebene kann man bekanntlich zu einer projektiven Ebene übergehen durch Hinzunahme der "uneigentlichen Punkte", die eineindeutig den Büscheln paralleler Geraden entsprechen und zusammen die sogenannte "uneigentliche Gerade" bilden. Geraden einer euklidischen Ebene sind damit parallel genau dann, wenn sie sich in einem uneigentlichen Punkt der zugehörigen projektiven Ebene schneiden. In 13.11 wird somit der Spezialfall behandelt, daß der betrachtete Kegelschnitt aus zwei Geraden besteht, und daß zwei der genannten Schnittpunkte (folglich auch der dritte) uneigentliche Punkte sind. Auch 13.10 läßt sich als Spezialfall des Pascalschen Satzes erhalten. Man kann nämlich schon in der absoluten Geometrie von einer Ebene (hier: einem Modell von A1 bis A9) durch Hinzunahme uneigentlicher Punkte zu einer projektiven Ebene übergehen (s. z.B. Hjelmslev 1907). Gewisse der uneigentlichen Punkte (nicht notwendig alle) entsprechen dabei eineindeutig den Büscheln von Geraden mit einer gemeinsamen Senkrechten; und solche uneigentlichen Punkte liegen auf einer Geraden genau dann, wenn sie sich erhalten lassen durch zugehörige Senkrechte, die durch einen Punkt gehen.

(ii) Für unseren Aufbau wird nur 13.11 benötigt. Tatsächlich liefert der Beweis die stärkere Version 13.10(a).

Beweis von 13.10: Wir betrachten die Verbindungsstrecken von o mit den Ecken des gegebenen Sechsecks und bezeichnen ihre Größen mit den entsprechenden großen Buchstaben (also $\langle o,a\rangle \in A$, $\langle o,a'\rangle \in A'$, ..., $\langle o,c'\rangle \in C'$). Außerdem betrachten wir die Lote von o auf fünf Seiten dieses Sechsecks; ihre Fußpunkte seien l für die Seite bc', l^* für cb', m für ca', m^* für ac' und n für ab' (Abb. 89). Nach Voraussetzung fallen diese Lote in zwei Fällen zusammen, genauer, es ist Col oll^* und Col omm^*. Es genügt zu zeigen, daß auch $on \perp ba'$ ist (so daß man einen entsprechenden Punkt n^* einführen könnte).

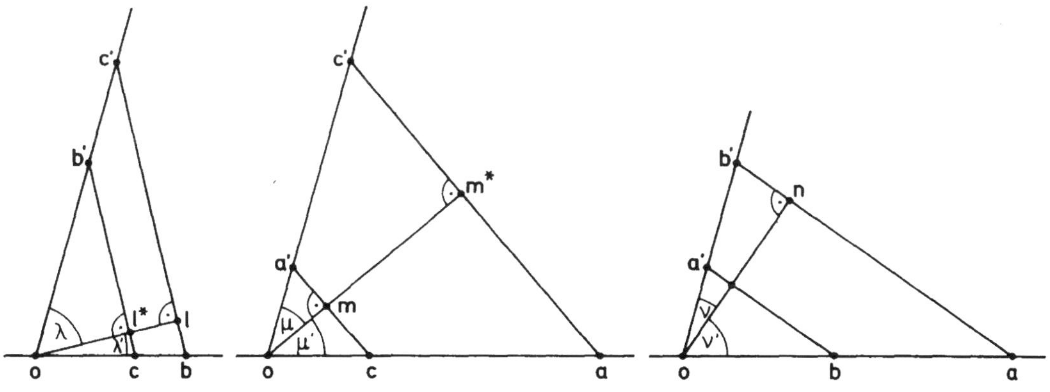

Abb. 89

Die Größen der spitzen Winkel, die die genannten Lote mit der Geraden L(oa) bzw. L(oa') bilden, seien mit entsprechenden griechischen Buchstaben mit bzw. ohne Strich bezeichnet, d.h., es sei (in naheliegender Kurzschreibweise) $lob \equiv l^*oc \in \lambda'$, $loc' \equiv l^*ob' \in \lambda$, $moc \equiv m^*oa \in \mu'$, $moa' \equiv m^*oc' \in \mu$, $noa \in \nu'$, $nob' \in \nu$ (die hier als kongruent genannten Winkel können zusammenfallen oder Scheitelwinkel zueinander sein, vgl. auch Abb. 90). Nach Konstruktion gilt

(1) $\lambda'B = \lambda C'$, (4) $\lambda'C = \lambda B'$,

(2) $\mu'C = \mu A'$, (5) $\mu'A = \mu C'$.

(3) $\nu'A = \nu B'$,

(Zur besseren Übersicht mache man sich klar, daß gewisse der genannten Bedingungen jeweils durch zyklische Vertauschung der Buchstaben auseinander hervorgehen.) Es soll zunächst gezeigt werden, daß auch die folgende (für die Behauptung jedenfalls notwendige) Gleichung gilt:

(6) $\nu'B = \nu A'$.

Das ergibt sich aus den "Rechengesetzen über die Projektionsoperation"; es ist nämlich

$\lambda'\nu'B = \nu'\lambda'B$ nach 13.7,

$= \nu'\lambda C'$ nach (1),

also

$\mu\lambda'\nu'B = \mu\nu'\lambda C'$

$= \nu'\lambda\mu C'$ nach 13.7 (zweimal angewendet),

$= \nu'\lambda\mu'A$ nach (5),

$= \lambda\mu'\nu'A$ nach 13.7,

$= \lambda\mu'\nu B'$ nach (3),

$= \mu'\nu\lambda B'$ nach 13.7,

$= \mu'\nu\lambda'C$ nach (4),

$= \nu\lambda'\mu'C$ nach 13.7,

$= \nu\lambda'\mu A'$ nach (2),

$= \mu\lambda'\nu A'$ nach 13.7,

nach 13.6 somit

$\lambda'\nu'B = \lambda'\nu A'$,

$\nu'B = \nu A'$, wie in (6) behauptet.

Seien nun n^* bzw. $n^{*\prime}$ die Fußpunkte der Lote von b bzw. a' auf L(on). Dann genügt es, $n^* = n^{*\prime}$ nachzuweisen. Auf Grund von (6) ist $on^* \equiv on^{*\prime}$, wegen der Eindeutigkeit der Streckenabtragung (6.11) genügt es also, $n^* \underset{o}{\approx} n^{*\prime}$ nachzuweisen. Dazu genügt die Bedingung $n^* \underset{o}{\approx} n \leftrightarrow n^{*\prime} \underset{o}{\approx} n$ und dafür nach Lemma 13.8 wiederum

(*) $a \underset{o}{\approx} b \leftrightarrow a' \underset{o}{\approx} b'$.

Durch zweimalige Anwendung von 13.8 (unter Benutzung der Lotfußpunkte)

erhält man

(7) $c \underset{o}{\approx} a \leftrightarrow c' \underset{o}{\approx} a'$

und ebenso

(8) $c \underset{o}{\approx} b \leftrightarrow c' \underset{o}{\approx} b'$.

Wegen Col *oabc* ist aber

$$a \underset{o}{\approx} c \vee \mathrm{B}aoc,$$

$$b \underset{o}{\approx} c \vee \mathrm{B}boc.$$

Die Kombination der beiden Möglichkeiten in der oberen Zeile und der in der unteren Zeile führt zu vier Fällen für die Lage von *a* und *b*, die sich nach (7) und (8) auf die entsprechenden "gestrichenen Punkte" übertragen, woraus sich (∗) ergibt. (In Abb. 90 ist einer der Fälle dargestellt, in denen die jeweils mit *o* kollinearen Ecken des gegebenen Sechsecks nicht alle auf derselben Seite von *o* liegen.) □

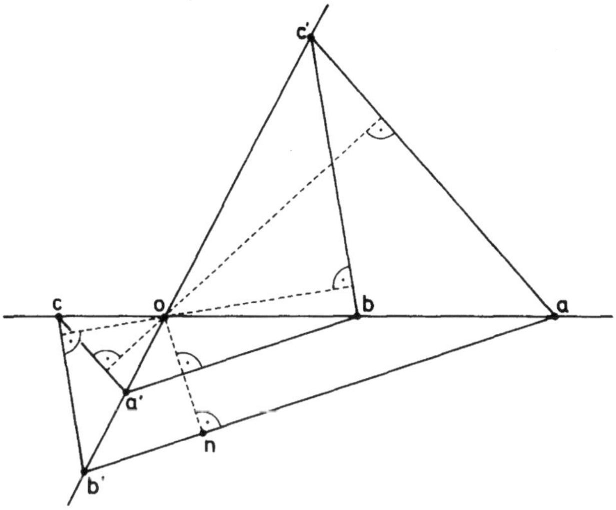

Abb. 90

<u>13.13 Anmerkung</u>. Von jetzt ab bis einschließlich 14.33 werden nur Sätze der euklidischen Geometrie benutzt, die sich mit der Kollinearität Col als einzigem Grundbegriff formulieren lassen, und zwar 13.11 sowie relativ einfache Sätze. Das bedeutet, daß diese Überlegungen - insbesondere die Einführung eines Körpers auf Grund des Satzes von Pappus-Pascal - weitgehend unabhängig von unserem speziell gewählten Axiomensystem sind und sich in vielen anderen Geometrien durchführen lassen, insbesondere schon in affinen Geometrien (mit geeigneten Axiomen, s. etwa Lingenberg 1969). Für den Rest von Abschnitt 14 (Einführung einer Anordnung im konstruierten Körper) werden lediglich Anordnungseigenschaften von Punkten, d.h. Sätze über die Zwischenbeziehung, hinzugenommen. Damit

läßt sich auch das in gewissen affinen Geometrien durchführen. Vgl. auch 13.17 und II.2.8, 13, 14(ii).

13.14 Satz (Satz von Pappus-Pascal für parallele Trägergeraden oder auch "kleiner Satz von Pappus-Pascal")

$oa \overset{\vee}{\parallel} o'a' \wedge \mathrm{Col}\ oabc \wedge \mathrm{Col}\ o'a'b'c' \wedge bc' \parallel cb' \wedge ca' \parallel ac' \rightarrow ab' \parallel ba'$.

(*Die Aussage von* 13.11 *gilt auch für Sechsecke, deren Ecken abwechselnd auf zwei zueiander parallelen Geraden liegen,* Abb. 91. Zur Bestimmung der Trägergeraden werden hier die Zusatzpunkte o, o' verwendet, um das Zusammenfallen von a, b, c und ebenso von a', b', c' nicht auszuschließen.)

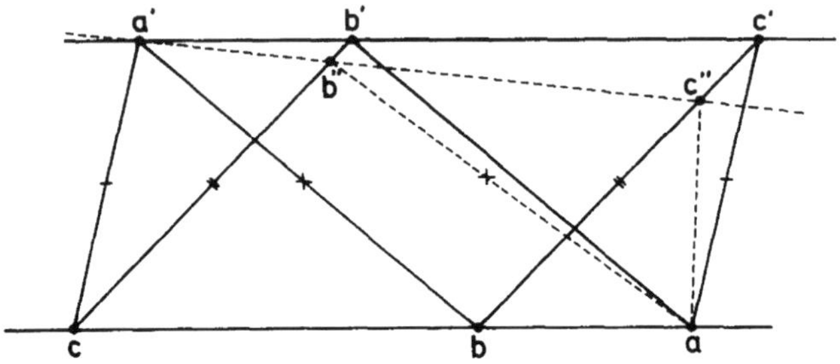

Abb. 91

Beweis: Falls irgendwelche der Punkte a, b, c zusammenfallen, so gilt das auf Grund der vorausgesetzten Parallelitäten auch für die entsprechenden "gestrichenen Punkte", und umgekehrt. Dann gilt die Behauptung; denn sie stimmt mit einer der Voraussetzungen überein (falls $a=c \vee b=c$) oder drückt die Reflexivität (12.4) aus (falls $a=b$).

Seien nun a, b, c, a', b', c' paarweise verschieden. Nehmen wir an, daß die Behauptung nicht gilt. Die Parallele G zu $L(ba')$ durch a muß dann die Gerade $L(cb')$ schneiden (sonst wäre, da alle betrachteten Punkte in einer Ebene liegen, $G \parallel L(cb')$, also auch $ba' \parallel bc'$ und damit $a'=c'$), und der Schnittpunkt b'' ist von b' verschieden. (Gleiche Markierungen in Abb. 91 kennzeichnen parallele Geraden.) Nach 12.16 schneidet die Gerade $L(a'b'')$ auch $L(bc')$ in einem von c' verschiedenen Punkt c'', und sie schneidet $L(oa)$ in einem gewissen Punkt o^*. Durch Anwendung von 13.11 auf o^* und das Sechseck mit den Ecken a, c'', b, a', c, b'' ergibt sich $ac'' \parallel ca'$, zusammen mit der Voraussetzung $ac' \parallel ca'$ also ein Widerspruch zur Eindeutigkeit der Parallelen. □

<u>13.15 Satz</u> (Satz von Desargues, Teil 1).

$$\neg \operatorname{Col} abc \wedge \operatorname{Cp} abca' \wedge ab \overset{\vee}{\parallel} a'b' \wedge ac \overset{\vee}{\parallel} a'c'$$
$$\wedge \operatorname{Col} oaa' \wedge \operatorname{Col} obb' \wedge \operatorname{Col} occ' \rightarrow bc \parallel b'c'.$$

(*In einer Ebene seien zwei nicht-entartete geordnete Dreiecke gegeben derart, daß zwei Seiten des einen Dreiecks im engeren Sinne parallel sind zu den entsprechenden (vgl.* 11.48*) Seiten des anderen Dreiecks,* Abb. 92. *Gehen dann die Verbindungsgeraden entsprechender Ecken durch einen Punkt o, so sind auch die dritten Seiten beider Dreiecke zueinander parallel.*)

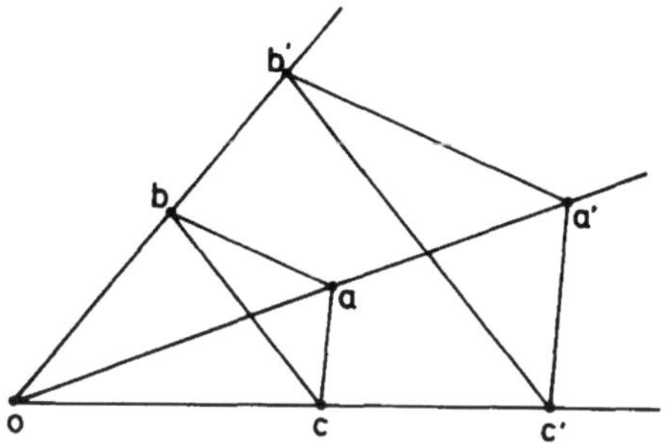

Abb. 92

<u>13.16 Anmerkung</u>. Satz 13.15 und die folgenden Teile (13.18) sind Spezialfälle (von trivialen Sonderfällen abgesehen) des folgenden Satzes von Desargues für die projektive Geometrie (vgl. Anmerkung 13.12(i)): *Sind in einer Ebene zwei nicht-entartete geordnete Dreiecke gegeben, so daß entsprechende Ecken und ebenso entsprechende Seiten niemals zusammenfallen, so gehen die Verbindungsgeraden entsprechender Ecken durch einen Punkt genau dann, wenn die Schnittpunkte entsprechender Seiten auf einer Geraden liegen.* Die genannten Schnittpunkte sind in allen hier behandelten Fällen uneigentlich, d.h. entsprechende Seiten zueinander parallel.

<u>Beweis von 13.15</u> (mittels des Satzes von Pappus-Pascal nach Hessenberg 1905, s.a. Hilbert 1977, § 35): Wegen der vorausgesetzten Parallelität im engeren Sinne liegt o nicht auf $L(ab)$ und nicht auf $L(ac)$. Wenn o auf $L(bc)$ liegt, erhält man leicht $L(bc)=L(b'c')$ und damit die Behauptung.

Sei also jetzt $\neg \operatorname{Col} obc$. Dann sind die Geraden $L(aa')$, $L(bb')$, $L(cc')$ paarweise verschieden und haben nur den Punkt o gemein.

<u>Fall 1</u>: $\neg ob \parallel ac$. Die Parallele zu $L(ob)$ durch a schneidet dann nach 12.16 die Gerade $L(a'c')$ in einem gewissen Punkt l. Sei m der Schnitt-

punkt von L(al) und L(oc), und sei n der Schnittpunkt von L($b'l$) und L(ab) (Abb. 93). Man kann nun den Satz von Pappus-Pascal (13.11 bzw. 13.14 je nach Art der Trägergeraden) dreimal hintereinander anwenden;

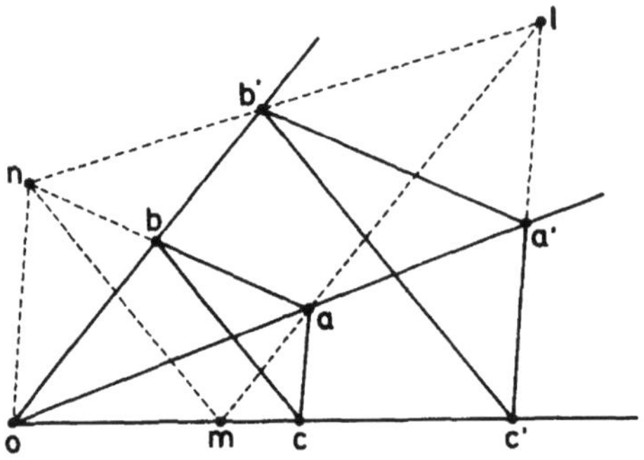

Abb. 93

die Anwendung auf das geordnete Sechseck ($onala'b'$) liefert $on \| a'l \| a'c' \| ac$, Anwendung auf ($nmacbo$) liefert $nm \| bc$, und Anwendung auf ($nmlc'b'o$) liefert $nm \| b'c'$ (man kann nachprüfen, daß die jeweiligen Voraussetzungen erfüllt sind). Aus den letzten beiden Parallelitäten ergibt sich die Behauptung.

Fall 2: $ob \| ac$. Nehmen wir an, die Behauptung gelte nicht, d.h. $\neg bc \| b'c'$. Dann ist wenigstens eine der beiden Geraden L(bc), L($b'c'$) nicht parallel zu L(oa), o.B.d.A. sei das L(bc), d.h. $\neg bc \| oa$ (sonst kann man die Rolle der "gestrichenen" und der "ungestrichenen" Punkte miteinander vertauschen). Die Parallele zu L(bc) durch b' schneidet nach 12.16 die Gerade L(oc) in einem Punkt c'', der nach Annahme von c' verschieden

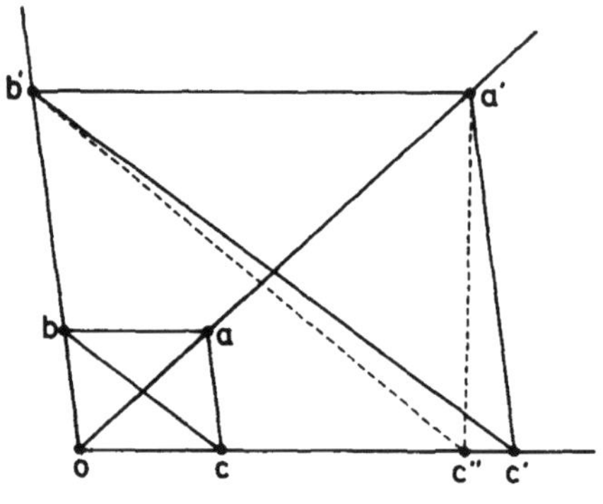

Abb. 94

ist (Abb. 94). Dann liegt die Situation von Fall 1 vor für die Punkte b, a, c, b', a', c'' an Stelle von a, b, c, a', b', c', woraus sich $ac \parallel a'c''$ ergibt im Widerspruch zur Eindeutigkeit der Parallelen. □

13.17 Anmerkung. Von jetzt ab bis einschließlich 14.28 (Einführung eines Schiefkörpers) wird an Stelle des Satzes von Pappus-Pascal nur die Folgerung 13.15 benutzt, so daß sich diese Überlegungen in noch allgemeineren ("desarguesschen") affinen Geometrien durchführen lassen (vgl. Anm. 13.13).

13.18 Satz (Satz von Desargues, Teile 2 bis 4).

$$\neg \mathrm{Col}\, abc \wedge \mathrm{Cp}\, abca' \wedge ab \overset{\vee}{\parallel} a'b' \wedge ac \overset{\vee}{\parallel} a'c'$$

$$\to [bc \overset{\vee}{\parallel} b'c' \wedge \mathrm{Col}\, oaa' \wedge \mathrm{Col}\, obb' \to \mathrm{Col}\, occ'] \qquad \text{(Teil 2)}$$

$$\wedge [bc \overset{\vee}{\parallel} b'c' \wedge aa' \parallel bb' \to cc' \parallel aa' \wedge cc' \parallel bb'] \qquad \text{(Teil 3)}$$

$$\wedge [aa' \parallel bb' \wedge aa' \parallel cc' \to bc \parallel b'c'] \qquad \text{(Teil 4).}$$

(*Seien zwei geordnete Dreiecke gegeben wie in* 13.15. *Dann gilt:* Teil 2: *Sind auch noch die dritten entsprechenden Seiten zueinander parallel und gehen zwei Verbindungsgeraden entsprechender Punkte durch einen Punkt o, so geht auch die dritte Verbindungsgerade durch diesen Punkt.* Teil 3: *Sind auch die dritten entsprechenden Seiten zueinander parallel und zwei Verbindungsgeraden zueinander parallel, so sind alle drei Verbindungsgeraden zueinander paarweise parallel,* Abb. 95. Teil 4: *Sind die Verbindungsgeraden entsprechender Ecken paarweise zueiander parallel, so sind auch die dritten entsprechenden Seiten zueinander parallel.*)

Die Teile 3 und 4 (zueinander parallele Verbindungsgeraden) werden auch als "kleiner Satz von Desargues" (affine Form) bezeichnet.

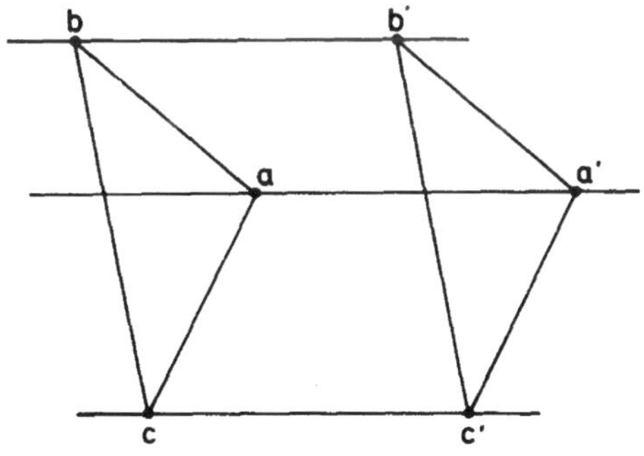

Abb. 95

Beweis: Mit der Methode, die im Beweis von 13.14 und von 13.15, Fall 2 verwendet wurde, läßt sich jeder dieser Teile auf den vorhergehenden Teil zurückführen; Einzelheiten können dem Leser überlassen bleiben.

□

13.19 Satz (Scherensatz).

$\mathrm{Cp}\,AB \wedge a,c,a',c' \in A-B \wedge b,d,b',d' \in B-A$
$\wedge ab \| a'b' \wedge ad \| a'd' \wedge bc \| b'c' \rightarrow cd \| c'd'$.

(*Sind (abcd) und (a'b'c'd') geordnete Vierecke, deren Ecken abwechselnd auf zwei komplanaren Geraden A, B - aber nicht auf beiden - liegen, und sind drei Seiten des einen parallel zu den entsprechenden Seiten des anderen Vierecks, so sind auch die vierten entsprechenden Seiten zueinander parallel*, Abb. 96.)

Der Fall paralleler Trägergeraden *A*, *B* wird auch als "kleiner Scherensatz" bezeichnet.

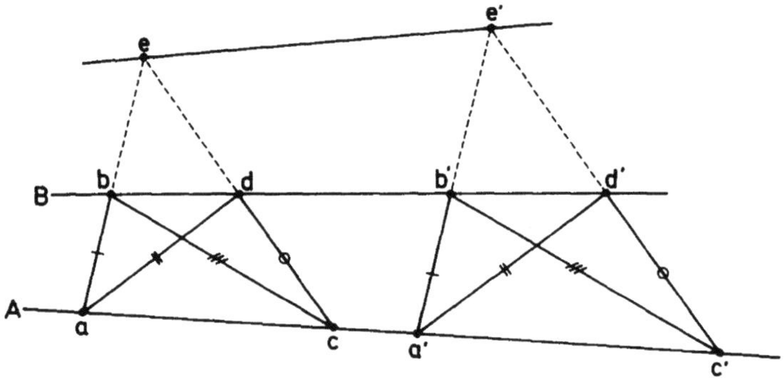

Abb. 96

Beweis: Seien o.B.d.A. die Punkte *a*, *b*, *c*, *d* (und damit auch *a'*, *b'*, *c'*, *d'*) paarweise verschieden und $a \neq a'$ (sonst ergibt sich die Behauptung unmittelbar). Nehmen wir an, die Behauptung gelte nicht. Dann ist wenigstens eine der Geraden L(*cd*), L(*c'd'*) nicht parallel zu den untereinander parallelen Geraden L(*ab*), L(*a'b'*), o.B.d.A. sei das L(*cd*) (sonst kann man die Rolle der "gestrichenen" und der "ungestrichenen" Punkte miteinander vertauschen). Dann schneiden sich L(*ab*) und L(*cd*) in einem gewissen Punkt *e*. Sei *C* diejenige Gerade durch *e*, die durch den Schnittpunkt von *A* und *B* geht (falls *A* und *B* sich schneiden) oder zu *A* und *B* parallel ist (falls $A \| B$), und sei *e'* der Schnittpunkt von *C* und L(*a'b'*). Durch Anwendungen des Satzes von Desargues (Teil 1 bzw. Teil 4) erhält man $cd \| ed \| e'd'$ und $cd \| ec \| e'c'$ (Abb. 96). Wegen der Eindeutigkeit der Parallelen ist also L(*e'd'*)=L(*e'c'*)=L(*c'd'*) und damit doch $cd \| c'd'$.

□

14. Einführung eines angeordneten Körpers

In weitgehender Analogie zum Aufbau der sog. "Hilbertschen Streckenrechnung" (Hilbert 1977, § 24) führen wir nun - auf Grund der Axiome A1 bis A8 und A10 oder der in Anm. 13.13, 13.17 genannten Voraussetzungen - einen angeordneten Körper ein. Er wird festgelegt durch zwei beliebig gewählte (verschiedene) Punkte o und e; seine Elemente sind dann die Punkte der Geraden durch o und e ("Zahlengerade"), insbesondere ist o das Nullelement und e das Einselement. Damit handelt es sich eigentlich um eine "Punktrechnung". Die für verschiedene Wahlen der Bezugspunkte o und e entstehenden Körper erweisen sich als zueinander isomorph.

14.1 Definition. $a_1, \ldots, a_n$ liegen *auf der durch o und e bestimmten Zahlengeraden*, oder kurz, $a_1, \ldots, a_n$ liegen *arithmetisch zu o, e*:

$$\mathrm{Ar}_{oe} a_1 \ldots a_n :\leftrightarrow o \neq e \wedge \bigwedge_{\nu=1}^{n} \mathrm{Col}\, oea_\nu \qquad (n=0,1,\ldots).$$

$a_1, \ldots, a_n$ liegen *arithmetisch zu o, e, e'* (e' ist ein geeigneter Hilfspunkt für die Definition der Körperoperationen):

$$\mathrm{Ar}_{oe}^{e'} a_1 \ldots a_n :\leftrightarrow \neg\, \mathrm{Col}\, oee' \wedge \bigwedge_{\nu=1}^{n} \mathrm{Col}\, oea_\nu \qquad (n=0,1,\ldots).$$

In beiden Fällen soll für $n=0$ die allgemeine Konjunktion ganz wegfallen; Ar_{oe} bedeutet, daß o und e eine Zahlengerade (einen Körper) bestimmen (für die hier behandelte Körperkonstruktion heißt das einfach $o \neq e$).

Zur Definition der Körperoperationen werden Parallelprojektionen benutzt. Dafür führen wir noch folgende Abkürzung ein.

14.2 Definition. d liegt auf der Parallelen zu $L(ab)$ durch c (entsteht durch *Parallelprojektion*) (falls $a \neq b$):

$$\mathrm{Pj}\, abcd :\leftrightarrow ab \parallel cd \vee c=d.$$

<u>14.3 Definition</u>. c ist eine *geometrische Summe* von a und b in bezug auf o, e, e' (Abb. 97):

$$\mathrm{Su}_{oe}^{e'}abc :\leftrightarrow \mathrm{Ar}_{oe}^{e'}abc \wedge \exists a' \exists c^{*}[\,\mathrm{Pj}\,ee'aa' \wedge \mathrm{Col}\,oe'a'$$

$$\wedge\, \mathrm{Pj}\,oea'c^{*} \wedge \mathrm{Pj}\,oe'bc^{*} \wedge \mathrm{Pj}\,e'ec^{*}c\,].$$

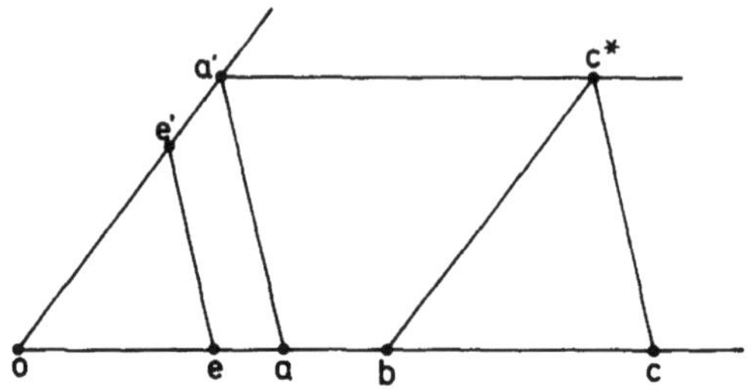

Abb. 97

<u>14.4 Definition</u>. c ist ein *geometrisches Produkt* von a und b in bezug auf o, e, e' (Abb. 98):

$$\mathrm{Pr}_{oe}^{e'}abc :\leftrightarrow \mathrm{Ar}_{oe}^{e'}abc \wedge \exists b'[\,\mathrm{Pj}\,ee'bb' \wedge \mathrm{Col}\,oe'b' \wedge \mathrm{Pj}\,e'ab'c\,].$$

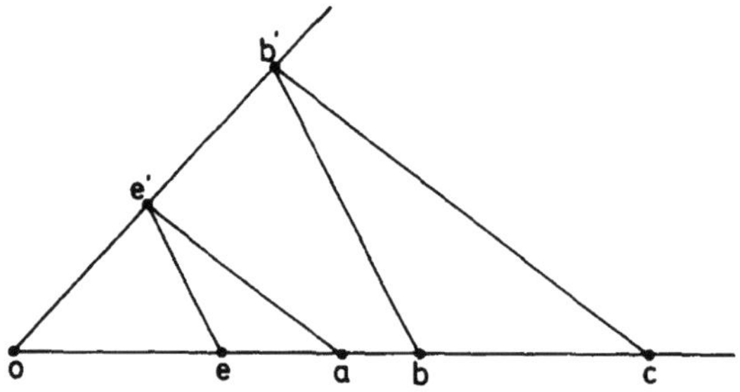

Abb. 98

<u>14.5 Anmerkung</u>. In anderen Darstellungen der Geometrie, in denen man bereits einen Zusammenhang zwischen Punkten und Zahlen (bzw. Körperelementen) zur Verfügung hat - z.B. auf Grund des bekannten Schulstoffes - kann man leicht beweisen, daß durch 14.3 und 14.4 die Summe bzw. das Produkt charakterisiert wird. Im hier durchgeführten Aufbau ist ein solcher Zusammenhang erst herzustellen und insbesondere ein Körper zu konstruieren, so daß 14.3 und 14.4 als Definitionen verwendet werden.

Man erhält nun sofort

<u>14.6 Satz.</u>

(1) $\mathrm{Ar}^{e'}_{oe}ab \to \overset{=1}{\exists} c\,\mathrm{Su}^{e'}_{oe}abc.$

(2) $\mathrm{Ar}^{e'}_{oe}ab \to \overset{=1}{\exists} c\,\mathrm{Pr}^{e'}_{oe}abc.$

(3) $\mathrm{Ar}^{e'}_{oe}a \to \overset{=1}{\exists} c\,\mathrm{Su}^{e'}_{oe}cao.$

(Es gibt genau eine geometrische Summe und genau ein geometrisches Produkt zu a, b und genau ein "geometrisch Inverses" zu a, falls die genannten Punkte arithmetisch zu den Bezugspunkten o, e, e' liegen.)

Wir führen nun entsprechende Operationen ein.

<u>14.7 Definition.</u>

(1) $\mathrm{Su}^{e'}_{oe}(ab) :=\iota c\{\mathrm{Su}^{e'}_{oe}abc \vee [\neg\mathrm{Ar}^{e'}_{oe}ab \wedge c=o]\}.$

(2) $\mathrm{Pr}^{e'}_{oe}(ab) :=\iota c\{\mathrm{Pr}^{e'}_{oe}abc \vee [\neg\mathrm{Ar}^{e'}_{oe}ab \wedge c=o]\}.$

(3) $\mathrm{Iv}^{e'}_{oe}(a) :=\iota c\{\mathrm{Su}^{e'}_{oe}cao \vee [\neg\mathrm{Ar}^{e'}_{oe}a \wedge c=o]\}.$

Hierbei wird zusätzlich festgelegt, daß die Operationen den Nullpunkt o liefern, falls die entsprechende Voraussetzung aus 14.6 nicht erfüllt ist. Dadurch wird erreicht, daß diese Operationen für beliebige Argumente definiert sind.

Zur Abkürzung benutzen wir auch die üblichen Bezeichnungen:

<u>14.8 Definition.</u>

(1) $a+b:=\mathrm{Su}^{e'}_{oe}(ab).$

(2) $a\cdot b:=\mathrm{Pr}^{e'}_{oe}(ab).$

(3) $-a:=\mathrm{Iv}^{e'}_{oe}(a).$

<u>14.9 Anmerkung.</u> Die so definierten Punkte werden jedoch nach wie vor in Abhängigkeit von den Bezugspunkten betrachtet; es wird lediglich die Angabe der Bezugspunkte weggelassen, wenn dafür die Buchstaben o, e, e' verwendet werden. In anderen Fällen (z.B. beim Vergleich verschiedener Körper) benutzen wir die ausführlicheren Bezeichnungen aus 14.7, allerdings auch ohne den oberen Index e' (ebenso in 14.3, 14.4), nachdem die Unabhängigkeit vom Hilfspunkt (14.15, 14.20) gezeigt ist.

Man erhält nun sofort die folgenden beiden Sätze.

14.10 Satz. $\mathrm{Ar}_{oe}^{e'} a \rightarrow o+a=a+o=a$.

14.11 Satz. $\mathrm{Ar}_{oe}^{e'} a \rightarrow -a+a=o$.

14.12 Satz. $\mathrm{Ar}_{oe}^{e'} ab \rightarrow a+b=b+a$.

Beweis: Seien a' und a'' (an Stelle von c^*) die gemäß 14.3 zur Bildung von $a+b$ benutzten Punkte und b', b'' die entsprechenden Punkte zur Bildung der Summe $b+a$ (Abb. 99, worin e, e' zur besseren Übersicht weggelassen sind). Die Behauptung ist dann gleichwertig mit $\mathrm{Pj}\, ee'a''b''$. Falls $a=b$ ist,

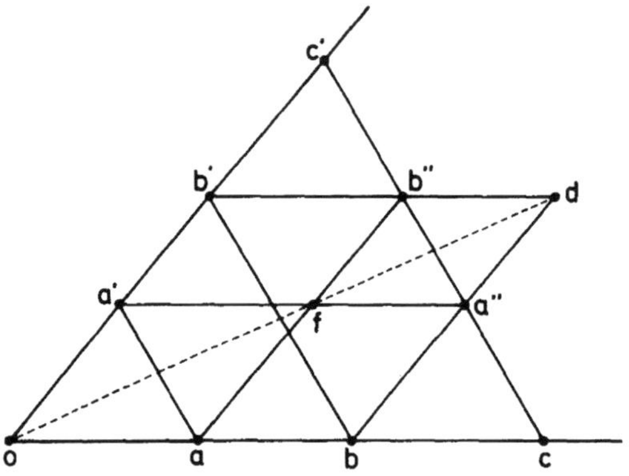

Abb. 99

gilt das wegen $a''=b''$. Falls $a=o \vee b=o$, gilt die Behauptung nach 14.10. Sei also jetzt $a \neq o \wedge b \neq o \wedge a \neq b$. Sei dann d der Schnittpunkt von $\mathrm{L}(ba'')$ und $\mathrm{L}(b'b'')$, und sei f der Schnittpunkt von $\mathrm{L}(ab'')$ und $\mathrm{L}(a'a'')$. Durch Anwendung des Satzes von Desargues, Teil 2, (13.18) auf die geordneten Dreiecke $(aa'f)$, $(bb'd)$ erhält man $\mathrm{Col}\, ofd$, und durch Anwendung von Teil 1 (13.15) auf (oaa'), $(db''a'')$ erhält man $aa' \| a''b''$ und mit $ee' \| aa'$ das Gewünschte. □

Die eben genannte Parallele zu $\mathrm{L}(ee')$, die die Punkte a'', b'' enthält, schneide $\mathrm{L}(oe)$ in $c=a+b$ und $\mathrm{L}(oe')$ in c' (Abb. 99). Damit ergibt sich

14.13 Korollar.

$\mathrm{Su}_{oe}^{e'} abc \wedge \mathrm{Ar}_{oe'}\, a'b'c' \wedge \mathrm{Pj}\, ee'aa' \wedge \mathrm{Pj}\, ee'bb' \wedge \mathrm{Pj}\, ee'cc' \rightarrow \mathrm{Su}_{oe'}^{e}\, a'b'c'$.

(Die Parallelprojektion in Richtung von L(ee') *liefert einen Isomorphismus zwischen den Zahlengeraden* L(oe) *und* L(oe') *bezüglich der Addition mit den angegebenen Bezugspunkten.)*

14.14 Satz. $\mathrm{Ar}_{oe}^{e'} abc \rightarrow a+(b+c)=(a+b)+c$.

Beweis (Skizze): Auf Grund des Kommutativgesetzes 14.12 genügt es, die Beziehung $(b+a)+c=(b+c)+a$ nachzuweisen. Die Konstruktion der hierin auftretenden Summen wird in Abb. 100 veranschaulicht. Die Bildpunkte von Punkten der Zahlengeraden bei der in 14.13 erwähnten Parallelprojektion

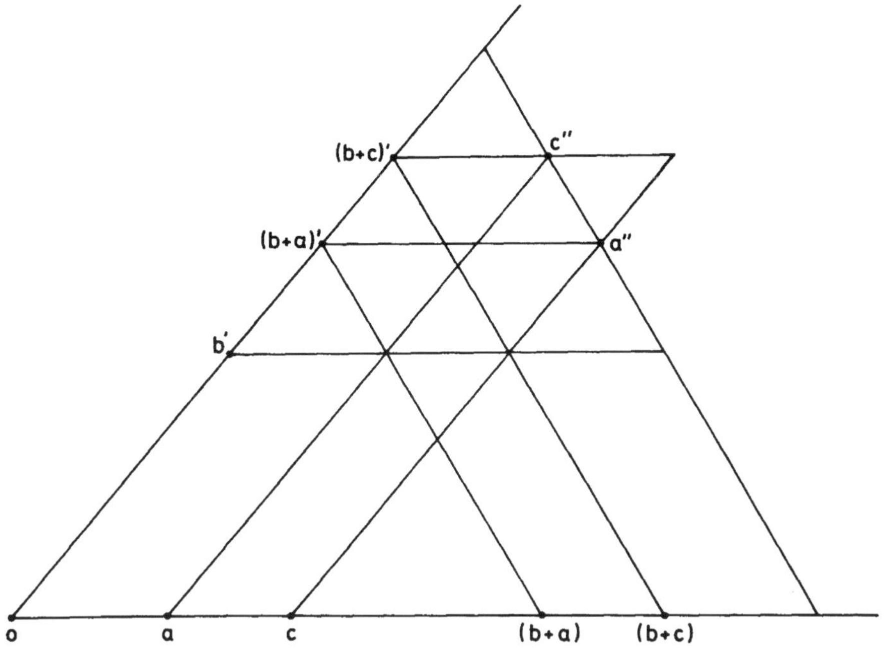

Abb. 100

sind dabei durch Anhängen eines Striches bezeichnet; der Punkt b ist nicht eingezeichnet, sondern nur der zugehörige Punkt b', der für die Bildung der Summen $b+a$ und $b+c$ verwendet wird; zur Konstruktion der Summen $(b+a)+c$ und $(b+c)+a$ werden die Hilfspunkte a'' bzw. c'' (statt c^* in Def. 14.3) benutzt. Dann genügt es wieder zu zeigen, daß $\mathrm{Pj}\, ee'a''c''$ ist. Das ergibt sich aber aus dem Beweis von 14.12; denn die in der Konstruktion benutzten Punkte außerhalb der Geraden $\mathrm{L}(oe)$ (oberer Teil von Abb. 100) liegen genauso wie die entsprechenden Punkte im Beweis von 14.12 (vgl. Abb. 99). □

14.15 Satz. $\mathrm{Su}_{oe}^{e'} abc \wedge \neg \mathrm{Col}\, oee'' \rightarrow \mathrm{Su}_{oe}^{e''} abc$.

(Die geometrische Summe ist unabhängig von der Wahl des Hilfspunktes e'.)

Beweis: Seien *a*' und *c*' (an Stelle von *c**) wie in Def. 14.3 gewählt und seien *a*" und *c*" die entsprechenden Punkte für die Konstruktion der geometrischen Summe $\mathrm{Su}_{oe}^{e''}(ab)$ (Abb. 101). Zu zeigen ist, daß diese Summe ebenfalls gleich *c* ist, d.h. Pj *e*"*ec*"*c*. Für $a=o \vee b=o$ ergibt sich die Behauptung sofort nach 14.10. Sei also jetzt $a \neq o \wedge b \neq o$ und o. B. d. A. $e' \neq e''$ (dann ist auch $a' \neq a''$).

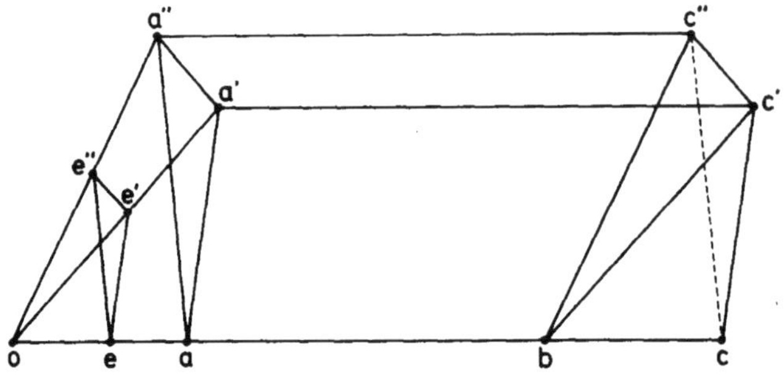

Abb. 101

Fall 1: Cp *oee*'*e*" (d.h. alle betrachteten Punkte liegen in einer Ebene).

Fall 1a: $\neg\, a'a'' \parallel oe$. Durch Anwendung des Satzes von Desargues, Teil 4, auf die geordneten Dreiecke (*oa*'*a*") und (*bc*'*c*") (bzw. trivialerweise für die Entartung Col *oa*'*a*") ergibt sich $a'a'' \parallel c'c''$; dieselbe Überlegung für (*a*'*a*"*a*) und (*c*'*c*"*c*) liefert $a''a \parallel c''c$ und damit das Gewünschte.

Falls 1b: $a'a'' \parallel oe$. Hier ist Col *a*'*a*"*c*'*c*", und der Scherensatz (13.19) liefert die gewünschte Parallelität $a''a \parallel c''c$.

Fall 2: ¬Cp *oee*'*e*". In diesem Fall wird der Satz von Desargues nicht zum Beweis gebraucht. Nach Def. 12.24 sind nämlich die Ebenen Pl(*oa*'*a*") und Pl(*bc*'*c*") zueinander parallel (man stelle sich Abb. 101 räumlich vor). Die Ebene Pl(*a*'*a*"*c*') enthält nach 12.14 auch die zu L(*a*'*c*') parallele Gerade L(*a*"*c*") und schneidet somit die vorher genannten Ebenen in den Geraden L(*a*'*a*") bzw. L(*c*'*c*"). Nach 12.26 sind also diese Geraden zueinander parallel. Die Wiederholung dieser Überlegung für die Ebene Pl(*aa*"*c*") und die zueinander parallelen Ebenen Pl(*a*'*a*"*a*) und Pl(*c*'*c*"*c*) liefert $a''a \parallel c''c$. □

14.16 Korollar. $\mathrm{Su}_{oe}^{e'}abc \wedge \mathrm{Col}\, oeu \wedge u \neq o \rightarrow \mathrm{Su}_{ou}^{e'}abc$.
(Die geometrische Summe ist auch unabhängig von der Wahl des Einheitspunktes auf der gegebenen Zahlengeraden.)

<u>Beweis</u>: In der Definition der geometrischen Summe (14.3, Abb. 97) werden die Punkte e und e' auf den gegebenen Geraden L(oe) und L(oe') offenbar nur zur Festlegung einer "Projektionsrichtung" gebraucht. Damit ist auch $\mathrm{Su}_{ou}^{u'}abc$, wobei u' der Punkt ist, in dem L(oe') von der Parallelen zu L(ee') durch u geschnitten wird. Nach 14.15 kann man von u' wieder zum Hilfspunkt e' (oder einem anderen Hilfspunkt) übergehen. □

Aus der Definition des geometrischen Produkts (14.4, Abb. 98) erhält man sofort die folgenden drei Sätze

<u>14.17 Satz</u>. $\mathrm{Ar}_{oe}^{e'}a \rightarrow o \cdot a = a \cdot o = o$.

<u>14.18 Satz</u>. $\mathrm{Ar}_{oe}^{e'}a \rightarrow e \cdot a = a \cdot e = a$.

<u>14.19 Satz</u>. $\mathrm{Ar}_{oe}^{e'}a \wedge a \neq o \rightarrow \exists b\ b \cdot a = e$.

(Zu jedem vom Nullpunkt verschiedenen Punkt der Zahlengeraden gibt es ein "reziprokes Element".)

<u>14.20 Satz</u>. $\mathrm{Pr}_{oe}^{e'}abc \wedge \neg \mathrm{Col}\, oee'' \rightarrow \mathrm{Pr}_{oe}^{e''}abc$.

(Das geometrische Produkt ist unabhängig von der Wahl des Hilfspunktes e'.)

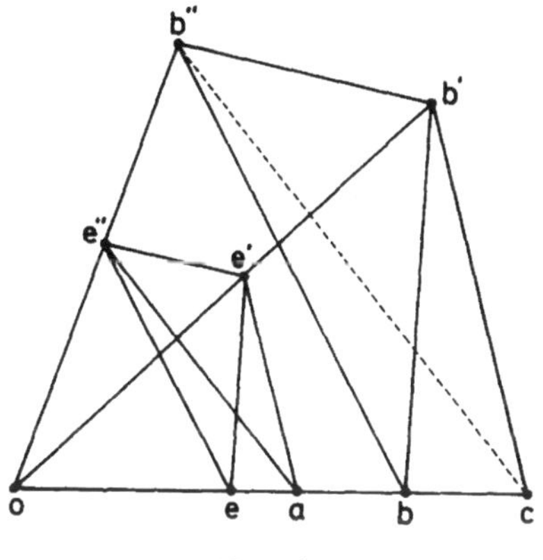

Abb. 102

<u>Zum Beweis</u>: Das ergibt sich mit den gleichen Methoden wie im Beweis von 14.15 und analogen Fallunterscheidungen (Abb. 102, der Fall 1b für die Anwendung des Scherensatzes ist hier natürlich $\mathrm{Col}\, oe'e''$). Auch hier wird im Fall 2 ($\neg \mathrm{Cp}\, oee'e''$) der Satz von Desargues (und überhaupt Abschnitt 13) nicht gebraucht. Einzelheiten können dem Leser überlassen bleiben. □

14.21 Anmerkungen. (i) Unter der Voraussetzung, daß nicht alle Punkte in einer Ebene liegen, läßt sich die Verwendung des Satzes von Desargues (und von Abschnitt 13) im Beweis von 14.20 (ebenso von 14.15) ganz vermeiden. Seien nämlich o, e, e', e'' Punkte in einer Ebene (Fall 1), und sei e''' ein Punkt außerhalb dieser Ebene. Dann gilt Fall 2 sowohl für die beiden Hilfspunkte e', e''' als auch für e''', e''. Zweimalige Anwendung des Ergebnisses von Fall 2 liefert also zunächst $\mathrm{Pr}_{oe}^{e'''} abc$ und dann die Behauptung. Der Satz von Desargues wird also nur für ebene Geometrien gebraucht.

(ii) Aus 14.20 kann man auf einfachem Wege auch umgekehrt den Satz von Desargues zurückerhalten. Es genügt dazu der Nachweis für 13.15, Fall 1 (auf den alle Teile zurückgeführt wurden). Seien dazu $(e'ee'')$ und $(b'bb'')$ Dreiecke der dort genannten Art (an Stelle von (abc) und $(a'b'c')$). Sei dann a der Schnittpunkt von $\mathrm{L}(oe)$ und $\mathrm{L}(e'e'')$, und sei $c=\mathrm{Pr}_{oe}^{e'}(ab)$ und damit $e'e''\| e'a\| b'c$. Nach 14.20 ist dann auch $\mathrm{Pr}_{oe}^{e''} abc$; das bedeutet aber (man stelle sich Abb. 102 mit kollinearen Punkten a, e', e'' vor) $e'e''\| e''a\| b''c$. Wegen der Eindeutigkeit der Parallelen ist also $\mathrm{L}(cb')$ $=\mathrm{L}(cb'')=\mathrm{L}(b'b'')$ und damit $e'e''\| b'b''$, wie zu zeigen.

(iii) Aus diesen beiden Überlegungen erhält man für Geometrien einer Dimension ≥ 3 einen Beweis des Satzes von Desargues aus einfacheren Sätzen über die Kollinearität (einschließlich Geraden und Ebenen). Dieser Beweis läßt sich wieder für affine Geometrien durchführen (vgl. Anm. 13.13 und 13.17). Damit ergibt sich das bekannte Resultat, daß die Gültigkeit des Satzes von Desargues notwendig ist für die Einbettbarkeit einer affinen Ebene in eine räumliche affine Geometrie (sie ist auch hinreichend, wie man aus der Charakterisierung der Modelle erhält). Es gibt tatsächlich affine Ebenen, in denen der Satz von Desargues nicht gilt; in Hilbert 1977, § 23 ist eine solche "Nicht-Desarguessche Geometrie" behandelt, die von Moulton stammt und in der sogar vier der fünf Hilbertschen Kongruenzaxiome gelten.

14.22 Satz. $\mathrm{Pr}_{oe}^{e'} abc \wedge \mathrm{Ar}_{oe'} a'b'c' \wedge \mathrm{Pj}\, ee'aa' \wedge \mathrm{Pj}\, ee'bb'$
$\wedge\, \mathrm{Pj}\, ee'cc' \rightarrow \mathrm{Pr}_{oe'}^{e} a'b'c'$.

(Die in 14.13 genannte Parallelprojektion ist auch ein Isomorphismus bezüglich der Multiplikation.)

Beweis: Sei o. B. d. A. $a \neq o$ (sonst gilt die Behauptung nach 14.17). Auf Grund der vorausgesetzten Projektionsbeziehungen (Abb. 103) ist auch $\mathrm{Pr}_{oe'}^{a} a'b'c'$, nach 14.20 gilt also die Behauptung. □

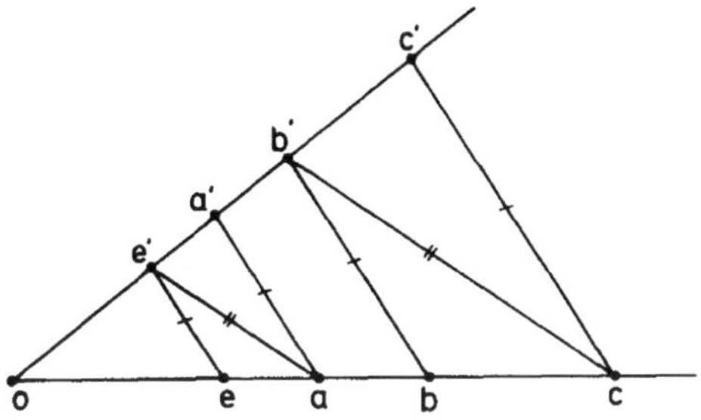

Abb. 103

<u>14.23 Satz</u>. $\mathrm{Ar}_{oe}^{e'} abc \to a\cdot(b\cdot c)=(a\cdot b)\cdot c$.

<u>Beweis</u>: Sei o. B. d. A. $b\neq o$, und sei $d=a\cdot b$, $f=b\cdot c$, $g=d\cdot c$ (Abb. 104, worin e und c weggelassen und dafür nur die entsprechenden Punkte e', c'

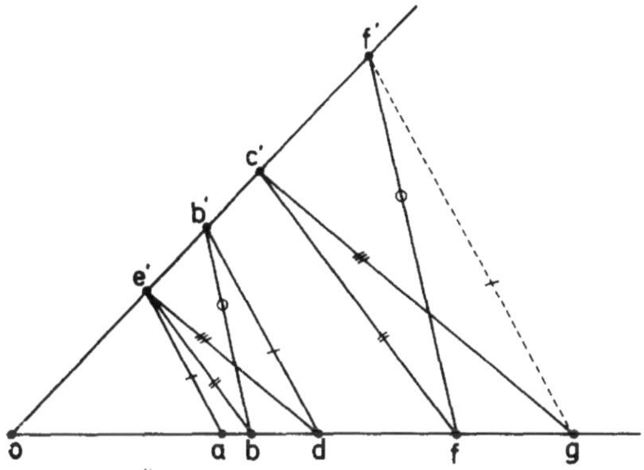

Abb. 104

eingezeichnet sind). Die in der Konstruktion der Produkte benutzten Projektionsbeziehungen liefern auch die Produktbeziehung $\mathrm{Pr}_{ob}^{e'} dfg$. Nach 14.20 ist nun $\mathrm{Pr}_{ob}^{b'} dfg$, woraus sich $\mathrm{Pj}\, b'df'g$ und damit auch $\mathrm{Pj}\, e'af'g$ ergibt. Damit ist $g=a\cdot f$, wie zu zeigen. □

<u>14.24 Satz</u>. $\mathrm{Ar}_{oe}^{e'} abc \to (a+b)\cdot c = a\cdot c+b\cdot c$.

<u>Beweis</u>: Sei o. B. d. A. $a,b,c\neq o$ (sonst gilt die Behauptung nach 14.17, 14.10). Sei $d=a+b$, $a^*=a\cdot c$, $b^*=b\cdot c$, $d^*=d\cdot c$. Nach 14.16 ist auch $\mathrm{Su}_{oa}^{e'} abd$, und es genügt, $\mathrm{Su}_{oa}^{e'} a^*b^*d^*$ nachzuweisen (Abb. 105, worin wieder e und c weggelassen und dafür nur e', c' eingezeichnet sind). Die zusätzlichen

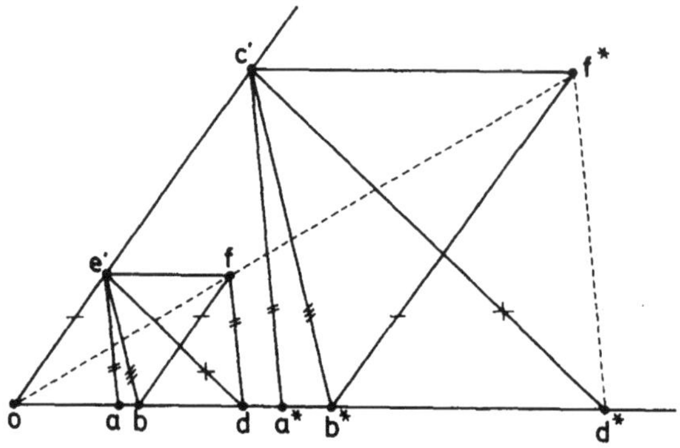

Abb. 105

Hilfspunkte für diese Summenbildungen (statt c^* in 14.3) seien mit f bzw. f^* bezeichnet. Die außer a und a^* in Abb. 105 angegebenen Punkte genügen nun denselben Bedingungen wie die entsprechenden Punkte in der Konstruktion zu 14.20 (Abb. 102) im Spezialfall $e'e''\parallel oe$, woraus sich Col off^* und die gewünschte Parallelität $fd\parallel f^*d^*$ ergibt; Einzelheiten können dem Leser überlassen bleiben. □

<u>14.25 Satz</u>. $\mathrm{Ar}_{oe}^{e'}abc \rightarrow a\cdot(b+c)=a\cdot b+a\cdot c$.

<u>Beweis</u>: Sei o. B. d. A. $a\neq o$. Sei $d=b+c$, und seien a', b', c', d' die Bildpunkte von a, b, c bzw. d bei der in 14.13 erwähnten Parallelprojektion; mit ihrer Hilfe werden die Produkte $b''=a\cdot b$, $c''=a\cdot c$ und $d''=a\cdot d$

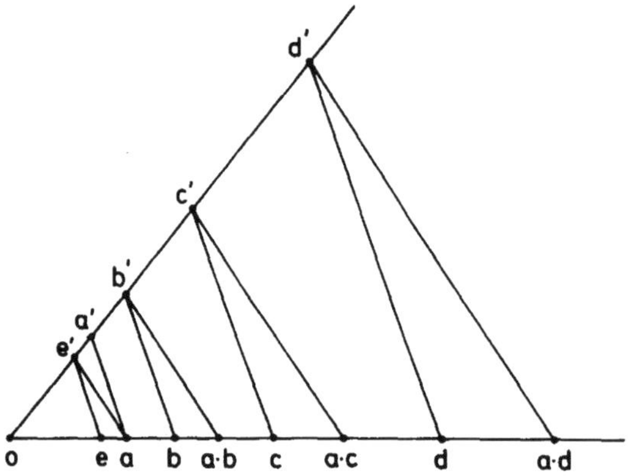

Abb. 106

konstruiert (Abb. 106). Aus $\mathrm{Su}_{oe}^{e'}bcd$ ergibt sich nach 14.13 die Beziehung

$\mathrm{Su}^{e}_{oe'}b'c'd'$ und nach 14.15 auch $\mathrm{Su}^{a}_{oe'}b'c'd'$, wieder nach 14.13 also $\mathrm{Su}^{e'}_{oa}b''c''d''$ und nach 14.16 somit $\mathrm{Su}^{e'}_{oe}b''c''d''$, wie behauptet. □

Die bisher bewiesenen Sätze über + und · lassen sich zusammenfassen in

<u>14.26 Satz</u>. *Die durch o und e bestimmte Zahlengerade bildet mit den Operationen + und · einen Schiefkörper.*

Zum Beweis der Isomorphie verschiedener Zahlengeraden benötigen wir noch

<u>14.27 Satz</u>. $oe \overset{\vee}{\parallel} o'e' \wedge \mathrm{Ar}_{oe}abc \wedge \mathrm{Ar}_{o'e'}a'b'c'$

$\wedge \mathrm{Pj}\, oo'ee' \wedge \mathrm{Pj}\, oo'aa' \wedge \mathrm{Pj}\, oo'bb' \wedge \mathrm{Pj}\, oo'cc'$

$\rightarrow (\mathrm{Su}^{e'}_{oe}abc \rightarrow \mathrm{Su}^{e}_{o'e'}a'b'c') \wedge (\mathrm{Pr}^{e'}_{oe}abc \rightarrow \mathrm{Pr}^{e}_{o'e'}a'b'c')$.

(Die Parallelprojektion von einer Geraden auf eine dazu parallele Gerade, die o und e in o' bzw. e' überführt, liefert einen Isomorphismus zwischen den durch diese Punkte bestimmten Zahlengeraden [Schiefkörpern].)

<u>Beweis</u>: <u>Teil 1 (Summen)</u>: Sei $\mathrm{Su}^{e'}_{oe}abc$ und o. B. d. A. $a \neq o$. Nach 14.15, 16 ist auch $\mathrm{Su}^{o'}_{oa}abc$ (Abb. 107). Auf Grund der damit festgestellten Projektionsbeziehungen ist auch $\mathrm{Su}^{a}_{o'a'}a'b'c'$, nach 14.15, 16 somit $\mathrm{Su}^{e}_{o'e'}a'b'c'$.

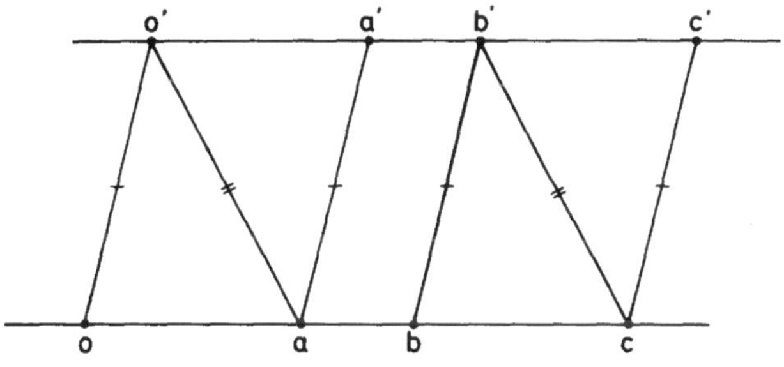

Abb. 107

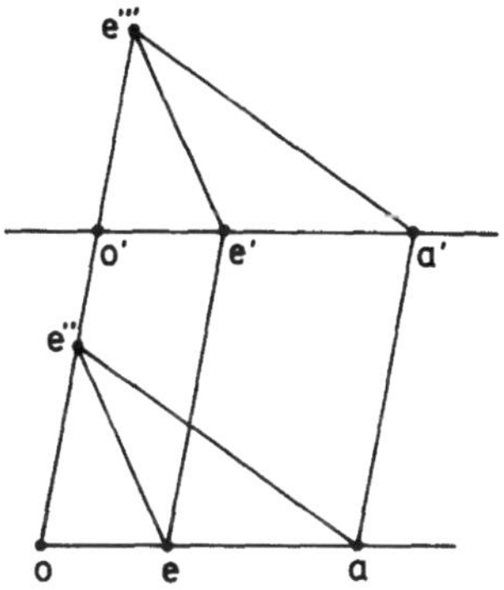

Abb. 108

<u>Teil 2 (Produkte)</u>: Sei $\mathrm{Pr}^{e'}_{oe}abc$ und o. B. d. A. $a,b,c \neq o$. Sei e'' ein von o verschiedener Punkt von $L(oo')$ (Abb. 108). Nach 14.20 ist $\mathrm{Pr}^{e''}_{oe}abc$; sei b'' der in der Definition dieser Produktbeziehung (gemäß 14.4) auftretende weitere Hilfspunkt. Sei nun $e'''=\mathrm{Su}^{e}_{oe''}(e''o')$; nach 14.15 ist dann auch $e'''=\mathrm{Su}^{a}_{oe''}(e''o')$. Die in der Definition dieser beiden Summen-

beziehungen (gemäß 14.3) auftretenden weiteren Hilfspunkte sind gerade e, e' bzw. a, a', so daß sich $ee''\parallel e'e'''$ und $ae''\parallel a'e'''$ ergibt. Die Wiederholung dieser Überlegung mit den Punkten b, b'', c an Stelle von e, e'', a liefert die Existenz eines Punktes b''' mit $bb''\parallel b'b'''$ und $cb''\parallel c'b'''$. Daraus ergibt sich $\mathrm{Pr}^{e'''}_{o'e'}a'b'c'$ und nach 14.20 auch $\mathrm{Pr}^{e}_{o'e'}a'b'c'$. □

<u>14.28 Satz</u>. *Für beliebige Paare $<o,e>$ und $<o',e'>$ von jeweils verschiedenen Punkten sind die dadurch bestimmten Zahlengeraden (Schiefkörper) zueinander isomorph, und ein Isomorphismus läßt sich durch Hintereinanderausführung von höchstens drei Parallelprojektionen der in* 14.13 *(und* 14.22*) und der in* 14.27 *genannten Art ("erster Art" und "zweiter Art") erhalten.*

<u>Beweis</u>: Sei $o\neq e$ und $o'\neq e'$. Es genügt, Projektionen zu finden, durch die man von $<o,e>$ zu $<o',e'>$ (einschließlich der dadurch bestimmten Zahlengeraden) übergehen kann.

<u>Fall 1</u>: $o=o'$. Fall o, e, e' nicht kollinear sind, leistet die Projektion gemäß 14.13 und 14.22 das Verlangte. Falls o, e, e' auf einer Geraden liegen, kann man einen Punkt e_1 außerhalb dieser Geraden wählen und durch zwei solche Projektionen von $<o,e>$ zu $<o,e_1>$ und dann zu $<o,e'>$ übergehen. In Fall 1 genügen also höchstens zwei Projektionen erster Art.

<u>Fall 2</u>: $o\neq o'$.

<u>Fall 2a</u>: $o'\notin L(oe)$. Eine Parallelprojektion zweiter Art liefert dann den Übergang von $<o,e>$ zu $<o',e_1>$ für einen geeigneten Bildpunkt e_1. Der Übergang von $<o',e_1>$ zu $<o',e'>$ läßt sich nach Fall 1 durchführen.

<u>Fall 2b</u>: $o\notin L(o'e')$. Dieser Fall ist analog zu 2a (unter Vertauschung der "gestrichenen" und der "ungestrichenen" Punkte).

<u>Fall 2c</u>: 2a und 2b gelten nicht. Dann ist $L(oe)=L(oo')=L(o'e')$. Sei e_1 ein Punkt außerhalb dieser Geraden (Abb. 109), und sei e_2 der (nach 12.16 existierende) Schnittpunkt der Parallelen zu $L(oe)$ durch e_1 und der Parallelen zu $L(oe_1)$ durch o'. Dann gelangt man durch eine Projektion erster Art von $<o,e>$ zu $<o,e_1>$, durch eine zweiter Art weiter zu $<o',e_2>$ und durch eine erster Art zu $<o',e'>$. □

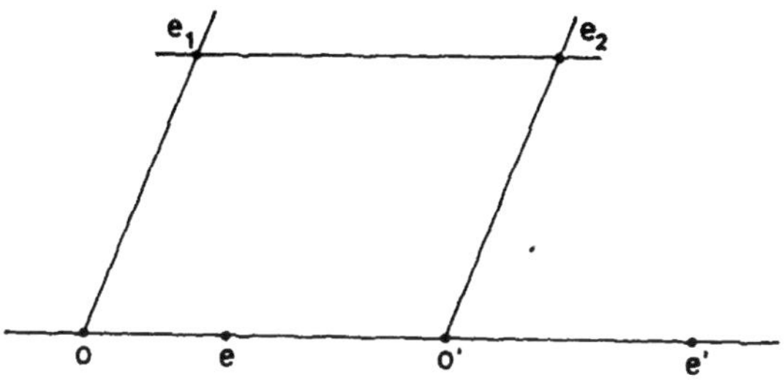

Abb. 109

Bisher wurde aus Abschnitt 13 nur der Satz von Desargues (mit Folgerungen) benutzt (vgl. Anm. 13.17). Mit Hilfe des Satzes von Pappus-Pascal erhalten wir nun

14.29 Satz. $\mathrm{Ar}_{oe}ab \to a\cdot b=b\cdot a$.

(Die geometrische Multiplikation ist kommutativ, d.h., der in 14.26 genannte Schiefkörper ist sogar ein Körper.)

Beweis: Sei $\mathrm{Pr}_{oe}^{e'}abc$ und o. B. d. A. $a,b\neq o$, und seien a', b' die Bilder von a bzw. b bei der Parallelprojektion gemäß 14.13 (Abb. 110). Durch Anwendung des Satzes von Pappus-Pascal (13.11) auf das geordnete Sechseck $(e'bb'ca'a)$ ergibt sich $e'b \parallel a'c$ und damit die gewünschte Beziehung $\mathrm{Pr}_{oe}^{e'}bac$. □

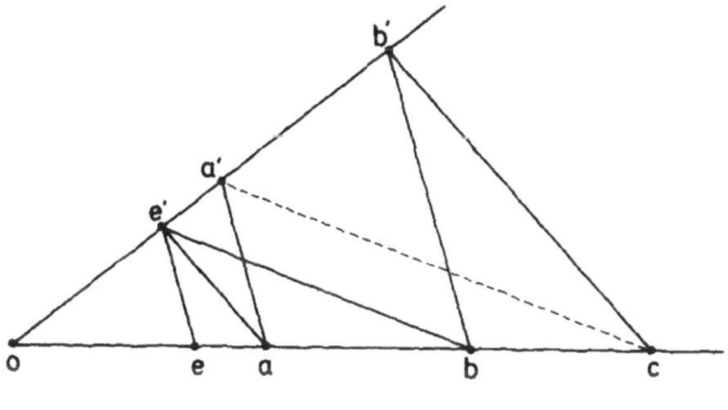

Abb. 110

14.30 Anmerkung. Auf Grund der bisherigen Überlegungen (allgemein in desarguesschen affinen Geometrien) ist 14.29 sogar gleichwertig mit dem Satz von Pappus-Pascal. Erfüllt nämlich das geordnete Sechseck $(e'bb'ca'a)$ zusammen mit dem Punkt o die Voraussetzung dieses Satzes, so braucht man nur e zu wählen mit $\mathrm{Col}\, oae \wedge \mathrm{Pj}\, a'ae'e$, um $\mathrm{Pr}_{oe}^{e'}abc$ und aus 14.29 die Parallelität $e'b \| a'c$ für das letzte Paar von Gegenseiten zu erhalten.

Als Hilfsmittel für einen speziellen Ähnlichkeitssatz (15.5) verwenden wir

14.31 Lemma. $\mathrm{Ar}_{oe}^{e'}abcd \wedge c \neq o \rightarrow [a \cdot b = c \cdot d \leftrightarrow \mathrm{Pr}_{oc}^{e'}abd]$.

Beweis: Sei o. B. d. A. auch $a,b,d \neq o$, und sei $a \cdot b = c \cdot d = p$ (Abb. 111), ausgezogene Linien). Die Parallele zu $L(e'a)$ durch d und die Parallele zu $L(e'c)$ durch b (gestrichelte Linien in Abb. 111) gehen nach dem Satz von Pappus-Pascal durch denselben Punkt b^* von $L(oe')$, woraus sich die

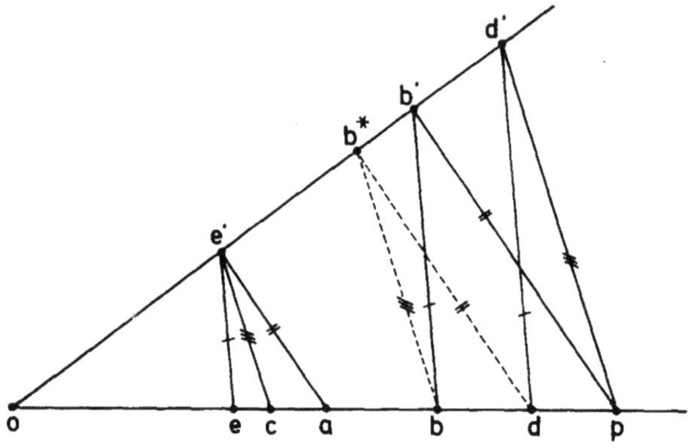

Abb. 111

gewünschte Produktbeziehung $\mathrm{Pr}_{oc}^{e'}abd$ ergibt. Die Umkehrung ergibt sich auf ähnlichem Wege oder durch Anwendung der schon bewiesenen Richtung auf den Punkt d_0 mit $a \cdot b = c \cdot d_0$ (wegen der Eindeutigkeit des geometrischen Produkts ist dann $d_0 = d$). □

Aus dieser Charakterisierung der Gleichheit von Produkten ergibt sich unmittelbar

14.32 Korollar.

$$\mathrm{Ar}_{oe}^{e'}abcd \wedge \mathrm{Col}\, oeu \wedge u \neq o \wedge a \cdot b = c \cdot d \rightarrow \mathrm{Pr}_{ou}^{e'}(ab) = \mathrm{Pr}_{ou}^{e'}(cd).$$

(*Die Gleichheit der Produkte von a, b und von c, d ist unabhängig von der Wahl des Einheitspunktes auf der gegebenen Zahlengeraden.*)

14.33 Anmerkung. Auch 14.31 und 14.32 erweisen sich als gleichwertig (wie in 14.30) mit dem Satz von Pappus-Pascal (Übung!).

Mit Hilfe der Zwischenbeziehung führen wir im konstruierten Körper noch eine Anordnung ein.

<u>14.34 Definition</u>. a ist *geometrisch positiv* in bezug auf o und e:

$$\mathrm{Ps}_{oe}a :\leftrightarrow a \underset{o}{\simeq} e.$$

<u>14.35 Lemma</u>.

$$\mathrm{Ln}\,A \wedge \mathrm{Ln}\,A' \wedge e\in A-A' \wedge e'\in A'-A$$
$$\wedge \bigwedge_{\nu=1}^{3} [a_\nu \in A \wedge a'_\nu \in A' \wedge \mathrm{Pj}\, ee'a_\nu a'_\nu]$$
$$\rightarrow [\mathrm{B}a_1a_2a_3 \leftrightarrow \mathrm{B}a'_1a'_2a'_3]$$
$$\wedge [a_1 \underset{a_2}{\simeq} a_3 \leftrightarrow a'_1 \underset{a'_2}{\simeq} a'_3].$$

(*Bei Parallelprojektionen von einer Geraden auf eine Gerade bleibt die Zwischenbeziehung erhalten und ebenso die Beziehung, auf derselben Seite eines Punktes zu liegen.*)

<u>14.36 Lemma</u>. $\mathrm{Su}_{oe}^{e'}abc \wedge a \underset{o}{\simeq} b \rightarrow \mathrm{B}oac \wedge \neq(oac)$.

<u>14.37 Satz</u>.

(1) $\neg\mathrm{Ps}_{oe}o$.

(2) $\mathrm{Ar}_{oe}^{e'}a \rightarrow \mathrm{Ps}_{oe}a \vee a=o \vee \mathrm{Ps}_{oe}(-a)$.

(3) $\mathrm{Ar}_{oe}^{e'}a \wedge \mathrm{Ps}_{oe}a \wedge \mathrm{Ps}_{oe}b \rightarrow \mathrm{Ps}_{oe}(a+b)$.

(4) $\mathrm{Ar}_{oe}^{e'}a \wedge \mathrm{Ps}_{oe}a \wedge \mathrm{Ps}_{oe}b \rightarrow \mathrm{Ps}_{oe}(a\cdot b)$.

((1) *Das Nullelement ist nicht positiv.* (2) *Für jedes Körperelement a ist $a=o$, a positiv oder $-a$ positiv.* (3) und (4) *Summe und Produkt positiver Elemente sind wieder positiv.*)

<u>Beweise von 14.35 bis 14.37</u>: Wir verwenden mehrfach die Sätze 9.19 und 9.18 (über den Zusammenhang zwischen den Relationen $\underset{p}{\simeq}$ und $\underset{A}{\simeq}$ sowie zwischen den entsprechenden Zwischenbeziehungen) und den Satz 12.6 (nach dem die Punkte einer Parallelen im engeren Sinne zu einer Geraden sämtlich auf derselben Seite dieser Geraden liegen).

Damit ergibt sich die zweite Behauptung von 14.35 unmittelbar und daraus die erste etwa nach 6.4.

Gelte die Voraussetzung von 14.36. Mit den Bezeichnungen von 14.3 (Abb. 97) ist $a \underset{a'o}{\overset{\frown}{\simeq}} b$, also auch $a \underset{a'o}{\overset{\frown}{\simeq}} c^*$ und außerdem $a \underset{a'c^*}{\overset{\frown}{\simeq}} o$. Nach 9.31

ergibt sich $\mathrm{Bo}_{a}^{a'} c^*$, also auch $\mathrm{Bo}_{a}^{a'} c$ und somit die Behauptung. Damit ist 14.36 bewiesen. Sind zusätzlich a und b (geometrisch) positiv, muß also auch c positiv sein, womit (3) gezeigt ist. Sind dagegen a und b zueinander geometrisch inverse Punkte, d.h. $c=a+b=o$, so können nach 14.36 a und b nicht auf derselben Seite von o liegen; daraus ergibt sich (2). (1) ist (nach Def. 6.1(1)) offensichtlich. Gilt die Voraussetzung von (4), so erhält man mit den Bezeichnungen von 14.4 (Abb. 98) aus $b \underset{o}{\simeq} e$ nach 14.35 zunächst $b' \underset{o}{\simeq} e'$ und dann $c \underset{o}{\simeq} a \underset{o}{\simeq} e$, wie gewünscht. □

<u>14.38 Definition</u>. Die *geometrische Differenz, Kleiner-Beziehung, Kleiner-Gleich-Beziehung*, bzw. das *Quadrat* (im Sinne der Punktrechnung) in bezug auf o, e, e' (nach 14.15 und 14.20 unabhängig von e') werden festgelegt durch

$$a-b:=a+-b,$$

$$a<b: \leftrightarrow \mathrm{Ps}_{oe}(b-a),$$

$$a\leq b: \leftrightarrow a<b \vee a=b,$$

$$a^2:=a\cdot a.$$

<u>14.39 Anmerkung</u>. In 14.37 wird eine Form der Axiome ausgedrückt, die man oft zusätzlich zu den Körperaxiomen zur Charakterisierung des Begriffs angeordneter Körper verwendet (mit der Positivität als zusätzlichem Grundbegriff). Die anderen üblichen Charakterisierungen mittels der Kleiner-Beziehung oder der Kleiner-Gleich-Beziehung erhält man daraus ohne weiteres mit Hilfe von 14.38.

Die früher (insbesondere in 14.28) genannten Parallelprojektionen erhalten nach 14.35 auch die Positivität und damit die (Körper-)Anordnung. Das hier Bewiesene läßt sich also zusammenfassen in

<u>14.40 Satz</u>. *Die durch o und e (mit $o\neq e$) bestimmte Zahlengerade bildet mit den Operationen + und · und der Relation* Ps_{oe} *(bzw. < oder ≤) einen angeordneten Körper* $\mathfrak{K}_{oe}$. *Für verschiedene Punktepaare sind die dadurch bestimmten Zahlengeraden (angeordneten Körper) zueinander isomorph, und ein Isomorphismus läßt sich durch Hintereinanderausführung von höchstens drei Parallelprojektionen erster und zweiter Art (vgl.* 14.28*) erhalten.*

Für kartesische Räume $\mathfrak{L}_n(\mathfrak{K})$ gilt außerdem

<u>14.41 Satz</u>. *Sei* $\mathfrak{K}$ *ein angeordneter Körper und* $n\geq 2$. *Dann sind die in* $\mathfrak{L}_n(\mathfrak{K})$ *konstruierten Zahlengeraden* $\mathfrak{K}_{oe}$ *auch isomorph zu* $\mathfrak{K}$.

<u>14.42 Anmerkung</u>. Der betrachtete Raum $\mathfrak{L}_n(\mathfrak{f})$ ist natürlich nur dann ein Modell von A1 bis A8 (und A10), wenn $\mathfrak{f}$ pythagoreisch ist (s. 1.6). Unsere Konstruktion von Zahlengeraden ist jedoch gemäß Anm. 13.13 in beliebigen kartesischen Räumen $\mathfrak{L}_n(\mathfrak{f})$ mit $n\geq 2$ sinnvoll.

<u>Zum Beweis von 14.41</u>: Nach 14.40 genügt es zu zeigen, daß eine Zahlengerade von $\mathfrak{L}_n(\mathfrak{f})$ zu $\mathfrak{f}$ isomorph ist. Sei $o=\langle 0,0,\ldots,0\rangle$, $e=\langle 1,0,\ldots,0\rangle$, $e'=\langle 0,1,0,\ldots,0\rangle$, und sei φ die Abbildung von $\mathfrak{f}$ auf die durch o und e bestimmte Zahlengerade mit $\varphi(x)=\langle x,0,\ldots,0\rangle$. Indem man geometrische Summen, Produkte und die Positivität (in bezug auf o, e und ggf. e') mit den Mitteln der analytischen Geometrie in $\mathfrak{L}_n(\mathfrak{f})$ beschreibt, erhält man ohne Schwierigkeiten, daß φ ein Isomorphismus der gewünschten Art ist. □

15. Längen von Strecken

Die Körperkonstruktion von Abschnitt 14 benutzen wir nun zur Einführung von Längen von Strecken und zum Beweis der grundlegenden Sätze darüber (auf Grund der Axiome A1 bis A8 und A10).

15.1 Definition. Die *Länge* der Strecke ab oder der *Abstand* der Punkte a, b in bezug auf o, e:

$$\mathrm{Lg}_{oe}(ab) := \iota c\{[\mathrm{Ar}_{oe}c \wedge o \leq c \wedge oc \equiv ab] \vee [\neg \mathrm{Ar}_{oe} \wedge c = o]\}$$

(d.i. der Punkt der durch o und e bestimmten Zahlengeraden, der durch Abtragen der Strecke ab von o aus nach der positiven Seite entsteht bzw. [damit die Bezeichnung stets erklärt ist] der Punkt o, falls o und e keine Zahlengerade bestimmen). Zur Abkürzung verwenden wir auch die übliche Bezeichnung (unter Weglassung von o und e analog zu 14.8, 9)

$$\overline{ab} := \mathrm{Lg}_{oe}(ab).$$

Aus bekannten Sätzen über die Streckenkongruenz (6.11, A3, 2.8, 2.7) erhält man unmittelbar, daß das obige c tatsächlich eindeutig bestimmt ist, sowie die Teile (1) bis (3) von

15.2 Satz.

(1) $\mathrm{Ar}_{oe} \rightarrow o \leq \overline{ab}$.

(2) $\mathrm{Ar}_{oe} \rightarrow [\overline{ab} = o \leftrightarrow a = b]$.

(3) $\mathrm{Ar}_{oe} \rightarrow [\overline{ab} = \overline{cd} \leftrightarrow ab \equiv cd]$.

(4) $\mathrm{Ar}_{oe} \rightarrow [\overline{ab} \leq \overline{cd} \leftrightarrow ab \leq cd]$.

Der in (4) ausgedrückte Zusammenhang zwischen den $\leq$-Beziehungen für Körperelemente (14.38) und für Strecken (5.4) läßt sich erhalten mit Hilfe von 14.36 (und geläufigen Sätzen über beide Beziehungen). Die Ausführung sei dem Leser überlassen.

15.3 Lemma. $\mathrm{Su}_{oe}abc \to ob\equiv ac$.

Beweis: Für $a=o$ bzw. $b=o$ ist $b=c$ bzw. $a=c$, woraus sich die Behauptung ergibt. Sonst seien e', a', c^* die Hilfspunkte für die Summenkonstruktion gemäß 14.3 (Abb. 97). Nach dem Parallelogramm-Satz 12.19 ist dann $ob\equiv a'c^*\equiv ac$. □

15.4 Satz. $\mathrm{Ar}_{oe} \wedge \mathrm{B}abc \to \overline{ab}+\overline{bc}=\overline{ac}$.

Beweis: Sei $a_1=\overline{ab}$, $b_1=\overline{bc}$ und $c_1=a_1+b_1$. Mittels Lemma 14.36 erhält man $\mathrm{B}oa_1c_1$. Nach Konstruktion ist $oa_1\equiv ab$ und nach 15.3 außerdem $a_1c_1\equiv ob_1\equiv bc$. Nach 2.11 ("Aneinanderlegen von Strecken") ist also $oc_1\equiv ac$, nach 14.37(3) auch $o\le c_1$ und damit $c_1=\overline{ac}$. □

15.5 Satz (spezieller Ähnlichkeitssatz, Abb. 112).

$\mathrm{Ar}_{oe} \wedge \mathrm{Col}\,pab \wedge \mathrm{Col}\,pcd \wedge \neg\mathrm{Col}\,pac \wedge \mathrm{Pj}\,acbd \to \overline{pa}\cdot\overline{pd}=\overline{pc}\cdot\overline{pb}$.

15.6 Anmerkung. Die Behauptung von 15.5 ist natürlich nur eine andere Ausdrucksweise für die (hier nicht erst eingeführten) "Proportionen" $\overline{pa}:\overline{pb}=\overline{pc}:\overline{pd}$ oder $\overline{pa}:\overline{pc}=\overline{pb}:\overline{pd}$ (abgesehen vom Fall $b=d=p$).

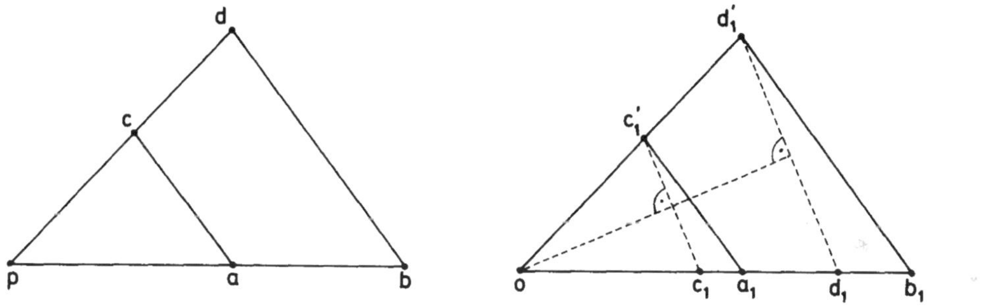

Abb. 112

Beweis von 15.5: Sei o. B. d. A. $b\neq p$ und damit $\neg\mathrm{Col}\,pbd$ (sonst ist auch $d=p$ und die Beh. trivial). Für $x=a,b,c,d$ sei $x_1=\overline{px}$ (Abb. 112). Durch Dreiecksantragung (10.16) erhält man einen Punkt c_1' mit $(pac)\equiv(oa_1c_1')$. Sei m der Mittelpunkt der Strecke c_1c_1' und $d_1'=\mathrm{S}_{om}d_1$. Da S_{om} eine Bewegung ist, ist $c_1'\underset{o}{\simeq}d_1' \wedge od_1'\equiv od_1\equiv pd$, wegen $om\perp c_1c_1'$ außerdem $\mathrm{Pj}\,c_1c_1'd_1d_1'$. Nach dem ersten Kongruenzsatz (11.49) und den Sätzen über Stufen- und Wechselwinkel an Parallelen (12.21, 22) ist $d_1'b_1o\equiv dbp\equiv cap\equiv c_1'a_1o$ und damit $a_1c_1'\parallel b_1d_1'$. Somit ist $\mathrm{Pr}^{c_1'}_{oc_1}a_1d_1b_1$, nach 14.31 und 14.20 also $a_1\cdot d_1=c_1\cdot b_1$. □

15.7 Satz (Euklid).

$\mathrm{Ar}_{oe} \wedge \mathrm{R}acb \wedge ch \underset{h}{\perp} ab \rightarrow \overline{ac}^2 = \overline{ah} \cdot \overline{ab}$.

(In einem rechtwinkligen Dreieck ist das Quadrat [der Länge] einer Kathete gleich dem Produkt [der Längen] des anliegenden Hypotenusenabschnitts und der Hypotenuse, Abb. 113.)

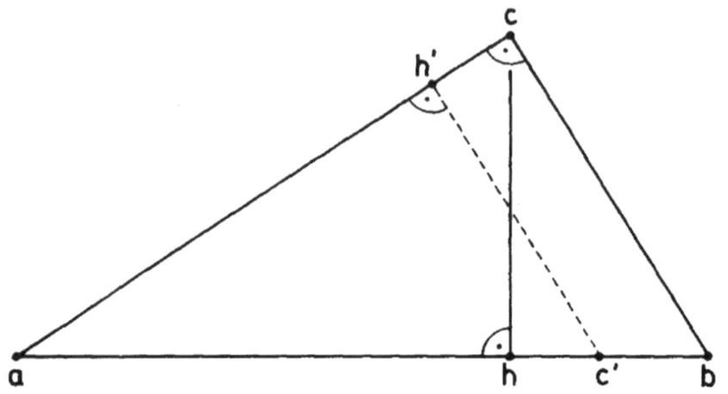

Abb. 113

Beweis: Durch Streckenabtragung erhält man Punkte c', h' mit $c' \underset{a}{\simeq} h \wedge ac' \equiv ac$, $h' \underset{a}{\simeq} c \wedge ah' \equiv ah$. Nach dem ersten Kongruenzsatz (11.49) sind dann die geordneten Dreiecke (ahc) und $(ah'c')$ zueinander kongruent, insbesondere ist $\mathrm{R}ah'c'$ und damit $cb \| h'c'$. Nach 15.5 ist also $\overline{ac'} \cdot \overline{ac} = \overline{ah'} \cdot \overline{ab}$, woraus sich nach 15.2(3) die Behauptung ergibt. □

15.8 Satz (Pythagoras).

$\mathrm{Ar}_{oe} \wedge \mathrm{R}acb \rightarrow \overline{ac}^2 + \overline{bc}^2 = \overline{ab}^2$.

(In einem rechtwinkligen Dreieck ist die Summe der Quadrate der [Längen der] Katheten gleich dem Quadrat der [Länge der] Hypotenuse.)

Beweis: Fall 1: $\mathrm{Col}\, abc$. Nach 8.9 ist dann $a=c$ oder $b=c$ und damit die Beh. trivial.

Fall 2: $\neg \mathrm{Col}\, abc$. Sei h der Fußpunkt des Lots von c auf $\mathrm{L}(ab)$. Nach 11.47 ist $\mathrm{B}ahb$, nach 15.4 also $\overline{ah} + \overline{bh} = \overline{ab}$. Nach 15.7 ist außerdem $\overline{ac}^2 = \overline{ah} \cdot \overline{ab}$ und ebenso $\overline{bc}^2 = \overline{bh} \cdot \overline{ab}$, also $\overline{ac}^2 + \overline{bc}^2 = (\overline{ah} + \overline{bh}) \cdot \overline{ab} = \overline{ab}^2$. □

16. Koordinaten

Um die am Ende von Abschnitt 1 angekündigten Darstellungssätze zu erhalten, führen wir noch Koordinaten von Punkten ein und beschreiben mit ihrer Hilfe die verwendeten Grundbegriffe. Für alle Punkte funktioniert das natürlich nur unter der Voraussetzung, daß der ganze Raum eine feste endliche Dimension n (mit $n\geq 2$) hat. Zunächst führen wir aber Koordinaten für beliebige endlichdimensionale Unterräume ein und legen nach wie vor nur die Axiome A1 bis A8 und A10 zugrunde (Sätze der absoluten Geometrie sind wieder mit "(a)" gekennzeichnet, vgl. Anm. 12.1).

16.1 Definition. Die Punkte s, $u_1,\ldots, u_n$ bilden in bezug auf o, e (auf die dadurch bestimmte Zahlengerade) ein *n-dimensionales kartesisches Koordinatensystem*:

$$\mathrm{Cs}_{oe}su_1\ldots u_n :\leftrightarrow \mathrm{Ar}_{oe} \wedge \bigwedge_{\nu=1}^{n} su_\nu \equiv oe \wedge \bigwedge_{\substack{\mu,\nu=1\\ \mu\neq\nu}}^{n} \mathrm{R}u_\mu su_\nu .$$

Dabei heißt s der *Ursprung(spunkt)* dieses Koordinatensystems (*Koordinatenursprung*), die Gerade $\mathrm{L}(su_\nu)$ die *ν-te Koordinatenachse* und der Punkt u_ν der *Einheitspunkt* auf dieser Achse. ($\mathrm{Cs}_{oe}su_1\ldots u_n$ bedeutet dann, daß die n Koordinatenachsen im Ursprung s paarweise aufeinander senkrecht stehen und daß die Einheitspunkte u_ν auf diesen Achsen vom Ursprung den Abstand e ["Eins"] haben.)

16.2 Satz (a). $\mathrm{Cs}_{oe}su_1\ldots u_n \rightarrow \neg\mathrm{Ah}_n su_1\ldots u_n \qquad (n\geq 1).$

(Jedes n-dimensionale kartesische Koordinatensystem spannt einen n-dimensionalen Unterraum auf.)

16.3 Satz (a). $\mathrm{Cs}_{oe}su_1\ldots u_n \wedge \mathrm{Sb}_{n+1}U \wedge s,u_1,\ldots,u_n\in U$

$\rightarrow \exists u_{n+1}[u_{n+1}\in U \wedge \mathrm{Cs}_{oe}su_1\ldots u_n u_{n+1}] \qquad (n\geq 1).$

(In einem (n+1)-dimensionalen Unterraum U läßt sich ein n-dimensionales kartesisches Koordinatensystem stets zu einem (n+1)-dimensionalen ergänzen.)

16.4 Satz (a).

$$Ar_{oe} \wedge Sb_n U \rightarrow \exists s \exists u_1 \dots \exists u_n [Cs_{oe} su_1 \dots u_n \wedge U = Sb_n(su_1 \dots u_n)] \quad (n \geq 1).$$

(In einem n-dimensionalen Unterraum U gibt es stets ein n-dimensionales kartesisches Koordinatensystem, das U aufspannt.)

Beweise von 16.2 bis 16.4: 16.2 ergibt sich durch Induktion über n aus 11.59, 60 (und daraus, daß eine Gerade nicht auf sich selbst senkrecht steht). 16.3 ergibt sich aus 11.63 und 16.4 durch Induktion über n aus 16.3. □

16.5 Definition. Der Punkt p hat in bezug auf s, $u_1, \dots, u_n$ (i.b.a. das von diesen Punkten gebildete kartesische Koordinatensystem) und o, e die *Koordinaten* $x_1, \dots, x_n$:

$$\begin{aligned} Cd_{oe}^{su_1 \dots u_n} px_1 \dots x_n :\leftrightarrow\ & Cs_{oe} su_1 \dots u_n \wedge p \in Sb_n(su_1 \dots u_n) \\ & \wedge \exists p_1 \dots \exists p_n \bigwedge_{\nu=1}^{n} [\, Col\, su_\nu p_\nu \wedge (p_\nu = p \vee pp_\nu \perp su_\nu) \\ & \wedge (oex_\nu) \equiv (su_\nu p_\nu)] \end{aligned}$$

(d.h., wenn p_ν die "senkrechte Projektion" von p auf die ν-te Achse ist [d.h. Lotfußpunkt oder, wenn p auf dieser Achse liegt, gleich p], so

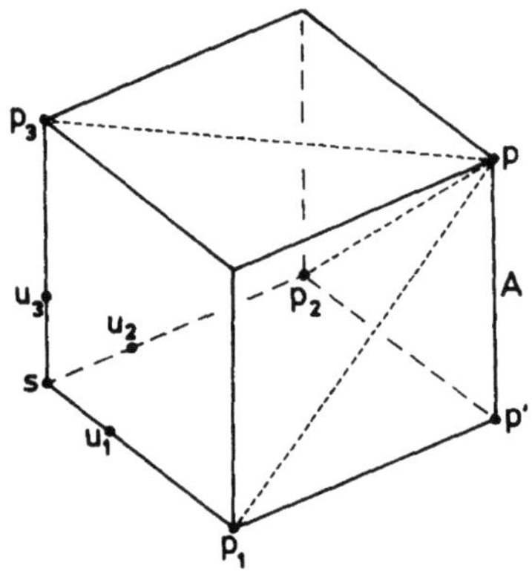

Abb. 114

entsteht x_ν aus p_ν durch kongruente Übertragung der ν-ten Achse auf die

Zahlengerade, wobei s, u_ν in o bzw. e übergehen ($\nu=1,\ldots,n$); s. Abb. 114 für $n=3$).

Zur Abkürzung schreiben wir dafür auch

$$\mathrm{Cd}\,\mathfrak{B}\,px_1\ldots x_n,$$

wenn $\mathfrak{B}=(su_1\ldots u_n oe)$ das System der Bezugspunkte ist.

Mittels 4.13 ergibt sich

16.6 Satz (a). $\mathrm{Cd}_{oe}^{su_1\ldots u_n}px_1\ldots x_n \to \mathrm{Ar}_{oe}x_1\ldots x_n \wedge p\in\mathrm{Sb}_n(su_1\ldots u_n)$.

16.7 Satz.

$$\mathrm{Cs}_{oe}su_1\ldots u_n \wedge U=\mathrm{Sb}_n(su_1\ldots u_n)$$
$$\to\forall p[p\in U\to\exists x_1\ldots\exists x_n\mathrm{Cd}_{oe}^{su_1\ldots u_n}px_1\ldots x_n]$$
$$\wedge\forall x_1\ldots\forall x_n[\mathrm{Ar}_{oe}x_1\ldots x_n\to\exists p\mathrm{Cd}_{oe}^{su_1\ldots u_n}px_1\ldots x_n]$$
$$\wedge[\mathrm{Cd}_{oe}^{su_1\ldots u_n}px_1\ldots x_n\wedge\mathrm{Cd}_{oe}^{su_1\ldots u_n}qy_1\ldots y_n\to(p=q\leftrightarrow\bigwedge_{\nu=1}^{n}x_\nu=y_\nu)].$$

(Wenn s, $u_1,\ldots,u_n$ ein n-dimensionales kartesisches Koordinatensystem bilden, so hat jeder Punkt p des davon aufgespannten Raumes Koordinaten $x_1,\ldots,x_n$, die eindeutig bestimmt sind, und zu beliebigen Punkten $x_1,\ldots,x_n$ der Zahlengeraden gibt es einen eindeutig bestimmten Punkt mit diesen Koordinaten.)

Beweis: 1. Sei $p\in U$ gegeben. Durch die Bedingungen in der zweiten Formelzeile von 16.5 sind die Punkte $p_1,\ldots,p_n$ eindeutig bestimmt. Ein Punkt x_ν mit der in der dritten Zeile genannten Eigenschaft existiert nach 4.14 und ist eindeutig bestimmt wegen 4.18. Somit hat p eindeutig bestimmte Koordinaten.

2. Sei $\mathrm{Ar}_{oe}x_1\ldots x_n$. Wie eben sind durch die Bedingung in der dritten Zeile zunächst Punkte $p_1,\ldots,p_n$ auf den Achsen eindeutig bestimmt. Sei nun U_ν der nach 11.67 eindeutig festgelegte $(n-1)$-dimensionale Unterraum von U, der in p_ν auf der ν-ten Achse senkrecht steht; nach 11.68 hat ein beliebiger Punkt p von U die ν-te Koordinate x_ν (d.h. $p=p_\nu \vee pp_\nu\perp su_\nu$) genau dann, wenn er in U_ν liegt ($\nu=1,\ldots,n$). Um die gewünschte Existenz und Eindeutigkeit eines Punktes mit den Koordinaten $x_1,\ldots,x_n$ zu erhalten, genügt es also zu zeigen, daß die Unterräume $U_1,\ldots,U_n$ genau einen gemeinsamen Punkt besitzen. Das wird u.a. ausgedrückt in

16.8 Lemma.

$$\mathrm{Cs}_{oe} su_1 \ldots u_n \wedge U=\mathrm{Sb}_n(su_1\ldots u_n) \wedge \bigwedge_{\nu=1}^{n} [\mathrm{Sb}_{n-1} U_\nu \wedge su_\nu \hat{\perp} U_\nu \wedge U_\nu \subseteq U]$$
$$\rightarrow [A=U \cap \bigcap_{\nu=1}^{n-1} U_\nu \rightarrow \mathrm{Ln} A \wedge A \hat{\perp} \mathrm{Sb}_{n-1}(su_1 \ldots u_{n-1}) \wedge A \hat{\perp} U_n \wedge A \parallel su_n]$$
$$\wedge \exists p \; p \in \bigcap_{\nu=1}^{n} U_\nu \qquad (n \geq 1).$$

Beweis (induktiv über n): Für $n=1$ ist das Lemma trivial (nur für diesen Fall wird U bei der Durchschnittsbildung in der zweiten Zeile gebraucht). Sei nun $n \geq 2$ und gelte das Lemma schon für $n-1$. Sei $U'=\mathrm{Sb}_{n-1}(su_1\ldots u_{n-1})$ und $U'_\nu = U_\nu \cap U'$ $(\nu=1,\ldots,n-1)$. Für diese Indizes erhält man leicht $U_\nu \cup U' = U$. Die Unterräume $U'_1,\ldots, U'_{n-1}$ sind nach 9.59 $(n-2)$-dimensional und erfüllen somit die Voraussetzung des Lemmas für das von s, $u_1,\ldots, u_{n-1}$ gebildete $(n-1)$-dimensionale kartesische Koordinatensystem. Damit haben sie genau einen gemeinsamen Punkt p' (Abb.114). Sei p_ν der Schnittpunkt von U_ν mit der ν-ten Achse. Für $\nu=1,\ldots,n-1$ enthält U_ν nach 11.68 die (auf der ν-ten Achse senkrecht stehende) Parallele zur n-ten Achse $\mathrm{L}(su_n)$ durch p_ν und damit auch die Parallele A' zu $\mathrm{L}(su_n)$ durch p'. Der Durchschnitt $A=\bigcap_{\nu=1}^{n-1} U_\nu$ ist aber nach 9.59 ein eindimensionaler Unterraum (da er mit U' den nulldimensionalen Durchschnitt $\{p'\}$ bildet), also ist $A=A'$ und somit $A \parallel su_n$ und nach 12.27 auch $A \hat{\perp} U'$ und $A \hat{\perp} U_n$. Damit leistet A das Verlangte. Außerdem ist $\bigcap_{\nu=1}^{n} U_\nu = A \cap U_n$ wegen $A \hat{\perp} U_n$ einelementig. Damit sind 16.8 und 16.7 bewiesen. □

16.9 Lemma.

(1) $\mathrm{Ar}_{oe} xy \wedge y \leq x \rightarrow \overline{xy} = x-y$.

(2) $\mathrm{Ar}_{oe} xy \rightarrow \overline{xy}^2 = (x-y)^2$.

Beweis: Zu (1): Sei $d=x-y$. Dann ist $y+d=x$, nach 15.3 also $od \equiv xy$, nach Voraussetzung außerdem $o \leq d$ und damit $d=\overline{xy}$.

Zu (2): Wegen (1) ist $\overline{xy}=x-y$ oder $\overline{xy}=y-x=-(x-y)$, woraus sich die Behauptung ergibt. □

16.10 Lemma. $\mathrm{Ar}_{oe} a_1 \ldots a_k \wedge \bigwedge_{\kappa=1}^{k} (oea_\kappa) \equiv (sup_\kappa)$
$\rightarrow (oea_1 \ldots a_k) \equiv (sup_1 \ldots p_k) \qquad (k \geq 1)$.

Beweis: Jede der zu zeigenden Kongruenzen $a_\kappa a_\lambda \equiv p_\kappa p_\lambda$ ergibt sich sofort aus dem Fünf-Strecken-Satz 4.16. □

16.11 Satz.

$$\mathrm{Cd}_{oe}^{su_1\ldots u_n}px_1\ldots x_n \wedge \mathrm{Cd}_{oe}^{su_1\ldots u_n}qy_1\ldots y_n \rightarrow \overline{pq}^2 = \sum_{\nu=1}^{n}(x_\nu - y_\nu)^2.$$

(Die "allgemeine [geometrische] Summe" von Punkten der Zahlengeraden sei hier mittels + induktiv erklärt, wie es in beliebigen Körpern [und allgemeineren Strukturen] üblich ist.)

Beweis (induktiv): Seien p_ν und die $(n-1)$-dimensionalen Unterräume U_ν $(\nu=1,\ldots,n)$, p' und A wie in 16.5 und den Beweisen zu 16.7, 16.8 für den Punkt p gewählt und analog q_ν, V_ν, q' und B für den Punkt q (Abb.115). Nach 16.10 ist dann $p_\nu q_\nu \equiv x_\nu y_\nu$ $(\nu=1,\ldots,n)$. Für $n=1$ ist $p=p_n$ und $q=q_n$, und die Behauptung ergibt sich sofort aus 16.9. Sei nun $n\geq 2$, und gelte

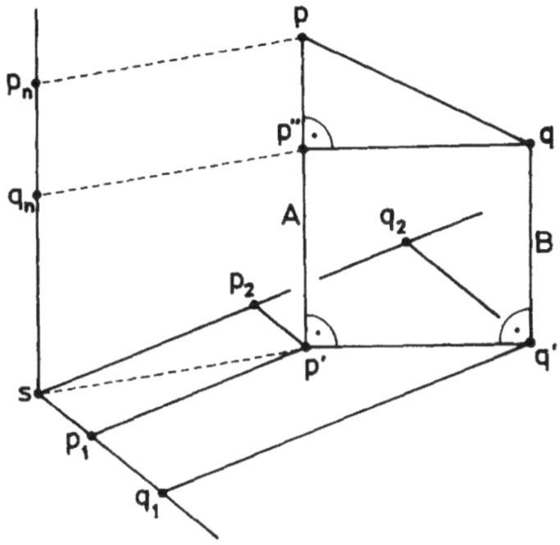

Abb. 115

16.11 für $(n-1)$-dimensionale Koordinatensysteme. Nach Konstruktion haben p' bzw. q' die Koordinaten $x_1,\ldots, x_{n-1}$ bzw. $y_1,\ldots, y_{n-1}$ in dem von s, $u_1,\ldots, u_{n-1}$ gebildeten Koordinatensystem, also ist $\overline{p'q'}^2 = \sum_{\nu=1}^{n-1}(x_\nu - y_\nu)^2$. Als Parallele zur n-ten Achse steht A auch senkrecht auf dem Unterraum V_n und schneidet diesen in einem gewissen Punkt p''. Wenn $A \neq \mathrm{L}(su_n)$, so sind $\mathrm{L}(p_n p)$ und $\mathrm{L}(q_n p'')$ Geraden, die in einer Ebene liegen und auf $\mathrm{L}(su_n)$ senkrecht stehen, also zueinander parallel. Somit bilden p_n, q_n, p'', p in jedem Falle ein - eventuell entartetes - Parallelogramm, und es ist $pp'' \equiv p_n q_n \equiv x_n y_n$. Ebenso erhält man, daß p'', p', q', q ein (eventuell entartetes) Parallelogramm bilden, und damit $p''q \equiv p'q'$. Wegen $A \hat{\perp} V_n$ ist schließlich auch $\mathrm{R}pp''q$, aus dem Satz des Pythagoras (15.8) (und 15.2(3)) ergibt sich also

$$\overline{pq}^2=\overline{p''q}^2+\overline{p''p}^2=\overline{p'q'}^2+\overline{p_nq_n}^2=\sum_{\nu=1}^{n-1}(x_\nu-y_\nu)^2+\overline{x_ny_n}^2$$

und aus 16.9(2) die Behauptung. □

Damit erhalten wir folgende Charakterisierung der Streckenkongruenz mit Hilfe von Koordinaten.

16.12 Satz.

$$\mathrm{Cd}\,\mathfrak{B}\,aa_1\ldots a_n \wedge \mathrm{Cd}\,\mathfrak{B}\,bb_1\ldots b_n \wedge \mathrm{Cd}\,\mathfrak{B}\,cc_1\ldots c_n \wedge \mathrm{Cd}\,\mathfrak{B}\,dd_1\ldots d_n$$

$$\rightarrow [ab\equiv cd \leftrightarrow \sum_{\nu=1}^{n}(a_\nu-b_\nu)^2=\sum_{\nu=1}^{n}(c_\nu-d_\nu)^2].$$

Beweis: Aus 15.2(3), (1) und den Rechengesetzen für angeordnete Körper ergibt sich die Gleichwertigkeit der Bedingungen $ab\equiv cd$, $\overline{ab}=\overline{cd}$ und $\overline{ab}^2=\overline{cd}^2$ und damit aus 16.11 die Behauptung. □

Die Zwischenbeziehung läßt sich folgendermaßen mit Hilfe von Koordinaten charakterisieren.

16.13 Satz.

$$\mathrm{Cd}\,\mathfrak{B}\,px_1\ldots x_n \wedge \mathrm{Cd}\,\mathfrak{B}\,qy_1\ldots y_n \wedge \mathrm{Cd}\,\mathfrak{B}\,rz_1\ldots z_n$$

$$\rightarrow \{\mathrm{B}pqr \leftrightarrow \exists t[\mathrm{Ar}_{oe}t \wedge o\leq t\leq e \wedge \bigwedge_{\nu=1}^{n} y_\nu-x_\nu=t\cdot(z_\nu-x_\nu)]\}.$$

Beweis: ($\rightarrow$): Sei $\mathrm{B}pqr$. Nach 15.4 ist $\overline{pq}+\overline{qr}=\overline{pr}$, nach 15.2(1) somit $o\leq\overline{pq}\leq\overline{pr}$. Sei t ein Punkt der Zahlengeraden mit $o\leq t\leq e$ und $\overline{pq}=t\cdot\overline{pr}$ (nach den Rechengesetzen für angeordnete Körper existiert ein solches t und ist eindeutig bestimmt, falls $p\neq r$; für $p=r$ hat jedes t zwischen o und e diese Eigenschaft). Seien die Punkte p_ν und die $(n-1)$-dimensionalen Unterräume U_ν ($\nu=1,\ldots,n$) für den Punkt p wie vorher gewählt und analog q_ν, V_ν für den Punkt q und r_ν, W_ν für den Punkt r (Abb. 116). Sei A_ν die Parallele zur ν-ten Achse durch p ($\nu=1,\ldots,n$). Dann steht auch A_ν senkrecht auf V_ν und W_ν und schneidet diese Unterräume in gewissen Punkten q'_ν bzw. r'_ν. Ist nun $\neg\,\mathrm{Col}\,prr'_\nu$, so ist $\overline{pr}\cdot\overline{pq'_\nu}=\overline{pr'_\nu}\cdot\overline{pq}=\overline{pr'_\nu}\cdot t\cdot\overline{pr}$ nach 15.5 und damit $\overline{pq'_\nu}=t\cdot\overline{pr'_\nu}$. Andernfalls ist $r=r'_\nu \wedge q=q'_\nu$ oder $p=q'_\nu=r'_\nu$ und damit ebenfalls $\overline{pq'_\nu}=t\cdot\overline{pr'_\nu}$. Wie im Beweis von 16.11 ergibt sich $(pq'_\nu r'_\nu)\equiv(p_\nu q_\nu r_\nu)$ und nach 16.10 auch $(p_\nu q_\nu r_\nu)\equiv(x_\nu y_\nu z_\nu)$. Somit ist $\overline{x_\nu y_\nu}=t\cdot\overline{x_\nu z_\nu}$, wegen 16.9(1) also

(1) $y_\nu - x_\nu = \pm t\cdot(z_\nu - x_\nu)$.

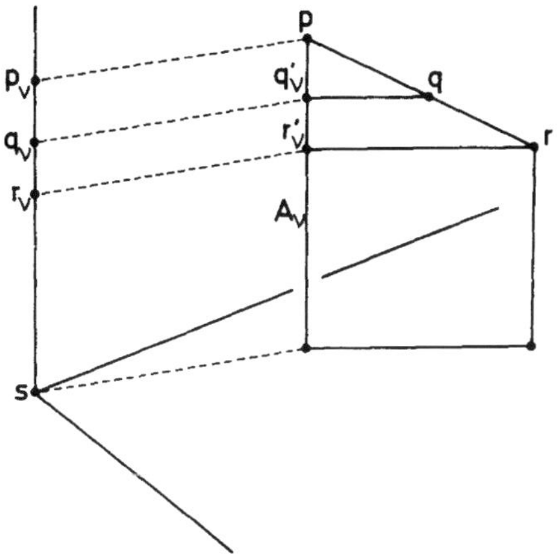

Abb. 116

Außerdem ergibt sich wie im Beweis von 16.11, daß q'_ν, r'_ν aus q bzw. r durch Parallelprojektion entstehen, und nach Lemma 14.35 somit $Bpq'_\nu r'_\nu$. Nach 4.6 ist dann auch $Bx_\nu y_\nu z_\nu$. Nach 14.35 bleibt die Zwischenbeziehung auch erhalten bei den Parallelprojektionen, die für Summenkonstruktionen

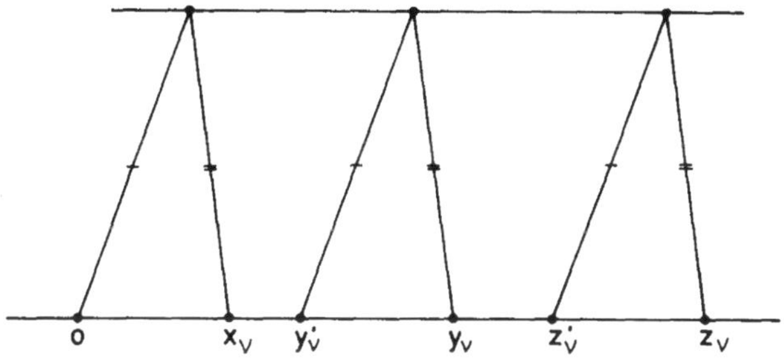

Abb. 117

verwendet werden; damit ist auch $Boy'_\nu z'_\nu$ für die Punkte $y'_\nu = y_\nu - x_\nu$ und $z'_\nu = z_\nu - x_\nu$ (Abb. 117). Somit ist $y'_\nu = o$, oder y'_ν und z'_ν haben dasselbe Vorzeichen. Aus (1) ergibt sich damit die gewünschte Beziehung $y_\nu - x_\nu = t\cdot(z_\nu - x_\nu)$.

($\leftarrow$): Sei t ein Punkt mit der auf der rechten Seite angegebenen Eigenschaft. Sei q^* der durch Streckenabtragung entstehende Punkt mit $\overline{pq^*} = t\cdot\overline{pr}$ und $q^* \underset{p}{=} r \vee q^* = p$. Wegen $o \leq t \leq e$ ist dann $\overline{pq^*} \leq \overline{pr}$, nach 15.2(4) also $pq^* \leq pr$ und nach 6.13 schließlich Bpq^*r. Für die Koordinaten y^*_ν von q^*

gelten nach dem schon bewiesenen Teil ebenfalls die Gleichungen $y^*_\nu - x_\nu = t\cdot(z_\nu - x_\nu)$. Somit ist $y^*_\nu = y_\nu$ $(\nu=1,\dots,n)$, nach 16.7 also $q^*=q$ und damit $Bpqr$. □

16.14 Anmerkung. In den Sätzen dieses Abschnitts kann der n-dimensionale Unterraum U, der vom Koordinatensystem aufgespannt wird, insbesondere die Menge aller Punkte sein, und zwar ist das genau dann der Fall, wenn das Dimensionsaxiom Dim_n gilt. Unter der Voraussetzung Dim_n kann man somit in 16.4 bis 16.7 die Bestandteile $\mathrm{Sb}_n U$, $U=\mathrm{Sb}_n(su_1\dots u_n)$, $p\in\mathrm{Sb}_n(su_1\dots u_n)$, $p\in U$ weglassen.

Damit erhalten wir nun den Darstellungssatz über die Charakterisierung der Modelle.

16.15 Darstellungssatz. (i) *Eine passende Relationalstruktur* $\mathfrak{A} = \langle A,D,B\rangle$ *(d.h. D ist eine vierstellige und B eine dreistellige Relation auf A) ist ein Modell von* A1,..., A7, Dim_n, A10 *genau dann, wenn sie isomorph ist zum n-dimensionalen kartesischen Raum* $\mathfrak{L}_n(\mathfrak{F})$ *über einem pythagoreischen angeordneten Körper* $\mathfrak{F}$ $(n\geq 2)$. *Statt* Dim_n *läßt sich dabei auch das gleichwertige Dimènsionsaxiom* $\mathrm{dim}_n := \mathrm{dim}_n^- \wedge \mathrm{dim}_n^+$ *verwenden (vgl.* 1.7.5, *Bem. zu* A8 *und* A9*).*

(ii) *Eine passende Relationalstruktur* $\mathfrak{A}$ *ist ein Modell von* A1 *bis* A10 *genau dann, wenn sie isomorph ist zur kartesischen Ebene (d.h. dem* 2*-dimensionalen kartesischen Raum)* $\mathfrak{L}_2(\mathfrak{F})$ *über einem pythagoreischen angeordneten Körper* $\mathfrak{F}$.

(iii) *In* (i) *und* (ii) *ist* $\mathfrak{F}$ *durch* $\mathfrak{A}$ *bis auf Isomorphie eindeutig bestimmt.*

(iv) *Seien* $\mathfrak{A}$ *und* $\mathfrak{F}$ *derart, daß beide Seiten von* (i) *oder* (ii) *gelten. Dann gelten in* $\mathfrak{A}$ *die Zusatzaxiome* A11, A11' *bzw.* CA *genau dann, wenn* $\mathfrak{F}$ *isomorph zu* $\mathbb{R}$, $\mathfrak{F}$ *reell-abgeschlossen bzw.* $\mathfrak{F}$ *euklidisch ist.*

Beweis: Zu (i): (←): Gelte die rechte Seite. Nach 1.6 und 11.75 ist $\mathfrak{L}_n(\mathfrak{F})$ ein Modell der genannten Axiome und damit auch die dazu isomorphe Struktur $\mathfrak{A}$.

(→): In $\mathfrak{A}$ seien o und e verschiedene Punkte und s, $u_1,\dots, u_n$ Punkte, die in bezug auf o und e ein n-dimensionales kartesisches Koordinatensystem bilden (solche Punkte existieren nach 3.13, 16.4). Sei $\mathfrak{F}$ die durch o und e bestimmte Zahlengerade, also nach 14.40 ein angeordneter

Körper. Sei φ die Abbildung, die jedem Punkt von $\mathfrak{A}$ das n-tupel seiner Koordinaten (in bezug auf das betrachtete Koordinatensystem) zuordnet. Nach 16.7 ist φ eine Bijektion von (der Trägermenge von) $\mathfrak{A}$ auf (die Trägermenge von) $\mathfrak{C}_n(\mathfrak{f})$. 16.12 und 16.13 besagen, daß sich die Streckenkongruenz und die Zwischenbeziehung durch die Koordinaten der jeweils betrachteten Punkte gerade so beschreiben lassen, wie es in der Definition kartesischer Räume (1.4) verlangt ist. Somit ist φ ein Isomorphismus von $\mathfrak{A}$ auf $\mathfrak{C}_n(\mathfrak{f})$. Mit $\mathfrak{A}$ ist auch $\mathfrak{C}_n(\mathfrak{f})$ ein Modell der genannten Axiome; nach 1.6 ist also $\mathfrak{f}$ pythagoreisch.

(ii) ergibt sich sofort aus (i) mit $n=2$ und der Gleichwertigkeit der benutzten Dimensionsaxiome (11.71, 72).

Zu (iii): Ist $\mathfrak{A}$ isomorph zu $\mathfrak{C}_n(\mathfrak{f})$ und zu $\mathfrak{C}_n(\mathfrak{f}')$, so sind $\mathfrak{f}$ und $\mathfrak{f}'$ nach 14.41 isomorph zu jeder Zahlengeraden in dem betreffenden Raum und damit auch isomorph zu jeder Zahlengeraden in $\mathfrak{A}$.

(iv) ergibt sich aus der vorausgesetzten Isomorphie und 1.6. □

Teil II: Metamathematische Betrachtungen

von Wolfram Schwabhäuser

1. Hilfsmittel aus der mathematischen Logik

1.1 Einführung. Schon beim axiomatischen Aufbau der Geometrie im ersten Teil dieses Buches wurde eine gewisse Formalisierung zum bequemen Aufschreiben geometrischer Sätze verwendet (vgl. I.0.5). Bei den nun folgenden metamathematischen Betrachtungen geht es darum, daß die Sätze einer gegebenen Theorie T (meist geometrische Sätze) zum Gegenstand einer neuen Theorie, der sog. *Metatheorie* zu T gemacht werden. Ein klassisches Beispiel (wesentlich älter als das Wort Metatheorie) ist das bekannte Dualitätstheorem für die projektive Geometrie. Dieses besagt: Wenn eine Aussage ein Satz der projektiven Geometrie ist, so ist eine weitere Aussage, die durch gewisse sprachliche Umformungen (Vertauschen der Worte "Punkt" und "Gerade" und dergleichen) aus der ersten Aussage entsteht, wieder ein Satz der projektiven Geometrie. Das Dualitätstheorem sagt also nicht etwas über Punkte und Geraden (wie die Sätze der projektiven Geometrie), sondern etwas über Aussagen, die sich im Rahmen der projektiven Geometrie formulieren lassen. In diesem Sinne ist es ein Satz der Metatheorie zur projektiven Geometrie. Zu metamathematischen Betrachtungen gehören insbesondere Untersuchungen über Widerspruchsfreiheit und Vollständigkeit von Axiomensystemen und Theorien, Entscheidbarkeit, Definierbarkeit, Axiomatisierbarkeit durch Axiomensysteme mit vorgegebenen Eigenschaften und vieles andere.

Die Sprache der gegebenen Theorie wird somit zum Gegenstand der Untersuchung in der Metatheorie, sie wird daher auch *Objektsprache* genannt. Sie ist zu unterscheiden von der *Metasprache*, d.h. der Sprache der Metatheorie, in der man Aussagen über Gebilde der Objektsprache machen kann. Um metamathematische Untersuchungen mit mathematischer Strenge durchführen zu können, muß natürlich die Objektsprache als Gegenstand der Metatheorie präzisiert werden. Das läuft darauf hinaus, genau abzugrenzen, welche sprachlichen Gebilde noch zur Objektsprache gehören und welche nicht. Diese Präzisierung erfolgt durch die in der mathematischen

Logik übliche Formalisierung. Die Formalisierung ist also in diesem Zusammenhang nicht nur eine bequeme Abkürzungstechnik, sondern ein unerläßliches Handwerkszeug zur Präzisierung des Gegenstandes der Metatheorie. Solange man keine Metatheorie treibt, ist dagegen die Abgrenzung der Sprache einer Theorie im allgemeinen nicht erforderlich und oft sogar unerwünscht, da man sich die Möglichkeit zur Formulierung neuer Ideen nicht von vornherein beschneiden möchte. Diese Situation liegt bei uns für die Metatheorie vor (da wir keine Metatheorie zur Metatheorie treiben). Als Metasprache verwenden wir daher einfach die Umgangssprache (hier Deutsch), wie üblich angereichert mit Formelzeichen, die von den Zeichen der Objektsprache zu unterscheiden sind.

1.2 Überblick. In diesem Abschnitt werden die verwendeten Begriffe und einige grundlegende Resultate aus der mathematischen Logik zusammengestellt. Für Leser ohne entsprechende Vorkenntnisse soll damit zumindest eine Übersicht gegeben werden, die es erlaubt, den weiteren Abschnitten zu folgen. Speziellere Resultate aus der mathematischen Logik werden dagegen erst später zitiert an den Stellen, wo sie jeweils benutzt werden. Ausgeführt wird hier vor allem die sogenannte "Standardformalisierung" (Tarski, Mostowski u. Robinson 1953) im Rahmen der Prädikatenlogik der ersten Stufe, die für die meisten der späteren Anwendungen verwendet wird, und die Interpretation der zugehörigen formalisierten (Objekt-)Sprachen in passenden Strukturen. Für die gelegentlich verwendeten anderen Formalisierungen wird angegeben, was abzuändern ist. Diese Zusammenstellung dient auch zur Festlegung der in der Literatur nicht ganz einheitlichen Bezeichnungen. Leser mit hinreichenden Kenntnissen in Prädikatenlogik können diesen Abschnitt überspringen und bei Bedarf zum Nachschlagen verwenden.

1.3 Formalisierte Sprachen im Rahmen der Prädikatenlogik der ersten Stufe. Die *Grundzeichen* einer solchen Sprache sind die folgenden unter 1. bis 3. angegebenen.

1.a) Die *Operationszeichen* (oder *Funktionszeichen*), die einer gegebenen Menge F angehören.

b) Die *Relationszeichen*, die einer gegebenen Menge R angehören.

Jedem Operationszeichen $f \in F$ und jedem Relationszeichen $R \in R$ sei eine *Stellenzahl* $r(f)$ bzw. $r(R)$ zugeordnet, die eine natürliche Zahl ist. (Eine entsprechende Stellenzahlfunktion r von $F \cup R$ in die Menge $\mathbb{N}$ der natürlichen Zahlen [einschließlich Null] sei ebenfalls gegeben.) Nullstellige Operationszeichen heißen auch *Individuenkonstanten*.

2. Die *Variablen* oder auch *Individuenvariablen*, die einer gegebenen abzählbaren Menge V angehören. Wir bezeichnen sie mit x, y,... (kleine Buchstaben in gerader Schrift).

3. Die endlich vielen weiteren Grundzeichen $\doteq$, ¬, ∧, ∨, →, ↔, ∀, ∃, (,) (*logische Zeichen* für "gleich", "nicht", "und", "oder", "wenn - so", "genau dann, wenn (gdw)", "für jedes", "es gibt ein" und die Klammern als *technische Zeichen*). Dabei heißt $\doteq$ das *Gleichheitszeichen* (der Objektsprache), die Zeichen ¬, ∧, ∨, →, ↔, werden *Junktoren* genannt und die Zeichen ∀, ∃ die *Quantoren*, und zwar ∀ der *Allquantor* (oder *Generalisator*) und ∃ der *Existenzquantor* (oder *Partikularisator*). In der Literatur finden sich häufig auch andere Bezeichnungen für die logischen Zeichen, insbesondere ~ statt ¬, & statt ∧, ⊃ statt →, ≡ oder ~ statt ↔, Λ statt ∀, V statt ∃ sowie = statt $\doteq$ (der Punkt wird auch hier bisweilen weggelassen, insbesondere in II.4).

In vielen Anwendungen sind die Mengen *F* und *R* endlich (evtl. leer). Andererseits werden für manche Betrachtungen (insbesondere in der Modelltheorie) auch Sprachen mit abzählbar oder überabzählbar vielen Operations- und Relationszeichen gebraucht. Wir machen daher keine einschränkenden Voraussetzungen über die Mächtigkeiten von *F* und *R* und über die Natur der Abbildung r.

Trotz des Namens "Zeichen" sind die Grundzeichen nicht zu verwechseln mit den Figuren, die wir dafür aufs Papier schreiben. Diese Figuren sind vielmehr Bezeichnungen für die (irgendwie abstrakt gedachten) Grundzeichen. Gleichgestaltete Figuren, wie etwa "→" und "→" sollen natürlich stets dasselbe Grundzeichen bezeichnen. In genügend einfachen Fällen kann man sich die Grundzeichen als die Formen dieser Figuren vorstellen. Diese Vorstellung versagt bei zu großen Mächtigkeiten von *F* und *R*. Es spielt aber für das Folgende gar keine Rolle, was für Dinge die Grundzeichen sind. Wir können sie daher ansehen als Elemente gegebener Mengen, und der weitere Aufbau ist eine mengentheoretische Konstruktion, für die wir nur folgende Voraussetzung benötigen:

(*) Die Mengen *F*, *R* und V sind paarweise disjunkt und enthalten nicht die unter 3. angegebenen endlich vielen weiteren (paarweise verschiedenen) Grundzeichen.

Wenn wir verschiedene Sprachen nebeneinander betrachten, halten wir nicht nur die Zeichen unter 3., sondern (im allgemeinen) auch die Menge V der Variablen fest. Eine Sprache L soll nun festgelegt sein durch ihren Zeichenvorrat mit der oben vorgenommenen Klassifikation, also schon durch die Angabe von F, R und r. Wir definieren daher eine *Sprache* (oder einen *Sprachrahmen*) als ein Tripel $L = \langle F,R,r\rangle$, wobei F, R, r von der angegebenen Art sind; diese Komponenten bezeichnen wir dann auch mit F_L, R_L, r_L. Man sieht leicht, daß eine Sprache eindeutig festgelegt ist durch die Menge ihrer Formeln (s. unten, 1.4), daher wird sie manchmal auch mit der Menge ihrer Formeln identifiziert. Sind F und R endlich, etwa $F = \{f_1,\ldots,f_k\}$, $R = \{R_1,\ldots,R_l\}$ (und r bekannt), so bezeichnen wir die so festgelegte Sprache zur Abkürzung auch mit $L(f_1,\ldots,f_k;R_1,\ldots,R_l)$. Ein Beispiel dafür ist die Sprache L(D,B) für die absolute (oder euklidische) Geometrie (mit $F = \emptyset$, $R = \{D,B\}$, $r(D)=4$, $r(B)=3$), die im wesentlichen schon in Teil I verwendet wurde (vgl. 1.17(ii), 28(v)).

<u>1.4 Definition</u>. Seien $L_i = \langle F_i,R_i,r_i\rangle$ ($i=1,2$) Sprachen der betrachteten Art. Dann heißt L_1 eine *Teilsprache* von L_2 oder L_2 eine *Erweiterungssprache* von L_1 (abgekürzt $L_1 \subseteq L_2$) falls gilt: $F_1 \subseteq F_2$, $R_1 \subseteq R_2$ und $r_1 = r_2|F_1 \cup R_1$ (d.h., r_1 ist die Einschränkung von r_2 auf den für L_1 gegebenen Definitionsbereich).

Unter *Zeichenreihen* verstehen wir endliche Folgen von Grundzeichen, m.a.W. Wörter über der (unendlichen) Menge der Grundzeichen als Alphabet; sie werden durch einfaches Hintereinanderschreiben der betreffenden Grundzeichen (Folgenglieder) bezeichnet.

Die sinnvollen sprachlichen Gebilde - Terme und Formeln - führen wir nun als spezielle Zeichenreihen ein. Die Terme sollen (bei einer Interpretation der unter 1. und 2. genannten Grundzeichen, s. 1.13ff.) Elemente einer gegebenen Menge oder Struktur bezeichen; die Formeln sollen etwas ausdrücken, was (bei geeigneter Interpretation) wahr oder falsch werden kann ("Aussageformen").

<u>1.5 Definition (Terme und Formeln)</u>. Als *Terme* der Sprache $L = \langle F,R,r\rangle$ bezeichnen wir diejenigen Zeichenreihen, die auf Grund der folgenden Regeln gebildet werden können:

(T1) Die Variablen sind Terme.

(T2) Wenn $f \in F$ und $\tau_1,\ldots,\tau_{r(f)}$ Terme sind, so ist auch $f\tau_1\ldots\tau_{r(f)}$ ein Term.

Als *Formeln* dieser Sprache bezeichnen wir diejenigen Zeichenreihen, die auf Grund der folgenden Regeln gebildet werden können:

(F1) Die Zeichenreihen der Form $R\tau_1 \dots \tau_{r(R)}$, wobei $R \in \mathcal{R}$ und $\tau_1, \dots, \tau_{r(R)}$ Terme, und die der Form $\sigma \doteq \tau$ (*Gleichungen*), wobei σ, τ Terme sind, sind Formeln. Diese speziellen Formeln heißen *atomare* oder *Primformeln*.

(F2) (a) Wenn α, β Formeln sind, so sind auch $\neg\alpha$, $(\alpha \wedge \beta)$, $(\alpha \vee \beta)$, $(\alpha \rightarrow \beta)$, $(\alpha \leftrightarrow \beta)$ Formeln.

(b) Wenn α eine Formel und $x \in V$ ist, so sind auch $\forall x \alpha$ und $\exists x \alpha$ Formeln.

Die gemäß (F2)(a) gebildeten Formeln heißen der Reihe nach die *Negation* von α, die *Konjunktion*, *Disjunktion*, *Implikation* bzw. *Äquivalenz* von α und β; die gemäß (F2)(b) gebildeten Formeln heißen die *Quantifizierungen*, und zwar die *Generalisierung* und *Partikularisierung* von α mit (oder bezüglich) der Variablen x.

Die Menge der Formeln von L bezeichnen wir mit Fml_L oder $Fml(\mathcal{F},\mathcal{R},r)$, im obigen Spezialfall für L auch mit $Fml(f_1,\dots,f_k;R_1,\dots,R_l)$. Entsprechend verwenden wir Tm_L und analoge andere Bezeichnungen für die Menge der Terme von L (die Angabe von $\mathcal{R}$ kann hier entfallen, da Tm_L davon offenbar nicht abhängt).

Beispiele. Seien $c, f \in \mathcal{F}$, $R \in \mathcal{R}$, $r(c)=0$, $r(f)=2$, $r(R)=2$, $x,y,z \in V$. Dann sind etwa die Zeichenreihen

x, c, fyc, $fxfyz$ und $ffxyz$

Terme (der Aufbau der letzten beiden aus einfacheren Termen ist durch die Bogen angedeutet); die Zeichenreihen

$\forall x(Rcx \rightarrow \exists y fyy \doteq x)$ und $\forall x \forall y \forall z fxfyz \doteq ffxyz$

sind Formeln.

Beim Aufschreiben einzelner Terme und Formeln werden wir in Anlehnung an den mathematischen Sprachgebrauch auch andere Bezeichnungen verwenden. Zweistellige Relationszeichen schreiben wir auch zwischen die betreffenden Terme, ebenso zweistellige Operationszeichen, gegebenenfalls unter Verwendung von Klammern beim Aufbau komplizierterer Terme. Schreiben wir zum Beispiel 0 statt c, $\sigma < \tau$ statt $R\sigma\tau$ und $\sigma \cdot \tau$ statt $f\sigma\tau$, so erhalten die Formeln des obigen Beispiels die Form $\forall x(0<x \longrightarrow \exists y\, y \cdot y \doteq x)$ und $\forall x \forall y \forall z\, x \cdot (y \cdot z) \doteq (x \cdot y) \cdot z$. Zur Ersparung von Klammern beim Aufschrei-

ben von Formeln lassen wir die Außenklammern weg. Außerdem verwenden wir Punkte in Verbindung mit Junktoren als Trennzeichen; ein mit mehr Punkten versehener Junktor soll dann stärker trennen als ein mit weniger Punkten versehener. Schließlich verabreden wir, daß ein späterer Junktor in der Reihenfolge $\wedge$, $\vee$, $\rightarrow$, $\leftrightarrow$ stets stärker trennt als ein früherer (der mit ebenso vielen Punkten versehen ist). Zum Beispiel bedeutet

$$\alpha\rightarrow\beta\cdot\rightarrow\cdot\alpha\rightarrow\gamma\rightarrow\cdot\alpha\rightarrow\beta\wedge\gamma$$

dasselbe wie

$$((\alpha\rightarrow\beta)\rightarrow((\alpha\rightarrow\gamma)\rightarrow(\alpha\rightarrow(\beta\wedge\gamma)))).$$

Zur besseren Übersicht verwenden wir auch andere Sorten von Klammern. Man kann das alles auffassen als Einführung von Abkürzungen oder bequemeren Bezeichnungen für Terme und Formeln, die nach den obigen (für eine systematische Behandlung besser geeigneten) Regeln gebildet sind.

Im folgenden sollen die Buchstaben α, β,... (kleine griechische Buchstaben vom Anfang des Alphabets) stets Formeln bedeuten, σ und τ Terme, f ein Operationszeichen, R ein Relationszeichen (ebenso mit angehängten Indizes). Mengen von Formeln werden mit großen griechischen Buchstaben (Σ, T, Γ u.a.) bezeichnet.

In der Literatur sind auch andere (präzisere und untereinander gleichwertige) Definitionen für die Begriffe Term und Formel üblich, auf die hier nicht näher eingegangen wird. In jedem Falle ergibt sich leicht das folgende Induktionsprinzip.

1.6 Satz (Beweise durch Induktion über den Term- bzw. Formelaufbau).

Sei L wie oben.

(IT) *Um zu beweisen, daß eine Behauptung* B(τ) *für jeden Term* τ *gilt, genügt es zu zeigen:*

(IT1) *Es gilt* B(x) *für jede Variable* x.

(IT2) *Für beliebiges* $f\in F$ *und beliebige Terme* $\tau_1,\ldots,\tau_{r(f)}$: *Wenn* $B(\tau_1),\ldots,B(\tau_{r(f)})$, *so gilt auch* $B(f\tau_1\ldots\tau_{r(f)})$.

(IF) *Um zu beweisen, daß eine Behauptung* B(α) *für jede Formel* α *gilt, genügt es zu zeigen:*

(IF1) *Es gilt* B(α) *für jede Primformel* α.

(IF2) *Für beliebige Formeln* α, β *und eine beliebige Variable* x: *Wenn* B(α) *und* B(β), *so gilt auch* $B(\neg\alpha)$, $B(\alpha\wedge\beta)$, $B(\alpha\vee\beta)$, $B(\alpha\rightarrow\beta)$, $B(\alpha\leftrightarrow\beta)$, $B(\forall x\alpha)$ *und* $B(\exists x\alpha)$.

Man nennt (IT1) bzw. (IF1) auch den *Anfangsschritt* und (IT2) bzw. (IF2) den *Induktionsschritt* eines nach diesem Prinzip geführten Induktionsbeweises.

<u>1.7 Induktive Definitionen</u>. Die Analogie von 1.6 zu Induktionsbeweisen für natürliche Zahlen ist offensichtlich. Ebenso läßt sich der Dedekindsche Rechtfertigungssatz für induktive Definitionen im Bereich der natürlichen Zahlen ("Rekursionssatz", vgl. etwa Oberschelp 1968, S. 18ff. oder Felscher 1978, S. 133ff.) auf den Bereich der Terme und Formeln übertragen, d.h., es läßt sich zeigen, daß man Funktionen von Tm_L bzw. Fml_L in eine beliebige Menge "durch Induktion über den Term- bzw. Formelaufbau" definieren kann. Wir verzichten hier auf die allgemeine Formulierung eines entsprechenden Rekursionssatzes, der besagt, daß es zu induktiven Definitionen, wie wir sie im folgenden verwenden (z.B. 1.15, 1.16), stets genau eine Funktion mit den in der Definition angegebenen Eigenschaften gibt. Es sei nur darauf hingewiesen, daß dieser Satz eines Beweises bedarf. Als ein wesentliches Hilfsmittel wird dazu noch gebraucht, daß sich ein Term bzw. eine Formel nur auf eine Weise aus einfacheren Termen bzw. Formeln aufbauen läßt (zum Beweis dazu vgl. etwa Hermes 1969, S. 57ff.). Das läuft auf folgende Feststellung hinaus: Die Menge der Formeln bildet mit den logischen Operationen (d.h. den Zusammensetzungen gemäß (F2)) eine "absolut freie Algebra" (vgl. etwa Grätzer 1968, S. 91) mit der Menge der Primformeln (F1) als Basis. Entsprechendes gilt für die Menge der Terme. Der erwähnte Rekursionssatz läßt sich für beliebige absolut freie Algebren formulieren und beweisen.

<u>1.8 Gebundene und freie Variablen</u>. Wenn in einer Formel α an einer gewissen Stelle ein Quantor vorkommt, so heißt die (notwendigerweise) unmittelbar darauf folgende Variable (nennen wir sie) x an dieser Stelle *quantifiziert*, und die unmittelbar danach beginnende (eindeutig bestimmte) Teilformel von α heißt der *Wirkungsbereich* dieses Quantors mit der Variablen x in α. Man sagt, daß eine Variable x an einer gewissen Stelle in einer Formel α *gebunden* vorkommt, falls sie dort quantifiziert oder im Wirkungsbereich eines Quantors mit der(selben) Variablen x vorkommt. Man sagt, daß x an einer Stelle in α *frei* vorkommt, falls sie dort vorkommt, aber dort nicht gebunden vorkommt.

<u>Beispiel</u>. Seien x, y, z paarweise verschiedene Variablen und α die Formel

$$\exists z\, \underbrace{x+z \doteq y} \wedge \forall y\, \underbrace{(1<y \wedge y<x \rightarrow \neg\exists z\, \underbrace{y\cdot z \doteq x})}$$

(einer geeigneten Sprache, vgl. Beisp. zu 1.5). Die Bogen markieren dann die Wirkungsbereiche sämtlicher Quantoren in α. Die Variable z kommt in α nur (an allen Stellen, wo sie vorkommt) gebunden vor, die Variable x kommt in α nur frei vor, die Variable y kommt in α sowohl frei (nämlich an genau einer Stelle im ersten Konjunktionsglied) als auch gebunden (nämlich an genau vier Stellen im zweiten Konjunktionsglied) vor.

Mit $Fr(\alpha)$ wird die (stets endliche) Menge der in der Formel α frei vorkommenden Variablen bezeichnet. $Fr(\alpha)$ läßt sich auch einfacher (ohne Berücksichtigung der Stelle des Vorkommens der Variablen) durch eine induktive Definition über den Formelaufbau von α charakterisieren (diese findet sich in vielen Darstellungen, z.B. [mit denselben Bezeichnungen] in Schwabhäuser 1972a, woraus hier einiges übernommen wurde).

Ist $K \subseteq F_L \cup R_L$, so bezeichnen wir mit Fml_K die Menge derjenigen Formeln, in denen nur Operations- und Relationszeichen aus K vorkommen, und mit $Fml_K(x_1,\ldots,x_n)$ die Menge derjenigen solchen Formeln, in denen außerdem nur (höchstens) $x_1,\ldots, x_n$ als freie Variablen vorkommen.

Unter *Aussagen* (engl. "sentences") oder *abgeschlossenen Formeln* verstehen wir Formeln α ohne freie Variablen (d.h. mit $Fr(\alpha)=\emptyset$); die Menge der Aussagen von L bezeichnen wir mit Aus_L.

1.9 Substitution. Sei α eine Formel, x eine Variable und τ ein Term. Man sagt, daß die *Substitution* (*Einsetzung*) des Terms τ für die Variable x in α *ausführbar* ist, falls folgende Bedingung (*) erfüllt ist.

(*) Für jede in τ vorkommende Variable z gilt: Die Variable x steht an allen Stellen, wo sie in α frei vorkommt, nicht im Wirkungsbereich eines Quantors mit der Variablen z.

Unter dem *Resultat* $\alpha^x/_\tau$ dieser Substitution versteht man dann die Formel (Zeichenreihe), die aus α dadurch entsteht, daß die Variable x an allen Stellen, wo sie in α frei vorkommt, durch den Term τ ersetzt wird. (Eine Verletzung der Bedingung (*) bei einer solchen formalen Ersetzung wird auch als eine *Variablenkonfusion* bezeichnet.)

Beispiel. Sei α wie im Beispiel zu 1.8. Für τ = x+y ist $\alpha^y/_\tau$ die Formel

$$\exists z\; x+z \doteq x+y \wedge \forall y (1<y \wedge y<x \rightarrow \neg \exists z\; y\cdot z \doteq x);$$

für τ = x+z+1 ist die Substitution von τ für y in α nicht ausführbar;

für jeden Term τ ist $\alpha^{z}/_{\tau} = \alpha$ und auch $\alpha^{u}/_{\tau} = \alpha$ für eine beliebige von x und y verschiedene Variable u.

Der Begriff Substitution läßt sich auch durch ein System von Regeln charakterisieren (analog zur Charakterisierung des Begriffs Formel durch die Regeln (F1), (F2) in 1.5); vgl. dazu Hermes 1969, S. 65ff. u. S. 158, wo auch strenge Beweise von Sätzen über die Substitution auf Grund einer solchen Charakterisierung ausgeführt sind.

Sind α, x, τ wie oben beliebig gegeben (ohne die Voraussetzung (*)), so kann man von α durch "Umbenennung gebundener Variablen" zu einer (dazu logisch äquivalenten, s. 1.30) Formel β übergehen, in der die Substitution von τ für x ausführbar ist. Das Resultat $\beta^{x}/_{\tau}$ heißt dann auch das Resultat einer *verallgemeinerten Substitution* von τ für x in α (die mithin stets ausführbar ist) und wird manchmal mit $\alpha^{x}//_{\tau}$ oder auch mit $\alpha^{x}/_{\tau}$ bezeichnet (vgl. Hermes 1969, S. 114ff.).

1.10 Argumentbezeichnung. Manchmal werden in der Bezeichnung von Formeln (freie) Variablen, die in weiteren Betrachtungen eine Rolle spielen, "wie Argumente" angegeben. Insbesondere wird die Bezeichnung $\alpha(x)$ für eine beliebige Formel α verwendet, um das Substitutionsresultat $\alpha^{x}/_{\tau}$ kürzer mit $\alpha(\tau)$ zu bezeichnen; analog mit mehreren Variablen. Häufig wird die Bezeichnung $\alpha(x_1,\ldots,x_n)$ auch verwendet zur Bezeichnung einer Formel, in der höchstens die Variablen $x_1,\ldots, x_n$ frei vorkommen.

1.11 Spezielle Formeln. Die schon in Teil I benutzten Bezeichnungen für allgemeine Konjunktionen und Disjunktionen, (paarweise) Verschiedenheit und Anzahlfeststellungen bleiben natürlich sinnvoll für den jetzt eingeführten Formelbegriff, und wir verwenden sie auch als Abkürzungen für entsprechende spezielle Formeln (s. I.0.5). Die Formeln $\overset{\geq k}{\exists} x\alpha(x)$, $\overset{\leq k}{\exists} x\alpha(x)$ bzw. $\overset{=k}{\exists} x\alpha(x)$ nennen wir dabei *Mindest-*, *Höchst-* bzw. *Anzahlformeln*; analog bilden wir *Mindest-*, *Höchst-* und *Anzahlaussagen* $\overset{\geq k}{\exists}$, $\overset{\leq k}{\exists}$, $\overset{=k}{\exists}$.

1.12 Semantik. Wir kommen nun zur "*Semantik*" (ursprünglich Wortbedeutungslehre) für die eingeführte Sprache, d.h. zur Einführung einer "Bedeutung" oder Interpretation der sprachlichen Gebilde. Dagegen hatte das Bisherige noch nichts mit einer Bedeutung zu tun und gehörte damit zur "*Syntax*" (ursprünglich Lehre von der richtigen [regelgerechten] Fügung der Wörter und Sätze).

1.13 Definition. Sei $L = \langle F,R,r\rangle$ eine Sprache der betrachteten Art. Eine *Struktur* $\mathfrak{A}$ für L soll festgelegt sein durch folgende Bestandteile:

1. Eine nichtleere Menge $U_{\mathfrak{A}}$; diese heißt die *Trägermenge* oder der *Individuenbereich* von $\mathfrak{A}$ (engl. "universe"), die Elemente dieser Menge heissen auch *Individuen* oder *Elemente der Struktur* $\mathfrak{A}$, dementsprechend schreiben wir auch $x \in \mathfrak{A}$ für $x \in U_{\mathfrak{A}}$.

2. Für jedes Operationszeichen $f \in F$ eine $r(f)$-stellige Operation (Funktion) $f_{\mathfrak{A}}$ auf $U_{\mathfrak{A}}$ (d.h. eine eindeutige Abbildung von $U_{\mathfrak{A}}^{r(f)}$ in $U_{\mathfrak{A}}$).

3. Für jedes Relationszeichen $R \in R$ eine $r(R)$-stellige Relation $R_{\mathfrak{A}}$ auf $U_{\mathfrak{A}}$ (d.h. eine Teilmenge von $U_{\mathfrak{A}}^{r(R)}$).

Eine solche Struktur kann also allgemein definiert werden als ein Paar $\langle U_{\mathfrak{A}}, s_{\mathfrak{A}}\rangle$, wobei $s_{\mathfrak{A}}$ eine ("Struktur"-)Funktion mit dem Definitionsbereich $F \cup R$ ist, so daß $U_{\mathfrak{A}}$ und die Bestandteile $f_{\mathfrak{A}} := s_{\mathfrak{A}}(f)$, $R_{\mathfrak{A}} := s_{\mathfrak{A}}(R)$ von der in 1. bis 3. angegebenen Art sind. In Spezialfällen sind auch andere Bezeichnungen und eventuell Definitionen üblich. Insbesondere wird eine Struktur für die Sprache am Schluß von 1.3 (mit endlichen Mengen F und R) oft angegeben in der Form

$$\mathfrak{A} = \langle U_{\mathfrak{A}}; (f_1)_{\mathfrak{A}}, \ldots, (f_k)_{\mathfrak{A}}; (R_1)_{\mathfrak{A}}, \ldots, (R_l)_{\mathfrak{A}}\rangle;$$

sie läßt sich als $(1+k+l)$-tupel auffassen (wodurch ebenfalls die in 1. bis 3. angegebenen Bestandteile festgelegt sind).

Beispiel. Die in I.1.4 eingeführten kartesischen Räume $\mathfrak{A} = \mathfrak{L}_n(\mathfrak{F})$ werden zu Strukturen für die Sprache L(D,B), indem wir setzen:

$$U_{\mathfrak{A}} = F^n,\ D_{\mathfrak{A}} = D_{\mathfrak{F}}^n,\ B_{\mathfrak{A}} = B_{\mathfrak{F}}^n.$$

1.14 Definition. Eine *Belegung* h (der Variablen) über der Struktur $\mathfrak{A}$ (oder auch über $U_{\mathfrak{A}}$) ist eine Funktion (eindeutige Abbildung) von V in $U_{\mathfrak{A}}$.

Im folgenden soll $\mathfrak{A}$ jeweils eine Struktur für die Sprache L bedeuten, deren Terme und Formeln betrachtet werden, und h eine Belegung über $\mathfrak{A}$.

1.15 Definition. Der (stets in $U_{\mathfrak{A}}$ liegende) *Wert* $w_{\mathfrak{A}}(\tau,h)$ eines Terms τ von L in der Struktur $\mathfrak{A}$ (für L) bei einer Belegung h (über $\mathfrak{A}$) wird gegeben durch folgende induktive Definition (über den Termaufbau) für die Wertfunktion $w_{\mathfrak{A}}$:

(1) $w_{\mathfrak{A}}(x,h) = h(x)$ für jede Variable $x \in V$.

(2) $w_{\mathfrak{A}}(f\tau_1 \ldots \tau_{r(f)}, h) = f_{\mathfrak{A}} w_{\mathfrak{A}}(\tau_1,h) \ldots w_{\mathfrak{A}}(\tau_{r(f)},h)$ für $f \in F$.

1.16 Definition. Für das Folgende sei $\{W,F\}$ eine zweielementige Menge, deren Elemente W ("wahr") und F ("falsch") wir *Wahrheitswerte* nennen (es spielt keine Rolle, was für Elemente das sind). Der *Wahrheitswert* $W_{\mathfrak{A}}(\alpha,h)$ *einer Formel* α von L in der Struktur $\mathfrak{A}$ bei einer Belegung h wird festgelegt durch die folgende induktive Definition über den Formelaufbau.

(1) $$W_{\mathfrak{A}}(R\tau_1\ldots\tau_{r(R)},h)=\begin{cases} W, & \text{falls } R_{\mathfrak{A}}w_{\mathfrak{A}}(\tau_1,h)\ldots w_{\mathfrak{A}}(\tau_{r(R)},h) \\ & \text{(d.h., die Relation } R_{\mathfrak{A}} \text{ trifft zu auf} \\ & \text{die Individuen } w_{\mathfrak{A}}(\tau_1,h),\ldots,\ w_{\mathfrak{A}}(\tau_{r(R)},h)), \\ F & \text{sonst.}\end{cases}$$

$$W_{\mathfrak{A}}(\sigma\doteq\tau,h) = \begin{cases} W, & \text{falls } w_{\mathfrak{A}}(\sigma,h)=w_{\mathfrak{A}}(\tau,h), \\ F & \text{sonst.}\end{cases}$$

(2) $$W_{\mathfrak{A}}(\neg\alpha,h) = \mathrm{non}(W_{\mathfrak{A}}(\alpha,h)).$$
$$W_{\mathfrak{A}}(\alpha\wedge\beta,h) = \mathrm{et}(W_{\mathfrak{A}}(\alpha,h),W_{\mathfrak{A}}(\beta,h)).$$
$$W_{\mathfrak{A}}(\alpha\vee\beta,h) = \mathrm{vel}(W_{\mathfrak{A}}(\alpha,h),W_{\mathfrak{A}}(\beta,h)).$$
$$W_{\mathfrak{A}}(\alpha\to\beta,h) = \mathrm{seq}(W_{\mathfrak{A}}(\alpha,h),W_{\mathfrak{A}}(\beta,h)).$$
$$W_{\mathfrak{A}}(\alpha\leftrightarrow\beta,h) = \mathrm{äq}(W_{\mathfrak{A}}(\alpha,h),W_{\mathfrak{A}}(\beta,h)).$$
$$W_{\mathfrak{A}}(\forall x\alpha,h) = \min_{x\in U_{\mathfrak{A}}} W_{\mathfrak{A}}(\alpha,h^x_x).$$
$$W_{\mathfrak{A}}(\exists x\alpha,h) = \max_{x\in U_{\mathfrak{A}}} W_{\mathfrak{A}}(\alpha,h^x_x).$$

Hierbei sind die Funktionen non, et, vel, seq, äq auf $\{W,F\}$ durch die üblichen "Wahrheitstafeln" erklärt:

non	
W	F
F	W

et	W	F
W	W	F
F	F	F

vel	W	F
W	W	W
F	W	F

seq	W	F
W	W	F
F	W	W

äq	W	F
W	W	F
F	F	W

(also für die Funktion seq z.B. seq(W,F)=F, seq(W,W)=seq(F,W)=seq(F,F)=W). Maximum und Minimum sind bezüglich der Ordnung von $\{W,F\}$ gebildet, in der F kleiner als W ist. Die Belegung h^x_x ist (wenn h Belegung, x Variable und $x\in U_{\mathfrak{A}}$) definiert durch

$$h^x_x(z) = \begin{cases} h(z), & \text{falls } z\neq x \text{ (d.h., x und z sind verschiedene Variablen)}, \\ x & \text{, falls } z=x.\end{cases}$$

Man sagt auch, daß eine Belegung h in der Struktur $\mathfrak{A}$ die Formel α *erfüllt* (Abkürzung $h \operatorname{Erf}_{\mathfrak{A}} \alpha$), falls $W_{\mathfrak{A}}(\alpha, h) = W$ ist. Bei Verwendung der Bezeichnung $\alpha(x_1, \ldots, x_n)$ (1.10) schreibt man für

$h^{x_1 \ldots x_n}_{x_1 \ldots x_n} \operatorname{Erf}_{\mathfrak{A}} \alpha(x_1, \ldots, x_n)$ manchmal kürzer $\alpha[x_1, \ldots, x_n]$ oder

$\models_{\mathfrak{A}} \alpha[x_1, \ldots, x_n]$ (bei Belegung der Variablen mit den in eckigen Klammern angegebenen Elementen von $\mathfrak{A}$ wird die Formel erfüllt).

<u>1.17 Anmerkungen</u>. (i) Die Bezeichnungen x, $x_1, \ldots, x_n$ gehören natürlich zur mathematischen Umgangssprache (Metasprache, s. 1.1); sie sind (metasprachliche) Variablen für Elemente der Trägermenge. Daher werden hier Kursivbuchstaben verwendet, wie es für Bezeichnungen mit variabler Bedeutung allgemein üblich ist. Im Vergleich dazu sind die entsprechenden objektsprachlichen Variablen x, $\mathrm{x}_1, \ldots, \mathrm{x}_n$ feste Objekte (Grundzeichen), die durch verschiedene Elemente der Trägermenge interpretiert werden können. Daher wird gerade Schrift verwendet, wie es üblich ist für Bezeichnungen mit einer (im betrachteten Zusammenhang) festen Bedeutung. Das entspricht zwar nicht ganz dem sonst üblichen Gebrauch der Schriftarten, da der Buchstabe x j e d e V a r i a b l e bedeuten kann. Es ist aber eine typographische Unterscheidung erforderlich, und diese ist wenigstens einigermaßen motiviert und hat sich weitgehend eingebürgert.

(ii) Da die geometrischen Sätze in Teil I in der mathematischen Umgangssprache formuliert wurden (vgl. I.0.5), erscheinen dort Kursivbuchstaben an Stelle der bei der jetzigen Formalisierung verwendeten Variablen in gerader Schrift (s.a. 1.28(v)).

<u>1.18 Satz</u> (Koinzidenztheorem). *Wenn* $\alpha \in \mathrm{Fml}_K(\mathrm{x}_1, \ldots, \mathrm{x}_n)$, $h_1(\mathrm{x}_\nu) = h_2(\mathrm{x}_\nu)$ $(\nu = 1, \ldots, n)$, $U_{\mathfrak{A}_1} = U_{\mathfrak{A}_2}$ *sowie* $f_{\mathfrak{A}_1} = f_{\mathfrak{A}_2}$ *und* $R_{\mathfrak{A}_1} = R_{\mathfrak{A}_2}$ *für jedes Operationszeichen* f *und jedes Relationszeichen* R *aus* K, *so ist* $W_{\mathfrak{A}_1}(\alpha, h_1) = W_{\mathfrak{A}_2}(\alpha, h_2)$. *Mit anderen Worten: Der Wahrheitswert einer Formel hängt nur davon ab, wie ihre freien Variablen belegt sind und wie die in ihr vorkommenden Operations- und Relationszeichen interpretiert sind.*

<u>1.19 Satz</u> (Überführungstheorem). *Wenn* $\gamma = \alpha^{\mathrm{x}}/_{\tau}$ *(Substitution oder verallgemeinerte Substitution gemäß* 1.9*) und* $w_{\mathfrak{A}}(\tau, h) = t$, *so ist* $W_{\mathfrak{A}}(\gamma, h) = W_{\mathfrak{A}}(\alpha, h^{t}_{\mathrm{x}})$.

Diese beiden anschaulich einleuchtenden Sätze (bzw. geeignete Hilfsbehauptungen) lassen sich durch Induktion über den Formelaufbau von α beweisen, vgl. etwa Hermes 1969, S. 81ff.

Die $\mathfrak{A}_i$ von 1.18 können auch Strukturen für verschiedene Sprachen L_i sein. Manchmal ist der folgende Spezialfall von Interesse.

<u>1.20 Definition</u>. Sei L_1 eine Teilsprache von L_2 (1.4) und $\mathfrak{A}_i$ Struktur für L_i (i=1,2). Dann heißt $\mathfrak{A}_1$ (das) *Redukt* von $\mathfrak{A}_2$ auf L_1 oder $\mathfrak{A}_2$ eine *Expansion* von $\mathfrak{A}_1$ auf L_2, falls gilt: $U_{\mathfrak{A}_1}=U_{\mathfrak{A}_2}$ sowie $f_{\mathfrak{A}_1}=f_{\mathfrak{A}_2}$ und $R_{\mathfrak{A}_1}=R_{\mathfrak{A}_2}$ für jedes Operationszeichen f und jedes Relationszeichen R von L_1.

<u>1.21 Definition</u>. Die Formel α ist in $\mathfrak{A}$ *gültig* (abgekürzt $\models_{\mathfrak{A}}\alpha$), falls $W_{\mathfrak{A}}(\alpha,h)=W$ ist für jede Belegung h über $\mathfrak{A}$. In $\mathfrak{A}$ gültige Aussagen heißen auch *wahre Aussagen* über die Struktur $\mathfrak{A}$ (der Wahrheitswert von Aussagen hängt nach 1.18 gar nicht von einer Belegung, sondern nur von der gegebenen Struktur ab).

<u>1.22 Anmerkung</u>. Die hier behandelte Präzisierung des Wahrheitsbegriffs (mit Hilfe der Begriffe Wahrheitswert bzw. Erfüllung) geht auf Tarski zurück, vgl. Tarski 1936. Sie trifft wohl das, was man sich anschaulich vorstellt. Ihre Bedeutung besteht vor allem darin, daß sie die Semantik unserer formalisierten Sprachen einer strengen Behandlung mit mathematischen Methoden zugänglich macht. Sie wird in der genannten Arbeit zum Beweis eines grundlegenden Nichtdefinierbarkeitsresultats über den Wahrheitsbegriff verwendet.

Auf Grund von 1.21 gilt offenbar

<u>1.23 Satz</u>. $\models_{\mathfrak{A}}\alpha$ *gdw* $\models_{\mathfrak{A}}\forall x\alpha$.

<u>1.24 Definition</u>. Unter einer *Generalisierten* einer Formel α verstehen wir eine Formel der Form $\forall x_1 \ldots \forall x_n\alpha$, wobei $x_1, \ldots, x_n$ die (paarweise verschiedenen) freien Variablen von α (in irgendeiner Reihenfolge) sind.

Dafür ergibt sich (mit 1.8, 1.23):

<u>1.25 Satz</u>. *Sei α* Generalisierte von α. Dann gilt:*

(1) α^* *ist Aussage.*

(2) $\models_{\mathfrak{A}}\alpha$ *gdw* $\models_{\mathfrak{A}}\alpha^*$.

1.26 Definition. Sei $\Sigma \subseteq \mathrm{Fml}_L$ und $\mathfrak{A}$ eine Struktur für L. Dann heißt $\mathfrak{A}$ ein *Modell* von Σ (in L) (abgekürzt $\mathfrak{A}\mathrm{Mod}_L\Sigma$), falls jede Formel von Σ in $\mathfrak{A}$ gültig ist; unter der *Modellklasse* $\mathrm{Mod}_L(\Sigma)$ von Σ in L versteht man die Klasse aller Modelle von Σ (in L).

1.27 Definition. Sei $\Sigma \subseteq \mathrm{Fml}_L$, $\alpha \in \mathrm{Fml}_L$. Wir sagen, aus Σ *folgt* α (in L) (abgekürzt $\Sigma \models_L \alpha$ oder $\Sigma \models \alpha$), falls α in jedem Modell von Σ (in L) gültig ist; unter der *Folgerungsmenge* $\mathrm{Cn}_L(\Sigma)$ von Σ in L versteht man die Menge aller Formeln α von L, die aus Σ folgen (diese heißen auch *Folgerungen* [engl. "consequences"] aus Σ).

1.28 Anmerkungen. (i) Mit diesen beiden Definitionen werden die geläufigen Begriffe Modell eines Axiomensystems Σ und der Folgerungsbegriff präzisiert für "Axiome" in der Objektsprache (das sind hier beliebige Formeln).

(ii) Es sei jedoch darauf hingewiesen, daß in der Literatur neben dieser auch eine andere Präzisierung des Folgerungsbegriffs betrachtet wird (z.B. in Hermes 1969): Dabei werden als Modelle statt der hier benutzten Strukturen sogenannte Interpretationen verwendet, diese sind jeweils festgelegt durch (mit unseren Bezeichnungen) eine Struktur und eine zugehörige Belegung. Dadurch wird für die Folgerungsbeziehung $\Sigma \models_L \alpha$ erreicht, daß nicht nur die Operations- und Relationszeichen, sondern auch die freien Variablen in Σ und α "dieselbe Bedeutung" erhalten. Damit werden grundlegende Betrachtungen vereinheitlicht und vereinfacht.

(iii) Dagegen ist unsere Folgerungsbeziehung nach Satz 1.25 invariant gegenüber dem Weglassen oder Hinzufügen von Allquantoren (mit Variablen) am Anfang von Formeln; wir können also zum Beispiel ein Axiom der Assoziativität auch in der üblichen Form $x\cdot(y\cdot z) \doteq (x\cdot y)\cdot z$ ohne Allquantoren schreiben, analog für andere Axiome und Sätze.

(iv) Diese beiden Folgerungsbegriffe werden beim weiteren Aufbau der Prädikatenlogik auch durch verschiedene Beweisbarkeitsbegriffe (vgl. 1.49) erfaßt.

(v) Die frühere Feststellung eines geometrischen Satzes in Teil I ist natürlich gleichwertig mit der Feststellung, daß die entsprechende Formel aus dem jeweils betrachteten Axiomensystem folgt. In den weiteren Abschnitten werden wir uns erlauben, solche Änderungen der Ausdrucksweise und der zugehörigen Schriftart (s. 1.17(ii)) ohne besondere Erwähnung vorzunehmen.

1.29 Definition. Eine Formel α von L heißt *logisch gültig*, *allgemeingültig*, eine *logische Identität*, *Tautologie* oder ein *Satz der Logik* (der Prädikatenlogik der ersten Stufe mit Identität in der Sprache L) (abgekürzt $\models\alpha$), falls α gültig ist in jeder Struktur für L. Das ist offenbar gleichbedeutend damit, daß α aus der leeren Menge $\emptyset$ folgt.

1.30 Definition. Formeln α, β von L heißen zueinander *äquivalent* in bezug auf eine Menge (ein Axiomensystem) $\Sigma \subseteq Fml_L$ (Bezeichnung $\alpha Äq_\Sigma \beta$) (oder auch *äquivalent* in der Theorie $T = Cn_L(\Sigma)$, vgl. 1.35(i)), falls $\Sigma \models \alpha \leftrightarrow \beta$ ist (d.h. $\alpha \leftrightarrow \beta$ Satz von T). Sie heißen *logisch äquivalent* (Bez. $\alpha Äq \beta$), falls $\alpha \leftrightarrow \beta$ ein Satz der Logik ist (Spezialfall mit $\Sigma=\emptyset$).

1.31 Definition. Ist Σ eine Formelmenge in einer beliebigen Sprache, so bezeichnen wir mit $L(\Sigma)$ (oder auch $\mathring{\Sigma}$) die kleinste Sprache (im Sinne der Halbordnung $\subseteq$ von 1.4), in der Σ eine Formelmenge ist; zu dieser Sprache gehören also genau die Operations- und Relationszeichen, die in Formeln aus Σ vorkommen. In Ergänzung zu 1.26, 1.27 setzen wir $Mod(\Sigma) := Mod_{L(\Sigma)}(\Sigma)$ und $Cn(\Sigma) := Cn_{L(\Sigma)}(\Sigma)$.

1.32 Definition (nach Tarski, Mostowski u. Robinson 1953, S. 11, vgl. 1.41). Eine Formelmenge T (Tau) heißt eine *Theorie*, falls $Cn(T) = T$ ist. Die Elemente von T heißen dann auch die *Sätze* der Theorie (im Sinne von Lehrsätze, engl. "theorems"), und $L(T)$ heißt die *Sprache* der Theorie.

1.33 Anmerkung. Es mag natürlicher erscheinen, eine Theorie einzuführen als ein Paar $\langle L,T\rangle$ aus der Sprache und der Satzmenge und dafür die (möglichst allgemeine) Forderung $Cn_L(T) = T$ zu stellen. Man erhält jedoch leicht den folgenden Satz (s. etwa Schwabhäuser 1971, S. 41), der besagt, daß man auf die Angabe der Sprache verzichten kann, da sie durch die Satzmenge mitbestimmt ist, und daß im übrigen die gerade genannte Forderung gleichwertig mit Def. 1.32 ist. (Zu anderen in der Literatur üblichen Definitionen s. 1.41.)

1.34 Satz. $Cn_L(T) = T$ *gdw* $Cn(T) = T$ *und* $L(T) = L$.

1.35 Definition (Beispiele für Theorien). (i) Sei $\Sigma \subseteq Fml_L$ beliebig. Dann heißt $T = Cn_L(\Sigma)$ die auf Σ *axiomatisch aufgebaute Theorie* oder die Theorie mit dem *Axiomensystem* Σ.

(ii) Sei K eine beliebige Klasse von Strukturen für die Sprache L. Unter der *Theorie der Klasse* K, Bezeichnung $Th(K)$, versteht man dann die Menge derjenigen Formeln, die in jeder Struktur aus K gültig sind.

Man erhält leicht, daß die so definierten Formelmengen tatsächlich Theorien im oben definierten Sinne sind. Es läßt sich sogar jede Theorie T in der Form (i) und in der Form (ii) darstellen; denn für jede gilt $T = Cn_{L(T)}(T) = Th(Mod(T))$. Unter der Voraussetzung (i) bzw. (ii) mit $T = Th(K)$ erhält man dagegen nur die Inklusionen $\Sigma \subseteq T$ bzw. $K \subseteq Mod(T)$ (wobei keineswegs Gleichheit eintreten muß).

(iii) Sei $\mathfrak{A}$ eine Struktur für L. Unter der *Theorie der Struktur* $\mathfrak{A}$, Bezeichnung $Th(\mathfrak{A})$, versteht man die Menge derjenigen Formeln, die in $\mathfrak{A}$ gültig sind. (Spezialfall von (ii) mit $K = \{\mathfrak{A}\}$.)

(iv) $Cn_L(\emptyset)$ ist nach 1.29 die Menge der Sätze der Prädikatenlogik in der Sprache L. Das ist ebenfalls eine Theorie (Spezialfall von (i) mit $\Sigma = \emptyset$), und zwar die kleinste Theorie mit der Sprache L.

(v) Fml_L ("die widerspruchsvolle Theorie mit der Sprache L", vgl. 1.38) ist die größte Theorie mit der Sprache L.

1.36 Definition. Seien T_1, T_2 Theorien. Dann heißt T_1 eine *Untertheorie* von T_2 oder T_2 eine *Obertheorie* oder *Erweiterung(stheorie)* von T_1, falls T_1 als Formelmenge in T_2 enthalten ist ($T_1 \subseteq T_2$). Die Untertheorie bzw. Erweiterung heißt *echt*, falls hierbei $T_1 \neq T_2$ ist. Die Sprachen L_1, L_2 beider Theorien können hierbei gleich oder voneinander verschieden sein, allgemein ist jedenfalls $L_1 \subseteq L_2$. Ist $T_2 = Cn(T_1 \cup \Sigma)$ (d.h. T_2 die kleinste

Theorie mit $T_1 \cup \Sigma \subseteq T_2$), so sagen wir, daß die Erweiterung T_2 aus T_1 durch *Adjunktion* (der Elemente) von Σ entsteht (analog zur Erweiterung von Strukturen in der Algebra, z.B. Körpererweiterungen). T_2 heißt *endliche* (bzw. *einfache*) Erweiterung von T_1, falls T_2 aus T_1 durch Adjunktion von endlich vielen Formeln (bzw. einer einzigen Formel) entsteht. (Beide Begriffe fallen hier zusammen; denn die Adjunktion von $\alpha_1 \wedge \ldots \wedge \alpha_n$ liefert offenbar dasselbe wie die von $\alpha_1, \ldots, \alpha_n$.)

<u>1.37 Definition</u>. Strukturen $\mathfrak{A}$, $\mathfrak{B}$ für L heißen zueinander *elementar äquivalent* (abgekürzt $\mathfrak{A} \equiv \mathfrak{B}$), falls $\mathrm{Th}(\mathfrak{A}) = \mathrm{Th}(\mathfrak{B})$ ist (d.h. in beiden Strukturen dieselben Formeln gültig sind). (Isomorphe Strukturen sind trivialerweise elementar äquivalent; die Umkehrung gilt nicht, s. etwa 3.22(ii).)

<u>1.38 Definition</u>. Eine Formelmenge Σ heißt *widerspruchsfrei*, falls $\mathrm{Cn}(\Sigma)$ verschieden von (und damit echte Teilmenge) der Menge $\mathrm{Fml}_{L(\Sigma)}$ aller Formeln der zugehörigen Sprache ist. Als gleichwertige Bedingungen erhält man (s. etwa Schwabhäuser 1972, wie auch für das Folgende):

(1) Für keine Aussage α von $L(\Sigma)$ ist $\Sigma \models \alpha \wedge \neg\alpha$.

(2) Für keine Aussage α von $L(\Sigma)$ ist $\Sigma \models \alpha$ und $\Sigma \models \neg\alpha$.

(3) Analoge Bedingungen mit "Formel" statt "Aussage".

(4) Σ besitzt ein Modell.

Andernfalls heißt Σ *widerspruchsvoll*.

<u>1.39 Definition</u>. Eine Formelmenge Σ heißt *vollständig*, falls gilt: Für jede Aussage α von $L(\Sigma)$ ist $\Sigma \models \alpha$ oder $\Sigma \models \neg\alpha$. Als gleichwertige Bedingungen erhält man:

(1) Je zwei Modelle von Σ in der zugehörigen Sprache $L(\Sigma)$ sind zueinander elementar äquivalent.

(2) Σ ist widerspruchsvoll oder $\mathrm{Cn}(\Sigma)$ ist eine maximale widerspruchsfreie Teilmenge von $\mathrm{Fml}_{L(\Sigma)}$.

Andernfalls heißt Σ *unvollständig*.

<u>Beispiele</u>. Die Theorie $\mathrm{Th}(\mathfrak{A})$ einer einzelnen Struktur $\mathfrak{A}$ (1.35(iii)) ist stets widerspruchsfrei und vollständig. Die Prädikatenlogik $\mathrm{Cn}_L(\emptyset)$ ist (im Sinne dieser Definition) eine unvollständige Theorie.

Aus 1.25, 1.32 ergibt sich

1.40 Satz. *Sei* Σ *eine Theorie und* α^* *Generalisierte von* α. *Dann ist* $\alpha \in T$ *gdw* $\alpha^* \in T$.

1.41 Anmerkung. Somit ist eine Theorie schon festgelegt durch die Menge derjenigen ihrer Sätze, die Aussagen sind. In der Literatur werden Theorien oft nur als Mengen von Aussagen eingeführt (insbesondere bei Verwendung des Folgerungsbegriffs nach 1.28(ii)). Manchmal werden Theorien auch als widerspruchsfrei vorausgesetzt. Schließlich werden bisweilen beliebige (evtl. widerspruchsfreie) Mengen Σ von Formeln oder Aussagen als Theorien bezeichnet (und in weiteren Definitionen wie die Axiomensysteme aus 1.35(i) verwendet). Der Übergang von einer zu einer anderen dieser Terminologien bereitet keine Schwierigkeiten.

Nach der bisher betrachteten Standardformalisierung in der ersten Stufe kommen wir nun zu anderen Formalisierungen und Spracherweiterungen.

1.42 Formalisierung in der ersten Stufe mit mehreren Sorten von Variablen. Diese Formalisierung unterscheidet sich von der ursprünglichen dadurch, daß mehrere Sorten von Individuenvariablen verwendet werden, sagen wir allgemein n Sorten. An Stelle der Variablenmenge V haben wir dann n paarweise disjunkte abzählbare Variablenmengen $V_1, \ldots, V_n$; die Elemente von V_i heißen auch (Individuen-)Variablen der i-ten Sorte. Bei den Operations- und Relationszeichen ist dann nicht nur die Stellenzahl festzulegen, sondern auch für jede Stelle, auf welche Sorte sie sich bezieht, bei den Operationszeichen (falls vorhanden) ist das auch für den Wert festzulegen. Bei der Definition der Terme hat man dann Terme jeder Sorte. Bei der Definition der Formeln hat man Gleichungen für jede Sorte von Termen als Primformeln; die Quantifizierungen (1.5(F2)(b)) sind für jede Sorte von Variablen erlaubt. Die Strukturen $\mathfrak{A}$ für eine so formalisierte Sprache in der e r s t e n Stufe enthalten dann nicht nur einen, sondern n Individuenbereiche $(U_1)_{\mathfrak{A}}, \ldots, (U_n)_{\mathfrak{A}}$ von "Individuen der Sorte i " ($i=1,\ldots,n$); die Operationen und Relationen beziehen sich in ihren Stellen und im Wert auf die richtige Sorte von Individuen entsprechend der Festlegung bei den Zeichen, denen sie (durch die Strukturfunktion s) zugeordnet sind. Belegungen über $\mathfrak{A}$ sind Abbildungen, die jeder Variablen aus V_i ein Element von $(U_i)_{\mathfrak{A}}$ zuordnen ($i=1,\ldots,n$). Die induktive

Definition der Werte von Termen und Formeln überträgt sich dann sinngemäß; bei der Festlegung des Wahrheitswertes einer Quantifizierung (analog zu 1.16(2)) mit Variablen der Sorte i erstreckt sich die Minimum- bzw. Maximumbildung auf Elemente von (U_i) . Alles Weitere läßt sich dann wie bisher einführen.

Beispiel. Eine Formalisierung der Überlegungen von Abschnitt I.6 läuft hinaus auf die Hinzunahme von Geradenvariablen (A, B,...) zu den schon vorhandenen Punktvariablen (a, b,...). Führt man noch ein Relationszeichen Iz für die Inzidenz zwischen Punkten und Geraden ein (und schreibt a Iz A statt der früheren Elementbeziehung a∈A), so gelangt man zu einer Formalisierung in der ersten Stufe mit zwei Sorten von Variablen. Genauer werden solche Formalisierungen behandelt in 4.58ff., 4.89f., 4.100. In beliebigen Modellen sind natürlich beliebige Individuenbereiche zugelassen, d.h., die Geraden sind dann irgendwelche Objekte und nicht notwendig Punktmengen.

Alle wichtigen Resultate der Prädikatenlogik der ersten Stufe könnte man von vornherein für mehrsortige Formalisierungen beweisen. Es genügt aber auch der bequemere Beweis für die einsortigen Standardformalisierungen. Zu jeder mehrsortigen Theorie kann man nämlich nach einer einheitlichen Methode eine "gleichwertige" einsortige Theorie konstruieren und damit die gewünschten Resultate übertragen. Diese Methode ist an Hand des obigen Beispiels skizziert in 4.100 (vgl. auch A. Schmidt 1938). Für spezielle Theorien können darüber hinaus andere (nicht allgemein anwendbare) Methoden zur Rückführung auf eine einsortige Theorie verfügbar sein; für das obige Beispiel s. dazu 4.59 und 4.89.

1.43 Zweite Stufe mit Mengenvariablen. Für eine "Standardformalisierung" dieser Art wird zu den Grundzeichen aus 1.3 folgendes hinzugenommen:
1. eine zweite abzählbare Menge V_M von *Mengenvariablen* (X, Y,...),
2. ein zweistelliges Relationszeichen $\dot{\in}$, das sich in der ersten Stelle auf Individuenvariablen und in der zweiten auf Mengenvariablen bezieht (und zwischen die Argumente geschrieben wird). Die Definition der Terme und Formeln erfolgt dann genauso wie in der erste Stufe mit zwei Sorten von Variablen (1.42). Die Unterscheidung erfolgt erst in der Semantik. Die betrachteten Strukturen $\mathfrak{A}$ sind jetzt nämlich dieselben wie für die Standardformalisierung in der ersten Stufe, sie haben insbesondere nur einen Individuenbereich $U_{\mathfrak{A}}$. Die Mengenvariablen werden dagegen stets als Mengen von Individuen und das Zeichen $\dot{\in}$ als die Elementbeziehung $\in$ inter-

pretiert. Belegungen h über $\mathfrak{A}$ sind also jetzt Abbildungen, die jedem $x \in V$ ein Element von $U_{\mathfrak{A}}$ und jedem $X \in V_M$ eine Teilmenge von $U_{\mathfrak{A}}$ zuordnen. Bei der Charakterisierung der Wahrheitswerte ist die zweite Bedingung von 1.16(1) sinngemäß auch für Gleichungen $X \doteq Y$ zwischen Mengenvariablen zu formulieren und hinzuzunehmen:

$$w_{\mathfrak{A}}(\tau \dot{\in} X, h) = \begin{cases} W, \text{ falls } w_{\mathfrak{A}}(\tau, h) \in h(X) , \\ \\ F \text{ sonst.} \end{cases}$$

Damit hat das Zeichen $\dot{\in}$ ebenso wie $\doteq$ eine feste (von den betrachteten Strukturen unabhängige) Bedeutung, und es ist in diesem Sinne zu den logischen Zeichen zu rechnen.

Dieselbe Semantik ergibt sich offenbar, wenn man in 1.42 nicht mehr beliebige zweisortige Strukturen zuläßt, sondern nur noch die Strukturen $\mathfrak{A}$ mit $(U_2)_{\mathfrak{A}} = \mathfrak{P}((U_1)_{\mathfrak{A}})$ (Potenzmenge) und $\dot{\in}_{\mathfrak{A}} = \in \cap ((U_1)_{\mathfrak{A}} \times (U_2)_{\mathfrak{A}})$ (Einschränkung der Elementbeziehung auf den betrachteten Bereich). Dementsprechend erhält man für die zweite Stufe einen stärkeren Folgerungsbegriff als für die erste, obwohl dieselben Formeln zugrunde gelegt werden: Für beliebige Mengen Σ von solchen Formeln ist stets $Cn_L(\Sigma) \subseteq Cn_L^2(\Sigma)$, wobei die rechte Seite die Folgerungsmenge im Sinne der zweiten Stufe (analog zu 1.27) bezeichnet (da in der zweiten Stufe weniger Modelle von Σ zugelassen werden, kann sich die Menge der Folgerungen höchstens vergrößern).

Alles übrige verläuft analog zu den bisherigen Betrachtungen. Die *Theorie* einer Strukturklasse K bzw. einer Struktur $\mathfrak{A}$ in der zweiten Stufe bezeichnen wir insbesondere mit $Th^2(K)$ bzw. $Th^2(\mathfrak{A})$.

Bei diesem ganzen Vorgehen wird natürlich (wie schon früher) für die Metasprache eine feste (Theorie der) Mengenlehre zugrunde gelegt; das kann irgendeine Mengenlehre sein, wie sie auch sonst für Anwendungen (z.B. in der Algebra) verwendet wird (das Auswahlaxiom wird häufig nicht gebraucht, aber bei Bedarf vorausgesetzt). Eventuelle metamathematische Betrachtungen über diese Mengenlehre (z.B. über verschiedene Modelle der Mengenlehre) würden dann in die Meta-Metasprache gehören.

1.44 Schwache zweite Stufe. Eine Formalisierung dieser Art mit Mengenvariablen unterscheidet sich von der in 1.43 lediglich dadurch, daß die Mengenvariablen nicht als beliebige, sondern als endliche Mengen von

Individuen interpretiert werden. Alles sonst in 1.43 Gesagte überträgt sich sinngemäß. Die entsprechenden Folgerungsmengen und Theorien in diesem Rahmen wollen wir mit $Cn_L^{S2}(\Sigma)$, $Th^{S2}(K)$ bzw. $Th^{S2}(\mathfrak{A})$ bezeichnen. In geometrischen Theorien dieser Art lassen sich z.B. Sätze über beliebige Polygone formulieren, während die Problematik beliebiger Mengen noch vermieden wird. Allgemeiner werden statt der Mengenvariablen auch manchmal Variablen für endliche Folgen von Individuen (und geeignete logische Operationszeichen) verwendet, insbesondere bei Tarski 1958, wo der Begriff schwache zweite Stufe eingeführt wurde.

1.45 Höhere Stufen. Zur Formalisierung nach 1.43 kann man in analoger Weise noch eine weitere Sorte von Variablen für "Mengen zweiter Stufe" (d.h. Mengen von Mengen von Individuen) hinzunehmen usw. Nimmt man Variablen für Mengen bis zur $(n-1)$-ten Stufe hinzu, so spricht man von einer Formalisierung in der n-ten Stufe. Diese Bezeichnung wird jedoch in der Literatur (schon für $n=2$) durchaus nicht einheitlich gebraucht. Manchmal werden auch Relationen beliebiger Stellenzahl zugelassen an Stelle von Mengen (die man mit einstelligen Relationen identifizieren kann) und manchmal Funktionen. Ein allgemeiner Rahmen für die Behandlung solcher Formalisierungen ist die sog. Typentheorie.

1.46 Anmerkung. Das wesentliche Merkmal der ersten Stufe ist also, daß nur Individuenvariablen zur Verfügung stehen, aber noch keine Variablen für Mengen von Individuen oder Ähnliches (Relationen, Folgen, Funktionen). Für diese Formalisierung und alles was sich darauf bezieht, wird auch die Bezeichnung *elementar* verwendet (z.B sagt man *elementare Theorien* für in der ersten Stufe formalisierte Theorien). In diesem Sinn ist auch die "elementare Äquivalenz" von Strukturen (1.37) zu verstehen. Für einen analogen Begriff in höheren Sprachen ist daher ein anderer Name sinnvoll (z.B. "Äquivalenz in der schwachen zweiten Stufe").

1.47 Der bestimmte Artikel (Kennzeichnungsoperator). Schon in Teil I wurde die Bezeichnung $\iota x \alpha(x)$ verwendet für das(jenige) x mit der Eigenschaft $\alpha(x)$. Wir betrachten nun eine Formalisierung, die aus der Standardformalisierung durch Hinzunahme solcher Bezeichnungen entsteht. Dazu wird die Definition der Terme und Formeln (1.5) durch folgende Regel erweitert.

(Tι) Wenn α eine Formel und $x \in V$ ist, so ist $\iota x \alpha$ ein Term.

Die Regeln (T1), (T2), (F1), (F2) und (Tι) liefern dann eine gemeinsame Charakterisierung (eine "verschränkte" induktive Definition) für die beiden Begriffe Term und Formel. Das (neue) Grundzeichen ι (jota) heißt dabei der *Kennzeichnungsoperator*, *ι-Operator* oder der *bestimmte Artikel*. Die Terme der Form $\iota x\alpha$ heißen *ι-Terme*. Tritt ein solcher Term innerhalb einer Formel auf, so heißt α der *Wirkungsbereich* des ι-Operators mit der Variablen x an der betreffenden Stelle (vgl. 1.8); dementsprechend wird die Variable x überall, wo sie im Term $\iota x\alpha$ vorkommt, als gebunden und hinter dem Zeichen ι insbesondere als quantifiziert betrachtet.

Der uneingeschränkte Gebrauch von ι-Termen in der formalisierten Sprache erscheint als Abweichung von dem üblichen Gebrauch in der Umgangssprache, bei dem der bestimmte Artikel nur dann verwendet wird, wenn es genau ein Ding mit der angegebenen Eigenschaft gibt. Eine solche Einschränkung würde jedoch für unsere Formalisierung zu unerwünschten Konsequenzen führen. Sie läuft ja darauf hinaus, daß die Bildung des ι-Terms $\iota x\alpha$ nur dann erlaubt wird, wenn die "Unitätsformel" $\overset{=1}{\exists} x\alpha$ (s. 1.11) wahr wird. Der Wahrheitswert $W_{\mathfrak{A}}(\overset{=1}{\exists} x\alpha, h)$ dieser Formel hängt aber von der gerade betrachteten Struktur $\mathfrak{A}$ und im allgemeinen (falls α noch andere freie Variablen außer x enthält) auch von der Belegung h ab. Die genannte Einschränkung wurde also bedeuten, daß die Syntax nicht mehr unabhängig von der wesentlich komplizierteren Semantik ist; es würde dadurch auch die für andere (nicht nur elementare) Formalisierungen selbstverständliche Eigenschaft verlorengehen, daß man für jede Zeichenreihe nachprüfen kann, ob sie ein Term bzw. eine Formel ist oder nicht (Rekursivität der Mengen der Terme und der Formeln), jedenfalls bei vernünftigen Zusatzvoraussetzungen über ein Bezeichnungssystem für die Sprache (vgl. 3.4, 3.6).

Eine geeignete Semantik für diesen uneingeschränkten Gebrauch von ι-Termen erhalten wir (nach Hilbert u. Bernays 1968, S. 432ff.), indem wir diesen Termen auch in den übrigen Fällen ("Unbestimmtheitsfällen") einen Wert zuordnen, und zwar den Wert einer für diesen Zweck ausgezeichneten Variablen u. Das geschieht durch eine entsprechend verschränkte induktive Definition der Werte von Termen und Formeln, die aus 1.15, 1.16 durch Hinzunahme der folgenden Bedingung entsteht:

$$w_{\mathfrak{A}}(\iota x\alpha, h) = \begin{cases} a, & \text{falls } a \text{ das einzige Element von } U_{\mathfrak{A}} \\ & \text{mit } h_x^a \operatorname{Erf}_{\mathfrak{A}} \alpha \text{ ist,} \\ h(u), & \text{falls es nicht genau ein solches } a \text{ gibt.} \end{cases}$$

Man könnte den betrachteten ι-Term genauer mit $\iota^u x\alpha$ bezeichnen oder die Bildung solcher Terme sogar für beliebige Variablen u zulassen; in jedem

Falle spielt u darin die Rolle einer freien Variablen (insbesondere für das Koinzidenztheorem 1.18), einerlei, ob sie in der Bezeichnung der Terme angegeben wird oder nicht.

Es läßt sich zeigen, (vgl. Hilbert u. Bernays 1968), daß man ι-Terme äquivalent eliminieren kann (sogar auf Grund eines Beweisbarkeitsbegriffs, vgl. 1.49, 60), d.h. zu jeder Formel gibt es eine dazu (beweisbar) logisch äquivalente Formel ohne ι-Terme. (Vgl. auch 3.38.)

Das Gesagte überträgt sich auf Formalisierungen mit mehreren Sorten von Variablen und in höheren Stufen.

1.48 Anmerkung. Eine andere für den uneingeschränkten Gebrauch von ι-Termen geeignete Semantik (und ein zugehöriger Beweisbarkeitsbegriff) wird behandelt bei Schröter 1956; dabei werden keine Werte für beliebige ι-Terme eingeführt, sondern erst Wahrheitswerte für die damit gebildeten Primformeln. Die oben genannte Variable u wird dann nicht gebraucht, aber der Wahrheitswert einer Formel ist i.a. nicht invariant gegenüber einem "Wechsel der Grundbegriffe" für eine Theorie. Damit ist diese Behandlung weniger geeignet für die späteren Betrachtungen in Abschnitt 4.

1.49 Beweisbarkeit für die erste Stufe. Ein wichtiges Resultat für Sprachen im Rahmen der Prädikatenlogik der ersten Stufe ist es, daß sich der Folgerungsbegriff gleichwertig ersetzen läßt durch einen Beweisbarkeitsbegriff. Ein solcher Beweisbarkeitsbegriff wird festgelegt durch ein System von logischen Axiomen und Schlußregeln. Eine Formel α heißt dann (in der Sprache L) aus Σ *beweisbar* (Bezeichnung $\Sigma\vdash_L\alpha$ oder kurz $\Sigma\vdash\alpha$), falls sie sich aus logischen Axiomen und Elementen von Σ (den sog. eigentlichen Axiomen) in endlich vielen Schritten durch Anwendung von Schlußregeln erhalten läßt. Als Präzisierung dieser anschaulichen Beschreibung kann man verwenden, daß es eine *Beweisfolge* oder einen *formalen Beweis* für α aus Σ in L gibt, das ist eine endliche Folge $\langle\alpha_1,\ldots,\alpha_k\rangle$ von Formeln von L mit dem letzten Glied $\alpha_k=\alpha$ derart, daß für jedes Folgenglied α_κ gilt: α_κ ist ein (logisches oder eigentliches) Axiom oder α_κ entsteht aus früheren Folgengliedern durch Anwendung einer Schlußregel.

Auf die Angabe des Systems der logischen Axiomen und Schlußregeln wird hier verzichtet. In der Literatur finden sich verschiedene solche Systeme (und damit Beweisbarkeitsbegriffe, zum Teil auch ohne Verwendung von logischen Axiomen). Ein für die hier betrachtete Standardformalisierung und den hier verwendeten Folgerungsbegriff (vgl. 1.28(ii), (iv)) geeignetes System wird z.B. angegeben in Schwabhäuser 1971 oder 1972a. Bei allen solchen Beweisbarkeitsbegriffen wird in der Formulierung der Schlußregeln und (eventuell) der logischen Axiome nur auf die Bauart der auftretenden Formeln und Terme Bezug genommen und nicht auf deren Bedeutung in Modellen (Semantik). Der Beweisbarkeitsbegriff wird daher - ebenso wie der Aufbau der formalisierten Sprache selbst - zur Syntax gerechnet. Insbesondere sind - unter einer Zusatzvoraussetzung über ein Bezeichnungssystem für die Sprache - die Schlußregeln und das System der logischen Axiome rekursiv (vgl. 3.4); das bedeutet anschaulich, daß sich ihre Anwendung in Beweisen nachprüfen läßt. Ist auch noch das System Σ der eigentlichen Axiome rekursiv (was für die meisten Anwendungen der Fall ist, vgl. 3.10(ii)), so werden formale Beweise aus Σ nachprüfbar. Die Feststellung einer Beweisbarkeit $\Sigma\vdash\alpha$ (ohne Angabe eines formalen Beweises) ist dagegen im allgemeinen nicht nachprüfbar (vgl. 3.14(iii)).

Die oben erwähnte Gleichwertigkeit wird in dem folgenden Satz ausgedrückt.

1.50 Satz. $\Sigma\vdash\alpha$ *gdw* $\Sigma\models\alpha$.

Hiervon wird die Richtung ($\rightarrow$) als *Korrektheit* des Beweisbarkeitsbegriffs bezeichnet. Diese ist etwas Selbstverständliches; man wird die logischen Axiome und Schlußregeln stets so wählen, daß der darauf beruhende Beweisbarkeitsbegriff korrekt ist. Das Hauptresultat ist also die Umkehrung ($\leftarrow$). Sie wird bezeichnet als die *Vollständigkeit* des Beweisbarkeitsbegriffs oder des Prädikatenkalküls der ersten Stufe oder auch als *Gödelscher Vollständigkeitssatz*. Die Vollständigkeit eines Beweisbarkeitsbegriffs (ursprünglich zur Charakterisierung der logischen Gültigkeit) wurde zuerst von Gödel im Jahre 1930 bewiesen.

Mit Hilfe von 1.50 und von Zusatzbetrachtungen im Beweis kann man die folgenden "modelltheoretischen" Resultate erhalten, in denen vom Beweisbarkeitsbegriff nicht mehr die Rede ist.

1.51 Satz (Endlichkeitssatz für das Folgern, 1. Stufe). *Wenn* $\Sigma\models\alpha$, *so existiert eine endliche Teilmenge* Σ' *von* Σ *mit* $\Sigma'\models\alpha$.

Zum Beweis: Das entsprechende Resultat für den Beweisbarkeitsbegriff $\vdash$ ist natürlich trivial; man braucht ja nur die in einer Beweisfolge für α aus Σ auftretenden Elemente von Σ zusammenzufassen zu Σ'. □

1.52 Satz (Endlichkeitssatz für Modelle, 1. Stufe). *Wenn jede endliche Teilmenge von* Σ *ein Modell besitzt, so besitzt auch* Σ *selbst ein Modell.*

1.53 Definition. Unter der *Mächtigkeit* $|\mathfrak{A}|$ *einer Struktur* $\mathfrak{A}$ verstehen wir die Mächtigkeit $|U_{\mathfrak{A}}|$ ihrer Trägermenge. Insbesondere nennen wir die Struktur $\mathfrak{A}$ endlich bzw. unendlich, falls ihre Trägermenge $U_{\mathfrak{A}}$ endlich bzw. unendlich (und damit $|\mathfrak{A}|$ eine endliche bzw. unendliche Kardinalzahl) ist. Unter der *Mächtigkeit* $|L|$ *einer Sprache* L verstehen wir die Mächtigkeit der Menge ihrer Grundzeichen; diese ist gleich der Mächtigkeit der Menge ihrer Formeln und gleich $\max(|F_L|,|R_L|,\aleph_0)$.

1.54 Satz (Satz von Löwenheim-Skolem-Tarski, 1. Stufe).
Vor.: Die Formelmenge Σ *besitzt ein unendliches Modell oder wenigstens zu jedem* $k\in\mathbb{N}$ *ein Modell mit einer Mächtigkeit* $\geq k$.
Beh.: Zu jeder Kardinalzahl $\kappa \geq |L(\Sigma)|$ *(Mächtigkeit der durch* Σ *bestimmten Sprache, s.* 1.31*) gibt es ein Modell von* Σ *mit der Mächtigkeit* κ.

Unter dem Satz von Löwenheim-Skolem verstand man ursprünglich den Spezialfall von 1.54 für $\kappa=|L(\Sigma)|=\aleph_0$ (wobei in der Voraussetzung nur ein überabzählbares Modell von Interesse ist). Diese Bezeichnung ist dann auch für Verschärfungen gebraucht worden. (Vgl. auch 7.53.)

Da Strukturen verschiedener Mächtigkeiten nicht zueinander isomorph sein können, ergibt sich aus 1.54 unmittelbar

1.55 Folgerung. *Eine unendliche Struktur* $\mathfrak{A}$ *läßt sich nicht bis auf Isomorphie charakterisieren durch ein Axiomensystem in der ersten Stufe.*

Natürlich kennt man *kategorische* Axiomensysteme, d.h. solche, die eine Struktur bis auf Isomorphie charakterisieren. Ein Beispiel ist das in Teil I angegebene System der Axiome A1 bis A11 für die kartesische

Ebene $\mathcal{L}_2(\mathbb{R})$ (s. I.16.15). Das "volle" Stetigkeitsaxiom A11 ist jedoch in der zweiten Stufe (mit Mengenvariablen) formuliert. Aus 1.55 ergibt sich nun das schon bei der Formulierung von A11 (I.1.2) erwähnte Resultat:

1.56 Satz. *Das Stetigkeitsaxiom* A11 *läßt sich nicht gleichwertig (auf Grund von* A1 *bis* A10*) ersetzen durch eine Menge von Axiomen in der ersten Stufe.*

1.57 Anmerkungen. (i) Aus der Existenz kategorischer Axiomensysteme ergibt sich, daß 1.54 nicht mehr für Sprachen im Rahmen der zweiten Stufe gelten kann. Für die schwache zweite Stufe gilt noch die folgende Abschwächung ("Abwärts-Richtung"): *Wenn* $\mathfrak{A}\mathrm{Mod}\,\Sigma$ *und* $|L(\Sigma)| \leq \kappa \leq |\mathfrak{A}|$, *so besitzt* Σ *auch ein Modell der Mächtigkeit* κ (s. Tarski 1958).

(ii) Tatsächlich gelten 1.51, 52, 54 weder für die schwache zweite, noch für die zweite oder höhere Stufen. Außerdem gibt es für diese Stufen keinen Beweisbarkeitsbegriff mit "nachprüfbaren formalen Beweisen" (vgl. 3.10(ii)), der mit dem Folgerungsbegriff gleichwertig ist. In der Literatur werden allerdings bisweilen andere Beweisbarkeitsbegriffe für diese Stufen und andere Formalisierungen betrachtet (unvollständig oder ohne nachprüfbare Beweise).

Der folgende *Satz von der pränexen Normalform* gilt dagegen für alle betrachteten Formalisierungen. Er wird häufig verwendet (u.a. schon zum Beweis des Vollständigkeitssatzes für die erste Stufe).

1.58 Definition. Eine Formel α heißt *pränex* oder eine *pränexe Normalform*, falls es ein π und ein α^* gibt derart, daß α^* eine *quantorenfreie* und ggf. (für die Formalisierung gemäß 1.47) ι-freie Formel ist (d.h. eine, in der die Grundzeichen $\forall$, $\exists$, ι nicht vorkommen), π eine aus Zeichenreihen der Form $\forall x$ und $\exists x$ (x Variable) zusammengesetzte Zeichenreihe (die auch leer sein kann) und $\alpha = \pi\alpha^*$. π heißt dann auch das *Präfix* und α^* der *Kern* (engl. "matrix") der pränexen Formel α.

1.59 Satz. *Zu jeder Formel* α *läßt sich eine dazu logisch äquivalente pränexe Normalform* β *angeben, die höchstens dieselben freien Variablen wie* α *enthält (d.h.* $\mathrm{Fr}(\beta) \subseteq \mathrm{Fr}(\alpha)$*).*

<u>1.60 Konstruktion</u> (zum Beweis). Ein Verfahren zur Konstruktion solcher Normalformen wird hier angegeben, da auch dieses später gebraucht wird. Es besteht in der schrittweisen Konstruktion von Formeln $\alpha_0=\alpha$, α_1, $\alpha_2,\ldots,$ $\alpha_m=\beta$ auf Grund der im folgenden angegebenen Äquivalenzen (1) bis (15). Im i-ten Schritt wird in α_{i-1} ein Bestandteil von der Form, wie sie auf der linken Seite der betreffenden Äquivalenz angegeben ist, ersetzt durch die zugehörige rechte Seite; dadurch entsteht aus der ganzen Formel α_{i-1} die dazu äquivalente Formel α_i. Alle genannten Äquivalenzen ergeben sich übrigens schon auf Grund der logischen Axiome und Schlußregeln eines Beweisbarkeitsbegriffs für die erste Stufe. Damit ergibt sich auch die sog. syntaktische oder "beweisbare" logische Äquivalenz $\emptyset\vdash\alpha\leftrightarrow\beta$, die zum Beweis des Vollständigkeitssatzes gebraucht wird.

Die Konstruktion besteht aus folgenden Teilen.

1. Ausschaltung des Junktors $\leftrightarrow$ (in endlich vielen Schritten) mittels der Äquivalenz

(1) $\gamma\leftrightarrow\delta$ Äq $(\gamma\rightarrow\delta)\wedge(\delta\rightarrow\gamma)$.

2. Ausschaltung von $\rightarrow$ mittels

(2) $\gamma\rightarrow\delta$ Äq $\neg\gamma\vee\delta$.

3. Verschiebung von $\neg$ nach innen mittels

(3) $\neg\neg\gamma$ Äq γ,

(4) $\neg(\gamma\wedge\delta)$ Äq $\neg\gamma\vee\neg\delta$,

(5) $\neg(\gamma\vee\delta)$ Äq $\neg\gamma\wedge\neg\delta$,

(6) $\neg\forall x\gamma$ Äq $\exists x\neg\gamma$,

(7) $\neg\exists x\gamma$ Äq $\forall x\neg\gamma$.

Dadurch läßt sich erreichen, daß das Negationszeichen $\neg$ nur noch unmittelbar vor Primformeln vorkommt.

4. Verschiebung der Quantoren nach vorn mittels der Äquivalenzen

(8) $\gamma\wedge\forall x\delta$ Äq $\forall x(\gamma\wedge\delta)$, falls x nicht frei in γ,

(9) $\gamma\vee\forall x\delta$ Äq $\forall x(\gamma\vee\delta)$, falls x nicht frei in γ,

(10) $\gamma\wedge\exists x\delta$ Äq $\exists x(\gamma\wedge\delta)$, falls x nicht frei in γ,

(11) $\gamma\vee\exists x\delta$ Äq $\exists x(\gamma\vee\delta)$, falls x nicht frei in γ

und analoger Sätze (12) bis (15) für den Quantor im ersten Konjunktions- bzw. Disjunktionsglied. Vor Anwendung einer solchen Äquivalenz kann man, wenn nötig, die gebundene Variable x in eine andere Variable umbenennen (wie in 1.9), um die geforderte Bedingung zu erfüllen.

Dadurch erhält man dann eine pränexe Normalform. (Es läßt sich zeigen, daß diese Konstruktion stets in endlich vielen Schritten zum Ziel führt.)

Manchmal ist es bequemer, auf die Ausschaltung von → zu verzichten und dafür noch folgende Sätze über die Quantorenverschiebung zu verwenden (die man aus den obigen Äquivalenzen erhalten kann).

(16) $\gamma \to \forall x\delta$ Äq $\forall x(\gamma \to \delta)$, falls x nicht frei in γ,

(17) $\gamma \to \exists x\delta$ Äq $\exists x(\gamma \to \delta)$, falls x nicht frei in γ,

(18) $\forall x\gamma \to \delta$ Äq $\exists x(\gamma \to \delta)$, falls x nicht frei in δ,

(19) $\exists x\gamma \to \delta$ Äq $\forall x(\gamma \to \delta)$, falls x nicht frei in δ.

1.61 Zusatz. *In 1.59 läßt sich zusätzlich erreichen, daß das Präfix π von β nur solche Variablen enthält, die auch im Kern auftreten, und daß jede dieser Variablen höchstens einmal in π vorkommt.*

Beweis: Im Präfix kann eine Variable x mit dem zugehörigen Quantor gestrichen werden, wenn sie an einer späteren Stelle im Präfix noch einmal vorkommt oder im Kern nicht vorkommt; das ergibt sich aus den Äquivalenzen

(1) $\forall x\gamma$ Äq γ, falls x nicht frei in γ,

(2) $\exists x\gamma$ Äq γ, falls x nicht frei in γ. □

2. Übersicht über betrachtete Geometrien.

Beim axiomatischen Aufbau der euklidischen Geometrie in Teil I wurde das System der Axiome A1 bis A8 (für die "dimensionsfreie absolute Geometrie") mit verschiedenen Erweiterungen betrachtet. Wir definieren nun in 2.1 die darauf beruhenden Geometrien als formalisierte Theorien (im Sinne von 1.32, 1.43) und führen Bezeichnungen dafür ein. Die metamathematischen Untersuchungen in den folgenden Abschnitten beziehen sich auch auf andere Geometrien, insbesondere auf verschiedene hyperbolische, elliptische, affine und projektive Geometrien. Darüber wird anschließend ein Überblick gegeben (die Durchführung eines axiomatischen Aufbaus wäre hier natürlich zu umfangreich). Wer sich zunächst nur mit den Geometrien von 2.1 beschäftigen möchte, kann den Rest dieses Abschnitts überspringen. Eine Zusammenfassung erfolgt in 2.16.

2.1 Euklidische Geometrien und Untertheorien. (i) Die *dimensionsfreie absolute Geometrie* ist die folgende Theorie erster Stufe mit der Sprache L(D,B):

(1) $\mathcal{A}$:= Cn({A1,...,A8}) (vgl. I.1.2).

Die *dimensionsfreie euklidische* (oder "parabolische") *Geometrie* $\mathcal{P}$ ist die daraus durch Hinzunahme ("Adjunktion") des euklidischen (Parallelen-) Axioms entstehende Erweiterung, d.h.

(2) $\mathcal{P}$:= Cn($\mathcal{A}$∪{A10}).

(ii) Durch Hinzunahme weiterer Axiome aus I.1.2 bilden wir für $T=\mathcal{A}$ und $T=\mathcal{P}$ die folgenden Erweiterungen.

(3) *Geometrien mit Kreisaxiom*:

T' := Cn(T∪{CA}).

(4) *Geometrien mit Schema der elementaren Stetigkeitsaxiome*:

T^* := Cn(T∪{A11' | für α(x), β(y) gilt (∗)}).

(5) *Geometrien zweiter Stufe mit (vollem) Stetigkeitsaxiom* ("volle dimensionsfreie absolute bzw. euklidische Geometrie"): $T^2 := Cn^2(T \cup \{A11\})$.

(iii) Ist T irgendeine der in (i), (ii) genannten (dimensionsfreien) Geometrien und $n \geq 2$ eine natürliche Zahl, so bezeichnen wir mit T_n ("*n-dimensionale Geometrie*"), $T_{\geq n}$ bzw. $T_{\leq n}$ die Theorie, die daraus durch Hinzunahme des Dimensionsaxioms, des unteren bzw. des oberen Dimensionsaxioms für die Dimension n (I.11.70) entsteht. Nach I.11.71, 72 entsteht insbesondere T_2 ("*ebene Geometrie*") auch durch Hinzunahme von A9.

Beispiele für die so erklärten Geometrien sind etwa P' (dimensionsfreie euklidische Geometrie mit Kreisaxiom), $A_{\geq 4}$ (absolute Geometrie von Dimension ≥ 4), $P_2^2 = Cn^2(\{A1, \ldots A11\})$ (volle ebene euklidische Geometrie).

Die Sätze in den Abschnitten I.2 bis I.11 sind dann, soweit sie in der ersten Stufe formuliert sind, schon Sätze von A im Sinne von (II.)1.32. Offenbar ist A eine Untertheorie (1.36) aller in 2.1 genannten Theorien.

Im Darstellungssatz I.16.15 wurden die Modelle der Theorien P_n ($n \geq 2$, $n \in \mathbb{N}$) und der hier betrachteten Obertheorien bis auf Isomorphie charakterisiert als die n-dimensionalen kartesischen Räume über pythagoreischen, euklidischen, reell-abgeschlossenen bzw. zu $\mathbb{R}$ isomorphen angeordneten Körpern.

2.2 Hyperbolische Geometrie. (i) Die *dimensionsfreie hyperbolische Geometrie* H werde aus der absoluten Geometrie A durch Hinzunahme des folgenden hyperbolischen Parallelenaxioms HP (nach Hilbert 1977, S. 162) gebildet.

$$\text{(HP)} \quad \neg \text{Col } acd \rightarrow \exists b_1 \exists b_2 \{\neg \text{Col } ab_1b_2$$
$$\wedge \forall u[\text{B}b_1ub_2 \wedge \neq(b_1b_2u) \rightarrow \exists x(\text{B}aux \wedge \text{Col } cdx)]$$
$$\wedge \neg \exists x[(\text{B}ab_1x \vee \text{B}ab_2x) \wedge \text{Col } cdx]\}$$

(zu beliebigen nicht-kollinearen Punkten a, c, d gibt es [in der dadurch bestimmten Ebene] stets zwei von a ausgehende Halbgeraden $H(ab_1)$, $H(ab_2)$, die nicht in einer Geraden enthalten sind und die Gerade $L(cd)$ nicht schneiden, während jede

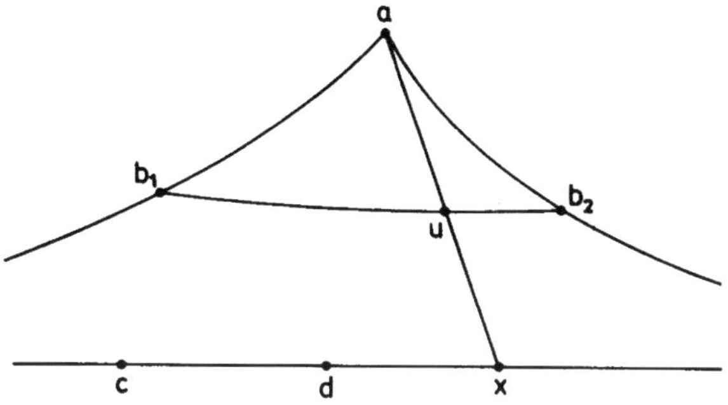

Abb. 118

echt innerhalb des Winkels $\sphericalangle b_1ab_2$ liegende von a ausgehende Halbgerade die Gerade L(cd) schneidet, Abb. 118).

(ii) Wir verwenden die Definitionen 2.1(4), (5) auch für T=H und 2.1(iii) für die so gebildeten Theorien T und erklären damit entsprechende Erweiterungen von H. (2.1(3) ist hier entbehrlich, da sich das Kreisaxiom CA bereits als Satz von H erweist.)

2.3 Definitionen und Sätze. (i) Einige grundlegende Begriffe und Sätze der hyperbolischen Geometrie seien hier ohne Beweis angegeben. Vgl. dazu etwa Hilbert 1977 (Anhang III) und Borsuk u. Szmielew 1960 (dort wird zwar jeweils ein anderer axiomatischer Aufbau und eine feste Dimension vorausgesetzt, auf Grund der in Teil I für A bewiesenen Sätze lassen sich die Beweise aber auch für unseren Aufbau verwenden).

(ii) Die Halbgeraden H(ab) und H(cd) sind zueinander *hyperbolisch parallel* oder *grenzparallel* (manchmal kurz "parallel") *im engeren Sinne* oder *echt grenzparallel*:

(1) $ab\check{\Vert}cd :\leftrightarrow \neg\exists x[Babx \wedge Bcdx]$
$\wedge \forall u\{Bbud \wedge u \neq b \rightarrow \exists x[Baux \wedge Bcdx]\}$

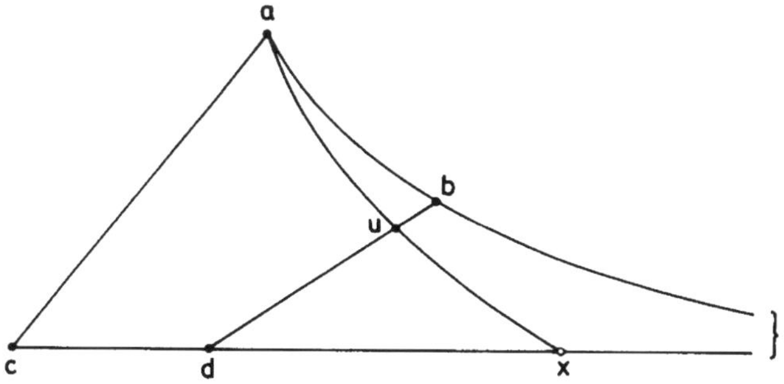

Abb. 119

(unter den von *a* ausgehenden Halbgeraden in der Halbebene [s. I.9.14] Hp(*ac*,*d*) [zu denen, wie sich zeigen läßt, H(*ab*) gehört] ist H(*ab*) die der Halbgeraden H(*cd*) "am nächsten gelegene", die H(*cd*) nicht schneidet, Abb. 119).

In Abbildungen wird dieser Sachverhalt durch eine geschweifte Klammer angedeutet, die die betreffenden Halbgeraden zusammenfaßt.

Die in der Formulierung von HP auftretenden Halbgeraden H(ab_1), H(ab_2) sind dann gerade diese hyperbolischen Parallelen zu H(*cd*) bzw. H(*dc*); sie heißen auch die beiden *hyperbolischen Parallelen* von *a* aus *zu der Geraden* L(*cd*).

Die Halbgeraden H(*ab*) und H(*cd*) entstehen auseinander durch *Verschiebung des Ursprungs* auf ihrer Trägergeraden oder die "gerichteten Strecken" (d.h. Punktepaare) <*a*,*b*>, <*c*,*d*> haben *gleiche Orientierung* (auf derselben Trägergeraden):

(2) $ab \underset{o}{=} cd \; :\leftrightarrow \exists u[a \underset{u}{\simeq} c \wedge a \neq b \wedge Babu \wedge c \neq d \wedge Bcdu]$.

Als gleichwertig (aber in einer komplizierteren Formalisierung) erhält man

(3) $ab \underset{o}{=} cd \leftrightarrow H(ab) \subseteq H(cd) \vee H(cd) \subseteq H(ab)$.

Die Halbgeraden H(*ab*) und H(*cd*) sind zueinander *(hyperbolisch) parallel* oder *grenzparallel im weiteren Sinne*:

(4) $ab \mathrel{\|\!\!-} cd \; :\leftrightarrow ab \overset{\vee}{\mathrel{\|\!\!-}} cd \vee ab \underset{o}{=} cd$.

(iii) *Geraden A* und *B* heißen dann *(hyperbolisch) parallel* oder *grenzparallel* im engeren bzw. weiteren Sinne (Bezeichnung $A \overset{\vee}{\mathrel{\|\!\!-}} B$ bzw. $A \mathrel{\|\!\!-} B$), falls sie hyperbolisch parallele Halbgeraden im engeren bzw. weiteren Sinne enthalten.

(iv) Dieser Begriff der Grenzparallelität ist natürlich auch in der absoluten Geometrie *A* sinnvoll und fällt in der euklidischen Geometrie *P* mit den früheren Begriffen $\overset{\vee}{\|}$ bzw. $\|$ (I.12.2, 3) zusammen. Schon in *A* läßt sich zeigen (s. Borsuk u. Szmielew 1960, S. 138f.), daß die Relation $\mathrel{\|\!\!-}$ eine Äquivalenzrelation zwischen Halbgeraden ist und daß sie invariant ist gegenüber Verschiebungen gemäß (2) (d.h. Übergang zu orientierungsgleichen Punktepaaren).

(v) In *H* (und in *P* und - auf Grund eines speziellen Stetigkeitsaxioms - auch in *A**) ergibt sich zusätzlich:

Zu jeder Halbgeraden gibt es von jedem Punkt aus genau eine dazu hyperbolisch parallele Halbgerade. (In A ergibt sich dagegen nur die Eindeutigkeit, aber nicht die Existenz solcher Grenzparallelen, vgl. 6.27ff.)

(vi) In A^* ist darüber hinaus das hyperbolische Parallelenaxiom HP äquivalent zur Negation des euklidischen Parallelenaxioms (A10 bzw. I.12.11). Das wurde (mit anderen Worten) für die Geometrien A_n^* festgestellt in Szmielew 1959a, S. 50f.; ein Beweis (für eine anders aufgebaute Geometrie, aber übertragbar im Sinne von (i)) ist enthalten in Schwabhäuser 1959, S. 143ff.

(vii) Die Äquivalenzklassen bezüglich $\nparallel$ werden (unter Voraussetzung der Existenz von Grenzparallelen gemäß (v)) *Enden* genannt. Die Enden lassen sich als eine Art von "uneigentlichen Punkten" auffassen; statt "*E* ist die Äquivalenzklasse von H(*ab*) " sagt man daher auch "die Halbgerade H(*ab*) " bzw. "die Gerade L(*ab*) *geht durch* das Ende *E* ".

Dann ergeben sich in *H* folgende Sätze.

(viii) *Zu zwei Geraden in einer Ebene, die sich weder schneiden noch zueinander hyperbolisch parallel sind, gibt es stets genau eine gemeinsame Senkrechte.*

(ix) *Durch zwei Enden geht stets genau eine Gerade.*

(x) *Von einem Ende gibt es (wie von einem Punkt) auf eine nicht durch dieses Ende gehende Gerade stets genau ein Lot.*

(xi) Damit kann man folgende Konstruktion einführen. In der Beziehung $ab\check{\nparallel}cd$ (Abb. 119) werde zusätzlich vorausgesetzt, daß der Winkel $\sphericalangle acd$ ein Rechter ist. Dann ist die Größe des (spitzen) Winkels $\sphericalangle cab$ (im Sinne von I.13.4) eindeutig festgelegt durch die Größe der Strecke *ac* und umgekehrt. Man kann somit eine Funktion Π einführen, die jeder solchen Streckengröße die zugehörige Winkelgröße zuordnet. Π ist dann eine eineindeutige Abbildung von der Menge der Größen von echten Strecken auf die Menge der Größen von spitzen Winkeln; außerdem erhält man, daß Π streng monoton fallend ist (im Sinne des Größenvergleichs für Strecken- und Winkelgrößen in I.5.13 bzw. I.11.37). Π heißt dann die *Parallelwinkelfunktion*. Ist $\Pi(A)=\alpha$, so heißt α *Parallelwinkel(größe)* von *A* und *A Paralleldistanz(größe)* von α. Die Bezeichnungen Parallelwinkel und Paralleldistanz sind auch für entsprechende (nur bis auf Kongruenz festgelegte) Strecken und Winkel üblich.

(xii) *In jedem nicht-entarteten Dreieck ist die Summe der Innenwinkel kleiner als ein gestreckter Winkel.*

2.4 (Hyperbolische) Kleinsche Räume. Unter dem n-dimensionalen (hyperbolischen) *Kleinschen Raum* $\mathfrak{K}_n(\mathfrak{F})$ über dem angeordneten Körper $\mathfrak{F}$ ($n\in\mathbb{N}$) verstehen wir die folgendermaßen festgelegte Struktur $\mathfrak{A}$ für die Sprache L(D,B):

(1) $U_{\mathfrak{A}} = \{x\in F^n \mid x^2 < 1\}$ (wobei $F=U_{\mathfrak{F}}$)
(d.h., geometrisch ausgedrückt, die Trägermenge ist das Innere der n-dimensionalen Einheitskugel um den Koordinatenursprung im kartesischen Raum $\mathfrak{L}_n(\mathfrak{F})$),

(2) $D_{\mathfrak{A}}xyuv$ gdw $\dfrac{(1-xy)^2}{(1-x^2)(1-y^2)} = \dfrac{(1-uv)^2}{(1-u^2)(1-v^2)}$,

(3) $B_{\mathfrak{A}}xyz$ gdw es gibt ein $\lambda\in F$ mit $0\leq\lambda\leq 1$ und $y-x=\lambda(z-x)$.

Hierbei sind natürlich wieder die üblichen Bezeichnungen aus der Vektorrechnung verwendet, insbesondere bedeutet xy das innere Produkt $\sum_{\nu=1}^{n} x_\nu y_\nu$ der Vektoren (n-tupel) x und y sowie x^2 das innere Produkt von x mit sich selbst. Zu einer weiteren Beschreibung von $\mathfrak{K}_n(\mathfrak{F})$ s. 2.10.

2.5 Darstellungssatz für H_n. *Sei* $n\geq 2$, $n\in\mathbb{N}$.

(i) *Eine Struktur* $\mathfrak{A}$ *für* L(D,B) *ist ein Modell von* H_n *genau dann, wenn sie isomorph ist zum n-dimensionalen Kleinschen Raum* $\mathfrak{K}_n(\mathfrak{F})$ *über einem euklidischen angeordneten Körper* $\mathfrak{F}$.

(ii) *In* (i) *ist* $\mathfrak{F}$ *durch* $\mathfrak{A}$ *bis auf Isomorphie eindeutig bestimmt.*

(iii) *Seien* $\mathfrak{A}$ *und* $\mathfrak{F}$ *derart, daß beide Seiten von* (i) *gelten. Dann ist* $\mathfrak{A}$ *sogar ein Modell von* H_n^* *bzw.* H_n^2 *genau dann, wenn* $\mathfrak{F}$ *reell-abgeschlossen bzw.* $\mathfrak{F}$ *isomorph zu* $\mathbb{R}$ *ist.*

2.6 Anmerkung. Zu verschiedenen Beweisen von 2.5 s. etwa Schwabhäuser 1959 und Szmielew 1959, 1959a, 1961, 1962. Ein wesentlicher Teil des Beweises ist natürlich in jedem Fall (wie für P_n in Teil I) die geometrische Konstruktion eines angeordneten Körpers. Der erste so konstruierte Körper ist die "Hilbertsche Endenrechnung" (1903)(s. Hilbert 1977, S. 170ff.). Verschiedene andere solche Körper ("Streckenkalküle") wurden in den genannten Arbeiten von Wanda Szmielew konstruiert, einige der Konstruktionen sind für die hyperbolische und die euklidische Geometrie gemeinsam anwendbar. Eine weitere Körperkonstruktion ist enthalten in Doraczyńska 1977. Unter Benutzung des konstruierten Körpers sind sodann Koordinaten einzuführen. Die zu den Kleinschen Räumen führenden Koordinaten heißen auch "Beltramische Koordinaten". In Szász 1958, 1959 wird ein Darstellungssatz mit sog. "Weierstraßschen homogenen Koordinaten" behandelt, was auf eine andere Definition der (zu den Kleinschen Räumen isomorphen) Modelle hinausläuft. Vgl. auch Gerretsen 1942, Szász 1962. Untersuchungen für den "klassischen Fall" (mit voller Stetigkeit, $\mathfrak{f} = \mathbb{R}$) sind wesentlich älter.

2.7 Weitere Geometrien. Die anschließend betrachteten Geometrien sind keine Erweiterungen von A. Für geeignete Axiomatisierungen lassen sich entsprechende Darstellungssätze beweisen. Wir verzichten hier auf die Angabe von Axiomensystemen (vgl. 2.14(ii)) und benutzen nur das Ergebnis des jeweiligen Darstellungssatzes, d.h., wir charakterisieren diese Geometrien als Theorien von Strukturklassen. Daher betrachten wir zuerst die zugehörigen Räume (Strukturen).

2.8 Affine Räume über Körpern. Sei n eine beliebige (endliche oder unendliche) Kardinalzahl und $\mathfrak{f}$ ein Körper (zunächst ohne Anordnung). Mit $\mathfrak{V}_n(\mathfrak{f})$ bezeichnen wir den (durch $\mathfrak{f}$ und n bis auf Isomorphie eindeutig bestimmten) n-dimensionalen *Vektorraum* über $\mathfrak{f}$. (Für viele Anwendungen werden nur endliche Dimensionen gebraucht; eine Darstellung unter Einbeziehung unendlicher Dimensionen findet sich z.B. in Day 1973, S. 1-5). Im hier betrachteten Zusammenhang ist es zweckmäßig, die affinen Räume als Strukturen für die Sprache L(Col,||) einzuführen, d.h. die (dreistellige) Kollinearität und die (als Relation zwischen Punkten vierstellige) Parallelität als Grundbegriffe zu verwenden (in der Literatur werden oft andere Grundbegriffe gewählt, zu einem Wechsel der Grundbegriffe vgl. 4.59, 60, bei Ausschluß von Körpern der Charakteristik 2 ist die Parallelität entbehrlich, vgl. 4.83, 5.3(i)).

(i) Unter dem *n-dimensionalen affinen Raum* $\mathfrak{A}_n(\mathfrak{f})$ über $\mathfrak{f}$ verstehen wir die folgendermaßen festgelegte Struktur $\mathfrak{A}$ für die Sprache L(Col,||):

(1) $U_{\mathfrak{A}}$ ist die Menge der Elemente ("Vektoren") von $\mathfrak{V}_n(\mathfrak{f})$, diese Elemente werden dann auch die *Punkte* von $\mathfrak{A}_n(\mathfrak{f})$ genannt,

(2) $Col_{\mathfrak{A}}\, xyz$ gdw die Vektoren $y-x$, $z-x$ sind linear abhängig,

(3) $xy\|_{\mathfrak{A}}\, uv$ gdw die Vektoren $y-x$ und $v-u$ sind vom Nullvektor ver- verschieden und linear abhängig.

(ii) Unter dem *n-dimensionalen affinen Raum* $\mathfrak{A}_n^B(\mathfrak{f})$ *(mit Anordnung)* über dem angeordneten Körper $\mathfrak{f}$ verstehen wir die Struktur $\mathfrak{A}$ für die Sprache L(B), die festgelegt ist durch (1) wie oben und

(2') $B_{\mathfrak{A}}\, xyz$ gdw es gibt ein $\lambda\in\mathfrak{f}$ mit $0\leq\lambda\leq 1$ und $y-x=\lambda(z-x)$.

<u>2.9 Projektive Räume über Körpern</u>. Sei jetzt n eine endliche Dimension ($n\in\mathbb{N}$) und $\mathfrak{f}$ ein Körper.

(i) Unter dem *n-dimensionalen projektiven Raum* $\mathfrak{P}_n(\mathfrak{f})$ über $\mathfrak{f}$ verstehen wir die folgendermaßen festgelegte Struktur $\mathfrak{P}$ für die Sprache L(Col):

(1) $U_{\mathfrak{P}}$ ist die Menge der eindimensionalen Unterräume von $\mathfrak{V}_{n+1}(\mathfrak{f})$, diese Unterräume werden dann auch die *Punkte* von $\mathfrak{P}_n(\mathfrak{f})$ genannt,

(2) $Col_{\mathfrak{P}}\, xyz$ gdw (die eindimensionalen Unterräume) x, y, z sind enthalten in einem zweidimensionalen Unterraum von $\mathfrak{V}_{n+1}(\mathfrak{f})$.

Zu den folgenden zusätzlichen Begriffsbildungen und Resultaten vgl. z.B. Schaal 1976a.

(ii) Eine Teilmenge M von $U_{\mathfrak{P}}$ (wie in (1)) heißt ein *k-dimensionaler Unterraum* von $\mathfrak{P}_n(\mathfrak{f})$ (für $k=1$ insbesondere eine *Gerade*, für $k=n-1$ eine *Hyperebene*), falls es einen $(k+1)$-dimensionalen Unterraum N von $\mathfrak{V}_{n+1}(\mathfrak{f})$ gibt, so daß M aus genau den in N enthaltenen eindimensionalen Unterräumen besteht ($0\leq k\leq n$).

(iii) Zeichnet man in $\mathfrak{P}=\mathfrak{P}_n(\mathfrak{f})$ eine beliebige Hyperebene H aus (sog. "*uneigentliche Hyperebene*") und nimmt diese aus $\mathfrak{P}$ heraus, so bleibt bekanntlich bis auf Isomorphie ein affiner Raum $\mathfrak{A}=\mathfrak{A}_n(\mathfrak{f})$ (gleicher Dimension über demselben Körper) übrig. Dabei sind Geraden in $\mathfrak{A}$ zueinander parallel genau dann, wenn sie (bzw. ihre Bilder in $\mathfrak{P}$) sich in einem Punkt von H schneiden. In analoger Weise läßt sich jeder affine Raum $\mathfrak{A}_n(\mathfrak{f})$ durch Hinzunahme einer "uneigentlichen Hyperebene" in den projektiven Raum $\mathfrak{P}_n(\mathfrak{K})$ isomorph einbetten. Die Punkte der uneigentlichen Hyperebene heißen dabei auch *uneigentliche Punkte* oder *Fernpunkte*. (Zum entsprechenden Zusammenhang für Räume mit Anordnung s. 6.60.)

(iv) Durch eine beliebige Basis $\langle b_o, b_1, \dots, b_n \rangle$ von $\mathfrak{V}_{n+1}(\mathfrak{K})$ wird ein *projektives Koordinatensystem* von $\mathfrak{P}_n(\mathfrak{K})$ festgelegt: $\langle x_o, x_1, \dots, x_n \rangle$ heißt ein *Koordinaten-*$(n+1)$*-tupel* des Punktes x in bezug auf dieses System, falls $\sum_{\nu=o}^{n} x_\nu b_\nu \in x-\{o\}$ ist (wobei o der Nullvektor ist); ein Koordinaten-$(n+1)$-tupel ist somit stets verschieden vom $(n+1)$-tupel aus lauter Nullen und bestimmt eindeutig den zugehörigen Punkt x, es ist aber durch diesen Punkt nur bis auf einen konstanten Faktor eindeutig festgelegt (Koordinaten[systeme] dieser Art nennt man auch *homogene* Koordinaten[systeme]). Ein solches Koordinatensystem läßt sich auch festlegen durch $n+1$ "*Grundpunkte*", die den Basisvektoren entsprechen, und einen weiteren "*Einheitspunkt*", zu dem das Koordinaten-$(n+1)$-tupel aus lauter Einsen gehört.

(v) Das *Doppelverhältnis* (engl. cross ratio) (ab,cd) von paarweise verschiedenen Punkten a, b, c, d auf einer Geraden läßt sich folgendermaßen einführen. Sei gemäß (iii) eine "uneigentliche Hyperebene" H so gewählt, daß sie keinen dieser vier Punkte enthält, und sei $x = p+t_x q$ eine Parameterdarstellung der betrachteten Geraden in dem verbleibenden affinen Raum $\mathfrak{A}$. Dann ist

$$(ab,cd) = \frac{t_a - t_c}{t_a - t_d} : \frac{t_b - t_c}{t_b - t_d} ;$$

das Doppelverhältnis ist also der Quotient der beiden "Teilverhältnisse", in denen die gerichtete Strecke $\langle c,d \rangle$ von den Punkten a bzw. b geteilt wird. Das Doppelverhältnis ist (im Unterschied zu den genannten Teilverhältnissen) invariant gegenüber "projektiven Abbildungen" und läßt sich - unabhängig von der Wahl von H - auch durch projektive homogene Koordinaten ausdrücken. Es kann (für paarweise verschiedene Punkte) alle Werte in $\mathfrak{K}$ außer 0 und 1 annehmen, durch diesen Wert und drei der betrachteten Punkte ist der vierte eindeutig bestimmt.

(vi) Analog zu 2.8 wollen wir auch projektive Räume mit Anordnung einführen. Zur Beschreibung der Anordnung von Punkten auf einer Geraden verwendet man üblicherweise die vierstellige Beziehung, daß ein Punktepaar $\langle a,b \rangle$ ein weiteres Paar $\langle c,d \rangle$ *trennt* (abgekürzt $ab|cd$). Unter dem *n-dimensionalen projektiven Raum* $\mathfrak{P}_n^{|}(\mathfrak{K})$ (*mit Anordnung*) über dem angeordneten Körper $\mathfrak{K}$ verstehen wir nun die Struktur $\mathfrak{P}$ für die Sprache L(|), die festgelegt ist durch (1) wie oben und

(2') $ab|_{\mathfrak{P}}cd$ gdw a, b, c, d sind paarweise verschiedene Punkte auf einer Geraden und $(ab,cd) < 0$.

Bei der "affinen Beschreibung" der Doppelverhältnisse aus (v) bedeutet (2'), daß die gerichtete Strecke $\langle c,d\rangle$ von einem der beiden Punkte a, b "innen" und von dem anderen "außen" geteilt wird (die Teilverhältnisse haben entgegengesetzte Vorzeichen). Das kann man sich (mit den Bezeichnungen von (iii)) auch folgendermaßen veranschaulichen. Eine Gerade von $\mathfrak{P}$ (die aus einer Geraden von $\mathfrak{A}$ durch Hinzunahme eines uneigentlichen Punktes entsteht) verhält sich topologisch wie eine Kreislinie (jedenfalls für $\mathfrak{F}=\mathbb{R}$). Daß Punktepaare sich trennen, bedeutet dann: Wenn man diese Kurve von einem gegebenen Punkt eines Paares aus in einem gewissen Umlaufssinn durchläuft, so stößt man stets zuerst auf einen Punkt des anderes Paares, bevor man den zweiten Punkt des ersten Paares erreicht. Damit wird auch anschaulich klar, daß eine dreistellige Zwischenbeziehung zur Beschreibung der Anordnung in projektiven Räumen ungeeignet ist.

In (vi) ist wieder (wie in 2.8(ii)) die Kollinearität nicht als Grundbegriff aufgenommen worden, da sie sich auch hier leicht durch die Anordnung charakterisieren läßt; aus (1), (2), (2') erhält man nämlich

(3) $\mathrm{Col}_{\mathfrak{P}}xyz$ gdw es gibt ein u mit $xy|_{\mathfrak{P}}zu$ oder x, y, z sind nicht paarweise verschieden.

2.10 Ergänzung zu 2.4. Betrachtet man den (hyperbolischen) Kleinschen Raum $\mathfrak{A}=\mathfrak{K}_n(\mathfrak{F})$ als eingebettet in den kartesischen Raum $\mathfrak{L}_n(\mathfrak{F})$, so ist die Zwischenbeziehung von $\mathfrak{A}$ nach 2.4(3) die Einschränkung der Zwischenbeziehung für kartesische Räume auf die Trägermenge $U_{\mathfrak{A}}$. Die Menge der Randpunkte von $U_{\mathfrak{A}}$, d.h. $\{x\in F^n \mid x^2=1\}$ heißt auch die *absolute Hyperfläche* oder *Eichhyperfläche* (für $n=3$ auch *absolute* oder *Eichfläche*, für $n=2$ auch *Eichkurve*). Unter der im Folgenden stets verwendeten Voraussetzung, daß $\mathfrak{F}$ euklidisch ist, lassen sich die Enden identifizieren mit den Punkten der absoluten Hyperfläche. Außerdem läßt sich diese Hyperfläche folgendermaßen zu einer anderen Charakterisierung der Streckenkongruenz verwenden. Bettet man den durch $\mathfrak{L}_n(\mathfrak{F})$ festgelegten affinen Raum noch in den projektiven Raum $\mathfrak{P}_n(\mathfrak{F})$ ein (vgl. 2.9(iii)), so sind die Bewegungen (s. I.4.8) genau die (Einschränkungen auf $U_{\mathfrak{A}}$ der) projektiven Abbildungen, die die absolute Hyperfläche in sich überführen; Strecken sind zueinander kongruent genau dann, wenn eine durch eine Bewegung in die andere übergeführt werden kann. Wegen der Invarianz von

Doppelverhältnissen unter projektiven Abbildungen (vgl. 2.9(v)) ist die Größe (und Orientierung) einer (gerichteten) Strecke <*a*,*b*> auch eindeutig festgelegt durch das Doppelverhältnis (ab,st), wobei s, t die Enden der Trägergeraden $L(ab)$ (in einer gegebenen Reihenfolge) sind. Im "Standardmodell" $\mathfrak{K}_n(\mathbb{R})$ definiert man dann üblicherweise die *Länge* der Strecke ab als die reelle Zahl $\frac{1}{2}|\ln(ab,st)|$. Zu weiteren Betrachtungen über dieses Modell vgl. etwa Baldus 1964.

2.11 Elliptische Kleinsche Räume. Sei $\mathfrak{f}$ ein angeordneter Körper und $n\in\mathbb{N}$. Sei $\mathfrak{P}_n(\mathfrak{f})$ der durch 2.9 erklärte projektive Raum, wobei $\mathfrak{V}_{n+1}(\mathfrak{f})$ jetzt der Vektorraum der $(n+1)$-tupel von Elementen von $\mathfrak{f}$ (mit den üblichen Operationen) sei. Für einen beliebigen Punkt $x\in\mathfrak{P}_n(\mathfrak{f})$ bedeute $\bar{x}$ jeweils einen Vektor von $\mathfrak{V}_{n+1}(\mathfrak{f})$, der x (als eindimensionalen Unterraum) aufspannt. (Damit ist ein projektives homogenes Koordinatensystem ausgezeichnet, in dem $\bar{x}$ ein Koordinaten-$(n+1)$-tupel ["Koordinatenvektor"] von x ist.)

(i) Unter dem *n-dimensionalen elliptischen Kleinschen Raum* $\mathfrak{E}_n(\mathfrak{f})$ über $\mathfrak{f}$ verstehen wir dann die folgendermaßen festgelegte Struktur $\mathfrak{A}$ für die Sprache $L(D;|)$:

(1) $U_{\mathfrak{A}}$ ist die Menge der Punkte von $\mathfrak{P}_n(\mathfrak{f})$,

(2) $D_{\mathfrak{A}}xyuv$ gdw $\frac{(\bar{x}\bar{y})^2}{\bar{x}^2\bar{y}^2} = \frac{(\bar{u}\bar{v})^2}{\bar{u}^2\bar{v}^2}$

(diese Bedingung ist offenbar unabhängig von der Wahl der Koordinatenvektoren, d.h. invariant unter Multiplikation der Vektoren $\bar{x}$, $\bar{y}$, $\bar{u}$, $\bar{v}$ mit Faktoren $\neq 0$ aus $\mathfrak{f}$),

(3) wie 2.9(2') (mit $\mathfrak{A}$ statt $\mathfrak{P}$).

(ii) Eine besonders einfache, anschaulichere Beschreibung ergibt sich für das "Standardmodell" $\mathfrak{E}_n(\mathbb{R})$. Veranschaulicht man sich die eindimensionalen Unterräume x, y,... als Geraden durch den Nullpunkt von $\mathfrak{V}_{n+1}(\mathbb{R})$, so ist die Gleichung auf der rechten Seite von (2) bekanntlich gleichwertig mit $\cos^2(x,y)=\cos^2(u,v)$, d.h., daß die von den jeweiligen Geraden eingeschlossenen (nichtstumpfen) Winkel oder die zugehörigen Bogenlängen auf der n-dimensionalen Einheitssphäre $S_n = \{\bar{x}\,|\,\bar{x}^2=1\}$ gleich groß sind. Diese Bogenlängen führt man dann üblicherweise als Längen von Strecken ein. Die Punkte x von $\mathfrak{E}_n(\mathbb{R})$ entsprechen eineindeutig den Zweiermengen $\{\bar{x}_1,\bar{x}_2\}$ von diametral gegenüberliegenden Punkten, in denen S_n von der "Geraden" x geschnitten wird. Die elliptische

Geometrie in $\mathfrak{E}_n(\mathbb{R})$ entsteht also aus der Geometrie auf der n-dimensionalen Sphäre (für $n=2$ insbesondere aus der Kugelgeometrie), indem diametral gegenüberliegende Punkte identifiziert werden. Dabei behalten übrigens auch Winkelmaße ihre Bedeutung, und Geraden von $\mathfrak{E}_n(\mathbb{R})$ werden durch Größtkreise von S_n dargestellt.

2.12 Definition. Zur bequemeren Schreibweise bezeichnen wir mit F die Klasse der Körper (engl. fields) und mit OF, PF, EF bzw. RF die Klasse der angeordneten Körper, der pythagoreischen, euklidischen bzw. reell-abgeschlossenen angeordneten Körper.

2.13 Zugehörige n-dimensionale Geometrien. Sei $n \geq 2$, $n \in \mathbb{N}$.

(i) In Analogie zu P_n charakterisieren wir die *n-dimensionale elliptische Geometrie* E_n als die Theorie der (Klasse der) entsprechenden Räume $\mathfrak{E}_n(\mathfrak{F})$ über pythagoreischen angeordneten Körpern, d.h.

(1) $E_n = \mathrm{Th}(\{\mathfrak{E}_n(\mathfrak{F}) \mid \mathfrak{F} \in PF\})$.

(ii) *n-dimensionale affine* und *projektive Geometrien* ohne und mit Anordnung charakterisieren wir als die Theorien der entsprechenden Räume über beliebigen Körpern bzw. angeordneten Körpern:

(2) $Af_n = \mathrm{Th}(\{\mathfrak{A}_n(\mathfrak{F}) \mid \mathfrak{F} \in F\})$,

(3) $Pr_n = \mathrm{Th}(\{\mathfrak{P}_n(\mathfrak{F}) \mid \mathfrak{F} \in F\})$,

(4) $Afo_n = \mathrm{Th}(\{\mathfrak{A}_n^B(\mathfrak{F}) \mid \mathfrak{F} \in OF\})$,

(5) $Pro_n = \mathrm{Th}(\{\mathfrak{P}_n^{|}(\mathfrak{F}) \mid \mathfrak{F} \in OF\})$.

(iii) Ist T eine beliebige der in (1) bis (5) genannten Theorien, so bilden wir die Erweiterungen T^* bzw. T^2 als Theorie der entsprechenden Räume über reell-abgeschlossenen (ggf. angeordneten) Körpern bzw. als Theorie zweiter Stufe des entsprechenden Raumes über $\mathbb{R}$, außerdem sei

(6) $E_n' = \mathrm{Th}(\{\mathfrak{E}_n(\mathfrak{F}) \mid \mathfrak{F} \in EF\})$.

2.14 Anmerkungen. (i) Für affine und projektive Geometrien ist eine Beschränkung auf pythagoreische oder euklidische Körper nicht geometrisch motiviert, da die Streckenkongruenz zur Formulierung eines Streckenabtragungs- oder Kreisaxioms nicht zur Verfügung steht. Im Unterschied zu diesen Geometrien werden die Geometrien, in denen die Streckenkongruenz oder ein gleichstarker Begriff (z.B. die Orthogonalität) zur Verfügung steht, auch als *metrische Geometrien* bezeichnet.

(ii) In der Literatur finden sich zahlreiche Axiomatisierungen für elliptische, affine und projektive Geometrien, meist allerdings mit anderen Grundbegriffen als den hier gewählten oder nur für eine feste Dimension, aber so, daß man die erforderlichen Abänderungen für die hier betrachteten Geometrien vornehmen kann (zum Wechsel der Grundbegriffe vgl. auch Abschnitt 4). Zu geeigneten Axiomen für die projektiven und die affinen Geometrien s. etwa Lingenberg 1969, für die Trennbeziehung | s. Heffter 1950 (S. 16ff.), für elliptische Geometrien s. etwa Podehl u. Reidemeister 1934 (ohne Trennbeziehung, mit geeigneten Formulierungen eines Axioms der Streckenabtragung für E_n und eines Kreisaxioms für E_n'), Schwabhäuser 1965 (elementare Stetigkeitsaxiome) und Kordos 1973 (mit einer einzigen binären Relation als Grundbegriff, vgl. 4.80). Im Beweis von Darstellungssätzen (vgl. 2.6) ist wieder die geometrische Konstruktion eines Körpers ein wesentlicher Bestandteil. Eine solche Konstruktion für affine Geometrien (unter Verwendung des Satzes von Pappus-Pascal als Axiom) wurde schon in Teil I mit behandelt (vgl. die Anmerkungen 13.13, 13.17 und 14.21 in Teil I). Auf Grund von 2.9(iii) läßt sich diese Konstruktion übertragen auf entsprechende projektive und damit auch auf elliptische Geometrien.

(iii) Unter den Axiomatisierungen in der Literatur sind viele auch stärker oder schwächer als die für die obige Darstellung geeigneten. Wesentlich allgemeiner sind insbesondere die *nicht-desarguesschen* (ebenen) affinen und projektiven Geometrien (vgl. Lingenberg 1969). Übrigens wird auch der Begriff *absolute Geometrie* oft wesentlich allgemeiner als hier gefaßt, nämlich als gemeinsame Untertheorie der metrischen (euklidischen, hyperbolischen und elliptischen) Geometrien, vgl. insbesondere Bachmann 1959, 1964a. Es wäre sicher interessant, auch solche Geometrien unter metamathematischen Gesichtspunkten zu untersuchen.

(iv) Für die "klassische" absolute Geometrie auf Grund der Hilbertschen Axiomgruppen I bis III und damit - nach den Ergebnissen von Teil I - auch für die hier betrachtete absolute Geometrie A_2 wurde ein Darstellungssatz bewiesen in Pejas 1961, vgl. auch Bachmann 1964; der Satz läßt sich auf die A_n $(n \geq 2)$ übertragen. Als Spezialfälle ergeben sich die Darstellungssätze für P_n, H_n und andere Geometrien; unter den Modellen sind auch solche mit "elliptischer Metrik", die sich in einen elliptischen Raum $\mathfrak{E}_n(\mathfrak{F})$ isomorph einbetten lassen (vgl. 6.29ff.).

2.15 Zugehörige dimensionsfreie Geometrien erster Stufe. Die bisherigen Darstellungssätze bezogen sich jeweils auf eine feste Dimension. Nach einem Resultat von D. Scott, das im nächsten Abschnitt behandelt wird (s. 3.70ff., 3.77), erweist sich die dimensionsfreie euklidische Geometrie P (mit Streckenabtragung!) als der Durchschnitt der P_n für alle endlichen Dimensionen $n \geq 2$. Eine Formel ist also ein Satz von P genau dann, wenn sie für jedes endliche $n \geq 2$ ein Satz von P_n ist, d.h., wenn sie in allen kartesischen Räumen beliebiger endlicher Dimension ≥ 2 über pythagoreischen Körpern gilt. (Die Klasse dieser Räume ist allerdings noch nicht die ganze Modellklasse von P, sondern P besitzt außerdem unendlichdimensionale Modelle, die nur für diese Charakterisierung von P nicht gebraucht werden, vgl. 3.67ff.). Entsprechendes ergibt sich für A, H und die betrachteten Erweiterungen. Mit derselben Methode läßt sich für die Geometrien in 2.13 zeigen, daß sich eine Axiomatisierung der in 2.14(ii) genannten Art aufspalten läßt in einen entsprechenden dimensionsfreien Teil und ein "Dimensionsaxiom" für die Dimension n. Damit ist die folgende Charakterisierung ein völliges Analogon zu den vorher betrachteten (axiomatisch aufgebauten) dimensionsfreien Geometrien.

(i) Die *dimensionsfreie elliptische Geometrie* E wird festgelegt durch

(1) $$E = \bigcap_{2 \leq n \in \mathbb{N}} E_n = \mathrm{Th}(\{\mathfrak{E}_n(\mathfrak{F}) \mid \mathfrak{F} \in PF,\ n \in \mathbb{N},\ n \geq 2\}).$$

(ii) Die *dimensionsfreien affinen* und *projektiven Geometrien* Af, Pr, Afo, Pro ohne und mit Anordnung charakterisieren wir mit Hilfe der Geometrien von 2.13(ii) durch eine analoge Durchschnittsbildung (und damit auch als Theorien entsprechender Strukturklassen).

(iii) Ist T eine beliebige der in (i), (ii) genannten Theorien, so charakterisieren wir auch die Erweiterungen T^* durch eine analoge Durchschnittsbildung mittels der Theorien T_n^* aus 2.13 und ebenso E' mittels der E_n'. Eine zu (1) analoge Charakterisierung als Theorien von Strukturklassen erhält man durch dieselben Beschränkungen für die zugrunde gelegten Körper bzw. angeordneten Körper wie in 2.13.

(iv) Die affinen Geometrien von (ii) und (iii) erhält man übrigens auch, wenn man in der Charakterisierung als Theorien von Strukturklassen die unendlichdimensionalen Räume von 2.8 mit einbezieht (vgl. den Beweis von 3.83, 3.84(i)).

(v) Geometrien $T_{\leq n}$ und $T_{\geq n}$ (analog zu 2.1(iii)) charakterisieren wir für alle betrachteten Fälle durch eine entsprechende Bildung des Durchschnitts über die erlaubten Dimensionen $\leq n$ bzw. $\geq n$.

2.16 Zusammenfassung. Für die betrachteten dimensionsfreien metrischen Geometrien kommen wir damit zu der folgenden Zusammenstellung (für die affinen und projektiven Geometrien s. 2.15, 2.13).

	absolute Geometrie	euklidische Geometrie	hyperbolische Geometrie	elliptische Geometrie
über pythagoreischen angeordneten Körpern (mit Streckenabtragung)	A	P		E
über euklidischen angeordneten Körpern	A'	P'	H	E'
über reell-abgeschl. angeordn. Körpern (mit Schema der elementaren Stetigkeitsaxiome)	A^*	P^*	H^*	E^*
("volle") Geometrie zweiter Stufe über $\mathbb{R}$ (mit Stetigkeitsaxiom)	A^2	P^2	H^2	E^2

Untere Indizes dienen zur Einschränkung oder Festlegung der Dimension (vgl. 2.1 (iii)).

Speziellere Geometrien werden auch später noch eingeführt, wo sie untersucht werden oder darüber berichtet wird (s. 3.52, 68, 84(iii); 4.59ff., 65, 80, 84, 89, 100, 103, 105, 108, 110; 5.16, 25, 32, 36; 6.29, 46, 69; 7.1, 2, 9(ii), 25, 33, 60; 8.2, 7, 19).

3. Entscheidbarkeit, Vollständigkeit, Finitisierbarkeit

3.1 Überblick. Zu den grundlegenden metamathematischen Problemen für eine gegebenen Theorie T gehören insbesondere die Fragen, ob diese Theorie entscheidbar bzw. unentscheidbar ist, ob sie vollständig bzw. unvollständig ist und ob sie finitisierbar (d.h. durch ein endliches Axiomensystem charakterisierbar) ist oder nicht. Diese Fragen sollen hier für die in Abschnitt 2 zusammengestellten Geometrien behandelt werden.

Für diese Geometrien gelingt das durch Rückführung auf entsprechende, bereits bekannte Resultate für die Theorien der zugehörigen Klassen von Körpern bzw. angeordneten Körpern. Das wichtigste Hilfsmittel für die Beweise sind daher geeignete Formelübersetzungen zwischen geometrischen und algebraischen Theorien, die hier durch die Konstruktion der sog. arithmetischen bzw. geometrischen Reduzierten vorgenommen werden (s. 3.25, 36, 64). (Diese Methode ist nicht immer anwendbar, vgl. den Unentscheidbarkeitsbeweis für eine andere Geometrie in 7.16f.)

Diese Methode wird - nach Bereitstellung der benötigten Hilfsmittel - ausführlich behandelt für euklidische Geometrien fester endlicher Dimension (ab 3.25). Die für andere Geometrien erforderlichen Abänderungen und die zugehörigen Resultate (ab 3.64) können dann kurz dargestellt werden. Danach (ab 3.67) werden Resultate für dimensionsfreie und für unendlichdimensionale Geometrien behandelt.

Auf präzise Definitionen der Begriffe Entscheidbarkeit, Berechenbarkeit, Berechnungsverfahren können wir hier verzichten (s. 3.5(ii)). Wir begnügen uns stattdessen mit der folgenden anschaulichen Beschreibung.

3.2 Berechnungsverfahren. Ein *Berechnungsverfahren (Algorithmus)* ist ein Verfahren, auf Grund dessen Umformungen von Bezeichnungen für irgendwelche Dinge (z.B. natürliche Zahlen, Formeln, Zeichenreihen) ausgeführt werden. Das Verfahren ist auf (abzählbar) unendlich viele Argumente anwendbar; für jedes einzelne Argument führt das Verfahren in endlich vielen Schritten zu einem Ergebnis. Das ganze Verfahren ist mitteilbar in einem Text endlicher Länge, durch den die Rechenschritte eindeutig festgelegt werden. Für die Ausführung des Verfahrens ist nur das Befolgen von Anweisungen des Textes und keine zusätzliche Intelligenz erforderlich, so daß die Ausführung auch maschinell vorgenommen werden kann.

Beispiele für Berechnungsverfahren sind etwa der euklidische Algorithmus zur Bestimmung des größten gemeinsamen Teilers zweier natürlicher Zahlen und das Verfahren zur Berechnung der n-ten Primzahl (für das Argument n).

3.3 Berechenbarkeit (Rekursivität) von Funktionen. Eine Funktion f von einer abzählbaren Menge Γ in eine Menge Δ, wobei Bezeichnungssysteme für diese Mengen vorausgesetzt werden, heißt *berechenbar* oder *rekursiv*, falls ein Berechnungsverfahren für f existiert, das ist ein Berechnungsverfahren, das zu jedem Argument x aus Γ den Funktionswert $f(x)$ liefert.

3.4 Entscheidbarkeit (Rekursivität) von Mengen. (i) Sei Γ eine abzählbare "Grundmenge", für die ein Bezeichnungssystem zur Verfügung steht. Eine Teilmenge Σ von Γ heißt dann *entscheidbar* oder *rekursiv*, falls ein *Entscheidungsverfahren* für Σ existiert, das ist ein Berechnungsverfahren, mit dem man für jedes Element von Γ feststellen ("entscheiden") kann, ob es in Σ liegt. Als gleichwertig erhält man, daß die charakteristische Funktion von Σ (mit dem Definitionsbereich Γ und den beiden Werten 1 und 0 für die Feststellungen "ja" und "nein") berechenbar ist. Andernfalls (falls kein solches Berechnungsverfahren existiert) heißt Σ *unentscheidbar*.

(ii) Wir benutzen diesen Begriff der Entscheidbarkeit hier für F o r m e l m e n g e n Σ, insbesondere für Theorien. Als Grundmenge können wir dabei die Menge Fml_L aller Formeln der betrachteten Sprache L verwenden (vgl. 3.7). In diesem Zusammenhang setzen wir über L (zusätzlich zu 1.3(*)) voraus, daß ein Bezeichnungssystem für die Menge der

Grundzeichen (und damit für die Mengen der Zeichenreihen und der Formeln) vorhanden ist. Diese Voraussetzung hat zur Folge, daß die Menge der Zeichenreihen abzählbar ist; sie ist insbesondere erfüllt für die im folgenden betrachteten Sprachen mit nur endlich vielen Operations- und Relationszeichen.

3.5 Anmerkungen. (i) In der Literatur finden sich eine ganze Reihe von untereinander gleichwertigen Präzisierungen für die Begriffe Berechnungsverfahren, Berechenbarkeit, Entscheidbarkeit; s. etwa Hermes 1978. Am engsten an anschauliche Vorstellungen angelehnt (aber nicht immer am einfachsten anwendbar) ist wohl die Präzisierung der Berechnungsverfahren als Turingmaschinen (nach A. M. Turing 1936).

(ii) Wenn man allerdings nur ein spezielles Verfahren angibt - etwa zum Nachweis der Entscheidbarkeit einer einzelnen Theorie - , so besteht meist kein Zweifel darüber, daß das ein ausführbares Verfahren im Sinne von 3.2 ist. In solchen Beweisen wird es daher meist nicht als notwendig angesehen, den Begriff des Berechnungsverfahrens überhaupt in seiner vollen Allgemeinheit zu präzisieren. Wichtig wird die Präzisierung vor allem, wenn gezeigt werden soll, daß es kein ("noch so allgemeines") Verfahren gibt, das die gewünschte Berechnung liefert; das ist gerade beim Nachweis der Unentscheidbarkeit einer Theorie der Fall. In diesem Buch wird jedoch die Unentscheidbarkeit einer Theorie T_1 immer nur zurückgeführt auf die schon (m i t Hilfe der Präzisierung) bewiesene Unentscheidbarkeit einer anderen Theorie T_2. Dafür ist die Präzisierung wieder nicht so wichtig; denn diese Rückführung beruht auf einfacheren Sätzen. Im allgemeinen wird wieder nur ein spezielles Verfahren angegeben, mit dem man aus einem als existent angenommenen Entscheidungsverfahren für T_1 auch ein Entscheidungsverfahren für T_2 und damit einen Widerspruch enthalten würde. Die in solchen Beweisen mehr oder weniger anschaulich beschriebenen Verfahren lassen sich jedoch stets umwandeln in Verfahren im Sinne einer Präzisierung nach 3.5(i).

(iii) Die Feststellung der Entscheidbarkeit einer Theorie sagt noch nichts über die Kompliziertheit oder "Komplexität" eines Entscheidungsverfahrens. Die Ausführung des Verfahrens ist zwar "prinzipiell" möglich, es kommt aber häufig vor, daß die Ausführung in interessanten Fällen mit den heutigen Rechenanlagen nicht oder nicht mit vertretbarem Aufwand möglich ist, da sie zu viel Speicherplatz oder Rechenzeit erfordern würde. Die Feststellung der Unentscheidbarkeit besagt dagegen, daß es überhaupt kein Entscheidungsverfahren gibt, also erst recht kein mit vertretbarem Aufwand ausführbares Verfahren.

Die folgende Entscheidbarkeitsfeststellung ist für beliebige Formalisierungen anschaulich einleuchtend (zur Angabe eines Verfahrens vgl. Hermes 1969, S. 56f., S. 62f.).

3.6 Satz (Entscheidbarkeit der Term-, Formel- bzw. Aussageneigenschaft). *Für die Sprachen L aus* 3.4(ii) *gilt bei Verwendung der Menge der Zeichenreihen als Grundmenge: Die Mengen der Terme, der Formeln und der Aussagen sind entscheidbar (bei allen Formalisierungen aus Abschnitt* 1*).*

3.7 Anmerkung. Daraus erhält man sofort, daß man in 3.4(ii) gleichwertig auch die Menge aller Zeichenreihen als Grundmenge verwenden könnte (aus einem auf Formeln anwendbaren Entscheidungsverfahren für Σ erhält man durch Vorschalten eines Verfahrens gemäß 3.6 das Gewünschte). Ebenso kann man sich bei Aussagenmengen Σ auf die Menge aller Aussagen als Grundmenge beschränken. Zum Nachweis der Entscheidbarkeit einer Theorie genügt nach 1.40 ebenfalls ein auf Aussagen anwendbares Entscheidungsverfahren.

3.8 Rekursive Aufzählbarkeit. Sei Γ wie in 3.4(i). Eine Teilmenge Σ von Γ (insbesondere eine Formelmenge wie in 3.4(ii)) heißt *rekursiv aufzählbar*, falls Σ leer ist oder eine berechenbare Funktion f mit dem Definitionsbereich $\mathbb{N}$ und dem Wertebereich Σ existiert. Im zweiten (nicht-trivialen) Fall ist also

$$\Sigma = \{f(n) \mid n\in\mathbb{N}\},$$

d.h., in der "Aufzählung" der Funktionswerte $f(0)$, $f(1)$, $f(2)$,... kommen genau die Elemente von Σ - mit oder ohne Wiederholungen - vor. Man nennt dann f selbst eine *rekursive Aufzählung* oder *Aufzählungsfunktion* für Σ. Ein Verfahren zur sukzessiven Berechnung der genannten Funktionswerte heißt ein *Aufzählungsverfahren* für Σ. Auch ein solches Verfahren ist "prinzipiell" maschinell durchführbar; es bricht zwar nicht ab (die Maschine würde unendlich lange rechnen), aber es liefert zu jedem $n\in\mathbb{N}$ den zugehörigen Funktionswert $f(n)$ "schon" nach endlich vielen Schritten.

3.9 Axiomatisierbarkeit. Sei L eine Sprache mit Bezeichnungssystem (wie in 3.4(ii)) und T eine Theorie mit dieser Sprache.

(i) T heißt *endlich axiomatisierbar* oder *finitisierbar*, falls es ein endliches Axiomensystem Σ für T (1.35(i)) gibt. Andernfalls heißt T *nicht-finitisierbar*.

(ii) T heißt *axiomatisierbar* oder (genauer) *rekursiv axiomatisierbar*, falls es ein entscheidbares Axiomensystem Σ für T gibt.

Beispiele. Die Geometrien P_n' und die in Abschnitt 2 betrachteten Untertheorien davon sind nach Definition "endlich axiomatisiert", also auch endlich axiomatisierbar. Die in der Definition von P^* und den P_n^* benutzten unendlichen Axiomensysteme sind entscheidbar; denn selbstverständlich kann man für jede Formel feststellen, ob sie von der Bauart des Schemas A11' der elementaren Stetigkeitsaxiome ist, und damit auch, ob sie zum gegebenen Axiomensystem gehört.

Interessanter ist die Frage nach der endlichen bzw. rekursiven Axiomatisierbarkeit natürlich, wenn sich die Antwort nicht einfach aus der Definition ergibt, also zum Beispiel, wenn Σ als Theorie einer Strukturklasse (1.35(ii)) definiert ist (vgl. 3.33, 57(ii), 52, 59, 84, 85).

3.10 Anmerkungen. (i) Würde man in 3.9(ii) beliebige Formelmengen als Axiomensysteme zulassen, so ergäbe sich ein trivialer Begriff, da jede Theorie (sich selbst als) ein Axiomensystem besitzt (1.35(i)).

(ii) Die Beschränkung auf entscheidbare Axiomensysteme Σ ist - zumindest in der ersten Stufe - dadurch motiviert, daß dann Beweise "nachprüfbar" werden (d.h., die Menge aller Beweisfolgen aus Σ ist dann rekursiv i.b.a. die Menge aller endlichen Folgen von Formeln als Grundmenge, vgl. 1.49).

Für die betrachteten Begriffe gelten die folgenden drei allgemeinen Sätze.

3.11 Satz. (i) *Jede endliche Menge ist entscheidbar.*

(ii) *Jede entscheidbare Menge ist rekursiv aufzählbar.*

3.12 Satz. *Ist T eine axiomatisierbare Theorie erster Stufe, so ist T rekursiv aufzählbar.*

3.13 Satz. *Ist T eine axiomatisierbare und vollständige Theorie erster Stufe, so ist T entscheidbar.*

Zum Beweis: 3.12 beruht auf dem Vollständigkeitssatz für die Prädikatenlogik der ersten Stufe, nach dem die Sätze von T gerade die aus den Axiomen beweisbaren Formeln sind (s. 1.50), und auf einer systematischen Aufzählung dieser Formeln; genauer und zu 3.11(ii) s. etwa Schwabhäuser 1972, S. 37ff. Zum Beweis von 3.13 sei o.B.d.A. T widerspruchsfrei (andernfalls ist $T=Fml_{L(T)}$ und damit trivialerweise entscheidbar). Ein (auf Aussagen anwendbares) Entscheidungsverfahren für T erhält man aus einem Aufzählungsverfahren (A) gemäß 3.12 folgendermaßen. Sei $\alpha \in Aus_{L(T)}$ beliebig. Dann berechne man der Reihe nach mittels (A) die Funktionswerte $f(0)$, $f(1)$, $f(2)$,... der Aufzählungsfunktion f und prüfe in jedem Schritt, ob a) $f(n)=\alpha$ oder b) $f(n)=\neg\alpha$ ist. Tritt für ein gewisses n einer dieser Fälle ein, so breche man das Verfahren ab; im Fall a) ist dann $\alpha \in T$, im Fall b) ist $\alpha \notin T$ (da T widerspruchsfrei). Da T vollständig (1.39) und α eine Aussage ist, ist $\alpha \in T$ oder $\neg\alpha \in T$, d.h., für ein gewisses n muß einer der beiden Fälle eintreten. Somit führt dieses Verfahren stets nach endlich vielen Schritten zum Ziel. □

3.14 Anmerkungen. (i) Das geschilderte Entscheidungsverfahren für 3.13 ist denkbar unökonomisch. Es arbeitet ja sozusagen "blind", d.h., die systematische Durchsicht der Sätze von T geschieht in einer festen Reihenfolge, die von der gerade untersuchten Aussage α völlig unabhängig ist. Dabei kommen normalerweise viele andere Sätze von T früher an die Reihe, deren Prüfung für das gewünschte Ergebnis bedeutungslos ist. Der Aufwand (3.5(iii)) ist dann schon in einfachen Fällen viel zu groß und wesentlich größer als bei den "gezielten" Verfahren, die für einzelne Theorien gefunden wurden (und häufig auch noch zu kompliziert für heutige Rechenanlagen sind).

(ii) Die Voraussetzung der Vollständigkeit wurde im Beweis von 3.13 wesentlich benutzt. Ohne diese Voraussetzung liefert das geschilderte Verfahren zwar immer noch das richtige Ergebnis, falls α oder $\neg\alpha$ ein Satz von T ist; ist aber weder α noch $\neg\alpha$ Satz von T, so tritt nie einer der Fälle a), b) ein, d.h., das geschilderte Verfahren liefert für solche α nicht in endlich vielen Schritten ein Ergebnis und ist damit für unvollständige Theorien kein Entscheidungsverfahren.

(iii) Tatsächlich ist die Voraussetzung der Vollständigkeit in 3.13 nicht entbehrlich, d.h., es gibt axiomatisierbare Theorien, die nicht entscheidbar sind (Beispiele s. 3.57(iii), 3.58ff.).

3.15 Entscheidbarkeit der elementaren euklidischen Geometrie. Das älteste und bekannteste geometrische Resultat in dieser Richtung ist die Entscheidbarkeit der elementaren Theorien $Th(\mathfrak{L}_n(\mathbb{R}))$ der kartesischen Räume über $\mathbb{R}$ und (damit) der euklidischen Geometrien P_n^*. Ein entsprechendes Verfahren wurde von Tarski 1930 zusammen mit seinem Entscheidungsverfahren für die elementare Theorie $Th(\mathbb{R})$ der reellen Zahlen und der reell-abgeschlossenen Körper gefunden (s. Tarski 1940 und 1948/51). Wir beginnen mit diesem Resultat und Folgerungen daraus nach Tarski 1959. Dafür werden zunächst die entsprechenden Resultate über reell-abgeschlossene Körper ohne Beweis zusammengestellt.

3.16 Definition. (i) Ein Körper $\mathfrak{F}$ heißt *formal-reell*, falls das Element -1 nicht Summe von endlich vielen Quadraten (von Elementen von $\mathfrak{F}$) ist (oder gleichwertig: falls jede "nicht-triviale" Quadratsumme [d.h. jede, bei der nicht alle Summanden Null sind] von Null verschieden ist).

(ii) Ein Körper $\mathfrak{F}$ heißt *reell-abgeschlossen*, falls $\mathfrak{F}$ formal-reell ist, aber keine echte algebraische Erweiterung besitzt, die auch noch formal-reell ist.

(iii) Die genannten Begriffe sind auch für angeordnete Körper sinnvoll. Offenbar ist jeder angeordnete Körper formal-reell.

Beispiele für reell-abgeschlossene Körper (ohne oder mit Anordnung) sind der Körper der reellen Zahlen und der Körper der reellen algebraischen Zahlen.

Die Theorie der reell-abgeschlossenen Körper wurde von Artin und Schreier entwickelt und ist z.B. in van der Waerden 1971, Kap.11, §§81, 82 dargestellt (Beweise für die Sätze in 3.17 s. dort).

<u>3.17 Sätze</u>. (i) *Ein Körper* $\mathfrak{K}$ *ist reell-abgeschlossen genau dann, wenn* $\mathfrak{K}$ *formal-reell und* $\mathfrak{K}$(i) *algebraisch abgeschlossen ist* (dabei bedeutet $\mathfrak{K}$(i) wie üblich den aus $\mathfrak{K}$ durch Adjunktion einer Nullstelle i des Polynoms x^2+1 entstehenden Körper).

(ii) *Ein Körper* $\mathfrak{K}$ *ist reell-abgeschlossen genau dann, wenn gilt:*

(1) $\mathfrak{K}$ *ist formal-reell,*

(2) $\mathfrak{K}$ *ist "euklidisch", d.h. für jedes* $a \in \mathfrak{K}$ *ist a ein Quadrat oder -a ein Quadrat, und*

(3) *jedes Polynom ungeraden Grades (in einer Unbestimmten) über* $\mathfrak{K}$ *besitzt in* $\mathfrak{K}$ *wenigstens eine Nullstelle.*

(iii) *Zu jedem formal-reellen Körper gibt es wenigstens eine algebraische Erweiterung, die reell-abgeschlossen ist (diese ist dann eine minimale reell-abgeschlossene Erweiterung).*

(iv) *Jeder Körper mit den Eigenschaften* (1) *und* (2) *aus* (ii) *läßt sich auf genau eine Weise anordnen (zu einem angeordneten Körper machen), und zwar so, daß für beliebige Körperelemente a gilt: a ist positiv gdw a ein von Null verschiedenes Quadrat ist.*

(v) *Zu jedem angeordneten Körper* $\mathfrak{K}$ *gibt es einen über* $\mathfrak{K}$ *algebraischen (und damit minimalen) reell-abgeschlossenen angeordneten Erweiterungskörper, und je zwei solche Erweiterungen* $\mathfrak{K}_1$, $\mathfrak{K}_2$ *sind über* $\mathfrak{K}$ *"äquivalent", d.h., es gibt einen Isomorphismus von* $\mathfrak{K}_1$ *auf* $\mathfrak{K}_2$, *der* $\mathfrak{K}$ *elementweise fest läßt.*

Man kann also von einer - bis auf Äquivalenz eindeutigen - "*reell-abgeschlossenen Hülle*" von $\mathfrak{K}$ sprechen in Analogie zur algebraisch abgeschlossenen Hülle von Körpern ohne Anordnung. Verschiedenen möglichen Anordnungen eines Körpers (zu deren Studium die Theorie von Artin und Schreier entwickelt wurde) entsprechen dann inäquivalente reell-abgeschlossene algebraische Erweiterungen.

<u>3.18 Definition</u>. (i) Unter den *natürlichen Elementen* eines Körpers (i.a. von Charakteristik 0) bzw. angeordneten Körpers $\mathfrak{K}$ verstehen wir die den natürlichen Zahlen entsprechenden Elemente, d.h. diejenigen, die sich vom Nullelement ausgehend erhalten lassen, indem man endlich oft das Einselement addiert. Unter den *rationalen Elementen* von $\mathfrak{K}$ verstehen wir dann die der Form $\pm \frac{p}{q}$, wobei p, q natürliche Elemente mit $q \neq 0$ sind.

(ii) Ein angeordneter Körper $\mathfrak{F}=\langle F,+,\cdot,<\rangle$ heißt bekanntlich *archimedisch angeordnet* und seine Anordnung < heißt *archimedisch*, falls es zu jedem Körperelement ein größeres natürliches Element gibt. Andernfalls heißt $\mathfrak{F}$ *nicht-archimedisch angeordnet* und seine Anordnung < *nicht-archimedisch*. Die Elemente, die größer als jedes natürliche Element sind, heissen dann *unendlich große Elemente*; die Reziproken von unendlich großen Elementen heißen *unendlich kleine Elemente* - das sind also diejenigen, die zwar positiv aber kleiner als jedes positive rationale Element sind.

3.19 Beispiele. Ist $\mathfrak{F}$ ein beliebiger reell-abgeschlossener Körper und $\mathfrak{F}(u)$ eine einfache transzendente Erweiterung, so ist $\mathfrak{F}(u)$ wieder formal-reell und läßt sich z.B. so anordnen, daß das adjungierte transzendente Element u größer ist als jedes Element von $\mathfrak{F}$. Eine reell-abgeschlossene Erweiterung $\mathfrak{F}'$ von $\mathfrak{F}(u)$ gemäß 3.17(iii) bzw. (v) ist dann eine echte Erweiterung von $\mathfrak{F}$. Mit der erwähnten Anordnung sind $\mathfrak{F}(u)$ und damit auch $\mathfrak{F}'$ nicht-archimedisch angeordnete Körper. In manchen Fällen (und zwar genau dann, wenn sich $\mathfrak{F}(u)$ isomorph in $\mathbb{R}$ einbetten läßt) kann die Anordnung von $\mathfrak{F}(u)$ auch archimedisch gewählt werden, und dabei wird $\mathfrak{F}'$ (gemäß (v)) ebenfalls archimedisch angeordnet. (Vgl. etwa van der Waerden 1971.) Auf diese Weise erhält man aus den Beispielen zu 3.16 unendlich viele (ineinander enthaltene) reell-abgeschlossene Körper.

3.20 Definition. Zur Festlegung entsprechender elementarer Theorien verwenden wir für Körper die Sprache $L_F:=L(+,\cdot,-,0,1)$ mit Operationszeichen in der üblichen Bedeutung (- einstellig für die Bildung des Inversen $-\tau$) und für angeordnete Körper die Sprache $L_{OF}:=L(+,\cdot,-,0,1;<)$ (mit den genannten "Grundbegriffen" werden weitere Begriffsbildungen wie $\sigma-\tau$, $\sigma\leq\tau$, $\sigma>\tau$ ggf. wie üblich definiert). Die Klasse der reell-abgeschlossenen angeordneten Körper (als Strukturen für die angegebene Sprache) bezeichen wir wieder mit RF, die Klasse der reell-abgeschlossenen Körper (ohne Anordnung) mit RF^-.

Die Theorien dieser Strukturklassen lassen sich nach 3.17(ii) auch durch rekursive Axiomensysteme charakterisieren.

Dazu führen wir noch *Terme* $\sum_{\nu=m}^{n}\tau_\nu$ *für endliche Summen* und τ^n *für Potenzen* ($m,n\in\mathbb{N}$, $m\leq n$) ein durch folgende induktive Definitionen:

(1) $\sum_{\nu=m}^{m} \tau_\nu = \tau_m$, $\sum_{\nu=m}^{n+1} \tau_\nu = \sum_{\nu=m}^{n} \tau_\nu + \tau_{n+1}$;

(2) $\tau^0 = 1$, $\tau^{n+1} = \tau^n \cdot \tau$.

Seien Σ_F bzw. Σ_{OF} die üblichen (endlichen) Axiomensysteme für Körper bzw. für angeordnete Körper in den angegebenen Sprachen. Weiter setzen wir

(3) $\alpha_n := \sum_{\nu=1}^{n} x_\nu^2 \doteq 0 \rightarrow \bigwedge_{\nu=1}^{n} x_\nu \doteq 0$ $\quad (n \in \mathbb{N} - \{0\})$
(zusammen: $\mathfrak{f}$ ist formal-reell),

(4) $\beta := \forall x \exists y [x \doteq y^2 \vee -x \doteq y^2]$
($\mathfrak{f}$ ist euklidisch),

(5) $\gamma_n := \forall u_0 \ldots \forall u_{n-1} \exists x \sum_{\nu=0}^{n-1} u_\nu \cdot x^\nu + x^n \doteq 0$ $\quad (n \in \mathbb{N} - \{0\})$
(jedes Polynom n-ten Grades [hier mit dem höchsten Koeffizienten 1 und den übrigen Koeffizienten u_ν] hat eine Nullstelle),

(6) $\beta' := \forall x [0 < x \rightarrow \exists y\, x \doteq y^2]$
(der angeordnete Körper $\mathfrak{f}$ ist euklidisch, vgl. 3.17(iv)),

(7) $\Sigma_{RF^-} := \Sigma_F \cup \{\alpha_n | n \geq 2\} \cup \{\beta\} \cup \{\gamma_n | n \geq 3,\ n \text{ ungerade}\}$,

(8) $\Sigma_{RF} := \Sigma_{OF} \cup \{\beta'\} \cup \{\gamma_n | n \geq 3,\ n \text{ ungerade}\}$.

Die in (3) bis (6) auftretenden (paarweise verschiedenen) Variablen x, y, x_ν, u_ν können irgendwie festgelegt werden, um die Definition eindeutig zu machen.

<u>3.21 Satz</u>. *Mit den Bezeichnungen von 3.20 gilt offenbar*

(1) $RF^- = Mod(\Sigma_{RF^-})$, $\qquad RF = Mod(\Sigma_{RF})$

und damit

(2) $Th(RF^-) = Cn(\Sigma_{RF^-})$, $\quad Th(RF) = Cn(\Sigma_{RF})$,

und die Axiomensysteme Σ_{RF^-} *und* Σ_{RF} *sind rekursiv.*

<u>3.22 Satz</u> (Tarski). (i) *Die elementare Theorie* $Th(RF)$ *der reell-abgeschlossenen angeordneten Körper ist entscheidbar und vollständig.*

(ii) (Folgerung, vgl. 1.39) *Je zwei reell-abgeschlossene angeordnete Körper sind zueinander und insbesondere zu* $\mathbb{R}$ *elementar äquivalent, und es ist* Th(RF)=Th($\mathbb{R}$) *(Theorie erster Stufe des angeordneten Körpers* $\mathbb{R}$ *der reellen Zahlen).*

(iii) *Für* Th(RF) *hat man ein Verfahren der "Quantorenelimination", d.h. ein (Berechnungs-)Verfahren, das zu jeder Formel* α *der zugehörigen Sprache eine dazu (in* Th(RF)*) äquivalente Formel* β *liefert, die quantorenfrei ist und keine anderen Variablen als die freien Variablen von* α *enthält.*

(iv) *Durch weitere äquivalente Umformungen läßt sich für* β *aus* (iii) *zusätzlich erreichen, daß es eine Disjunktion von endlich vielen Formeln folgender Bauart wird:*

(1) $\pi \doteq 0 \wedge \rho_1 > 0 \wedge \ldots \wedge \rho_m > 0.$

Hierbei ist auch $m=0$ *zugelassen, d.h., die "Ungleichungen"* $\rho_\mu > 0$ *können ganz wegfallen. Andererseits kann* π *der Term* 0 *sein, was darauf hinausläuft, daß die Gleichung in* (1) *äquivalent weggelassen werden kann. Im allgemeinen können in den Termen* π *und* ρ_μ *alle Variablen aus* Fr(α) *auftreten. Für eine vorgegebene Variable* x *lassen sich diese Terme schließlich noch "nach Potenzen von* x *entwickeln" und damit als "formale Polynome" in* x *darstellen, d.h. als Terme der speziellen Bauart*

(2) $\sum_{\nu=0}^{n} \tau_\nu \cdot x^\nu$,

wobei die Variable x *in den "Koeffiziententermen"* τ_ν *nicht mehr vorkommt.*

(v) (Folgerung) *Ist* $\mathfrak{f} \in$ RF *und* α(x) *eine Formel mit der einzigen freien Variablen* x*, so ist die durch* α(x) *definierte Menge, d.h.* $\{x \mid \models_{\mathfrak{f}} \alpha[x]\}$ (s. 1.16), *eine Vereinigung von endlich vielen Intervallen in* $\mathfrak{f}$*, deren Endpunkte reelle algebraische Zahlen sind (die Intervalle können dabei offen, halboffen oder abgeschlossen sein, auch uneigentlich oder nur aus einem Punkt bestehend). Enthält* α *weitere freie Variablen, die mit Elementen* $a_1, \ldots, a_n$ *fest belegt sind, so gilt für die entsprechende Menge (d.h.* $\{x \mid \models_{\mathfrak{f}} \alpha[x, a_1, \ldots, a_n]\}$*) dasselbe mit Endpunkten, die in der reell-abgeschlossenen Hülle von* $\mathbb{Q}(a_1, \ldots, a_n)$ *liegen. (Diese Hülle und insbesondere der Körper der reellen algebraischen Zahlen werden hierbei natürlich als Unterkörper von* $\mathfrak{f}$ *aufgefaßt, was zumindest bis auf Isomorphie der Falle ist analog zu* 3.18(i)*.)*

(vi) *In der elementaren Theorie* Th(RF^-) *der reell-abgeschlossenen Körper ohne Anordnung läßt sich die Kleiner-Beziehung ausdrücken durch*

(3) $\sigma < \tau \leftrightarrow \exists y[\tau - \sigma \doteq y^2 \wedge y \neq 0]$.

Damit übertragen sich (i) *und* (ii) *auch auf diese Theorie. Dagegen besitzt diese Theorie keine Quantorenelimination: Auf den in* (3) *benutzten Quantor kann man nicht verzichten, z.B. gibt es keine zu* $0 < x$ *äquivalente quantorenfreie Formel der Sprache* L_F.

3.23 Anmerkungen. (i) Das Kernstück des Beweises bei Tarski 1951 ist das erwähnte Verfahren zur Quantorenelimination. Seine Anwendung auf eine Formel α besteht in der sukzessiven Elimination einzelner Quantoren durch äquivalente Umformungen. Dabei wird (nach Anwendung der Umformungen gemäß 3.22(iv)) eine Verallgemeinerung der Sätze von Sturm über Nullstellen von Polynomen benutzt. Das Entscheidungsverfahren ergibt sich durch Spezialisierung des Eliminationsverfahrens auf Aussagen α; dabei erhält man nämlich zum Schluß in den Bestandteilen 3.22(1) variablenfreie Terme, deren Werte (ganze Zahlen sind und) ausgerechnet werden können. Die Vollständigkeit von Th(RF) ergibt sich daraus, daß alle äquivalenten Umformungen des (zunächst für Th($\mathbb{R}$) aufgestellten) Verfahrens schon auf Grund des Axiomensystems Σ_{RF} (3.20(8)) durchgeführt werden können.

(ii) Inzwischen wurden mehrere andere Beweise für die Resultate aus 3.22, insbesondere 3.22(i), gefunden. Besonders einfach ist ein modelltheoretischer Beweis von A. Robinson (s. auch Schwabhäuser 1972, S. 26ff.) für die (Modellvollständigkeit und die) Vollständigkeit von Th(RF). Daraus und aus der Axiomatisierbarkeit (3.21) ergibt sich nach 3.13 auch die Entscheidbarkeit. Der Preis für die Einfachheit dieses Beweises ist, daß das darauf beruhende Entscheidungsverfahren besonders kompliziert ist (vgl. 3.14(i)). Das kürzeste zur Zeit bekannte Verfahren zur Quantorenelimination und Entscheidungsverfahren sowie eine Abschätzung der Komplexität dieses Verfahrens wurden von G.E. Collins angegeben, s. Collins 1975.

3.24 Korollar. (i) *Ein geeignetes (rekursives) Axiomensystem für* Th(RF) *erhält man auch, indem man zum Axiomensystem* Σ_{OF} *für angeordnete Körper das folgende Axiomenschema* DSN' *hinzunimmt (elementare Axiome vom Dedekindschen Schnitt mit nicht notwendig erschöpfender Einteilung, vgl. die Bemerkungen zu* A11 *in* I.1.2*).*

(DSN') $\exists u\, \alpha(u) \wedge \exists v\, \beta(v) \wedge \forall u \forall v[\alpha(u) \wedge \beta(v) \rightarrow u < v]$
$\rightarrow \exists t\{\forall u[\alpha(u) \rightarrow u \leq t] \wedge \forall v[\beta(v) \rightarrow t \leq v]\}$,

wobei gilt

(*) t, v *nicht frei in* $\alpha(u)$, t, u *nicht frei in* $\beta(v)$.

(ii) *Statt* DSN' *genügt in* (i) *auch das Schema* DS', *das aus* DSN' *durch Hinzunahme der jeweiligen Voraussetzung* $\forall x[\alpha(x) \vee \beta(x)]$ *entsteht ("elementare Axiome vom Dedekindschen Schnitt").*

(iii) *Statt* Σ_{OF} *genügt in* (i) *und* (ii) *ein Axiomensystem für angeordnete Schiefkörper.*

Beweis: Zu (ii): Jedes Axiom aus DS' ist ein Spezialfall des bekannten (mit Mengenvariablen formulierten) Axioms vom Dedekindschen Schnitt und gilt damit zunächst in $\mathbb{R}$, nach 3.22(ii) also auch in jedem reell-abgeschlossenen Körper. Andererseits erhält man aus DS' leicht die Axiome von Σ_{RF} (3.20(5), (6), (8)); die üblicherweise für die Sätze β' und γ_n (*n* ungerade) geführten Beweise auf Grund des Axioms vom Dedekindschen Schnitt benutzen nämlich nur Unter- und Oberklassen, die sich in der ersten Stufe (durch Formeln der betrachteten Sprache) definieren lassen.

Zu (i): Gilt für $\alpha(u)$ und $\beta(v)$ nur die Voraussetzung von DSN', so kann man mittels $\bar{\alpha}(u) := \exists u'[\alpha(u') \wedge u \leq u']$ und $\bar{\beta}(v) := \neg\bar{\alpha}(v)$ zu Formeln übergehen, die eine erschöpfende Einteilung bestimmen, und damit das Gewünschte aus DS' erhalten. Somit ist das Schema DSN' gleichwertig mit dem Schema DS'.

Zu (iii): Ein Beweis des Kommutativgesetzes der Multiplikation aus den übrigen Axiomen ist ausgeführt in Schwabhäuser 1956, S. 152ff. □

Die Übertragung von Resultaten für Th(RF) auf entsprechende euklidische Geometrien läßt sich nun mit der folgenden Formelübersetzung erreichen (vgl. 3.1).

3.25 Definition (Arithmetische Reduzierte, nach Tarski 1951, S. 44).
(i) Sei $n \geq 1$ irgendeine natürliche Zahl (Dimension), die für das Folgende festgehalten wird. Sei V_P die Menge der "Punktvariablen" (a, b, c,...) für die Sprache L(D,B) der Geometrie, und sei V die Menge der "Zahlenvariablen" oder Variablen für Körperelemente (u, v,...) für die

Sprache L_{OF} (3.20) der "Arithmetik" der angeordneten Körper. Es spielt hierbei keine Rolle, ob für beide Sprachen dieselbe Variablenmenge verwendet wird oder nicht, wir benutzen jedoch verschiedene Bezeichnungen, um die jeweilige Verwendung der Variablen deutlicher zu machen. Sei ξ eine (für das Folgende festgehaltene) berechenbare Abbildung von V_P in V^n mit folgender Eigenschaft:

(*) Wenn $\xi(a)=\langle a',a'',\ldots,a^{(n)}\rangle$, $\xi(b)=\langle b',b'',\ldots,b^{(n)}\rangle$ und $a\neq b$, so sind die Zahlenvariablen $a',\ldots,a^{(n)}, b',\ldots,b^{(n)}$ paarweise verschieden.

Allgemein erklären wir Komponenten-Abbildungen ξ_ν durch $\xi(a)=\langle\xi_1(a),\ldots,\xi_n(a)\rangle$ und bezeichnen die einer Punktvariablen zugeordneten Zahlenvariablen zur Abkürzung (wie in (*)) durch Anhängen von Strichen (d.h. $a^{(\nu)}=\xi_\nu(a)$, analog für andere Buchstaben). Sind die Variablenmengen mit einer Numerierung gegeben, etwa $V_P=\{p_i \mid i\in\mathbb{N}\}$, $V=\{v_i \mid i\in\mathbb{N}\}$, so können wir als ξ z.B. die Abbildung mit $\xi_\nu(p_i) = v_{n\cdot i+\nu-1}$ verwenden.

(ii) Jeder Formel γ der Sprache L(D,B), wird nun eine Formel Rda(γ) der Sprache L_{OF}, die "*arithmetische Reduzierte* (für n-dimensionale kartesische Räume)" von γ zugeordnet durch die folgende induktive Definition über den Formelaufbau von γ:

(1) $$\mathrm{Rda}(a\doteq b) = \bigwedge_{\nu=1}^{n} a^{(\nu)}\doteq b^{(\nu)}.$$

$$\mathrm{Rda}(Dabcd) = \sum_{\nu=1}^{n} (a^{(\nu)}-b^{(\nu)})^2 \doteq \sum_{\nu=1}^{n} (c^{(\nu)}-d^{(\nu)})^2.$$

$$\mathrm{Rda}(Babc) = \exists u[0\leq u\leq 1 \wedge \bigwedge_{\nu=1}^{n} (b^{(\nu)}-a^{(\nu)})\doteq u\cdot(c^{(\nu)}-a^{(\nu)})].$$

(2) $$\mathrm{Rda}(\neg\gamma) = \neg\mathrm{Rda}(\gamma).$$

$$\mathrm{Rda}(\gamma\circ\delta) = \mathrm{Rda}(\gamma)\circ\mathrm{Rda}(\delta) \quad \text{für } \circ = \wedge,\vee,\rightarrow,\leftrightarrow.$$

$$\mathrm{Rda}(\forall a\gamma) = \forall a'\ldots\forall a^{(n)}\,\mathrm{Rda}(\gamma).$$

$$\mathrm{Rda}(\exists a\gamma) = \exists a'\ldots\exists a^{(n)}\,\mathrm{Rda}(\gamma).$$

(iii) Sei $\mathfrak{f}$ ein angeordneter Körper. Eine Belegung h über dem kartesischen Raum $\mathfrak{C}_n(\mathfrak{f})$ und eine Belegung j über $\mathfrak{f}$ nennen wir dann (bezüglich $\mathfrak{f}$) *assoziiert*, falls für jede Punktvariable a gilt

(3) $h(a) = \langle j(a'),j(a''),\ldots,j(a^{(n)})\rangle$.

<u>3.26 Satz</u>. *Sei $\mathfrak{F}$ ein angeordneter Körper.*

(i) *Wenn* γ *eine Formel der Sprache* L(D,B) *ist, so gilt für jedes Paar von (bezüglich $\mathfrak{F}$) assoziierten Belegungen h, j:*

h Erf$_{\mathfrak{L}_n(\mathfrak{F})}$ γ *genau dann, wenn* j Erf$_{\mathfrak{F}}$ Rda(γ)

(ii) *Wenn* γ *eine Aussage von* L(D,B) *ist, so gilt*

$\models_{\mathfrak{L}_n(\mathfrak{F})} \gamma$ *genau dann, wenn* $\models_{\mathfrak{F}}$ Rda(γ).

<u>Beweis</u>: Für Primformeln γ ergibt sich (i) unmittelbar aus der Definition der kartesischen Räume (I.1.4) und der dieser nachgebildeten Definition 3.25(1) zusammen mit (3). Für beliebige Formeln γ erhält man dann (i) ohne Schwierigkeiten durch Induktion über den Formelaufbau auf Grund von 3.25(2); für die Quantifizierungen im Induktionsschritt wird dabei die Voraussetzung 3.25(*) gebraucht (Übung!). Ist insbesondere γ eine Aussage, so ist es auch Rda(γ), und man erhält (ii) als Spezialfall von (i) (vgl. 1.16). □

<u>3.27 Satz</u>. *Für je zwei reell-abgeschlossene angeordnete Körper* $\mathfrak{F}_1$, $\mathfrak{F}_2$ *sind die kartesischen Räume* $\mathfrak{L}_n(\mathfrak{F}_1)$ *und* $\mathfrak{L}_n(\mathfrak{F}_2)$ *zueinander und insbesondere zu* $\mathfrak{L}_n(\mathbb{R})$ *elementar äquivalent.*

<u>Beweis</u>: Seien $\mathfrak{F}_1, \mathfrak{F}_2 \in$ RF. Nach 1.37, 1.41 genügt es, für beliebige Aussagen γ von L(D,B) die folgende Bedingung nachzuweisen:

(1) $\models_{\mathfrak{L}_n(\mathfrak{F}_1)} \gamma$ gdw $\models_{\mathfrak{L}_n(\mathfrak{F}_2)} \gamma$.

Nach 3.26(ii) ist (1) jedoch gleichwertig mit

(2) $\models_{\mathfrak{F}_1}$ Rda(γ) gdw $\models_{\mathfrak{F}_2}$ Rda(γ),

und diese Bedingung gilt nach 3.22(ii). □

<u>3.28 Ergänzung zum Beweis des Darstellungssatzes für P_n^*</u>. Der Darstellungssatz in I.16.15 (Charakterisierung der Modelle von P_n^*, s. 2.1) war bereits zurückgeführt worden auf den früheren Satz

I.1.6(vi): *Wenn* $n \geq 2$, *so gelten die elementaren Stetigkeitsaxiome* (A11')
in $\mathfrak{L}_n(\mathfrak{F})$ *gdw* $\mathfrak{F}$ *reell-abgeschlossen ist.*

Die im damaligen Beweis noch ausstehende "genauere Untersuchung der auftretenden Formeln" können wir mit den jetzt bereitgestellten Hilfsmitteln durchführen und den Beweis noch vereinfachen.

Beweis: Zu (←): Wir können als bekannt voraussetzen, daß das volle Stetigkeitsaxiom A11 in $\mathcal{L}_n(\mathbb{R})$ gilt (I.1.6(v)). Die Axiome des Schemas A11' sind Spezialfälle davon, also gelten sie ebenfalls in $\mathcal{L}_n(\mathbb{R})$ und, da sie in der ersten Stufe formuliert sind, nach 3.27 auch in $\mathcal{L}_n(\mathfrak{F})$, wenn $\mathfrak{F}$ reell-abgeschlossen ist.

Zu (→): Sei $\mathfrak{F}_{oe}$ ein im Modell $\mathcal{L}_n(\mathfrak{F})$ gemäß I.14 geometrisch konstruierter angeordneter Körper ("Zahlengerade", I.14.40). (Für die Konstruktion wird die Voraussetzung $n \geq 2$ benutzt!) Nach I.14.41 ist $\mathfrak{F}_{oe}$ isomorph zu $\mathfrak{F}$. Somit genügt es zu zeigen, daß $\mathfrak{F}_{oe}$ reell-abgeschlossen ist, d.h., daß in $\mathfrak{F}_{oe}$ die Sätze β' und γ_m für ungerades $m \geq 3$ (3.20(6), (5)) gelten. Die Gültigkeit dieser Sätze in $\mathfrak{F}_{oe}$ wird - mit Hilfe geometrischer Summen und Produkte und der geometrischen Positivitätsrelation - durch Formeln von L(B) ausgedrückt, deren Gültigkeit im gegebenen Raum zu zeigen ist. Diese Formeln lassen sich aber als Folgerungen aus Axiomen des Schemas A11' in derselben Weise erhalten, wie es für die ursprünglichen Sätze β' und γ_m auf Grund der Axiome des Schemas DS' bzw. DSN' (3.24) geschieht (Übung!). □

3.29 Anmerkungen. (i) Im Sinne des ursprünglich skizzierten Beweises für I.1.6(vi) kann man auch direkt (ohne Benutzung von 3.17 bis 3.24) beweisen, daß die Gültigkeit von A11' in $\mathcal{L}_n(\mathfrak{F})$ gleichwertig ist mit der Gültigkeit von DS' bzw. DSN' in $\mathfrak{F}$. Für die Richtung (←) kann man nämlich von einem beliebigen Axiom γ von A11' ausgehen, die arithmetische Reduzierte Rda(γ) bilden und eine Parameterdarstellung für die Punkte der Geraden, in der der Schnitt enthalten ist, verwenden; dadurch erhält man γ nach 3.26(i) aus einem Spezialfall von DSN'. Für die Umkehrung kann man in ähnlicher Weise die später noch eingeführten geometrischen Reduzierten (3.36) benutzen. Die Ausführung sei dem Leser überlassen.

(ii) Die Voraussetzung $n \geq 2$ ist in I.1.6(vi) nicht entbehrlich, man kann sogar zeigen:

3.30 Satz. *Für jeden angeordneten Körper* $\mathfrak{F}$ *gelten die elementaren Stetigkeitsaxiome* (A11') *im eindimensionalen kartesischen Raum* $\mathcal{L}_1(\mathfrak{F})$.

Beweis: Die Streckenkongruenz und die Zwischenbeziehung des eindimensionalen kartesischen Raumes $\mathcal{L}_1(\mathfrak{F})$ lassen sich nämlich schon mit Hilfe der Addition und der Anordnung von $\mathfrak{F}$ beschreiben (ohne Verwendung der Multiplikation). Damit erhält man, daß sich für ein beliebiges Axiom γ aus A11' die Reduzierte Rda(γ) äquivalent umformen läßt in ein Axiom der Form DSN', wobei $\alpha(u)$ und $\beta(v)$ Formeln der Sprache $L(+,-,0;<)$ sind. Nun hat man schon in der Theorie der dividierbaren abelschen nicht-trivialen angeordneten Gruppen (die in dieser Sprache formalisiert ist und deren Axiome in $\mathfrak{f}$ gelten) eine Quantorenelimination (s. A. Robinson 1956, S. 36), d.h., $\alpha(u)$ und $\beta(v)$ lassen sich äquivalent umformen in quantorenfreie Formeln derselben Sprache. Die durch solche Formeln definierbaren Mengen sind Vereinigungen von endlich vielen Intervallen, deren Endpunkte in $\mathfrak{f}$ liegen. Damit existiert ein Schnittelement t der in DSN' angegebenen Art in $\mathfrak{f}$, wie zu zeigen. □

3.31 Satz (Tarski). *Die n-dimensionale euklidische Geometrie P_n^* ist entscheidbar.*

Beweis (Entscheidungsverfahren): Für eine beliebige Aussage γ von L(D,B) erhält man, daß die folgenden Bedingungen untereinander gleichwertig sind:

(1) $\gamma \in P_n^*$,

(2) $\models \mathfrak{L}_n(\mathfrak{f})^\gamma$ für jedes $\mathfrak{f} \in$ RF (nach dem Darstellungssatz),

(3) $\models \mathfrak{L}_n(\mathbb{R})^\gamma$ (nach 3.27),

(4) $\models_{\mathbb{R}}$ Rda(γ) (nach 3.26(ii)).

Ein Entscheidungsverfahren der gewünschten Art (d.h. zur Nachprüfung der Bedingung (1)) besteht also darin, daß man aus γ die arithmetische Reduzierte Rda(γ) bildet und darauf ein (nach 3.22, 23 existierendes) Entscheidungsverfahren für Th($\mathbb{R}$) (d.h. zur Nachprüfung von (4)) anwendet. (Die Definition 3.25 liefert tatsächlich nicht nur die Existenz, sondern auch ein einfaches Verfahren zur Angabe der arithmetischen Reduzierten für beliebige γ.) □

3.32 Satz (Tarski). *P_n^* ist vollständig.*

Beweis (semantische Methode, vgl. 3.35, 3.46(i)): Nachdem der Darstellungssatz zur Verfügung steht, ergibt sich die Vollständigkeit ("Je zwei Modelle von P_n^* sind zueinander elementar äquivalent") sofort aus 3.27. □

3.33 Satz (Tarski). *Die Theorien* Th(RF) *und* P_n^* ($n \geq 2$, $n \in \mathbb{N}$) *sind nicht endlich axiomatisierbar.*

Zum Beweis verwenden wir

3.34 Lemma. *Ist* Σ *ein Axiomensystem für die Theorie* T *erster Stufe (1.35(i)) und* T *endlich axiomatisierbar, so gibt es eine endliche Teilmenge* Σ' *von* Σ, *die ebenfalls ein Axiomensystem von* T *ist.*

Beweis: Nach Voraussetzung gibt es überhaupt ein endliches Axiomensystem $\Gamma = \{\gamma_1, \ldots, \gamma_k\}$ für T. Da Σ ein Axiomensystem ist, folgt die Formel $\gamma := \bigwedge_{\kappa=1}^{k} \gamma_\kappa$ aus Σ und nach dem Endlichkeitssatz schon aus einer endlichen Teilmenge Σ' von Σ. Dieses Σ' leistet das Verlangte; denn jeder Satz von T folgt aus $\{\gamma\}$ und damit auch aus Σ'. □

Zum Beweis von 3.33: Zur Behauptung für Th(RF) genügt nach dem Lemma zu zeigen: Keine endliche Teilmenge Σ' des Axiomensystems Σ_{RF} (3.20(8)) ist selbst ein Axiomensystem für Th(RF). Ist Σ' eine solche endliche Teilmenge, so kann man eine ungerade Primzahl wählen, die größer ist als die Indizes m der in Σ' noch vorhandenen Axiome γ_m aus Σ_{RF}. Somit erhält man das Gewünschte aus dem folgenden Satz der Algebra (Tarski 1959): *Ist* p *eine ungerade Primzahl, so gibt es einen euklidischen angeordneten Körper, in dem jedes Polynom ungeraden Grades* $< p$, *aber nicht jedes Polynom vom Grade* p *eine Nullstelle besitzt.* Eine Konstruktion für solche Körper ist z.B. ausgeführt in Schwabhäuser 1971, S. 109f.

Für P_n^* läßt sich der Beweis analog führen. Sei nämlich Σ_0 ein endliches Axiomensystem für P_n (vgl. 2.1). Ein geeignetes Axiomensystem Σ für P_n^* erhält man daraus durch Hinzunahme von Axiomen $\overline{\beta'}$ und $\overline{\gamma_m}$, die die Gültigkeit der entsprechenden Axiome β' und γ_m aus Σ_{OF} im geometrisch konstruierten angeordnenten Körper $\mathfrak{F}_{oe}$ ausdrücken (vgl. den Beweis zu (→) in 3.28). Für ein endliches Teilsystem Σ' kann man dann einen angeordneten Körper $\mathfrak{F}$ wie oben konstruieren und erhält mit $\mathfrak{L}_n(\mathfrak{F})$ ein Modell von Σ', das kein Modell von Σ ist. □

3.35 Anmerkung (Semantische und syntaktische Methode). Die Vollständigkeit von P_n^* wurde auf dem angegebenen Wege bewiesen in Tarski 1959, das Resultat wurde aber schon in Tarski 1951 angekündigt. Dadurch angeregt wurde ein anderer Beweis für die Vollständigkeit in Schwabhäuser 1956, ursprünglich für eine andere Sprache und Axiomatisierung in stärkerer Anlehnung an Hilbert 1977. Dieser Beweis wird im folgenden für P_n^*

behandelt. Die verwendete Methode können wir im Unterschied zur vorigen als "syntaktisch" bezeichnen, da sie sich auf den Beweisbarkeitsbegriff bezieht (vgl. auch 3.46). Die dabei eingeführten "geometrischen Reduzierten" werden auch noch zum Beweis der Unvollständigkeit und der Unentscheidbarkeit anderer Geometrien verwendet.

Ist α eine Formel der Sprache L_{OF} (der "Arithmetik", 3.20), so soll, anschaulich gesagt, die geometrische Reduzierte von α den durch α ausgedrückten Sachverhalt übersetzen in eine entsprechende Feststellung für die geometrisch eingeführten angeordneten Körper $\mathfrak{F}_{oe}$ (I.14.40).

<u>3.36 Definition (Geometrische Reduzierte)</u>. (i) Wir verwenden wieder die Bezeichnungen von 3.25(i). Seien zwei Punktvariablen o und e für das Folgende fest gewählt, und sei η eine berechenbare eineindeutige Abbildung von der Menge V der Zahlenvariablen in die Menge $V_P-\{o,e\}$ der übrigen Punktvariablen. Unter Benutzung der Numerierung aus 3.25(i) können wir z.B. $o=p_0$, $e=p_1$ setzen und als η die Abbildung mit $\eta(v_i)=p_{i+2}$ verwenden. Zur Abkürzung bezeichnen wir die einer Zahlenvariablen x zugeordnete Punktvariable $\eta(x)$ mit $\bar{x}$. Mit $L^\iota(D,B)$ bzw. $L^\iota(B)$ bezeichnen wir die aus L(D,B) bzw. L(B) durch Hinzunahme des bestimmten Artikels (1.47) entstehende Sprache erster Stufe (analog für andere Systeme von "Grundbegriffen" [Operations- und Relationszeichen]).

(ii) Jedem Term τ der Sprache L_{OF} wird ein Term Rdt(τ) der Sprache $L^\iota(B)$, die *Termreduzierte* von τ, zugeordnet durch die folgende induktive Definition über den Aufbau von τ:

(1) $\mathrm{Rdt}(0) = o.$
$\mathrm{Rdt}(1) = e.$
$\mathrm{Rdt}(x) = \bar{x}.$

(2) $\mathrm{Rdt}(\sigma+\tau) = \mathrm{Rdt}(\sigma) \underset{o,e}{+} \mathrm{Rdt}(\tau).$
$\mathrm{Rdt}(\sigma\cdot\tau) = \mathrm{Rdt}(\sigma) \underset{o,e}{\cdot} \mathrm{Rdt}(\tau).$
$\mathrm{Rdt}(-\sigma) = \underset{o,e}{-} \mathrm{Rdt}(\sigma).$

Dabei seien die Terme für geometrische Summen, Produkte und Inverse auf den rechten Seiten von (2) (gemäß I.14.7 bis 9) folgendermaßen eingeführt unter Verwendung der Definitionen I.14.1, 3, 4:

(3) $a \underset{o,e}{+} b = Su_{oe}(ab) := \iota c\{Su_{oe}abc \vee [\neg Ar_{oe}ab \wedge c\doteq o]\}$,
wobei $Su_{oe}abc := \exists e' Su_{oe}^{e'}abc$,
analog für $\cdot$, Pr statt +, Su,
$\underset{o,e}{-} a = Iv_{oe}(a) := \iota c\{Su_{oe}cao \vee [\neg Ar_{oe}a \wedge c\doteq 0]\}$;
dabei dürfen an Stelle der (freien) Variablen a, b auch Terme von

$L^1(B)$ stehen.

(iii) Jeder Formel α von L_{OF} wird eine Formel Rdgv(α) der Sprache $L^1(B)$, die *vorläufige geometrische Reduzierte* von α, zugeordnet durch die folgende induktive Definition über den Aufbau von α:

(4) $\mathrm{Rdgv}(\sigma \doteq \tau) = \mathrm{Rdt}(\sigma) \doteq \mathrm{Rdt}(\tau)$.

$\mathrm{Rdgv}(\sigma < \tau) = \mathrm{Rdt}(\sigma) \;{}_{o,e}\!\!< \mathrm{Rdt}(\tau)$.

(5) $\mathrm{Rdgv}(\neg\alpha) = \neg\mathrm{Rdgv}(\alpha)$.

$\mathrm{Rdgv}(\alpha \circ \beta) = \mathrm{Rdgv}(\alpha) \circ \mathrm{Rdgv}(\beta)$ für $\circ = \wedge, \vee, \rightarrow, \leftrightarrow$.

$\mathrm{Rdgv}(\forall x\alpha) = \forall\bar{x}[\mathrm{Ar}_{oe}\bar{x} \rightarrow \mathrm{Rdgv}(\alpha)]$.

$\mathrm{Rdgv}(\exists x\alpha) = \exists\bar{x}[\mathrm{Ar}_{oe}\bar{x} \wedge \mathrm{Rdgv}(\alpha)]$.

Dabei ist die Formel für die geometrische Kleiner-Beziehung in (4) (gemäß I.14.38) unter Verwendung von I.14.34 eingeführt durch

(6) $a \;{}_{o,e}\!\!< b := \mathrm{Ps}_{oe}(b \;{}_{o,e}\!\!- a)$,

wobei $b \;{}_{o,e}\!\!- a := b \;{}_{o,e}\!\!+ ({}_{o,e}\!\!- a)$;

analog mit Termen an Stelle der freien Variablen a, b.

(iv) Die *geometrische Reduzierte* Rdg(α) einer beliebigen Formel α von L_{OF} wird nun erklärt durch

(7) $\mathrm{Rdg}(\alpha) := \forall o \forall e[\mathrm{Ar}_{oe} a_1 \ldots a_k \rightarrow \mathrm{Rdgv}(\alpha)]$,

wobei $a_1, \ldots, a_k$ die außer o und e in Rdgv(α) frei vorkommenden Variablen (in irgendeiner Reihenfolge) sind.

3.37 Anmerkungen. (i) (Festlegung der gebundenen Variablen) In den Termen in 3.36(2), (3) treten gebundene Variablen auf, die natürlich von den jeweils verwendeten freien Variablen verschieden sein sollen (z.B. in (3) verschieden von a, b oder von den freien Variablen der dafür stehenden Terme und von o, e). Mehr wird für Anwendungen im allgemeinen nicht gebraucht. Nur um die Definition eindeutig zu machen, hat man (in Abhängigkeit von den gegebenen freien Variablen!) festzulegen, welche gebundenen Variablen tatsächlich verwendet werden sollen. Das kann mit Hilfe der betrachteten Numerierung der Variablen nach folgendem Prinzip geschehen.

(*) Eine an einer Stelle quantifizierte (festzulegende) Variable wird stets so gewählt, daß ihr Index der kleinste Index ist, der verschieden ist von den Indizes der übrigen im zugehörigen Wirkungsbereich frei vorkommenden Variablen.

Dieses Prinzip läßt sich allgemein anwenden bei der Einführung von Abkürzungen für Formeln oder ι-Terme, wenn die auftretenden gebundenen Variablen in der Abkürzung nicht mehr angegeben werden.

Beispiel. Der Term $a_o +_{o,e} b$ aus 3.36(3) (mit $o=p_o$, $e=p_1$) werde betrachtet für den Fall $a=p_3$, $b=p_6$. Dann ergibt sich für die genannten gebundenen Variablen folgende Festlegung: $c=p_2$, $e'=p_4$ und mit den Bezeichnungen aus I.14.3 weiter $a'=p_5$, $c^*=p_7$. (Die in I.14.3 noch auftretenden Formeln für die Parallelität lassen sich ohne Verwendung von Geradenvariablen definieren, vgl. I.12.7; die dabei noch vorkommenden gebundenen Punktvariablen lassen sich dann nach demselben Prinzip (*) festlegen.)

(ii) (Vermeidung von Konstanten) Zur Beschreibung von "arithmetischen Feststellungen" in der Geometrie wählt man üblicherweise feste Punkte *o* und *e* (eventuell noch weitere Hilfspunkte) und betrachtet den dadurch bestimmten angeordneten Körper (wie hier $\mathfrak{k}_{oe}$ in I.14.40). Eine dementsprechende Formelübersetzung (vgl. 3.35) würde darauf hinauslaufen, daß man Individuenkonstanten zur Bezeichnung dieser Bestimmungspunkte in die Sprache der Geometrie aufnimmt und bei der Bildung der geometrischen Reduzierten verwendet. Die Beschreibung ist jedoch unabhängig davon, wie man die Bestimmungspunkte (in einem gegebenen Raum) wirklich wählt, da für verschiedene Wahlen die zugehörigen angeordneten Körper zueinander isomorph sind (für unsere Konstruktion s. I.14.40). Daher kann man für die Formelübersetzung statt der zusätzlichen Individuenkonstanten auch Variablen mit dem Allquantor (hier in 3.36(7)) verwenden. Auf diese Weise gelingt eine Übersetzung in die ursprüngliche Sprache der Geometrie ohne zusätzliche Konstanten.

(iii) (Elimination des bestimmten Artikels) Die geometrischen Reduzierten wurden hier als Formeln der Sprache $L^1(B)$ mit bestimmtem Artikel definiert. Der bestimmte Artikel läßt sich nach dem in 1.47 erwähnten Verfahren eliminieren, so daß man äquivalente Formeln der Sprache L(B) erhält (es genügt L(D,B)). Im hier betrachteten Spezialfall läßt sich der bestimmte Artikel jedoch auf einfachere Weise eliminieren. In I.14.7 wurden nämlich die (ι-Terme für die) geometrischen Summen, Produkte und Inversen so definiert, daß es stets (sogar in den geometrisch uninteressanten Fällen) genau einen Punkt mit der genannten Eigenschaft gibt. Für solche Fälle gilt der folgende Satz über eine vereinfachte Elimination. (Zu einer anderen Konstruktion von geometrischen Reduzierten, bei der ι gar nicht erst eingeführt wird, vgl. 6.66, 71ff.).

3.38 Satz. *Vor.:* $\exists^{=1} x\alpha(x)$ *ist ein Satz der Theorie* T *(dabei darf* $\alpha(x)$ *außer* x *noch weitere freie Variablen enthalten).*

Beh.: In T gelten die Äquivalenzen

(1) $\beta(\iota x\alpha(x))$ $Äq_T$ $\exists x[\alpha(x) \wedge \beta(x)]$,

(2) $\beta(\iota x\alpha(x))$ $Äq_T$ $\forall x[\alpha(x) \rightarrow \beta(x)]$,

und damit auch

(3) $\exists x[\alpha(x) \wedge \beta(x)]$ $Äq_T$ $\forall x[\alpha(x) \rightarrow \beta(x)]$.

Man kann also zur Elimination nach Belieben den Existenz- oder den Allquantor verwenden; außerdem kann man in Formeln der in (3) *auftretenden Bauart vom Existenzquantor zum Allquantor übergehen und umgekehrt.*

Etwas allgemeiner als (3) *gilt schon in der Prädikatenlogik (ohne obige Voraussetzung über T):*

(4) $\models \exists^{=1} x\alpha(x) \rightarrow \{\exists x[\alpha(x) \wedge \beta(x)] \leftrightarrow \forall x[\alpha(x) \rightarrow \beta(x)]\}$.

<u>3.39 Satz</u>. *Sei* Σ *ein Axiomensystem für (irgendeine) Klasse von angeordneten Körpern in der Sprache* L_{OF}. *Sei* T *eine Geometrie in der Sprache* L(D,B) *mit einem Axiomensystem* Σ_T, *aus dem die geometrischen Reduzierten der Axiome von* Σ *beweisbar sind. Dann gilt für jede Formel* α *von* L_{OF}:

(1) *Wenn* $\Sigma \vdash \alpha$, *so* $\Sigma_T \vdash \mathrm{Rdg}(\alpha)$.

<u>Beweis</u> (Skizze): Der Beweis geschieht durch eine Induktion entsprechend der Definition des Beweisbarkeitsbegriffs (nicht durch Induktion über den Formelaufbau!). Für den Anfangsschritt der Induktion ("α ist ein Axiom") wird gerade die Voraussetzung benötigt. Im Induktionsschritt ist zu zeigen: Wenn α aus anderen Formeln $\alpha_1, \ldots, \alpha_k$ durch Anwendung einer Schlußregel entsteht, so läßt sich auch Rdg(α) aus Rdg(α_1), ..., Rdg(α_k) durch (i.a. mehrfache) Anwendungen von Schlußregeln und (evtl.) Axiomen erhalten. Das läßt sich für die einzelnen Schlußregeln leicht durchführen. (Für die Regeln des in 1.49 erwähnten Beweisbarkeitsbegriffs kommen nur die Fälle $k=1$ und $k=2$ in Frage. Da die Regeln hier nicht angegeben wurden, wird auf die Ausführung verzichtet.)

Auf diese Weise werden formale Beweise aus dem Axiomensystem Σ sozusagen übersetzt in geometrische Beweise aus Σ_T. Die beschriebene Methode liefert auch ein Verfahren, mit dem man zu jedem vorgelegten formalen Beweis aus Σ einen entsprechenden Beweis aus Σ_T effektiv angeben kann.

□

3.40 Definition. Ist T irgendeine der in 2.1 betrachteten Geometrien, so sei Σ_T das dort dafür angegebene Axiomensystem. Außerdem seien Σ_{PF} bzw. Σ_{EF} (endliche) Axiomensysteme für die Klassen PF bzw. EF der pythagoreischen bzw. euklidischen angeordneten Körper.

3.41 Korollar. *Für jede Formel* α *von* L_{OF}:
(1) *Wenn* $\Sigma_{RF} \vdash \alpha$, *so* $\Sigma_{P*} \vdash \mathrm{Rdg}(\alpha)$.
Entsprechendes gilt mit Σ_{PF} *und* Σ_P *sowie mit* Σ_{EF} *und* $\Sigma_{P'}$.

Zum Beweis: Es genügt in jedem der Fälle zu zeigen, daß die Voraussetzung von 3.39 erfüllt ist. An Hand der Überlegungen im Beweis von 3.28(→) erhält man auch formale Beweise für die dort betrachteten geometrischen Sätze (über die angeordneten Körper $\mathfrak{F}_{oe}$) aus dem Axiomensystem Σ_{P*}; diese Sätze sind aber gerade die geometrischen Reduzierten der (nicht schon in Σ_{OF} liegenden) Axiome von Σ_{RF}. Für Σ_{PF} und Σ_{EF} ergibt sich das Gewünschte ("die $\mathfrak{F}_{oe}$ sind pythagoreisch bzw. euklidisch") unmittelbar aus dem Darstellungssatz (I.16.15) und der Isomorphie in I.14.41. □

Für eine beliebige Aussage γ von L(D,B) kann man nun beide Formelübersetzungen hintereinander ausführen. Der folgende Satz besagt, daß die so entstehende (von γ verschiedene) Aussage zu γ beweisbar äquivalent ist.

3.42 Satz. *Sei n die Dimension, für die die arithmetischen Reduzierten gebildet werden (3.25), und $n \geq 2$. Dann gilt für jede Aussage γ von* L(D,B):

$$\Sigma_{P_n} \vdash [\gamma \leftrightarrow \mathrm{Rdg}(\mathrm{Rda}(\gamma))].$$

Der Beweis beruht darauf, daß eine geeignete Hilfsbehauptung für beliebige Formeln formuliert wird, die sich durch Induktion über den Formelaufbau beweisen läßt.

3.43 Lemma. *Sei n wie vorher. Dann gilt für jede Formel γ von* L(D,B):
Wenn $\mathrm{Fr}(\gamma)=\{a_1,\dots,a_k\}$, *so*

$$\Sigma_{P_n} \vdash \mathrm{Cs}_{oe}su_1\dots u_n \wedge \bigwedge_{\kappa=1}^{k} \mathrm{Cd}_{oe}^{su_1\dots u_n} a_\kappa \overline{a'_\kappa\dots a_\kappa^{(n)}} \rightarrow [\gamma \leftrightarrow \mathrm{Rdgv}(\mathrm{Rda}(\gamma))].$$

Die anschauliche Bedeutung des Lemmas liegt auf der Hand: Bei der doppelten Reduziertenbildung wird jeder freien Punktvariablen a_κ von γ ein n-tupel von Punktvariablen $\overline{a_\kappa^{(\nu)}}$ ($\nu=1,\dots,n$) zugeordnet. Aus der Voraussetzung, daß die durch diese Variablen bezeichneten Punkte jeweils die Koordinaten des durch a_κ bezeichneten Punktes in einem gegebenen kartesischen Koordinatensystem (I.16.1, 5) sind, ergibt sich die Gleichwertigkeit von γ mit der "Übersetzung" Rdgv(Rda(γ)).

Zum Beweis von 3.43: Wir betrachten zunächst die Formeln γ, die der folgenden Zusatzvoraussetzung genügen:

(*) Für jede in γ quantifiziert vorkommende Variable a gilt:

($*_a$) a und die zugehörigen $\overline{a^{(\nu)}}$ ($\nu=1,\ldots,n$) sind verschieden von den Variablen o, e, s, $u_1,\ldots, u_n$.

Für diese Formeln γ läßt sich die Behauptung durch Induktion über den Aufbau von γ beweisen (wobei die Variablen o,..., u_n für die Bezugspunkte des Koordinatensystems festgehalten werden können). Ist γ eine Primformel der Form Dabcd bzw. Bpqr bzw. $p \doteq q$, so ist die zu beweisende Formel gerade der Satz I.16.12 bzw. 13 bzw. die Eindeutigkeitsbehauptung in der letzten Formelzeile von I.16.7. Im Induktionsschritt ist die Verknüpfung mit Junktoren trivial, und es genügt die Behandlung der Quantifizierung mit $\exists$ (da sich $\forall$ durch $\exists$ und $\neg$ ausdrücken läßt). Die zu beweisende Formel ist dann von der Form

(1) $\pi \rightarrow [\exists a\gamma \leftrightarrow \varepsilon]$, wobei ε = Rdgv(Rda($\exists a\gamma$)).

Nach 3.25(2), 3.36(5) ist ε logisch äquivalent zu

(2) $\varepsilon^* = \exists\overline{a'}\ldots\exists\overline{a^{(n)}}[Ar_{oe}\overline{a'}\ldots\overline{a^{(n)}} \wedge \bar{\gamma}]$, wobei $\bar{\gamma}$ = Rdgv(Rda(γ)).

Nach Induktionsvoraussetzung ist schon aus Σ_{P_n} beweisbar:

(3) $\pi \wedge \pi_o \rightarrow [\gamma \leftrightarrow \bar{\gamma}]$, wobei $\pi_o = Cd_{oe}^{su_1\ldots u_n}a\overline{a'}\ldots\overline{a^{(n)}}$.

Ein inhaltlicher Beweis für (1) (aus dem man ohne weiteres einen formalen Beweis konstruieren kann) ist nun ganz einfach. Gelte nämlich die Voraussetzung π. Ist dann ein Punkt a mit der Eigenschaft γ gegeben, so wähle man $\overline{a'},\ldots, \overline{a^{(n)}}$ (gemäß I.16.7) als seine Koordinaten im gegebenen System (d.h. mit der Eigenschaft π_o), diese Punkte leisten nach I.16.6 und (3) das Verlangte. Umgekehrt kann man zu Punkten $\overline{a'},\ldots, \overline{a^{(n)}}$ der in (2) angegebenen Art einen Punkt a mit diesen Koordinaten (d.h. mit der Eigenschaft π_o) wählen, der nach (3) das Verlangte leistet.

Von der Zusatzvoraussetzung (*) kann man sich nachträglich leicht befreien durch Umbenennungen von gebundenen Variablen in γ und entsprechende Umbenennungen in Rdgv(Rda(γ)). □

Zum Beweis von 3.42: Sei γ jetzt eine Aussage. Dann fällt in 3.43 die verallgemeinerte Konjunktion ganz weg, und als Prämisse genügt Ar_{oe} (da man zugehörige Punkte s, $u_1,\ldots, u_n$ nach I.16.4 wählen kann), d.h. es ist in P_n beweisbar:

(1) $Ar_{oe} \to [\gamma \leftrightarrow \bar{\gamma}]$, wobei $\bar{\gamma} = Rdgv(Rda(\gamma))$.

Nach 3.36(7) ist dann

(2) $\bar{\bar{\gamma}} := Rdg(Rda(\gamma)) = \forall o \forall e[Ar_{oe} \to \bar{\gamma}]$,

und $\bar{\bar{\gamma}}$ ist ebenfalls eine Aussage. Die Richtung $\gamma \to \bar{\bar{\gamma}}$ der Behauptung ergibt sich nun unmittelbar aus (1). Die Umkehrung $\bar{\bar{\gamma}} \to \gamma$ erhält man (inhaltlich), indem man Punkte *o*, *e* mit Ar_{oe} wählt (nach I.3.13) und (1) anwendet.

Die hier und im Beweis zu 3.43 beschriebene Methode liefert wieder ein Verfahren, mit dem man zu jeder Formel bzw. Aussage der angegebenen Art einen formalen Beweis aus Σ_{P_n} effektiv angeben kann. □

3.44 Satz (Syntaktische Version der Vollständigkeit). *Für jede Aussage* γ *von* L(D,B) *gilt:*

Wenn $\models_{\mathfrak{L}_n(\mathbb{R})} \gamma$, *so* $\Sigma_{P_n^*} \vdash \gamma$.

Beweis: Das ergibt sich durch Rückführung auf den entsprechenden Satz für $\mathbb{R}$:

(*) *Wenn* $\models_{\mathbb{R}} \alpha$, *so* $\Sigma_{RF} \vdash \alpha$ *(für Aussagen* α *von* L_{OF}*).*

Von der Voraussetzung ausgehend, erhält man nämlich folgende Bedingungen:

(1) $\models_{\mathfrak{L}_n(\mathbb{R})} \gamma$ (Voraussetzung),

(2) $\models_{\mathbb{R}} Rda(\gamma)$ (nach 3.26(ii)),

(3) $\Sigma_{RF} \vdash Rda(\gamma)$ (nach (*)),

(4) $\Sigma_{P_n^*} \vdash Rdg(Rda(\gamma))$ (nach 3.41(1)),

(5) $\Sigma_{P_n^*} \vdash \gamma$ (nach 3.42).

Aus (5) ergibt sich natürlich auch wieder (1), da $\mathfrak{L}_n(\mathbb{R})$ ein Modell von $\Sigma_{P_n^*}$ ist. Somit sind die Bedingungen (1) bis (5) untereinander gleichwertig. □

3.45 Korollar. *Für jede Aussage* γ *von* L(D,B) *gilt:*

(1) $\Sigma_{P_n^*} \vdash \gamma$ *oder* $\Sigma_{P_n^*} \vdash \neg\gamma$;

(2) $\Sigma_{P_n^*} \models \gamma$ *oder* $\Sigma_{P_n^*} \models \neg\gamma$;

(3) $\gamma \in P_n^*$ *oder* $\neg\gamma \in P_n^*$ *(Vollständigkeit von* P_n^**).*

Beweis: Für eine beliebige Aussage γ von L(D,B) ist trivialerweise γ oder $\neg\gamma$ in der Struktur $\mathcal{L}_n(\mathbb{R})$ gültig (1.21), somit ergibt sich (1) sofort aus 3.44. (2) ergibt sich daraus auf Grund der Korrektheit des Beweisbarkeitsbegriffs (s. 1.50), und (3) ist mit (2) gleichbedeutend.

□

3.46 Anmerkungen. (i) Im Vollständigkeitsbeweis nach der "semantischen Methode" (bis 3.32) wurde der (zur Syntax gerechnete) Begriff der (formalen) Beweisbarkeit überhaupt nicht gebraucht. Aus der so bewiesenen Vollständigkeit von P_n^* erhält man natürlich auch die Resultate 3.45(1) und 3.44 sofort mit Hilfe des Vollständigkeitssatzes der Prädikatenlogik. Mit den Methoden aus dem Beweis von 3.13 erhält man dann sogar ein Verfahren, einen formalen Beweis entsprechend der Behauptung von 3.44 effektiv anzugeben (mit der Menge der beweisbaren Formeln ist auch eine Menge von zugehörigen formalen Beweisen rekursiv aufzählbar). Dieses Verfahren ist allerdings wieder denkbar unökonomisch; es gilt sinngemäß das in 3.14(i) Gesagte.

(ii) Die danach geschilderte "syntaktische Methode" zum Beweis von 3.44 ist zwar wesentlich aufwendiger als die semantische. Sie macht aber keinen Gebrauch von der Betrachtung anderer Modelle und vom Vollständigkeitssatz der Prädikatenlogik und liefert sozusagen "direkt" ein "gezieltes" Verfahren zur effektiven Angabe eines entsprechenden formalen Beweises. Falls nämlich 3.44(1) (und damit (2)) gilt, liefert das Tarskische (oder ein anderes) Entscheidungsverfahren für Th($\mathbb{R}$) effektiv einen formalen Beweis für Rda(γ) aus Σ_{RF}. Der Beweis von 3.39 (s. dort) liefert daraus (effektiv) einen formalen Beweis für Rdg(Rda(γ)) aus Σ_{P^*} und der Beweis von 3.42 daraus einen formalen Beweis für γ aus $\Sigma_{P_n^*}$.

Allerdings ist auch dieses "gezielte" Verfahren kaum für praktische Anwendungen geeignet; weder das Entscheidungsverfahren für Th($\mathbb{R}$) noch die Rückführung darauf liefert besonders einfache formale Beweise - für geometrische Sätze hat man ja im allgemeinen wesentlich einfachere Beweise als durch Übersetzung in die Arithmetik.

(iii) In diesem Zusammenhang sei erwähnt, daß heute auch automatische Beweisverfahren untersucht und zum Teil maschinell durchgeführt werden, die der jeweiligen (evtl. sogar unentscheidbaren) Theorie angepaßt sind und relativ einfache formale Beweise (falls solche überhaupt existieren) liefern.

(iv) In Schwabhäuser 1956 wurden (nach einem älteren Sprachgebrauch) noch andere Bezeichnungen für die verschiedenen Arten von Vollständigkeit verwendet. Dort stand der Beweisbarkeitsbegriff im Vordergrund, und dementsprechend wurde 3.45(1) als "klassische" Vollständigkeit bezeichnet. Im Unterschied zu diesem - nur syntaktische Begriffsbildungen enthaltenden - Satz wurde 3.44 als "inhaltliche" (d.h. soviel wie "semantische") Vollständigkeit bezeichnet.

<u>3.47 Beispiel eines interessanten Satzes von P_3^*</u>. Für eine gegebene Kugel im dreidimensionalen Raum $\mathfrak{L}_3(\mathbb{R})$ kann man fragen, wieviele Kugeln gleicher Größe an die gegebene "angelegt" werden können, so daß sie die gegebene berühren und sich gegenseitig nicht durchdringen (man denke an eine Realisierung des Problems durch Billardkugeln). Im Unterschied zum analogen Problem für die Ebene (wo sich sechs Kreise anlegen lassen) ist die Situation im Raum unübersichtlich. Man sieht noch relativ leicht, daß sich zwölf Kugeln im genannten Sinne anlegen lassen. Das Problem, ob das auch noch für dreizehn Kugeln möglich ist, wurde schon von Newton und Gregory diskutiert. Ein Beweis für die Unmöglichkeit wurde aber erst durch Schütte u. van der Waerden 1953 gegeben "by an ingenious combination of trigonometry and topology" (Coxeter, Math. Reviews 14(1953), 787).

Man beachte, daß der damit bewiesene Satz in der Sprache L(D,B) formalisiert werden kann und damit zu P_3^* gehört. Das genannte Problem hätte sich also prinzipiell auch durch Anwendung des Tarskischen Entscheidungsverfahrens lösen lassen, das zu diesem Zeitpunkt schon bekannt war (Tarski 1948).

Zur Vereinfachung können wir dabei noch annehmen, daß der Mittelpunkt der gegebenen Kugel der Koordinatenursprung ist und alle Kugeln den Radius $\frac{1}{2}$ haben. Damit erhalten wir folgende Form (einer "vereinfachten arithmetischen Reduzierten") des genannten Satzes:

$$\neg\exists a_1'\exists a_1''\exists a_1''' \ldots \exists a_{13}'\exists a_{13}''\exists a_{13}'''[\bigwedge_{k=1}^{13} \sum_{\nu=1}^{3} (a_k^{(\nu)})^2 \doteq 1 \ \wedge$$

$$\wedge \bigwedge_{1\le k<l\le 13} \sum_{\nu=1}^{3} (a_k^{(\nu)} - a_l^{(\nu)})^2 \ge 1]$$

(a_k', a_k'', a_k''' bedeuten dabei die Koordinaten des Mittelpunkts der k-ten anzulegenden Kugel). Bei Anwendung des Tarskischen Verfahrens (jetzt für Th($\mathbb{R}$)) auf diese Aussage wären also insbesondere 39 Quantoren zu eliminieren. Der Aufwand für eine tatsächliche Durchführung ist natürlich viel zu groß.

3.48 Geometrische Reduzierten und Semantik, Unentscheidbarkeitsbeweise. Im folgenden führen wir noch eine semantische Überlegung für geometrische Reduzierten aus, die sich an 3.36 anschließt und analog ist zu der Überlegung für arithmetische Reduzierten in 3.25(iii), 3.26. Sie wird dann zum Beweis von Unentscheidbarkeitsresultaten (3.58ff.) verwendet.

3.49 Definition. Sei $n \geq 2$, $n \in \mathbb{N}$. Sei $\mathfrak{f}$ ein angeordneter Körper, seien o, e verschiedene Punkte des Raumes $\mathfrak{L}_n(\mathfrak{f})$, und sei φ ein Isomorphismus von $\mathfrak{f}$ auf die "Zahlengerade" $\mathfrak{f}_{oe}$ gemäß I.14.40, 41. Eine Belegung j über $\mathfrak{f}$ und eine Belegung h über $\mathfrak{L}_n(\mathfrak{f})$ heißen dann bezüglich o, e *assoziiert*, falls für jede Zahlenvariable x gilt:

(1) $h(\bar{x}) = \varphi(j(x))$, $h(o) = o$, $h(e) = e$.

3.50 Anmerkung. Für den Isomorphismus φ läßt sich zeigen, daß er durch die oben angegebene Zusatzeigenschaft (wegen der Beschränkung auf Parallelprojektionen in I.14.40) eindeutig bestimmt ist. Man kann jedoch auch auf den Nachweis dafür verzichten und statt dessen "bezüglich φ assoziierte" Belegungen betrachten.

3.51 Satz. *Seien n, $\mathfrak{f}$, o, e, φ wie in* 3.49.

(i) *Wenn τ ein Term der Sprache* L_{OF} *ist, so gilt für jedes Paar von bezüglich o, e assoziierten Belegungen j, h:*

$$\varphi(w_{\mathfrak{f}}(\tau, j)) = w_{\mathfrak{L}_n(\mathfrak{f})}(\mathrm{Rdt}(\tau), h).$$

(ii) *Wenn α eine Formel der Sprache* L_{OF} *ist, so gilt für jedes Paar von bezüglich o, e assoziierten Belegungen j, h:*

$$j\,\mathrm{Erf}_{\mathfrak{f}}\ \alpha \text{ genau dann, wenn } h\,\mathrm{Erf}_{\mathfrak{L}_n(\mathfrak{f})}\mathrm{Rdgv}(\alpha).$$

(iii) *Wenn α eine Aussage von* L_{OF} *ist, so gilt:*

$$\models_{\mathfrak{f}} \alpha \text{ genau dann, wenn } \models_{\mathfrak{L}_n(\mathfrak{f})}\mathrm{Rdg}(\alpha).$$

Beweis: (i) ergibt sich ohne weiteres durch Induktion über den Termaufbau, wobei verwendet wird, daß die Terme in 3.36(3) gerade die Operationen von $\mathfrak{f}_{oe}$ (I.14.7, 8) beschreiben (sofern nur Argumente in $\mathfrak{f}_{oe}$ betrachtet werden). (ii) ergibt sich dann durch Induktion über den Formelaufbau (wobei für die Primformeln $\sigma < \tau$ ein entsprechender Zusammenhang zwischen den Formeln 3.36(6) und der Anordnung von $\mathfrak{f}_{oe}$ benutzt wird). Ist insbesondere α eine Aussage, so ist auch $\mathrm{Rdg}(\alpha) = \forall o \forall e[\mathrm{Ar}_{oe} \rightarrow \mathrm{Rdgv}(\alpha)]$ eine Aussage, und (iii) ergibt sich dann daraus, daß man zu beliebigen voneinander verschiedenen Punkten *o*, *e* von $\mathfrak{L}_n(\mathfrak{F})$ Belegungen *j*, *h* gemäß 3.49 wählen kann. □

Wir können nun einen allgemeinen Zusammenhang zwischen der Entscheidbarkeit einer Geometrie und der einer zugehörigen Theorie von angeordneten Körpern herstellen.

3.52 Definition. Sei K eine Klasse von angeordneten Körpern. Unter der (elementaren) *n-dimensionalen euklidischen Geometrie* $\Gamma_n(K)$ *über (Strukturen aus) K* verstehen wir dann die Theorie der Klasse der n-dimensionalen kartesischen Räume über Strukturen aus K, d.h.

$$\Gamma_n(K) = \mathrm{Th}(\{\mathfrak{L}_n(\mathfrak{F}) \mid \mathfrak{F}\in K\}).$$

Unter der entsprechenden *dimensionsfreien euklidischen Geometrie* $\Gamma(K)$ verstehen wir (in Analogie zu 2.15) den Durchschnitt

$$\Gamma(K) = \bigcap_{2\leq n\in\mathbb{N}} \Gamma_n(K) = \mathrm{Th}(\{\mathfrak{L}_n(\mathfrak{F}) \mid \mathfrak{F}\in K,\ n\in\mathbb{N},\ n\geq 2\}).$$

Ebenso definieren wir $\Gamma_{\leq n}(K)$ und $\Gamma_{\geq n}(K)$ (analog zu 2.15(v)) durch eine entsprechende Bildung des Durchschnitts über die erlaubten Dimensionen $\leq n$ bzw. $\geq n$.

Beispiel. Für K = RF, EF bzw. PF (2.12) ist insbesondere $\Gamma_n(K) = P_n^*$, P_n' bzw. P_n $(n\geq 2)$.

3.53 Satz. *Sei K wie vorher,* $n\geq 2$, $n\in\mathbb{N}$. *Dann ist* $\Gamma_n(K)$ *entscheidbar genau dann, wenn* Th(*K*) *entscheidbar ist.*

Beweis: Zu (←): Sei γ eine beliebige Aussage von L(D,B). Dann sind (nach Definition und 3.26(ii)) die folgenden Bedingungen untereinander gleichwertig:

(1) $\gamma\in\Gamma_n(K)$,

(2) $\models_{\mathfrak{L}_n(\mathfrak{F})}\gamma$ für jedes $\mathfrak{F}\in K$,

(3) $\models_{\mathfrak{F}} \mathrm{Rda}(\gamma)$ für jedes $\mathfrak{F}\in K$,

(4) $\mathrm{Rda}(\gamma)\in\mathrm{Th}(K)$.

Zur Nachprüfung von (4) gibt es nach Voraussetzung ein Entscheidungsverfahren; daraus erhält man auch eins zur Nachprüfung von (1).

Zu (→): Sei α eine beliebige Aussage von L_{OF}. Dann sind (nach Definition und 3.51(iii)) die folgenden Bedingungen untereinander gleichwertig:

(5) $\alpha \in Th(K)$,

(6) $\models_{\mathfrak{F}} \alpha$ für jedes $\mathfrak{F} \in K$,

(7) $\models_{\mathfrak{L}_n(\mathfrak{F})} Rdg(\alpha)$ für jedes $\mathfrak{F} \in K$,

(8) $Rdg(\alpha) \in \Gamma_n(K)$.

Zur Nachprüfung von (8) gibt es nach Voraussetzung ein Entscheidungsverfahren; daraus erhält man eins zur Nachprüfung von (5) (da man die geometrischen Reduzierten zu gegebenen Formeln - ebenso wie die arithmetischen Reduzierten [vgl. 3.31] - stets effektiv angeben kann). □

Mittels 3.53(←) ergibt sich aus 3.22 natürlich noch einmal die Entscheidbarkeit von P_n^* (die in 3.31 in ähnlicher Weise bewiesen wurde).

Mit Hilfe der Umkehrung 3.53(→) ergibt sich nun sofort die U n e n t s c h e i d b a r k e i t geeigneter Geometrien $\Gamma_n(K)$ aus der Unentscheidbarkeit entsprechender Theorien von angeordneten Körpern. Einige dafür und später verwendete Unentscheidbarkeitsresultate werden zunächst ohne Beweis zusammengestellt.

<u>3.54 Satz</u>. *Vor.: Die Theorie T besitzt ein Modell $\mathfrak{A}$ mit folgender Eigenschaft: Es gibt eine Teilmenge M von $U_{\mathfrak{A}}$, die sich durch eine Formel α(x) der Sprache von T definieren läßt (wie in 3.22(v)) und dreistellige Relationen S, P auf M, die sich in derselben Weise durch Formeln β(x,y,z), γ(x,y,z) definieren lassen, so daß die Struktur ⟨M,S,P⟩ isomorph ist zur Struktur der natürlichen Zahlen mit Addition und Multiplikation (hier aufgefaßt als dreistellige Relationen).*

Beh.: T ist unentscheidbar.

<u>3.55 Satz</u>. *Vor.: T_2 ist eine sog. "unwesentliche Erweiterung" von T_1, d.h., die Sprache L_2 von T_2 entsteht aus der Sprache L_1 von T_1 durch Hinzunahme von (höchstens) Individuenkonstanten, und es ist $T_2 = Cn_{L_2}(T_1)$ (anschaulich gesagt, für T_2 werden keine zusätzlichen Axiome zugrunde gelegt).*

Beh.: T_1 ist unentscheidbar genau dann, wenn T_2 unentscheidbar ist.

<u>3.56 Satz</u>. *Vor.: T_2 ist eine endliche Erweiterung von T_1 (1.36) mit derselben Sprache.*

Beh.: Wenn T_2 unentscheidbar ist, so auch T_1.

Beweise für diese Sätze werden gegeben in Tarski, Mostowski u. Robinson 1953: 3.55 und 3.56 sind einfach und werden dort direkt angegeben (S.16 bzw. 17), 3.54 ergibt sich durch Zusammenfassung von Sätzen in den Abschnitten I und II (die stärkere Resultate enthalten) und beschreibt eine besonders häufig angewendete Methode für Unentscheidbarkeitsbeweise (durch Rückführung auf die Unentscheidbarkeit der Theorie der natürlichen Zahlen).

3.57 Satz (Julia Robinson). (i) *Der Körper* $\mathbb{Q}$ *der rationalen Zahlen hat die in* 3.54 *für* $\mathfrak{A}$ *angegebene Eigenschaft.*

(ii) (Folgerung) *Die elementare Theorie des Körpers der rationalen Zahlen - mit oder ohne Anordnung - ist unentscheidbar und (damit nach* 3.13 *und Beispiel zu* 1.39*) nicht axiomatisierbar.*

(iii) (Folgerung) *Die elementaren Theorien* Th(F) *und* Th(OF) *aller Körper bzw. aller angeordneten Körper sind unentscheidbar (aber natürlich endlich axiomatisierbar).*

Das wurde gezeigt in J. Robinson 1949 unter Benutzung der Zahlentheorie zum Beweis von (i).

Aus 3.57(iii) ergibt sich mittels 3.53(→):

3.58 Satz. *Die n-dimensionale euklidische Geometrie* Γ_n(OF) *über beliebigen angeordneten Körpern* ($2 \leq n \in \mathbb{N}$) *ist unentscheidbar.*

3.59 Anmerkung. Dieses Ergebnis ist für eine in einer anderen Sprache formalisierte Geometrie enthalten in Rautenberg 1962, wo auch eine (endliche) Axiomatisierung für diese Geometrie behandelt wurde. Ein endliches Axiomensystem für Γ_n(OF) in der hier betrachteten Sprache L(D,B) wurde angegeben in Gupta 1965, 1965a, für $n=2$ s. auch Gupta u. Piesyk 1965.

Die Frage, ob die Theorien Th(PF) und Th(EF) der pythagoreischen bzw. euklidischen angeordneten Körper entscheidbar sind, war lange Zeit offen. Die Unentscheidbarkeit wurde erst kürzlich bewiesen durch Ziegler 1982 unter Verwendung von Ideen aus Hauschild 1974, 1977. Daraus ergibt sich mittels 3.53(→) sofort:

3.60 Satz. *Die euklidischen Geometrien* P_n *und* P'_n *(ohne bzw. mit Kreisaxiom)* $(2 \leq n \in \mathbb{N})$ *sind unentscheidbar.*

Natürlich gilt auch:

3.61 Satz. *Die betrachteten Geometrien* $\Gamma_n(\mathrm{OF})$, P_n *und* P'_n $(2 \leq n \in \mathbb{N})$ *sind (als echte Untertheorien von* P^*_n*) unvollständig und (nach Anmerkung 3.59 bzw. nach Definition) endlich axiomatisierbar.*

Bei den Entscheidbarkeitsresultaten 3.22 und 3.31 ist die Beschränkung auf die erste Stufe wesentlich; denn Theorien zweiter Stufe sind fast immer unentscheidbar. Insbesondere gilt

3.62 Satz. (i) *Für eine beliebige nichtleere Klasse* K *von angeordneten Körpern ist die zugehörige Theorie* $\mathrm{Th}^2(K)$ *zweiter Stufe unentscheidbar.*

(ii) (Folgerung) *Die Theorie* $\mathrm{Th}^2(\mathbb{R})$ *zweiter Stufe der reellen Zahlen ist unentscheidbar.*

(iii) *Die n-dimensionale euklidische Geometrie* P^2_n *zweiter Stufe (2.1) ist unentscheidbar.*

Beweis: Zu (i): Sei $\mathfrak{f} \in K$ beliebig. Nach 3.54 genügt es zu zeigen, daß $\mathfrak{f}$ die dort für $\mathfrak{A}$ angegebene Eigenschaft besitzt. Der Bereich M der natürlichen Elemente von $\mathfrak{f}$ (3.18(i)) wird offenbar definiert durch die Formel

(1) $\alpha(x) = \forall X\{0\dot{\in}X \wedge \forall z[z\dot{\in}X \to z+1\dot{\in}X] \to x\dot{\in}X\}$.

M leistet zusammen mit der auf M eingeschränkten Addition und Multiplikation von $\mathfrak{f}$ das Verlangte; man kann also setzen

(2) $\beta(x,y,z) = \alpha(x) \wedge \alpha(y) \wedge x+y\doteq z$,

(3) $\gamma(x,y,z) = \alpha(x) \wedge \alpha(y) \wedge x\cdot y\doteq z$.

Zu (iii): Statt P^2_n betrachten wir gemäß 3.55 die unwesentliche Erweiterung, die daraus durch Hinzunahme von zwei Individuenkonstanten für Punkte entsteht, und ein Modell $\mathfrak{A}$, das aus $\mathfrak{L}_n(\mathbb{R})$ dadurch entsteht, daß diesen Konstanten verschiedene Punkte o, e von $\mathfrak{L}_n(\mathbb{R})$ zugeordnet werden. Dann läßt sich für den geometrisch eingeführten Körper $\mathfrak{f}_{oe}$ eine zur vorigen analoge Konstruktion durchführen. □

3.63 Anmerkung. Aus der Entscheidbarkeit von $\mathrm{Th}(\mathbb{R})$ ergibt sich, daß sich die Überlegung zu 3.62(i) nicht in die erste Stufe übertragen läßt, d.h., daß die Menge der natürlichen Zahlen innerhalb $\mathbb{R}$ nicht durch eine

Formel der betrachteten Sprache erster Stufe definiert werden kann. Das steht in Einklang mit der Charakterisierung aller durch solche Formeln definierten Mengen in 3.22(v).

3.64 Reduziertenbildungen für andere Geometrien. Die Konstruktion der arithmetischen Reduzierten in 3.25 war im wesentlichen der Definition der kartesischen Räume nachgebildet. Man sieht nun, daß sich analoge Konstruktionen von arithmetischen Reduzierten auch für die anderen in Abschnitt 2 betrachteten Räume durchführen lassen und daß sich damit entsprechende Sätze für hyperbolische, affine, projektive und elliptische Geometrien ergeben. Die Ausführung dieser Konstruktionen kann dem Leser überlassen bleiben. Es sei nur auf die folgenden beiden Unterschiede hingewiesen.

Für die n-dimensionalen hyperbolischen Kleinschen Räume $\mathfrak{K}_n(\mathfrak{F})$ ist die Trägermenge (2.4(1)) nicht die Menge aller n-tupel; dementsprechend sind die letzten beiden Rekursionsgleichungen von 3.25(2) zu ersetzen durch

(1) $\mathrm{Rda}(\forall a\gamma) = \forall a' \ldots \forall a^{(n)} [\varepsilon \rightarrow \mathrm{Rda}(\gamma)]$,

(2) $\mathrm{Rda}(\exists a\gamma) = \exists a' \ldots \exists a^{(n)} [\varepsilon \wedge \mathrm{Rda}(\gamma)]$, wobei

(3) $\varepsilon = \sum_{\nu=1}^{n} (a^{(\nu)})^2 < 1$.

Für n-dimensionale projektive oder elliptische Räume sind die $(n+1)$-tupel von homogenen Koordinaten (2.9(iv)) nicht eindeutig bestimmt; dementsprechend kann man für Gleichungen die folgenden arithmetischen Reduzierten einführen (wobei jeder Punktvariablen a jetzt $n+1$ Zahlenvariablen $a^{(0)}, \ldots, a^{(n)}$ zugeordnet werden):

(4) $\mathrm{Rda}(a \doteq b) = \exists u [u \neq 0 \wedge \bigwedge_{\nu=0}^{n} a^{(\nu)} \doteq u \cdot b^{(\nu)}]$.

Für die Quantifizierungen verwendet man dann (1) und (2), wobei

(5) $\varepsilon = \neg \bigwedge_{\nu=0}^{n} a^{(\nu)} \doteq 0$.

Die Konstruktion der geometrischen Reduzierten in 3.36 war ebenso der geometrischen Definition der angeordneten Körper $\mathfrak{F}_{oe}$ nachgebildet. Für hyperbolische, affine, projektive und elliptische Geometrien lassen sich geometrische Reduzierten in analoger Weise konstruieren auf Grund der in 2.6 und 2.14(ii) erwähnten geometrischen Definitionen von Körpern bzw. angeordneten Körpern (Durchführung für die hyperbolische Geometrie s. Schwabhäuser 1959). Dann ergeben sich auch zu den bisherigen analoge

Sätze über geometrische Reduzierten für diese Geometrien.

Damit erhält man dann die folgenden (zu 3.31, 32, 58, 60, 61, 62 analogen) Resultate.

3.65 Satz. *Sei* $2 \le n \in \mathbb{N}$. *Für* $T = H_n, E_n, Af_n, Afo_n, Pr_n, Pro_n$ *(n-dimensionale hyperbolische, elliptische, affine bzw. projektive Geometrie ohne und mit Anordnung, s.* 2.16, 13*) gilt:*

(i) T^* *ist vollständig und entscheidbar.*

(ii) (Tarski 1949 für eine anders aufgebaute projektive Geometrie) T *und* E_n' *sind unentscheidbar (und unvollständig).*

(iii) T^2 *(Geometrie zweiter Stufe) ist unentscheidbar.*

3.66 Anmerkung. Verschiedenen Möglichkeiten zur Beschreibung der betrachteten Modelle bzw. zur geometrischen Einführung eines Körpers (vgl. 2.6) entsprechen natürlich auch verschiedene Möglichkeiten zur Konstruktion der arithmetischen bzw. geometrischen Reduzierten und damit verschiedene Beweismöglichkeiten für 3.65. Für die arithmetischen Reduzierten wird (für unsere Anwendungen) gebraucht, daß sie in der Sprache L_{OF} liegen, das bedeutet im wesentlichen, daß sich die geometrischen Grundbegriffe r a t i o n a l in den Koordinaten der Punkte ausdrücken lassen. In der Literatur werden allerdings auch Koordinatenarten (und damit Beschreibungen von Modellen) benutzt, für die das nicht der Fall ist (s. etwa Liebmann 1923 für hyperbolische Geometrie).

3.67 Dimensionsfreie und unendlichdimensionale Geometrien. Die bisherigen Ergebnisse bezogen sich auf Geometrien mit einer (beliebigen, aber) festen endlichen Dimension n. Für die zugehörigen dimensionsfreien Geometrien erster Stufe erhalten wir zunächst, daß sie auch unendlichdimensionale Modelle besitzen.

3.68 Definition. Sei T eine der bisher betrachteten dimensionsfreien Geometrien erster Stufe (s. 2.16, 2.15, 3.52). Allgemein sei dann das *untere Dimensionsaxiom für die Dimension* n (wie für A in I.11.70, $n \ge 1$) eingeführt als eine gewisse Aussage Dim_n^- derart, daß diese in einem Raum $\mathfrak{A}$ von der jeweiligen für T betrachteten Art gültig ist genau dann, wenn $\mathfrak{A}$ wenigstens die Dimension n hat. (Ohne Ausführung der Konstruktion sei nur erwähnt, daß man solche Formeln in der jeweils für T betrachteten Sprache angeben kann.) Das Dimensionsaxiom für die Dimension n sei dann wieder die Aussage $\mathrm{Dim}_n := \mathrm{Dim}_n^- \wedge \neg \mathrm{Dim}_{n+1}^-$.

Unter der zugehörigen *unendlichdimensionalen Geometrie* T_∞ verstehen wir dann die Theorie, die aus T durch Hinzunahme aller dieser Axiome entsteht, d.h.

(1) $T_\infty = \mathrm{Cn}(T \cup \Delta_\infty)$, wobei

(2) $\Delta_\infty = \{\mathrm{Dim}_n^- \mid n \in \mathbb{N} - \{0\}\}$.

Die Modelle von T_∞ heißen dann die *unendlichdimensionalen Modelle* von T.

3.69 Satz. (i) *Solche unendlichdimensionalen Modelle existieren.*

(ii) T_∞ *ist keine endliche Erweiterung von* T.

Beweis: Zu (i) genügt es nach dem Endlichkeitssatz (1.52) zu zeigen, daß jede endliche Teilmenge Σ' von $T \cup \Delta_\infty$ ein Modell besitzt. Ein beliebiges solches Σ' enthält aber insbesondere nur für endlich viele n die Axiome Dim_n^-; wählt man eine natürliche Zahl n_0 größer-gleich allen diesen n und ≥ 2, so hat T ein Modell der endlichen Dimension n_0, das auch Modell von Σ' ist.

Zu (ii): Nehmen wir das Gegenteil an. Dann ist T_∞ schon eine einfache Erweiterung (1.36), also von der Form $T_\infty = \mathrm{Cn}(T \cup \alpha)$. Nach 3.68(1) und dem Endlichkeitssatz für das Folgern (1.51) folgt α schon aus einer gewissen endlichen Teilmenge Σ' von $T \cup \Delta_\infty$. Wie in (i) hat aber $T \cup \Sigma'$ ein Modell einer endlichen Dimension n_0, das dann auch Modell von $T \cup \alpha$ und damit von T_∞ ist im Widerspruch zur Definition von T_∞. □

3.70 Anmerkung. Natürlich kann man die unendlichdimensionalen Modelle für die betrachteten Geometrien genauer untersuchen. Jedes solche Modell läßt sich dann charakterisieren als ein Raum mit einer Dimension, die eine eindeutig bestimmte unendliche Kardinalzahl ist (wie bei den affinen Räumen in 2.8). Eine solche Charakterisierung wird jedoch für die folgenden Überlegungen nicht gebraucht.

Von D. Scott (1959) wurde nämlich mit Hilfe einer modelltheoretischen Überlegung gezeigt, daß sich die dimensionsfreie Geometrie P^* schon als Theorie der kartesischen Räume über $\mathbb{R}$ von endlicher Dimension ≥ 2 charakterisieren läßt. Die genannte Arbeit war damit grundlegend für die Untersuchung elementarer dimensionsfreier Geometrien. Die wichtigsten Resultate daraus werden anschließend behandelt und auf Modelle der absoluten Geometrie A verallgemeinert. Eine entsprechende Verallgemeinerung für die euklidische Geometrie $\mathcal{P}$ ist bereits enthalten in Gupta 1965 (vgl. I.1.7.4).

3.71 Definition (Scott). Seien $\mathfrak{A}$ und $\mathfrak{B}$ Strukturen für dieselbe Sprache L. Dann heißt $\mathfrak{A}$ eine *m-gradig elementare Unterstruktur* von $\mathfrak{B}$ oder $\mathfrak{B}$ eine *m-gradig elementare Erweiterung* von $\mathfrak{A}$, Bezeichnung $\mathfrak{A} \underset{m}{\prec} \mathfrak{B}$, falls die folgenden beiden Bedingungen gelten.

(i) $\mathfrak{A}$ ist eine *Unterstruktur* von $\mathfrak{B}$ (Bezeichnung $\mathfrak{A} \subseteq \mathfrak{B}$), d.h. $U_{\mathfrak{A}} \subseteq U_{\mathfrak{B}}$, und die Operationen und Relationen von $\mathfrak{A}$ sind die Einschränkungen der entsprechenden Operationen und Relationen von $\mathfrak{B}$ auf die Trägermenge $U_{\mathfrak{A}}$ ($\mathfrak{B}$ wird dann auch eine *Erweiterung* oder *Oberstruktur* von $\mathfrak{A}$ genannt).

(ii) Für jede Formel α der Sprache L (erster Stufe), in der höchstens m verschiedene Variablen auftreten, und jede Belegung h über $\mathfrak{A}$ gilt: h erfüllt α in $\mathfrak{A}$ gdw h erfüllt α in $\mathfrak{B}$.

Es sei besonders darauf hingewiesen, daß hierbei über die Art des Auftretens der Variablen keinerlei zusätzliche Voraussetzungen gemacht werden; es können also einige von diesen m Variablen frei und andere (nur) gebunden vorkommen. Der eingeführte Begriff ist eine Abschwächung des folgenden Begriffs, den wir später noch benötigen.

3.72 Definition (Tarski u. Vaught 1957). Seien $\mathfrak{A}$, $\mathfrak{B}$ wie vorher. Dann heißt $\mathfrak{A}$ eine *elementare Unterstruktur* von $\mathfrak{B}$ oder $\mathfrak{B}$ eine *elementare Erweiterung* von $\mathfrak{A}$, Bezeichnung $\mathfrak{A} \prec \mathfrak{B}$, falls $\mathfrak{A}$ Unterstruktur von $\mathfrak{B}$ ist und jede Formel α von L die in 3.71(ii) angegebene Eigenschaft hat (d.h., daß 3.71(ii) für jedes $m \in \mathbb{N}$ gilt).

3.73 Satz (Scott). *Vor.:* (i) $\mathfrak{A}$ *ist eine Unterstruktur von* $\mathfrak{B}$.

(ii') *Zu jeder Teilmenge M von* $U_{\mathfrak{A}}$ *mit weniger als m Elementen und jedem* $b \in U_{\mathfrak{B}}$ *existiert ein Automorphismus* φ *von* $\mathfrak{B}$, *der M elementweise fest läßt und b in ein Element von* $\mathfrak{A}$ *überführt.*

Beh.: $\mathfrak{A} \underset{m}{\prec} \mathfrak{B}$.

Das ist eine Verallgemeinerung eines entsprechenden Satzes über elementare Unterstrukturen aus Tarski u. Vaught 1957 und läßt sich durch Induktion über den Formelaufbau beweisen (s. etwa Schwabhäuser 1972, S. 57).

3.74 Satz. *Vor.:* $2 \le n \in \mathbb{N}$. $\mathfrak{B}$ *ist ein Modell der dimensionsfreien absoluten Geometrie* $\mathcal{A}$ *mit der Trägermenge* $B = U_{\mathfrak{B}}$ *("Menge aller Punkte").*

$\mathfrak{A}$ *(bzw. die zugehörige Trägermenge* $A=U_{\mathfrak{A}}$*) ist ein Unterraum von* $\mathfrak{B}$ *mit einer Dimension* $\geq n$. *(Ohne Präzisierung der hier zugelassenen unendlichen Dimensionen* [3.70] *genügt als Voraussetzung über A, daß es ein Unterraum ist* [I.9.53] *und einen n-dimensionalen Unterraum* [I.9.44] *enthält.)*
Beh.: $\mathfrak{A}$ *ist eine* $(n+1)$*-gradig elementare Unterstruktur von* $\mathfrak{B}$.

Beweis: Es genügt der Nachweis der Bedingung 3.73(ii'), wobei $m=n+1$. Sei also M eine beliebige Teilmenge von A mit höchstens n Punkten und $b \in B$ beliebig. Ist $b \in A$, so leistet schon die identische Abbildung von $\mathfrak{B}$ auf sich das für φ Verlangte. Sei nun $b \notin A$. Die Punkte von M spannen (I.9.36, 44) einen höchstens $(n-1)$-dimensionalen Unterraum auf; dementsprechend wählen wir einen $(n-1)$-dimensionalen Unterraum U mit $M \subseteq U \subseteq A$. Sei (gemäß I.11.62, 63) q der Fußpunkt des Lots von b auf U und L eine in q auf U errichtete Senkrechte innerhalb A. Sei b' ein Punkt auf L mit $qb' \equiv qb$. Sei U^* der von U, b', b aufgespannte $(n+1)$-dimensionale Unterraum und V der (n-dimensionale) Mittellotraum zu b' und b in U^* (I.11.65, 66), dann ist $U \subseteq V$. Sei schließlich φ die Spiegelung an V, d.h. die Abbildung S_V mit $y=S_V(x) \leftrightarrow M(xy) \in V \wedge (xy \hat{\perp} V \vee x=y)$. (Aus I.11.62, I.10.10 ergibt sich, daß diese Spiegelung eindeutig bestimmt ist und eine involutorische Bewegung von $\mathfrak{B}$ mit der Fixpunktmenge V ist.) Dann ist insbesondere $\varphi(b)=b' \in A$, und wegen $M \subseteq U \subseteq V$ läßt φ die Punkte von M fest. Damit leistet φ das Verlangte. □

3.75 Satz. *Vor.:* T *ist eine der in Abschnitt* 2 *betrachteten dimensionsfreien Obertheorien erster Stufe von* A. γ *ist eine Formel der zugehörigen Sprache* L(D,B), *in der höchstens* $n+1$ *verschiedene (Punkt-)Variablen auftreten.* $2 \leq n < m$, $n,m \in \mathbb{N}$.

Beh.: (i) *Folgende Bedingungen sind untereinander gleichwertig:*

(1) γ *ist gültig in allen Modellen von* T, *deren Dimension* $\geq n$ *ist (vgl.* 3.74*).*

(2) γ *ist gültig in allen Modellen von* T *der Dimension* n.

(ii) (Beweis nur für P⊆T und für H⊆T) *Diese Bedingungen sind auch gleichwertig mit*

(3) γ *ist gültig in allen Modellen von* T *der Dimension* m.

Beweis: Sei o.B.d.A. γ eine Aussage (sonst kann man zu einer Generalisierten γ^* übergehen [1.24, 25], die dieselben Variablen enthält).

Zu (i): Gelte (2), und sei $\mathfrak{B}$ ein beliebiges Modell von T mit einer Dimension $\geq n$. Sei $\mathfrak{A}$ ein n-dimensionaler Unterraum von $\mathfrak{B}$. Nach I.11.74 ist $\mathfrak{A}$ ein Modell von A, in dem die eventuell für T verwendeten Zusatzaxiome gelten (das gilt natürlich auch für das hyperbolische Parallelenaxiom HP, da sich dieses als Forderung an beliebige zweidimensionale Unterräume auffassen läßt). Wegen (2) ist γ in $\mathfrak{A}$ gültig. Nach 3.74 ist $\mathfrak{A} \prec_{n+1} \mathfrak{B}$, also γ auch in $\mathfrak{B}$ gültig. Damit ist (2) $\rightarrow$ (1) gezeigt; die Umkehrung ist trivial.

Zu (ii): Gelte (3). Der Übergang von m zu größeren Dimensionen ergibt sich wie in (i). Sei nun $\mathfrak{A}$ ein beliebiges Modell von T mit einer Dimension k, wobei $n \leq k < m$. Dann genügt der Nachweis von $\models_{\mathfrak{A}} \gamma$. Dazu wiederum genügt es zu zeigen:

(*) $\mathfrak{A}$ ist Unterraum eines geeigneten m-dimensionalen Modells $\mathfrak{B}$ von T.

Nach 3.74 ist dann nämlich wieder $\mathfrak{A} \prec_{n+1} \mathfrak{B}$, und aus $\models_{\mathfrak{B}} \gamma$ ergibt sich das Gewünschte. (*) ergibt sich nun aus den Darstellungssätzen für P, H und die betrachteten Obertheorien. Danach ist ja $\mathfrak{A}$ isomorph zu einem kartesischen bzw. Kleinschen Raum $\mathfrak{C}_k(\mathfrak{f})$ bzw. $\mathfrak{K}_k(\mathfrak{f})$, und der entsprechende m-dimensionale Raum $\mathfrak{C}_m(\mathfrak{f})$ bzw. $\mathfrak{K}_m(\mathfrak{f})$ über demselben angeordneten Körper $\mathfrak{f}$ leistet bis auf Isomorphie das Verlangte. □

<u>3.76 Anmerkungen</u>. (i) Allgemein ergibt sich (*) und damit 3.75(ii) in analoger Weise mit Hilfe des (hier nicht formulierten) Darstellungssatzes von Pejas für die absolute Geometrie und seiner Verallgemeinerung auf beliebige endliche Dimensionen ≥ 2 (Pejas 1961, Bachmann 1964, vgl. 2.14(iv)): Ist $\mathfrak{A}$ ein k-dimensionales Modell von A, so leistet ein m-dimensionales Modell mit demselben angeordneten Grundkörper, derselben metrischen Konstanten und derselben Abszissenmenge das für $\mathfrak{B}$ Verlangte (bis auf Isomorphie). Der Beweis von 3.75(i) und insbesondere I.11.74 macht dagegen keinen Gebrauch von den Darstellungssätzen. Für die folgenden Beweise wird 3.75(ii) nicht benötigt.

(ii) Tatsächlich läßt sich die Dimensionsschranke n in der Beh. von 3.75 nicht unterbieten. Mit bekannten Mitteln erhält man nämlich - jedenfalls für Modelle von A' - , daß die $n+1$ Variablen enthaltende Aussage (Scott 1959)

(1) $\exists a_0 \ldots \exists a_n [a_0 \neq a_1 \wedge \bigwedge_{0 \leq \mu < \nu \leq n} a_\mu a_\nu \equiv a_0 a_1]$

(es gibt $n+1$ paarweise verschiedene Punkte, deren Verbindungsstrecken alle gleich lang sind, d.h., daß die Punkte einen "gleichseitigen n-dimensionalen Simplex" bilden) gültig ist genau dann, wenn das Modell eine Dimension $\geq n$ hat. Diese Aussage läßt sich damit ebenfalls als unteres Dimensionsaxiom (3.68) verwenden.

3.77 Satz. *Sei* T *wie in* 3.75. *Dann ist*

$$T = \bigcap_{2\leq n\in \mathbb{N}} T_n ,$$

d.h., die dimensionsfreie Geometrie T *besteht aus denjenigen Sätzen, die für jede endliche Dimension* $n \geq 2$ *Sätze der zugehörigen n-dimensionalen Geometrie* T_n *sind.*

Beweis: Es genügt zu zeigen, daß eine beliebige Formel γ von L(D,B) in allen Modelle von T gültig ist genau dann, wenn sie in den Modellen mit endlicher Dimension ≥ 2 gültig ist. Die Richtung $(\rightarrow)$ ist trivial. Zur Umkehrung sei n so gewählt, daß $n \geq 2$ ist und γ höchstens $n+1$ verschiedene Variablen enthält. Dann ergibt sich das Gewünschte aus 3.75(i). □

3.78 Satz. *Die folgenden dimensionsfreien bzw. unendlichdimensionalen Geometrien sind entscheidbar:*

(1) (Scott 1959) P^*, P^*_∞ *(euklidisch);*
(2) H^*, H^*_∞ *(hyperbolisch);*
(3) A^*, A^*_∞ *(absolut); außerdem auch die* A^*_n $(2\leq n\in \mathbb{N})$.

Beweis: Zu (1): Der Beweis von 3.31 liefert nicht nur für jede der Geometrien P^*_n $(2\leq n\in \mathbb{N})$ ein Entscheidungsverfahren, sondern auch eine für alle betrachteten n einheitliche Beschreibung dieser Verfahren und damit ein Verfahren, mit dem man für jedes solche n und jede Aussage γ feststellen kann, ob γ ein Satz von P^*_n ist. Das führt zu folgendem Verfahren für (1).

Für eine beliebig gegebene Aussage γ von L(D,B) bestimme man zunächst die Anzahl t der verschiedenen Variablen in γ und wähle $n=\max(2,t-1)$. Dann gilt:

(1a) $\gamma \in P^*$ gdw $\gamma \in P^*_2, \ldots, \gamma \in P^*_{n-1}$ und $\gamma \in P^*_n$.

(1b) $\gamma \in P^*_\infty$ gdw $\gamma \in P^*_n$.

Aus 3.75(i) ergibt sich nämlich sofort (1a) und außerdem die Richtung $(\leftarrow)$ von (1b). Dieselbe Richtung ergibt sich auch für $\neg\gamma$ an Stelle von γ. Die Umkehrung $(\rightarrow)$ ergibt sich dann aus der Vollständigkeit von P^*_n (3.32) und der Widerspruchsfreiheit von P_∞^* (3.69(i)). Man prüfe nun nach, ob die rechte Seite von (1a) bzw. (1b) gilt, was nach dem vorher genannten Verfahren möglich ist. Damit erhält man die gewünschte Antwort.

Zu (2): Das ergibt sich natürlich in völlig analoger Weise aus der Entscheidbarkeit der hyperbolischen Geometrien H_n^* (3.65) und der auf Grund der Konstruktion möglichen einheitlichen Beschreibung der zugehörigen Verfahren.

Die Entscheidbarkeit der absoluten Geometrien (3) ergibt sich sofort aus ihrer Charakterisierung als Durchschnitte von entscheidbaren Theorien in dem folgenden Lemma ((ii) und (iii)). □

3.79 Lemma. (i) $\mathrm{Mod}(A^*) = \mathrm{Mod}(P^*) \cup \mathrm{Mod}(H^*)$,
d.h., eine Struktur $\mathfrak{A}$ *(für* L(D,B)*) ist Modell von* A^* *genau dann, wenn sie Modell von* P^* *oder von* H^* *ist.*

(ii) $A^* = P^* \cap H^*$.

(iii) *Die Behauptungen* (i) *und* (ii) *gelten analog für* A_n^*, P_n^*, H_n^* $(2 \le n \in \mathbb{N})$ *und für* A_∞^*, P_∞^*, H_∞^*.

(iv) (Szmielew 1959a, vgl. 2.3(vi)) *Die einzigen echten widerspruchsfreien Erweiterungen von* A_n^* *mit derselben Sprache sind* P_n^* *und* H_n^* $(2 \le n \in \mathbb{N})$.

3.80 Anmerkung. Für die Feststellung in 3.79(iv) sagt man bisweilen, daß A_n^* "gabelbar" ist. Ausführlicher nennt man A_n^* "gabelbar am Parallelenaxiom (in die genannten Erweiterungen)" (die eine dieser Erweiterungen entsteht aus A_n^* durch Hinzunahme des Parallelenaxioms, die andere durch Hinzunahme seiner Negation).

Beweis von 3.79: Zu (i): Sei $\mathfrak{A}$ ein beliebiges Modell von A^*. Dann ist in $\mathfrak{A}$ das euklidische Parallelenaxiom (als Aussage formuliert) oder seine Verneinung gültig; die Verneinung ist aber äquivalent zum hyperbolischen Parallelenaxiom (2.3(vi)). Im ersten Fall ist also $\mathfrak{A}$ Modell von P^*, im zweiten Modell von H^*. Die Umkehrung ($\supseteq$ bzw. $\leftarrow$) ist trivial.

Zu (ii): Die Richtung $\subseteq$ ist klar (da A^* Untertheorie der rechts genannten Theorien). Zur Umkehrung sei $\gamma \in P^* \cap H^*$. Es genügt zu zeigen, daß $\gamma \in \mathrm{Cn}(A^*)$ ist, d.h., daß γ in jedem Modell von A^* gültig ist. Das ergibt sich aber aus (i).

Zu (iii): Dasselbe wie in (i) und (ii) ergibt sich, wenn zu allen drei Geometrien das Zusatzaxiom Dim_n bzw. das zusätzliche Axiomensystem Δ_∞ (3.68(2)) hinzugenommen wird.

Zu (iv): Sei T eine beliebige widerspruchsfreie Erweiterung von A_n^* mit derselben Sprache. Auf Grund von (i), (iii) sind dann die folgenden drei Fälle möglich.

Fall 1: Alle Modelle von T liegen in $Mod(P_n^*)$. Die zu einem solchen Modell elementar äquivalenten Strukturen sind dann ebenfalls Modelle von T. Das sind aber wegen der Vollständigkeit von P_n^* schon alle Modelle von P_n^*. Damit haben T und P_n^* dieselben Modelle, also ist $T=P_n^*$.

Fall 2: Alle Modelle von T liegen in $Mod(H_n^*)$. Analog zu Fall 1 ergibt sich dann $T=H_n^*$.

Fall 3: T besitzt sowohl Modelle aus $Mod(P_n^*)$ als auch solche aus $Mod(H_n^*)$. Zu den dazu elementar äquivalenten Strukturen, die ebenfalls Modelle von T sein müssen, gehören dann wie vorher sämtliche Modelle von P_n^* und sämtliche von H_n^*, wegen (i), (iii) also sämtliche Modelle von A_n^*. Damit ist $T=A_n^*$. □

3.81 Korollar. *Von den entscheidbaren Geometrien aus* 3.78 *sind* P_∞^*, H_∞^* *vollständig und alle anderen unvollständig.*

Beweis: Für jede Aussage γ von $L(D,B)$ ergibt sich aus der Charakterisierung 3.78(1b) und der Vollständigkeit des entsprechenden P_n^*, daß $\gamma \in P_\infty^*$ oder $\neg\gamma \in P_\infty^*$ ist. Somit ist P_∞^* vollständig. Der Beweis für H_∞^* verläuft analog. Die Unvollständigkeit der übrigen Geometrien ergibt sich unmittelbar daraus, daß jeweils eine echte widerspruchsfreie Erweiterung mit derselben Sprache vorhanden ist (für die absoluten Geometrien (3) etwa die entsprechende euklidische Geometrie, für P^*, H^* eine entsprechende n-dimensionale Geometrie). □

3.82 Satz. (i) (Scott) *Die* P_n^* $(2 \leq n \in \mathbb{N})$ *und* P_∞^* *sind die einzigen widerspruchsfreien Vervollständigungen von* P^* *(d.h. Erweiterungen mit derselben Sprache, die widerspruchsfrei und vollständig sind).*

(ii) *Dasselbe gilt für die* H_n^*, H_∞^* *und* H^*.

Beweis: Sei T eine beliebige Vervollständigung von P^*. Dann gilt einer der folgenden beiden Fälle.

Fall 1: Es ist $\mathrm{Dim}_n \in T$ für ein n mit $2 \leq n \in \mathbb{N}$. (Das gilt dann für nur ein solches n, da die Dimensionsaxiome für verschiedene Dimensionen einander widersprechen.) Für dieses n ist dann $P_n^* \subseteq T$, wegen der Vollständigkeit von P_n^* also $P_n^* = T$.

Fall 2: Für kein solches n ist $\mathrm{Dim}_n \in T$. Wegen der Vollständigkeit von T ist dann $\neg\mathrm{Dim}_n \in T$ für jedes solche n. Offenbar ist $\bigwedge_{i=2}^{n} \neg\mathrm{Dim}_n$ (schon in A^*) äquivalent zu Dim_{n+1}^-, also ist $\Delta_\infty \subseteq T$ (3.68) und damit $P_\infty^* \subseteq T$. Wegen der Vollständigkeit von P_∞^* ist wieder $P_\infty^* = T$.

Der Beweis von (ii) verläuft analog. □

Die bisherigen Ergebnisse (ab 3.67) für euklidische Geometrien sind schon in Scott 1959 enthalten. Die verwendete Methode (Satz 3.73) reicht jedoch weiter und ist nicht nur auf Erweiterungen von A, sondern auch auf viele andere Geometrien anwendbar. Darüber sei als Zusammenfassung nur der folgende Satz formuliert.

3.83 Satz. *Satz 3.74 und die daraus erzielten Folgerungen gelten analog für die dimensionsfreien Geometrien* $\mathfrak{Af}$, $\mathfrak{Afo}$, $\mathfrak{Pr}$, $\mathfrak{Pro}$, $\mathfrak{E}$ *mit den in Abschnitt 2 betrachteten Erweiterungen (2.15, 16). Ist T irgendeine der genannten Geometrien, so sind insbesondere* T^* *(dimensionsfrei, über reell-abgeschlossenen Körpern) und die zugehörige unendlichdimensionale Geometrie* T_∞^* *(3.68) entscheidbar.*

Zum Beweis: Für die Übertragung von Satz 3.74 auf die jeweils betrachtete Geometrie ist natürlich wieder die Existenz von Automorphismen der in 3.73(ii') verlangten Art zu zeigen. Wir beschränken uns auf eine Beweisskizze für die affinen Geometrien. Seien $\mathfrak{B}$, A, M, $b \notin A$, U wie im Beweis von 3.74. Sei ein "Nullpunkt" in U gewählt, so daß sich U als $(n-1)$-dimensionaler Unterraum (Unterraum des zu $\mathfrak{B}$ gehörigen Vektorraums) auffassen läßt. Sei U_1 ein n-dimensionaler Vektorraum mit $U \subseteq U_1 \subseteq A$, und sei b' ein Vektor, der eine Basis von U zu einer Basis von U_1 ergänzt. Durch Hinzunahme von b entsteht eine Basis eines $(n+1)$-dimensionalen Vektorraums U^*. Diese werde schließlich zu einer Basis C des ganzen Raumes ergänzt. Da sich lineare Abbildungen durch die Bilder der Basisvektoren festlegen lassen, kann man insbesondere eine lineare Abbildung φ wählen mit $\varphi(b) = b'$, $\varphi(b') = b$, $\varphi(a) = a$ für jeden anderen Basisvektor $a \in C$. Diese ist dann auch eine affine Abbildung und ein Automorphismus der gewünschten Art. □

<u>3.84 Anmerkungen</u>. (i) Das Analogon zu 3.77 ist natürlich für die in 3.83 betrachteten Geometrien trivial, da es als Definition für diese Geometrien verwendet wurde (2.15). Es gilt aber auch für entsprechende axiomatisch eingeführte Geometrien. Umgekehrt gibt das bewiesene Resultat die Möglichkeit, aus einer geeigneten Axiomatisierung aller T_n ($2 \leq n \in \mathbb{N}$) ein Axiomensystem für die dimensionsfreie Geometrie T gemäß 2.15 zu erhalten, indem man die Dimensionsaxiome wegläßt.

(ii) Im ursprünglichen Beweis von 3.74 für A wurde bei der Konstruktion von b' das Axiom der Streckenabtragung benutzt. Tatsächlich ist dieses Axiom nicht entbehrlich für den Nachweis der Existenz von Automorphismen der angegebenen Art. In Gupta 1965 (S. 408), 1965a wurde nämlich festgestellt, daß Unterräume gleicher Dimension eines kartesischen Raumes über einem (nicht-pythagoreischen) angeordneten Körper nicht immer zueinander isometrisch und nicht immer kartesisch sind und daß man durch Weglassen der Dimensionsaxiome aus den betrachteten endlichen Axiomensystemen für die $\Gamma_n(\mathrm{OF})$ (vgl. 3.59) jedenfalls kein Axiomensystem für $\Gamma(\mathrm{OF})$ erhält. Auch 3.74 selbst gilt nicht für die Geometrie $\Gamma(\mathrm{OF})$; ein einfaches Gegenbeispiel (mit n=2) liefert die 3 Variablen enthaltende Aussage 3.76(1) ("es gibt ein gleichseitiges Dreieck") (als α gemäß 3.71(ii)), die wahr ist z.B. in $\mathfrak{B} = \mathfrak{L}_3(\mathbb{Q})$, aber nicht in einem Unterraum $\mathfrak{A} \cong \mathfrak{L}_2(\mathbb{Q})$.

(iii) In Schwabhäuser 1970 wird für beliebiges $k \in \mathbb{N} - \{0\}$ die dimensionsfreie Geometrie $\Gamma(\mathrm{PF}_k)$ über der Klasse PF_k der k-pythagoreischen angeordneten Körper untersucht. Dabei heißt ein Körper *k-pythagoreisch*, falls jede endliche Quadratsumme von Körperelementen sich schon als Summe von k Quadraten darstellen läßt. Insbesondere ist dann $\mathrm{PF}_1 = \mathrm{PF}$. Für $k \geq 2$ ergibt sich wenigstens noch eine Abschwächung von 3.74 mit der Dimensionsschranke $k \cdot n$ an Stelle von n (für dieselbe Behauptung $\mathfrak{A} \prec_{n+1} \mathfrak{B}$). Zu diesem Zweck werden auch unendlichdimensionale Modelle etwas genauer untersucht (vgl. 3.70); als Hilfsmittel dazu wird benutzt, daß das innere Produkt von Vektoren sich durch eine geeignete geometrische Konstruktion beschreiben läßt (ebenso wie die Operationen des Grundkörpers). Dann ergeben sich folgende Resultate. *Jedes* $\Gamma(\mathrm{PF}_k)$ *ist (rekursiv) axiomatisierbar, aber für* $k \geq 2$ *nicht mehr endlich axiomatisierbar.* $\Gamma(\mathrm{OF})$ *ist jedenfalls nicht endlich axiomatisierbar.* Korrektur: Zu den in Schwabhäuser 1970 angegebenen Axiomen E1 bis E12 für endlichdimensionale Modelle über beliebigen angeordneten Körpern ist übrigens Satz I.11.47 (oder etwas Gleichwertiges) als Axiom E13 hinzuzufügen. Das stellte sich heraus im Zusammenhang mit der Durchführung eines axiomatischen Aufbaus der Geometrie auf Grund dieser Axiome in der Staatsexamensarbeit Hecht 1973.

Bisher ungelöst ist dagegen folgendes Problem.

3.85 Problem. Ist die dimensionsfreie euklidische Geometrie Γ(OF) über beliebigen angeordneten Körpern rekursiv axiomatisierbar?

Zum Abschluß übertragen wir noch die Unentscheidbarkeitsresultate, die früher für die Geometrien fester Dimension behandelt wurden, auf andere Geometrien.

3.86 Satz. (i) *Die dimensionsfreien Geometrien* A, A', P, P', H, E, E', Aƒ, Aƒo, Pr, Pro, Γ(OF) *(2.16, 2.15, 3.52) sowie die* A_n, A'_n $(2 \leq n \in \mathbb{N})$ *sind unentscheidbar.*

(ii) *Die zugehörigen unendlichdimensionalen Geometrien (3.68) sind unentscheidbar.*

Beweis: Zu (i) genügt nach 3.56 die Feststellung, daß jede der genannten Geometrien eine unentscheidbare endliche Erweiterung mit derselben Sprache besitzt. Tatsächlich entsteht eine solche Erweiterung, deren Unentscheidbarkeit schon früher festgestellt wurde (3.58, 60, 65), indem ein Dimensionsaxiom Dim_n und bei den absoluten Geometrien außerdem etwa das Parallelenaxiom hinzugenommen wird. (Das gilt sogar für Γ(OF); eine Axiomatisierung dafür wird in diesem Zusammenhang nicht gebraucht!)

Zu (ii) (Ausführung für euklidische Geometrien): Sei T eine der in (i) betrachteten euklidischen Geometrien (Γ(OF), P oder P') und K die zugehörige Klasse von angeordneten Körpern (OF, PF oder EF). Für ein beliebiges Modell $\mathfrak{A}$ von T gilt nun:

(*) Die geometrische Konstruktion aus I.14 liefert einen angeordneten Körper $\mathfrak{f}_{oe}$ aus K.

Für $T \supseteq P$ ergibt sich das direkt aus I.14. Für T=Γ(OF) ergibt es sich daraus, daß (*) jedenfalls für die endlichdimensionalen Modelle $\mathfrak{L}_n(\mathfrak{f})$ gilt und sich als Aussage der Sprache L(D,B) formulieren läßt; nach Definition 3.52 ist diese Aussage also ein Satz von Γ(OF) und gilt somit in allen Modellen. Nun waren die Konstruktion der geometrischen Reduzierten und der Beweis von 3.51 unabhängig von der Dimension des betrachteten Raumes, sie beruhten nur auf der Konstruktion von $\mathfrak{f}_{oe}$. Ebenso wie 3.51(iii) ergibt sich also die allgemeinere Feststellung

(**) Ist $\mathfrak{A}$ ein beliebiges Modell von T und das darin gemäß (*) konstruierte $\mathfrak{F}_{oe}$ isomorph zu $\mathfrak{F}$, so gilt für jede Aussage α von L_{OF}:

$\models_{\mathfrak{F}} \alpha$ gdw $\models_{\mathfrak{A}} Rdg(\alpha)$.

Wie in 3.53(→) ergibt sich dann die Unentscheidbarkeit von T_∞ aus der von $Th(K)$ auf Grund der Gleichwertigkeit der folgenden Bedingungen für beliebige Aussagen α von L_{OF}:

(1) $\alpha \in Th(K)$,

(2) $\models_{\mathfrak{F}} \alpha$ für jedes $\mathfrak{F} \in K$,

(3) $\models_{\mathfrak{A}} Rdg(\alpha)$ für jedes Modell $\mathfrak{A}$ von T_∞,

(4) $Rdg(\alpha) \in T_\infty$.

Hiervon sind (1) ↔ (2) und (3) ↔ (4) trivial. (2) → (3) ergibt sich sofort aus (*) und (**). Die Umkehrung (3) → (2) ergibt sich analog mit Hilfe der folgenden Feststellung:

(***) zu jedem $\mathfrak{F} \in K$ gibt es ein Modell $\mathfrak{A}$ von T_∞, in dem das zugehörige $\mathfrak{F}_{oe}$ zu $\mathfrak{F}$ elementar äquivalent ist.

Für jedes solche $\mathfrak{F}$ ergibt sich dann nämlich aus (3) und (**), daß $\models_{\mathfrak{F}_{oe}} \alpha$ und damit auch $\models_{\mathfrak{F}} \alpha$ ist. (***) läßt sich nun wieder ohne genauere Untersuchung der unendlichdimensionalen Modelle beweisen. Dazu bilden wir zur Theorie $Th(\mathfrak{F})$ die Menge der zugehörigen Reduzierten $\Sigma_{\mathfrak{F}} = \{Rdg(\alpha) \mid \alpha \in Th(\mathfrak{F}),\ \alpha \text{ Aussage}\}$ und betrachten die Menge

$$\Theta := T \cup \Sigma_{\mathfrak{F}} \cup \Delta_\infty .$$

Jede endliche Teilmenge von Θ enthält nur endlich viele der unteren Dimensionsaxiome aus Δ_∞ (3.68) und besitzt damit einen geeigneten kartesischen Raum $\mathfrak{C}_n(\mathfrak{F})$ als Modell. Nach dem Endlichkeitssatz besitzt auch Θ ein gewisses Modell $\mathfrak{A}$. Dieses ist zunächst Modell von T_∞. Sei $\mathfrak{F}_{oe}$ darin gemäß (*) eingeführt. Wegen der Gültigkeit der Elemente von $\Sigma_{\mathfrak{F}}$ in $\mathfrak{A}$ sind nach (**) die Aussagen $\alpha \in Th(\mathfrak{F})$ in $\mathfrak{F}_{oe}$ gültig, d.h., daß $\mathfrak{F}_{oe}$ ein Modell von $Th(\mathfrak{F})$ ist. Wegen der Vollständigkeit von $Th(\mathfrak{F})$ (Beispiel zu 1.39) ist $\mathfrak{F}_{oe} \equiv \mathfrak{F}$, wie zu zeigen.

Für die übrigen betrachteten Geometrien läßt sich der Beweis analog führen mit Hilfe der für diese Geometrien zur Verfügung stehenden Konstruktionen von (ggf. angeordneten) Körpern und der entsprechenden Konstruktionen von geometrischen Reduzierten (vgl. 3.64). Statt der absoluten Geometrien A_∞, A'_∞ genügt wieder die Betrachtung der endlichen Erweiterungen P_∞ bzw. P'_∞. □

<u>3.87 Wesentliche Unentscheidbarkeit</u>. Nach Tarski, Mostowski u. Robinson 1953 wird eine Theorie *wesentlich unentscheidbar ("essentially undecidable")* genannt, falls jede widerspruchsfreie Obertheorie mit derselben Sprache ebenfalls unentscheidbar ist. Dieser Begriff ist in anderen Zusammenhängen von großer Bedeutung. Dagegen gilt:

<u>3.88 Satz</u>. *Von den bisher betrachteten unentscheidbaren Geometrien erster Stufe ist keine wesentlich unentscheidbar.*

Zu jeder dieser Geometrien wurde ja eine geeignete entscheidbare Erweiterung T^* (über reell-abgeschlossenen Körpern) angegeben.

4. Definierbarkeitsfragen

4.1 Überblick. Für den axiomatischen Aufbau der Geometrie in Teil I haben wir die Grundbegriffe D und B (Streckenkongruenz und Zwischenbeziehung) verwendet. Vielfach werden andere Grundbegriffe (z.B. Rechtwinkligkeit und Zwischenbeziehung oder die Grundbegriffe des Hilbertschen Axiomensystems) für den Aufbau "derselben" Geometrie verwendet. Ist eine Geometrie mit Hilfe einer Menge K von Grundbegriffen aufgebaut (z.B. $K = \{D,B\}$), so werden wir eine zweite Menge K' von geometrischen Begriffen ein "geeignetes System von Grundbegriffen" für diese Geometrie nennen, falls sich die Begriffe von K' mit Hilfe der Begriffe von K und auch die von K mit Hilfe derer von K' ausdrücken (oder "definieren", Präzisierung s. Def. 4.2) lassen. Allgemeiner erhebt sich die Frage, ob ein gewisser Begriff sich mit Hilfe von gewissen anderen Begriffen ausdrücken läßt. (Die Antwort hängt nicht nur von den gegebenen Begriffen, sondern i.a. auch von der betrachteten Theorie [z.B. von der Stärke ihrer Axiome] ab.) Fragen dieser Art sollen in diesem Abschnitt behandelt werden.

Nach allgemeinen Betrachtungen über die Definierbarkeit (4.2ff.) werden Resultate für die in Abschnitt 2 betrachteten Geometrien behandelt, und zwar zunächst für euklidische und absolute Geometrien (ab 4.10), dann für hyperbolische (ab 4.57), elliptische (ab 4.73) und affine (ab 4.81) Geometrien. Danach wird noch eine Geometrie mit Geraden statt Punkten als geometrischen Objekten behandelt (ab 4.89) (die Hinzunahme und Beseitigung von Geraden als Objekten wird schon in 4.59f.) betrachtet). Zum Schluß wird hingewiesen auf Geometrien mit Operationen als Grundbegriffen (4.103f.), auf die Einbeziehung der bisher ausgeschlossenen Dimension 1 (4.105ff.) und auf die Verwendung anderer geometrischer Objekte, insbesondere Bälle, an Stelle von Punkten und Geraden (4.110).

4.2 Definition (nach Tarski, Mostowski u. Robinson 1953). Sei K eine Menge von Operations- und Relationszeichen, sei R ein weiteres Relationszeichen der Stellenzahl r, und sei T eine in der ersten Stufe formalisierte Theorie, in deren Sprache mindestens R und die Zeichen aus K vorkommen. Dann heißt R *definierbar mittels* (oder *mit Hilfe von*) *K in* T, falls folgendes gilt:

(i) Es gibt eine Formel $\delta \in \mathrm{Fml}_K(x_1,\ldots,x_r)$ (d.h. eine Formel, deren Operations- und Relationszeichen sämtlich in K liegen und in der als freie Variablen höchstens $x_1,\ldots, x_r$ vorkommen, s. 1.8), so daß die Aussage

(1) $\forall x_1\ldots\forall x_r[Rx_1\ldots x_r \leftrightarrow \delta]$

ein Satz von T ist.

Ist dabei $K=\{\zeta_1,\ldots,\zeta_n\}$ (in den folgenden Anwendungen haben wir das meist mit Relationszeichen ζ_ν), so nennen wir R auch *definierbar mittels $\zeta_1,\ldots, \zeta_n$ in* T.

Unter analogen Voraussetzungen für ein r-stelliges Operationszeichen f an Stelle von R heißt f *definierbar mittels* (der Zeichen von) *K in* T, falls

(ii) es gibt ein $\delta \in \mathrm{Fml}_K(x_1,\ldots,x_r,y)$, so daß die Aussage

(2) $\forall x_1\ldots\forall x_r\forall y[fx_1\ldots x_r \doteq y \leftrightarrow \delta]$

ein Satz von T ist.

Die Aussage (1) bzw. (2) oder auch nur der Bestandteil in eckigen Klammern heißt dabei eine *formale Definition* für R bzw. für f mittels (der Zeichen von) K; $Rx_1\ldots x_r$ bzw. $fx_1\ldots x_r \doteq y$ heißt das *definiendum* (das "zu definierende") und δ das *definiens* (das "Definierende") in dieser Definition. Selbstverständlich sollen $x_1,\ldots, x_r$ und evtl. y hierbei paarweise verschiedene Variablen sein; diese können im übrigen (etwa auf Grund der Sätze 4.4 und 4.5, s.u.) beliebig vorgegeben werden (z.B. die ersten in einer gegebenen Numerierung der Variablen).

Der Leser, der sich sofort mit geometrischen Anwendungen beschäftigen möchte, kann zu 4.10 übergehen und die hier folgenden allgemeinen Betrachtungen bei Bedarf später nachschlagen.

4.3 Anmerkungen. (i) In 4.2 geht es nicht um die Definierbarkeit einer einzelnen Relation oder Operation. Der zu definierende "Begriff" läßt sich mit unseren Mitteln gerade charakterisieren durch das Zeichen R bzw. f, das in verschiedenen Modellen von T verschiedene (Relationen bzw. Operationen als) Bedeutungen haben kann. Wenn man mehr Bezug auf die Bedeutung nehmen will, kann man den "Begriff" auch auffassen als die (durch das Zeichen und die Theorie schon festgelegte) Abbildung mit $\mathfrak{A} \mapsto R_{\mathfrak{A}}$ bzw. $\mathfrak{A} \mapsto f_{\mathfrak{A}}$, die auf der Klasse aller Modelle von T definiert ist.

(ii) Manchmal wird daneben die *Definierbarkeit einer einzelnen Relation oder Operation* in folgendem Sinne betrachtet. Sei eine Struktur $\mathfrak{A}$ für eine Sprache L vorgegeben. Eine r-stellige Relation R auf der Trägermenge $U_{\mathfrak{A}}$ heißt dann *definierbar* mit Hilfe (der Zeichen von) K bzw. mit Hilfe der zugehörigen Operationen und Relationen von $\mathfrak{A}$ - im Falle $K=F_L \cup R_L$ auch einfach definierbar in L - , falls es eine Formel $\delta(x_1,\ldots,x_r) \in Fml_K(x_1,\ldots,x_r)$ gibt, so daß für jedes $x_1,\ldots,x_r \in U$ gilt: $Rx_1\ldots x_r$ gdw $\models_{\mathfrak{A}} \delta[x_1,\ldots,x_r]$; R heißt dann auch die durch $\delta(x_1,\ldots,x_r)$ *definierte* Relation. Beispiele sind die in 3.22(v) betrachteten, durch Formeln von L_{OF} definierten Mengen von reellen Zahlen - aufgefaßt als einstellige Relationen.

Diese Definierbarkeit läßt sich als Spezialfall von 4.2 behandeln. Ist nämlich $\hat{L}$ die aus L durch Hinzunahme eines entsprechenden r-stelligen Relationszeichens R entstehende Erweiterungssprache und $\hat{\mathfrak{A}}$ die durch Hinzunahme von $R_{\mathfrak{A}}=R$ entstehende Expansion von $\mathfrak{A}$ auf $\hat{L}$ (1.20), so liegt die genannte Definierbarkeit vor genau dann, wenn R definierbar ist mittels K in der Theorie $Th(\hat{\mathfrak{A}})$ im Sinne von 4.2.

Das hier für eine Relation Gesagte gilt sinngemäß für eine Operation.

(iii) Die in 4.2 genannte Aussage (1) bzw. (2) hat offenbar die Bauart einer Definition (daher die Bezeichnung "formale Definition", in Tarski, Mostowski u. Robinson 1953 wird dafür die Bezeichnung "possible definition" verwendet). Es spielt jedoch in 4.2 keine Rolle, ob diese Aussage beim Aufbau der formalisierten Theorie T tatsächlich als Definition verwendet worden ist. Wichtig ist nur, daß man sie jedenfalls als einen Satz der Theorie erhält (sei es durch einen Beweis oder trivialerweise auf Grund der Definition). So ist z.B. in Teil I jede der Formeln 6.1(1), 6.3 und 6.4 eine formale Definition des dreistelligen Relationszeichens $\simeq$ mittels B in A. Somit kommt es vielleicht der

Anschauung näher, wenn man statt "definierbar" eins der Worte "charakterisierbar", "beschreibbar", "ausdrückbar" verwendet. Die Bezeichnung "definierbar" ist jedoch inzwischen allgemein gebräuchlich. Wenn (1) tatsächlich als Definition verwendet worden ist, könnte man zur Unterscheidung sagen, daß R mittels (der Zeichen von) *K* in *T* (nämlich durch (1)) "definiert" ist (und damit natürlich auch "definierbar").

(iv) Zu den beim Aufbau einer Theorie (z.B. in Teil I) tatsächlich verwendeten (expliziten) Definitionen ist (spätestens) in diesem Zusammenhang noch etwas zu sagen. Die Einführung eines neuen Begriffs durch eine solche Definition läßt sich stets auffassen als die Einführung einer Abkürzung für ein komplizierteres sprachliches Gebilde. Für formalisierte Theorien gibt es jedoch zwei Möglichkeiten zur Präzisierung dieser Auffassung, die grundsätzlich zu unterscheiden sind.

Die erste Möglichkeit besteht darin, daß die Abkürzung lediglich in der Metasprache eingeführt wird (z.B. ist dann $a \underset{p}{=} b$ eine Abkürzung für die Formel auf der rechten Seite von 6.1(1)). Die Objektsprache (im Beispiel L(D,B)) wird dabei nicht verändert.

Bei der zweiten Möglichkeit dagegen wird die Objektsprache um ein Zeichen für den neuen Begriff erweitert, d.h., die Abkürzung in die Objektsprache aufgenommen. Die verwendete Definition ist dann nichts anderes als ein Zusatzaxiom ϑ; durch seine Hinzunahme entsteht aus der bisherigen Theorie *T'* (ohne das neue Zeichen) eine neue Theorie *T*, die als eine *definitorische Erweiterung* von *T'* (auf Grund von ϑ) bezeichnet wird. Definitionen der Form 4.2(2) werden dabei nur verwendet, wenn die Aussage $\forall x_1 \ldots \forall x_r \overset{=1}{\exists} y\delta$ ein Satz der bisherigen Theorie *T'* ist (nur dann handelt es sich um eine "definitorische" Einführung eines neuen Funktionszeichens); statt 4.2(2) verwendet man dann auch

(2') $fx_1 \ldots x_r = \iota y\delta$ ("dasjenige y mit der Eigenschaft δ ").

Allgemein werden definitorische Erweiterungen durch Hinzunahme einer Menge von neuen Zeichen mit zugehörigen formalen Definitionen gebildet. Es ist wohl (zumindest anschaulich) klar, daß eine definitorische Erweiterung "im wesentlichen gleichwertig" mit der ursprünglichen Theorie ist (Genaueres dazu s. etwa Schwabhäuser 1971, S. 114ff.). Insbesondere besagen die folgenden Sätze 4.4 und 4.5, daß man den unter Benutzung definitorischer Erweiterungen vorgenommenen Aufbau einer Theorie auch in der ursprünglichen Objektsprache durchführen kann.

Damit spielt die Unterscheidung der genannten beiden Möglichkeiten in vielen Fällen (insbesondere in Teil I) keine Rolle.

(v) Für das Folgende legen wir natürlich die zweite Möglichkeit zugrunde, um möglichst viele Mengen *K* im Sinne von 4.2 bilden zu können. Wir erlauben uns jedoch, eine geometrische Theorie und verschiedene ihrer definitorischen Erweiterungen jeweils mit demselben Buchstaben (A, A_n, P_2', H^* usw.) zu bezeichnen. (Aus dem Zusammenhang geht stets hervor, welche Zeichen mindestens in der Objektsprache liegen müssen; auf eine genauere Festlegung kommt es für die Resultate nicht an.)

(vi) Definition 4.2 bleibt sinnvoll für Theorien T, die in anderen Sprachen - z.B. höherer Stufe - formalisiert sind. Auch von der zur Verfügung stehenden Sprache kann es abhängen, ob R mittels *K* definierbar ist (ein Beispiel für die schwache zweite Stufe s. 4.84(ii), (iii)). Solange nichts anderes gesagt ist, wird jedoch im Folgenden stets eine Formalisierung in der ersten Stufe zugrundegelegt.

(vii) In 4.2 wurde insbesondere gefordert, daß das definiens δ keine anderen freien Variablen enthält als diejenigen, die im definiendum auftreten. Diese Forderung kann man äquivalent weglassen, d.h. die Bedingung $\delta \in \mathrm{Fml}_K(\ldots)$ ersetzen durch $\delta \in \mathrm{Fml}_K$. Sei nämlich δ ein derartiges definiens mit den zusätzlichen freien Variablen $u_1, \ldots, u_k$. Dann gibt es auch ein definiens δ' der in 4.2 genannten Art, z.B. $\delta' = \forall u_1 \ldots \forall u_k \delta$ oder $\delta' = \exists u_1 \ldots \exists u_k \delta$. Schließlich kann man als δ' auch eine Formel (ohne zusätzliche Quantoren) verwenden, die aus δ entsteht, indem man für $u_1, \ldots, u_k$ Variablen substituiert, die im definiendum auftreten - bei formalen Definitionen gemäß 4.2(1) unter der Voraussetzung $r \geq 1$. (In jedem dieser Fälle ergibt sich leicht [z.B. aus bekannten Schlußregeln der Prädikatenlogik], daß mit 4.2(1) auch $Rx_1 \ldots x_r \leftrightarrow \delta'$ ein Satz von T ist, analog für 4.2(2).)

Wir formulieren noch einige allgemeine Sätze über den betrachteten Definierbarkeitsbegriff.

4.4 Satz. *Vor.: R ist definierbar mittels K in T.*

Beh.: Zu jeder Formel α der Sprache von T läßt sich eine bezüglich T äquivalente Formel α' mit höchstens denselben freien Variablen angeben, in der als Operations- und Relationszeichen höchstens die Zeichen aus K und die von R verschiedenen Operations- und Relationszeichen der Formel α vorkommen.

4.5 Satz. *Entsprechendes gilt für ein beliebiges Operationszeichen* f *an Stelle von* R.

Zum Beweis: Die beiden Sätze besagen anschaulich, daß man ein durch eine formale Definition charakterisiertes Zeichen (unter Verwendung der Zeichen aus K) "in jedem Zusammenhang" (der in der Objektsprache ausgedrückt ist) "eliminieren" kann. Daß das möglich ist, ist wohl zumindest anschaulich klar; man braucht ja nur das, was man mit Hilfe des Zeichens gesagt wurde, durch das zugehörige definiens auszudrücken.

Ein strenger Beweis von 4.4 läßt sich durch Induktion über den Formelaufbau von α führen. Hierbei ist alles Routine bis auf den zum Anfangsschritt gehörenden Fall, daß α von der Form $R\tau_1 \ldots \tau_r$ ist mit Termen (oder auch nur Variablen), die i.a. von den Variablen aus 4.2(1) verschieden sind. In diesem Falle erhält man das Gewünschte aus dem Bestand-$Rx_1 \ldots x_r \leftrightarrow \delta$ von 4.2(1) durch geeignete verallgemeinerte Substitutionen (d.h. ggf. mit Umbenennungen von gebundenen Variablen von δ, s. 1.9).

In einem strengen Beweis von 4.5 ist noch zu berücksichtigen, daß das Funktionszeichen f in Termen verschachtelt auftreten kann. Die Ausführung sei dem Leser überlassen. Sie findet sich z.B. auch in Schwabhäuser 1971, S. 117ff. (Aus der dort genannten allgemeineren Behauptung (v) von Satz 2.3.2 ergeben sich unmittelbar unsere beiden Sätze 4.4 und 4.5). □

4.6 Satz. *Vor.:* (i) *Das (Operations- oder Relations-)Zeichen* ζ *ist definierbar mittels* K' *in* T.

(ii) *Jedes Zeichen aus* K' *ist definierbar mittels* K *in* T.

Beh.: ζ *ist definierbar mittels* K *in* T.

Beweis: Sei z.B. ζ=R (r-stelliges Relationszeichen), und sei

(1) $Rx_1 \ldots x_r \leftrightarrow \delta'$

eine (nach Vor. (i) existierende) formale Definition für ζ mittels K' in T (der Beweis funktioniert genauso mit einem definiendum der Form $fx_1 \ldots x_r \doteq y$). Seien $\zeta_1', \ldots, \zeta_m'$ die im definiens δ' vorkommenden Zeichen aus K'. Nach Vor. (ii) kann man für jedes dieser Zeichen sukzessive einen der Sätze 4.4, 4.5 anwenden und erhält damit (nach m Anwendungen) schließlich eine in T zu δ' äquivalente Formel δ (mit höchstens denselben freien Variablen), in der nur noch Operations- und Relationszeichen aus K vorkommen. Also ist

(2) $Rx_1 \ldots x_r \leftrightarrow \delta$

ein Satz von T und damit eine formale Definition von der in der Behauptung gewünschten Art. □

Trivialerweise gelten auch noch die folgenden beiden Sätze.

<u>4.7 Satz</u>. *Wenn* ζ *definierbar mittels* K_1 *in* T *und* $K_1 \subseteq K_2$, *so ist* ζ *auch definierbar mittels* K_2 *in* T.

<u>4.8 Satz</u>. *Wenn* ζ *definierbar mittels K in* T_1 *und* $T_1 \subseteq T_2$ *(d.h.* T_1 *Untertheorie von* T_2*), so ist* ζ *auch definierbar mittels K in* T_2.

<u>4.9 Anmerkungen</u>. (i) Wenn in späteren Sätzen die Definierbarkeit eines speziellen Begriffs festgestellt wird, ist es somit zweckmäßig, diese Feststellung für eine möglichst schwache Theorie T und evtl. eine möglichst kleine Konstantenmenge *K* zu treffen, um ein möglichst starkes Resultat zu erhalten. Wird dagegen festgestellt, daß ein Begriff nicht definierbar ist, so ist es ebenso zweckmäßig, das für eine möglichst starke Theorie und eine möglichst große Konstantenmenge zu tun (in einer schwächeren Theorie und evtl. mittels einer kleineren Konstantenmenge ist der Begriff dann erst recht nicht definierbar). Folgerungen, die sich aus solchen Resultaten auf Grund der Sätze 4.6 bis 4.8 ergeben, werden i.a. nicht besonders formuliert.

(ii) Die F r a g e , ob ein gewisser Begriff mit Hilfe von anderen definierbar ist, ist natürlich metamathematischer Natur - das ist ja die Frage nach der Existenz einer formalen Definition, d.h. einer Formel mit gewissen Eigenschaften. Zum Auffinden einer solchen formalen Definition hat man übrigens keine praktisch brauchbare Methode (s. jedoch (iii)); zur Beantwortung der genannten Frage ist man daher in jedem Einzelfall auf eine geeignete Idee angewiesen. Die B e w e i s e für die Feststellung einer Definierbarkeit bzw. Nichtdefinierbarkeit lassen sich dagegen (wenn man die Idee erst einmal hat) meist mit den Mitteln der gegebenen Theorie T führen, sie sind also in den folgenden Anwendungen (s. dort) meist rein geometrischer Natur. Nur in wenigen Fällen werden für solche Beweise Hilfsmittel aus der mathematischen Logik benutzt (die über die übliche Formalisierungstechnik hinausgehen), z. B. in 4.56, 67, 68, 71, 72, 84 und 88.

(iii) Wenn die betrachtete Theorie T rekursiv aufzählbar (s. 3.8ff.) ist (z.B. rekursiv axiomatisierbar und von erster Stufe, wie es für die hier betrachteten Geometrien meist der Fall ist), so gibt es allerdings ein "prinzipiell ausführbares" Verfahren (vgl. 3.2, 3.5(i), (iii)), das zu jedem "Begriff" ζ und jedem "Begriffssystem" K in der Sprache von T eine formale Definition von ζ mittels K liefert, falls ζ überhaupt mittels K in T definierbar ist (andernfalls bricht das Verfahren nicht ab). Das Verfahren besteht - anschaulich gesagt - darin, daß man 1. die Sätze von T sukzessive aufzählt (wofür nach Voraussetzung ein Verfahren existiert), 2. bei jedem Satz prüft, ob er die Bauart einer formalen Definition von ζ mittels K hat, und 3., wenn das der Fall ist, das Verfahren abbricht. Angesichts der Kompliziertheit von Verfahren zur Aufzählung aller ("beweisbaren") Sätze (vgl. die Beweisidee für 3.12) kann das aber nicht als eine praktisch brauchbare Methode angesehen werden.

Wenn die Theorie T sogar entscheidbar ist, ist kein einfacheres Verfahren zum Finden formaler Definitionen zu erwarten, da man vorher nicht weiß, für welche Formel man nachprüfen soll, ob sie ein Satz von T ist.

(iv) Wenn ζ als Grundbegriff in einem Axiomensystem Σ für T verwendet wird und ζ in T mittels K definierbar ist, so kann man natürlich (durch Elimination von ζ gemäß 4.4 oder 4.5 auch ein neues A x i o m e n - s y s t e m Σ' bilden, das an Stelle von ζ "nur noch" die Begriffe von K enthält. Dieses Axiomensystem ist jedoch im allgemeinen für den Aufbau der Theorie "unbefriedigend"; denn die neuen Axiomen sind gewöhnlich lang und unübersichtlich und drücken nicht mehr einfache Eigenschaften der jetzt verwendeten Grundbegriffe aus, wie man es sich normalerweise wünschen würde. Die Aufgabe, ein "befriedigendes" Axiomensystem in den neuen Grundbegriffen zu finden, ist wesentlich komplizierter als die bloße Feststellung einer Definierbarkeit und wird im folgenden nicht behandelt. Sie scheint nicht allgemein angreifbar zu sein. Auch in Spezialfällen kann man nicht erwarten, daß es ein "ebenso befriedigendes" oder "ebenso einfaches" Axiomensystem in den neuen Grundbegriffen überhaupt gibt (vgl. z.B. 4.24 und die Bemerkung in I.1.7.2 über die Kompliziertheit des Axiomensystems von Pieri 1908 im Grundbegriff P).
Nach diesen allgemeinen Betrachtungen kommen wir nun zu geometrischen Anwendungen.

<u>4.10 Geometrische Begriffe</u>. Beim Aufbau der dimensionsfreien absoluten Geometrie A in Teil I (Abschnitte 1 bis 11) wurden neben den Grundbegriffen D und B (Streckenkongruenz und Zwischenbeziehung) insbesondere schon die folgenden Begriffe eingeführt (die wir jetzt als Relationszeichen in einer definitorischen Erweiterung im Sinne von Anmerkung 4.3(iv), (v) auffassen):

Col (Kollinearität, Def. I.4.10 mittels B),

$\leq$ (kleiner-Gleich-Beziehung für Strecken, Def. I.5.4),

$\approx$ (mit $a \underset{p}{\approx} b$ gdw a und b liegen auf derselben Seite von p, Def. I.6.1(1) mittels B),

$\underset{A}{\equiv}$ (Winkelkongruenz, Def. I.11.2, kurz $\equiv$),

M (Mittelpunktsbeziehung, Def. I.7.1),

R (Rechtwinkelbeziehung),

Cp (Komplanarität).

Für R kann statt Def. I.8.1 die folgende formale Definition (ohne Einschaltung eines Operationszeichens) verwendet werden:

(1) $\text{Rabc} \leftrightarrow \exists d[\text{Mcbd} \wedge \text{ac} \equiv \text{ad}]$.

Cp werde hier als Zeichen für eine vierstellige Relation zwischen Punkten verwendet; eine einfache formale Definition mittels Col ist Satz I.9.33.

4.11 Anmerkungen. (i) In allgemeineren Betrachtungen (z.B. 4.2) wurde der Buchstabe R für ein beliebiges Relationszeichen verwendet; Verwechslungen sind wohl nicht zu befürchten.

(ii) An die Stelle der genannten (metasprachlichen) Definitionen und Sätze aus Teil I sollen jetzt natürlich entsprechende Formeln (mit Variablen in gerader Schrift) treten (vgl. Anmerkung 1.17(ii)).

(iii) Rein geometrische Beweise für derart formalisierte Sätze (vgl. 4.9(ii)) führen wir dagegen nach wie vor in der Metasprache und verwenden dabei die eingeführten Relationszeichen auch zur Bezeichnung der entsprechenden Relationen (wir schreiben also z.B. kurz Col statt $\text{Col}_{\mathfrak{A}}$ im Sinne von 1.13).

4.12 Definition. Darüberhinaus betrachten wir die Begriffe $\underset{H}{\equiv}$, P, P', E, S, die (im selben Sinne) durch folgende formale Definitionen in A eingeführt werden.

(1) $\text{abc} \underset{H}{\equiv} \text{def} \leftrightarrow \text{abc} \underset{A}{\equiv} \text{def} \wedge \neg\, \text{Col abc}$

(Winkelkongruenz im Sinne des Hilbertschen Axiomensystems für nicht entartete Winkel),

(2) $\text{Pabc} \leftrightarrow \text{ab} \equiv \text{ac}$ ("Pierische Relation"),

(3) $\text{P'abc} \leftrightarrow \text{ab} \leq \text{ac}$,

(4) $\text{Eabc} \leftrightarrow \text{ab} \equiv \text{ac} \equiv \text{bc}$

(a, b, c bilden ein gleichseitiges Dreieck [engl. equilateral triangle] oder fallen zusammen),

(5) $Sabc \leftrightarrow Rabc \vee Rbca \vee Rcab$
("vollsymmetrische Rechtwinkelbeziehung": a, b, c bilden in irgendeiner Reihenfolge ein rechtwinkliges Dreieck oder sind nicht paarweise verschieden).

P wurde von Pieri 1908 als Grundbegriff beim Aufbau der euklidischen Geometrie verwendet. P' wurde - im Zusammenhang mit Definierbarkeitsfragen in der euklidischen Geometrie - betrachtet von Tarski 1956, E von Beth u. Tarski 1956, S von Scott 1956.

Auf Grund von (2) ist P trivialerweise definierbar mittels D. In den folgenden Sätzen soll gezeigt werden, daß mittels P auch D in A definierbar ist. Wir beginnen mit einer häufig benutzten Charakterisierung der Kollinearität.

4.13 Satz. Col *ist definierbar mittels* P *in der zweidimensionalen absoluten Geometrie* A_2; *eine geeignete formale Definition ist*

(1) $Col\,abc \leftrightarrow \exists p \exists q [p \neq q \wedge Papq \wedge Pbpq \wedge Pcpq]$ (Abb. 120).

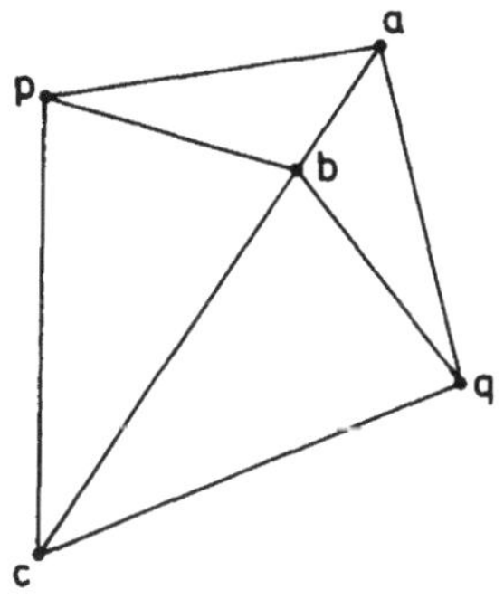

Abb. 120

Beweis: In A_2 läßt sich jede Gerade als Mittellot von geeigneten voneinander verschiedenen Punkten p, q charakterisieren; damit erhält man leicht (1) als Satz von A_2. □

Die angegebene Charakterisierung 4.13(1) versagt für höhere Dimension. In der n-dimensionalen Geometrie A_n bedeutet die Bedingung $Pxpq$ - für voneinander verschiedene Punkte p, q - nämlich, daß x in dem ($(n-1)$-dimensionalen) Mittellotraum Ml(pq) zu p und q liegt (vgl. I.11.65, 66). Man kann jedoch Geraden als Schnittgebilde von Unterräumen höherer Dimension charakterisieren. Für A_3 erhält man z.B.

4.14 Satz. Col *ist definierbar mittels* P *in* A_3; *und zwar mit der formalen Definition*

$$\text{(1)}\quad \mathrm{Col}\,abc \leftrightarrow \exists p\exists q\exists p'\exists q'[p\neq q \wedge Papq \wedge Pbpq \wedge Pcpq \wedge p'\neq q' \wedge Pap'q' \wedge Pbp'q' \wedge Pcp'q' \wedge \exists d(Pdpq \wedge \neg Pdp'q')].$$

Zum Beweis: Die Bedingung in der letzten Zeile von (1) besagt, daß die Mittelebenen Ml(pq) und Ml($p'q'$) voneinander verschieden sind (da es einen Punkt d gibt, der in der ersten, aber nicht in der zweiten liegt) und ihr Durchschnitt (in dem a, b und c liegen) somit eine Gerade ist (I.9.30, 58). Alles andere ist leicht zu sehen. □

4.15 Anmerkung. In analoger Weise kann man für beliebiges natürliches $n\geq 2$ erhalten, daß Col mittels P in A_n definierbar ist. Nachdem man nämlich erst einmal die $(n-1)$-dimensionalen Unterräume wie in 4.13(1) charakterisiert hat, kann man auch die $(n-i)$-dimensionalen Unterräume für $i=2,...,n-1$ als Schnittgebilde charakterisieren (vgl. etwa I.9.59) oder stattdessen die Begriffe Ah_{n-i+1} (Def. I.9.46) definieren. Die Ausführung sei dem Leser überlassen.

4.16 Anmerkung. 4.13 bis 4.15 liefern für verschiedene Dimensionen verschiedene formale Definitionen von Col mittels P. Damit ist noch nicht gesagt, ob Col auch in der dimensionsfreien absoluten Geometrie A mittels P definierbar ist (d.h. insbesondere mit einer einheitlichen [dimensionsunabhängigen] formalen Definition). Aber auch diese Frage (die vorher anscheinend noch nicht in der Literatur aufgeworfen wurde), läßt sich - sogar wesentlich einfacher - in dem folgenden Satz beantworten. (4.13 bis 4.15 werden damit als Definierbarkeitsresultate entbehrlich; die verwendete Methode wird jedoch später noch einmal benutzt.)

4.17 Satz. Col *ist definierbar mittels* P *in* A ; *nämlich mit*

$$\text{(1)}\quad \mathrm{Col}\,abc \leftrightarrow a=b \vee \forall p\forall q[Papq \wedge Pbpq \rightarrow Pcpq].$$

Nach dem vorher Gesagten erkennt man leicht die folgende anschauliche Bedeutung der rechten Seite von (1): Wenn $a\neq b$ ist, so liegt c in jeder Hyperebene, in der a und b liegen. Für einen strengen Beweis werden jedoch nicht einmal die Begriffe Hyperebene oder Unterraum gebraucht.

Beweis von 4.17: Es genügt, (1) als Satz von A zu beweisen. Die Richtung (→) ist eine unmittelbare Folgerung aus Satz I.4.17. Die Umkehrung zeigen wir indirekt, nehmen also ¬Col *abc* an. Dann ist zunächst $a \neq b$, und *c* liegt nicht auf der Verbindungsgeraden L(*ab*) von *a* und *b*. Wir setzen nun $p=c$ und $q=S_{ab}(p)$ (Abb. 121).

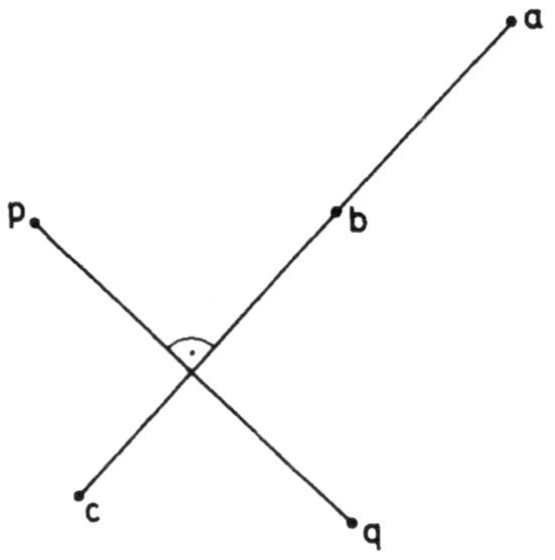

Abb. 121

Da *p* nicht auf der Spiegelungsachse liegt, ist $p \neq q$ (Satz I.10.8), also die Nullstrecke *cp* nicht kongruent zu *cq*, d.h. ¬P*cpq*. Nach Konstruktion ist aber P*apq* und P*bpq*. Somit wird die rechte Seite von (1) falsch, wie zu zeigen. (Um das zu erreichen, hätte man als *p* auch einen beliebigen Punkt der Ebene Pl(*abc*) außerhalb L(*ab*) wählen können, vgl. den Beweis von 4.19(3).)

4.18 Satz. *In* A *gelten (die folgenden Formeln) als Sätze:*

(1) $\text{Mabc} \leftrightarrow \text{Pbac} \wedge \text{Col abc} \wedge (a \neq c \vee b=c)$,

(2) $\text{Rabc} \leftrightarrow \exists d[\text{Mcbd} \wedge \text{Pacd}]$,

(3) $\text{Dabcd} \leftrightarrow \exists e \exists f[\text{Mbec} \wedge \text{Maef} \wedge \text{Pcfd}]$;

folglich sind auch M, R *und* D *definierbar mittels* P *in* A *(vgl.* 4.6*).*

Beweis: (1) ergibt sich leicht aus Abschnitt I.7 (die Richtung (←) insbesondere aus Satz I.7.20). (2) ergibt sich unmittelbar aus 4.10(1) und 4.12(2).

Zu (3): Seien *a*, *b*, *c*, *d* beliebig, und seien *e* und *f* die eindeutig bestimmten Punkte mit der Eigenschaft M*bec* ∧ M*aef* (Abb. 122). Nach I.7.13 (Spiegelung an *e*) ist dann $ab \equiv cf$. Damit ist die Streckenkongruenz D*abcd* gleichwertig mit $cf \equiv cd$, d.h. $Pcfd$. □

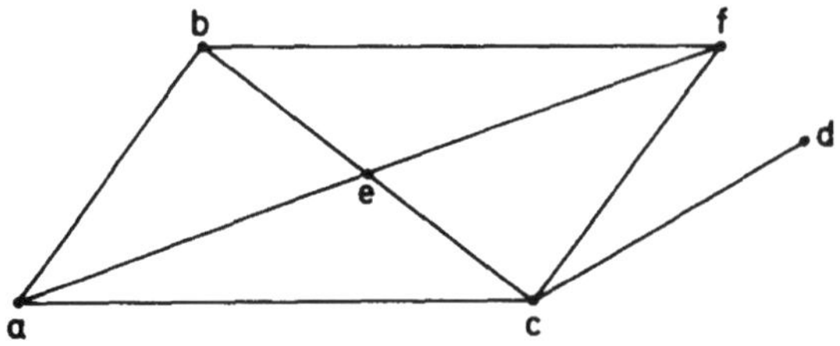

Abb. 122

Als nächstes soll gezeigt werden, daß mittels R dieselben Begriffe definierbar sind wie mittels P.

4.19 Satz. *In* A *sei* P^* *durch folgende formale Definition eingeführt:*

(1) $P^*abc \leftrightarrow Pabc \wedge \neg \mathrm{Col}\, abc$.

Dann gelten in A *als Sätze:*

(2) $Pabc \leftrightarrow \exists d[P^*abd \wedge P^*acd] \vee [a=b \wedge a=c]$ (Abb. 123),

(3) $\mathrm{Col}\, abc \leftrightarrow a=b \vee \forall p[Rabp \rightarrow Rcbp]$ (Abb. 124),

(4) $P^*abc \leftrightarrow \exists d \exists e \exists b' \exists c'[\ddagger(b'ec') \wedge Rab'd \wedge Rac'd \wedge \mathrm{Col}\, aed \wedge \mathrm{Col}\, b'ec' \wedge Raeb' \wedge \mathrm{Col}\, abb' \wedge \mathrm{Col}\, acc' \wedge \mathrm{Col}\, bdc \wedge Radb]$ (Abb. 126).

Darüberhinaus gilt in A_2

(3') $\mathrm{Col}\, abc \leftrightarrow \exists p[p \neq b \wedge Rabp \wedge Rcbp]$ (Abb. 124),

und in der euklidischen Geometrie P gilt

(4') $P^*abc \leftrightarrow \exists d \exists e[\ddagger(bec) \wedge Rabd \wedge Racd \wedge \mathrm{Col}\, aed \wedge \mathrm{Col}\, bec \wedge Raeb]$ (Abb. 125).

Folglich sind Col, P^*, P *definierbar mittels* R *in* A *(in stärkeren Theorien zum Teil mit einfacheren formalen Definitionen).*

4.20 Korollar. *Jeder der Begriffe* D, P, P^*, R *ist mittels eines jeden dieser Begriffe definierbar in* A. Col *und* M *sind mittels eines jeden dieser Begriffe definierbar in* A.

Beweis von 4.19 und 4.20: Es genügt der Beweis der angegebenen geometrischen Sätze. (2) ist offensichtlich. (3)(→) ergibt sich sofort aus I.8.3. Zum Beweis der Umkehrung nehmen wir wieder $\neg\mathrm{Col}\, abc$ an und wählen einen von b verschiedenen Punkt p auf der Senkrechten, die in b auf $L(ab)$ in der Ebene $Pl(abc)$ errichtet ist. Dann ist $Rcbp$, und aus der Eindeutigkeit der Senkrechten (auf $L(pb)$ in b) in einer Ebene (I.11.20)

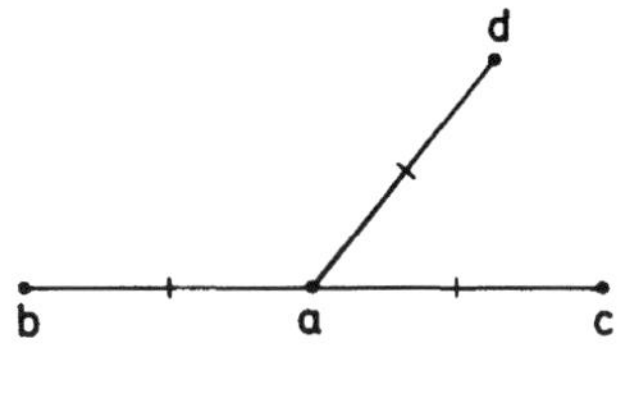

Abb. 123

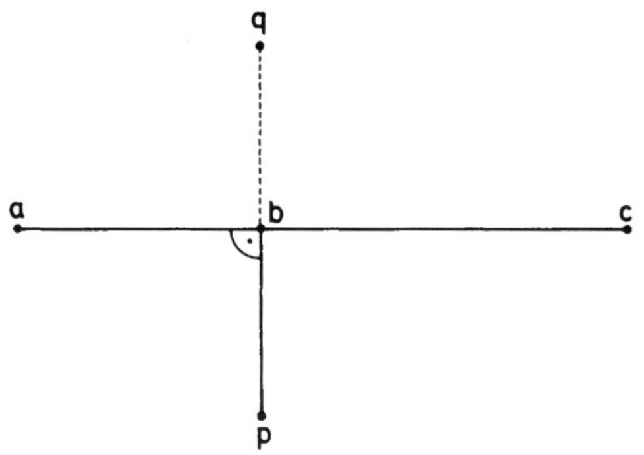

Abb. 124

ergibt sich doch Col abc. Mit denselben Hilfsmitteln ergibt sich (3'). Setzt man noch $q=S_{ab}(p)$, so erkennt man die Analogie von (3) bzw. (3') zu 4.17(1) bzw. 4.13(1).

Zu (4')(←)(Beweis in A): Seien d, e Punkte der angegebenen Art (Abb. 125). Dann ist zunächst $a \neq e$; denn sonst wäre $Rebd \wedge Recd$ und

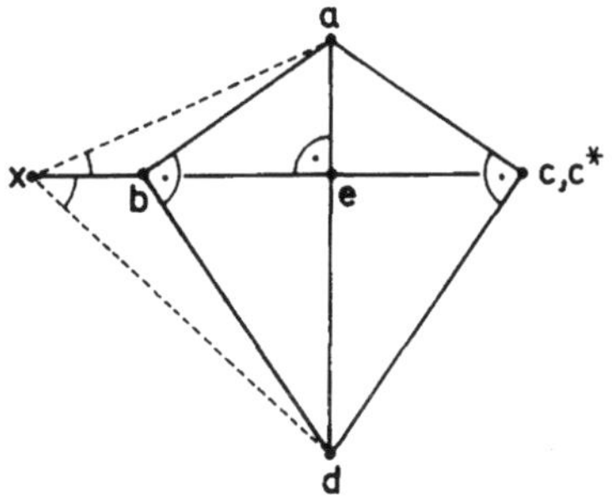

Abb. 125

(nach I.8.3) auch $Rcbd \wedge Rbcd$, nach I.8.7 also $c=b$ im Widerspruch zur Voraussetzung. Mit I.8.9 (Nicht-Kollinearität der Punkte in einer Rechtwinkelbeziehung) ergibt sich dann ¬Col aeb und $a \neq d$ und damit $bc \underset{e}{\perp} ad$. Nach I.11.47 liegt e (als Höhenfußpunkt im rechtwinkligen Dreieck abd) echt zwischen a und d. Sei nun $c^*=S_{ad}(b)$. Es genügt $c=c^*$ zu zeigen. Aus bekannten Eigenschaften der Geradenspiegelungen (I.10.3 und I.10.11) ergibt sich zunächst $Mbec^*$ und Rac^*d. Ist nun x ein von b verschiedener Punkt der Halbgeraden H(eb), so ist $Bebx$ oder $Bexb$ und dementsprechend (nach Lemma I.11.53) $axe < abe \wedge dxe < dbe$ bzw. $axe > abe \wedge dxe > dbe$; aus den Sätzen über das Aneinanderlegen von Winkeln (I.11.22) und über die

Kleiner-Gleich-Beziehung erhält man leicht, daß $\sphericalangle axd$ im Falle Bebx ein spitzer und im Falle Bexb ein stumpfer Winkel ist, also jedenfalls kein rechter Winkel. Dasselbe ergibt sich für einen von c^* verschiedenen Punkt der Halbgeraden H(ec^*). Damit sind b und c^* die einzigen beiden Punkte x auf der Geraden L(be) mit der Eigenschaft Raxd. Nach Voraussetzung ist aber auch c ein solcher Punkt und $c \neq b$, also $c = c^*$, wie zu zeigen.

Zu (4')(→)(Beweis in P): Gelte P^*abc, und sei e der Mittelpunkt von bc. Nach Definition von R ist dann Raeb. Nach Voraussetzung ist ¬ Col abc. Damit sind insbesondere die Punkte a, b, e paarweise verschieden, und L(ae) steht nicht senkrecht auf L(ab). Sei nun G die Senkrechte auf L(ab) in b in der Ebene Pl(abc). Dann ist G (nach I.12.9 und 12.13 oder 12.21) nicht parallel zu L(ae), liegt aber mit L(ae) in einer Ebene und schneidet daher L(ae) in einem gewissen Punkt d. Nach Konstruktion ist Rabd und $c = S_{ad}(b)$ und damit auch Racd. Somit leisten d und e das Verlangte.

In diesem Beweis wurde das Parallelenaxiom (oder Folgerungen daraus) benutzt, um zu erhalten, daß G und L(ae) sich schneiden. In der absoluten Geometrie A läßt sich das tatsächlich nicht zeigen; Gegenbeispiele liefert die hyperbolische Geometrie H (wo G und L(ae) sich nicht schneiden, falls $\sphericalangle bae$ größer-gleich dem Parallelwinkel von ab ist).

Zu (4)(←): Seien d, e, b', c' Punkte der angegebenen Art (Abb. 126). Aus dem (in A geführten) Beweis von (4')(←) ergibt sich $P^*ab'c'$ und

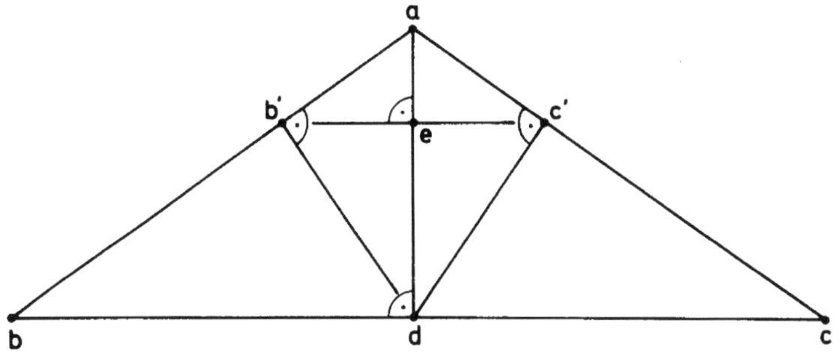

Abb. 126

$S_{ad}(b') = c'$. Da die Geradenspiegelung S_{ad} ein Automorphismus ist, ist $S_{ad}(b)$ der Schnittpunkt von L(ac') und der Senkrechten auf L(ad) in d in der Ebene Pl$(ab'c')$, d.h. $S_{ad}(b) = c$. Somit ist $ab \equiv ac$ und wegen $a \neq d$

auch $\neg \mathrm{Col}\, abc$, d.h. $\mathrm{P}^* abc$.

Zu (4)(→): Gelte $\mathrm{P}^* abc$, und sei d der Mittelpunkt von bc. Dann ist $\mathrm{S}_{ad}(b)=c$ und $\neg \mathrm{Col}\, abc$. Weiter sei b' der Fußpunkt des Lots von d auf $\mathrm{L}(ab)$, $c'=\mathrm{S}_{ad}(b')$ und e der Mittelpunkt von $b'c'$. Dann leisten d, e, b', c' das Verlangte. □

4.21 Definition. Eine Menge K' von Relations- und (evtl.) Operationszeichen heißt ein *geeignetes System von Grundbegriffen für* A, falls gilt:

(i) Jedes Zeichen aus K' ist definierbar mittels D und B,

(ii) D und B sind definierbar mittels K'.

Analog definiert man geeignete Systeme von Grundbegriffen für andere Theorien T mit Bezug auf eine Menge K von "ursprünglich gegebenen Begriffen" für diese Theorie (für A ist $K=\{\mathrm{D},\mathrm{B}\}$). Ist hierbei $K'=\{\zeta\}$, so sagen wir auch, daß ζ *als einziger Grundbegriff für* T *geeignet* ist. Die Bedingung (i) ist natürlich trivial, wenn die Begriffe von K' durch Bildung einer definitorischen Erweiterung (4.3(iv), (v)) eingeführt sind.

Wir zeigen zunächst, daß die beiden Winkelkongruenzen für A jeweils als einziger Grundbegriff geeignet sind.

4.22 Satz. *In* A *gelten als Sätze:*

(1) $a \underset{b}{\simeq} c \leftrightarrow aba \underset{A}{\equiv} abc$,

(2) $\mathrm{Col}\, abc \leftrightarrow a=c \vee b \underset{a}{\simeq} c \vee b \underset{c}{\simeq} a$;

(3) $\mathrm{P}^* abc \leftrightarrow abc \underset{H}{\equiv} acb$ (Abb. 127, P^* s. 4.19(1)),

(4) $\mathrm{B}abc \leftrightarrow b=a \vee b=c \vee [b \underset{a}{\simeq} c \wedge b \underset{c}{\simeq} a]$,

(5) $\mathrm{Col}\, abc \leftrightarrow \neg abc \underset{H}{\equiv} abc$,

(6) $\mathrm{B}abc \leftrightarrow b=a \vee b=c \vee \{\mathrm{Col}\, abc \wedge \exists d[\neg \mathrm{Col}\, abd \wedge \neg abd \underset{H}{\equiv} cbd]\}$.

Folglich (4.12(1)) *ist* $\underset{H}{\equiv}$ *definierbar mittels* $\underset{A}{\equiv}$ *in* A; *weiter sind* P^*, Col *und* B *mittels* $\underset{A}{\equiv}$ *und auch mittels* $\underset{H}{\equiv}$ *definierbar und damit* (4.19(2), 18) *auch* P, R, D. *Somit ist jede der beiden Winkelkongruenzen* $\underset{A}{\equiv}$, $\underset{H}{\equiv}$ *als einziger Grundbegriff für* A *geeignet.*

Beweis: Es genügt wieder der Beweis der angegebenen geometrischen Sätze. Zu (1) beachte man, daß die rechte Seite besagt, daß $\sphericalangle abc$ ein Nullwinkel ist.

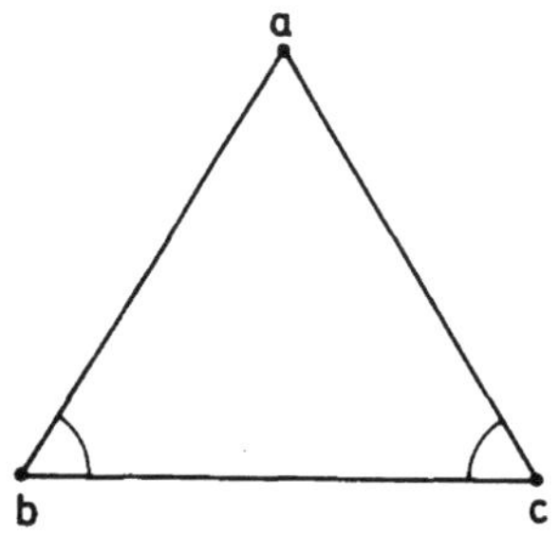

Abb. 127

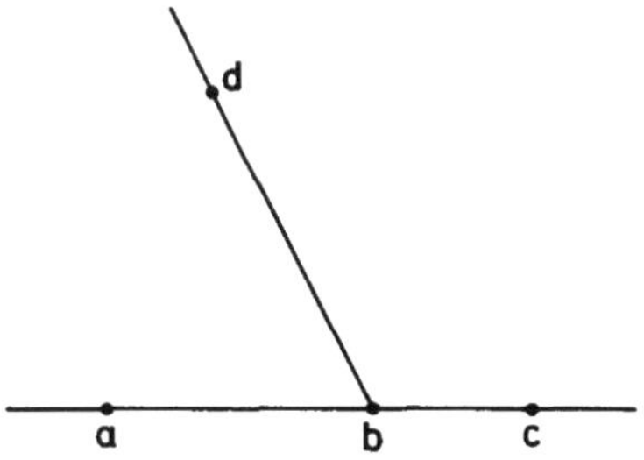

Abb. 128

Zu (6) für den Fall $a,c \neq b$ (Abb. 128): (←): Wegen der ausgedrückten Inkongruenz von Winkeln mit dem gemeinsamen Schenkel H(bd) können die verbleibenden Schenkel nicht zusammenfallen, müssen also wegen Col$\,abc$ zueinander entgegengesetzte Halbgeraden sein. (→): ein Punkt d außerhalb L(ab) mit $\neg ab \perp db$ leistet das Verlangte.

Die übrigen Sätze gelten offensichtlich (vgl. I.6, I.11). □

Die Winkelkongruenzen $\underset{A}{\equiv}$, $\underset{H}{\equiv}$, sind hier sechsstellige Relationszeichen. Andererseits zeigt 4.20, daß für A neben dem ursprünglichen System {D,B} auch Systeme von dreistelligen Grundbegriffen geeignet sind, nämlich {P,B} und {R,B}. Wir zeigen nun, daß schon der dreistellige Begriff P' (4.12(3)) als einziger Grundbegriff geeignet ist.

4.23 Satz. *In* A *gelten als Sätze:*

(1) $\text{Pabc} \leftrightarrow \text{P'abc} \wedge \text{P'acb}$,

(2) $\text{Babc} \leftrightarrow \text{Col abc} \wedge \text{P'abc} \wedge \text{P'cba}$,

(2') $\text{Babc} \leftrightarrow \forall x[\text{P'axb} \wedge \text{P'cxb} \rightarrow x=b]$ (Abb. 129);

folglich sind P, B *und damit auch* D *und* R *definierbar mittels* P' *in* A *(vgl.* 4.17, 18*), und* P' *ist als einziger Grundbegriff für* A *geeignet.*

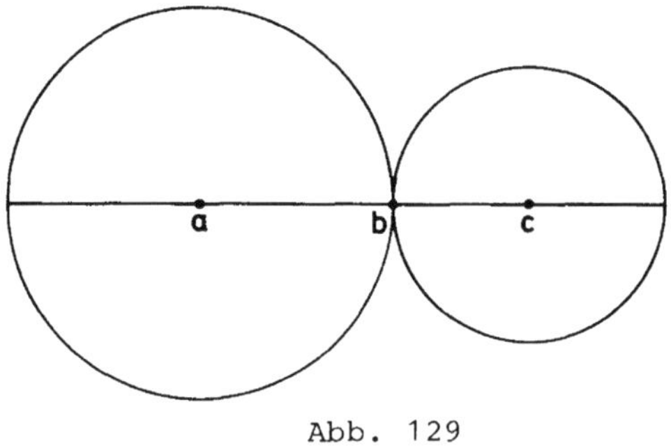

Abb. 129

Beweis: (1) und (2) ergeben sich sofort aus den Sätzen über die Kleiner-Gleich-Beziehung für Strecken (insbesondere I.5.9 und I.5.12). Die Ausführung des Beweises für den (statt (2) verwendbaren) Satz (2')(etwa mit Hilfe von I.11.46) kann dem Leser überlassen bleiben. □

4.24 Satz. *In* A' *(der dimensionsfreien absoluten Geometrie mit Kreisaxiom) gilt als Satz* (R.M. Robinson 1959)*:*

(1) $P'abc \leftrightarrow \forall d\{Pdac \rightarrow \exists e[Peba \wedge Paed]\}$ (Abb. 130);

folglich ist für A' *auch* P *als einziger Grundbegriff geeignet (und damit nach* 4.20 *auch jeder der Begriffe* D, R, P**). Außerdem gilt in* P' *(entsprechende euklidische Geometrie):*

(2) $Babc \leftrightarrow Col\,abc \wedge \exists d[Rabd \wedge Radc]$ (Abb. 131).

Beweis: Zu (1): Seien b', c' die Mittelpunkte von ab bzw. ac (Abb. 130). Aus den Sätzen über die $\leq$-Beziehung (I.5) erhält man, daß dann $P'abc$ (d.h. $ab \leq ac$) gleichwertig ist mit $ab' \leq ac'$.

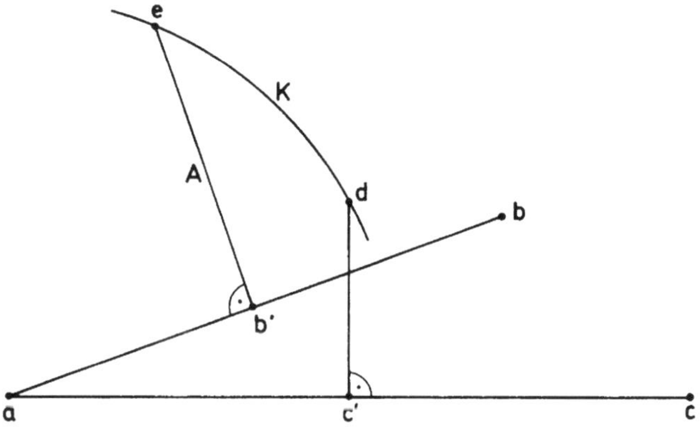

Abb. 130

Zu ($\rightarrow$): Gelte $P'abc$ und $Pdac$. Dann ist $Rac'd$, also $ab' \leq ac' \leq ad$. Sei E eine Ebene durch a und b' (in der d nicht zu liegen braucht), sei K der Kreis in E um a mit dem Radius ad, und sei A die Senkrechte auf $L(ab)$ in b' in der Ebene E bzw., falls $a=b$ ist, irgendeine Gerade durch b' in E. Nach Konstruktion ist dann b' ein Punkt im Inneren des Kreises K oder liegt auf diesem Kreis. Nach dem Kreisaxiom CA haben also A und K einen gemeinsamen Punkt e. Wegen $e \in A$ ist $Peba$ und wegen $e \in K$ auch $Paed$, d.h., e leistet das Verlangte.

Zu ($\leftarrow$): Wir setzen speziell $d=c'$. Dann ist $Pdac$. Sei dann e gemäß der rechten Seite von (1) gewählt. Dann ist $Rab'e$ und damit $ab' \leq ae \equiv ad \equiv ac'$, also P'$abc$, wie zu zeigen.

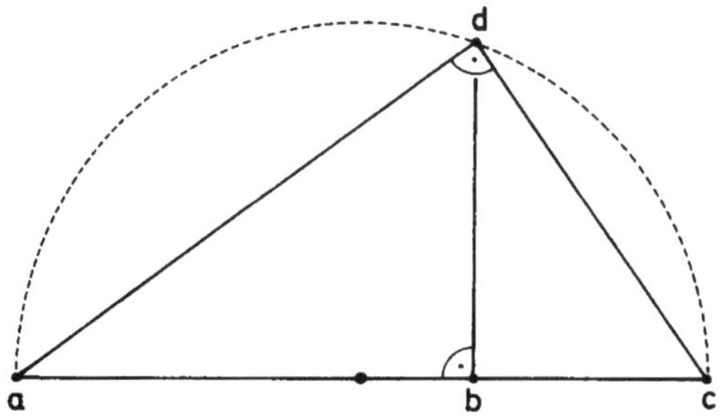

Abb. 131

Zu (2)(←)(Beweis in A): Falls ¬Col acd ist, liegt b als Höhenfußpunkt im rechtwinkligen Dreieck adc zwischen a und c (I.11.47). Falls Col acd ist, ergibt sich aus Radc nach I.8.9 zunächst $d=a \vee d=c$ und damit jedenfalls Col abd; aus Rabd ergibt sich dann ebenso $b=a \vee b=d$, also $b=a \vee b=c$ und damit Babc.

Zu (2)(→)(Beweis in P'): Falls $b=a$ oder $b=c$ ist, leistet $d=b$ das Verlangte. Seien nun a, b, c paarweise verschiedene Punkte mit Babc. Sei E eine Ebene durch diese Punkte. Nach dem Kreisaxiom schneidet die in b auf L(ac) in der Ebene E errichtete Senkrechte einen in E geschlagenen Halbkreis mit dem Durchmesser ac in einem gewissen Punkte d. Nach dem bekannten Satz von Thales (Beweise finden sich in Büchern über Elementargeometrie und lassen sich in P durchführen) ist der "Sehnenwinkel" $\sphericalangle adc$ im betrachteten Halbkreis ein Rechter. Somit leistet d das Verlangte. □

4.25 Anmerkung. Der Satz von Thales ist kein Satz der absoluten Geometrie; in der hyperbolischen Geometrie H läßt sich zeigen, daß der Sehnenwinkel im Halbkreis stets ein spitzer Winkel ist. Der angegebene Beweis von 4.24(2) versagt somit in A'. Aus der Charakterisierung der Modelle von A von Pejas (vgl. 2.14(iv)) kann man jedoch erhalten, daß 4.24(2) sogar ein Satz von A' ist.

4.26 Problem. Man gebe einen möglichst einfachen geometrischen Beweis für 4.24(2) in A'.

4.27 Nichtdefinierbarkeit. Als nächstes wollen wir zeigen, daß das Kreisaxiom für die Definierbarkeitsfeststellungen in 4.24 nicht entbehrlich ist. Dazu genügt es, zu zeigen, daß z.B. die Zwischenbeziehung B nicht mittels D in A definierbar ist. Im Sinne von Anmerkung 4.9(i) zeigen wir dieses erste Nichtdefinierbarkeitsresultat für die stärkere Theorie P_n. Wir verwenden dazu die in dem folgenden Satz beschriebene Methode, die auf Padoa zurückgeht und häufig für Nichtdefinierbarkeitsbeweise verwendet wird.

4.28 Definition. Seien $\mathfrak{A}$, $\mathfrak{B}$ Strukturen für ein und dieselbe Sprache L (einer Theorie T). Eine eineindeutige Abbildung φ von $U_{\mathfrak{A}}$ auf $U_{\mathfrak{B}}$ (Trägermengen der Strukturen) heißt dann ein *Isomorphismus* bezüglich des r-stelligen Relationszeichens R von L, falls gilt:

(i) Für jedes $a_1,\ldots,a_r \in U_{\mathfrak{A}}$ ist $R_{\mathfrak{A}}a_1\ldots a_r$ gdw $R_{\mathfrak{B}}\varphi(a_1)\ldots\varphi(a_r)$.

Analog dazu heißt φ ein Isomorphismus bezüglich eines r-stelligen Operationszeichens f von L, falls gilt:

(ii) Für jedes $a_1,\ldots a_r, b \in U_{\mathfrak{A}}$ ist

$$f_{\mathfrak{A}}a_1\ldots a_r = b \quad \text{gdw} \quad f_{\mathfrak{B}}\varphi(a_1)\ldots\varphi(a_r) = \varphi(b).$$

Statt (ii) kann man offenbar auch folgende gleichwertige Bedingung verwenden:

(ii') Für jedes $a_1,\ldots,a_r \in U_{\mathfrak{A}}$ ist $\varphi(f_{\mathfrak{A}}a_1\ldots a_r) = f_{\mathfrak{B}}\varphi(a_1)\ldots\varphi(a_r)$.

Statt "φ ist Isomorphismus bezüglich R bzw. f " sagen wir auch, daß φ den Begriff R bzw. f *erhält* oder (von $\mathfrak{A}$ nach $\mathfrak{B}$) *überträgt*; im Falle $\mathfrak{A}=\mathfrak{B}$ sagen wir dafür auch, daß R bzw. f oder auch $R_{\mathfrak{A}}$ bzw. $f_{\mathfrak{A}}$ unter (oder gegenüber) φ *invariant* ist.

4.29 Satz (Methode von Padoa für Nichtdefinierbarkeitsbeweise).

Vor.: Es gibt Modelle $\mathfrak{A}$ *und* $\mathfrak{B}$ *von* T *und eine eineindeutige Abbildung* φ *von* $U_{\mathfrak{A}}$ *auf* $U_{\mathfrak{B}}$, *so daß gilt:*

(i) *Für jedes Zeichen* $\zeta' \in K$ *ist* φ *ein Isomorphismus bezüglich* ζ'.

(ii) φ *ist kein Isomorphismus bezüglich* ζ.

Beh.: ζ *ist nicht definierbar mittels* K *in* T.

Zum Beweis: Seien $\mathfrak{A}$, $\mathfrak{B}$, φ wie in der Voraussetzung ohne Bedingung (ii), und sei ζ doch definierbar mittels K in T. Es genügt zu zeigen, daß dann φ auch den Begriff ζ erhält. Das ist - wegen der Gültigkeit der gegebenen formalen Definition in $\mathfrak{A}$ und in $\mathfrak{B}$ - anschaulich sofort klar: das, was im definiens δ für ζ (4.2) über die Struktur $\mathfrak{A}$ ausgedrückt wird, überträgt sich wegen Bedingung (i) auch auf $\mathfrak{B}$ und umgekehrt. Das läßt sich präzisieren (wenn $\delta \in \mathrm{Fml}_K(x_1,\ldots,x_r)$) zu

(1) $\models_{\mathfrak{A}} \delta[a_1,\ldots,a_r]$ gdw $\models_{\mathfrak{B}} \delta[\varphi(a_1),\ldots,\varphi(a_r)]$.

Um zu einem strengen Beweis zu kommen, kann man zeigen, daß (1) sogar für jede Formel $\delta \in \mathrm{Fml}_K$ (mit einer beliebigen Anzahl r von freien Variablen und für beliebige Belegungen dieser Variablen) gilt. Das erhält man ohne Schwierigkeiten durch Induktion über den Aufbau von δ. □

4.30 Anmerkungen. (i) 4.29 ergibt sich auch für Theorien höherer Stufe durch einen genauso einfachen Rückgang auf die Definitionen. Für Theorien erster Stufe gilt auch die Umkehrung von 4.29, wie von Beth (1953) gezeigt wurde (s.a. Chang u. Keisler 1973, S. 87 u. 92). Für Theorien erster Stufe ist also die Voraussetzung von 4.29 sogar gleichwertig mit der Behauptung. Bisweilen sagt man dafür, daß man Nichtdefinierbarkeitsbeweise für solche Theorien stets nach der Methode von Padoa "führen kann". Genau genommen besagt der Satz von Beth natürlich nicht, daß man Modelle $\mathfrak{A}$, $\mathfrak{B}$ und eine Abbildung φ der genannten Art stets finden kann, sondern nur, daß sie jedenfalls existieren, wenn die Behauptung von 4.29 gilt.

(ii) Manchmal wird in der Voraussetzung von 4.29 zusätzlich gefordert, daß die Strukturen $\mathfrak{A}$ und $\mathfrak{B}$ die gleiche Trägermenge haben. Das ist keine Beschränkung der Allgemeinheit, da man statt der Struktur $\mathfrak{A}$ stets ein isomorphes Bild mit der Trägermenge $U_{\mathfrak{B}}$ verwenden kann. Ebenso kann man fordern, daß φ die identische Abbildung ist.

4.31 Satz. *Sei n eine beliebige natürliche Zahl mit $n \geq 2$. Dann ist* B *nicht definierbar mittels* D *in* P_n *(der n-dimensionalen euklidischen Geometrie ohne Kreisaxiom).*

Der Beweis beruht im wesentlichen auf dem folgenden Lemma.

4.32 Lemma. *Es gibt einen pythagoreischen Körper $\mathfrak{F}$, der zwei verschiedene Anordnungen $<_1$ und $<_2$ besitzt.*

Beweis (Konstruktion nach Pejas 1961, S. 233): Sei $\mathfrak{F}_0$ ein beliebiger Körper, der zwei verschiedene Anordnungen $<_1$ und $<_2$ besitzt (z.B. $\mathbb{Q}(\sqrt{2})$), und sei $\mathfrak{A}$ die algebraisch abgeschlossene Hülle von $\mathfrak{F}_0$. Für $i=1,2$ sei $\mathfrak{L}_i$ der in $\mathfrak{A}$ liegende reell-abgeschlossene algebraische Erweiterungskörper von $\mathfrak{F}_0$, dessen (durch + und · eindeutig festgelegte) Anordnung die Anordnung $<_i$ fortsetzt (dieser existiert nach den Sätzen über reell-abgeschlossene Körper, s. 3.17), seine Anordnung bezeichnen wir wieder mit $<_i$. Sei schließlich $\mathfrak{F} = \mathfrak{L}_1 \cap \mathfrak{L}_2$ (Durchschnitt beider Körper, d.h. mit der Trägermenge $F = U_{\mathfrak{L}_1} \cap U_{\mathfrak{L}_2}$ und den Operationen von $\mathfrak{A}$, eingeschränkt auf F), und sei $\mathfrak{F}_i$ der angeordnete Körper, der aus $\mathfrak{F}$ durch Hinzunahme der Anordnung $<_i$ (eingeschränkt auf F) entsteht ($i=1,2$). Für jedes Element c von $\mathfrak{f}$, das in beiden Anordnungen positiv ist (d.h. $c>_1 0$ und $c>_2 0$), liegen die Wurzeln $\pm\sqrt{c}$ in $\mathfrak{L}_1$ und in $\mathfrak{L}_2$ (da $\mathfrak{L}_i$ reell-abgeschlossen) und damit in F, d.h., c ist ein Quadrat in $\mathfrak{f}$. Das gilt insbesondere für von Null verschiedene Quadratsummen c, die ja in jeder Anordnung positiv sind. Damit ist $\mathfrak{f}$ pythagoreisch. □

Beweis von 4.31: Seien $\mathfrak{f}$, $<_i$ wie in 4.32 und $\mathfrak{f}_i$ die entsprechenden angeordneten Körper (i=1,2). Da in einem angeordneten Körper die Anordnung schon durch die Menge der positiven Elemente festgelegt ist, sind diese Mengen in $\mathfrak{f}_1$ und $\mathfrak{f}_2$ voneinander verschieden; es gibt also ein Element von $F=\bigcup\mathfrak{f}$, das o.B.d.A. in $\mathfrak{f}_1$ positiv und in $\mathfrak{f}_2$ negativ ist, sei c ein solches. Zur Anwendung der Methode von Padoa setzen wir nun $\mathfrak{A}=\mathfrak{L}_n(\mathfrak{f}_1)$, $\mathfrak{B}=\mathfrak{L}_n(\mathfrak{f}_2)$ (das sind nach I.16.15 Modelle von P_n) und φ gleich der identischen Abbildung von F^n (der Trägermenge von $\mathfrak{A}$ und $\mathfrak{B}$) auf sich. Nach Definition der kartesischen Räume (I.1.4) ist die Streckenkongruenz eines solchen Raumes schon durch Addition und Multiplikation des Grundkörpers festgelegt, also in beiden Modellen $\mathfrak{A}$, $\mathfrak{B}$ dieselbe. Die Zwischenbeziehungen beider Modelle sind dagegen voneinander verschieden. Betrachten wir nämlich die Punkte $o=\langle 0,0,\ldots,0\rangle$, $e=\langle 1,0,\ldots,0\rangle$ und $x=\langle c,0,\ldots,0\rangle$ von F^n (auf der x_1-Achse), so ist $B_{\mathfrak{B}}xoe$ (da $c<_2 0$), aber nicht $B_{\mathfrak{A}}xoe$ (da $c>_1 0$). Somit ist φ ein Isomorphismus von $\mathfrak{A}$ auf $\mathfrak{B}$ bezüglich D, aber nicht bezüglich B. Satz 4.29 liefert also die gewünschte Behauptung. □

4.33 Minimale Stellenzahl von Grundbegriffen. Für die hier betrachteten Geometrien haben wir mit P' bzw. P oder R jeweils einen dreistelligen Begriff gefunden, der als einziger Grundbegriff geeignet ist. Die Stellenzahl 3 läßt sich dabei tatsächlich nicht weiter herabdrücken, solange man nur Punkte als geometrische Objekte und Relation(szeich)en als Grundbegriffe verwendet (vgl. aber 4.73, 76, 93(ii), 93(iii), 101, 104, 110).

Wir beschäftigen uns zunächst mit dem klassischen Beweis dieses Ergebnisses für (möglichst starke, vgl. 4.9(i)) euklidische Geometrien, wobei wesentlich die Existenz von (echten) Ähnlichkeitstransformationen benutzt wird. Später (4.67) wird etwas Analoges auch für andere Geometrien gezeigt.

4.34 Satz (Lindenbaum u. Tarski 1936). *Sei n eine beliebige natürliche Zahl mit $n\geq 2$. Für P_n^2 (die n-dimensionale volle euklidische Geometrie [zweiter Stufe]) gibt es kein geeignetes System von Grundbegriffen, das* *nur aus zweistelligen Relationszeichen (für Relationen zwischen Punkten) besteht.*

Der Beweis beruht auf den folgenden beiden Lemmata.

4.35 Lemma. *Sei K ein geeignetes System von Grundbegriffen für P_n^2, sei $\mathfrak{A}$ das Modell $\mathfrak{L}_n(\mathbb{R})$, und sei φ eine eineindeutige Abbildung von $\mathbb{R}^n$ auf sich. Dann gilt: Alle Begriffe aus K sind invariant unter φ gdw φ eine Ähnlichkeitstransformation ist.*

Beweis: Wie aus der analytischen Geometrie bekannt ist, sind die Ähnlichkeitstransformationen diejenigen Abbildungen des $\mathbb{R}^n$ auf sich, die die Streckenkongruenz (als vierstellige Relation) (und damit auch die Zwischenbeziehung) invariant lassen. Aus 4.29 (in der ursprünglich bewiesenen kontraponierten Form) und der Voraussetzung (s. 4.21) ergibt sich aber sofort, daß eine Abbildung die Begriffe des Systems *K* invariant läßt genau dann, wenn das für das System {D,B} der Fall ist. □

4.36 Lemma. *Auf der Menge $\mathbb{R}^n$ gibt es nur vier zweistellige Relationen, die invariant sind gegenüber Ähnlichkeitstransformationen, und zwar die leere Relation, die Allrelation, die Gleichheitsrelation und die Verschiedenheit.*

Beweis: Sei *R* eine zweistellige Relation auf $\mathbb{R}^n$, die gegenüber Ähnlichkeitstransformationen invariant ist. Zu zwei Paaren von untereinander verschiedenen Punkten gibt es bekanntlich stets eine Ähnlichkeitstransformation, die das erste in das zweite Paar überführt (für $n \geq 2$ sogar mehrere solche Transformationen). Aus der Invarianzeigenschaft ergibt sich somit: Wenn die Relation *R* auf ein Paar von untereinander verschiedenen Punkten zutrifft, so trifft sie auf jedes solche Paar zu. Ebenso ergibt sich: Wenn die Relation *R* auf ein Paar von untereinander gleichen Punkten zutrifft, so trifft sie auf jedes solche Paar zu.

Somit gilt genau einer der folgenden beiden Fälle:
Fall 1: *R* trifft auf kein Paar von untereinander verschiedenen Punkten zu.
Fall 2: *R* trifft auf jedes solche Paar zu.

Unabhängig davon gilt auch genau einer der folgenden beiden Fälle A, B.
Fall A: *R* trifft auf kein Paar von untereinander gleichen Punkten zu.
Fall B: *R* trifft auf jedes solche Paar zu.

Die Fallkombinationen A1, B2, B1, A2 (in dieser Reihenfolge) liefern die im Lemma genannten vier Relationen, und diese vier sind tatsächlich invariant gegenüber Ähnlichkeitstransformationen. □

Beweis von 4.34: Annahme: Sei K ein System der genannten Art. Als zugehörige Relationen im Modell $\mathfrak{A} = \mathcal{L}_n(\mathbb{R})$ (d.h. als die $R_{\mathfrak{A}}$ mit $R \in K$) kommen dann nur die vier in 4.36 genannten Relationen in Frage; diese sind aber nicht nur gegenüber Ähnlichkeitstransformationen invariant, sondern gegenüber beliebigen eineindeutigen Abbildungen des $\mathbb{R}^n$ auf sich im Widerspruch zu 4.35. □

4.37 Parameter in den elementaren Stetigkeitsaxiomen. Aus der (auf andere Modelle von P_n übertragbaren) Charakterisierung der unter Ähnlichkeitstransformationen invarianten zweistelligen Relationen in 4.36 kann man leicht das in I.1.2 angekündigte Resultat erhalten, daß die Parameter in den elementaren Stetigkeitsaxiomen des Schemas A11' nicht entbehrlich sind. Wir zeigen jedoch gleich ein stärkeres Resultat für die absolute Geometrie A.

4.38 Satz. *Vor.: α(x) und β(x) sind für die Bildung eines elementaren Stetigkeitsaxioms zugelassene Formeln der Sprache* L(D,B) *(s. Bedingung (*) vor* A11' *in* I.1.2*), die außer* x *bzw.* y *höchstens eine weitere freie Variable* p *enthalten.*
Beh.: Das (höchstens den Parameter p *enthaltende) zugehörige Stetigkeitsaxiom*
(1) $\exists a \forall x \forall y[\alpha(x) \wedge \beta(y) \rightarrow Baxy] \rightarrow \exists b \forall x \forall y[\alpha(x) \wedge \beta(y) \rightarrow Bxby]$
ist schon ein Satz von A *(und damit als zusätzliches Axiom überflüssig).*

Beweis: Sei $\mathfrak{A}$ ein beliebiges Modell von A. Dann brauchen in $\mathfrak{A}$ keine "echten" Ähnlichkeitstransformationen zu existieren (d.h. keine, die nicht schon Bewegungen sind). In $\mathfrak{A}$ gilt aber:
(2) Ist $pq \equiv rs$, so gibt es eine Bewegung, die p in r und q in s überführt. Das ergibt sich sofort aus dem (in A geführten) Beweis von Satz I.10.12, wo gezeigt wurde, daß sich sogar rechtwinklige Dreiecke mit kongruenten entsprechenden Katheten durch eine Bewegung ineinander überführen lassen.

Wir schreiben die gegebenen Formeln jetzt in der Form $\alpha(p,x)$, $\beta(p,y)$. Sei p ein beliebiger Punkt von $\mathfrak{A}$, und seien X, Y die durch diese Formeln "für den Parameterwert p definierten Punktmengen", d.h. $X=\{x \mid \alpha[p,x]\}$, $Y=\{y \mid \beta[p,y]\}$. Da Streckenkongruenz und Zwischenbeziehung jedenfalls unter Bewegungen invariant sind, gilt dasselbe nach 4.29

auch für die durch α bzw. β (gemäß 4.3(ii)) definierten zweistelligen Relationen. Insbesondere ergibt sich aus $px \equiv px'$ und $x \in X$ auch $x' \in X$. Mit einem Punkt $x \neq p$ enthält X also eine ganze Hyperkugel. Somit ist X höchstens einelementig oder nicht enthalten in einer Geraden. Dasselbe gilt für Y. Somit wird das Stetigkeitsaxiom für diese Mengen trivial (vgl. die Bemerkung hinter A11 in I.1.2). □

4.39 Anmerkung. Tatsächlich erhält man mit zwei Parametern die ersten nicht-trivialen Stetigkeitsaxiome. Für die in P eingeführte Punktrechnung (I.14) läßt sich zum Beispiel die Existenz der Wurzeln $\sqrt[m]{k}$ ($m, k \in \mathbb{N} - \{0\}$) jeweils beweisen mit Hilfe eines elementaren Stetigkeitsaxioms mit den zwei Parametern o und e (für Null- und Einheitspunkt); dagegen ergibt sich diese Existenz i.a. (schon für $m=3$, $\sqrt[3]{k} \notin \mathbb{N}$) nicht in P oder P' allein (vgl. den Beweis von 3.33).

Aus der Existenz von echten affinen Abbildungen in der euklidischen Geometrie ergibt sich

4.40 Satz. *Die Streckenkongruenz* D *ist nicht definierbar mittels* B *(also erst recht nicht mittels* Col*) in* P_n^2.

Beweis: Sei $\mathfrak{A} = \mathfrak{B} = \mathcal{L}_n(\mathbb{R})$, und sei φ eine affine Abbildung von $\mathfrak{A}$ auf sich, die keine Ähnlichkeitstransformation ist. Dann erhält φ bekanntlich die Zwischenbeziehung (und damit die Kollinearität), aber nicht die Streckenkongruenz. Damit ergibt sich nach 4.29 die Behauptung. □

4.41 Vollsymmetrische Rechtwinkelbeziehung. Nachdem gezeigt ist, daß zweistellige Grundbegriffe nicht ausreichen, bleibt noch die Möglichkeit, nach einfacheren dreistelligen Grundbegriffen zu suchen. Als einen solchen kann man die vollsymmetrische Rechtwinkelbeziehung S (4.12(5)) ansehen, da sie invariant ist gegenüber Permutationen der Argumente. Scott (1956) zeigte für die ebene euklidische Geometrie und Henkin (1962) für die euklidische Geometrie beliebiger Dimension $n \geq 2$ (ursprünglich über $\mathbb{R}$), daß man S als einzigen Grundbegriff verwenden kann. Wir formulieren ihre Ergebnisse hier für möglichst schwache Geometrien.

4.42 Satz. *In* A *gilt als Satz* (Scott)*:*
(1) $\mathrm{R}abc \leftrightarrow a=b \vee \forall d[\mathrm{Col}\,abd \rightarrow \mathrm{S}dbc]$,
folglich ist R *(und damit auch* D, P, P^*, M *s.* 4.20*) definierbar mittels* S *und* Col *in* A.

Beweis: Die Richtung (→) ist klar (Abb. 132, vgl. I.8.3).

Zu (←): Gelte die rechte Seite von (1). Falls $a=b \vee c=b$ ist, gilt die Behauptung Rabc trivialerweise (I.8.5, 8.2). Gelte also jetzt $a \neq b \wedge c \neq b$. Durch Anwendung der Voraussetzung auf den speziellen Punkt $d=a$ erhält man zunächst Sabc.

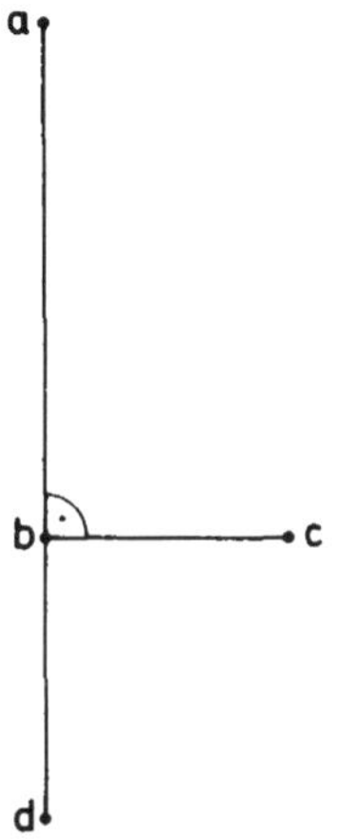

Abb. 132

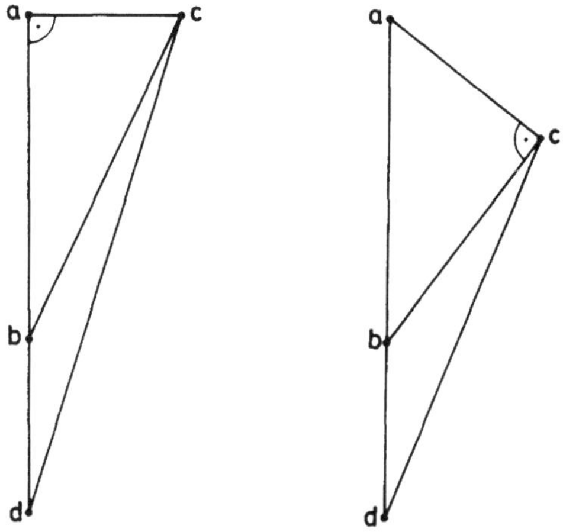

Abb. 133

Annahme: Rcab ∨ Rbca (d.h., es gilt eine der beiden Rechtwinkelbeziehungen, die nach Definition von Sabc außer der Behauptung möglich sind, Abb. 133). Sei dann d ein Punkt mit Babd ∧ $d \neq b$. Falls $a \neq c$ ist, bilden a, b, c ein nicht-entartetes rechtwinkliges Dreieck mit dem rechten Winkel bei a oder bei c, somit ist der Winkel $\sphericalangle abc$ spitz (I.11.43). Falls $a=c$ ist, ist $\sphericalangle abc$ als Nullwinkel natürlich ebenfalls spitz. Der zugehörige Nebenwinkel $\sphericalangle dbc$ ist damit stumpf, und das Dreieck dbc enthält keine rechten Innenwinkel, d.h., es ist ¬Sdbc im Widerspruch zur Voraussetzung. □

4.43 Satz. *In* A_2 *gilt als Satz* (Scott)*:*

(1) Col abc ↔ ∀u∃v[u≠v ∧ Suva ∧ Suvb ∧ Suvc],

folglich sind Col *und damit auch* R *(4.42),* D*,* P*,* P**,* M *(4.20) definierbar mittels* S *in* A_2.

4.44 Lemma. *In* A *gilt: Die Punktmenge*

(1) $\hat{S}(ab) := \{x \mid Sabx\}$

(Abb. 134) besteht, falls $a \neq b$ ist, aus den beiden "Hyperebenen", die in a bzw. in b auf L(ab) *senkrecht stehen, und gewissen Punkten zwischen diesen Hyperebenen (d.h. Punkten, die für jede dieser beiden Hyperebenen auf derselben Seite von ihr liegen wie die Punkte der anderen Hyperebene).*

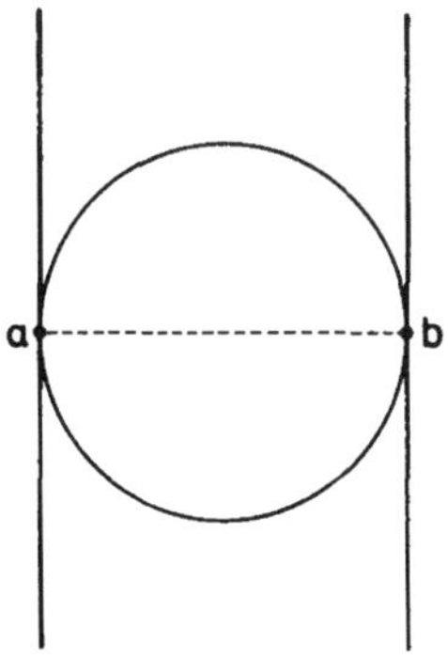

Abb. 134

Beweis und Anmerkung: Die genannten Hyperebenen seien hierbei einfach definiert als die Menge der Punkte x mit Rbax bzw. Rabx. Nach dem Satz I.11.66 über Mittelloträume sind diese Hyperebenen jedenfalls affine Unterräume, deren Kompositum mit L(ab) der ganze Raum ist; falls der ganze Raum die endliche Dimension n hat, sind es $(n-1)$-dimensionale Unterräume (für $n=2$ also Geraden). Die übrigen Elemente von $\hat{S}(ab)$ sind die Punkte x mit Raxb. In der euklidischen Geometrie sind das natürlich die Punkte der "Hypersphäre" mit dem Durchmesser ab (d.h. $\{x \mid mx \equiv ma\}$ wobei m der Mittelpunkt von ab), für $n=2$ insbesondere die Punkte des Thales-Kreises. Aber auch in der absoluten Geometrie A ergibt sich aus Raxb und (o.B.d.A.) $x \neq a,b$, daß der Höhenfußpunkt im rechtwinkligen Dreieck axb zwischen a und b liegt (I.11.47) und damit das Gewünschte. □

Beweis von 4.43: Zu $(\rightarrow)$: Sei A eine Gerade durch a, b, c, und sei u ein beliebiger Punkt. Falls u nicht auf A liegt, sei v der Fußpunkt des Lots

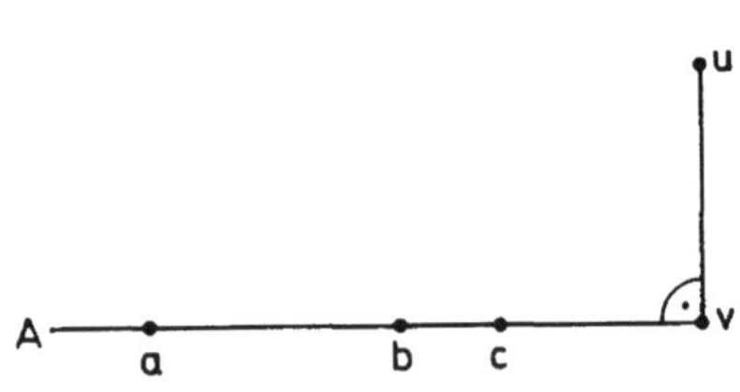

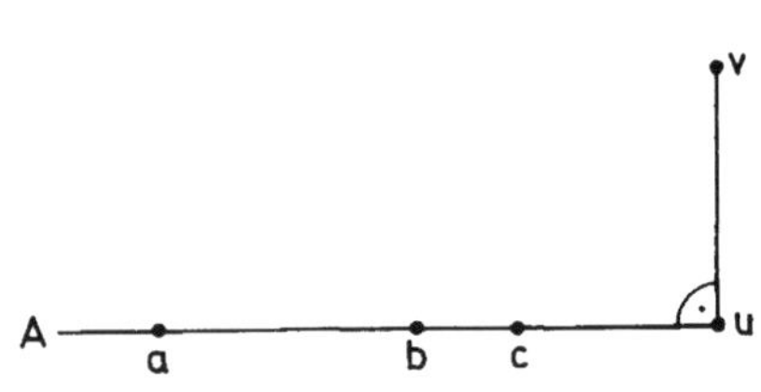

Abb. 135

von u auf A; falls u auf A liegt, sei v ein von u verschiedener Punkt auf der in u auf A errichteten Senkrechten (Abb. 135). Dann leistet v offenbar das Verlangte.

Zu (←): Annahme: ¬ Col abc (Abb. 136). Sei dann u ein Punkt innerhalb des Dreiecks abc (d.h. für jede Ecke des Dreiecks liegt u mit dieser auf derselben Seite der Verbindungsgeraden der anderen beiden Ecken).

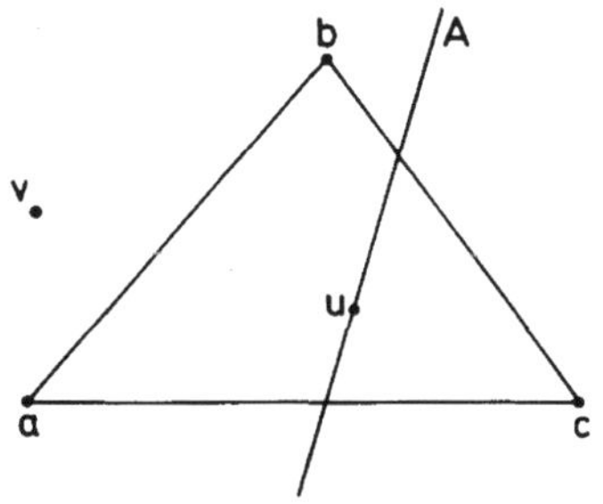

Abb. 136

Für jede durch u gehende Gerade A gibt es dann (wenigstens ein Paar von) zwei Dreiecksecken, die auf entgegengesetzten Seiten von A liegen. (Das erhält man leicht mittels I.9; die Ausführung sei dem Leser überlassen, s.a. 4.48.) Ist nun v ein beliebiger von u verschiedener Punkt, so sei A die in $\hat{S}(uv)$ enthaltene Gerade, die durch u geht. Nach 4.44 liegen alle Punkte von $\hat{S}(uv)$ auf A oder mit v auf derselben Seite von A. Demnach können die Dreiecksecken a, b, c nicht alle in $\hat{S}(uv)$ liegen. Zu dem gewählten u gibt es somit keinen Punkt v der in 4.43(1) genannten Art, d.h., die rechte Seite von 4.43(1) wird falsch, wie zu zeigen. □

4.45 Satz. *Sei* Str *durch folgende formale Definition eingeführt:*

(1) $\mathrm{Str}_{ab}x \leftrightarrow a \neq b \wedge \exists x'[\mathrm{B}ax'b \wedge (ab \perp xx' \vee x = x')]$

(x liegt in dem durch a und b bestimmten "Streifen", d.h. auf einer der in 4.44 genannten Hyperebenen oder dazwischen; Abb. 137).

Dann gelten in P' (euklidische Geometrie mit Kreisaxiom) als Sätze (Henkin)*:*

(2) $\mathrm{Str}_{ab}x \leftrightarrow \forall u\{u \neq x \rightarrow \exists v[\mathrm{S}abv \wedge \mathrm{S}xuv]\}$,

(3) $\mathrm{R}abc \leftrightarrow a = b \vee c = b \vee \{\mathrm{S}abc \wedge \neg\exists u \forall x[\mathrm{Str}_{ac}x \leftrightarrow \mathrm{Str}_{bu}x]\}$.

Folglich sind Str *und* R *definierbar mittels* S *in* P'*, und damit* (4.24, 20) *ist* S *als einziger Grundbegriff für* P' *geeignet.*

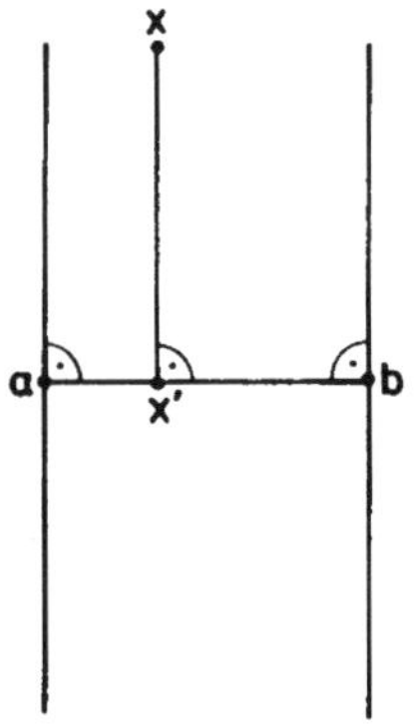

Abb. 137

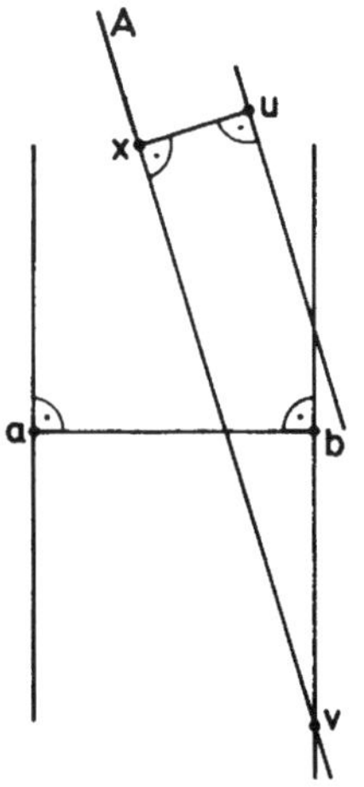

Abb. 138

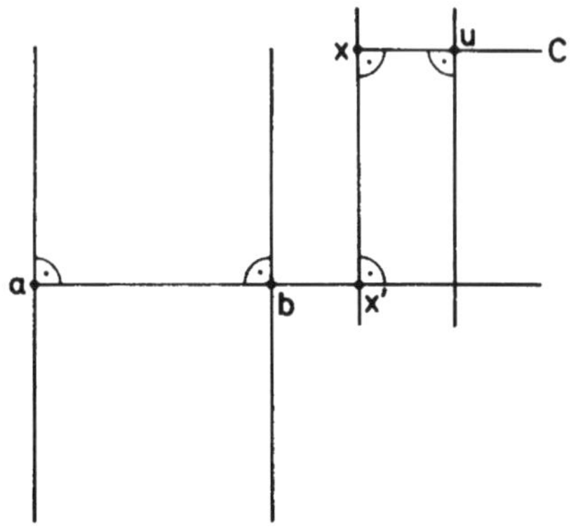

Abb. 139

Beweis: Zu (2)(→): Gelte $\mathrm{Str}_{ab}x$, und sei u ein beliebiger von x verschiedener Punkt. Sei A die durch x gehende Hyperebene von $\hat{S}(xu)$. Falls $L(ab)$ nicht senkrecht auf A steht (Abb. 138), schneidet A die Hyperebenen von $\hat{S}(ab)$, hat also mit $\hat{S}(ab)$ wenigstens einen gemeinsamen Punkt v. Falls $L(ab)$ senkrecht auf A steht (Abb. 137), hat A nach dem Kreisaxiom wenigstens einen gemeinsamen Punkt v mit der Hypersphäre von $\hat{S}(ab)$. Dann leistet v das Verlangte. (Zur bequemeren Anwendung der Sätze aus I.12 kann man statt A auch den Durchschnitt dieser Hyperebene mit einer Ebene durch a, b und x betrachten.)

Zu (2)(←) (Beweis in A): Sei x ein Punkt außerhalb des durch a und b bestimmten Streifens (Abb. 139). Es genügt zu zeigen, daß die rechte Seite von (2) falsch wird. Falls x auf $L(ab)$ liegt, sei $C=L(ab)$. Sonst sei x' der Fußpunkt des Lots von x auf $L(ab)$ und C die im Punkte x auf der Geraden $L(xx')$ in der Ebene $Pl(abx)$ errichtete Senkrechte. Sei dann u ein von x verschiedener Punkt von C derart, daß u und a auf entgegengesetzten Seiten von x bzw. von $L(xx')$ liegen. Man erkennt dann leicht, daß $L(ab)$ sogar senkrecht steht auf der durch x gehenden Hyperebene von $\hat{S}(xu)$ (vgl. I.11.59 und die folgenden Sätze, Abb. 75). Damit haben die durch a, b und durch x, u bestimmten Streifen und erst recht die Mengen $\hat{S}(ab)$ und $\hat{S}(xu)$ keinen gemeinsamen Punkt v, wie zu zeigen.

Zu (3)(Beweis in P): Falls $a=b \vee c=b$ ist, gelten beide Seiten. Sei also jetzt $a \neq b \wedge c \neq b$.

Zu (→): Wegen $Rabc$ gilt zunächst $Sabc$, und b ist ein "innerer Punkt" des durch a, c bestimmten Streifens (Abb. 140 mit $b=b_1$). Ist u ein beliebiger Punkt, so besteht der durch b und u bestimmte Streifen, falls $u \neq b$ ist, nur aus Punkten, die auf einer durch b gehenden Hyperebene oder

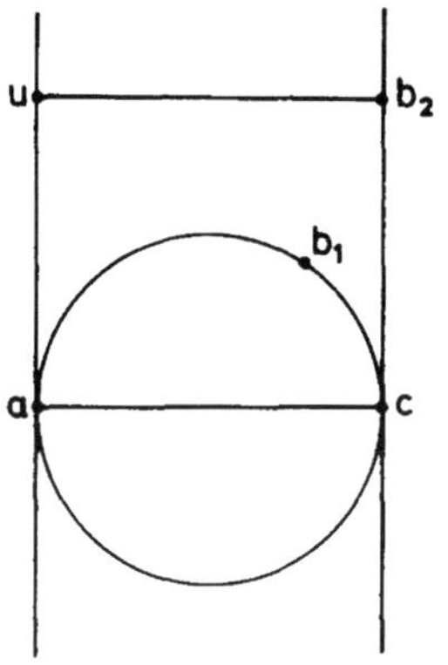

Abb. 140

auf einer gegebenen Seite dieser Hyperebene liegen, falls $u=b$ ist, ist dieser "Streifen" die leere Menge ($\mathrm{Str}_{bu}x$ trifft nie zu). Der genannte Streifen ist also in jedem Fall verschieden von dem durch a, c bestimmten Streifen, wie zu zeigen.

Zu ($\leftarrow$): Gelte der Bestandteil von (3) in geschweiften Klammern, Dann ist zunächst auch $a \neq c$ (sonst würde die $\leftrightarrow$-Beziehung mit $u=b$ gelten).

Annahme: $\neg Rabc$. Aus $Sabc$ ergibt sich dann, daß b auf einer der beiden Hyperebenen von $\hat{S}(ac)$ liegt (Abb. 140 mit $b=b_2$). Sei dann u der Punkt auf der anderen Hyperebene, so daß $bu \| ac$ ist. Man erhält dann leicht, daß b, u denselben Streifen bestimmen wie a, c im Widerspruch zur Voraussetzung. □

4.46 Anmerkungen. (i) In Henkin 1962 wird zusätzlich eine formale Definition der Pierischen Relation P mittels S (kürzer als auf Grund unserer bisherigen Sätze) gegeben. Außerdem wird gezeigt, daß man auch einen anderen vollsymmetrischen Grundbegriff H als einzigen Grundbegriff der euklidischen Geometrie verwenden kann, der eingeführt wird durch

(1) $Habc \leftrightarrow Pabc \vee Pbca \vee Pcab$

(a, b, c bilden in irgendeiner Reihenfolge ein gleichschenkliges Dreieck oder sind nicht paarweise verschieden).

Beide Resultate ergeben sich wiederum schon für P'.

(ii) In 4.45 wurde das Parallelenaxiom (oder etwas dazu Äquivalentes) wesentlich benutzt, ebenso in 4.43 das Dimensionsaxiom Dim_2 (I.11.70). Die Betrachtungen aus 4.43 lassen sich jedoch folgendermaßen auf die absolute Geometrie beliebiger Dimension n ausdehnen.

4.47 Satz. *Sei n eine beliebige natürliche Zahl mit n≥2. Dann gelten in* A_n *(der n-dimensionalen absoluten Geometrie) als Sätze (vgl. Def.* I.9.46*):*

(1) $Ah_n a_0 a_1 \dots a_n \leftrightarrow \forall u \exists v [u \neq v \wedge \bigwedge_{\nu=0}^{n} Suva_\nu]$,

(2) $Col\, abc \leftrightarrow \forall a_3 \dots \forall a_n Ah_n abca_3 \dots a_n$

(für n=2 sollen die Variablen $a_3, \dots, a_n$ *mit den zugehörigen Quantoren ganz wegfallen).*

Folglich sind die affine Abhängigkeit Ah_n *von n+1 Punkten, die Kollinearität und damit (nach* 4.42*) auch* R, D, P, P*, M *definierbar mittels* S *in* A_n.

Als Hilfsmittel aus der absoluten Geometrie verwenden wir

4.48 Lemma. *In* A *gilt: Sei k eine natürliche Zahl mit k≥*1*. Dann gilt für jeden "k-dimensionalen Simplex"* $a_0 a_1 \dots a_k$ *(d.h. für jede Menge von k+*1 *affin unabhängigen Punkten* $a_0, a_1, \dots, a_k$*):*

(i) $a_0 \dots a_k$ *hat "innere Punkte", d.h. Punkte u derart, daß u mit jeder "Ecke"* a_i *des Simplex auf derselben Seite des von den übrigen Ecken* a_j (0≤*j*≤*k*, *j*≠*i*) *aufgespannten Unterraumes liegt.*

(ii) *Wenn jede der Ecken* $a_0, \dots, a_k$ *auf einer vorgegebenen Seite eines Unterraumes A oder bis auf eine Ecke in diesem Unterraum liegt, so liegt jeder innere Punkte des Simplex auf der vorgegebenen Seite von A.*

Zum Beweis, Anmerkung: Mit den Hilfsmitteln von Abschnitt I.9 erhält man dieses Lemma durch Induktion über *k*. Die Existenz eines *k*-dimensionalen Simplex ist natürlich (nach Def. I.11.70) gleichbedeutend mit der Gültigkeit des unteren Dimensionsaxioms Dim_k^- ("die Dimension des ganzen Raumes ist wenigstens *k* "). Die Existenz eines Simplex wie in (ii) bedeutet dann, daß außerdem *A* wenigstens (*k*-1)-dimensional ist.

Beweis von 4.47: Zu (1)(→): Das ergibt sich wie in 4.43, wobei *A* jetzt eine Hyperebene durch die Punkte $a_0, \dots, a_n$ ist.

Zu (1)(←): Annahme: $\neg Ah_n a_0 a_1 \dots a_n$. Sei dann *u* (gemäß 4.48(i)) ein innerer Punkt des *n*-dimensionalen Simplex $a_0 a_1 \dots a_n$. Ist nun *v* ein beliebiger von *u* verschiedener Punkt, so sei *A* die in S(*uv*) enthaltene Hyperebene, die durch *u* geht. Dann ist *A* ein (*n*-1)-dimensionaler Unterraum, und nach Annahme liegen nicht alle Ecken des betrachteten Simplex in *A*. Nach 4.48(ii) (dessen Behauptung für den inneren Punkt *u* falsch wird) gibt es also (wenigstens ein Paar von) zwei Simplexecken, die auf entgegengesetzten Seiten von A liegen. Andererseits liegen nach 4.44 alle Punkte von $\hat{S}(uv)$ in A oder mit v auf derselben Seite von A. Demnach können die Simplexecken $a_0, \ldots, a_n$ nicht alle in $\hat{S}(uv)$ liegen. Somit wird wieder die rechte Seite von (1) falsch, wie zu zeigen.

Zu (2): Col abc ist gleichbedeutend mit $\mathrm{Ah}_2 abc$ (I.9.46(3)). Für $n=2$ ist damit (2) trivial, und (1) fällt mit 4.43(1) zusammen.

Zu (→): Das ergibt sich daraus, daß die affine Abhängigkeit bei Hinzunahme weiterer Punkte erhalten bleibt (I.9.49).

Zu (←): Annahme: ¬ Col abc. Dann spannen a, b, c einen zweidimensionalen Unterraum (eine Ebene) auf. Aus den Sätzen über affine Unterräume (I.9.) und der Voraussetzung Dim_n^- ergibt sich, daß man sukzessive weitere Punkte $a_3, \ldots, a_n$ hinzunehmen kann, so daß sich jedesmal die Dimension des von den betrachteten Punkten aufgespannten Raumes um eins vergrößert bis zur Dimension n. Damit wird die rechte Seite von (2) falsch. □

4.49 Anmerkung. Man erkennt eine Analogie zwischen 4.47 und der Methode aus Anmerkung 4.15 (s.a. 4.16): Zuerst werden in A_n die $(n-1)$-dimensionalen Unterräume bzw. der Begriff Ah_n charakterisiert und danach - mit Hilfe dieser Charakterisierung - die Kollinearität. Damit erhält man für verschiedene Dimensionen wieder verschiedene formale Definitionen für Col. Der Versuch, auch 4.17 vom Grundbegriff P auf S zu übertragen, ist meines Wissens bisher nicht gelungen. Damit erhebt sich

4.50 Problem. Sind die Begriffe Col, R, D, P, P^* schon in A definierbar mittels S, mit anderen Worten, gibt es eine einheitliche - d.h. d i m e n s i o n s u n a b h ä n g i g e - formale Definition für einen dieser Begriffe mittels S in A? Wenn einer dieser Begriffe in diesem Sinne definierbar ist, sind es nach 4.42, 4.20 auch die anderen. Um zu zeigen, daß eine als formale Definition geeignete Formel ein Satz von A ist, genügt es nach 3.75 (für $T=A$) zu zeigen, daß sie in allen endlichdimensionalen Modellen von A gilt (obwohl A auch unendlichdimensionale Modelle besitzt); es genügt sogar der entsprechende Nachweis für alle Dimensionen bis zu einer gewissen Schranke n, die allerdings von der Formel abhängt. Ein eventueller Nichtdefinierbarkeitsbeweis nach der Methode von Padoa (4.29, 30) könnte dagegen nur geführt werden mit einem unendlichdimensionalen Modell (wegen 4.47), das nicht Modell von P' ist (wegen 4.45).

4.51 Dimensionsabhängigkeit bei der Gleichseitigkeit. In allen bisherigen Resultaten erfolgte die Feststellung, daß ein gewisser Begriff mittels anderer definierbar bzw. nicht definierbar ist, noch einheitlich für alle Dimensionen $n\geq 2$ (auch bei unterschiedlichen formalen Definitionen). Zu einer anderen Situation kommen wir jetzt bei dem ebenfalls vollsymmetrischen Begriff E der Gleichseitigkeit (4.12(4)).

4.52 Satz (Beth u. Tarski 1956). (i) *In* $P'_{\geq 3}$ *(der euklidischen Geometrie von Dimension ≥3 mit Kreisaxiom) gilt als Satz*

(1) $\text{Pabc} \leftrightarrow \exists d \exists e[\text{Eabd} \wedge \text{Eace} \wedge \text{Eade}]$;

folglich ist P *definierbar mit Hilfe von* E *in* $P'_{\geq 3}$, *und damit (4.24) ist* E *als einziger Grundbegriff für* $P'_{\geq 3}$ *geeignet.*

(ii) *Die Begriffe* Col, P, R, D *sind nicht definierbar mittels* E *in* P_2^2 *(der zweidimensionalen euklidischen Geometrie zweiter Stufe; dasselbe gilt dann auch* [4.9(i)] *für die Theorie* P_2^* *erster Stufe und die Untertheorien davon, ebenso für die absolute Geometrie* A_2^2*).*

Beweis von (i): Zu (←)(Beweis in *A*): Seien *d*, *e* Punkte der angegebenen Art. Aus den mit E ausgedrückten Kongruenzen ergibt sich unmittelbar das Gewünschte.

Zu (→): Zur Veranschaulichung betrachten wir zunächst ein "ebenes Netz" der Figur, die von *a*, *b*, *c* und den gesuchten Punkten *d*, *e* gebildet werden soll (Abb. 141, die entsprechenden Punkte des Netzes sind durch Anhängen eines Striches bezeichnet): Das Netz besteht aus drei gleich-

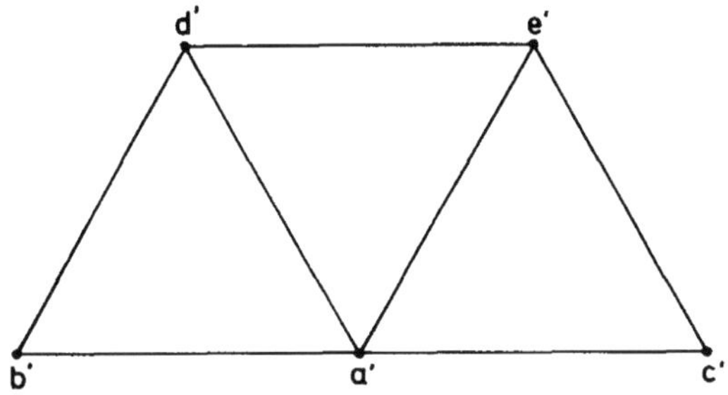

Abb. 141

seitigen Dreiecken, deren Seiten alle kongruent zu *ab* (und damit auch zu *ac*) sind. Nach dem Satz über die Winkelsumme im Dreieck (I.12.23) sind *a'*, *b'*, *c'* kollinear, damit ist B*b'a'c'*. Es ist anschaulich klar, daß man durch "Falten" des Netzes entlang der Kanten L(*a'd'*) und L(*a'e'*) die Punkte *b'* und *c'* überführen kann in Punkte mit einem beliebig vorgegebenen Abstand kleiner-gleich $b'c'$, insbesondere also in Punkte b'', c'' mit dem Abstand bc; am einfachsten ist dabei wohl eine "symmetrische Faltung" in einem dreidimensionalen affinen Unterraum des ganzen Raumes (d.h., daß b'', c'' symmetrisch zur Mittelebene von $d'e'$ liegen). Zu den gegebenen Punkten a, b, c und einem sie enthaltenden dreidimensionalen Unterraum U kann man dann Punkte d, e der gewünschten Art in U erhalten durch kongruente Übertragung der Figur aus den Punkten a', b'', c'', d', e'. Man kann sie auch direkt konstruieren als Schnittpunkte von geeigneten Kreisen mit Geraden oder - auf Grund des Darstellungssatzes für U (I.16.15(i), (iv)) - durch Bestimmung ihrer Koordinaten in einem geeigneten Koordinatensystem erhalten. Eine genaue Durchführung sei dem Leser überlassen. □

4.53 Anmerkung. Es sei noch darauf hingewiesen, daß 4.52(1) kein Satz der absoluten Geometrie ist. Gegenbeispiele liefert wieder die hyperbolische Geometrie H, in der folgendes gilt: Die Summe der Innenwinkel im Dreieck ist kleiner als ein gestreckter Winkel, somit sind im oben betrachteten Netz die Punkte a', b', c' nicht kollinear (für genügend große Seiten ab lassen sich sogar die untereinander kongruenten Innenwinkel im gleichseitigen Dreieck und damit auch $\sphericalangle b'a'c'$ beliebig klein machen); es existieren also keine Punkte der gewünschten Art, wenn $b'a'c' < bac$ ist.

Beweis von 4.52(ii): Nach 4.20 genügt der Beweis für die Nichtdefinierbarkeit von Col. Zur Anwendung der Methode von Padoa (4.29) betrachten wir als Modell $\mathfrak{A}$ von P_2^2 die Gaußsche Zahlenebene mit der üblichen geometrischen Interpretation (d.h. das isomorphe Bild von $\mathfrak{L}_2(\mathbb{R})$ unter der Abbildung χ von $\mathbb{R}^2$ auf die Menge $\mathbb{C}$ der komplexen Zahlen mit $\chi(x_1,x_2) = x_1+x_2 i$). Für ein gleichseitiges Dreieck mit den Ecken 0 und 1 berechnet man leicht als Möglichkeiten für die dritte Ecke die beiden komplexen Zahlen $\frac{1}{2} \pm \frac{\sqrt{3}}{2} i$. An Hand der bekannten Beschreibung von Drehungen durch Multiplikation mit Einheitsvektoren erhält man dann die folgende Charakterisierung der Relation der Gleichseitigkeit in $\mathfrak{A}$:

$$(2) \quad \mathrm{E}_{\mathfrak{A}}xyz \quad \text{gdw} \quad z = x\cdot(\tfrac{1}{2} - \tfrac{\sqrt{3}}{2} i) + y\cdot(\tfrac{1}{2} + \tfrac{\sqrt{3}}{2} i) \quad \text{oder}$$
$$z = x\cdot(\tfrac{1}{2} + \tfrac{\sqrt{3}}{2} i) + y\cdot(\tfrac{1}{2} - \tfrac{\sqrt{3}}{2} i) \quad .$$

Gesucht ist eine Abbildung φ, die E aber nicht Col erhält. Damit wird die folgende Konstruktion motiviert.

Sei $\mathfrak{K}$ der Unterkörper $\mathbb{Q}(\sqrt{3},i)$ von $\mathbb{C}$. Dann bildet $\mathbb{C}$ über $\mathfrak{K}$ einen (unendlichdimensionalen, vgl. 2.8) Vektorraum $\mathfrak{V}$. Sei c eine reelle Zahl, die nicht in $\mathfrak{K}$ liegt, und d eine komplexe Zahl, die nicht reell ist und nicht in $\mathfrak{K}(c)$ liegt. Dann sind 1, c, d linear unabhängige Vektoren von $\mathfrak{V}$. Sei B eine (auf Grund des Auswahlaxioms existierende) Basis von $\mathfrak{V}$, die 1, c und d als Elemente enthält. Sei ψ eine eineindeutige Abbildung von B auf sich mit

(3) $\psi(c)=d,\ \psi(d)=c,\ \psi(u)=u$ für jedes $u\in B$ mit $u\neq c,d$.

Sei schließlich φ die lineare Abbildung von $\mathfrak{V}$ auf sich, die die Abbildung ψ der Basis fortsetzt, d.h.

(4) $\varphi(\sum_{\mu=0}^{m} c_\mu \cdot u_\mu) = \sum_{\mu=0}^{m} c_\mu \cdot \psi(u_\mu)$ $(m\in\mathbb{N},\ c_\mu\in\mathfrak{K},\ u_\mu\in B$ für $\mu\leq m)$.

Dann ist φ eine eineindeutige Abbildung von $\mathbb{C}=U_{\mathfrak{A}}$ auf sich. Die in (2) auftretenden Bestandteile in Klammern sind Elemente von $\mathfrak{K}$; wegen der Linearität ist φ also ein Isomorphismus bezüglich E. φ führt jedoch die in $\mathfrak{A}$ kollinearen Punkte 0, 1, c über in die nicht kollinearen Punkte 0, 1, d und ist damit kein Isomorphismus bezüglich Col. Nach der Methode von Padoa (4.29 mit $\mathfrak{B}=\mathfrak{A}$) ergibt sich also das Gewünschte. □

Eine Verallgemeinerung der gerade angestellten Überlegung führt zu dem folgenden Satz.

4.54 Satz (Tarski 1956). *Sei R ein dreistelliges Relationszeichen, das als einziger Grundbegriff für P_2^2 (oder eine Untertheorie davon, vgl. 4.9(i)) geeignet ist. Sei $\mathfrak{A}$ das Modell von P_2^2 aus dem Beweis von 4.52(ii) (Gaußsche Zahlenebene). Seien a, b verschiedene komplexe Zahlen, und sei M die Menge der komplexen Zahlen z mit $R_{\mathfrak{A}}abz$. Dann ist der von $M\cup\{a,b\}$ erzeugte Körper der Körper $\mathbb{C}$ aller komplexen Zahlen.*

Beweis: Für beliebige komplexe Zahlen x, y mit $x\neq y$ gibt es eine Ähnlichkeitstransformation $\vartheta_{x,y}$ von $\mathfrak{A}$, die a in x und b in y überführt; eine solche wird gegeben durch

(1) $\vartheta_{x,y}(u) = x+\frac{u-a}{b-a}\cdot(y-x) = \frac{b-u}{b-a}\cdot x+\frac{u-a}{b-a}\cdot y$.

Nach Lemma 4.35 ist $R_{\mathfrak{A}}$ invariant gegenüber Ähnlichkeitstransformationen. Somit gilt:

(i) Für beliebige x, y, z mit $x \neq y$ ist $R_{\mathfrak{A}}xyz$ gdw es ein $c \in M$ gibt mit

$$z = \frac{b-c}{b-a} \cdot x + \frac{c-a}{b-a} \cdot y \,.$$

(Somit wird die Relation $R_{\mathfrak{A}}$, abgesehen von den Tripeln $\langle x,y,z \rangle$ mit $x=y$, schon durch die Menge M festgelegt.)

Sei nun $\mathfrak{f}$ der von $M \cup \{a,b\}$ erzeugte Körper. Dann bildet $\mathbb{C}$ über $\mathfrak{f}$ einen (wenigstens eindimensionalen) Vektorraum $\mathfrak{V}$. Sei B eine Basis von $\mathfrak{V}$, die 1 als Element enthält. Sei ψ die eineindeutige Abbildung von B in $\mathfrak{V}$ mit

(2) $\psi(1)=2$, $\psi(u)=u$ für jedes $u \in B$ mit $u \neq 1$.

Sei wieder φ die lineare Abbildung von $\mathfrak{V}$ auf sich, die die Abbildung ψ fortsetzt (wie in (4) im Beweis von 4.52(ii)). Wir zeigen nun, daß $R_{\mathfrak{A}}$ auch invariant gegenüber φ ist, d.h.

(ii) für beliebige x, y, $z \in \mathbb{C}$ ist $R_{\mathfrak{A}}xyz$ gdw $R_{\mathfrak{A}}\varphi(x)\varphi(y)\varphi(z)$.

Falls $x \neq y$ ist, ergibt sich das schon aus (i), da die dort auftretenden Brüche in $\mathfrak{f}$ liegen. Zur Behandlung des Falles $x=y$ genügt es, an Stelle von $R_{\mathfrak{A}}$ die zweistellige Relation $Q_{\mathfrak{A}}$ mit $Q_{\mathfrak{A}}xz$ gdw $R_{\mathfrak{A}}xxz$ zu betrachten. Mit $R_{\mathfrak{A}}$ ist auch $Q_{\mathfrak{A}}$ invariant gegenüber Ähnlichkeitstransformationen, also nach 4.36 gleich einer der vier dort genannten Relationen und damit invariant gegenüber beliebigen eineindeutigen Abbildungen, insbesondere auch gegenüber φ. Somit gilt (ii). Nach 4.35 ist also auch φ eine Ähnlichkeitstransformation.

Annahme: $\mathfrak{f} \neq \mathbb{C}$. Dann ist die Dimension von $\mathfrak{V}$ wenigstens 2, und es gibt ein von 1 verschiedenes Element u in B. Aus der Linearität von φ und (2) ergibt sich

(3) $\varphi(0)=0$, $\varphi(1)=2$, $\varphi(u)=u$ für ein u mit $u \neq 0,1$.

Das bedeutet aber, daß φ keine Ähnlichkeitstransformation sein kann, im Widerspruch zum Obigen. □

4.55 Korollare (Tarski 1956). (i) *Unter den Voraussetzungen von Satz* 4.54 *hat M die Mächtigkeit des Kontinuums.*

(ii) *Die folgenden Begriffe sind jeweils nicht geeignet als einziger Grundbegriff für* P_2^2: E, Col *und der Begriff* K *mit* $K_{\mathfrak{A}}xyz$ *gdw* z *aus* x *und* y *durch Konstruktionen mittels Zirkel und Lineal erhalten werden kann.*

Beweis: (i) ergibt sich daraus, daß der von einer unendlichen Menge erzeugte Körper stets dieselbe Mächtigkeit hat wie diese Menge.

Zu (ii): Für jeden der genannten Begriffe betrachten wir die zugehörige Menge M aus 4.54.

Für E ergibt sich das Gewünschte sofort aus (i); die Menge M besteht nämlich stets (für beliebig gewählte a, b) nur aus zwei Punkten.

Für Col: Für $a=0$, $b=1$ besteht M und damit auch der erzeugte Körper nur aus den reellen Zahlen; damit ist die Behauptung von 4.54 verletzt.

Für K: Wir benutzen das bekannte Resultat, daß sich aus endlich vielen Punkten nur abzählbar viele weitere mittels Zirkel und Lineal konstruieren lassen. Damit ist M stets abzählbar und die Behauptung von (i) verletzt. □

4.56 Anmerkungen. (i) Die obigen Resultate für E und Col sind natürlich schon enthalten in 4.52(ii) bzw. 4.40. Bemerkenswert an diesem zweiten Beweis ist die gemeinsame Methode, mit der sich Resultate ergeben, die sonst auf recht unterschiedliche Art bewiesen werden. In Tarski 1956 wird noch bemerkt, daß sich 4.54 und 4.55(i) in geeigneter Form auf eine beliebige Dimension $n\geq 1$ und auf Systeme von Begriffen mit Stellenzahlen ≥ 3 ausdehnen lassen. Auf diesem Wege ergibt sich dann auch ein weiterer Beweis für das Resultat von Lindenbaum, daß die Pierische Relation P nicht als einziger Grundbegriff für die eindimensionale Geometrie geeignet ist (in diesem Falle besteht M wieder nur aus zwei Punkten).

(ii) Das in den Beweisen von 4.52(ii) und 4.54 verwendete Auswahlaxiom wird nach Beth u. Tarski 1956 (S. 466, Fußnote [4])) nicht benötigt, um für die genannten Theorien erster Stufe dieselben Resultate zu erhalten. Dann kann man nämlich statt $\mathbb{C}$ die Ebene der algebraischen Zahlen als Modell von P_2^* betrachten; der entsprechende Vektorraum $\mathfrak{V}$ ist dann abzählbar, und die Existenz einer Basis B mit den angegebenen Eigenschaften läßt sich ohne Auswahlaxiom zeigen. Für geeignete "syntaktische Theorien" höherer Stufe genügt statt des Auswahlaxioms selbst die von Gödel 1940 gezeigte relative Verträglichkeit dieses Axioms. In solchen syntaktischen Theorien sind die "Sätze" (vgl. Def. 4.2 mit Anm. 4.3(vi)) festgelegt als diejenigen Formeln, die sich aus gegebenen logischen, mengentheoretischen und eigentlich geometrischen Axiomen mit Hilfe gegebener Schlußregeln ableiten (formal beweisen) lassen. Das 4.52(ii) bzw. 4.54, 55 entsprechende Resultat für solche Theorien besagt also, daß sich formale Definitionen der betrachteten Art jedenfalls nicht beweisen lassen, falls die Axiome widerspruchsfrei sind.

4.57 Andere Geometrien, Unterschiede. Die bisher in diesem Abschnitt betrachteten Geometrien waren jeweils Untertheorien (Abschwächungen) einer euklidischen Geometrie P_n^2 (für geeignete Dimension *n*; häufig dimensionsfrei). Im folgenden beschäftigen wir uns mit anderen, insbesondere mit nichteuklidischen Geometrien. Dabei wird sich zeigen, daß die euklidische, die hyperbolische und die elliptische Geometrie sich wesentlich voneinander unterscheiden in den Möglichkeiten für die Wahl der Grundbegriffe (im Gegensatz zu den einheitlichen Resultaten über Vollständigkeit und Entscheidbarkeit).

4.58 Satz (Menger 1938). *Für die hyperbolische Geometrie H ist die Kollinearität* Col *als einziger Grundbegriff geeignet.*

Zur bequemeren Formulierung des Beweises (hinter 4.61) verwenden wir (wie in der ursprünglichen Darstellung von Menger, die hier auf den dimensionsfreien Fall verallgemeinert wird) einen "zweisortigen" Aufbau (vgl. das Beispiel in 1.42) der hyperbolischen Geometrie, in dem neben den Punkten auch Geraden als geometrische Objekte und der Begriff Iz der Inzidenz von Punkten und Geraden vorkommen (*a*Iz*X* bedeute, daß der Punkt *a* mit der Geraden *X* "inzidiert", d.h. auf *X* liegt). Wir zeigen zunächst allgemein, daß man in einem formalisierten Aufbau dieser Art die Geradenvariablen (unter Verwendung von Col) eliminieren kann.

4.59 Satz (Elimination von Geradenvariablen).

Vor.: (i) *Die (geometrische) Theorie* T_L *habe eine Sprache erster Stufe mit zwei Sorten von Variablen - Punktvariablen* (a, b,...) *und Geradenvariablen* (X, Y,...) *- ; die Grundbegriffe der Sprache seien* Iz, Col *und (eventuell) die (Relations- und Operations-)Zeichen aus einer Menge K, die sich sämtlich nur auf Punktvariablen beziehen (K kann auch leer sein).*

(ii) *In* T_L *gelten als Sätze:*

(L1) $a \neq b \rightarrow \exists^{=1} X[aIzX \wedge bIzX]$,

(L2) $\forall X \exists a \exists b[aIzX \wedge bIzX \wedge a \neq b]$,

(L3) $\mathrm{Col}\, abc \leftrightarrow \exists X[aIzX \wedge bIzX \wedge cIzX]$.

Beh.: Zu jeder Formel α *ohne freie Geradenvariablen aus der genannten Sprache läßt sich eine dazu bezüglich* T_L *äquivalente Formel* α' *mit höchstens denselben freien (Punkt-)Variablen angeben, in der keine Geradenvariablen vorkommen und in der als Operations- und Relationszeichen höchstens* Col *und die in* α *auftretenden Zeichen aus K vorkommen.*

Beweis: Als Folgerung aus (L1) und (L2) erhält man zunächst

(L4) $X{=}Y \leftrightarrow \forall a[aIzX \rightarrow aIzY]$.

Somit kann man Gleichungen zwischen Geradenvariablen X, Y äquivalent ersetzen durch die rechte Seite von (L4) (wobei a eine beliebige Punktvariable ist) und sich für das Weitere auf Formeln α beschränken, in denen keine solchen "Geradengleichungen" mehr vorkommen. Die einzigen Primformeln, in denen noch Geradenvariablen vorkommen, sind dann von der Form aIzX. Insbesondere enthält jede solche Primformel nur eine Geradenvariable. Aus (L1) bis (L3) erhält man auch

(L5) $x{\neq}y \wedge xIzX \wedge yIzX \rightarrow [aIzX \leftrightarrow \mathrm{Col}\,axy]$

als Satz von T_L (anschaulich bedeutet das, daß man die Inzidenz eines beliebigen Punktes mit X beschreiben kann durch seine Kollinearität mit zwei Punkten x, y, die die Gerade X bestimmen). Ist nun eine Geradenvariable X beliebig gegeben und ist γ eine Formel, in der keine Geradengleichungen und keine quantifizierten Geradenvariablen vorkommen (dagegen dürfen X und auch andere Geradenvariablen frei vorkommen), so bedeute $\gamma_{X|xy}$ die Formel, die aus γ entsteht, indem darin jede Primformel der Form aIzX (für beliebige Punktvariablen a) ersetzt wird durch Col axy; wir bilden diese Formel für Punktvariablen x, y, die nicht in γ vorkommen (solche Punktvariablen kann man stets wählen).
Dann ist auch

(L6) $x{\neq}y \wedge xIzX \wedge yIzX \rightarrow (\gamma \leftrightarrow \gamma_{X|xy})$

ein Satz von T_L, wie man leicht aus (L5) durch Induktion über den Formelaufbau von γ erhält. Daraus ergibt sich als Satz von T_L:

(L7) $\exists X\gamma \leftrightarrow \exists x \exists y[x{\neq}y \wedge \gamma_{X|xy}]$.

Ist nämlich X eine Gerade mit der links genannten Eigenschaft, so erhält man Punkte x, y mit der rechts genannten Eigenschaft auf Grund von (L2). Sind umgekehrt x, y solche Punkte, so erhält man eine entsprechende Gerade X auf Grund von (L1). Analog oder durch Kontraposition von (L7) ergibt sich

(L8) $\forall X\gamma \leftrightarrow \forall x \forall y[x{\neq}y \rightarrow \gamma_{X|xy}]$.

Betrachten wir nun in α die letzte Quantifizierung, die sich auf eine Geradenvariable bezieht, und sei X diese Variable. Der zugehörige Wirkungsbereich enthält dann keine weitere quantifizierte Geradenvariable, er ist also eine Formel γ der betrachteten Art. Das betreffende Teilstück $\forall X\gamma$ bzw. $\exists X\gamma$ von α läßt sich also äquivalent durch die rechte Seite von (L8) bzw. (L7) ersetzen, d.h., die Geradenvariable X läßt sich in diesem Stück in der gewünschten Weise eliminieren.

Durch Wiederholung dieses Eliminationsschritts (Induktion über die Länge oder den Aufbau von α) gelangt man schließlich zu einer Formel α' der verlangten Art. □

4.60 Anmerkungen. (i) Ist T_L' eine Theorie ohne den Grundbegriff Col, in der (L1) und (L2) gelten (sonst wie in 4.59, Vor.(i)), so kann man natürlich mit (L3) als formaler Definition zu einer definitorischen Erweiterung T_L der obengenannten Art übergehen.

(ii) Betrachten wir andererseits die Sprache, die aus der von T_L durch Weglassen der Geradenvariablen und des Begriffs Iz entsteht. Die dieser Sprache angehörenden Sätze von T_L bilden für sich eine Theorie T. Als Folgerungen aus (L1) bis (L3) ergeben sich insbesondere folgende Sätze von T:

(1) $\overset{\geq 2}{\exists}$ (*es gibt wenigstens zwei Punkte*, das ist Satz I.3.13),

(2) $\mathrm{Col}\,abc \rightarrow \mathrm{Col}\,bca \wedge \mathrm{Col}\,bac$
([hinreichend für die] Invarianz von Col gegenüber Permutationen der Argumente, I.4.11),

(3) $\mathrm{Col}\,aab$ (I.4.12),

(4) $p \neq q \wedge \mathrm{Col}\,pqa \wedge \mathrm{Col}\,pqb \rightarrow \mathrm{Col}\,pab \wedge \mathrm{Col}\,qab$ (I.6.26(2)).

(iii) Sei umgekehrt eine Theorie T mit dieser eingeschränkten Sprache gegeben, so daß (1) bis (4) Sätze von T sind. (Alle bisher betrachteten Geometrien mit nur Punktvariablen sind von dieser Art.) Dann kann man Geraden und einen Inzidenzbegriff wie in I.6 einführen, d.h. durch

(5) $\mathrm{Ln}A \leftrightarrow \exists p \exists q [p \neq q \wedge A = \{x \mid \mathrm{Col}\,pqx\}]$,

(6) $a\mathrm{Iz}A \leftrightarrow \mathrm{Ln}A \wedge a \in A$

(man sieht, daß dafür nur die Sätze (1) bis (4) gebraucht werden). Führt man noch eine neue Sorte von Variablen (X, Y, ...) für Geraden ein, so erhält man aus T eine Theorie T_L mit der ursprünglichen Sprache, die der Voraussetzung von 4.59 genügt und die sich zu T so ähnlich wie eine definitorische Erweiterung verhält (die Behauptung von 4.59 entspricht den Sätzen 4.4, 4.5). Natürlich sind (5) und (6) keine Sätze von T_L (da sie in der zweiten Stufe formalisiert sind); auf Grund von T_L sind Geraden nur irgendwelche Objekte und nicht notwendig Punktmengen (was sich in der ersten Stufe gar nicht formulieren läßt). Die Hinzunahme der Forderungen (5) und (6) ändert jedoch nichts an der Menge der (in der ersten Stufe formulierbaren) Sätze von T_L. Man erhält nämlich zu jedem Modell von T_L ein isomorphes Modell, in welchem zusätzlich (5) und (6) gelten, indem man jede Gerade durch die Menge der mit ihr inzidierenden Punkte ersetzt.

(iv) Die Hintereinanderausführung der beiden in (ii) und (iii) angegebenen Prozesse - Einschränkung und anschließende Erweiterung der Sprache oder umgekehrt - führt wieder zur usprünglich gegebenen Theorie T_L bzw. T. Die Theorien T der hier genannten Art und die zugehörigen T_L (wie in 4.59) entsprechend sich also eineindeutig.

4.61 Anmerkung. Analog zu 4.59, 60 kann man Variablen für Ebenen - wie sie etwa im Hilbertschen Aufbau der Geometrie vorkommen - eliminieren oder einführen. Die Elimination beruht darauf, daß man eine Ebene E durch drei nicht-kollineare Punkte x, y, z, die mit ihr inzidieren, beschreiben kann; die Inzidenz eines vierten Punktes mit E wird dann durch seine Komplanarität mit x, y und z ausgedrückt (vgl. (L5)). Die Methode läßt sich auf Unterräume höherer Dimension übertragen.

Beweis von 4.58: Es genügt, die Definierbarkeit der ursprünglichen Grundbegriffe D und B mittels Iz in der hyperbolischen Geometrie H_L zu zeigen, da man dann das jeweilige definiens nach 4.59 auf die gewünschte Form bringen kann. Somit genügt der Beweis von

4.62 Satz (nach Menger 1938). *In der hyperbolischen Geometrie* H_L *(gemäß* 4.60(iii) *mit* $K=\{D,B\}$*) seien die Begriffe* Is, $\check{H}$, $\check{H}$, Mt, $\ddagger$ *folgendermassen durch formale Definitionen (mittels* D, B, Iz*) eingeführt.*

(1) $XY \,\mathrm{Is}\, a \leftrightarrow X \neq Y \wedge a\mathrm{Iz}X \wedge a\mathrm{Iz}Y$

(*X* und *Y* *schneiden sich* in *a* [analog zu I.6.22, ausgedrückt in der Sprache von H_L]).

Analog seien die hyperbolischen Parallelitäten $\check{H}$, $\check{H}$ *im engeren Sinne von Halbgeraden bzw. Geraden gemäß* 2.3(1) *bzw.* 2.3(iii) *eingeführt.*

(2) $$\mathrm{Mt}X_1X_2X_3 \leftrightarrow \exists x_1\exists y_1\exists x_2\exists y_2\exists x_3\exists y_3\{\bigwedge_{\nu=1}^{3}[x_\nu \mathrm{Iz} X_\nu \wedge y_\nu \mathrm{Iz} X_\nu]$$
$$\wedge\, x_1y_1\check{H}y_2x_2 \wedge x_2y_2\check{H}y_3x_3 \wedge x_3y_3\check{H}y_1x_1\}$$

(die Geraden X_1, X_2, X_3 bilden ein *maximales Dreieck* ["maximal triangle"], d.h., je zwei von ihnen, aber nicht alle drei, haben ein gemeinsames Ende, Abb. 142).

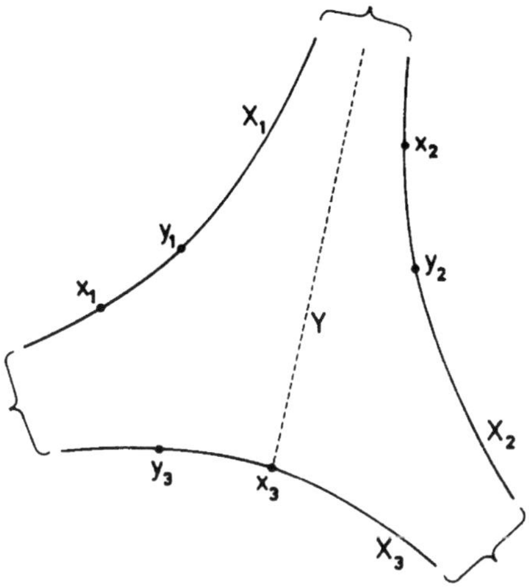

Abb. 142

(3) $ab \overset{\uparrow}{=} cd \leftrightarrow ab \equiv cd \wedge [ab \overset{\vee}{\parallel} cd \vee ba \overset{\vee}{\parallel} dc]$
(die "gerichteten Strecken" [Paare] $\langle a,b \rangle$ und $\langle c,d \rangle$ sind *gerichtet kongruent*, d.h., sie bilden kongruente Strecken, und die Halbgeraden H(*ab*), H(*cd*) oder ihre rückwärtigen Verlängerungen sind zueinander parallel, Abb. 143; insbesondere haben die Strecken also voneinander verschiedene Endpunkte).

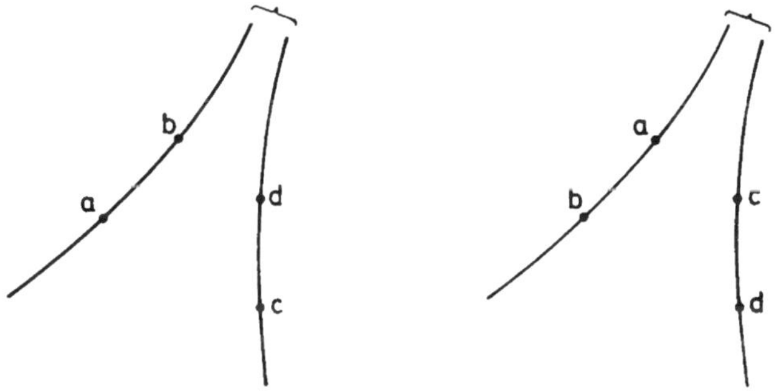

Abb. 143

Der n-stellige Begriff der Komplanarität von Geraden (I.9.32(ii)) *werde hier genauer mit* Cp_n *bezeichnet im Unterschied zu der gemäß* 4.10(I.9.33) *eingeführten Komplanarität* Cp *von Punkten (vierstellig); er sei einge-*

führt durch die (einfachere, aber mit I.9.32(ii) *gleichwertige) formale Definition*

(4) $\mathrm{Cp}_n X_1 \ldots X_n \leftrightarrow \exists a \exists b \exists c[\neg \mathrm{Col}\, abc \wedge \bigwedge_{\nu=1}^{n} \forall x(x\mathrm{Iz}X_\nu \to \mathrm{Cp}\, abcx)]$ $(n \geq 2)$.

Dann gelten in H_L *als Sätze:*

(5) $$\mathrm{Mt}X_1X_2X_3 \leftrightarrow \mathrm{Cp}_3X_1X_2X_3 \wedge \bigwedge_{\substack{\mu,\nu=1\\ \mu\neq\nu}}^{3} \neg\exists x[x\mathrm{Iz}X_\mu \wedge x\mathrm{Iz}X_\nu]$$
$$\wedge \bigwedge_{\lambda=1}^{3} \forall a\{a\mathrm{Iz}X_\lambda \to \exists^{=1} Y[a\mathrm{Iz}Y \wedge Y\neq X_\lambda \wedge \mathrm{Cp}_4YX_1X_2X_3$$
$$\wedge \bigwedge_{\substack{\nu=1\\ \nu\neq\lambda}}^{3} \neg\exists x(x\mathrm{Iz}Y \wedge x\mathrm{Iz}X_\nu)]\}$$

(Abb. 142 mit $a=x_3$),

(6) $X_1 \overset{\vee}{\nparallel} X_2 \leftrightarrow \exists X_3 \mathrm{Mt}X_1X_2X_3$,

(7) $$x_1y_1 \overset{\uparrow}{\ddagger} x_2y_2 \leftrightarrow \exists X_1\exists X_2\exists Y_1\exists Y_2\exists Z_0\exists Z_1\exists Z_2\exists x_0\exists y_0\{Z_1 \overset{\vee}{\nparallel} Z_2$$
$$\wedge \bigwedge_{\nu=1}^{2}[X_\nu Z_\nu \mathrm{Is}\, x_\nu \wedge Y_\nu Z_\nu \mathrm{Is}\, y_\nu \wedge x_\nu \neq y_\nu$$
$$\wedge X_\nu \overset{\vee}{\nparallel} Z_{3-\nu} \wedge Y_\nu \overset{\vee}{\nparallel} Z_{3-\nu}$$
$$\wedge x_0\mathrm{Iz}X_\nu \wedge y_0\mathrm{Iz}Y_\nu \wedge Z_0 \overset{\vee}{\nparallel} Z_\nu]$$
$$\wedge x_0\mathrm{Iz}Z_0 \wedge y_0\mathrm{Iz}Z_0\}$$ (Abb. 144),

(8) $ab \equiv cd \leftrightarrow [a=b \wedge c=d] \vee \exists e\exists f\{ab \overset{\uparrow}{\ddagger} ef \wedge [cd \overset{\uparrow}{\ddagger} ef \vee dc \overset{\uparrow}{\ddagger} ef]\}$,

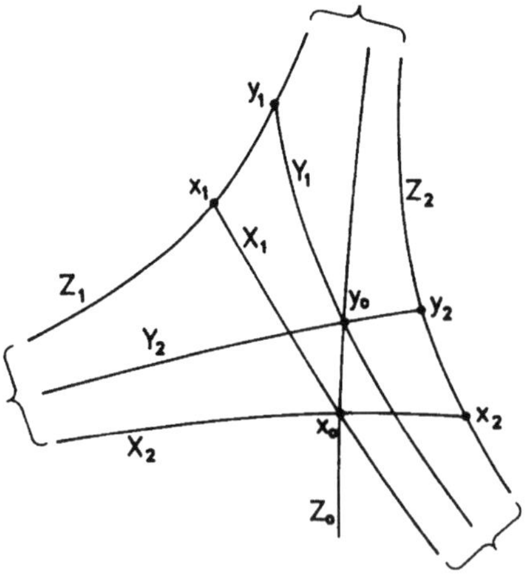

Abb. 144

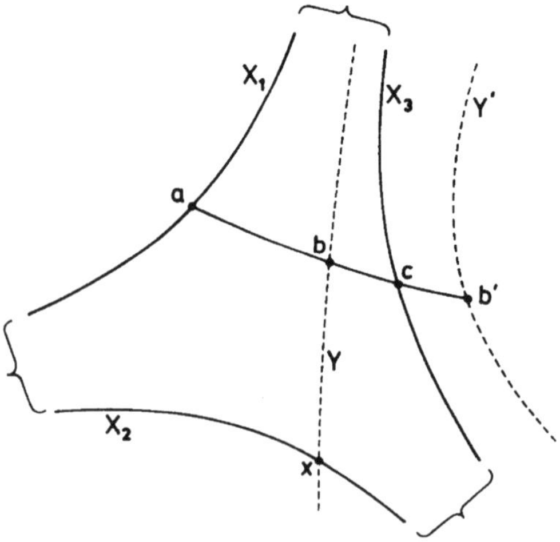

Abb. 145

(9) $\text{Babc} \leftrightarrow a=b=c$

$$\vee \{\text{Col}\,abc \wedge \exists X_1 \exists X_2 \exists X_3 [\text{Mt}X_1X_2X_3 \wedge a\text{Iz}X_1 \wedge c\text{Iz}X_3 \wedge \forall Y\langle b\text{Iz}Y \rightarrow \exists x(x\text{Iz}Y \wedge \bigvee_{\nu=1}^{3} x\text{Iz}X_\nu)\rangle]\},$$

(9') $\text{Babc} \leftrightarrow a=b=c$

$$\vee\{\text{Col}\,abc \wedge a\neq c \wedge \forall X_1 \forall X_2 \forall X_3 [\text{Mt}X_1X_2X_3 \wedge a\text{Iz}X_1 \wedge c\text{Iz}X_3 \rightarrow \forall Y\langle b\text{Iz}Y \rightarrow \exists x(x\text{Iz}Y \wedge \bigvee_{\nu=1}^{3} x\text{Iz}X_\nu)\rangle]\}$$

(Abb. 145).

Folglich sind die in (5) *bis* (9) *links genannten Begriffe mittels* Col *und* Iz *in* H_L *definierbar, nach* 4.59(L3) *also auch mittels* Iz *allein. Insbesondere ist die Inzidenz* Iz *als einziger Grundbegriff für* H_L *geeignet.*

Beweis: Es genügt wieder der Beweis der angegebenen geometrischen Sätze.

Zu (5)(→): Gelte $\text{Mt}X_1X_2X_3$. Nach Definition (2) liegen die "Seiten" X_1, X_2, X_3 des gegebenen maximalen Dreiecks zunächst in einer Ebene E und sind (als Punktmengen) paarweise disjunkt (Zeile 1 der Beh.). Sei nun a ein beliebiger Punkt einer Seite X_λ. Dann leistet die Gerade Y_o, die a mit dem gemeinsamen Ende der anderen beiden Seiten verbindet, das für Y Verlangte. Jede andere von X_λ verschiedene Gerade durch a in E schneidet dagegen (auf Grund der Definition der hyperbolischen Parallelität) eine der anderen beiden Seiten; also ist Y_o auch die einzige solche Gerade.

Zu (5)(←): Sei eine der Geraden X_λ und ein Punkt a darauf beliebig gewählt, und sei Y die dazu gemäß (5)(rechts) eindeutig bestimmte Gerade. Die Ebene E, in der alle betrachteten Geraden liegen, wird durch X_λ und Y in vier "Winkelgebiete" zerlegt. Dann ist jede der anderen beiden Geraden X_μ, X_ν (mit $\{\lambda,\mu,\nu\}=\{1,2,3\}$) in genau einem dieser Winkelgebiete enthalten (da sie mit $X_\lambda \cup Y$ keinen Punkt gemein hat), nennen wir es G_μ bzw. G_ν. Die Winkelgebiete G_μ und G_ν liegen nicht auf derselben Seite von Y; denn sonst könnte man auf der entgegengesetzten Seite einen Punkt a' auf X_λ wählen (Abb. 146), und die hyperbolischen Parallelen Y_1, Y_2 zu Y durch a' wären dann zwei verschiedene Geraden zu a' mit der in (5) für Y genannten Eigenschaft. G_μ und G_ν liegen aber auf derselben Seite von X_λ; denn sonst wären sie Winkelgebiete von Scheitelwinkeln,

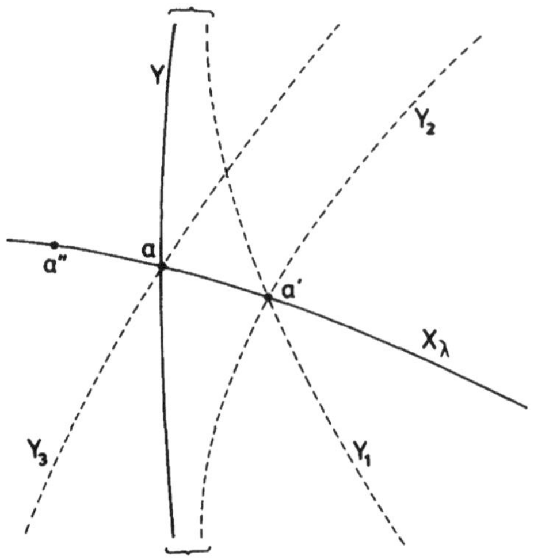

Abb. 146

und eine Gerade Y_3 durch a, die innere Punkte der anderen beiden Winkelgebiete enthält, wäre eine weitere Gerade mit der für Y genannten Eigenschaft. Schließlich ist X_μ hyperbolisch parallel zu den beiden Schenkeln (Halbgeraden von X_λ und von Y) des Winkels mit dem Gebiet G_μ; denn sonst enthielte X_μ eine zu keinem dieser Schenkel parallele Halbgerade H, und die hyperbolische Parallele zu H durch a wäre eine weitere Gerade mit der in (5) für Y genannten Eigenschaft. Dasselbe erhält man für X_ν und G_ν. Damit bilden X_1, X_2, X_3 tatsächlich ein maximales Dreieck. Aus dem Beweis sieht man noch, daß es genügt hätte, die Forderung in den Zeilen 2 und 3 von (5) für nur ein λ und für drei Punkte a', a, a'' auf X_λ zu stellen; mit einer leichten Modifikation des Beweises (Übung!) sieht man auch, daß man mit zwei Punkten a_1, a_2 auf X_λ auskommt.

(6) ergibt sich sofort daraus, daß man zwei Enden stets durch eine Gerade verbinden kann (hier die vom gemeinsamen Ende verschiedenen Enden von X_1 und X_2).

Zu (7): Wir führen den Beweis im Kleinschen Modell, was auf Grund des Darstellungssatzes für die hyperbolische Geometrie (2.5, hier für die Ebene durch die Punkte x_ν, y_ν, siehe auch I.11.74) genügt (Abb. 147). Für beide Richtungen können wir voraussetzen, daß die Z_ν, X_ν, Y_ν (ν=1,2) Geraden sind, die die in den ersten drei Zeilen von

(7)(rechts) angegebenen Eigenschaften besitzen; für die Richtung (→) lassen sie sich nämlich so wählen, außerdem sind sie durch diese Eigenschaften eindeutig bestimmt (Z_ν als $L(x_\nu y_\nu)$, X_ν und Y_ν als entsprechende Parallelen durch x_ν bzw. y_ν). Sei dann s das gemeinsame Ende von Z_1 und

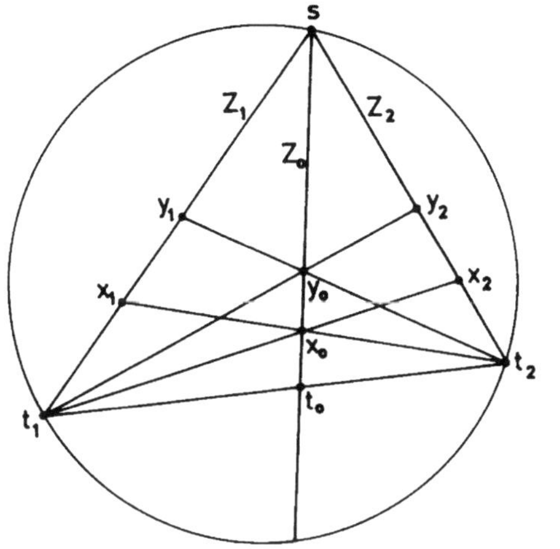

Abb. 147

Z_2 und t_ν das von s verschiedene Ende von Z_ν ($\nu=1,2$). Weiter sei (wie durch (7) rechts festgelegt) x_0 der Schnittpunkt von X_1 und X_2, und sei Z_0 die Verbindungsgerade von s und x_0. Sei außerdem t_0 der Schnittpunkt von Z_0 mit der Verbindungsgeraden von t_1 und t_2, und sei $y_{0\nu}$ der Schnittpunkt von Z_0 mit Y_ν ($\nu=1,2$). Da Abstand und Orientierung von zwei Punkten auf einer Geraden eindeutig festgelegt sind durch das Doppelverhältnis dieser Punkte mit den "Enden" dieser Geraden (und umgekehrt), ist die gerichtete Kongruenz in (7) links gleichwertig mit der Gleichheit der entsprechenden Doppelverhältnisse, d.h. mit

(7.1) $(x_1y_1,st_1) = (x_2y_2,st_2)$.

Da Doppelverhältnisse unter Projektionen invariant bleiben, ist

(7.2) $(x_\nu y_\nu,st_\nu) = (x_0y_{0\nu},st_0)$ $(\nu=1,2)$.

Da auf einer Geraden zu drei gegebenen Punkten (hier x_0, s, t_0) ein vierter Punkt eindeutig festgelegt wird durch das Doppelverhältnis mit ihnen (und umgekehrt), ist (7.1) schließlich gleichwertig mit $y_{01}=y_{02}$. Mit dieser Bedingung ist aber auch die rechte Seite von (7) gleichwertig.

Zu (8): Die Richtung (←) gilt offensichtlich. Sei nun $ab\equiv cd$, und seien o.B.d.A. ab und cd keine Nullstrecken (sonst ist auch die Umkehrung klar).

Es genügt, zu zeigen, daß es stets eine Gerade Z gibt, auf der sich eine gerichtete Strecke $\langle e,f \rangle$ der angegebenen Art wählen läßt. Sei X=L(ab), Y=L(cd) (Z muß zu diesen Geraden im engeren Sinne parallel sein).

<u>Fall 1</u>: $X \neq Y$. Wir zeigen, daß in diesem Falle das letzte Disjunktionsglied $dc \overset{\uparrow}{\neq} ef$ in (8) entbehrlich ist. Die gewünschte Orientierung von $\langle e,f \rangle$ läßt sich dann nämlich dadurch erreichen, daß von X und von Y ein geeignetes Ende ausgewählt wird, durch das Z gehen soll.

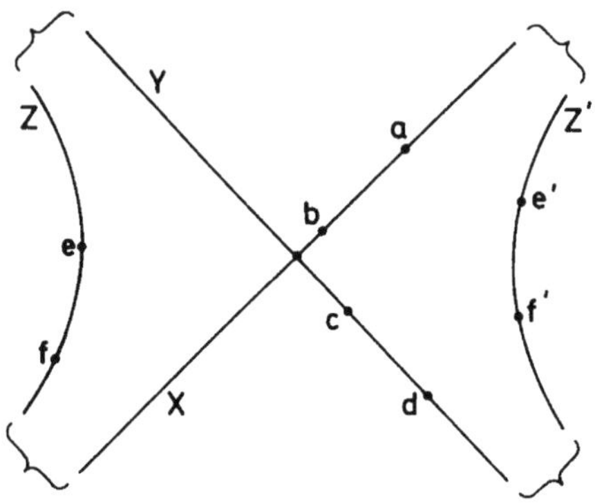

Abb. 148

<u>Fall 1a)</u>: X und Y sind nicht hyperbolisch parallel (d.h., sie haben zusammen vier paarweise verschiedene Enden, einerlei, ob sie sich schneiden oder nicht). Die Verbindung des Endes der Halbgeraden H(ab) und des Endes von H(dc) leistet in diesem Falle das für Z Verlangte, ebenso die Verbindung der Enden von H(ba) und von H(cd) (Abb. 148).

<u>Fall 1b)</u>: $X \overset{\vee}{\Vert} Y$. Falls schon $ab \overset{\uparrow}{\neq} cd$ ist, leistet eine weitere Gerade durch das gemeinsame Ende von X und Y das für Z Verlangte, andernfalls die Ergänzung von X und Y zu einem maximalen Dreieck (Abb. 149).

<u>Fall 2</u>: $X=Y$. Hier leistet eine beliebige hyperbolische Parallele zu X im engeren Sinne das Verlangte. (Von der letzten Disjunktion in (8) gilt das erste oder das zweite Glied je nachdem, ob $\langle a,b \rangle$ und $\langle c,d \rangle$ auf X dieselbe oder entgegengesetzte Orientierung besitzen.)

Zu (9) und (9'): Falls $a=c$ ist, gilt die Behauptung offensichtlich. Sei nun $a \neq c$ (Abb. 145). Betrachten wir ein beliebiges maximales Dreieck der angegebenen Art (d.h., X_1 und X_3 sind beliebige Geraden durch a bzw. c mit $X_1 \overset{\vee}{\Vert} X_3$, und X_2 ist ihre Ergänzung zu einem maximalen Dreieck). Man sieht leicht, daß jede Gerade Y durch einen Punkt b zwischen a und c wenigstens eine Seite X_λ des maximalen Dreiecks trifft, während es durch andere Punkte b' von $L(ac)$ stets eine Gerade Y' gibt, die mit der "nächstgelegenen" Seite und somit mit allen Seiten X_λ keine gemeinsamen Punkte hat. Daraus ergeben sich (9) und (9'). □

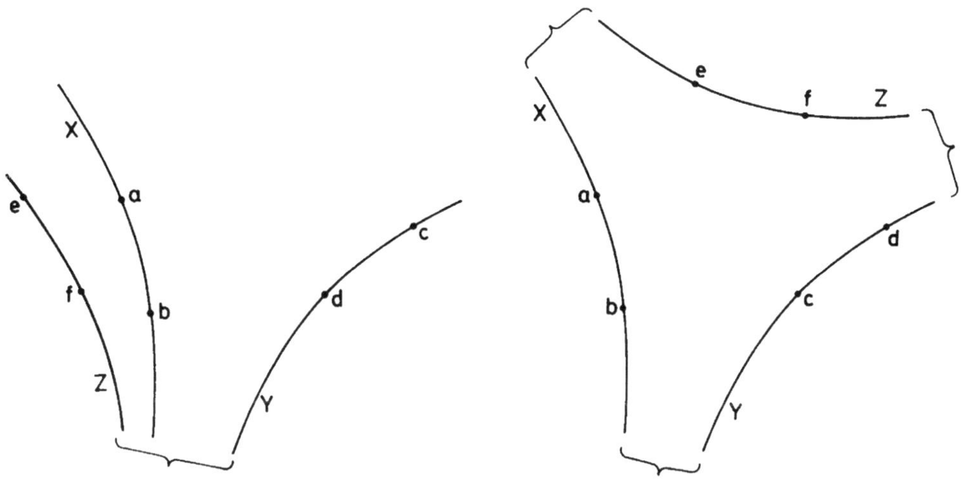

Abb. 149

4.63 Anmerkung. Der letzte Satz besagt insbesondere, daß es in (Modellen) der hyperbolischen Geometrie H - im Gegensatz zur euklidischen Geometrie (vgl. 1.10) - keine echten affinen Abbildungen gibt; jede Abbildung, die die Kollinearität erhält, erhält schon die Streckenkongruenz und die Zwischenbeziehung.

4.64 Weitere Geometrien, für die zweistellige Grundbegriffe nicht ausreichen. In H gibt es bekanntlich auch keine echten Ähnlichkeitstransformationen (da man [einige] feste Längen definieren kann). Wir kommen nun zu dem Resultat von R.M. Robinson 1959, daß es auch dafür kein geeignetes System von zweistelligen Grundbegriffen gibt (natürlich ist der Beweis des entsprechenden Satzes 4.34 für euklidische Geometrien hier nicht mehr anwendbar). Das Resultat bezieht sich auf verschiedene Geometrien ohne echte Ähnlichkeitstransformationen. Dazu gehört als wichtiges Beispiel auch eine euklidische Geometrie, in der neben den bisherigen Grundbegriffen ein "Einheitsabstand" gegeben ist, d.h. (ein Zeichen für) eine zweistellige Relation, die auf Punkte a, b zutrifft genau dann, wenn diese den Abstand Eins haben (dann gibt es als Automorphismen nur noch Bewegungen). Allgemein werden Geometrien betrachtet, die aus H^* oder P^* durch Hinzunahme beliebig vieler fester Abstände (ausgedrückt durch zweistellige "Abstandsrelationen") entstehen. Im Beweis wird als Hilfsmittel aus der mathematischen Logik das Tarskische Resultat über die Quantorenelimination für Th($\mathbb{R}$) benutzt.

4.65 Definition. Seien $d_1,\ldots, d_r$ positive reelle Zahlen ("Abstände"). Wir betrachten eine zugehörige Sprache $L_r := L(D,B,D_1,\ldots,D_r)$ erster Stufe mit zweistelligen Relationszeichen D_ρ $(1\le\rho\le r)$. Unter dem *n-dimensionalen kartesischen Raum* $\mathfrak{L}^n_{d_1,\ldots,d_r}$ (über $\mathbb{R}$) *mit den festen Abständen* $d_1,\ldots, d_r$ verstehen wir die Relationalstruktur $\mathfrak{A}$ für L_r, die aus dem üblichen kartesischen Raum $\mathfrak{L}_n(\mathbb{R})$ (s. I.1.4) dadurch entsteht, daß die entsprechenden Abstandsrelationen $(D_\rho)_\mathfrak{A}$ hinzugenommen werden, d.h.

(1) $\langle U_\mathfrak{A}, D_\mathfrak{A}, B_\mathfrak{A}\rangle = \mathfrak{L}_n(\mathbb{R})$,

(2) $(D_\rho)_\mathfrak{A}\, ab$ gdw $\sum_{\nu=1}^{n} (a_\nu - b_\nu)^2 = d_\rho^2$ $(1\le\rho\le r)$.

Unter dem *n-dimensionalen Kleinschen Raum* $\mathfrak{K}^n_{d_1,\ldots,d_r}$ (über $\mathbb{R}$) *mit den festen Abständen* $d_1,\ldots, d_r$ verstehen wir die in analoger Weise aus dem Kleinschen Raum $\mathfrak{K}_n(\mathbb{R})$ (s. 2.4) gebildete Struktur $\mathfrak{A}$, d.h.

(1') $\langle U_\mathfrak{A}, D_\mathfrak{A}, B_\mathfrak{A}\rangle = \mathfrak{K}_n(\mathbb{R})$,

(2') $(D_\rho)_\mathfrak{A}\, ab$ gdw $\frac{1}{2}|\ln(ab,st)| = d_\rho$,
wobei s und t die Schnittpunkte der Geraden $L(ab)$ mit der absoluten (Hyper-)Fläche (2.10) sind. (Bei Verwendung der üblichen "natürlichen Längeneinheit" heißt das gerade, daß a und b den Abstand d_ρ haben.)

Unter der *n-dimensionalen euklidischen* bzw. *hyperbolischen Geometrie* $P^*_{n,d_1,\ldots,d_r}$ bzw. $H^*_{n,d_1,\ldots,d_r}$ *mit den festen Abständen* $d_1,\ldots, d_r$ verstehen wir die Theorie mit der Sprache L_r der jeweils genannten Struktur, also

(3) $P^*_{n,d_1,\ldots,d_r} = \mathrm{Th}(\mathfrak{L}^n_{d_1,\ldots,d_r})$,

(3') $H^*_{n,d_1,\ldots,d_r} = \mathrm{Th}(\mathfrak{K}^n_{d_1,\ldots,d_r})$.

Die genannte Struktur bezeichnen wir dann als "Standardmodell" der betreffenden Theorie.

Hierbei ist der Fall $r=0$ zugelassen; die betrachtete Struktur fällt dann zusammen mit $\mathfrak{L}_n(\mathbb{R})$ bzw. $\mathfrak{K}_n(\mathbb{R})$ und die betrachtete Theorie mit P_n^* bzw. H_n^* (wegen der Vollständigkeit, s. 3.32 oder 44, 3.65(i)).

4.66 Anmerkung. Ist $c>0$ und $d'_\rho = c\cdot d_\rho$ $(1\le\rho\le r)$, so ist offenbar

$$\mathfrak{L}^n_{d_1,\ldots,d_r} \cong \mathfrak{L}^n_{d'_1,\ldots,d'_r}$$

(eine Ähnlichkeitstransformation von $\mathfrak{L}_n(\mathbb{R})$, die Abstände in ihr c-faches überführt, ist ein Isomorphismus der gewünschten Art) und damit

$$P^*_{n,d_1,\ldots,d_r} = P^*_{n,d'_1,\ldots,d'_r} .$$

Im euklidischen Fall kommt es also nicht auf die gegebenen Abstände selbst, sondern nur auf ihre Verhältnisse untereinander an. Für $r=1$ ist insbesondere stets $P^*_{n,d} = P^*_{n,1}$, d.h., man kann $P^*_{n,d}$ als die euklidische Geometrie mit festem Einheitsabstand ansehen.

4.67 Satz (R. M. Robinson 1959). *Seien $d_1,\ldots, d_r$ wie oben und $n\ge 2$. Für $P^*_{n,d_1,\ldots,d_r}$ und für $H^*_{n,d_1,\ldots,d_r}$ gibt es kein geeignetes System von Grundbegriffen, das nur aus zweistelligen Relationszeichen (für Relationen zwischen Punkten) besteht.*

Der Beweis beruht auf den folgenden drei Lemmata.

4.68 Lemma. *Sei T eine der in* 4.67 *genannten Theorien und R_λ ein darin mit den Grundbegriffen von L_r definierbares (d.h. durch definitorische Erweiterung eingeführtes) zweistelliges Relationszeichen. Dann sind die folgende Formel* (1) *und* e i n e *der beiden Formeln* (2)(a), (2)(b) *Sätze von T:*

(1) $ab\equiv cd \rightarrow [R_\lambda ab \leftrightarrow R_\lambda cd]$,

(2)(a) $\exists u\exists v\forall a\forall b[ab>uv \rightarrow R_\lambda ab]$

(R_λ trifft auf alle hinreichend großen Strecken zu),

(b) $\exists u\exists v\forall a\forall b[ab>uv \rightarrow \neg R_\lambda ab]$

(R_λ trifft auf alle hinreichend großen Strecken nicht zu).

Beweis: Zu (1): Die Grundbegriffe von L_r sind invariant gegenüber Bewegungen. Dasselbe muß also auch (4.29) für R_λ gelten; das wird aber gerade durch (1) ausgedrückt.

Zu (2): Die rechten Seiten der Bedingungen (2) bzw. (2') aus 4.65 sind gleichwertig mit (unter Benutzung der Vektorschreibweise für die n-tupel a, b)

(*) $(a-b)^2 = e_\rho$,

wobei $e_\rho = d_\rho^2$, bzw.

(*') $\dfrac{(1-ab)^2}{(1-a^2)(1-b^2)} = e_\rho$,

wobei $e_\rho = \cos h^2 d_\rho$.

Die Zahl e_ρ, die man statt des Abstandes d_ρ zur Charakterisierung der Abstandsrelation $(D_\rho)_{\mathfrak{A}}$ verwenden kann, wollen wir hier als die entsprechende *Abstandscharakteristik* bezeichnen; analog auch für andere Abstände. Wesentlich an dieser Charakterisierung ist, daß sie sich (auch im hyperbolischen Fall) rational ausdrücken läßt. Die früher behandelte Konstruktion der arithmetischen Reduzierten (s. 3.25 und 3.64) läßt sich damit von L(D,B) auf die Sprache L_r ausdehnen, indem man den zusätzlichen Primformeln der Form $D_\rho pq$ (mit beliebigen Punktvariablen) Reduzierten zuordnet, die durch Formalisierung von (*) (im euklidischen Fall) bzw. (*') (im hyperbolischen Fall) (mit entsprechenden Zahlenvariablen $p^{(1)}, \ldots, q^{(n)}$) entstehen, dabei sollen statt der Zahlen e_ρ zunächst entsprechende Zahlenvariablen y_ρ verwendet werden (die Zahlen e_ρ brauchen in Th($\mathbb{R}$) keine Bezeichnungen zu besitzen). Insbesondere erhalten wir aus dem definiens δ von R_λ eine Reduzierte δ^* in der Sprache $L_{OF} = L(+,-,\cdot,0,1;<)$ von Th($\mathbb{R}$), die mit Hilfe der Koordinaten von a, b ausdrückt, daß die Relation $(R_\lambda)_{\mathfrak{A}}$ (im Standardmodell $\mathfrak{A}$) auf a und b zutrifft, vorausgesetzt, daß die zusätzlichen freien Variablen y_ρ von δ^* mit den Zahlen e_ρ belegt sind $(1 \leq \rho \leq r)$. Damit erhalten wir auch eine Formel $\alpha = \alpha(x, y_1, \ldots, y_r)$ von L_{OF} mit den angegebenen freien Variablen, die - bei derselben Belegung der y_ρ - ausdrückt, daß die Relation $(R_\lambda)_{\mathfrak{A}}$ auf alle Strecken ab mit der Abstandscharakteristik x zutrifft; wir setzen nämlich etwa

$$\alpha = \forall a^{(1)} \ldots \forall b^{(n)} [\gamma \rightarrow \delta^*];$$

dabei entstehe γ durch Formalisierung von (*) bzw. von $a^2<1 \wedge b^2<1 \wedge$ (*') mit der freien Variablen x statt e_ρ. Wegen (1) hängt das Zutreffen von $(R_\lambda)_{\mathfrak{A}}$ auf eine Strecke nur von ihrer Abstandscharakteristik ab. Die Negation von α drückt also aus, daß (R_λ) auf keine Strecke mit der Abstandscharakteristik x zutrifft.

Nach den Resultaten von Tarski aus 3.22 (Elimination der Quantoren für Th($\mathbb{R}$) und weitere Umformungen) ist α äquivalent zu einer gewissen quantorenfreien disjunktiven Normalform

$$\bigvee_{\kappa=1}^{k} \bigwedge_{\mu=1}^{m} \alpha_{\kappa,\mu} ,$$

wobei für jedes Indexpaar $\langle\kappa,\mu\rangle$ die Primformel $\alpha_{\kappa,\mu}$ die Form $\pi_{\kappa,\mu}=0$ oder $\pi_{\kappa,\mu}>0$ hat mit formalen Polynomen (Termen) $\pi_{\kappa,\mu}$ in x, in deren Koeffizienten höchstens die Variablen y_ρ vorkommen. Für unsere Beschreibung interessiert nur die Belegung der y_ρ mit den Zahlen e_ρ, dabei beschreiben die Terme $\pi_{\kappa,\mu}$ gewisse Polynome über $\mathbb{R}$ in einer Unbestimmten. Da sich das Vorzeichen (des Wertes) eines Polynoms für hinreichend große Argumente nicht mehr ändert, kann man zu jedem Indexpaar $\langle\kappa,\mu\rangle$ eine Zahl $t_{\kappa,\mu}$ wählen, so daß gilt:

$\alpha_{\kappa,\mu}$ wird wahr für alle $x > t_{\kappa,\mu}$

oder $\alpha_{\kappa,\mu}$ wird falsch für alle $x > t_{\kappa,\mu}$.

Dasselbe gilt dann jeweils erst recht für alle $x > t$, wobei t das Maximum der $t_{\kappa,\mu}$ ist. Für diese x haben die $\alpha_{\kappa,\mu}$ also feste Wahrheitswerte, aus denen sich der Wahrheitswert der obigen disjunktiven Normalform ergibt. Somit gilt (bei der genannten Belegung der y_ρ) genau einer der folgenden beiden Fälle:

(a) α wird wahr für alle $x > t$,

(b) α wird falsch für alle $x > t$.

Zu größeren Abstandscharakteristiken gehören natürlich größere Strecken (im Sinne von Def. I.5.14). Auf Grund der oben genannten Bedeutung von α sind also die Fälle (a), (b) gleichwertig damit, daß (2)(a) bzw. (2)(b) ein Satz von T ist. □

<u>Beweis von 4.67</u> (Anfang): Nehmen wir an, es gäbe ein geeignetes System K von Grundbegriffen der genannten Art. Dann ist insbesondere die Streckenkongruenz D mittels K definierbar in der betreffenden Theorie T. Seien $R_1, \ldots, R_l$ die in einem entsprechenden definiens tatsächlich vorkommenden Relationszeichen. Für jedes solche R_λ gilt dann wegen $R_\lambda \in K$ (4.21) die Behauptung von 4.68. Da man von Begriffen der einen Art (Gültigkeit von (2)(a)) durch Negationsbildung zu solchen der anderen Art (Gültigkeit von (2)(b)) übergehen kann und umgekehrt, erhält man aus der Annahme auch die

Beh. (i): D *ist in* T *definierbar mit Hilfe von gewissen zweistelligen Relationszeichen* $R_1, \ldots, R_l$ *derart, daß für jedes* R_λ *die Formeln* (1) *und* (2)(b) *von* 4.68 *Sätze von* T *sind.*

Damit wird die folgende Definition motiviert.

4.69 Definition. Für jedes natürliche m wird in der Sprache L(D,B) der $(2m+2)$-stellige Begriff Pat^m der *stückweisen Kongruenz* ("patch-wise congruence" in R. M. Robinson 1959) durch folgende formale Definition eingeführt:

(1) $(a_1 \ldots a_m)\mathrm{Pat}^m_{uv}(b_1 \ldots b_m)$

$$\longleftrightarrow \bigvee_{\mathrm{Pt}_m \mathfrak{P}} \Big\{ \bigwedge_{M \in \mathfrak{P}} \ \bigwedge_{\mu,\nu \in M} a_\mu a_\nu \equiv b_\mu b_\nu \wedge \bigwedge_{\substack{M,N \in \mathfrak{P} \\ M \neq N}} \ \bigwedge_{\substack{\mu \in M \\ \nu \in N}} [a_\mu a_\nu > uv \wedge b_\mu b_\nu > uv] \Big\}$$

(die m-tupel $\langle a_1,\ldots,a_m\rangle$ und $\langle b_1,\ldots,b_m\rangle$ sind *stückweise kongruent mit dem Trennabstand uv*). Hierbei bedeute $\mathrm{Pt}_m \mathfrak{P}$, daß $\mathfrak{P}$ eine Klasseneinteilung ("partition") der Indexmenge $\{1,\ldots,m\}$ ist (d.h. von der Form $\mathfrak{P}=\{M_1,\ldots,M_k\}$ mit $M_1 \cup \ldots \cup M_k = \{1,\ldots,m\}$, $M_\kappa \neq \emptyset$ und $M_\kappa \cap M_\lambda = \emptyset$ für $1 \le \kappa, \lambda \le k$ und $\kappa \neq \lambda$, insbesondere ist $k \le m$); die auftretende Disjunktion erstreckt sich also über alle (endlich vielen) solchen Klasseneinteilungen. Im Falle $m=0$ ("leere Konjunktionen") werde die rechte Seite von (1) durch etwas immer Gültiges ersetzt, etwa $uv \equiv uv$.

Durch (1) wird ausgedrückt, daß sich die Indexmenge und damit auch die Punkte beider m-tupel in derselben Weise so in "Stücke" einteilen lassen (Abb. 150), daß die Punkte a_μ in einem Stück und die entsprechenden Punkte b_μ im entsprechenden Stück kongruente Figuren (im Sinne von I.4.4)

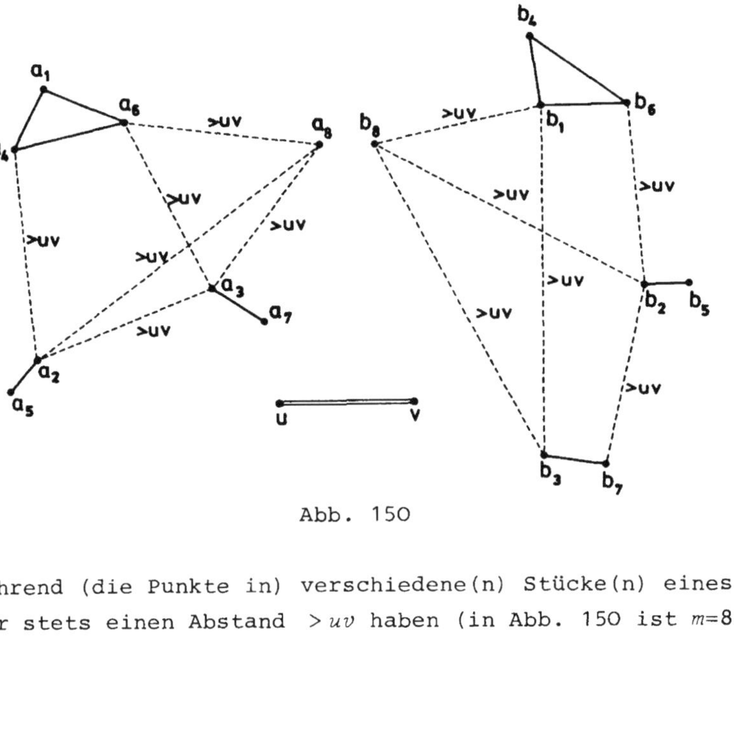

Abb. 150

bilden, während (die Punkte in) verschiedene(n) Stücke(n) eines m-tupels voneinander stets einen Abstand $> uv$ haben (in Abb. 150 ist $m=8$ und

$\mathfrak{P}=\{\{1,4,6\},\{2,5\},\{3,7\},\{8\}\}$, die Punkte eines Stücks sind jeweils durch ausgezogene Linien verbunden, zwischen verschiedenen Stücken eines m-tupels sind nur kürzeste Verbindungen [gestrichelt] eingezeichnet). Die Forderung der Existenz einer solchen Einteilung wird durch die Verwendung der angegebenen endlichen Disjunktion tatsächlich (wenn auch kompliziert) in der ersten Stufe ausgedrückt.

Wir verwenden auch die Bezeichnung $2\cdot uv$ für eine Strecke (Trennabstand), die durch "Verdoppelung" von uv entsteht (etwa die Strecke $uS_v(u)$), und (induktiv) weiter $2^k\cdot uv$ für $2\cdot(2^{k-1}\cdot uv)$; ebenso entsprechende objektsprachliche Bezeichnungen.

4.70 Lemma. *In der absoluten Geometrie* A *(und damit erst recht in den Theorien* T *von* 4.67*) gilt als Satz:*

(1) $(a_1\dots a_m)\mathrm{Pat}^m_{2\cdot uv}(b_1\dots b_m) \to \forall a\exists b(a_1\dots a_m a)\mathrm{Pat}^{m+1}_{uv}(b_1\dots b_m b)$.

Beweis: Wir betrachten eine Einteilung gemäß der Voraussetzung und einen beliebigen Punkt a. Wenn a von allen Punkten a_μ einen Abstand $>uv$ hat, so kann man ein neues Stück bilden, das nur aus dem Punkt a besteht, und dazu einen entsprechenden Punkt b wählen, der von allen b_μ ebenfalls einen Abstand $>uv$ hat. Andernfalls gibt es ein eindeutig bestimmtes Stück $\mathfrak{a}$ (eine Teilfolge) von $\langle a_1,\dots,a_m\rangle$, so daß a von (wenigstens) einem Punkt von $\mathfrak{a}$ einen Abstand $\leq uv$ hat (gäbe es nämlich zwei solche Stücke mit entsprechenden Punkten a_μ, a_ν, so hätten diese Punkte voneinander einen Abstand $\leq 2\cdot uv$ im Widerspruch zur Voraussetzung). Sei dann $\mathfrak{b}$ das entsprechende (zu $\mathfrak{a}$ kongruente) Stück und b ein Punkt derart, daß auch die erweiterten Stücke $(\mathfrak{a},a)$ und $(\mathfrak{b},b)$ zueinander kongruent sind. Ebenso wie a hat dann auch b von den Punkten b_μ anderer Stücke Abstände $>uv$ und leistet somit das Verlangte. □

4.71 Lemma: *Sei* T *eine Theorie, in der mindestens die Axiome von* A *zur Verfügung stehen und in deren Sprache die zweistelligen Relationszeichen* $R_1,\dots,R_l$ *vorkommen. Seien* α *und* β *die Formeln*

(1) $\bigwedge_{\lambda=1}^{l} \forall a\forall b\forall c\forall d\{ab\equiv cd \to [R_\lambda ab \leftrightarrow R_\lambda cd]\}$,

(2) $\bigwedge_{\lambda=1}^{l} \forall a\forall b[ab>uv \to \neg R_\lambda ab]$.

Dann gilt für jede Formel $\delta=\delta(a_1,\dots,a_m)$ *der Sprache* $L(R_1,\dots,R_l)$*: Wenn* δ *höchstens die freien Variablen* $a_1,\dots,a_m$ *enthält und* $b_1,\dots,b_m$ *darin nicht vorkommen und* δ *höchstens den Quantorenrang* k *besitzt (d.h.,* *daß Quantoren höchstens k-fach ineinandergeschachtelt auftreten), so gilt als Satz von T:*

(3) $\alpha \wedge \beta \wedge (a_1 \ldots a_m)\,\mathrm{Pat}^m_{2^k \cdot uv}(b_1 \ldots b_m) \rightarrow [\delta(a_1,\ldots,a_m) \leftrightarrow \delta(b_1,\ldots,b_m)]$.

Beweis: Das Lemma wird induktiv über den Formelaufbau von δ (mit beliebigen m und k) bewiesen. Die Primformeln sind hier von einer der beiden Formen $a_\mu = a_\nu$, $R_\lambda a_\mu a_\nu$. Dafür ergibt sich die Behauptung von (3) direkt aus der Voraussetzung (aus α oder aus β je nachdem, ob die entsprechenden Punkte a_μ und a_ν im selben oder in verschiedenen Stücken liegen). Bei aussagenlogischen Zusammensetzungen überträgt sich das Lemma unmittelbar. Bei den Quantifizierungen genügt die Behandlung des Existenzquantors (da sich $\forall$ mit $\neg$ und $\exists$ ausdrücken läßt). Sei also jetzt δ von der Form

$$\delta(a_1,\ldots,a_m) = \exists a\delta'(a_1,\ldots,a_m,a),$$

und gelte für δ' die Induktionsvoraussetzung, außerdem gelte die Voraussetzung von (3). Für die Richtung ($\rightarrow$) gelte auch noch $\delta[a_1,\ldots,a_m]$, und sei a dementsprechend gewählt. Sei schließlich b gemäß 4.70(1) (mit $2^{k-1}\cdot uv$ statt uv) gewählt. Damit läßt sich die Induktionsvoraussetzung anwenden, diese besagt ja gerade

$$\alpha \wedge \beta \wedge (a_1 \ldots a_m a)\,\mathrm{Pat}^{m+1}_{2^{k-1}\cdot uv}(b_1 \ldots b_m b)$$

$$\rightarrow (\delta'[a_1,\ldots,a_m,a] \leftrightarrow \delta'[b_1,\ldots,b_m,b]).$$

Somit leistet b das Verlangte. Die Umkehrung ($\leftarrow$) ergibt sich analog (unter Verwendung der offensichtlichen Symmetrie von Pat). □

Beweis von 4.67 (Fortsetzung): Sei

$$a_1a_2 \equiv a_3a_4 \leftrightarrow \delta(a_1,a_2,a_3,a_4)$$

eine formale Definition gemäß Behauptung (i) aus dem ersten Teil des Beweises, und sei k der Quantorenrang des definiens δ. Wir kommen nun folgendermaßen zu einem Widerspruch in T. Zu jedem R_λ sei $u_\lambda v_\lambda$ eine Strecke gemäß 4.68(2)(b), und sei uv eine Strecke mit $uv \geq u_\lambda v_\lambda$ ($\lambda=1,\ldots,l$). Weiter wählen wir Punkte a_μ, b_μ ($\mu=1,\ldots,4$), so daß

$$a_\mu a_\nu > 2^k \cdot uv, \quad b_\mu b_\nu > 2^k \cdot uv \ (\mu \neq \nu), \quad \text{aber}$$

$$a_1a_2 \equiv a_3a_4 \quad \text{und} \quad \neg b_1b_2 \equiv b_3b_4$$

(solche gibt es offenbar). Nun sind alle Voraussetzungen von 4.71 erfüllt. Somit gilt der Satz 4.71(3), dessen Voraussetzung für die betrachteten Punkte ebenfalls erfüllt ist (jedes a_μ und jedes b_μ bildet für sich ein Stück). Daraus ergibt sich

$$a_1a_2 \equiv a_3a_4 \leftrightarrow b_1b_2 \equiv b_3b_4$$

im Widerspruch zur Konstruktion. □

4.72 Anmerkung. Im vorangehenden Beweis wurde wesentlich benutzt, daß es beliebig große Abstände gibt. Für die elliptische Geometrie ist das nicht der Fall. Tatsächlich kommt man für die elliptische Geometrie mit einem zweistelligen Grundbegriff aus (s. 4.73ff.). Darüber hinaus wird in R. M. Robinson 1959 für die euklidische und die hyperbolische Geometrie (als Gegenstück zum vorigen Beweis) gezeigt, daß man mit Hilfe eines zweistelligen Relationszeichens (für einen festen Abstand) definieren kann, daß die Streckenkongruenz zutrifft auf Punkte, deren Abstände unterhalb einer gegebenen Schranke liegen; die Schranke kann dabei sogar beliebig groß gemacht werden. Außerdem wird die Definierbarkeit von Abständen untersucht. Die Resultate lassen sich folgendermaßen zusammenfassen.

Mit D^d *werde die Abstandsrelation für den festen Abstand* d *bezeichnet (d.h.* $D^d = D_\rho$, *falls* $d = d_\rho$ *in* 4.65*). Seien die Begriffe* $D^{\leq h}$ *und* $\overset{h}{\equiv}$ *durch folgende formale Definitionen eingeführt:*

(1) $D^{\leq h}ab \leftrightarrow \exists a' \exists b' [D^h a'b' \wedge ab \leq a'b']$

(die Punkte a, b haben einen Abstand kleiner-gleich h),

(2) $a_1a_2 \overset{h}{\equiv} a_3a_4 \leftrightarrow a_1a_2 \equiv a_3a_4 \wedge \bigwedge_{\substack{\mu,\nu=1 \\ \mu \neq \nu}}^{4} D^{\leq h}a_\mu a_\nu$

("lokale Streckenkongruenz mit der Abstandsbeschränkung h").

Sei T *eine der in* 4.67 *genannten Geometrien mit* $d, 2d, h \in \{d_1, \ldots, d_r\}$. *Dann gelten die folgenden Behauptungen.*

(i) ("Lokale Definierbarkeit der Streckenkongruenz") *In* T *ist* $\overset{h}{\equiv}$ *mittels* D^d *definierbar gdw* D^h *mittels* D^d *definierbar ist.*

(ii) D^{2d} *ist definierbar mittels* D^d *in* T.

(iii) D^h *ist in* T *definierbar mittels* D^d *gdw gilt:*

(a) *im euklidischen Fall* ($T = P^*_{n,d_1,\ldots,d_r}$): $\frac{h}{d}$ *ist algebraisch,*

(b) *im hyperbolischen Fall* ($T = H^*_{n,d_1,\ldots,d_r}$): e^h *ist algebraisch in* e^d.

Die Resultate bleiben dieselben für die Definierbarkeit mit Hilfe von D^d *u n d der Streckenkongruenz* ≡.

(iv) D^h *ist in* T *definierbar mittels der Streckenkongruenz allein gdw gilt:*

(b) *im hyperbolischen Fall:* e^h *ist algebraisch.*

(Im euklidischen Fall ist natürlich keine Abstandsrelation mittels ≡ *definierbar, s.* 4.36.*)*

(iii) und (iv) (b) werden außerdem auf eine analog eingeführte elliptische Geometrie übertragen, dabei treten cos *h* und cos *d* an die Stelle von e^h und e^d.

Als Korollar aus (iv) (b) ergibt sich noch:

(v) *Die natürliche Längeneinheit (d.h. hier* D^1*) und die Multiplikation von Abständen sind in der hyperbolischen Geometrie* H_n^* *nicht (mittels der Grundbegriffe* D, B*) definierbar.*

Die Behauptungen (iii) bis (v) beruhen wieder auf der Tarskischen Quantorenelimination für Th(ℝ) .

In ähnlicher Weise ergibt sich (Schwabhäuser 1962, ursprünglich für eine andere Formalisierung der Geometrie):

(vi) *Die Abstandsfunktion, die jeder Strecke ihre Länge zuordnet, und eine entsprechende Größenfunktion für Winkel sind nicht in* H_n^* *definierbar. Die Längen und Größen - in bezug auf eine beliebige Einheit der Messung - seien dabei im Standardmodell eingeführt als Elemente des (zu* ℝ *isomorphen) Körpers der Hilbertschen Endenrechnung (oder eines anderen geometrisch definierten Körpers, wie er im Beweis des Darstellungssatzes verwendet wird).*

(vii) *Für einen anderen Körper, dessen Addition durch Aneinanderlegen von Strecken erklärt ist, ist die Multiplikation nicht in* H_n^* *definierbar.* (Dieser ebenfalls zu ℝ isomorphe Körper läßt sich als Körper der Abstände im Sinne von (v) auffassen.)

4.73 Elliptische Geometrie. Wir betrachten nun die elliptische Geometrie *E* (dimensionsfrei, mit den Grundbegriffen ≡ [Streckenkongruenz] und | [Trennen von Punktepaaren], s. 2.11, 13, 15) und erwähnen für das Folgende einige grundlegende Sätze.

In einer elliptischen Ebene F (d.h. in einem zweidimensionalen Unterraum eines Modells von E) heißt bekanntlich ein Punkt a der *Pol* der Geraden A oder A die *Polare* von a, falls wenigstens zwei - und damit alle - auf A senkrechten Geraden (in F) durch a gehen. Das ist auch genau dann der Fall, wenn a von wenigstens drei - und damit von allen - Punkten von A denselben Abstand hat. In F hat jede Gerade genau einen Pol und jeder Punkt genau eine Polare. Der genannte Abstand ist von a und F unabhängig und heißt die *Polardistanz*. Er ist der größtmögliche Abstand von Punkten in der elliptischen Geometrie; bei der üblichen Längenmessung im "Standardmodell" $\mathfrak{E}_n(\mathbb{R})$ wird er gleich $\frac{\pi}{2}$ gesetzt. Punkte a, b in F haben Polardistanz genau dann, wenn a auf der Polaren in F von b liegt, und auch genau dann, wenn b auf der Polaren in F von a liegt. Eine Strecke ab mit $a \neq b$ hat genau zwei Mittelpunkte (d.h. mit a, b kollineare Punkte x mit $ax \equiv bx$), diese haben voneinander Polardistanz und trennen das Paar $\langle a,b \rangle$. In einer Ebene besteht die Menge der Punkte x mit $xa \equiv xb$ (für $a \neq b$) aus zwei Geraden, die in den Mittelpunkten von ab auf $L(ab)$ senkrecht stehen (die Senkrechte in einem Punkt auf einer Geraden in einer Ebene ist - wie in A - stets vorhanden und eindeutig bestimmt).

Wir verwenden Qab als Abkürzung dafür, daß a und b Polardistanz haben, und $Q'ab$ dafür, daß a und b einen kleineren Abstand als die halbe Polardistanz haben.

<u>4.74 Anmerkung zur Literatur</u>. Der Begriff Q wurde von R. M. Robinson und der Begriff Q' von H. L. Royden gefunden als zweistelliger Begriff (zwischen Punkten), der jeweils als einziger Grundbegriff einer ebenen elliptischen Geometrie geeignet ist. In R. M. Robinson 1959 wird gezeigt, daß Q als einziger Grundbegriff für die Theorie des Standardmodells geeignet ist; die Beweise lassen sich praktisch ohne Änderung übertragen auf die Geometrie $E_2^!$ (mit Kreisaxiom). In Royden 1959 wird festgestellt (und zum Teil bewiesen), daß Q und $\equiv$ in E_2 (ohne Kreisaxiom) gegenseitig definierbar sind, und daß Q', aber nicht Q, als einziger Grundbegriff für E_2 geeignet ist.

Wir benutzen hier einige neue formale Definitionen und kommen damit zu Vereinfachungen und zu einer Ausdehnung der genannten Resultate auf die dimensionsfreie elliptische Geometrie.

<u>4.75 Satz</u>. *In E seien die Begriffe* Col, P, P^*, E, Q, ⊥, M, Pc *durch folgende formale Definitionen eingeführt (vgl.* 4.10, 12, 19*):*

(1) $\mathrm{Col}\, abc \leftrightarrow a{=}b \vee b{=}c \vee c{=}a \vee \exists d\, ab|cd$
(*a*, *b*, *c* sind kollinear),

(2) $\mathrm{P}abc \leftrightarrow ab{\equiv}ac$ (Pierische Relation),

(3) $\mathrm{P}^*abc \leftrightarrow \mathrm{P}abc \wedge \neg\, \mathrm{Col}\, abc$,

(4) $\mathrm{E}abc \leftrightarrow \mathrm{P}abc \wedge \mathrm{P}bca$
(*a*, *b*, *c* bilden ein gleichseitiges Dreieck oder fallen zusammen),

(5) $\mathrm{Q}ab \leftrightarrow \exists u \exists v[\neq(buv) \wedge \mathrm{Col}\, buv \wedge \mathrm{P}abu \wedge \mathrm{P}abv]$
(*a*, *b* haben Polardistanz),

(6) $ab \perp cd \leftrightarrow a{\neq}b \wedge c{\neq}d$
$\wedge\ \exists u \exists x\, [\mathrm{Col}\, abu \wedge \mathrm{Col}\, cdu \wedge \mathrm{Col}\, cdx \wedge \mathrm{Q}ax \wedge \mathrm{Q}bx]$
(die Verbindungsgerade L(*ab*) steht senkrecht auf L(*cd*)),

(7) $\mathrm{M}abc \leftrightarrow \mathrm{Col}\, abc \wedge \mathrm{P}bac \wedge [a{=}c \rightarrow b{=}c \vee \mathrm{Q}ab]$
(durch Spiegelung an *b* geht *a* in *c* über, für $a{\neq}c$ heißt das, daß *b* ein[er der] Mittelpunkt[e] von *ac* ist),

(8) $\mathrm{Pc}\, abc \leftrightarrow \mathrm{Q}ab \wedge \neq(abc) \wedge \mathrm{Col}\, abc$
(*a*, *b* haben Polardistanz, und *a*, *b*, *c* sind paarweise verschiedene kollineare Punkte).

Dann gelten in E als Sätze:

(9) $\mathrm{Pc}\, abc \leftrightarrow \neq(abc) \wedge \exists x \mathrm{E}abx \wedge \forall x[\mathrm{E}abx \rightarrow \mathrm{P}xca]$,

(10) $\mathrm{Q}ab \leftrightarrow \exists c \mathrm{Pc}\, abc$;

(11) $\mathrm{Col}\, abc \leftrightarrow a{=}b \vee \forall x[\mathrm{Q}ax \wedge \mathrm{Q}bx \rightarrow \mathrm{Q}cx]$,

(12) $\mathrm{P}^*abc \leftrightarrow \neg\, \mathrm{Col}\, abc \wedge \exists d[ad \perp bc \wedge ab \perp bd \wedge ac \perp cd]$ (Abb. 125),

(13) $\mathrm{P}abc \leftrightarrow \exists d[\mathrm{P}^*abd \wedge \mathrm{P}^*acd] \vee [a{=}b \wedge a{=}c]$,

(14) $ab{\equiv}cd \leftrightarrow \exists e \exists f[\mathrm{M}bec \wedge \mathrm{M}aef \wedge \mathrm{P}cfd]$ (Abb. 122);

(15) $\mathrm{Q}ab \leftrightarrow \mathrm{M}aba \wedge b{\neq}a$;

(16) $\mathrm{Col}\, abc \leftrightarrow a{=}b \vee c{=}b \vee \forall p\, [ab \perp bp \rightarrow cb \perp bp]$.

Folglich ist jeder der Begriffe ≡, P, P^*, Q, Pc, M, ⊥ *mittels eines jeden dieser Begriffe definierbar in E, und* Col *ist mittels eines jeden dieser Begriffe definierbar in E.*

<u>Beweis</u>: Zu (9)(←): Sei *F* eine Ebene, in der *a*, *b*, *c* liegen. Nach Voraussetzung gibt es einen Punkt *x* mit E*abx*. Dann gibt es auch in jeder Ebene durch L(*ab*) einen solchen Punkt (man erhält ihn durch Fällen des

Lots von x auf L(ab) und Abtragen seiner Länge vom Fußpunkt aus auf der Senkrechten zu L(ab) in der Ebene). Sei x_1 ein Punkt mit Eabx_1 in F und x_2 das Bild von x_1 bei der Spiegelung an L(ab).

Annahme: ¬Qab. Dann ist x_1 nicht Pol von L(ab) und damit $x_1 \neq x_2$. Sei K_ν der Kreis um x_ν in F mit dem Radius $ab \equiv xa$ (ν=1,2). Durch Anwendung der Voraussetzung auf x_1 und x_2 ergibt sich, daß außer a und b auch c auf beiden Kreisen liegt. Das steht im Widerspruch zu dem (auch in E gültigen) Satz, daß zwei Kreise höchstens zwei gemeinsame Punkte haben.

Somit ist Qab und damit x_1 der Pol von L(ab) in F (und $x_2=x_1$). Nach Voraussetzung ist nun $x_1c \equiv x_1a \equiv ab$, also Q$x_1c$ und damit $c \in$L(ab).

Zu (9)(→): Gelte die Voraussetzung Pcabc (siehe (8)). Dann ist die Bedingung Eabx gleichwertig damit, daß x der Pol von L(ab) in der Ebene durch a, b und x ist. Damit ergeben sich alle Teile der Behauptung unmittelbar.

Der Bestandteil ∃xEabx in (9) ist natürlich entbehrlich (beweisbar), wenn das Kreisaxiom zur Verfügung steht.

Zu (11): Die Bedingung Qax ∧ Qbx bedeutet für verschiedene Punkte a, b, daß x der Pol von L(ab) in der Ebene durch a, b und x ist (nämlich: a und b liegen auf der Polaren von x). Damit gilt die Richtung (→) offensichtlich; die Umkehrung (←) ergibt sich durch Anwendung der rechten Seite auf den Pol x in einer Ebene durch a, b, c.

Zu (12): Der "Sonderfall", daß a Pol von L(bc) ist, bereitet keine Schwierigkeiten (er ist gleichwertig mit $d \in$L(bc), wenn d von der angegebenen Art ist). Im sonst eintretenden "Normalfall" verläuft der Beweis wie für 4.19(4'), wobei wieder benutzt wird, daß es höchstens zwei - zueinander symmetrische - Punkte x auf L(bc) mit der Eigenschaft $ax \perp xd$ gibt (das gilt auch in E und läßt sich z.B. in den Kleinschen Modellen leicht nachprüfen) (vgl. Abb. 125; im Sonderfall braucht diese Figur dagegen nicht symmetrisch zu sein).

(10), (13) und (15) gelten trivialerweise auf Grund der vorangehenden Definitionen. (14) und (16) lassen sich analog zu 4.18(3) und 4.19(3) beweisen mit dem Unterschied, daß der Mittelpunkt e in (14) nicht mehr eindeutig bestimmt ist. □

<u>4.76 Satz</u>. *In E seien die Begriffe* Q', B^*, Ss, $\leq$, P' *(zusätzlich zu denen von* 4.75*) durch folgende formale Definitionen eingeführt:*

(1) $Q'ab \leftrightarrow a=b \vee \exists c \exists d[ac|bd \wedge Qad \wedge Pcad]$

(a, b haben kleineren Abstand als die halbe Polardistanz [die hier durch a, c gegeben wird]),

(2) $B^*abc \leftrightarrow \exists d[ac|bd \wedge Qad]$

("b liegt *echt zwischen* a und c ", d.h., a und c haben nicht Polardistanz, und der davon verschiedene Punkt b liegt auf der [in diesem Falle eindeutig bestimmten] kürzesten Verbindung von a und c),

(3) $ab \ Ss \ cd \leftrightarrow ab|cd \wedge \neg Qab$

("das Paar $\langle a,b \rangle$ *trennt speziell* ["separates specially"] das Paar $\langle c,d \rangle$", d.h. so, daß a, b nicht Polardistanz haben),

(4) $ab \leq cd \leftrightarrow Qcd \vee \exists y\{[B^*cyd \vee y=c \vee y=d] \wedge ab \equiv cy\}$,

(5) $P'abc \leftrightarrow ab \leq ac$ *(vgl.* 4.12(3)*).*

Dann gelten in E als Sätze:

(6) $Qab \leftrightarrow \neg\exists c[Q'ac \wedge Q'bc]$ (Royden 1959),

(7) $B^*abc \leftrightarrow \neq(abc) \wedge Col\ abc \wedge \neg Qac$
$\wedge \forall u[Col\ acu \wedge Q'au \wedge Q'cu \rightarrow Q'bu]$,

(8) $ab \ Ss \ cd \leftrightarrow \neq(abcd) \wedge Col\ abc \wedge Col\ abd \wedge \neg Qab$
$\wedge \{[B^*acb \wedge \neg B^*adb] \vee [B^*adb \wedge \neg B^*acb]\}$,

(9) $ab|cd \leftrightarrow ab \ Ss \ cd \vee cd \ Ss \ ab$
$\vee [\neq(abcd) \wedge Col\ abc \wedge Col\ abd \wedge Qab \wedge Qcd]$;

(10) $Qab \leftrightarrow a \neq b \wedge \neg\exists c \ B^* acb$;

(11) $Qab \leftrightarrow a \neq b \wedge \neg\exists c \exists d \ ab \ Ss \ cd$;

(12) $Qab \leftrightarrow \forall x P'axb$,

(13) $B^*abc \leftrightarrow \neq(abc) \wedge \forall x[P'axb \wedge P'cxb \rightarrow x=b]$ (R. M. Robinson 1959).

Folglich (unter Verwendung von 4.75*) ist jeder der Begriffe* Q', B^*, Ss, P', $\leq$ *als einziger Grundbegriff für E geeignet.*

<u>Beweis</u> (Skizze): Die Beweise für (6), (8) bis (13) und einige andere anschaulich einleuchtende Sätze, insbesondere über B^*, können dem Leser überlassen bleiben.

Zu (7)($\rightarrow$): Die erste Zeile der Behauptung ergibt sich unmittelbar aus (2). Sei nun u ein beliebiger Punkt der angegebenen Art. Sei u' der Punkt mit Polardistanz von u auf der Geraden $L(ac)$. Dann ist

$\nparallel(acu') \wedge u \neq u'$. Somit gilt einer der folgenden drei Fälle (falls $\nparallel(acuu')$ ist, gilt eine der in diesen Fällen angegebenen Trennungsbeziehungen).

Fall 1: $uu'|ac$ (Abb. 151; darin wird die Situation veranschaulicht in einem Modell, in dem die Punkte von $\mathfrak{P}_n(\mathfrak{F})$ dargestellt werden durch Geraden durch den Nullpunkt von $\mathfrak{L}_{n+1}(\mathfrak{F})$, die Punkte von $L(ac)$ insbesondere durch Geraden in der Zeichenebene, vgl. 2.11). Wegen $Q'au \wedge Q'cu$ liegen a und c echt zwischen u und den Punkten, die von u halbe Polardistanz haben (in Abb. 151 gestrichelt); daraus ergibt sich B^*auc. Da auch B^*abc ist, gilt einer der folgenden beiden Unterfälle.

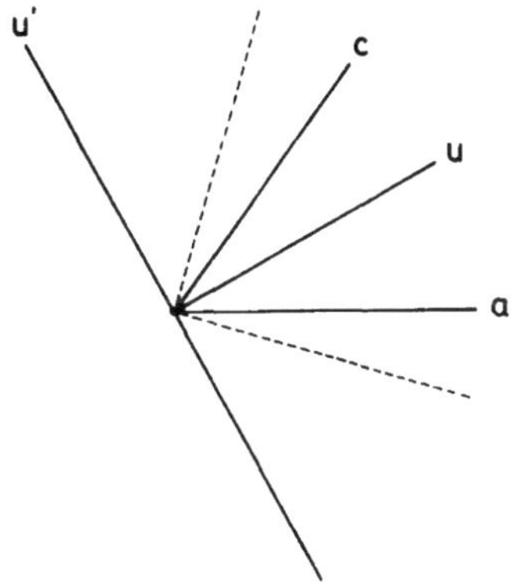

Abb. 151

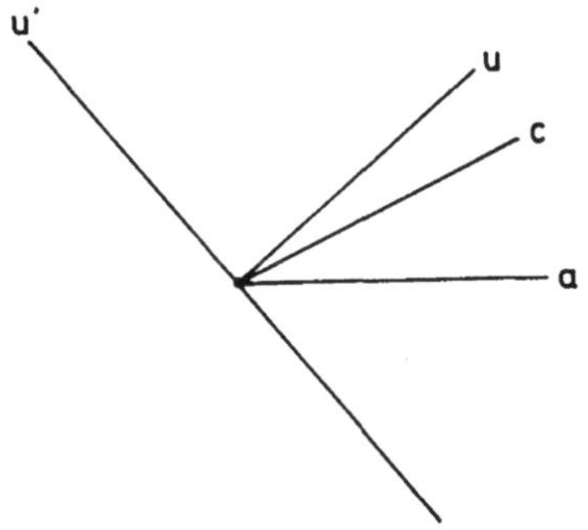

Abb. 152

Fall 1a): $B^*uba \vee b=u$. Wegen $Q'au$ ist dann erst recht $Q'bu$.

Fall 1b): $B^*ubc \vee b=u$. Das ist analog zu 1a) unter Vertauschung von a mit c.

Fall 2: $ua|u'c \vee u=c$ (Abb. 152 mit derselben Darstellung der Punkte). Nach (2) ergibt sich hier $B^*uca \vee u=c$. Wegen B^*abc ist dann B^*uba. Damit gilt die Bedingung von Fall 1a), woraus sich $Q'bu$ ergibt.

Fall 3: $uc|u'a \vee u=a$. Hier ist alles analog zu Fall 2 unter Vertauschung von a mit c.

Zu (7)(←) (Abb. 153 mit derselben Darstellung): Sei u_o der "innere Mittelpunkt" von ac, d.h. derjenige mit B^*au_oc. Das ist ein Punkt u der angegebenen Art, also ist $Q'bu_o$. Seien a' und c' die beiden Punkte auf $L(ac)$, die von u_o halbe Polardistanz haben, und zwar in der Reihenfolge, daß $B^*u_oaa' \wedge B^*u_occ'$ gilt (a', c' haben dann Polardistanz). Wegen $Q'bu_o$ gilt dann $b=u_o \vee B^*u_oba' \vee B^*u_obc'$.

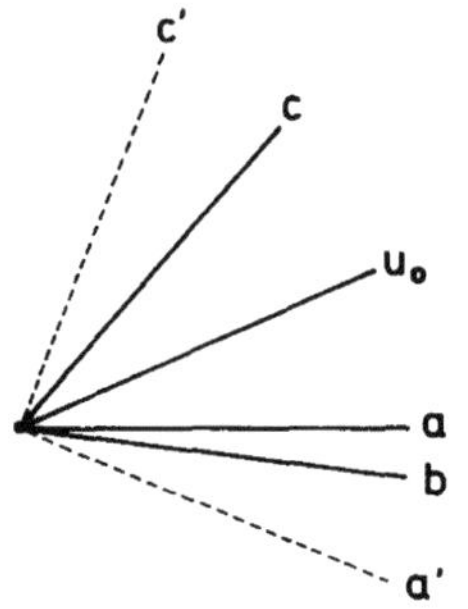

Abb. 153

Annahme: $\neg B^*abc$. Dann ist $b \neq u_o$, und es bleiben die folgenden beiden Fälle.

Fall 1: B^*u_oba'. Dann ist sogar B^*aba'. Sei dann u derjenige Punkt auf $L(ac)$ mit halber Polardistanz von b, für den B^*bu_ou gilt. Wegen B^*u_oab ist dann $Q'au$. Nach Konstruktion ist außerdem $Q'cu$. Anwendung der Voraussetzung auf u liefert also, daß auch $Q'bu$ sein müßte im Widerspruch zur Wahl von u. Somit ist die Annahme falsch.

Fall 2: B^*u_obc'. Hier ist alles analog zu Fall 1 unter Vertauschung von a mit c und a' mit c'. □

4.77 Satz. *In E' (der dimensionsfreien elliptischen Geometrie mit Kreisaxiom) gilt als Satz* (R.M. Robinson 1959):

(1) $P'abc \leftrightarrow \forall d\{Pdac \rightarrow \exists e[Peba \wedge Paed]\}$.

Folglich ist für E' auch P als einziger Grundbegriff geeignet und damit (nach 4.75*) auch jeder der Begriffe* Q, ≡, P^*, Pc, M, ⊥.

Zum Beweis: Der Beweis verläuft analog zu dem von 4.24(1) mit folgenden Unterschieden. Die Mittelpunkte b' bzw. c' seien speziell die "inneren Mittelpunkte" der Strecken ab bzw. ac (d.h. die auf der kürzeren Verbindung), falls die Endpunkte dieser Strecken nicht Polardistanz haben. Im

Beweis der Richtung (→) (für $a \neq c$) kann der Fußpunkt des Lots von d auf L(ac) auch der andere Mittelpunkt c" sein (der "äußere Mittelpunkt", falls ¬Qac, vgl. 4.73); für diesen gilt aber erst recht die Ungleichung $ab' \leq ac' \leq ad$. Der genannte Kreis K kann auch die Polare von a in der betrachteten Ebene sein. □

4.78 Satz. *Sei n eine beliebige natürliche Zahl mit $n \geq 2$. Dann ist die Trennbeziehung | nicht definierbar mittels ≡ in E_n (der n-dimensionalen elliptischen Geometrie ohne Kreisaxiom). Folglich sind die in* 4.75 *genannten Begriffe nicht als einzige Grundbegriffe für E_n geeignet.*

Zum Beweis: Der Beweis verläuft analog zu dem von 4.31 und beruht wieder darauf, daß auch im Kleinschen Modell $\mathfrak{E}_n(\mathfrak{f})$ die Streckenkongruenz schon durch die Körperoperationen des Grundkörpers $\mathfrak{f}$ festgelegt ist, während für die Festlegung der Trennbeziehung die Anordnung von $\mathfrak{f}$ benötigt wird. Wählt man einen (nach 4.32 existierenden) pythagoreischen Körper, der zwei verschiedene Anordnungen besitzt, so erhält man wieder zwei Modelle, auf die sich die Methode von Padoa anwenden läßt. □

Analog zu 4.40 ergibt sich auch

4.79 Satz. *Die Streckenkongruenz ≡ ist nicht definierbar mittels | (also erst recht nicht mittels* Col*) in E_n^2 (der n-dimensionalen vollen elliptischen Geometrie [zweiter Stufe]).*

Beweis: Sei $\mathfrak{A} = \mathfrak{B} = \mathfrak{E}_n(\mathbb{R})$, und sei φ eine projektive Abbildung von $\mathfrak{A}$ auf sich, die zwei zueinander orthogonale Geraden in nicht-orthogonale Geraden überführt. Dann erhält φ als (reguläre) projektive Abbildung bekanntlich (negative Doppelverhältnisse und damit) die Trennbeziehung, aber nicht den Begriff ⊥ und somit (4.75) auch nicht die Streckenkongruenz. Nach 4.29 ergibt sich dann die Behauptung. □

4.80 Anmerkung. Aus unseren Sätzen ergeben sich, wie schon früher gesagt, keine "befriedigenden" Axiomensysteme in den als geeignet gefundenen Grundbegriffen (vgl. 4.9(iv)). Daher sei hier noch hingewiesen auf besonders schöne Axiomensysteme in Kordos 1973 für eine ebene elliptische Geometrie, die aus E_2 durch Weglassen des Begriffs der Trennbeziehung entsteht (für diese Geometrie ist also nach 4.75 jeder der dort genannten untereinander gegenseitig definierbaren Begriffe ≡,..., ⊥ als einziger Grundbegriff geeignet). Das erste Axiomensysten besteht aus 12 Axiomen in einer zweisortigen Sprache (1.42) mit Punktvariablen und Geradenvariablen und den beiden Grundbegriffen ρ der Inzidenz (hier Iz, vgl. 4.59, 60) und σ der Polardistanzeigenschaft (hier Q). Dabei bilden die ersten 5 Axiome für sich ein selbstduales Axiomensystem für projektive Ebenen über beliebigen Körpern von Charakteristik $\neq 2$ mit dem Grundbegriff ρ. Man kann das ganze System auch "dual auffassen", d.h. die Rolle von Punkten und Geraden vertauschen; dann wird σ zum Begriff der Orthogonalität von Geraden. Die Modelle lassen sich (bis auf Isomorphie) so darstellen, daß die beiden Trägermengen U_1, U_2 (der Punkte und der Geraden) zusammenfallen und auch die Begriffe ρ und σ zusammenfallen. Eine entsprechende Identifikation führt dann zu einer elliptischen Geometrie mit einer einsortigen Sprache und einem einzigen zweistelligen Grundbegriff, die sich wahlweise auffassen läßt als Geometrie mit Punktvariablen und dem Grundbegriff der Polardistanzeigenschaft oder als Geometrie mit Geradenvariablen und dem Grundbegriff der Orthogonalität. Für diese Geometrie wird ein vereinfachtes Axiomensystem von 7 Axiomen aufgestellt. Die Modelle der betrachteten Geometrien sind in allen Fällen geeignete elliptische Kleinsche Ebenen (mit geeigneten Relationen zur Interpretation der Grundbegriffe) über formal-reellen pythagoreischen Körpern. Die Eigenschaft, formal-reell zu sein, bedeutet aber gerade die Existenz einer Anordnung (vgl. 3.17). An Stelle der angeordneten Körper in 2.13 hat man also hier anordenbare Körper (dasselbe gilt auch für die elliptische Geometrie in Podehl u. Reidemeister 1934, vgl. 2.14(ii)).

4.81 Die Mittelpunktsbeziehung. Als nächstes betrachten wir die Rolle der Mittelpunktsbeziehung M (hinsichtlich der Definierbarkeit) in euklidischen und affinen Geometrien. Dabei zeigt sich (4.87), daß M hier ein besonders "schwacher" Begriff ist (im Gegensatz zur elliptischen Geometrie, vgl. 4.75). Das folgende Resultat aus Schwabhäuser 1967 geht zurück auf die von H.N. Gupta aufgeworfene Frage, ob die Kollinearität Col mittels M definierbar ist in einem beliebig vorgegebenen kartesischen Raum $\mathfrak{C}_n(\mathfrak{F})$ (im Sinne von 4.3(ii)). Die Antwort wird allgemeiner gegeben für affine Räume über beliebigen Körpern (ohne Anordnung) und unter Einschluß von unendlichen Dimensionen. Im Beweis wird neben M noch eine vierstellige Parallelogrammbeziehung Pg (s.u.) behandelt. Der Einfachheit halber betrachten wir daher die affinen Räume an dieser Stelle von vornherein als Strukturen für eine entsprechende Sprache L(Col,M,Pg).

4.82 Definition. Sei n eine beliebige (endliche oder unendliche) Kardinalzahl und $\mathfrak{F}$ ein beliebiger Körper. Mit $\mathfrak{A}_n(\mathfrak{F})$ bezeichnen wir (wie in 2.8) den (bis auf Isomorphie eindeutig bestimmten) n-dimensionalen

Vektorraum über $\mathfrak{F}$ und benutzen die üblichen Bezeichnungen aus der Vektorrechnung und auch $\overrightarrow{ab}$ für $b-a$ ("der Vektor von a nach b "). Unter dem *n-dimensionalen affinen Raum* $\mathfrak{A}_n(\mathfrak{F})$ *über* $\mathfrak{F}$ (mit Mittelpunkts- und Parallelogrammbeziehung) verstehen wir die Struktur $\mathfrak{A}$ für L(Col,M,Pg), die folgendermaßen festgelegt ist:

(1) $U_{\mathfrak{A}}$ ist die Menge der Vektoren aus $\mathfrak{V}_n(\mathfrak{F})$,

(2) $Col_{\mathfrak{A}}\ abc$ gdw die Vektoren $b-a$, $c-a$ sind linear abhängig,

(3) $M_{\mathfrak{A}}abc$ gdw $2b=a+c$,

(4) $ab\ Pg_{\mathfrak{A}}\ cd$ gdw $b-a=d-c$
(die gerichteten Strecken $\langle a,b\rangle$ und $\langle c,d\rangle$ bestimmen denselben Vektor, d.h., sie sind [gleich orientierte] Gegenseiten eines - eventuell entarteten - Parallelogramms).

4.83 Übungsaufgabe. Wenn man Körper der Charakteristik 2 ausschließt, so sind M, Pg, $\|$ (s. 2.8(i)) definierbar mittels Col in der entsprechenden dimensionsfreien affinen Geometrie, d.h. in der Theorie der (Klasse der) zugehörigen affinen Räume $\mathfrak{A}_n(\mathfrak{F})$ mit $n\geq 2$ (hierbei können auch unendliche n zugelassen werden).

4.84 Satz (Schwabhäuser 1967). *Sei n eine (endliche oder unendliche) Kardinalzahl mit* $n\geq 2$ *und* $\mathfrak{F}$ *ein Körper. Dann gelten die folgenden Behauptungen.*

(i) *Sei* $\mathfrak{F}$ *kein Primkörper. Dann gibt es eine eineindeutige Abbildung* φ *von* $\mathfrak{A}_n(\mathfrak{F})$ *auf sich, die die Addition von Vektoren (von* $\mathfrak{V}_n(\mathfrak{F})$*) und (damit) die Begriffe* M *und* Pg *invariant läßt, aber nicht den Begriff* Col. *Folglich (*4.29*) ist* Col *nicht definierbar mittels* M *oder mittels* Pg *in* Th($\mathfrak{A}_n(\mathfrak{F})$) *(der Theorie erster Stufe des Raumes* $\mathfrak{A}_n(\mathfrak{F})$*). Dasselbe gilt (wie* 4.29*) auch für Formalisierungen der Theorie von* $\mathfrak{A}_n(\mathfrak{F})$ *in anderen Systemen der Logik; dazu gehören die Formalisierungen in höheren Stufen, in der schwachen zweiten Stufe und in den "infinitären Sprachen"* $L_{\kappa\lambda}$ *(s. etwa* Keisler 1971).

(ii) *Sei* $\mathfrak{F}$ *der Körper* $\mathbb{Q}$ *der rationalen Zahlen. Dann ist* Col *nicht definierbar mittels* M *oder mittels* Pg *in* Th($\mathfrak{A}_n(\mathbb{Q})$) *(Theorie erster Stufe).*

(iii) *In* $T = Th^{S2}(\mathfrak{A}_n(\mathbb{Q}))$ *(Theorie in der schwachen zweiten Stufe) sei der Begriff* N *durch folgende formale Definition eingeführt:*

(1) $Nabc \leftrightarrow \exists S\{a\in S \wedge c\in S \wedge \forall r\forall s[r\in S \wedge r\neq c \wedge ab\ Pg\ rs \rightarrow s\in S]\}$
(hierbei variiert S *über endliche Punktmengen).*

In $\mathfrak{A} = \mathfrak{A}_n(\mathbb{Q})$ *(und allgemeiner in* $\mathfrak{A}_n(\mathfrak{f})$*, wobei* $\mathfrak{f}$ *von Charakteristik 0) gilt dann für die zugehörige Relation* $N_{\mathfrak{A}}$ *:*

(2) $N_{\mathfrak{A}}$ *abc gdw a=b oder* $\overrightarrow{ac}$ *ist ein "natürliches Vielfaches"* $k\cdot\overrightarrow{ab}$ *(für ein* $k \in \mathbb{N}$*) des vom Nullvektor verschiedenen Vektors* $\overrightarrow{ab}$*.*

In T *gilt dann als Satz:*

(3) $\text{Col}\,abc \leftrightarrow \exists d\{d \neq a \wedge \text{N}acd \wedge [\text{N}abd \vee \text{N}bad]\}$

und außerdem (sogar in jedem $\mathfrak{A}_n(\mathfrak{f})$ *mit* $\text{char}(\mathfrak{f}) \neq 2$*)*

(4) $ab\,\text{Pg}\,rs \leftrightarrow \exists t[\text{M}ats \wedge \text{M}btr]$.

Folglich ist Col *in* T *definierbar mittels* Pg *und mittels* M.

(iv) *In* $T = \text{Th}^2(\mathfrak{A}_n(\mathbb{Q}))$ *(Theorie in der zweiten Stufe) sei der Begriff* N *eingeführt durch*

(1') $\text{N}abc \leftrightarrow a=b \vee \forall S\{a \in S \wedge \forall r \forall s[r \in S \wedge ab\,\text{Pg}\,rs \rightarrow s \in S] \rightarrow c \in S\}$
(hierbei variiert S *über beliebige Punktmengen).*

Dann gilt alles Weitere wie in (iii).

(v) *Sei p eine Primzahl und* $\mathbb{Z}_p$ *der Primkörper mit p Elementen. In* $T = \text{Th}(\mathfrak{A}_n(\mathbb{Z}_p))$ *(Theorie in der ersten Stufe) gilt dann als Satz:*

(5) $\text{Col}\,abc \leftrightarrow a=b \vee \exists r_0 .. \exists r_{p-1}\{\vartheta \wedge \bigvee_{i=0}^{p-1} c=r_i\}$,

wobei ϑ *folgende Formel ist*

$$r_0=a \wedge r_1=b \wedge \bigwedge_{i=2}^{p-1} \text{M}r_{i-2}\,r_{i-1}\,r_i .$$

Für p=2 soll hierbei die "leere Konjunktion" weggelassen werden, d.h., (5) *vereinfacht sich in diesem Falle zu*

(5') $\text{Col}\,abc \leftrightarrow a=b \vee a=c \vee b=c$.

Für p=2 gilt außerdem (sogar in jedem $\mathfrak{A}_n(\mathfrak{f})$ *mit* $\text{char}(\mathfrak{f})=2$*)*

(6) $\text{M}abc \leftrightarrow a=c$.

Schließlich gilt (in T *und allgemein in jedem affinen Raum* $\mathfrak{A}_n(\mathfrak{f})$*)*

(7) $\text{M}abc \leftrightarrow ab\,\text{Pg}\,bc$.

Folglich ist Col *in* T *definierbar mittels* M *und mittels* Pg. *Für p=2 ist* Col *(ebenso wie* M*) sogar "logisch" definierbar, d.h. ohne Verwendung von anderen Relations- oder Operationszeichen (das bedeutet, daß im definiens nur Gleichungen zwischen Variablen als Primformeln vorkommen).*

4.85 Anmerkung. Für die Dimensionen 0 und 1 ist Col ebenfalls logisch definierbar (da beliebige Punkte stets kollinear sind). Die Voraussetzung $n \geq 2$ ist also für die Nichtdefinierbarkeitsfeststellungen in 4.84 nicht entbehrlich. Die Definierbarkeitsfeststellungen in (iii) bis (v) übertragen sich natürlich auf die entsprechende Untertheorie der Räume beliebiger Dimension n über dem betrachteten Körper (d.h. auf den Durchschnitt der Theorien $\mathrm{Th}^{S2}(\mathfrak{A}_n(\mathbb{Q}))$ bzw. $\mathrm{Th}^{2}(\mathfrak{A}_n(\mathbb{Q}))$ bzw. $\mathrm{Th}(\mathfrak{A}_n(\mathbb{Z}_p))$ für beliebige n), da die angegebenen formalen Definitionen von der Dimension unabhängig sind. (Selbstverständlich übertragen sich auch die Nichtdefinierbarkeitsfeststellungen auf Untertheorien gemäß 4.9(i).)

Der Beweis von 4.84(i) beruht auf einer Überlegung, die später noch mehrfach verwendet wird. Deshalb formulieren wir sie als

4.86 Lemma. *Vor.: $\mathfrak{V}$ ist ein Vektorraum über dem Körper $\mathfrak{f}$ und $\mathfrak{f}_0$ ein Unterkörper von $\mathfrak{f}$. Die Struktur $\mathfrak{V}_0$ entsteht aus $\mathfrak{V}$ durch Einschränkung der Multiplikation von Vektoren mit Skalaren auf (die Skalare von) $\mathfrak{f}_0$. (Im Gegensatz zur Bildung von Unterräumen wird hier also nicht die Menge der Vektoren, sondern der Skalarbereich eingeschränkt.)*

Beh.: (i) *$\mathfrak{V}_0$ ist ein Vektorraum über $\mathfrak{f}_0$, und die Menge der Vektoren und die Vektoraddition in $\mathfrak{V}_0$ sind dieselben wie in $\mathfrak{V}$.*

(ii) *Wenn $\mathfrak{V}$ die Dimension n hat und $\mathfrak{f}$ über $\mathfrak{f}_0$ den Körpergrad $(\mathfrak{f} : \mathfrak{f}_0)=m$ (d.h. die Dimension m als Vektorraum über $\mathfrak{f}_0$, hierbei lassen wir für m und n auch unendliche Kardinalzahlen zu), so hat $\mathfrak{V}_0$ die Dimension $m \cdot n$.*

Zum Beweis: (i) ist trivial (wie man beim Durchgehen der Axiome sofort sieht). Zum Beweis von (ii) sei B eine Basis (der Mächtigkeit n) von $\mathfrak{V}$ und A eine Basis (der Mächtigkeit m) von $\mathfrak{f}$ über $\mathfrak{f}_0$. Weiter sei $B_0 = \{a \cdot b \mid a \in A,\ b \in B\}$, also von der Mächtigkeit $m \cdot n$. Man sieht dann leicht, daß B_0 eine Basis von $\mathfrak{V}_0$ ist, d.h., daß jeder Vektor von $\mathfrak{V}_0$ sich eindeutig als Linearkombination von (endlich vielen Vektoren aus) B_0 (mit Skalarfaktoren aus $\mathfrak{f}_0$) darstellen läßt. □

Beweis von 4.84: Zu (i): Sei $\mathfrak{V} = \mathfrak{V}_n(\mathfrak{f})$, sei $\mathfrak{f}_0$ ein echter Unterkörper von $\mathfrak{f}$ und $\mathfrak{V}_0$ wie in Lemma 4.86 konstruiert. Weiter seien c, d Vektoren, die in $\mathfrak{V}$ linear unabhängig sind (solche existieren wegen $n \geq 2$), sei α ein nicht in $\mathfrak{f}_0$ liegendes Element von $\mathfrak{f}$, und sei $e = \alpha \cdot c$. Dann sind c und e linear u n a b h ä n g i g im Vektorraum $\mathfrak{V}_0$ (wegen $\alpha \notin \mathfrak{f}_0$); nach Konstruktion sind dann auch c, d, e linear unabhängig in $\mathfrak{V}_0$. Sei nun B_0 eine (auf Grund des Auswahlaxioms existierende) Basis von $\mathfrak{V}_o$, die c, d, e als Elemente enthält. Sei ψ die eineindeutige Abbildung von B_o auf sich mit

(8) $\psi(d)=e$, $\psi(e)=d$, $\psi(u)=u$ für jedes $u\in B_o$ mit $u\neq d,e$.

Sei schließlich φ die lineare Abbildung von $\mathfrak{V}_o$ auf sich, die die Abbildung ψ der Basis fortsetzt (vgl. (4) im Beweis von 4.52(ii)). Dann erhält φ als lineare Abbildung die Vektoraddition in $\mathfrak{V}_o$, also auch die Vektoraddition in $\mathfrak{V}$ und damit die Begriffe M und Pg im affinen Raum $\mathfrak{A}_n(\mathfrak{f})$ (4.82)). φ führt jedoch die kollinearen Punkte o (Nullvektor), c, e von $\mathfrak{A}_n(\mathfrak{f})$ in die nicht kollinearen Punkte o, c, d über. Somit leistet φ das Verlangte.

Zu (ii): Nehmen wir das Gegenteil an. Sei dann ϑ von der Form

(9) $\forall a \forall b \forall c[\mathrm{Col}\, abc \leftrightarrow \delta(a,b,c)]$

eine in $\mathfrak{A}_n(\mathbb{Q})$ gültige formale Definition mittels M bzw. Pg (als einzigem Relationszeichen in δ).

<u>Fall 1</u>: n ist endlich. Sei dann ϑ^* eine "arithmetische Reduzierte" von ϑ (auf Grund der Definition 4.82 analog zu 3.25, 3.64 gebildet), d.h. eine Formel der Körpersprache $L_F:=L(+,\cdot,-,0,1)$ (die ϑ mit Hilfe der Koordinaten der auftretenden Punkte beschreibt), so daß für jeden Körper $\mathfrak{f}$ (analog zu 3.26(ii)) gilt

(*) ϑ ist gültig in $\mathfrak{A}_n(\mathfrak{f})$ gdw ϑ^* in $\mathfrak{f}$ gültig ist.

Nach Annahme ist ϑ gültig in $\mathfrak{A}_n(\mathbb{Q})$, also ϑ^* gültig in $\mathbb{Q}$. Sei nun $\mathfrak{f}$ ein Nichtstandardmodell (d.h. ein nicht zu $\mathbb{Q}$ isomorphes Modell) von $\mathrm{Th}(\mathbb{Q})$, ein solches erhält man z.B. nach dem Satz von Löwenheim-Skolem-Tarski (s. 1.54). Damit ist $\mathfrak{f}$ ein Körper der Charakteristik Null, in dem ϑ^* gültig ist, aber kein Primkörper. Wegen (*) ist dann ϑ auch gültig im affinen Raum $\mathfrak{A}_n(\mathfrak{f})$; das bedeutet, daß auch in diesem Raum Col mittels M bzw. Pg definierbar ist. Das widerspricht Beh.(i).

<u>Fall 2</u>: n ist unendlich. Aus dem Satz 3.73 von Scott ergibt sich ein Analogon zu 3.74 für die jetzt betrachteten affinen Räume (da Automorphismen der gewünschten Art existieren). Damit erhält man eine endliche Dimension m (die von der Anzahl der Variablen in ϑ abhängt, vgl. 3.75), so daß die Formel ϑ in $\mathfrak{A}_n(\mathbb{Q})$ gültig ist genau dann, wenn sie in $\mathfrak{A}_m(\mathbb{Q})$ gültig ist. Das bedeutet, daß Col auch im Raum $\mathfrak{A}_m(\mathbb{Q})$ mittels M bzw. Pg definierbar ist, und widerspricht dem Ergebnis von Fall 1.

Zu (iii): Es genügt der Beweis von (2) und (3) ((4) gilt offensichtlich).

Zu (2): Sei r_κ der Punkt mit $\overrightarrow{ar_\kappa}=\kappa\cdot\overrightarrow{ab}$ ($\kappa=0,1,2,\ldots$). Gelte zunächst die rechte Seite von (2). Falls $a=b$ ist, so sei $S=\{a,c\}$. Falls $a\neq b$ und $\overrightarrow{ac}=k\cdot\overrightarrow{ab}$ ist, so sei $S=\{r_0,\ldots,r_k\}$. Dann genügt S der Bedingung auf der rechten Seite von (1), d.h., es ist $N_{\mathfrak{A}}abc$. Gelte nun die rechte Seite von (2) nicht, d.h., es sei $a\neq b$ und c verschieden von allen Punkten r_κ. Eine Punktmenge S, die der Bedingung auf der rechten Seite von (1) genügt, muß dann alle r_κ als Elemente enthalten und damit unendlich sein. Somit gibt es keine endliche Punktmenge dieser Art, d.h., es ist nicht $N_{\mathfrak{A}}abc$.

Zu (3): Abgesehen von den Trivialfällen, daß a mit einem der Punkte b, c zusammenfällt, besagt die rechte Seite, daß es einen Punkt d gibt, so daß $\overrightarrow{ad}=l\cdot\overrightarrow{ac}$ und $\overrightarrow{ad}=\pm k\cdot\overrightarrow{ab}$ für gewisse natürliche Zahlen k, l mit $l\neq 0$. Das bedeutet aber gerade, daß $\overrightarrow{ac}$ ein "rationales Vielfaches" von $\overrightarrow{ab}$ ist (von der Form $\pm\frac{k}{l}\cdot\overrightarrow{ab}$), wie es in der Definition der Kollinearität des Raumes $\mathfrak{A}_n(\mathbb{Q})$ verlangt wird.

Zu (iv): Im Falle $a\neq b$ besagt die rechte Seite von (1'), daß c im Durchschnitt aller Mengen S liegt, die a als Element enthalten und abgeschlossen sind in bezug auf das Antragen des Vektors $\overrightarrow{ab}$. Dieser Durchschnitt ist aber die Menge der r_κ aus dem Beweis von (iii). Also gilt wieder (2) und damit auch alles Weitere wie in (iii).

Zu (v): Abgesehen vom Trivialfall $a=b$ drückt die Formel ϑ offenbar aus, daß $r_0,\ldots, r_{p-1}$ die p Punkte der Geraden durch a und b sind (und zwar r_i der Punkt mit dem Parameter $i\in\mathbb{Z}_p$ in einer Vektorparameterdarstellung dieser Geraden, in der a und b die Parameter 0 bzw. 1 haben). Somit gilt (5). Das Weitere ergibt sich unmittelbar aus den Definitionen. □

4.87 Korollar. *Für* $n\geq 2$ *ist* Col *nicht definierbar mittels* M *in* P_n^2 *(n-dim. euklidische Geometrie zweiter Stufe) und (nach* 4.9(i)*) in Untertheorien davon.*

Beweis: Im Modell $\mathfrak{L}_n(\mathbb{R})$ sind Kollinearität und Mittelpunktsbeziehung (durch definitorische Erweiterung) ebenso festgelegt wie im affinen Raum $\mathfrak{A}_n(\mathbb{R})$. Damit ergibt sich das Gewünschte aus 4.84(i). □

4.88 Anmerkung. Für die Verwendung des Auswahlaxioms im Beweis von 4.84(i), (ii) und 4.87 gilt analog das in Anmerkung 4.56(ii) Gesagte. Insbesondere ist das Auswahlaxiom entbehrlich für die betrachteten Theorien erster Stufe (die Existenz eines höchstens abzählbaren Modells ergibt sich in jedem Fall ohne Auswahlaxiom aus einem geeigneten Beweis des Vollständigkeits- und des Endlichkeitssatzes für abzählbare Sprachen, s. etwa Schwabhäuser 1971, S. 73). Außerdem ist das Auswahlaxiom entbehrlich für Räume $\mathfrak{A}_n(\mathfrak{f})$, für die sich der Unterkörper $\mathfrak{f}_0$ in geeigneter Weise wählen läßt (z.B. so, daß das Produkt $m \cdot n$ von 4.86(ii) höchstens abzählbar wird), und insbesondere für höchstens abzählbare Strukturen.

<u>4.89 Elimination der Punktvariablen</u>. (i) Für das folgende sei T wieder (wie in 4.59) eine (geometrische) Theorie erster Stufe mit zwei Sorten von Variablen - Punktvariablen (a, b,..., x, y,...) und Geradenvariablen (A, B,..., X, Y,...) - ; die Grundbegriffe der zugehörigen Sprache seien Iz, Cpt (s. die folgende Formel (P3)), Isl (s. (P4)) und die (Relations- und Operations-)Zeichen aus der Menge K, die sich sämtlich nur auf Geradenvariablen beziehen; außerdem mögen in T als Sätze gelten:

(P1) $A \neq B \rightarrow \overset{\leq 1}{\exists} x[xIzA \wedge xIzB]$,

(P2) $\forall x \exists A \exists B[A \neq B \wedge xIzA \wedge xIzB]$,

(P3) $CptABC \leftrightarrow \exists x[xIzA \wedge xIzB \wedge xIzC]$
(die Geraden A, B, C sind *kompunktual*, d.h., sie gehen durch einen Punkt).

Der Begriff Cpt kann dabei auch durch (P3) als formale Definition eingeführt sein. Schließlich sei noch Isl eingeführt durch

(P4) $IslAB \leftrightarrow A \neq B \wedge CptABB$
(die Geraden A, B schneiden sich ["intersecting lines"]);

wegen (P3) ist das gleichwertig mit

(P5) $IslAB \leftrightarrow A \neq B \wedge \exists x[xIzA \wedge xIzB]$.

Dann kann man analog zu 4.59 die Punktvariablen eliminieren, indem man zunächst Gleichungen zwischen Punktvariablen ausschaltet und dann die Inzidenz einer beliebigen Geraden mit einem Punkt x beschreibt durch ihre Kompunktualität mit zwei Geraden X, Y, die den Punkt x bestimmen, d.h. sich in x schneiden. Den Sätzen (L4) bis (L8) aus dem Beweis von 4.59 entsprechen dabei analoge Sätze (P6) bis (P10), die Folgerungen aus (P1) bis (P4) sind, dem Satz (L7) entspricht z.B.

(P9) $\exists x\gamma \leftrightarrow \exists X \exists Y[IslXY \wedge \gamma_{x|XY}]$.

Die Durchführung der Einzelheiten sei dem Leser überlassen.

Mit T^L bezeichnen wir dann die Theorie, die aus T durch diese Elimination der Punktvariablen entsteht; d.h., T^L besteht aus denjenigen Sätzen von T, die formuliert sind in der Sprache L^L, die aus der Sprache L

von T durch Weglassen der Punktvariablen und des Begriffs Iz entsteht. (Diese Bildung von T^L aus T ist analog zur Bildung von T aus T_L in 4.60(ii).) Als Folgerungen aus (P1) bis (P4) ergeben sich insbesondere folgende Sätze von T^L:

(1) $\exists A \exists B\, \mathrm{Isl}\, AB$,

(2) $\mathrm{Cpt}ABC \to \mathrm{Cpt}BCA \wedge \mathrm{Cpt}BAC$
([hinreichend für die] Invarianz von Cpt gegenüber Permutationen der Argumente),

(3) $\mathrm{Cpt}ABC \to \mathrm{Cpt}ABB$,

(4) $\mathrm{Cpt}AAA \to \exists B\, \mathrm{Isl}\, AB$,

(5) $A \neq B \wedge \mathrm{Cpt}ABX \wedge \mathrm{Cpt}ABY \to \mathrm{Cpt}AXY \wedge \mathrm{Cpt}BXY$
(analog zu 4.60(4), I.6.26(2)).

Als Folgerungen aus (5) bzw. aus (2) und (5) ergeben sich noch

(6) $A \neq B \wedge \bigwedge_{\nu=1}^{3} \mathrm{Cpt}ABX_\nu \to \mathrm{Cpt}X_1X_2X_3$,

(7) $A \neq B \wedge X \neq Y \wedge \mathrm{Cpt}ABX \wedge \mathrm{Cpt}ABY \to \forall Z[\mathrm{Cpt}ABZ \leftrightarrow \mathrm{Cpt}XYZ]$
(analog zu I.6.26(1)).

(ii) Sei nun umgekehrt T^L eine beliebige Theorie mit der obengenannten Sprache L^L derart, daß (1) bis (5) Sätze von T^L sind, wobei Isl durch (P4) eingeführt ist. Analog zu I.6 bzw. 4.60(iii) - unter Vertauschung der Rollen der Punkte und der Geraden - kann man dann in (den Modellen von) T^L Punkte als Mengen von kompunktualen Geraden ("*Geradenbüschel*", vgl. 4.90) und einen zugehörigen Inzidenzbegriff einführen, indem man folgende Definitionen benutzt:

(8) $\mathrm{Pt}\, x \leftrightarrow \exists A \exists B[\mathrm{Isl}\, AB \wedge x = \{X \mid \mathrm{Cpt}ABX\}]$
("x ist ein Punkt"),

(9) $x\mathrm{Iz}X \leftrightarrow \mathrm{Pt}\, x \wedge X \in x$.

Damit erhält man (als Theorie der konstruierten Modelle) wieder eine Theorie T mit der ursprünglichen Sprache L, in der (P1) bis (P4) als Sätze gelten.

(iii) Für diese Konstruktionen gilt das in 4.60(iii), (iv) Gesagte analog; insbesondere entsprechen sich die Theorien T^L und T der hier genannten Arten eineindeutig.

<u>4.90 Anmerkung</u>. Die Bezeichnung "Geradenbüschel" wird oft nur bei Dimension 2 verwendet, bei Dimension 3 spricht man dann von "Geradenbündeln". Wir brauchen hier jedoch keine Dimensionsbeschränkungen.

4.91 Definition (Grundbegriffe, die sich auf Geraden beziehen). Sei T eine geometrische Theorie, in der nur Punktvariablen (als einzige Variablen) vorkommen, von der in 4.60(iii) genannten Art. Sei T_L aus T durch Hinzunahme von Geradenvariablen und der Inzidenz Iz gemäß 4.60(iii) gebildet, und gelte darin auch 4.89(P2) ((P1) folgt trivialerweise aus 4.59(L1)). (Für alle bisher betrachteten geometrischen Theorien mit nur Punktvariablen gelten diese Voraussetzungen über T.) Weiter sei K eine Menge von (nicht notwendig allen) "punktgeometrischen Begriffen" (d.h. Zeichen für Relationen zwischen Punkten) von T, so daß Col mittels K definierbar ist. Schließlich sei K' eine Menge von "geradengeometrischen Begriffen" (Zeichen für Relationen zwischen Geraden), die in (einer definitorischen Erweiterung von) T_L durch formale Definitionen mittels $K \cup \{Iz\}$ eingeführt sind, und sei der Begriff Cpt aus 4.89(P3) auch mittels K' definierbar.

Sind dann auch die Begriffe aus K mittels $K' \cup \{Iz\}$ in T_L definierbar, so sagen wir, daß K' ein *mit K gleichstarkes System von geradengeometrischen Begriffen* für T ist.

Insbesondere nennen wir K' ein *geeignetes System von geradengeometrischen Begriffen* für T, falls hierbei K ein geeignetes System von Grundbegriffen für T ist (vgl. 4.21). In diesem Falle läßt sich also T_L auch als (definitorische Erweiterung einer) Theorie mit den Grundbegriffen aus $K' \cup \{Iz\}$ auffassen, und die daraus durch Elimination der Punktvariablen gemäß 4.89(i) entstehende Theorie $(T_L)^L$ ist "gleichwertig" mit T (die Modelle entsprechen sich eineindeutig, und die Übergänge in der umgekehrten Richtung gemäß 4.89(ii) und 4.60(ii) führen von $(T_L)^L$ wieder zur ursprünglichen Theorie T). Dieses $(T_L)^L$ nennen wir dann die *zu* T *gehörige Geradengeometrie*; für sie ist K' ein geeignetes System von Grundbegriffen.

4.92 Anmerkung zur Literatur. Geeignete Systeme von geradengeometrischen Begriffen wurden zunächst für euklidische Geometrien mit Kreisaxiom untersucht in Schwabhäuser u. Szczerba 1975. Für solche geeigneten Systeme wurde dabei eine andere, kürzere Definition unter Benutzung von kartesischen Räumen verwendet, die aber im betrachteten Fall zu 4.91 äquivalent ist. Durch Verwendung von 4.91 und der neuen Charakterisierung 4.93(3) ergibt sich hier eine Verallgemeinerung auf die absolute Geometrie.

4.93 Satz (nach Schwabhäuser u. Szczerba 1975). *In der absoluten Geome-*

trie A_L *(4.60(iii)) seien* Cpt *und* Isl *eingeführt durch* 4.89(P3), (P4) *und die Orthogonalität* $\perp$ *von Geraden durch*

(1) $A \perp B \leftrightarrow \exists p \exists q \exists r[\neq(pqr) \wedge Rpqr \wedge pIzA \wedge qIzA \wedge qIzB \wedge rIzB]$.

(i) *In* A_L *gilt (offensichtlich) als Satz*

(2) $Rabc \leftrightarrow \exists A \exists B[A \perp B \wedge aIzA \wedge bIzA \wedge bIzB \wedge cIzB]$.

Folglich (nach 4.20*) ist* {Cpt,⊥} *ein mit der Streckenkongruenz* D *gleichstarkes System von geradengeometrischen Begriffen für die absolute Geometrie* A, *und es ist insbesondere (nach* 4.24*) ein geeignetes System von geradengeometrischen Begriffen für* A' *(mit Kreisaxiom).*

(ii) *Außerdem gilt als Satz in* $(A_{\geq 3})_L$

(3) $$CptA_1A_2A_3 \leftrightarrow A_1=A_2=A_3 \vee \bigvee_{\substack{1\leq\mu<\nu\leq 3\\ \kappa=6-\mu-\nu}} [A_\mu=A_\nu \wedge Isl\, A_\nu A_\kappa]$$
$$\vee \{\neq(A_1A_2A_3) \wedge \bigwedge_{1\leq\mu<\nu\leq 3} Isl\, A_\mu A_\nu$$
$$\wedge \forall B[B\perp A_1 \wedge B\perp A_2 \rightarrow Isl\, BA_3 \vee B=A_3]\}$$

und in $(A_{\geq 4})_L$

(4) $$Isl\, A_1A_2 \leftrightarrow A_1 \neq A_2 \wedge \exists B_1 \exists B_2[\bigwedge_{\mu=1}^{2} \bigwedge_{\nu=1}^{2} B_\mu \perp A_\nu \wedge B_1 \perp B_2]$$

(A_1*,* A_2 *sind voneinander verschieden und besitzen zwei gemeinsame Senkrechten* B_1*,* B_2*, die auch aufeinander senkrecht stehen).*

Folglich ist {Isl,⊥} *für* $A_{\geq 3}$ *und* {⊥} *für* $A_{\geq 4}$ *ein mit* D *gleichstarkes System von geradengeometrischen Begriffen; für* $A'_{\geq 3}$ *bzw.* $A'_{\geq 4}$ *sind diese Systeme geeignete Systeme von geradengeometrischen Begriffen.*

(iii) *Für* P_2^2 *(ebene euklidische Geometrie zweiter Stufe) - und folglich auch für jede Untertheorie davon - gibt es kein geeignetes System von zweistelligen geradengeometrischen Begriffen.*

Für die dreidimensionale Geometrie P_3^2 haben wir dagegen mit {Isl,⊥} ein geeignetes System von zwei zweistelligen geradengeometrischen Begriffen gefunden. Es ist aber meines Wissens nicht bekannt, ob dafür die Orthogonalität ⊥ allein als Grundbegriff genügt.

<u>4.94 Problem</u>. (i) Ist {⊥} ein mit D gleichstarkes System von geradengeometrischen Begriffen für A_3 oder wenigstens für eine geeignete Obertheorie davon? (Das bedeutet, daß Isl mittels ⊥ definierbar ist in der

zugehörigen Geradengeometrie.)

(ii) Gibt es überhaupt ein System mit derselben Eigenschaft, das nur aus einem einzigen zweistelligen Begriff besteht?

Beweis von 4.93: Zu (ii) genügt der Beweis von (3) und (4) in den genannten Geometrien.

Zu (3): Falls die Geraden A_1, A_2, A_3 nicht paarweise verschieden sind (erste Zeile von (3)), ist alles klar. Seien sie nun paarweise verschieden, dann ergibt sich aus beiden Seiten von (3), daß sich je zwei von ihnen schneiden. Sei x der Schnittpunkt von A_1 und A_2. Eine gemeinsame Senkrechte B von A_1 und A_2 kann nicht in der von A_1 und A_2 aufgespannten Ebene $\mathrm{Pl}(A_1,A_2)$ liegen, sie steht somit senkrecht auf dieser Ebene im Punkte x. Cpt $A_1A_2A_3$ ist nun gleichwertig damit, daß auch A_3 durch x geht. Daraus ergibt sich die rechte Seite von (3). Geht dagegen A_3 nicht durch x, so schneidet sie A_1 und A_2 in verschiedenen Punkten, liegt also ganz in der Ebene $\mathrm{Pl}(A_1,A_2)$ und hat damit keinen gemeinsamen Punkt mit einer solchen (auf Grund der Dimensionsvoraussetzung vorhandenen) Senkrechten B, d.h., daß die letzte Bedingung von (3) (dritte Zeile) verletzt wird.

Zu (4)(→): Sei x der Schnittpunkt von A_1 und A_2. Sei B_1 eine Senkrechte in x auf $\mathrm{Pl}(A_1,A_2)$ und B_2 eine Senkrechte in x auf dem von A_1, A_2, B_1 aufgespannten dreidimensionalen Unterraum U. Diese (nach I.11.63 existierenden) Senkrechten leisten dann das Verlangte.

Zu (4)(←): Nehmen wir an, daß die voneinander verschiedenen Geraden A_1, A_2 sich nicht schneiden, also keine gemeinsamen Punkte haben. Seien B_1, B_2 beliebige gemeinsame Senkrechten von A_1, A_2, die voneinander verschieden sind. Dann können B_1, B_2 die Gerade A_1 nicht im selben Punkt schneiden, da das Lot von einem solchen Punkt auf A_2 eindeutig bestimmt ist. Damit haben B_1, B_2 aber keinen gemeinsamen Punkt (falls sie in einer Ebene liegen, etwa nach I.12.9, sonst trivialerweise), also können sie nicht aufeinander senkrecht stehen. Das widerspricht der Voraussetzung ((4) rechts).

Zum Beweis von (iii) formulieren wir zunächst ein Lemma.

4.95 Definition. Für ein beliebiges Modell $\mathfrak{A}$ der euklidischen Geometrie P sei Ami ("angle measures interval") die Menge der Größen von spitzen und von rechten Winkeln. Diese Winkelgrößen seien dabei als Äquivalenzklassen von kongruenten Winkeln bzw. den sie bestimmenden Punktetripeln eingeführt (vgl. I.13.4); bei der üblichen Winkelmessung in den Modellen $\mathcal{L}_n(\mathbb{R})$ entspricht Ami dem Intervall $<0,\frac{\pi}{2}>$. Für $\alpha \in$ Ami betrachten wir die folgenden zweistelligen Relationen Ag_α, Agd_α zwischen Geraden von $\mathfrak{A}$ (die gemäß 4.60(iii) eingeführt seien):

(1) $\mathrm{Ag}_\alpha A_1 A_2 \leftrightarrow \exists b \exists a_1 \exists a_2 [\bigwedge_{\nu=1}^{2} (b \mathrm{Iz} A_\nu \wedge a_\nu \mathrm{Iz} A_\nu \wedge a_\nu \neq b) \wedge <a_1, b, a_2> \in \alpha]$

(A_1, A_2 *bilden* miteinander *einen Winkel der Größe* α),

(2) $\mathrm{Agd}_\alpha A_1 A_2 \leftrightarrow \neg \exists x \bigwedge_{\nu=1}^{2} x \mathrm{Iz} A_\nu \wedge \exists A_2' [A_2' \| A_2 \wedge \mathrm{Ag}_\alpha A_1 A_2']$

(A_1, A_2 sind [als Punktmengen] *disjunkt* und *bilden "im weiteren Sinne"* [bis auf Parallelverschiebung] *einen Winkel der Größe* α).

Für $U, V \subseteq$ Ami werde die zweistellige Relation W_{UV} zwischen Geraden von $\mathfrak{A}$ definiert durch

(3) $\mathrm{W}_{UV} AB \leftrightarrow \exists \alpha [\alpha \in U \wedge \mathrm{Ag}_\alpha AB] \vee \exists \beta [\beta \in V \wedge \mathrm{Agd}_\beta AB]$.

Man beachte:

(i) Für die Größe 0 der Nullwinkel ist Ag_0 die Gleichheitsrelation und Agd_0 die Parallelität $\check{\|}$ im engeren Sinne (I.12.2).

(ii) Für beliebige Geraden A_1, A_2 von $\mathfrak{A}$ trifft genau eine der Relationen (1), (2) mit einem eindeutig bestimmten α zu.

(iii) Wenn $\mathfrak{A}$ die Dimension 2 hat, so kann die Relation Agd_β für $\beta \neq 0$ niemals zutreffen, also sind für V in (3) nur die beiden Fälle $V=\emptyset$ und $V=\{0\}$ von Interesse.

4.96 Lemma. *Sei* $\mathfrak{A}$ *wie in* 4.95. *Die zweistelligen Relationen zwischen Geraden von* $\mathfrak{A}$, *die invariant sind gegenüber Ähnlichkeitstransformationen, sind genau die Relationen* W_{UV} *mit* $U, V \subseteq$ Ami.

Beweis: In beliebigen Modellen der euklidischen Geometrie sind die Ähnlichkeitstransformationen gerade die Automorphismen, d.h. diejenigen eineindeutigen Abbildungen des Modells auf sich, die die Grundbegriffe (D und B) invariant lassen. Sie lassen sich auch in bekannter Weise mit den Mitteln der analytischen Geometrie charakterisieren.

Wenn die Relation Ag_α ($\alpha \in$ Ami) auf irgendwelche Geraden A_1, A_2 zutrifft, so trifft sie auf weitere Geraden A_1', A_2' offenbar genau dann zu, wenn sich diese aus A_1, A_2 durch eine Ähnlichkeitstransformation erhalten lassen. Entsprechendes gilt für die Relationen Agd_β mit $\beta \in$ Ami (für $\beta \neq 0$

kann man das daraus erhalten, daß zwei nicht in einer Ebene liegende Geraden stets genau eine gemeinsame Senkrechte besitzen).

Damit sind auch die W_{UV} invariant gegenüber Ähnlichkeitstransformationen. Sei andererseits R eine beliebige gegenüber Ähnlichkeitstransformationen invariante zweistellige Relation zwischen Geraden von $\mathfrak{A}$. Wir setzen

(1) $U:=\{\alpha \mid \exists A_1 \exists A_2 [\mathrm{Ag}_\alpha A_1 A_2 \wedge R\, A_1 A_2]\}$,

(2) $V:=\{\beta \mid \exists A_1 \exists A_2 [\mathrm{Agd}_\beta A_1 A_2 \wedge R\, A_1 A_2]\}$.

Aus dem Vorangehenden ergibt sich dann ohne weiteres, daß $R = W_{UV}$ ist. □

Beweis von 4.93(iii): Sei $\mathfrak{A}$ ein beliebiges Modell der ebenen euklidischen Geometrie P_2. Sei C eine feste Gerade von $\mathfrak{A}$ und σ eine eineindeutige Abbildung derart, daß $A \| \sigma(A)$ für alle Geraden A und $\sigma(C) \neq C$ (z.B. eine Parallelverschiebung). Sei dann die Abbildung φ für Geraden von $\mathfrak{A}$ definiert durch

(5) $$\varphi(A) = \begin{cases} \sigma(A), & \text{falls } A \| C, \\ A & \text{sonst.} \end{cases}$$

Man sieht dann, daß unter der Abbildung φ alle Relationen der Form W_{UV} invariant sind (vgl. 4.95(iii)), aber die Kompunktualität $\mathrm{Cpt}_{\mathfrak{A}}$ nicht invariant ist. Wäre nun K' ein geeignetes (oder auch nur ein mit D gleichstarkes) System von zweistelligen geradengeometrischen Begriffen, so müßten diese Begriffe unter Ähnlichkeitstransformationen invariant, also die zugehörigen Relationen in $\mathfrak{A}$ nach 4.96 von der Form W_{UV} sein, es müßte aber auch Cpt mittels K' definierbar sein. Das widerspricht dem vorherigen Ergebnis über die Abbildung φ. Ist $\mathfrak{A}$ insbesondere das Modell $\mathfrak{L}_2(\mathbb{R})$ von P_2^2, so ergibt sich das Gewünschte. □

Aus diesem Beweis ergibt sich unmittelbar die folgende Verschärfung von 4.93(iii).

4.97 Korollar. *Sei* T *irgendeine widerspruchsfreie Erweiterung (Obertheorie) der ebenen euklidischen Geometrie* P_2 *(d.h. eine, die ein Modell* $\mathfrak{A}$ *besitzt). Dann gibt es für* T *- und folglich auch für jede Untertheorie davon - kein geeignetes und kein mit* D *gleichstarkes System von zweistelligen geradengeometrischen Begriffen.*

4.98 Anmerkung. Verwendet man im obigen Beweis ein beliebiges Modell $\mathfrak{A}$ von P (ohne Dimensionsforderung), so ergibt sich ebenso, daß jedenfalls die Relationen der Form W_{UU} unter φ invariant sind (beliebige Geraden

A, B mit $\mathrm{Ag}_\alpha AB \vee \mathrm{Agd}_\alpha AB$ werden durch φ in ebensolche Geraden übergeführt). Als Spezialfall erhalten wir

4.99 Korollar. *Sei* T *irgendeine widerspruchsfreie Erweiterung von P. Dann ist* Cpt *in* T *(und Untertheorien) nicht definierbar mittels* $\{\mathrm{I},\|\}$, *wobei* I *eingeführt ist durch*

(1) $A\,\mathrm{I}\,B \leftrightarrow \exists B'[B'\|B \wedge A \perp B']$
(A, B bilden "im weiteren Sinne" [bis auf Parallelverschiebung] *einen rechten Winkel).*

4.100 Punkte und Geraden als einheitliche geometrische Objekte. Für das Folgende sei T_L wieder eine "zweisortige" geometrische Theorie mit Punktvariablen (a, b, c,...) und Geradenvariablen (A, B, C,...) als den beiden Sorten von Variablen. Unter einfachen geometrischen Voraussetzungen an die Theorie kann man nach 4.59 bzw. 4.89(i) die Geradenvariablen bzw. die Punktvariablen eliminieren und damit zu einer einsortigen Theorie übergehen.

Wir betrachten jetzt noch einen anderen Übergang zu einer einsortigen Theorie T_U mit einer Sorte von Variablen (x, y, z,...) für ("unifizierte") geometrische Objekte, das sind Punkte u n d Geraden. Über die Sprache L von T_L setzen wir voraus, daß sie kein Operationszeichen enthält und daß bei den Relationszeichen für jede Stelle festgelegt ist, auf welche Sorte sie sich bezieht. Die Sprache L_U von T_U enthalte dieselben Relationszeichen und zwei zusätzliche einstellige Relationszeichen ("Sortenprädikate") Pt und Ln; Pt x bzw. Ln x bedeute, daß das zugehörige Objekt x ein Punkt bzw. eine Gerade ist.

Aus einem beliebigen Modell von T_L entsteht eine Struktur für L_U dadurch, daß die beiden (o.B.d.A. disjunkten) Individuenbereiche der Punkte bzw. der Geraden zu einem neuen Individuenbereich vereinigt werden, daß Pt und Ln die angegebene Bedeutung erhalten und daß die übrigen Relationen (als Mengen von n-tupeln) ungeändert bleiben. Die Sätze von T_U seien dann gerade diejenigen, die in allen so (aus Modellen von T_L) konstruierten Strukturen gelten. Ein solcher Satz ist z.B.

(1) $\forall x[(\mathrm{Pt}\,x \vee \mathrm{Ln}\,x) \wedge \neg(\mathrm{Pt}\,x \wedge \mathrm{Ln}\,x)]$.

Man sieht, daß diese Konstruktion der einsortigen Theorie T_U aus T_L von speziellen geometrischen Voraussetzungen unabhängig ist und damit allgemein für mehrsortige Theorien durchführbar ist.

Wir verwenden die Bezeichnung T_U insbesondere, wenn T gegeben (z.B. A oder A') und T_L daraus gemäß 4.60(iii) gebildet ist.

Für Geometrien dieser Art kann man nun ohne Dimensionsvoraussetzungen einen zweistelligen Grundbegriff einführen.

4.101 Satz. *In der (dimensionsfreien) absoluten Geometrie* A_U *sei der zweistellige Begriff* I *(nach Bachmann* 1959*) eingeführt durch*

(1) $I\,xy \leftrightarrow x\,Iz\,y \vee x \perp y$

(oder ausführlicher

(1') $I\,xy \leftrightarrow Ln\,y \wedge [(Pt\,x \wedge x\,Iz\,y) \vee (Ln\,x \wedge x \perp y)]$ *)*.

Dann gelten in A_U *als Sätze:*

(2) $Ln\,y \leftrightarrow \exists x\, I\,xy$,

(3) $Pt\,x \leftrightarrow \neg Ln\,x$,

(4) $x\,Iz\,y \leftrightarrow Pt\,x \wedge I\,xy$,

(5) $x \perp y \leftrightarrow Ln\,x \wedge I\,xy$.

Damit ist I *als einziger Grundbegriff für* A'_U *(mit Kreisaxiom) geeignet. Für* A_U *ist* I *"gleichstark" (in einem Sinne analog zu* 4.91*) mit der Streckenkongruenz* D.

Beweis: Die Sätze (2) bis (5) gelten offensichtlich. Als System der "ursprünglich gegebenen Begriffe" (4.21) für A_U erhält man zunächst auf Grund der Definitionen (4.60(iii) und 4.100) das System {D,B,Iz,Pt,Ln}. Die gewünschten Definierbarkeiten mittels I bzw. mittels D ergeben sich dann aus (2) bis (5), 4.93(1), (2), 4.20 und 4.24. □

4.102 Anmerkung. In Bachmanns "Aufbau der Geometrie aus dem Spiegelungsbegriff" (Bachmann 1959), S. 32ff. wird der Begriff I (dort bezeichnet mit |) im Zusammenhang mit einem gruppentheoretischen Aufbau von allgemeineren ebenen Geometrien eingeführt. In diesem Aufbau werden geometrische Sätze formuliert als Sätze über die Gruppe der Bewegungen und das Erzeugendensystem der Geradenspiegelungen. Punkte und Geraden werden dann aufgefaßt (dargestellt) als spezielle Bewegungen, nämlich Punkt- bzw. Geradenspiegelungen; I xy bedeutet, daß das Produkt der involutorischen Gruppenelemente x, y ebenfalls involutorisch ist.

4.103 Geometrien mit Operationen als Grundbegriffen. In diesem Buch werden fast ausschließlich Relationen bzw. Relationszeichen (s. 4.3(i)) als

Grundbegriffe von Geometrien betrachtet. Inzwischen sind aber auch Geometrien mit Operationen (insbesondere Konstruktionsoperationen) als Grundbegriffen untersucht worden, vgl. 8.24. In diesem Zusammenhang sei das folgende Definierbarkeitsresultat von Engeler erwähnt.

4.104 Satz (Engeler 1983, S. 77, 80). (i) *Für die ebene (volle) euklidische Geometrie* P_2^2 *gibt es kein geeignetes System von Grundbegriffen, das nur aus zweistelligen Operationszeichen für Punkte (Zeichen für Operationen, die Punkten wieder Punkte zuordnen) besteht.*

(ii) *Dagegen ist ein zweistelliges Operationszeichen für Geraden als einziger Grundbegriff für* P_2' *geeignet.*

Zum Beweis (Hinweise): Zu (i): Für die Ebene $\mathbb{C}$ der komplexen Zahlen als Modell wird gezeigt, daß sich jede Operation ρ_i der angegebenen Art (auf Grund der Invarianz unter Ähnlichkeitstransformationen) darstellen läßt in der Form

(1) $\rho_i(x,y) = s_i\cdot(x-y)+y$,

wobei s_i eine geeignete komplexe Zahl ist. Darauf werden ähnliche Methoden wie im Beweis von 4.52(ii) angewendet. Für den Spezialfall einer einzigen zweistelligen Operation (die sich als dreistellige Relation auffassen läßt) kann man das Ergebnis auch als Folgerung aus dem Tarskischen Resultat 4.55(i) erhalten.

Zu (ii): Für die Operation, die beliebigen Geraden a, b das Bild von b bei der Spiegelung an a zuordnet, wird gezeigt, daß sie das Verlangte leistet (vgl. 4.102); dazu wird eine geeignete Charakterisierung (formale Definition) der Pierischen Relation gegeben. □

4.105 Einbeziehung der Dimension 1 (nach Notizen von Alfred Tarski). Wenn man bei der Betrachtung n-dimensionaler Geometrien die Dimension 1 einbeziehen will, liegt es nahe, die auch sonst verwendeten Grundbegriffe D und B beizubehalten (vgl. aber 4.108). In diesem Sinne definieren wir die *elementare* bzw. *volle eindimensionale euklidische Geometrie* P_1^* bzw. P_1^2 als die Theorie erster bzw. zweiter Stufe des eindimensionalen kartesischen Raumes $\mathcal{L}_1(\mathbb{R})$, d.h. (vgl. 1.35(iii), 1.43, I.1.4)

(1) $P_1^* = \mathrm{Th}(\mathcal{L}_1(\mathbb{R}))$,

(2) $P_1^2 = \mathrm{Th}^2(\mathcal{L}_1(\mathbb{R}))$.

Die Definition I.1.4 der kartesischen Räume $\mathcal{L}_n(\mathfrak{F})$ läßt sich im Falle $n=1$ noch vereinfachen; für die Relationen von $\mathcal{L}_1(\mathfrak{F})$ gilt nämlich statt I.1.4(2), (3) offenbar auch folgende Charakterisierung

(3) $D^1_{\mathfrak{f}}xyuv$ gdw $|x-y|=|u-v|$
gdw $x-y=u-v$ oder $x-y=v-u$,

(4) $B^1_{\mathfrak{f}}xyz$ gdw $x\leq y\leq z$ oder $z\leq y\leq x$.

Somit wird die Multiplikation von $\mathfrak{f}$ in dieser Charakterisierung nicht gebraucht, und $\mathfrak{L}_1(\mathfrak{f})$ wird schon eindeutig festgelegt durch die angeordnete additive Gruppe von $\mathfrak{f}$. Diese ist stets nicht-trivial (d.h. wenigstens zweielementig), abelsch und dividierbar. Nun sind aber alle solchen angeordneten Gruppen untereinander elementar äquivalent (Robinson 1956, vgl. auch Beweis von 3.30). Die Modelle von P^*_1 sind daher bis auf Isomorphie gerade die eindimensionalen kartesischen Räume über solchen angeordneten Gruppen. (Damit entfallen die sonst vorhandenen Unterschiede zwischen P^*_n, P'_n und P_n.) Dagegen ist P^2_1 kategorisch, d.h., alle Modelle sind untereinander isomorph. Auf Grund der Gültigkeit des Axioms vom Dedekindschen Schnitt läßt sich nämlich in einer angeordneten Gruppe der betrachteten Art - nach Wahl eines beliebigen positiven Elements zum Einselement - eine Multiplikation (mit den Mitteln der zweiten Stufe) definieren, so daß eine zu $\mathbb{R}$ isomorphe Struktur entsteht.

Wir verzichten auf die Angabe eines Axiomensystems und erwähnen nur, daß man für P^2_1 ein endliches, aber für P^*_1 nur ein unendliches (rekursives) Axiomensystem angeben kann (an die Stelle der jetzt entbehrlichen elementaren Stetigkeitsaxiome [s. 3.30] treten jetzt unendlich viele andere Axiome, die für jedes $n\in\mathbb{N}-\{0,1\}$ oder wenigstens jede Primzahl n die Dividierbarkeit durch n ausdrücken).

In den folgenden Definierbarkeitsresultaten zeigen sich einige Besonderheiten für die Dimension 1.

<u>4.106 Satz</u> (nach Tarski). (i) *In* P^*_1 *sind die Begriffe* D *und* P *(Streckenkongruenz und Pierische Relation) wechselseitig definierbar; als formale Definition von* D *mittels* P *in* P^*_1 *hat man (einfacher als die nach wie vor gültigen Sätze* 4.18(1), (3)*)*

(1) Mabc $\leftrightarrow$ Pbac $\wedge$ (a$\neq$c $\vee$ b=c) *(Mittelpunktsbeziehung),*

(2) Dabcd $\leftrightarrow$ $\exists$e{[Maed $\wedge$ Mbec] $\vee$ [Maec $\wedge$ Mbed]}.

(ii) *Für den wie früher definierten Begriff* P' *(*P'abc $\leftrightarrow$ ab$\leq$ac*) gelten in* P^*_1 *als Sätze (analog zu* 4.23*)*

(3) Babc $\leftrightarrow$ P'abc $\wedge$ P'cba,

(4) $Pabc \leftrightarrow P'abc \wedge P'acb$;
folglich ist P' *auch für* P_1^* *als einziger Grundbegriff geeignet.*

(iii) *In* P_1^2 *(und entsprechenden Formalisierungen höherer Stufe) ist* D *nicht definierbar mittels* B.

(iv) *In* P_1^* *ist* B *nicht definierbar mittels* D.

Zum Beweis: (i) und (ii) sind offensichtlich ((2) beruht darauf, daß eine Strecke in eine beliebige dazu kongruente Strecke auf derselben Geraden übergeführt werden kann durch Spiegelung an einem geeigneten Punkt e dieser Geraden).

Zu (iii): Diese Feststellung ist analog zu dem entsprechenden Resultat 4.40 für Dimensionen $n \geq 2$, der Beweis verläuft jedoch anders. Hier leistet z.B. die Bijektion φ von $\mathfrak{A} = \mathcal{L}_1(\mathbb{R})$ auf sich mit

$$(5) \quad \varphi(x) = \begin{cases} x & \text{, falls } x \geq 0, \\ x+x, & \text{falls } x < 0 \end{cases}$$

das Verlangte (sie erhält die Zwischenbeziehung, aber nicht die Streckenkongruenz). Dagegen wäre eine affine Abbildung (wie sie im Beweis von 4.40 verwendet wurde) für diesen Zweck ungeeignet.

Zu (iv): Hier unterscheidet sich schon das Resultat von dem entsprechenden für höhere Dimensionen (4.24). Zum Beweis genügt es (auf Grund der Methode der Reduziertenbildung) zu zeigen, daß in der Struktur der reellen Zahlen $\leq$ nicht (in der ersten Stufe) definierbar ist mittels + und 1. Das wiederum ergibt sich aus einer Charakterisierung derjenigen Mengen von reellen Zahlen, die sich durch Formeln $\alpha(x)$ der Sprache L(+,1) mit einer einzigen freien Variablen x definieren lassen (im Sinne von 4.3(ii)). Mit Hilfe einer Quantorenelimination für die Theorie der reellen Zahlen in der erweiterten Sprache L(+,-,0,1) (vgl. auch 5.18, 23) erhält man, daß sich lediglich die endlichen Mengen von rationalen Zahlen und deren Komplemente (in $\mathbb{R}$) in dieser Weise definieren lassen. Insbesondere ist also die Menge der nicht-negativen reellen Zahlen nicht in dieser Weise definierbar, während sie selbstverständlich mittels + und $\leq$ definierbar ist. □

4.107 Anmerkung (nach Tarski). Als kompliziert erweist sich das Problem, ob in P_1^2 (zweite Stufe) und Formalisierungen höherer Stufe der Begriff B mittels D definierbar ist oder - gleichwertig - ob $\leq$ definierbar ist mittels + und 1 in der Theorie der reellen Zahlen entsprechender Stufe.

Die Antwort hängt nämlich ab von dem System der Mengenlehre, das für die Metatheorie zugrunde gelegt wird.

Auf Grund einer Überlegung von Lindenbaum (s. Tarski u. Lindenbaum 1927) ergibt sich die entsprechende Nichtdefinierbarkeit aus der Existenz einer nicht Lebesgue-meßbaren Lösung der "Cauchyschen Funktionalgleichung" $g(u+v)=g(u)+g(v)$ mit (o.B.d.A.) $g(1)=1$, die eine Bijektion von $\mathbb{R}$ ist. Eine solche kann nämlich nicht monoton sein; sie erhält also die Begriffe + und 1, aber nicht ≤ (und B) und leistet damit das für die Methode von Padoa (4.29) Verlangte. Die Existenz eines solchen g ergibt sich aus dem Auswahlaxiom (AC).

Setzt man andererseits statt AC das sog. Determiniertheitsaxiom (AD) voraus, so ergibt sich, daß jede reelle Funktion Lebesgue-meßbar ist, und daraus ergibt sich die oben genannte Definierbarkeit in einer "zweiten Stufe mit Funktionsvariablen". Als formale Definition von ≤ mittels + und 1 kann man verwenden (Tarski)

(1) $x \leq y \leftrightarrow \exists g[\forall u \forall v\, g(u+v)=g(u)+g(v) \wedge x+g(g(1))=y]$
(hierbei variiert g über Funktionen von $\mathbb{R}$ in $\mathbb{R}$).

Aus der ersten in (1) geforderten Eigenschaft von g ergibt sich dann nämlich (s. etwa Sierpiński 1958, S.442), daß g monoton und genauer $g(x)=x\cdot g(1)$ ist für jedes $x\in\mathbb{R}$, somit $g(g(1))=g(1)\cdot g(1)\geq 0$. Natürlich läßt sich auch jede nicht-negative reelle Zahl in der Form $g(g(1))$ mit einem solchen g darstellen. Damit besagt die rechte Seite von (1), daß y aus x durch Addition einer nicht-negativen Zahl entsteht.

Eine analoge Abhängigkeit vom zugrunde gelegten System der Mengenlehre ergibt sich übrigens für das Problem der Abhängigkeit des Axioms von Pasch von den übrigen Axiomen eines geeigneten Axiomensystems, vgl. 8.18 .

4.108 Stärkere Grundbegriffe für Dimension 1. Die Behandlung der eindimensionalen Geometrie im Sinne von 4.105 unter Beibehaltung der üblichen geometrischen Grundbegriffe hat zur Folge, daß man als zugehörige algebraische Strukturen gewisse angeordnete Gruppen erhält, also "einfachere Strukturen" als für die höheren Dimensionen. Man kann aber auch umgekehrt die geometrischen Grundbegriffe so stark wählen, daß man für die elementare eindimensionale Geometrie ebenfalls die üblichen angeordneten Körper als zugehörige algebraische Strukturen erhält.

Dieser Weg - bei dem sich sozusagen die Geometrie an der Algebra orientiert - wird eingeschlagen in den Untersuchungen von Makowiecka 1965,

1975a bis e, 1976, 1977. Da dabei die Dimension 1 keine Ausnahme mehr bildet, kann sie auch (besser) einbezogen werden in eine geeignete Definition von dimensionsfreien Geometrien. In den genannten Arbeiten werden solche dimensionsfreien und n-dimensionale ($n \geq 1$) euklidische Geometrien (mit Modellen) über reell-abgeschlossenen, euklidischen, pythagoreischen und beliebigen Körpern mit und ohne Anordnung betrachtet. Dafür werden verschiedene (fünf- und) vierstellige Grundbegriffe eingeführt; als die wichtigsten erwähnen wir nur die Begriffe K und T, die sich für kartesische Räume $\mathfrak{A}$ (geeigneter Art) folgendermaßen charakterisieren lassen:

(1) $K_{\mathfrak{A}}xyuv$ gdw $\overrightarrow{xu}^2 = \overrightarrow{xy} \cdot \overrightarrow{xv}$,

(2) $T_{\mathfrak{A}}xyuv$ gdw $K_{\mathfrak{A}}xyuv$ und $B_{\mathfrak{A}}uxy$,

dabei bedeutet $\cdot$ das innere Produkt von Vektoren und $B_{\mathfrak{A}}$ die wie üblich eingeführte Zwischenbeziehung.

4.109 Satz (nach Makowiecka). (i) K *ist als einziger Grundbegriff geeignet für die in* 4.108 *betrachteten n-dimensionalen und dimensionsfreien Geometrien ohne Anordnung und für diejenigen mit Anordnung, in denen das Kreisaxiom gilt (d.h. über euklidischen und über reell-abgeschlossenen angeordneten Körpern); für die übrigen betrachteten Geometrien mit Anordnung ist* T *(aber nicht* K*) als einziger Grundbegriff geeignet.* T *ist darüber hinaus geeignet für sog. "Pasch-freie" Geometrien (vgl.* 8.1*ff.).*

(ii) *Für die betrachteten eindimensionalen und dimensionsfreien elementaren Geometrien gibt es kein geeignetes System von Grundbegriffen, das nur aus dreistelligen Relationszeichen (für Relationen zwischen Punkten) besteht.*

(iii) *Für die betrachteten Geometrien zweiter Stufe mit Stetigkeitsaxiom (auch für die eindimensionale und die dimensionsfreie) ist dagegen der dreistellige Begriff* P' *als einziger Grundbegriff geeignet (vgl.* 4.106 *und die Bemerkung in* 4.105 *über die Definierbarkeit der Multiplikation von* $\mathbb{R}$*).*

4.110 Verwendung anderer geometrischer Objekte. Bisher haben wir Punkte und Geraden als geometrische Objekte betrachtet. Daneben lassen sich auch andere Figuren (Punktmengen) als Objekte und Relationen zwischen ihnen als Grundbegriffe verwenden. Im ursprünglichen Hilbertschen Axiomensystem (Hilbert 1977) werden z.B. Punkte, Geraden, Ebenen, Strecken und Winkel als Objekte verwendet.

Ganz andere Wege für einen Aufbau der euklidischen Geometrie werden eingeschlagen mit der Verwendung von Bällen bei Tarski 1929 und Jaśkowski 1949 (s. auch 1948 und 1949a). Bei Tarski werden offene Bälle, bei Jaśkowski abgeschlossene Bälle als Objekte verwendet, einziger Grundbegriff ist in beiden Fällen die (mengentheoretische) Inklusion $\subseteq$. (Unter dem *offenen* bzw. *abgeschlossenen Ball* mit dem Mittelpunkt c und dem Radius cp versteht man bekanntlich die Menge der Punkte x mit $cx<cp$ bzw. $cx\leq cp$, für den offenen Ball werde dabei $c\neq p$ vorausgesetzt; die (Hyper-)Kugel mit demselben Mittelpunkt und Radius ist dann die Menge der Randpunkte dieses Balls.)

Für den zweiten Weg sei hier ein Beweis der "Gleichwertigkeit" mit dem früheren Aufbau angedeutet (mit Vereinfachungen auf Grund einer Skizze von L. W. Szczerba). Sei $n\geq 2$ und T die n-dimensionale euklidische Geometrie P_n. Zunächst ist klar, wie man in T abgeschlossene (oder auch offene) Bälle einführen kann durch eine geeignete Spracherweiterung (analog zu 4.60(iii) und 4.89(ii)). Dadurch entstehe die Theorie $T_{\mathcal{B}}$. Die Theorie $(T_{\mathcal{B}})^{\mathcal{B}}$ der oben erwähnten Art bestehe dann aus den Sätzen von $T_{\mathcal{B}}$, die in der Sprache $L(\subseteq)$ mit Variablen für abgeschlossene Bälle (A, B, C,...) als den einzigen Variablen formuliert sind. In dieser Theorie lassen sich weitere Begriffe folgendermaßen durch formale Definitionen einführen.

(1) $\mathrm{Pt}\,X \leftrightarrow \forall B[B\subseteq X \rightarrow B=X]$
(X ist einpunktig).

Die einpunktigen Bälle spielen dann natürlich die Rolle der Punkte.

(2) $B^*ABC \leftrightarrow \forall D[A\subseteq D \wedge C\subseteq D \rightarrow B\subseteq D]$
(der Ball B liegt "zwischen" den Bällen A und C, d.h., daß er in der konvexen Hülle von $A\cup C$ enthalten ist).

Für "Punkte" A, B, C ist das insbesondere die übliche Zwischenbeziehung.

(3) $A\,\mathrm{Ti}\,B \leftrightarrow A\subseteq B \wedge \forall X_1 \forall X_2 \bigwedge_{\nu=1}^{2} [A\subseteq X_\nu \subseteq B \rightarrow X_1\subseteq X_2 \vee X_2\subseteq X_1]$
(A berührt B innen).

(4) $A\,\mathrm{Te}\,B \leftrightarrow \neg\mathrm{Pt}\,A \wedge \neg\mathrm{Pt}\,B \wedge \exists^{=1} C[C\subseteq A \wedge C\subseteq B]$
(die echten [d.h. nicht einpunktigen] Bälle A, B berühren sich außen).

(5) $X\,\mathrm{Bo}\,A \leftrightarrow \exists B[B\,\mathrm{Te}\,A \wedge X\subseteq B \wedge X\subseteq A] \vee [\mathrm{Pt}\,A \wedge X=A]$
(der "Punkt" X liegt auf dem Rand ["boundary"] von A).

(6) $X \,\mathrm{Di}\, U_1U_2A \leftrightarrow \exists B_1\exists B_2\{\bigwedge_{\nu=1}^{2}[B_\nu \,\mathrm{Ti}\, A \wedge B_\nu \neq A \wedge U_\nu \,\mathrm{Bo}\, A \wedge U_\nu \subseteq B_\nu \wedge X \subseteq B_\nu]$

$\wedge\, B_1 \,\mathrm{Te}\, B_2 \wedge B^{*}U_1XU_2\}$

(der innere Punkt X liegt auf dem die Randpunkte U_1, U_2 verbindenden Durchmesser ["diameter"] von A).

(7) $X \,\mathrm{Ce}\, A \leftrightarrow [\mathrm{Pt}\, A \wedge X=A]$

$\vee\, \exists U_1\exists U_2\exists V_1\exists V_2[X \,\mathrm{Di}\, U_1U_2A \wedge X \,\mathrm{Di}\, V_1V_2A \wedge \neq(U_1U_2V_1V_2)]$

(X ist der Mittelpunkt von A).

(8) $P\,XYZ \leftrightarrow \exists A[X \,\mathrm{Ce}\, A \wedge Y \,\mathrm{Bo}\, A \wedge Z \,\mathrm{Bo}\, A]$
(Pierische Relation).

Man erhält, daß diese Begriffe die in den jeweiligen Beschreibungen angegebene Bedeutung haben. Damit ergibt sich die gewünschte Gleichwertigkeit. Man kann die Beweise sogar so anlegen, daß sich diese Gleichwertigkeit auch für die dimensionsfreie absolute Geometrie A (als T) ergibt.

Für den ersten Weg, bei dem offene Bälle als Objekte verwendet werden, sei hier nur auf einen wesentlichen Unterschied hingewiesen: Bei Tarski 1929 werden Punkte eingeführt als Äquivalenzklassen von konzentrischen Bällen (die einfache Charakterisierung (1) ist hier natürlich nicht möglich).

5. Modellvollständigkeit

5.1 Überblick (s. auch 5.2, 3). Aus dem Satz 4.84, wo affine Räume einzeln behandelt wurden, ergab sich insbesondere (4.87), daß die Kollinearität Col nicht mit Hilfe der Mittelpunktsbeziehung M definierbar ist in den betrachteten euklidischen und affinen Geometrien (2.1, 13, 15). Dasselbe ergab sich für die Parallelogrammbeziehung Pg an Stelle von M. In diesem Abschnitt soll im wesentlichen das folgende Resultat bewiesen werden (s.a. 5.3(iv), 5.38), das als eine gewisse Ergänzung zu 4.84 angesehen werden kann.

5.2 Satz (Schwabhäuser 1979a). *Sei p eine gegebene Charakteristik (Null oder Primzahl), und sei η eine Aussage der Sprache* L(M) *bzw.* L(Pg) *(Sprache erster Stufe mit* M *bzw.* Pg *als einzigem Relationszeichen). Wenn dann η in einem affinen Raum mit unendlich vielen Punkten über einem Körper der Charakteristik p gültig ist, so ist η gültig in jedem solchen Raum.*

5.3 Anmerkungen. (i) Hält man hierbei einen unendlichen Grundkörper $\mathfrak{f}$ fest, so ergibt sich, daß in den Räumen $\mathfrak{A}_m(\mathfrak{f})$ und $\mathfrak{A}_n(\mathfrak{f})$ verschiedener Dimensionen $m,n\geq 1$ dieselben Aussagen von L(M) oder L(Pg) gelten. Insbesondere läßt sich also auch die Dimension eines Raumes nicht mit Hilfe von M oder Pg charakterisieren, während das - jedenfalls für endliche Dimension n und Räume mit Anordnung - mit Hilfe von Col möglich war (vgl. die Dimensionsaxiome Dim_n, I.11.70; diese lassen sich übrigens auch für Räume ohne Anordnung verwenden, wenn nur $p\neq 2$ ist; allgemein ist eine Charakterisierung der endlichen Dimension n mittels Col und $\parallel$ und auch mittels Col und Pg möglich). Die Nichtdefinierbarkeit von Col mittels M oder Pg könnte man auch daraus erhalten, wenigstens für Geometrien, in denen mehr als eine endliche Dimension ≥ 1 zugelassen ist.

(ii) Neben dem Übergang zu einer anderen Dimension ist in 5.2 auch der Übergang zu einem anderen Grundkörper derselben Charakteristik erlaubt. Im Unterschied zu 4.84 spielen jetzt die Primkörper und die Dimension 1 keine Ausnahmerolle. Ist der Grundkörper $\mathfrak{f}$ endlich (Primkörper oder nicht), so muß nur die Dimension n unendlich sein, damit der Raum $\mathfrak{A}_n(\mathfrak{f})$ unendlich viele Punkte besitzt und somit für 5.2 in Betracht kommt.

(iii) Satz 5.2 ist eine Verallgemeinerung einer Vermutung von H. N. Gupta.

(iv) Zum Beweis werden der Begriff der Modellvollständigkeit und Sätze darüber benutzt. Diese Hilfsmittel aus der Modelltheorie werden hier zunächst zitiert. Dann wird eine geeignete "Mittelpunktsgeometrie" bzw. "Parallelogrammgeometrie" als modellvollständig nachgewiesen, und zwar auf dem Weg über die Modellvollständigkeit der Theorie der "Vektorgruppen", d.h. der additiven Gruppen von Vektorräumen, mit unendlich vielen Elementen.

5.4 Definition. Sei $L=\langle F,R,r\rangle$ eine formalisierte Sprache erster Stufe (1.3). Sei $\mathfrak{A}$ eine Struktur für L und M eine beliebige Teilmenge ihrer Trägermenge $U_{\mathfrak{A}}$. Mit L_M werde "die" Erweiterungssprache (1.4) von L bezeichnet, die aus L dadurch entsteht, daß für jedes Element a von M eine neue (d.h. nicht zu L gehörende) Individuenkonstante $\bar{a}$ hinzugenommen wird. D.h., die Erweiterungssprache L_M ist von der Form $\langle F\cup C_M,R,r'\rangle$, wobei $C_M=\{\bar{a}\,|\,a\in M\}$ disjunkt zur Menge der Grundzeichen von L und die Abbildung $a\mapsto\bar{a}$ eine Bijektion ist sowie $r'(\bar{a})=0$ für jedes $a\in M$. (Eine solche Menge C_M von neuen Individuenkonstanten mit einer zugehörigen Bijektion kann dabei beliebig gewählt werden; auf eine genauere Festlegung kommt es aber für das Folgende nicht an, insofern ist L_M "im wesentlichen" eindeutig bestimmt.) Mit $\mathfrak{A}_M$ werde dann die Expansion (1.20) von $\mathfrak{A}$ auf L_M mit $(\bar{a})_{\mathfrak{A}_M}=a$ bezeichnet; sie entsteht also aus $\mathfrak{A}$ durch Hinzunahme der festen Elemente (nullstelligen Funktionen) $a\in M$ zur Interpretation der entsprechenden Individuenkonstanten $\bar{a}$.

Ist M Trägermenge einer Struktur $\mathfrak{L}$ (mit $M=U_{\mathfrak{L}}\subseteq U_{\mathfrak{A}}$), so werden auch die Bezeichnungen $L_{\mathfrak{L}}:=L_{U_{\mathfrak{L}}}$ und $\mathfrak{A}_{\mathfrak{L}}:=\mathfrak{A}_{U_{\mathfrak{L}}}$ verwendet. Insbesondere heißt $L_{\mathfrak{A}}$ die *Diagrammsprache* von $\mathfrak{A}$.

Unter dem *Diagramm* $D(\mathfrak{A})$ der Struktur $\mathfrak{A}$ verstehen wir (nach Ronbinson 1963) die folgende Menge von quantorenfreien Aussagen von $L_{\mathfrak{A}}$:

$$D(\mathfrak{A}) := \{f\overline{a_1}\ldots\overline{a_{r(f)}}\doteq\overline{a} \mid f\in F,\ a_1,\ldots,a_{r(f)}\in U_{\mathfrak{A}},\ f_{\mathfrak{A}}a_1\ldots a_{r(f)}=a\}$$
$$\cup\{R\overline{a_1}\ldots\overline{a_{r(R)}} \mid R\in R,\ a_1,\ldots,a_{r(R)}\in U_{\mathfrak{A}},\ R_{\mathfrak{A}}a_1\ldots a_{r(R)}\}$$
$$\cup\{\neg R\overline{a_1}\ldots\overline{a_{r(R)}} \mid R\in R,\ a_1,\ldots,a_{r(R)}\in U_{\mathfrak{A}},\ \text{nicht } R_{\mathfrak{A}}a_1\ldots a_{r(R)}\}$$
$$\cup\{\overline{a}\ne\overline{b} \mid a,b\in U_{\mathfrak{A}},\ a\ne b\}.$$

5.5 Anmerkung. Durch das Diagramm wird sozusagen der Bereich der mit den neuen Konstanten $\overline{a}$ bezeichneten Elemente als isomorph zu $\mathfrak{A}$ beschrieben. Für beliebige Modelle von $D(\mathfrak{A})$ wird jedoch nicht ausgeschlossen, daß sie auch andere Elemente enthalten. (Für unendliches $\mathfrak{A}$ läßt sich das auch nicht ausschließen, da dann eine Charakterisierung bis auf Isomorphie durch ein Axiomensystem in der ersten Stufe unmöglich ist, vgl. 1.55.) Die Modelle von $D(\mathfrak{A})$ lassen sich somit im wesentlichen dadurch charakterisieren, daß der genannte Teilbereich zu $\mathfrak{A}$ isomorph ist. Genauer wird das in dem folgenden Satz ausgedrückt.

5.6 Satz. *Seien* $\mathfrak{A}$, L *wie in* 5.4. *Sei* $\mathfrak{B}$ *eine Struktur für die Diagrammsprache* $L_{\mathfrak{A}}$ *und* φ *die Abbildung von* $U_{\mathfrak{A}}$ *in* $U_{\mathfrak{B}}$ *mit* $\varphi(a)=w_{\mathfrak{B}}(\overline{a},h)$ *(wobei es auf die Wahl der Belegung h nicht ankommt). Dann sind die folgenden drei Bedingungen untereinander äquivalent.*

(i) $\mathfrak{B}$ *ist ein Modell von* $D(\mathfrak{A})$.

(ii) φ *ist ein Isomorphismus von* $\mathfrak{A}_{\mathfrak{A}}$ *auf eine Unterstruktur von* $\mathfrak{B}$.

(iii) φ *ist ein Isomorphismus von* $\mathfrak{A}$ *auf eine Unterstruktur des Redukts von* $\mathfrak{B}$ *auf* L.

Insbesondere ist $\mathfrak{A}_{\mathfrak{A}}$ *Modell von* $D(\mathfrak{A})$.

5.7 Anmerkung. In der Literatur werden manchmal auch größere Mengen als in 5.4 als Diagramm von $\mathfrak{A}$ bezeichnet. An Stelle von 5.6(ii), (iii) treten dann ggf. stärkere Bedingungen.

5.8 Definition. Eine Formelmenge Σ heißt *modellvollständig* (A. Robinson), falls gilt: Für jedes Modell $\mathfrak{A}$ von Σ in der zugehörigen Sprache $L(\Sigma)$ ist $\Sigma\cup D(\mathfrak{A})$ vollständig (1.39).

5.9 Definition. Seien $\mathfrak{A}$, L wie in 5.4. Die Menge $\exists_{\mathfrak{A}}$ der *für* $\mathfrak{A}$ *primitiven Aussagen* ist erklärt als die Menge der Aussagen α von $L_{\mathfrak{A}}$ von folgender Bauart:

(1) $\alpha = \exists x_1\ldots\exists x_k \bigwedge_{\lambda=1}^{l} \varepsilon_\lambda,$

wobei $k,l\in\mathbb{N}$ und jedes Konjunktionsglied ε_λ eine "spezielle Grundformel" ist, d.h. von einer der Formen $f\sigma_1\ldots\sigma_{r(f)}\doteq\sigma$, $R\sigma_1\ldots\sigma_{r(R)}$, $\neg R\sigma_1\ldots\sigma_{r(R)}$, $\sigma_1\neq\sigma_2$ ($f\in\mathcal{F}$, $R\in\mathcal{R}$, jedes σ_ρ eine Variable oder eine der neuen Konstanten $\bar{a}$).

5.10 Satz (Modellvollständigkeitstest von A. Robinson). *Sei Σ eine Menge von Formeln ("Axiomen") und L=L(Σ). Dann sind die folgenden drei Bedingungen untereinander äquivalent.*

(i) *Σ ist modellvollständig.*

(ii) *Für je zwei Modelle $\mathfrak{A}$, $\mathfrak{B}$ von Σ (in L) mit $\mathfrak{A}\subseteq\mathfrak{B}$ ($\mathfrak{A}$ Unterstruktur von $\mathfrak{B}$) ist sogar $\mathfrak{A}\prec\mathfrak{B}$ ($\mathfrak{A}$ elementare Unterstruktur von $\mathfrak{B}$,* s. 3.71, 72).

(iii) *Für je zwei Modelle $\mathfrak{A}$, $\mathfrak{B}$ von Σ (in L) mit $\mathfrak{A}\subseteq\mathfrak{B}$ gilt: Jede in $\mathfrak{B}_{\mathfrak{A}}$ gültige Aussage von $\exists_{\mathfrak{A}}$ ist auch in $\mathfrak{A}_{\mathfrak{A}}$ gültig.*

Die Beziehung $\mathfrak{A}\prec\mathfrak{B}$ läßt sich auch folgendermaßen charakterisieren.

5.11 Satz. *$\mathfrak{A}\prec\mathfrak{B}$ genau dann, wenn $\mathfrak{A}\subseteq\mathfrak{B}$ und in den Strukturen $\mathfrak{A}_{\mathfrak{A}}$ und $\mathfrak{B}_{\mathfrak{A}}$ dieselben Aussagen von $L_{\mathfrak{A}}$ gültig sind.*

5.12 Anmerkungen. (i) Damit bedeutet die Bedingung 5.10(ii), daß sich die Gültigkeit beliebiger Aussagen (mit Zeichen für feste Elemente der jeweils kleineren Struktur) "nach oben" und "nach unten" (d.h. auf größere und auf kleinere Modelle von Σ) überträgt. Zum Nachweis dafür genügt es nach 5.10(iii), dasselbe für recht spezielle Aussagen in der Richtung "nach unten" zu zeigen. In diesem Nachweis besteht der "Test" für die Modellvollständigkeit.

(ii) Für Formelmengen ergibt sich weder aus der Vollständigkeit die Modellvollständigkeit noch umgekehrt. Ein Zusammenhang wird jedoch durch den folgenden Primmodelltest hergestellt.

5.13 Definition. Eine Struktur $\mathfrak{A}$ für die Sprache L=L(Σ) heißt ein *Primmodell* von Σ, falls $\mathfrak{A}$ Modell von Σ ist und jedes Modell $\mathfrak{B}$ von Σ (in L) eine zu $\mathfrak{A}$ isomorphe Unterstruktur besitzt.

Beispiel. Für die Theorie der Körper gegebener Charakteristik p (Null oder Primzahl) sind Primmodelle dasselbe wie Primkörper. Die Theorie aller Körper besitzt dagegen keine Primmodelle.

5.14 Satz (Primmodelltest von A. Robinson).

Vor.: (i) Σ *ist modellvollständig.*

(ii) Σ *besitzt ein Primmodell.*

Beh.: Σ *ist vollständig.*

5.15 Betrachtete Räume und Geometrien. Die hier betrachteten Mittelpunkts- bzw. Parallelogrammräume werden eingeführt als Redukte der in 4.82 betrachteten affinen Räume, d.h., sie entstehen aus diesen durch Weglassen von Relationen.

5.16 Definition. Mit $\mathfrak{V}_n(\mathfrak{F})$ (wobei $\mathfrak{F}$ Körper und n eine [endliche oder unendliche] Kardinalzahl) sei wieder der (bis auf Isomorphie eindeutig bestimmte) n-dimensionale Vektorraum über $\mathfrak{F}$ bezeichnet, mit char($\mathfrak{F}$) die Charakteristik von $\mathfrak{F}$. Unter dem *n-dimensionalen Mittelpunktsraum* $\mathfrak{M}_n(\mathfrak{F})$ über $\mathfrak{F}$ verstehen wir die folgendermaßen festgelegte Struktur $\mathfrak{A}$ für die Sprache L(M):

(1) $U_{\mathfrak{A}}$ ist die Menge der Vektoren aus $\mathfrak{V}_n(\mathfrak{F})$,

(2) $M_{\mathfrak{A}}abc$ gdw $2b=a+c$.

Unter dem *n-dimensionalen Parallelogrammraum* $\mathfrak{Pg}_n(\mathfrak{F})$ über $\mathfrak{F}$ verstehen wir die Struktur $\mathfrak{A}$ für L(Pg), die festgelegt ist durch (1) wie vorher und

(3) $ab\ Pg_{\mathfrak{A}}\ cd$ gdw $b-a=d-c$

(vgl. 4.82).

Unter der *Mittelpunktsgeometrie* M bzw. der *Parallelogrammgeometrie* Pg verstehen wir die Theorie (der Klasse) aller Mittelpunktsräume bzw. aller Parallelogrammräume. Nach Konstruktion lassen sich diese Geometrien auch charakterisieren als die Menge ("Satzmenge") der Formeln der eingeschränkten Sprache L(M) bzw. L(Pg), die in allen affinen Räumen gültig sind. (Im Unterschied zur früher betrachteten dimensionsfreien affinen Geometrie Af sind jetzt auch die Dimensionen 0 und 1 zugelassen.)

5.17 Anmerkung. Für die Definition der Räume $\mathfrak{M}_n(\mathfrak{F})$ und $\mathfrak{Pg}_n(\mathfrak{F})$ wurde der Vektorraum $\mathfrak{V}_n(\mathfrak{F})$ nicht ganz benötigt, sondern nur seine additive Gruppe. Umgekehrt läßt sich diese Gruppe bis auf Isomorphie aus $\mathfrak{Pg}_n(\mathfrak{F})$ und, falls char($\mathfrak{F}$)$\neq 2$ ist, auch aus $\mathfrak{M}_n(\mathfrak{F})$ zurückgewinnen. Nach Wahl eines beliebigen Punktes o als Nullpunkt läßt sich nämlich die Addition $+_o$ in bezug auf diesen Nullpunkt offenbar charakterisieren durch die formale Definition

(1) $a +_o b \doteq c \leftrightarrow$ oa Pg bc;

und für char($\mathfrak{f}$)$\neq$2 läßt sich Pg mittels M definieren (wie in 4.84(4)) durch

(2) oa Pg bc $\leftrightarrow \exists e$[Moec $\wedge$ Maeb].

Es liegt also nahe, statt der eingeführten Räume zunächst die additiven Gruppen von Vektorräumen zu betrachten, die wir zur Abkürzung als *Vektorgruppen* bezeichnen.

5.18 Definition. Zur Behandlung abelscher Gruppen verwenden wir die formalisierte Sprache L_G:=L(+,-,0) mit Operationszeichen in der üblichen Bedeutung (- einstellig für die Inversenbildung). Für Terme τ dieser Sprache und natürliche Zahlen k seien die Terme $k\cdot\tau$ induktiv definiert durch $0\cdot\tau$=0, $(k+1)\cdot\tau = k\cdot\tau + \tau$. Analog seien die entsprechenden Elemente $k\cdot x$ einer abelschen Gruppe $\mathfrak{G}$ ($k\in\mathbb{N}$, $x\in\mathfrak{G}$) eingeführt. Ein Gruppenelement x heißt bekanntlich von der *(endlichen) Ordnung* n ($n\in\mathbb{N}-\{0\}$), falls $n\cdot x$=0, aber $k\cdot x\neq 0$ für alle k mit $1\leq k<n$; es heißt von *unendlicher Ordnung*, falls es kein solches n gibt. Eine abelsche Gruppe $\mathfrak{G}$ heißt eine *elementar abelsche p-Gruppe*, falls alle von Null verschiedenen Elemente dieselbe Ordnung p besitzen (p Primzahl); sie heißt *torsionsfrei*, falls alle von Null verschiedenen Elemente unendliche Ordnung besitzen, sie heißt darüber hinaus *dividierbar*, falls die Behauptungen der Axiome $\gamma_{3,k}$ (s.u.) gelten.

Sei Σ_{VG} das System der folgenden Axiome.

γ_1: (Konjunktion der) Axiome für abelsche Gruppen,

$\gamma_{2,k}$:= $x\neq 0 \wedge k\cdot x \doteq 0 \rightarrow \forall y\, k\cdot y \doteq 0$ $(k\geq 2)$,

$\gamma_{3,k}$:= $\exists z\, k\cdot z \neq 0 \rightarrow \forall x \exists y\, k\cdot y \doteq x$ $(k\geq 2)$.

5.19 Satz. *Für Strukturen* $\mathfrak{G}$ *für* L_G *sind die folgenden fünf Bedingungen untereinander äquivalent.*

(i) $\mathfrak{G}$ *ist die additive Gruppe eines Vektorraums ("Vektorgruppe").*

(ii) $\mathfrak{G}$ *ist die additive Gruppe eines Vektorraums über einem Primkörper.*

(iii) $\mathfrak{G}$ *ist Modell von* Σ_{VG}.

(iv) $\mathfrak{G}$ *ist eine torsionsfreie dividierbare abelsche Gruppe oder eine elementar abelsche Gruppe.*

(v) $\mathfrak{G}$ *ist die additive Gruppe eines Körpers oder besteht nur aus dem neutralen Element.*

<u>Zum Beweis</u>: (i) $\leftrightarrow$ (ii): Das ergibt sich sofort aus Lemma 4.86.

(i) $\rightarrow$ (iii): Man sieht unmittelbar, daß die Axiome von Σ_{VG} in jeder Vektorgruppe gelten.

Für das Folgende sei $\mathfrak{G}$ o.B.d.A. verschieden von der "trivialen" (nur aus dem neutralen Element Null bestehenden) Gruppe (für diese gelten (i) bis (v)).

(iii) $\rightarrow$ (iv): Aus den Axiomen $\gamma_{2,k}$ ergibt sich zunächst, daß die von Null verschiedenen Elemente alle von derselben Ordnung sind.

<u>Fall 1</u>: Diese Ordnung ist eine natürliche Zahl k (endlich). (In diesem Fall werden die Axiome $\gamma_{3,k}$ nicht gebraucht.) Sind m, n Zahlen mit $1 \leq m, n < k$, so ergibt sich für Elemente $x \neq 0$, daß auch $n \cdot x \neq 0$ ist, und ebenso, daß $m \cdot (n \cdot x) = (m \cdot n) \cdot x \neq 0$ ist. Somit kann sich die Ordnung k nicht als Produkt solcher Faktoren m, n darstellen lassen, d.h., k ist eine Primzahl p und $\mathfrak{G}$ eine elementar abelsche p-Gruppe.

<u>Fall 2</u>: Die erwähnte Ordnung ist unendlich, d.h., $\mathfrak{G}$ ist torsionsfrei. Dann gelten die Voraussetzungen und damit auch die Behauptungen aller Axiome $\gamma_{3,k}$, also ist $\mathfrak{G}$ auch dividierbar.

(iv) $\rightarrow$ (ii): Ist $\mathfrak{G}$ eine elementar abelsche p-Gruppe, so läßt sich die in 5.18 beschriebene Multiplikation von Gruppenelementen mit natürlichen Zahlen in natürlicher Weise auf die Elemente des Primkörpers $\mathbb{Z}_p$ (Restklassen modulo p) übertragen. Ist dagegen $\mathfrak{G}$ torsionsfrei und dividierbar, so läßt sich diese Multiplikation auf rationale Zahlen ausdehnen mittels der Festsetzungen

(1) $\frac{m}{n} \cdot x =$ dasjenige y mit $n \cdot y = m \cdot x$,

(2) $(-r) \cdot x = -(r \cdot x)$.

Man prüft leicht nach, daß $\mathfrak{G}$ mit dieser Multiplikation zu einem Vektorraum über dem Primkörper $\mathbb{Z}_p$ bzw. $\mathbb{Q}$ wird.

(ii) $\rightarrow$ (v): Der in (ii) betrachtete Vektorraum (und damit auch $\mathfrak{G}$) ist bis auf Isomorphie eindeutig bestimmt durch seine Dimension n und die Charakteristik p (Null oder Primzahl) des betreffenden Primkörpers. Somit genügt es, einen Körper der Charakteristik p anzugeben, der als

Vektorraum über seinem Primkörper die Dimension n besitzt. Das geschieht durch folgende Fallunterscheidung für die in Frage kommenden n und p.

Fall 1: n ist endlich. 1.a): $p=0$. Das Polynom x^n-q, wobei q Primzahl, ist bekanntlich irreduzibel über $\mathbb{Q}$ (s. etwa van der Waerden 1971, S. 97), Adjunktion einer Nullstelle liefert also eine Erweiterung der gewünschten Art. 1.b): p Primzahl. Das Galoisfeld mit p^n Elementen leistet das Verlangte.

Fall 2: n ist unendlich. Sei $\mathfrak{f}$ der betrachtete Primkörper. Da dieser endlich oder abzählbar ist, ergibt sich mit bekannten Mächtigkeitsabschätzungen:

(3) $|\mathfrak{V}_n(\mathfrak{f})| = n$
(d.h., die Menge der Vektoren des gewünschten Raumes hat die Mächtigkeit n).

Bedeute $\mathfrak{T}_m(\mathfrak{f})$ die rein transzendente Körpererweiterung von $\mathfrak{f}$ vom Transzendenzgrad m (die bis auf Isomorphie eindeutig bestimmt ist und durch Adjunktion einer Menge von m Unbestimmten erhalten werden kann). Dann ergibt sich auch

(4) $|\mathfrak{T}_n(\mathfrak{f})| = n$.

2.a): $n>\aleph_0$. Dann leistet $\mathfrak{T}_n(\mathfrak{f})$ das Verlangte; der zugehörige Vektorraum, den diese Erweiterung über $\mathfrak{f}$ bildet, kann nämlich wegen (3) ebenfalls nur die Dimension n haben. 2.b): $n=\aleph_0$. Dann leistet $\mathfrak{T}_m(\mathfrak{f})$ das Verlangte, wobei m beliebig mit $1\leq m\leq\aleph_0$. Der zugehörige Vektorraum hat nämlich wegen (3) höchstens die Dimension $\aleph_0$. Andererseits hat er auch mindestens diese Dimension; denn ist x ein über $\mathfrak{f}$ transzendentes Element, so bilden die abzählbar vielen Elemente x^i ($i\in\mathbb{N}$) eine linear unabhängige Menge.

(v) $\rightarrow$ (i): Der eindimensionale Vektorraum über dem gegebenen Körper leistet das Verlangte. □

5.20 Anmerkung. Falls $\mathfrak{g}$ nicht die triviale Gruppe ist, ist der Primkörper gemäß 5.19(ii) (bis auf Isomorphie) eindeutig bestimmt. Seine Charakteristik bezeichnen wir dann auch als die *Charakteristik von* $\mathfrak{g}$.

5.21 Definition. Sei $\Sigma_{VG\infty}$ das Axiomensystem für unendliche Vektorgruppen, das aus Σ_{VG} (5.18) durch Hinzunahme der folgenden Axiome entsteht:

$$\gamma_{4,k} = \overset{\geq k}{\exists} \quad (k\geq 2).$$

5.22 Satz. $\Sigma_{VG\infty}$ *(und damit die Theorie der unendlichen Vektorgruppen) ist modellvollständig.*

Beweis (nach A. Robinson 1956, S. 32f., wo der Spezialfall der torsionsfreien dividierbaren abelschen Gruppen behandelt wurde): Wegen 5.10 genügt der Beweis des folgenden Lemmas.

5.23 Lemma. *Vor.:* $\mathfrak{A}$, $\mathfrak{B}$ *sind unendliche Vektorgruppen,* $\mathfrak{A} \subseteq \mathfrak{B}$, $\eta \in \exists_{\mathfrak{A}}$, η *ist in* $\mathfrak{B}_{\mathfrak{A}}$ *gültig.*

Beh.: η *ist in* $\mathfrak{A}_{\mathfrak{A}}$ *gültig.*

Beweis: $\mathfrak{A}$ und $\mathfrak{B}$ haben offenbar dieselbe Charakteristik; seien sie gemäß 5.20, 19(ii) als Vektorräume über dem Primkörper $\mathfrak{f}$ dieser Charakteristik ($\mathbb{Q}$ oder ein $\mathbb{Z}_p$) dargestellt. Nach Vor. ist η von der Form $\eta = \exists x_1 \ldots \exists x_k \varepsilon$, dabei ist $\varepsilon = \bigwedge_{\lambda=1}^{l} \varepsilon_\lambda$ nach 5.9 ein System von Gleichungen und Ungleichungen in den Variablen $x_1, \ldots, x_k$ (da die betrachtete Sprache L_G keine Relationszeichen enthält). Sei $\langle x_1, \ldots, x_k \rangle$ eine (nach Vor. existierende) Lösung dieses Systems in $\mathfrak{B}$. Dann liegen $x_1, \ldots, x_k$ schon in einer Vektorgruppe $\mathfrak{A}' = \mathfrak{A}(d_1, \ldots, d_m)$, die aus $\mathfrak{A}$ durch "Adjunktion" endlich vieler Elemente $d_1, \ldots, d_m$ von $\mathfrak{B}$ entsteht (darunter verstehen wir [analog zu Körpern] die kleinste Vektorgruppe $\mathfrak{A}'$ mit $\mathfrak{A} \subseteq \mathfrak{A}' \subseteq \mathfrak{B}$, die $d_1, \ldots, d_m$ als Elemente enthält; diese besteht offenbar genau aus den Elementen der Form $a + \sum_{\mu=1}^{m} r_\mu \cdot d_\mu$ mit $a \in \mathfrak{A}$, $r_\mu \in \mathfrak{f}$). Sei nämlich $d_\mu = x_\kappa$, wobei κ minimal mit $x_\kappa \notin \mathfrak{A}(d_1, \ldots, d_{\mu-1})$ (falls vorhanden). Dann ist $m \leq k$. Sei außerdem $\mathfrak{A}_\mu = \mathfrak{A}(d_1, \ldots, d_\mu)$ $(0 \leq \mu \leq m)$. Dann ist $\mathfrak{A} = \mathfrak{A}_0 \subsetneqq \mathfrak{A}_1 \subsetneqq \ldots \subsetneqq \mathfrak{A}_m = \mathfrak{A}'$ und $\mathfrak{A}_\mu = \mathfrak{A}_{\mu-1}(d_\mu)$ $(1 \leq \mu \leq m)$.

Nach Konstruktion besitzt ε eine Lösung in $\mathfrak{A}_m$. Die Existenz einer Lösung in den Vektorgruppen $\mathfrak{A}_{m-j}$ $(j \leq m)$ (und damit, wie gewünscht in $\mathfrak{A}$) ergibt sich durch Induktion über j. Dazu genügt es, das Lemma 5.23 unter der Zusatzvoraussetzung zu beweisen, daß $\mathfrak{B}$ von der Form $\mathfrak{A}(d)$ ist mit $d \notin \mathfrak{A}$.

Seien dazu $\mathfrak{f}$, ε, $x_1, \ldots, x_k$ wie vorher. Dann hat jedes x_κ die Form $x_\kappa = b_\kappa + r_\kappa \cdot d$ für geeignete Elemente $b_\kappa \in \mathfrak{A}$, $r_\kappa \in \mathfrak{f}$. Wir setzen nun

(1) $x_\kappa(u) := b_\kappa + r_\kappa \cdot u$ $(u \in \mathfrak{B}$ beliebig, $\kappa = 1, \ldots, k)$

und wollen zeigen, daß diese Elemente schon für ein $u \in \mathfrak{A}$ eine Lösung von ε bilden. Die Gültigkeit der Gleichungen und Ungleichungen von ε für $x_1(u), \ldots, x_k(u)$ ist gleichwertig mit Bedingungen der Form

(2) $c_\lambda + s_\lambda \cdot u = 0 \qquad (\lambda=1,\ldots,i)$,

(3) $c_\lambda + s_\lambda \cdot u \neq 0 \qquad (\lambda=i+1,\ldots,l)$,

wobei $c_\lambda \in \mathfrak{A}$, $s_\lambda \in \mathfrak{f}$ $(\lambda=1,\ldots,l)$. Das erhält man, indem man alles auf eine Seite bringt und ordnet; die s_λ lassen sich dann ausdrücken durch die r_κ, die c_λ lassen sich ausdrücken durch die b_κ und die endlich vielen Elemente a_ν von $\mathfrak{A}$, die durch Konstanten in ε bezeichnet werden.

Nach Vor. bilden die Elemente $x_\kappa(d)$ $(\kappa=1,\ldots,k)$ eine Lösung von ε; d.h., die Bedingungen (2) und (3) gelten für $u=d$. Wegen $c_\lambda \in \mathfrak{A}$ und $d \notin \mathfrak{A}$ gilt statt (2) sogar stärker

(4) $c_\lambda = 0,\ s_\lambda = 0 \qquad (\lambda=1,\ldots,i)$

(andernfalls wäre $d = -s_\lambda^{-1} \cdot c_\lambda \in \mathfrak{A}$). Somit gelten die Bedingungen (2) sogar für jedes u. Betrachten wir nun die Bedingungen der Form (3). Diejenigen davon, in denen $s_\lambda = 0$ ist, gelten natürlich ebenfalls für jedes u. Diejenigen, in denen $s_\lambda \neq 0$ ist, lassen sich umformen zu

(5) $u \neq -s_\lambda^{-1} \cdot c_\lambda$.

Diese Bedingungen besagen also zusammen, daß u verschieden ist von endlich vielen gegebenen Elementen von $\mathfrak{A}$. Da $\mathfrak{A}$ aber unendlich viele Elemente besitzt, gibt es ein solches u (sogar unendlich viele) in $\mathfrak{A}$. Das war zu zeigen. □

5.24 Geometrische Axiome. Für die Mittelpunktsgeometrie (5.16) wird zunächst ebenfalls eine Axiomatisierung angegeben.

5.25 Definition. In der Sprache L(M) verwenden wir folgende Abkürzungen.

(1) χ_2 := ∀a∀bMaba.

Diese Aussage drückt offenbar aus, daß die Charakteristik gleich zwei ist (sie gilt in einem Mittelpunktsraum $\mathfrak{M}_n(\mathfrak{f})$ mit $n \geq 1$ gdw char($\mathfrak{f}$)=2 ist, vgl. 5.16(2); sie gilt natürlich auch im "einpunktigen" Mittelpunktsraum [$n=0$, $\mathfrak{f}$ beliebig]).

(2) $F_k(abc)$:=

$$\exists a_0 \exists a_1 \ldots \exists a_k [a_0 \doteq a \wedge a_1 \doteq b \wedge a_k \doteq c \wedge \bigwedge_{\kappa=0}^{k-2} M a_\kappa a_{\kappa+1} a_{\kappa+2}] \qquad (k \geq 2)$$

("$\overrightarrow{ac}$ ist das *k-fache* von $\overrightarrow{ab}$ "; insbesondere ist $F_2(abc)$ äquivalent mit Mabc.)

Schließlich sei der Begriff Pg definiert wie in 5.17(2).

Sei Σ_M das System der folgenden Axiome.

$\alpha_1 := \text{Maaa}.$

$\alpha_2 := \text{Mabc} \rightarrow \text{Mcba}.$

$\alpha_3 := \forall a \forall b \overset{=1}{\exists} c\ \text{Mabc}.$

$\alpha_4 := \neg\chi_2 \rightarrow \forall a \forall c \overset{=1}{\exists} b\ \text{Mabc}.$

$\alpha_5 := \neg\chi_2 \wedge ab\ \text{Pg}\ cd \wedge cd\ \text{Pg}\ ef \rightarrow ab\ \text{Pg}\ ef.$

$\alpha_{6,k} := F_k(aba) \wedge a \neq b \rightarrow \forall x \forall y \forall z [F_k(xyz) \rightarrow x \doteq z] \qquad (k \geq 2).$

$\alpha_{7,k} := F_k(abc) \wedge a \neq c \rightarrow \forall x \forall z \exists y F_k(xyz) \qquad (k \geq 2).$

$\alpha_{8,n,k} := \chi_2 \wedge \overset{\geq k}{\exists} \rightarrow \overset{\geq 2^{n+1}}{\exists} \qquad (2^n < k < 2^{n+1}).$

Sei $\Sigma_{M\infty}$ das System, das hieraus entsteht durch Ersetzung der $\alpha_{8,n,k}$ durch die (zusammen stärkeren) Axiome

$\alpha'_{8,k} := \overset{\geq k}{\exists} \qquad (k \geq 2),$

und sei M_∞ die darauf aufgebaute Theorie (d.h. $M_\infty = Cn(\Sigma_{M\infty})$).

5.26 Anmerkungen. (i) Die Existenzforderung in α_4 ist gleichwertig mit $\alpha_{7,2}$; eine der beiden Forderungen könnte also weggelassen werden.

(ii) Unter der Voraussetzung χ_2 (d.h. für die Theorie $Cn(\Sigma_M \cup \{\chi_2\})$) erhält man mittels α_3, daß M folgendermaßen "logisch" definierbar ist (vgl. 4.84(6)ff.):

(1) $\forall a \forall b \forall c [\text{Mabc} \leftrightarrow a \doteq c].$

Auch in diesem Falle hat die Formel $F_k(abc)$ noch die in 5.25(2) angegebene Bedeutung (sie ist äquivalent mit $c \doteq b$ für ungerades k und mit $c \doteq a$ für gerades k).

5.27 Satz. *Eine Relationalstruktur $\mathfrak{A}$ für die Sprache L(M) ist Modell von Σ_M bzw. von $\Sigma_{M\infty}$ gdw sie isomorph ist zu einem Mittelpunktsraum bzw. zu einem Mittelpunktsraum mit unendlich vielen Punkten.*

Zum Beweis: Zunächst sieht man, daß die angegebenen Axiome in den genannten Räumen gelten. Sei nun $\mathfrak{A}$ ein beliebiges Modell von Σ_M (in L(M)). Falls χ_2 in $\mathfrak{A}$ gilt, ist die Struktur $\mathfrak{A}$ wegen 5.26(1) schon festgelegt durch ihre Trägermenge; falls die Anzahl ihrer Elemente endlich ist, wird noch gebraucht, daß diese Anzahl eine Zweierpotenz ist, was durch die $\alpha_{8,n,k}$ gesichert wird. Gelte nun $\neg\chi_2$. Sei ein beliebiger Punkt o von $\mathfrak{A}$ als "Nullpunkt" gewählt, und sei die Addition $+_o$ gemäß 5.17(1) in $\mathfrak{A}$ eingeführt. Auf Grund von α_1 bis α_5 erhält man ohne

Schwierigkeiten, daß die Punkte (Elemente von $\mathfrak{A}$) in bezug auf diese Addition eine abelsche Gruppe bilden und daß sich die Mittelpunktsrelation damit wie in 5.16(2) charakterisieren läßt, außerdem, daß für verschiedene Punkte o, o' die entsprechend gebildeten Gruppen zueinander isomorph sind. Die Axiome $\alpha_{6,k}$ und $\alpha_{7,k}$ sind gerade die Übersetzungen der Zusatzaxiome $\gamma_{2,k}$ und $\gamma_{3,k}$. Somit ist die eingeführte Gruppe eine Vektorgruppe, wie zu zeigen. □

Aus dieser Axiomatisierung ergibt sich insbesondere

5.28 Korollar. *Jedes Modell der Mittelpunktsgeometrie M ist (bis auf Isomorphie) ein Mittelpunktsraum.*

5.29 Anmerkung. Dieses Korollar ergibt sich keineswegs unmittelbar aus der Definition 5.16. Ist nämlich T die Theorie einer Klasse K von Strukturen, so braucht i.a. nicht jedes Modell von T in K zu liegen; vgl. etwa 3.69(i) oder das Beispiel $K=\{\mathbb{R}\}$ (s. 3.22(ii)).

5.30 Satz. Σ_{M_∞} *(und damit die Mittelpunktsgeometrie* M_∞ *mit unendlich vielen Punkten) ist modellvollständig.*

Beweis: Es genügt wieder der Nachweis der Bedingung 5.10(iii). Dementsprechend seien $\mathfrak{A}$, $\mathfrak{B}$ beliebige Modelle von Σ_{M_∞} in L(M) mit $\mathfrak{A} \subseteq \mathfrak{B}$.

Fall 1: In $\mathfrak{B}$ gilt $\neg\chi_2$. Dann gilt $\neg\chi_2$ auch in $\mathfrak{A}$ (sonst hätten Strecken der Form aa mehrere Mittelpunkte in $\mathfrak{A}$ und damit auch in $\mathfrak{B}$ im Widerspruch zur Eindeutigkeitsforderung von α_4). Sei ein Punkt o von $\mathfrak{A}$ beliebig gewählt. Wir betrachten zunächst eine Erweiterungssprache L'-L(o,+;M) mit den zusätzlichen Grundzeichen o und +. Seien $\mathfrak{A}'$, $\mathfrak{B}'$ die Strukturen für L' (Expansionen), die aus $\mathfrak{A}$ bzw. $\mathfrak{B}$ entstehen, indem zusätzlich o als der gewählte Punkt und + als die Addition $+_o$ bezüglich o interpretiert wird (vgl. den vorigen Beweis). Dann ist sogar $\mathfrak{A}' \prec \mathfrak{B}'$, d.h. (5.11), in beiden Strukturen sind dieselben Aussagen von $L'_{\mathfrak{A}'}$ gültig. Für diejenigen Aussagen, die das Zeichen M nicht enthalten, ergibt sich das nämlich aus der Modellvollständigkeit in 5.22; für beliebige Aussagen von $L'_{\mathfrak{A}'}$ ergibt es sich daraus, daß der Begriff M mittels + definierbar ist (Mabc $\leftrightarrow$ b+b$\doteq$a+c). Durch Einschränkung auf die Sprache L(M) ergibt sich insbesondere $\mathfrak{A} \prec \mathfrak{B}$. Damit ist im Fall 1 für $\mathfrak{A}$ und $\mathfrak{B}$ die stärkere Forderung 5.10(ii) gezeigt (s. 5.12(i)).

Fall 2: In $\mathfrak{B}$ (und damit auch in $\mathfrak{A}$) gilt χ_2. Hier läßt sich die Forderung 5.10(iii) leicht direkt nachweisen. Sei nämlich $\eta = \exists x_1 \ldots \exists x_k \varepsilon \in \exists_{\mathfrak{A}}$

beliebig. Wegen 5.26(1) besagt ε, daß von den Elementen $x_1, \ldots, x_k$ einige untereinander gleich, voneinander verschieden, gleich gegebenen Elementen von $\mathfrak{A}$ bzw. verschieden von gegebenen Elementen von $\mathfrak{A}$ sind. Wenn solche x_κ in $\mathfrak{B}$ existieren, so gibt es solche Elemente auch schon in $\mathfrak{A}$, da $\mathfrak{A}$ unendlich viele Elemente besitzt. □

5.31 Anmerkungen. (i) Die Modellvollständigkeit der Mittelpunktsgeometrie M_∞ wurde hier im wesentlichen zurückgeführt auf die Modellvollständigkeit der Theorie der unendlichen Vektorgruppen. Dabei wurde benutzt, daß - abgesehen von der Ausnahme für die Charakteristik 2 - jede der beiden Theorien in der anderen "interpretierbar" ist, d.h., daß sie sich in geeigneter Weise (unter Benutzung formaler Definitionen) innerhalb der anderen Theorie beschreiben läßt. Es sei jedoch ausdrücklich darauf hingewiesen, daß die Feststellung der wechselseitigen Interpretierbarkeit allein nicht ausreicht, um aus der Modellvollständigkeit einer Theorie die der anderen zu erhalten (für ein Gegenbeispiel s. etwa Chang u. Keisler 1973, Example 3.4.4).

(ii) Ganz allgemein ist die Feststellung solcher Interpretierbarkeiten ein häufig benutztes Hilfsmittel, um gewisse Resultate von einer Theorie auf eine andere zu übertragen. Ein Überblick über derartige Übertragungssätze wird (für einen geeigneten Interpretierbarkeitsbegriff) gegeben in Pinter 1978. In der Literatur werden verschiedene Interpretierbarkeitsbegriffe betrachtet. Besonders weit gefaßt ist der Interpretierbarkeitsbegriff in Szczerba 1977, und zwar so, daß sich auch die Übertragungen mit Hilfe von arithmetischen und geometrischen Reduzierten aus Abschnitt 3 als Spezialfälle einordnen lassen. Für diesen Begriff wurden Methoden zur Klassifikation von Theorien bezüglich der Interpretierbarkeit entwickelt; s. auch Sette u. Szczerba 1978.

5.32 Definition. In Ergänzung zu 5.25(1) setzen wir

(1) $\chi_p := \forall a \forall b\, F_p(aba)$ (p Primzahl, $p>2$).

Für beliebige Primzahlen p ist damit χ_p eine Aussage von L(M), die ausdrückt, daß die Charakteristik p ist. Sei χ'_p eine entsprechende Aussage für Vektorgruppen, etwa

(2) $\chi'_p := \forall x\, p\cdot x \doteq 0$.

Die Mittelpunktsgeometrien M^p_∞ der Charakteristik p (Null oder Primzahl) *mit unendlich vielen Punkten* lassen sich dann durch folgende Axiomensysteme $\Sigma^p_{M\infty}$ charakterisieren:

(3) $\Sigma^p_{M\infty} := \Sigma_{M\infty} \cup \{\chi_p\}$ (p Primzahl)

(4) $\Sigma^0_{M\infty} := \Sigma_{M\infty} \cup \{\neg\chi_p \mid p \text{ Primzahl}\}$.

In analoger Weise seien Axiomensysteme $\Sigma^p_{VG\infty}$ für die Theorie der unendlichen Vektorgruppen der Charakteristik p gebildet.

5.33 Satz. *Die Axiomensysteme* $\Sigma^p_{VG\infty}$ *und* $\Sigma^p_{M\infty}$ *(p Null oder Primzahl) sind vollständig.*

Beweis: Jedes der genannten Axiomensysteme besitzt ein Primmodell (5.13). Primmodelle von $\Sigma^p_{VG\infty}$ sind nämlich (vgl. 5.19): falls $p=0$, die additive Gruppe von $\mathbb{Q}$ (des eindimensionalen Vektorraums über $\mathbb{Q}$); falls p Primzahl, die additive Gruppe des Vektorraums der Dimension $\aleph_0$ über $\mathbb{Z}_p$. Die mit diesen Gruppen gebildeten Mittelpunktsräume sind Primmodelle des entsprechenden Axiomensystems $\Sigma^p_{M\infty}$ (vgl. den Beweis von 5.27). Nach 5.14 ergibt sich somit das Gewünschte. □

5.34 Anmerkungen. (i) Falls p Primzahl ist, ist $\Sigma^p_{VG\infty}$ natürlich ein Axiomensystem für unendliche elementar abelsche p-Gruppen, das man vereinfachen kann durch Weglassen der Axiome $\gamma_{2,k}$ und $\gamma_{3,k}$; sei $\bar{\Sigma}^p_{VG\infty}$ das so gebildete vereinfachte Axiomensystem. $\Sigma^0_{VG\infty}$ ist ein Axiomensystem für nicht-triviale (und damit unendliche) torsionsfreie dividierbare abelsche Gruppen; es ist gleichwertig mit der Vereinfachung $\bar{\Sigma}^0_{VG\infty}$, die aus γ_1 und den folgenden Axiomen besteht:

$\bar{\gamma}_{2,k} := \forall x[x \neq 0 \rightarrow k \cdot x \neq 0]$ ($k \geq 2$) ("Torsionsfreiheit"),

$\bar{\gamma}_{3,k} := \forall x \exists y\, k \cdot y \doteq x$ ($k \geq 2$) ("Dividierbarkeit"),

$\bar{\gamma}_4 := \exists^{\geq 2}$.

In analoger Weise kann man die Axiomensysteme $\Sigma^p_{M\infty}$ vereinfachen.

(ii) Die Vollständigkeit der Systeme $\bar{\Sigma}^p_{VG\infty}$ wurde auf anderem Wege gezeigt in Bell u. Slomson 1969, S. 180. In Doraczyńska 1974 wurden "Mittelpunktsalgebren" (mit einer zweistelligen Mittelpunktsoperation) eingeführt, die den torsionsfreien dividierbaren abelschen Gruppen entsprechen, und die Vollständigkeit ihrer Theorie gezeigt.

Auf Grund von 3.13 ergibt sich noch

5.35 Korollar. *Die Theorien mit den in* 5.33 *(oder* 5.34*) genannten Axiomensystemen sind entscheidbar.*

5.36 Satz. (i) *Jedes Modell der Parallelogrammgeometrie Pg ist (bis auf Isomorphie) ein Parallelogrammraum.*

(ii) *Die Parallelogrammgeometrie mit unendlich vielen Punkten*

$$Pg_{\infty} := Cn(Pg \cup \{\exists^{\geq k} \mid k \geq 2\})$$

ist modellvollständig. Die daraus durch Festlegung einer Charakteristik p (Null oder Primzahl) entstehende Theorie Pg_{∞}^{p} *ist vollständig und entscheidbar.*

Zum Beweis: Es ist leicht zu sehen, daß man analog zu den bisherigen Überlegungen vorgehen kann, wenn man statt der Mittelpunktsbeziehung M die Parallelogrammbeziehung Pg als geometrischen Grundbegriff verwendet. Man erhält ebenso eine Übertragung der Resultate für unendliche Vektorgruppen auf die Parallelogrammgeometrie Pg_{∞}. Der Unterschied ist lediglich, daß die Charakteristik 2 keine Ausnahmerolle spielt (s. 5.16(3) und 5.17(1)). Einzelheiten können dem Leser überlassen bleiben. □

5.37 Anmerkung. Für Parallelogrammräume $\mathfrak{A}$ der Charakteristik 2 läßt sich noch zeigen (s. Schwabhäuser 1979a), daß die Parallelogrammbeziehung Pg logisch definierbar ist gdw $\mathfrak{A}$ höchstens vier Punkte besitzt.

5.38 Schlußbemerkungen. (i) Mit der Vollständigkeit der Geometrien in 5.33, 36 und der Charakterisierung der zugehörigen Modelle ist insbesondere der Satz 5.2 bewiesen.

(ii) In den Betrachtungen dieses Abschnitts spielt es durchweg keine Rolle, ob man unter Körpern nur kommutative Körper oder - allgemeiner - Schiefkörper versteht; denn die Primkörper bleiben bei beiden Auffassungen dieselben.

6. Präfixtypen

6.1 Überblick (s. auch 6.6). Nach 1.59, 60 läßt sich jede Formel logisch äquivalent umformen in eine pränexe Formel (1.58), d.h. eine Formel, in der alle Quantoren mit den zugehörigen Variablen im sog. Präfix am Anfang der Formel zusammengefaßt sind. Je nach Art (∀ oder ∃) und Stellung der Quantoren zueinander kann man verschiedene Präfixtypen unterscheiden und als einen Ausdruck für die logische Kompliziertheit einer Formel auffassen (vgl. I.1.7.3). Aufeinanderfolgende Quantoren derselben Art mit den zugehörigen Variablen faßt man dabei zu sog. Quantorenblöcken zusammen. Meist klassifiziert man die möglichen Präfixe nur nach den auftretenden Quantorenblöcken, bisweilen aber auch nach der Anzahl der in den einzelnen Blöcken vorkommenden Variablen.

6.2 Definition. Sei L eine Sprache in Standardformalisierung (s. 1.3, 5).

(i) Eine Zeichenreihe der Form $\forall x_1 \ldots \forall x_k$ bzw. $\exists x_1 \ldots \exists x_k$ (wie sie als Bestandteil des Präfixes einer pränexen Formel auftreten kann) heißt ein *Quantorenblock*, und zwar ein *∀-Block* bzw. ein *∃-Block* der *Länge* k. (Hierbei ist k=0 zugelassen.) Zur Abkürzung schreiben wir dafür auch $\forall \mathfrak{x}$ bzw. $\exists \mathfrak{x}$, wobei $\mathfrak{x} = \langle x_1, \ldots, x_k \rangle$.

(ii) Ein *Präfixtyp* ist eine Zeichenverbindung (Zeichenreihe, Wort) aus Grundzeichen der Form $\forall$, $\exists$, $\forall^i$, $\exists^i$ ($i \in \mathbb{N}$). Die *Formeln vom (Präfix-)Typ* Ξ oder kurz *Ξ-Formeln* (von L) werden induktiv über die Länge des Präfixtyps Ξ definiert durch folgende Bedingungen:

(1) Ist Ξ das leere Wort, so sind Formeln vom Typ Ξ dasselbe wie quantorenfreie Formeln.

(2) α ist vom Typ $\zeta\Xi$ gdw α aus einer Formel vom Typ Ξ entsteht durch Vorsetzen

(a) eines ∀-Blocks (beliebiger Länge), falls $\zeta = \forall$,

(b) eines $\exists$-Blocks (beliebiger Länge), falls $\zeta = \exists$,

(c) eines $\forall$-Blocks mit einer Länge $\leq i$, falls $\zeta = \forall^i$,

(d) eines $\exists$-Blocks mit einer Länge $\leq i$, falls $= \exists^i$.

Bisweilen nennen wir auch eine nicht pränexe Formel α vom Typ Ξ, falls aus α nach der Konstruktion 1.60 eine pränexe Normalform von diesem Typ entsteht.

Beispiel. Aus dem als Aussage geschriebenen euklidischen Axiom A10 (I.1.2) erhält man durch Quantorenverschiebung für $\rightarrow$ (1.60(17)) die pränexe Normalform

$$\forall a \forall b \forall c \forall d \forall t \exists x \exists y (Badt \wedge Bbdc \wedge a \neq d \rightarrow Babx \wedge Bacy \wedge Bxty).$$

Das ist eine $\forall\exists$-Aussage oder Aussage vom Typ $\forall\exists$; sie ist (genauer) auch von jedem der Typen $\forall\exists^2$, $\forall^5\exists$, $\forall^5\exists^2$. Außerdem ist sie (da Blöcke der Länge 0 zugelassen wurden) auch von jedem komplizierteren Typ wie z.B. $\forall\exists\forall$, $\exists\forall\exists$, $\forall\exists^2\forall$ usw.

Axiomensysteme, die nur aus Ξ-Aussagen bestehen, nennen wir auch *Ξ-Axiomensysteme*.

Ist der Präfixtyp Ξ ein Wort der Länge m und besteht nur aus den Grundzeichen $\forall$ und $\exists$ in abwechselnder Reihenfolge, so wird die Menge der Ξ-Formeln auch mit A_m oder E_m bezeichnet je nachdem, ob Ξ (falls $m \geq 1$) mit dem Zeichen $\forall$ oder $\exists$ beginnt (für $m=0$ ist $A_0 = E_0$ die Menge der quantorenfreien Formeln). Diese Mengen lassen sich natürlich auch direkt durch eine induktive Definition charakterisieren. Die Formel des obigen Beispiels liegt dann in A_2.

6.3 Anmerkung. In der Literatur werden die für die Quantoren üblichen anderen Bezeichnungen wie Π oder Λ für $\forall$ und Σ oder V für $\exists$ in entsprechender Weise auch zur Angabe von Präfixtypen verwendet. Die Mengen A_m bzw. E_m werden auch mit Π_m bzw. Σ_m bezeichnet.

6.4 Axiome für P_2'. Betrachten wir nun einmal das System der Axiome A1 bis A10 und CA aus I.1.2 (zu den elementaren Stetigkeitsaxiomen aus A11' s. 6.85). Wir fassen jetzt die Axiome als Aussagen auf, d.h., wir verwenden die Bezeichnungen A1,... für die Generalisierten (s. 1.24) der in I.1.2 formulierten Axiome. (Soweit diese noch nicht pränex sind, erhält man daraus pränexe Normalformen schon mittels der Quantorenverschiebung für $\rightarrow$ wie im obigen Beispiel.) Wie im Beispiel ergibt sich unmittelbar:

6.5 Satz. *Die Axiome* A1 *bis* A10 *und* CA *sind* ∀∃*-Aussagen; darüberhinaus sind* A1 *bis* A3, A5, A6, A9 *schon vom Typ* ∀ *und* A8 *vom Typ* ∃. *Somit besitzt jede der Geometrien*

A, A', A_2, A_2' *(absolut) und* P, P', P_2, P_2' *(euklidisch)*

(nach Definition, s. 2.1*) ein Axiomensystem, das aus endlich vielen* ∀∃*-Aussagen besteht (kurz, ein endliches* ∀∃*-Axiomensystem).*

6.6 Überblick (Fortsetzung). In diesem Abschnitt beschäftigen wir uns vor allem mit der Frage, welche Geometrien ∀∃-Axiomensysteme besitzen. (Zu Resultaten über einige andere Präfixtypen s. auch 7.37ff., außerdem 7.26ff. zu Resultaten über ∀∃ und Spezialfälle.)

Es sei darauf hingewiesen, daß ∀∃-Axiomensysteme besonders häufig vorkommen. Zum Beispiel bestehen auch die üblicherweise formulierten Axiomensysteme für Gruppen, für Körper, für angeordnete Körper sowie die Axiomensysteme Σ_{RF^-} und Σ_{RF} aus 3.20 für reell-abgeschlossene Körper (ohne und mit Anordnung) aus ∀∃-Aussagen.

Nach allgemeinen Sätzen über die Präfixtypen ∀∃ und ∀ (bis 6.19) werden zunächst (als Ergänzung zu 6.5) die unteren und die oberen Dimensionsaxiome betrachtet. Dann werden Erweiterungen der absoluten Geometrie A, u.a. die Nicht-Legendresche und die semi-euklidische Geometrie in Hinblick auf den Präfixtyp von Axiomensystemen behandelt (6.26 bis 37). Es folgen weitere Betrachtungen über die Rolle des Axioms der Grenzparallelen (6.38ff.), des Parallelenaxioms (6.42 und 6.58ff.) und der Dimensionsaxiome (6.43 bis 57). Danach wird eine Bildung von "geometrischen Reduzierten" in einer möglichst allgemeinen sog. "konvexen Geometrie" unter Erhaltung des Präfixtyps betrachtet (6.65 bis 81). Damit ergibt sich (Szczerba), daß das Schema der elementaren Stetigkeitsaxiome in manchen (6.85), aber nicht in allen (6.86) betrachteten Geometrien äquivalent zu einer Menge von ∀∃-Aussagen ist. Zum Schluß wird noch ein Entscheidbarkeitsresultat für ∀-Aussagen behandelt (6.87ff.).

6.7 Allgemeine Sätze über die Präfixtypen ∀∃ und ∀. Zunächst werden einige Sätze aus der Modelltheorie angegeben. Die darin auftretenden Ketten von Strukturen werden im üblichen Sinne eingeführt unter Verwendung der Unterstruktur-Beziehung $\subseteq$ (3.71(i)) als Halbordnung in der Klasse aller Strukturen für eine gegebene Sprache L. Als ω-Ketten bezeichnen wir insbesondere die Ordnungen vom Ordnungstyp ω der natürlichen Zahlen. D.h., es wird folgende Definition zugrunde gelegt.

6.8 Definition. Sei K eine Menge von Strukturen für die Sprache L (in Standardformalisierung).

(i) K ist eine *Kette*, falls für jedes $\mathfrak{A},\mathfrak{B}\in K$ gilt: $\mathfrak{A}\subseteq\mathfrak{B}$ oder $\mathfrak{B}\subseteq\mathfrak{A}$.

(ii) K ist eine ω-*Kette*, falls K von der Form $\{\mathfrak{A}_i \mid i\in\mathbb{N}\}$ ist, wobei stets $\mathfrak{A}_i\subseteq\mathfrak{A}_{i+1}$. (Auch die Folge $\langle\mathfrak{A}_i \mid i\in\mathbb{N}\rangle$ wird ω-Kette genannt.)

(iii) Die *Vereinigung* $\mathfrak{V}=\bigcup K$ der Kette K wird komponentenweise gebildet, d.h.

$$U_{\mathfrak{V}} = \bigcup_{\mathfrak{A}\in K} U_{\mathfrak{A}},\quad f_{\mathfrak{V}} = \bigcup_{\mathfrak{A}\in K} f_{\mathfrak{A}},\quad R_{\mathfrak{V}} = \bigcup_{\mathfrak{A}\in K} R_{\mathfrak{A}}\quad (f\in F_L,\ R\in R_L);$$

die $f_{\mathfrak{A}}$ bzw. $R_{\mathfrak{A}}$ seien dabei wie üblich als Mengen von r-tupeln von Individuen aufgefaßt mit $r=r(f)+1$ bzw. $r=r(R)$. (Man erhält leicht, daß dann $\mathfrak{V}$ wieder eine Struktur für L und [kleinste] gemeinsame Erweiterung der $\mathfrak{A}$ aus K ist.)

6.9 Satz. *Vor.:* $\mathfrak{V}$ *ist die Vereinigung der Kette* K *von Strukturen.*

Beh.: (i) *Ist* α *eine* $\forall\exists$-*Formel und gültig in jedem* $\mathfrak{A}\in K$, *so ist* α *auch gültig in* $\mathfrak{V}$.

(ii) *Ist* Σ *eine Menge von* $\forall\exists$-*Formeln und jedes* $\mathfrak{A}\in K$ *ein Modell von* Σ, *so ist auch* $\mathfrak{V}$ *Modell von* Σ.

6.10 Satz (Łoś, Suszko, Chang). *Sei* T *eine Theorie mit der Sprache* L *(wie oben). Dann sind die folgenden Bedingungen untereinander äquivalent.*

(i) T *besitzt ein* $\forall\exists$-*Axiomensystem.*

(ii) *Für jede* ω-*Kette von Modellen von* T *(in* L*) ist auch ihre Vereinigung ein Modell von* T.

(iii) *Für jede Kette von Modellen von* T *(in* L*) ist auch ihre Vereinigung ein Modell von* T.

Dafür, daß sich ein einzelnes Axiom äquivalent (auf Grund der übrigen Axiome) durch eine $\forall\exists$-Aussage ersetzen läßt, ergibt sich folgende Charakterisierung.

6.11 Satz. *Sei* Σ *ein Axiomensystem für die Theorie* T *(wie oben),* $\Sigma=\Gamma\cup\{\alpha\}$, α *Aussage. Dann sind die folgenden Bedingungen untereinander äquivalent.*

(i) *Es gibt eine* $\forall\exists$-*Aussage* β, *so daß* $\alpha\ddot{\mathrm{A}}\mathrm{q}_{\Gamma}\beta$.

(ii) *Für jede* ω-*Kette* K *von Modellen von* T *(in* L*) gilt:*

Ist die Vereinigung $\bigcup K$ *Modell von* Γ, *so ist sie auch Modell von* $\{\alpha\}$ *(und damit von* T*).*

(iii) *Analog zu* (ii) *für beliebige Ketten statt* ω*-Ketten.*

Den Sätzen 6.9 bis 6.11 entsprechen folgende Sätze für den Präfixtyp $\forall$.

6.12 Satz. *Vor.:* $\mathfrak{A}$ *ist Unterstruktur von* $\mathfrak{B}$ $(\mathfrak{A} \subseteq \mathfrak{B})$.

Beh.: (i) *Ist* α *eine* $\forall$*-Formel und gültig in* $\mathfrak{B}$, *so ist* α *auch gültig in* $\mathfrak{A}$.

(ii) *Ist* Σ *eine Menge von* $\forall$*-Formeln und* $\mathfrak{B}$ *Modell von* Σ, *so ist auch* $\mathfrak{A}$ *Modell von* Σ.

6.13 Satz (Łoś, Tarski). *Sei* T *wie oben. Dann sind die folgenden Bedingungen untereinander äquivalent.*

(i) T *besitzt ein* $\forall$*-Axiomensystem.*

(ii) *Für jedes Modell* $\mathfrak{A}$ *von* T *(in* L*) ist auch jede Unterstruktur von* $\mathfrak{A}$ *ein Modell von* T *(m.a.W.: Die Modellklasse* $\mathrm{Mod}_L(T)$ *ist abgeschlossen in bezug auf Bildung von Unterstrukturen).*

6.14 Satz. *Sei* Σ *ein Axiomensystem für die Theorie* T *(wie oben),* $\Sigma=\Gamma\cup\{\alpha\}$, α *Aussage. Dann sind die folgenden Bedingungen untereinander äquivalent.*

(i) *Es gibt eine* $\forall$*-Aussage* β, *so daß* $\alpha \mathrm{\ddot{A}q}_\Gamma \beta$.

(ii) *Für je zwei Modelle* $\mathfrak{A}, \mathfrak{B}$ *von* Γ *(in* L*) mit* $\mathfrak{A} \subseteq \mathfrak{B}$ *gilt: Wenn* $\models_{\mathfrak{B}} \alpha$, *so* $\models_{\mathfrak{A}} \alpha$.

6.15 Anmerkung (zu den Beweisen von 6.9 bis 14). Die Sätze 6.9 und 6.12 und damit auch die Richtungen (i) $\to$ (iii) $\to$ (ii) bzw. (i) $\to$ (ii) der übrigen Sätze ergeben sich ganz einfach durch Rückgang auf die benutzten Definitionen. Die nicht-triviale Richtung in diesen vier Sätzen ist jeweils die Umkehrung (ii) $\to$ (i). Zum Beweis für die genannten Behauptungen s. etwa Schwabhäuser 1972, S. 47, 104, 110.

6.16 Beispiel. Als Anwendung (der trivialen Richtung) von 6.10 betrachten wir die elementare hyperbolische Geometrie H. Für diese ist nach 4.58 die Zwischenbeziehung B (sogar die Kollinearität Col) als einziger Grundbegriff geeignet.

6.17 Satz (nach Tarski und Szmielew, mündlich weitergegeben). *Die dimensionsfreie und die n-dimensionalen hyperbolischen Geometrien* H *bzw.*

H_n $(2 \le n \in \mathbb{N})$ *mit der Sprache* L(B) *besitzen kein* ∀∃*-Axiomensystem.*

Beweis: Im n-dimensionalen kartesischen Raum $\mathfrak{L}_n(\mathfrak{f})$, wobei $\mathfrak{f}$ euklidisch (z.B. $\mathfrak{f} = \mathbb{R}$), betrachten wir die offenen Hyperkugeln $U_i = \{x \mid x^2 < i^2\}$ um den Ursprung mit natürlichen Zahlen $i \ge 1$ als Radien. Sei $\mathfrak{A}_i$ die Struktur mit der Trägermenge U_i und der Zwischenbeziehung B_i, die durch Einschränkung der Zwischenbeziehung von $\mathfrak{L}_n(\mathfrak{f})$ auf U_i entsteht. Dann ist B_i auch die Einschränkung von B_{i+1} auf U_i, also $\mathfrak{A}_i$ Unterstruktur von $\mathfrak{A}_{i+1}$. Außerdem ist jedes $\mathfrak{A}_i$ isomorph zum hyperbolischen Kleinschen Raum $\mathfrak{K}_n(\mathfrak{f})$ (s. 2.4, 5), jetzt aufgefaßt als Struktur für L(B). Damit ist $K = \{\mathfrak{A}_{i+1} \mid i \in \mathbb{N}\}$ eine ω-Kette von Modellen von H_n und (damit auch) von H. Die Vereinigung $\cup K$ ist dagegen der affine Raum $\mathfrak{A}_n^B(\mathfrak{f})$ (mit Anordnung, s. 2.8(ii)). Darin gilt das euklidische Parallelenaxiom, also ist $\cup K$ kein Modell von H und von H_n. Somit ist für diese Theorien die Bedingung 6.10(ii) und damit auch 6.10(i) verletzt. □

6.18 Anmerkungen. (i) In 6.17 ist die Beschränkung auf B als einzigen Grundbegriff ganz wesentlich. Führt man nämlich eine analoge Konstruktion der Strukturen $\mathfrak{A}_i$ für die Sprache L(D,B) aus, so ergibt sich, daß die Streckenkongruenz D_i von $\mathfrak{A}_i$ (die gemäß 4.58, 62 durch B_i eindeutig bestimmt ist) keine Einschränkung der entsprechenden Streckenkongruenz von $\mathfrak{A}_{i+1}$, also $\mathfrak{A}_i$ nicht Unterstruktur von $\mathfrak{A}_{i+1}$ ist. Die konstruierte Klasse K ist somit keine Kette, d.h., 6.10 ist nicht mehr anwendbar. In der Sprache L(D,B) besitzen H und die H_n tatsächlich ∀∃-Axiomensysteme (s. 6.39).

(ii) Dieses Beispiel zeigt, daß Resultate über mögliche Präfixtypen von der Wahl der Grundbegriffe abhängen. Beim Übergang zu anderen Grundbegriffen ist (im Gegensatz zu den Untersuchungen in Abschnitt 4) auch die Stellung der Quantoren in den benutzten formalen Definitionen zu berücksichtigen.

(iii) Auf Grund des folgenden Satzes läßt sich die Existenz von ∀∃-Axiomensystemen sogar stets (trivialerweise) erreichen, wenn man beliebige definitorische Erweiterungen zuläßt.

(iv) Die weiteren Ergebnisse beziehen sich also immer auf ein fest vorgegebenes System von geometrischen Grundbegriffen. Bezeichnungen für zusätzliche Begriffe dienen lediglich zur (metasprachlichen) Abkürzung von Formeln mit den gegebenen Grundbegriffen (vgl. 4.3(iv)).

6.19 Satz. *Sei* $T = Cn_L(\Sigma)$. *Dann gibt es eine definitorische Erweiterung* T^+ *von* T *und ein* Σ^+, *so daß gilt:*

(i) Σ^+ *ist ein* $\forall\exists$-*Axiomensystem für* T^+,

(ii) *wenn* Σ *endlich, so ist auch* Σ^+ *endlich.*

Beweis: Sei $\alpha\in\Sigma$ beliebig und o.B.d.A. α eine pränexe Aussage, in deren Präfix keine Variable mehrfach auftritt (1.59, 61, 40). Falls α noch keine $\forall\exists$-Aussage ist, sei das Präfix in sich abwechselnde $\forall$- und $\exists$-Blöcke zerlegt, und sei $Q\mathfrak{z}$ (d.h. $\forall\mathfrak{z}$ bzw. $\exists\mathfrak{z}$) davon der letzte Quantorenblock. Dann ist α von der Form

(1) $\alpha = \pi Q\mathfrak{z}\beta$,

wobei der Präfix-Bestandteil π die Variablen $x_1,\ldots, x_r$ enthalte. Wir bilden nun eine definitorische Erweiterung (4.3(iv)) durch Hinzunahme des neuen (bisher nicht verwendeten) r-stelligen Relationszeichens R mit der formalen Definition

(2) $\forall x_1\ldots\forall x_r[Rx_1\ldots x_r \leftrightarrow Q\mathfrak{z}\beta]$.

Nach 1.60(1), (16) bis (19) ist das "Zusatzaxiom" (2) logisch äquivalent zur Konjunktion der beiden Aussagen

(3) $\forall x_1\ldots\forall x_r Q\mathfrak{z}[Rx_1\ldots x_r \rightarrow \beta]$,

(4) $\forall x_1\ldots\forall x_r \bar{Q}\mathfrak{z}[\beta \rightarrow Rx_1\ldots x_r]$,

wobei $\bar{Q}$ den von Q verschiedenen der beiden Quantoren $\forall$ und $\exists$ bedeutet. Hierin sind die Bestandteile in eckigen Klammern quantorenfrei, also sind (3) und (4) vom Präfixtyp $\forall\exists$ (eine sogar vom Präfixtyp $\forall$). Auf Grund von (2) ist α äquivalent zur Aussage

(5) $\alpha' = \pi Rx_1\ldots x_r$,

die einen Quantorenblock weniger enthält. In der genannten definitorischen Erweiterung kann man (3) bis (5) als neue Axiome an Stelle von α und (2) verwenden.

Durch Iteration dieses Prozesses kann man weitere Quantorenblöcke abbauen und gelangt nach endlich vielen Schritten zu einer definitorischen Erweiterung, in der statt α eine endliche Menge Σ_α von Axiomen vom Typ $\forall\exists$ verwendet wird.

Diese Konstruktion der Σ_α sei nun für alle $\alpha\in\Sigma$ durchgeführt, wobei für die verschiedenen α paarweise disjunkte Mengen von (jeweils endlich vielen) neuen Relationszeichen verwendet werden; für Axiome α vom Typ $\forall\exists$ sei $\Sigma_\alpha=\{\alpha\}$ gesetzt. Durch Hinzunahme aller dieser Relationszeichen entstehe aus L die Sprache L^+. Dann leisten

(6) $\Sigma^+ := \bigcup_{\alpha\in\Sigma} \Sigma_\alpha$ und

(7) $T^+ := Cn_{L^+}(\Sigma^+)$

das Verlangte. □

6.20 Dimensionsaxiome. In Ergänzung zu 6.5 betrachten wir noch beliebige Dimensionen $n\geq 2$.

6.21 Satz. *Für die unteren und oberen Dimensionsaxiome (*I.11.70*, formuliert in der Sprache* L(B)*) gilt:*

(i) *Der Bestandteil* $Ah_n a_o \ldots a_n$ *(für die affine Abhängigkeit von Punkten)* $(n\geq 2)$ *ist eine* ∃*-Formel, für* $n=2$ *(äquivalent zu einer) quantorenfrei(en Formel).*

(ii) *Dieser Bestandteil ist logisch äquivalent zu einer Formel vom Typ* $\exists^{n-2}$.

(iii) *Das untere Dimensionsaxiom* Dim_n^- *("die Dimension ist mindestens* n*")* $(n\geq 2)$ *ist eine* ∃∀*-Aussage, für* $n=2$ *sogar vom Typ* ∃.

(iv) *Das obere Dimensionsaxiom* Dim_n^+ *("die Dimension ist höchstens* n*")* $(n\geq 2)$ *ist eine* ∀∃*-Aussage*.

Beweis: Zu (i): Der genannte Bestandteil (jetzt aufgefaßt als Abkürzung für eine Formel) läßt sich gemäß 1.9.50(1), 46(3) induktiv definieren durch

(1) $Ah_2 a_o a_1 a_2 = Col\, a_o a_1 a_2$,

(2) $Ah_n a_o \ldots a_n = \bigvee_{Perm\langle i_o,\ldots,i_n\rangle} \exists x[Col\, a_{i_o} a_{i_1} x \wedge Ah_{n-1} a_{i_2} \ldots a_{i_n} x]$.

Aus den Sätzen 1.60(10), (11), (15) über die Verschiebung des Quantors ∃ ergibt sich die Behauptung ohne weiteres durch Induktion über n.

Zu (ii): Für $n=2$ ist das wieder trivial. Zur Ausführung des Induktionsschritts seien $x, x_1, \ldots, x_{n-3}$ paarweise verschiedene und von $a_o, \ldots, a_n$ verschiedene Variablen. In jedem Disjunktionsglied auf der rechten Seite von (2) läßt sich dann der Bestandteil $Ah_{n-1}\ldots$ nach Induktionsvoraussetzung ersetzen durch eine äquivalente Formel mit dem Präfix $\exists x_1 \ldots \exists x_{n-3}$ (die ggf. durch Umbenennung der gebundenen Variablen entsteht). Nach 1.60(10) läßt sich dieses Präfix zunächst wieder vor die eckigen Klammern verschieben. Das so entstehende Präfix $\exists x \exists x_1 \ldots \exists x_{n-3}$ läßt sich dann vor die ganze Disjunktion ziehen auf Grund der Äquivalenz

(3) $\exists x\gamma \vee \exists x\delta$ Äq $\exists x(\gamma \vee \delta)$,

die statt 1.60(11), (15) (mehrfach) angewendet werden kann.

Zu (iii) und (iv): Das ergibt sich sofort aus (i) an Hand der Definitionen

(4) $\mathrm{Dim}_n^- = \exists a_o \ldots \exists a_n \neg \mathrm{Ah}_n a_o \ldots a_n$,

(5) $\mathrm{Dim}_n^+ = \neg \mathrm{Dim}_{n+1}^+$

(s. I.11.70) und der Äquivalenzen 1.60(6), (7). □

6.22 Anmerkungen. (i) Nach I.9.48, 50(i) könnte man zur Charakterisierung der affinen Abhängigkeit statt 6.21(2) auch eine Disjunktion mit nur drei Gliedern verwenden. Der Präfixtyp in 6.21(ii) wird aber dadurch nicht weiter vereinfacht.

(ii) In der Sprache L(B) läßt sich das untere Dimensionsaxiom Dim_n^- für $n \geq 3$ tatsächlich nicht durch eine $\forall\exists$-Aussage ersetzen (s. 6.50). In der Sprache L(D,B) läßt sich dagegen die folgende Vereinfachung des Präfixtyps (s.a. 6.24(iii)) erreichen.

6.23 Satz. (i) *Das untere Dimensionsaxiom* Dim_n^- $(n \geq 1)$ *ist in der absoluten Geometrie* A *(und allen Obertheorien) äquivalent zu jeder der beiden* $\exists$*-Aussagen (vgl.* 6.24(iii) *bis* (v)*)*

(1) $$\exists s \exists u_1 \ldots \exists u_n \left[\bigwedge_{\nu=1}^{n} u_\nu \neq s \wedge \bigwedge_{\substack{\mu,\nu=1 \\ \mu\neq\nu}}^{n} \mathrm{R} u_\mu s u_\nu \right]$$

(es gibt ein n-dimensionales rechtwinkliges Koordinatensystem),

(2) $$\exists s \exists u_1' \exists u_1 \ldots \exists u_n [s \neq u_1' \wedge \mathrm{B} u_1 s u_1' \wedge \bigwedge_{\nu=1}^{n} su_\nu \equiv su_1' \wedge \bigwedge_{1\leq\mu<\nu\leq n} u_\mu u_\nu \equiv u_1' u_2]$$

(es gibt ein n-dimensionales kartesisches Koordinatensystem, dim_n^-, *vgl. die Bem. zu* A8 *und* A9 *in* I.1.7.5*).*

(ii) *Folglich* (1.60(7)) *ist das obere Dimensionsaxiom* Dim_n^+ *in* A *äquivalent zu einer* $\forall$*-Aussage.*

(iii) *Somit besitzt auch jede der Geometrien*

$$A_n,\ A_n',\ P_n,\ P_n' \qquad (n \geq 2)$$

(vgl. 6.5*) ein* $\forall\exists$*-Axiomensystem.*

Beweis: (i) ergibt sich ohne weiteres aus dem (schon in A durchführbaren) Beweis der Sätze I.16.2 bis 4. Zu (2) ist lediglich zu bemerken, daß - im Falle $n\geq 2$ - nur für die eine Rechtwinkelbeziehung Ru_1su_2 die übliche Definition (4.10(1) oder auch 6.24(1)) in den Forderungen über s, u_1, u_2, u_1' enthalten ist und die weiteren Rechtwinkelbeziehungen in (1) sich aus den Kongruenzen $(u_\mu s u_\nu)\equiv(u_1' s u_2)$ von geordneten Dreiecken (I.4.4) ergeben; für diese Charakterisierung werden gleich lange Einheitsstrecken auf den Achsen benötigt. (ii) und (iii) gelten dann offensichtlich. □

6.24 Anmerkungen. (i) In Gupta 1965 wurde 6.23(1) als unteres Dimensionsaxiom verwendet für schwächere Geometrien ohne Streckenabtragung (vgl. I.0.3), in denen gleich lange Strecken auf den Achsen nicht zu existieren brauchen. Auch in diesen Geometrien gilt die Äquivalenz mit Dim_n^-.

(ii) Die Formel für die Rechtwinkelbeziehung sei in der Sprache L(D,B) (in Standardformalisierung, ohne bestimmten Artikel) gemäß 4.10(1) definiert als die ∃-Formel

(1) Rabc := ∃d[Mcbd ∧ ac≡ad],

wobei

(2) Mcbd := Bcbd ∧ bc≡bd.

Bei Bedarf läßt sich auch die folgende Äquivalenz von (1) mit einer ∀-Formel verwenden:

(3) Rabc Äq_A ∀d[Mcbd → ac≡ad].

(Vgl. den Satz 3.38 über die Elimination des bestimmten Artikels.)

(iii) Mit (1) wird 6.23(1) zu einer ∃-Aussage der Sprache L(D,B), die allerdings wesentlich mehr Variablen enthält, da für jede Rechtwinkelbeziehung in der letzten Konjunktion (wobei man sich auf $\mu<\nu$ beschränken kann) eine Variable hinzuzunehmen ist.

(iv) Interessanter ist die Auffassung von 6.23(1) als ∃-Aussage der Sprache L(R) (mit R als Grundbegriff). Dann enthält sie nämlich nur $n+1$ paarweise verschiedene Variablen; das ist die Anzahl, die nach dem auf Scott zurückgehenden Resultat 3.75(ii) (s.a. 3.76(i)) mindestens in einer zu Dim_n^- äquivalenten Aussage auftreten muß. (Dieses Resultat ist nicht nur unabhängig vom Präfixtyp, sondern auch unabhängig von der Wahl der Grundbegriffe in einem geeigneten Rahmen, es gilt insbesondere auch für die Sprachen L(R), L(R,B) u.a.)

(v) Dagegen ist 6.23(2) eine ∃-Aussage der Sprache L(D,B), die nur eine Variable mehr enthält, als nach dem Scottschen Resultat erforderlich ist. Damit erhebt sich folgendes Problem.

6.25 Problem. Gibt es eine in *P*, in *A* oder in schwächeren Geometrien (vgl. 6.24(i)) zu Dim_n^- äquivalente Aussage oder sogar ∃-Aussage der Sprache L(D,B), in der nur $n+1$ paarweise verschiedene Variablen auftreten? (Zu *A'* vgl. 3.76(ii).)

6.26 Weitere Geometrien. Als nächstes behandeln wir Resultate aus Kordos u. Szczerba 1969 über die Existenz von ∀∃-Axiomensystemen für weitere Geometrien, die Erweiterungen der absoluten Geometrie *A* sind. Die ursprünglich für ebene Geometrien formulierten Ergebnisse werden gleich auf den dimensionsfreien Fall und beliebige Dimension $n \geq 2$ verallgemeinert.

Zunächst werden die Definitionen und einige Sätze für die betrachteten Geometrien angegeben. Diese Geometrien entstehen durch Unterscheidung der möglichen Fälle für die Winkelsumme (Summe der Innenwinkel) im Dreieck und für das Verhalten von Punkten zu Geraden in Hinblick auf das Parallelenaxiom.

6.27 Definition. (i) Sei Em(abc), Hm(abc) bzw. Pm(abc) irdendeine Formel, die (mit entsprechenden Punktvariablen) ausdrückt, daß die Winkelsumme im nicht-entarteten Dreieck *abc* größer, kleiner bzw. gleich einem gestreckten Winkel ist (wie in der elliptischen, hyperbolischen bzw. euklidischen Geometrie). Diese Bedingung läßt sich in naheliegender Weise durch Aneinanderlegen von Winkeln beschreiben, was in L(D,B) (ohne Einführung einer Winkelmessung) möglich ist (für den euklidischen Fall vgl. I.12.23, s.a. 6.33(i)).

Die Forderung, daß die betreffende Bedingung für jedes nicht-entartete Dreieck gilt, wird ausgedrückt durch die folgenden *Axiome der elliptischen, hyperbolischen* bzw. *euklidischen (oder parabolischen) Metrik*:

(1) Em := ∀a∀b∀c[¬ Col abc → Em(abc)],

(2) Hm := ∀a∀b∀c[¬ Col abc → Hm(abc)],

(3) Pm := ∀a∀b∀c[¬ Col abc → Pm(abc)].

Für das Zutreffen eines dieser Axiome sagen wir auch, daß die betreffende Metrik vorliegt; Modelle von *A*, in denen das Axiom gilt, nennen wir auch Modelle mit der betreffenden Metrik.

(ii) Entsprechend bedeute Eukl(a,cd) bzw. Hyp(a,cd), daß es durch den nicht auf L(*cd*) liegenden Punkt *a* höchstens eine bzw. wenigstens zwei (und damit unendlich viele) Parallelen zur Geraden L(*cd*) (im euklidischen Sinne, I.12.2, 3) gibt ("*a* verhält sich zu L(*cd*) *euklidisch bzw. hyperbolisch*"). Gp(a,cd) bedeute, daß es von dem nicht auf L(*cd*) liegenden Punkt *a* aus eine Grenzparallele zur Halbgeraden H(*cd*) (2.3) gibt. (Wegen ¬ Col *acd* handelt es sich dabei jeweils um echte Parallelen bzw. Grenzparallelen. Bekanntlich gilt in A, daß es für solche Punkte stets wenigstens eine Parallele [im euklidischen Sinne] und höchstens eine Grenzparallele gibt [I.12.10, II.2.3(v)].)

Die Forderung, daß die betreffende Bedingung für beliebige nicht-kollineare Punkte gilt, wird ausgedrückt durch die folgenden Axiome:

(4) ("euklidisches Punkt-Geraden-Verhalten")
Eukl := ∀a∀c∀d[¬ Col acd → Eukl(a,cd)],

(5) ("hyperbolisches Punkt-Geraden-Verhalten")
Hyp := ∀a∀c∀d[¬ Col acd → Hyp(a,cd)],

(6) ("Grenzparallelenaxiom")
Gp := ∀a∀c∀d[¬ Col acd → Gp(a,cd)].

Entsprechende Redeweisen für die Geometrie und für Modelle werden analog zu (i) verwendet.

6.28 Satz. (i) *In* A *gelten folgende Sätze.*

(1) ∃a∃b∃c Em(abc) ↔ Em ,
(2) ∃a∃b∃c Hm(abc) ↔ Hm ,
(3) ∃a∃b∃c Pm(abc) ↔ Pm
(wenn in irgend e i n e m nicht-entarteten Dreieck die Winkelsumme größer, kleiner bzw. gleich einem gestreckten Winkel ist, so ist sie es in j e d e m nicht-entarteten Dreieck).

(4) ∃a∃c∃d Eukl(a,cd) ↔ Eukl ,
(5) ∃a∃c∃d Hyp(a,cd) ↔ Hyp
(analog für das Punkt-Geraden-Verhalten).

(6) ∃a∃c∃d Gp(a,cd) ↔ Gp
(analog für die Existenz von Grenzparallelen).

Somit gilt auch

(7) Em v Hm v Pm ,
(8) Eukl v Hyp ,
(9) Gp v ∀a∀c∀d[¬ Col acd → ¬Gp(a,cd)].

(ii) (Folgerung) Eukl *ist bekanntlich (in A) äquivalent zum euklidischen Parallelenaxiom und damit (*I.12.11, 12*) auch zu unserem Euklidischen Axiom* A10. Hyp *ist somit äquivalent zur Negation von* Eukl *und zu der von* A10 *(als Aussage aufgefaßt).* Hyp *und* Gp *zusammen sind offenbar äquivalent zum hyperbolischen Parallelenaxiom* HP *(2.2).*

(iii) *Aus dem Archimedischen Axiom (das sich in einer geometrischen Version, aber nicht in der ersten Stufe formulieren läßt, s. etwa* Hilbert 1977, §8, *und das aus dem Stetigkeitsaxiom* A11 *folgt) ergibt sich, daß die Metrik nicht elliptisch sein kann und daß* A10 *gleichwertig ist mit der Forderung, daß die Metrik euklidisch ist (*Legendre 1833*). Statt des Archimedischen Axioms läßt sich dabei nach* (ii) *auch das Grenzparallelenaxiom verwenden (dieses folgt aus einem Spezialfall des Schemas* A11' *der elementaren Stetigkeitsaxiome).*

Zum Beweis (Literaturangaben): (1) bis (3) (in *A*) wurden bewiesen von Dehn 1900 (vgl. 6.30). Zu anderen Beweisen s. Bonola u. Liebmann 1921 und für (1) auch Fladt 1920. Verschiedene Beweise der (untereinander gleichwertigen) Sätze (4) und (5) wurden gegeben in Strommer 1960 und Piesyk 1961. Verschiedene Beweise von (6) finden sich in Strommer 1962 und Baumann u. Schwabhäuser 1970. Zu (iii) s. Hilbert 1977, §10. Alle diese Beweise wurden elementargeometrisch für die absolute Geometrie einer festen Dimension (und mit anderen Grundbegriffen) ausgeführt. Durch Anwendung von I.11.74 und kongruente Übertragung in andere Unterräume erhält man dasselbe als Sätze der dimensionsfreien Geometrie *A*.

Außerdem kann man alle genannten Resultate aus der Charakterisierung der Modelle der A_n (vgl. 2.14(iv)) nach Pejas 1961 erhalten (zu (1) bis (3) s. dort S. 231).

6.29 Definition. (i) Durch Hinzunahme von Axiomen aus 6.27 werden entsprechende Geometrien folgendermaßen eingeführt.

(1) *Em* := Cn(*A*∪{Em})
("*Nicht-Legendresche Geometrie*", vgl. 6.30, d.h. mit elliptischer Metrik).

(2) *Hm* := Cn(*A*∪{Hm})
(Geometrie mit hyperbolischer Metrik).

(3) *Pm* := Cn(*A*∪{Pm})
(Geometrie mit euklidischer Metrik).

(4) *Ps* := Cn(A∪{Pm,Hyp})
("*semi-euklidische Geometrie*", vgl. 6.30, mit euklidischer Metrik und hyperbolischem Punkt-Geraden-Verhalten).

(5) *Ag* := Cn(A∪{Gp})
(absolute Geometrie mit [echten] Grenzparallelen).

(ii) Ist T eine dieser Geometrien, so bezeichnen wir mit T_n wieder die zugehörige n-dimensionale Geometrie, die durch Hinzunahme des Dimensionsaxioms Dim_n (oder der beiden Axiome Dim_n^-, Dim_n^+) entsteht ($2 \leq n \in \mathbb{N}$).

<u>6.30 Anmerkung</u>. Die Nicht-Legendresche und die semi-euklidische Geometrie wurden (für Dimension 3 und mit anderen Grundbegriffen) schon von Dehn 1900 eingeführt (s. Hilbert 1977, §§12 und 10) und durch Angabe von Modellen als widerspruchsfrei nachgewiesen. (Diese Modelle sind natürlich enthalten in der Klassifikation von Pejas, s.o.)

Aus A10, d.h. in *P* (2.1), ergibt sich bereits, daß die Metrik euklidisch ist und daß echte Grenzparallelen existieren. Damit erhalten wir mittels 6.28

<u>6.31 Satz</u>. (i) *Jedes Modell der absoluten Geometrie* A *ist Modell von genau einer der Geometrien Em, Hm, Ps, P.*

(ii) *In Em, Hm, Ps gilt hyperbolisches Punkt-Geraden-Verhalten (d.h.* Hyp*).*

(iii) *In den Geometrien Em und Ps gibt es keine echten Grenzparallelen (d.h.,* Gp *gilt nicht).*

(iv) *Jedes Modell von Ag ist Modell von genau einer der Geometrien P, H.*

Außerdem hat man

<u>6.32 Satz</u>. (i) (Dehn 1900) *Jede der genannten Geometrien (*6.29, 31*) ist widerspruchsfrei, d.h., sie besitzt Modelle (für die* Ps_n *s.a.* 6.36*).*

(ii) *Hm ist eine echte Untertheorie von H, d.h., es gibt Modelle von Hm, in denen keine echten Grenzparallelen existieren.*

Zu beiden Resultaten s.a. Pejas 1961 und Bachmann 1964.

Über den Präfixtyp in Axiomen erhalten wir nun folgende Sätze.

6.33 Satz (nach einer Idee von Tarski und Szmielew, s. Kordos u. Szczerba 1969 für Dimension 2). (i) *Als Abkürzung (einer quantorenfreien Formel) werde verwendet:*

(1) $\text{Ta}(abcdefgh) := \neg\,\text{Col}\,abc$

$\wedge\ ab \equiv cd \wedge cb \equiv ad \wedge \text{B}afc \wedge \text{B}bfd$

$\wedge\ ac \equiv de \wedge dc \equiv ae \wedge \text{B}agd \wedge \text{B}cge$

$\wedge\ \text{B}bhe \wedge \text{Col}\,cah$

((*abc*), (*cda*) und (*ead*) sind kongruente nicht-entartete geordnete Dreiecke, die in der [mittels *f* und *g*] beschriebenen Weise aneinanderstoßen ["adjacent triangles"], und *h* ist der Schnittpunkt von L(*ca*) und L(*be*), Abb. 154 und 155).

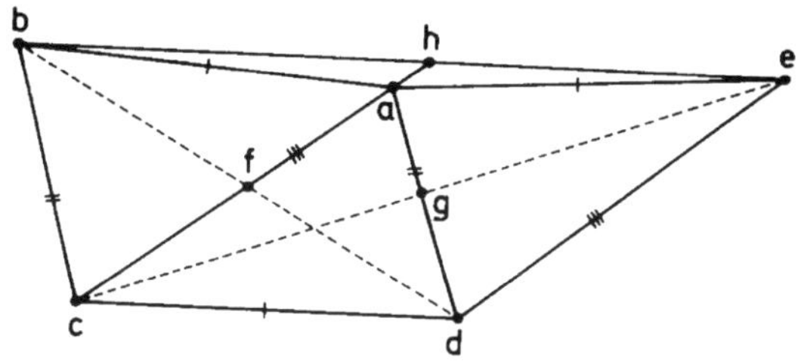

Abb. 154

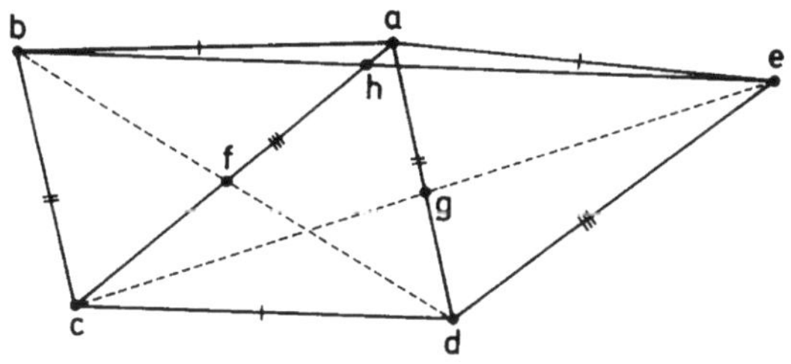

Abb. 155

Dann gelten in A *folgende Äquivalenzen.*

(2) $\text{Em}(abc)\ \text{Äq}_A\ \exists d \exists e \exists f \exists g \exists h[\text{Ta}(abcdefgh) \wedge \text{B}cah \wedge h \neq a]$.

(3) $\text{Hm}(abc)\ \text{Äq}_A\ \exists d \exists e \exists f \exists g \exists h[\text{Ta}(abcdefgh) \wedge \text{B}cha \wedge h \neq a]$.

(4) $\text{Pm}(abc)\ \text{Äq}_A\ \exists d \exists e \exists f \exists g \exists h[\text{Ta}(abcdefgh) \wedge h \doteq a]$.

(ii) *Damit sind die Zusatzaxiome* Em, Hm *und* Pm *(nach* 6.28(1) *bis* (3)*) äquivalent zu* ∃*-Aussagen.*

(iii) *Folglich besitzen die Geometrien* Em, Hm, Pm *(6.29) und (nach 6.23(i), (ii) auch)* Em_n, Hm_n, Pm_n $(2 \le n \in \mathbb{N})$ *jeweils ein* $\forall\exists$*-Axiomensystem.*

<u>Zum Beweis</u>: Aus der in (1) formulierten Bedingung erhält man zunächst, daß alle genannten Punkte in einer Ebene liegen und daß *b* und *e* auf entgegengesetzten Seiten der Geraden L(*ca*) liegen. Damit ist klar, daß durch die in (2), (3), (4) rechts angegebenen Formeln die gewünschte Bedingung über die Winkelsumme ausgedrückt wird. (Man könnte auch Em(abc), Hm(abc) bzw. Pm(abc) in 6.27(i) gleich als diese Formel definieren.) □

<u>6.34 Korollar</u>. (i) *In* A *gilt: Durch beliebige nicht-kollineare Punkte a, b, c sind weitere Punkte d, e, f, g, h mit* Ta(*abcdefgh*) *eindeutig festgelegt. Somit kann man (vgl. die Bemerkungen in* 3.38 *über die Elimination des bestimmten Artikels, insbesondere* 3.38(4)*) statt der* $\exists$*-Formeln auf den rechten Seiten von* 6.33(2) *bis* (4) *auch* $\forall$*-Formeln wie folgt verwenden.*

(1) $\mathrm{Em(abc)}\ \ddot{A}q_A\ \neg \mathrm{Col\ abc} \wedge \forall d \forall e \forall f \forall g \forall h[\mathrm{Ta(abcdefgh)} \rightarrow \mathrm{Bcah} \wedge h \neq a]$.

(2) $\mathrm{Hm(abc)}\ \ddot{A}q_A\ \neg \mathrm{Col\ abc} \wedge \forall d \forall e \forall f \forall g \forall h[\mathrm{Ta(abcdefgh)} \rightarrow \mathrm{Bcha} \wedge h \neq a]$.

(3) $\mathrm{Pm(abc)}\ \ddot{A}q_A\ \neg \mathrm{Col\ abc} \wedge \forall d \forall e \forall f \forall g \forall h[\mathrm{Ta(abcdefgh)} \rightarrow h \doteq a]$.

(ii) *Damit sind die Zusatzaxiome über die Metrik (nach Def.* 6.27(i)*) auch äquivalent zu* $\forall$*-Aussagen, nämlich*

(4) $\mathrm{Em}\ \ddot{A}q_A\ \forall a \forall b \ldots \forall h[\mathrm{Ta(ab\ldots h)} \rightarrow \mathrm{Bcah} \wedge h \neq a]$,

(5) $\mathrm{Hm}\ \ddot{A}q_A\ \forall a \forall b \ldots \forall h[\mathrm{Ta(ab\ldots h)} \rightarrow \mathrm{Bcha} \wedge h \neq a]$,

(6) $\mathrm{Pm}\ \ddot{A}q_A\ \forall a \forall b \ldots \forall h[\mathrm{Ta(ab\ldots h)} \rightarrow h \doteq a]$.

<u>6.35 Anmerkung</u>. In allen Charakterisierungen von Pm(abc) und Pm gemäß 6.33, 34 (d.h. bei euklidischer Metrik) kann man natürlich auf die Benutzung der Variablen h verzichten und stattdessen eine geeignete Bedingung mittels a formulieren.

<u>6.36 Satz</u> (Kordos u. Szczerba 1969 für Dimension 2). *Die dimensionsfreie und die n-dimensionalen semi-euklidischen Geometrien* Ps *bzw.* Ps_n *(mit der Sprache* L(D,B)*) besitzen kein* $\forall\exists$*-Axiomensystem.*

<u>Beweis</u>: Sei $\mathfrak{f}$ zunächst ein beliebiger pythagoreischer, nicht-archimedisch angeordneter Körper (3.18). Für positive Elemente *r*, *s* von $\mathfrak{f}$ bedeute $r \ll s$, daß $kr < s$ ist für jedes natürliche Element (d.h. jedes

natürliche Vielfache des Einselements) k von $\mathfrak{f}$. Im n-dimensionalen kartesischen Raum $\mathfrak{L}_n(\mathfrak{f})$ betrachten wir für beliebiges $t>0$ die Menge U_t derjenigen Punkte $\langle x_1,\ldots,x_n\rangle$, für die $\sqrt{x_1^2+\ldots+x_n^2} \ll t$ gilt (oder gleichwertig: $|x_\nu| \ll t$ für $\nu=1,\ldots,n$). Dann ist U_t eine unendliche Menge (da die Anordnung von $\mathfrak{f}$ nicht-archimedisch ist). Sei dann $\mathfrak{M}_t$ die Unterstruktur von $\mathfrak{L}_n(\mathfrak{f})$, die durch Einschränkung auf die Trägermenge U_t entsteht. Dann ergibt sich, daß $\mathfrak{M}_t$ ein Modell von Ps_n ist. Man prüft nämlich leicht nach, daß die Streckenabtragung nicht aus der Menge U_t herausführt. Außerdem wird das euklidische Parallelenaxiom falsch; denn jede Gerade L von $\mathfrak{M}_t$ ist eine echte Teilmenge einer entsprechenden Geraden $\bar{L}$ von $\mathfrak{L}_n(\mathfrak{f})$, und es gibt solche Geraden K, L, so daß sich $\bar{K}$ und $\bar{L}$ in einem nicht zu U_t gehörenden Punkt schneiden. Die Gültigkeit aller anderen Axiome von Ps_n erhält man schon daraus, daß diese in $\mathfrak{L}_n(\mathfrak{f})$ gelten (vgl. den Beweis von I.11.74).

Wir setzen nun spezieller über $\mathfrak{f}$ voraus, daß darin eine monoton wachsende Folge $\mathfrak{t}=\langle t_i\rangle_{i\in\mathbb{N}}$ von positiven Elementen vorhanden ist, die in $\mathfrak{f}$ unbeschränkt ist (d.h., zu jedem $x\in\mathfrak{f}$ gibt es ein i mit $x<t_i$). Ein solches $\mathfrak{f}$ mit einem solchen $\mathfrak{t}$ existiert; z.B. leistet der reell-abgeschlossene Körper $\mathfrak{f}'$ aus dem ersten Beispiel in 3.19 mit $t_i=u^i$ das Verlangte. Wir setzen nun $\mathfrak{A}_i = \mathfrak{M}_{t_i}$. Dann ist $K:=\{\mathfrak{A}_i \mid i\in\mathbb{N}\}$ eine ω-Kette von Modellen von Ps_n und (damit auch) von Ps. Die Vereinigung $\bigcup K$ ist dagegen der ganze kartesische Raum $\mathfrak{L}_n(\mathfrak{f})$, in dem das euklidische Parallelenaxiom gilt, also ist $\bigcup K$ kein Modell von Ps_n und von Ps. Aus (der trivialen Richtung von) 6.10 (oder 6.9(ii), vgl. 6.15) ergibt sich also das Gewünschte. □

In dieser Konstruktion ist aber $\bigcup K$ ein Modell von Pm_n. Damit ergibt sich mittels 6.11

<u>6.37 Korollar</u>. *Die* $\exists\forall$*-Aussage* $\neg$A10 *(wobei das Euklidische Axiom* A10 *als Aussage formuliert sei, s.a.* 6.28(ii)*) ist auf Grund der übrigen Axiome (d.h. in* Pm_n*) nicht äquivalent zu einer* $\forall\exists$*-Aussage (und damit erst recht nicht in Untertheorien wie* A_n, A*).*

<u>6.38 Lambertsche Vierecke</u>. Im folgenden Resultat 6.39 wird der Begriff des Lambertschen Vierecks verwendet; darunter versteht man ein ebenes Viereck mit drei rechten Winkeln. Nach 6.28(1) bis (3) ist dann der vierte Winkel stumpf, spitz bzw. ein Rechter genau dann, wenn die Metrik elliptisch, hyperbolisch bzw. euklidisch ist. Diese Vierecke wurden schon 1766 betrachtet von J. H. Lambert im Zusammenhang mit Untersuchungen, die zu einem Beweis des Parallelenaxioms führen sollten (vgl. etwa Baldus 1964, Abschn. I).

<u>6.39 Satz</u> (Kordos u. Szczerba 1969 für Dimension 2). (i) *Als Abkürzung werde verwendet:*

(1) $\mathrm{Lq}(abcds) := \neg\,\mathrm{Col}\,bcd \wedge \mathrm{R}abc \wedge \mathrm{R}bcd \wedge \mathrm{R}cda$
$\wedge\ \mathrm{B}asc \wedge \mathrm{B}bsd$

(*a*, *b*, *c*, *d* bilden ein nicht-entartetes *Lambertsches Viereck* ["quadrangle"] mit rechten Winkeln bei *b*, *c*, *d* mit dem Diagonalenschnittpunkt *s*, Abb. 156).

Das Grenzparallelenaxiom Gp *(*6.27(6)*) ist dann in* A *äquivalent zur Konjunktion der folgenden drei* ∀∃*-Aussagen:*

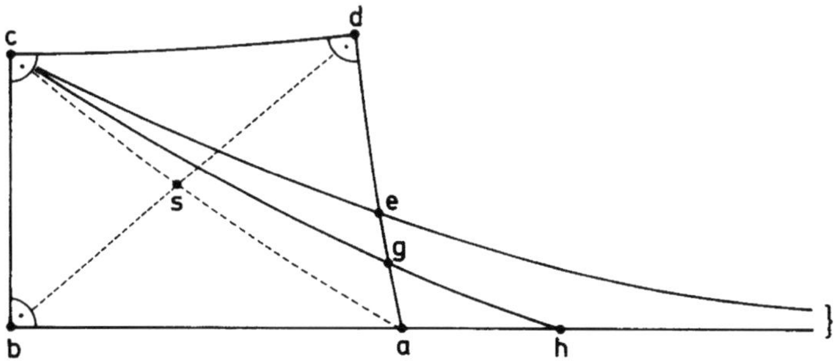

Abb. 156

(2) $\mathrm{L1} := \forall a \forall b \forall c \forall d \forall s\{\mathrm{Lq}(abcds) \to \exists e[\mathrm{B}aed \wedge ce \equiv ba]\}$,

(3) $\mathrm{L2} := \forall a \forall b \forall c \forall d \forall s \forall e \forall f\{\mathrm{Lq}(abcds) \wedge \mathrm{B}aed \wedge ce \equiv ba \wedge \mathrm{B}cef \to \neg \mathrm{B}baf\}$,

(4) $\mathrm{L3} := \forall a \forall b \forall c \forall d \forall s \forall e \forall g\{\mathrm{Lq}(abcds) \wedge \mathrm{B}aed \wedge ce \equiv ba \wedge \mathrm{B}ega \wedge g \neq e$
$\to \exists h[\mathrm{B}cgh \wedge \mathrm{B}bah]\}$.

(ii) *Folglich besitzen die Geometrien* Ag*,* H *und (nach* 6.23(i), (ii) *auch)* Ag_n, H_n ($2 \leq n \in \mathbb{N}$) *jeweils ein* ∀∃*-Axiomensystem.*

Zum Beweis sei zunächst ein Resultat aus der hyperbolischen Geometrie zitiert (zu dessen Beweis s. etwa Liebmann 1923, §10, wo die Lambertschen Vierecke als "Spitzecke" bezeichnet werden).

<u>6.40 Lemma</u>. (i) *In* H *gilt:*

(1) $\mathrm{Lq}(abcds) \to cd < ba$.

(2) $\mathrm{Lq}(abcds) \wedge \mathrm{B}aed \to [ce \overset{\vee}{\nparallel} ba \leftrightarrow ce \equiv ba]$
(unter der genannten Voraussetzung ist die Halbgerade H(*ce*) *grenzparallel zu* H(*ba*) *genau dann, wenn e auf dem Kreis um c mit dem Radius ba [in der Ebene des Lambertschen Vierecks] liegt).*

(ii) (Folgerung für *H*) *Eine Grenzparallele von einem Punkt c aus zu einer Halbgeraden läßt sich stets konstruieren durch Konstruktion eines passenden Lambertschen Vierecks, Schlagen des genannten Kreises und Bestimmung seines Schnittpunkts mit der Seite ad.*

Beweis von 6.39: Zu (i): Aus 6.40 erhält man zunächst, daß L1 bis L3 Sätze von *H* sind (die Existenz von *e* in L1 ergibt sich aus dem Grenzparallelenaxiom). Offenbar sind diese Formeln aber auch Sätze von *P* (die in L1 für *e* verlangte Eigenschaft ist dann gleichwertig mit *e=d*, s.a. 2.3(iv)). Nach 6.31(iv) gelten diese Formeln also in jedem Modell von *Ag*, d.h., sie folgen in *A* aus Gp. Umgekehrt erhält man aus L1 bis L3 ohne weiteres Gp. Ist nämlich ¬ Col *cb'a'*, so kann man Punkte *a*, *b*, *d*, *s* mit $ba \equiv_0 b'a'$ (s. 2.3(2)) und Lq(*abcds*) konstruieren und *e* gemäß L1 wählen; dann besagen L2 und L3, daß H(*ce*) die gewünschte Grenzparallele zu H(*b'a'*) ist.

Zu (ii): *Ag* entsteht somit aus *A* durch Hinzunahme der Axiome L1 bis L3; *H* entsteht daraus durch Hinzunahme des Axioms Hm. □

6.41 Anmerkungen. (i) In 6.39(2) bis (4) sind nicht alle in der Sprache L(D,B) verwendeten Variablen aufgeschrieben. In der Formel Lq(abcds) enthalten ja die mit R gebildeten Bestandteile nach 6.24(1) je eine Variable mit dem Existenzquantor. Bei der Bildung der pränexen Normalformen von L1 bis L3 entstehen daraus im Präfix drei zusätzliche Variablen mit dem Allquantor (1.60(19)). Für die mit R gebildeten Bestandteile kann man auch die Charakterisierung 6.24(3) (mit Allquantor) verwenden und erhält dann drei zusätzliche Variablen mit dem Existenzquantor im Präfix entsprechender Axiome. In jedem Falle erhält man Axiome vom Präfixtyp ∀∃.

(ii) Die Variable s in der Formel Lq(abcds) wird nur verwendet, um auszudrücken, daß die Ecken des Lambertschen Vierecks in einer Ebene liegen. In den ebenen Geometrien Ag_2 und H_2 kann man daher auf die Variable s und die Forderung in der zweiten Zeile von 6.39(1) verzichten; man kommt also mit einer Variablen weniger im Präfix aus (so wurden die Axiome ursprünglich bei Kordos u. Szczerba 1969 formuliert).

(iii) Im Gegensatz zu (ii) wird in der Formel Ta(abcdefgh) (6.33(1)) durch die Verwendung der Variablen f und g nicht nur ausgedrückt, daß die betrachteten Dreiecke in einer Ebene liegen, sondern auch, daß sie sich nicht überlappen, d.h., daß die betrachteten Winkel aneinander und nicht ineinander gelegt werden. Auch in ebenen Geometrien kann man nicht ohne weiteres (und tatsächlich nicht in allen Fällen) auf die Verwendung dieser Variablen verzichten.

6.42 Satz (Antwort auf eine Frage von Alfred Tarski, vgl. die Bem. zu A10 in I.1.7.5). *Sei n beliebig mit* $2 \le n \in \mathbb{N}$. *Im Axiomensystem für die volle n-dimensionale euklidische Geometrie* P_n^2 *(2.1) läßt sich das Euklidische Axiom* A10 *nicht äquivalent ersetzen durch eine Formel der Sprache* L(B) *vom Präfixtyp* $\forall$. *D.h., es gibt keine Formel der angegebenen Art, zu der* A10 *auf Grund der übrigen Axiome - also in (der entsprechenden absoluten Geometrie)* A_n^2 *- äquivalent ist. (Dasselbe gilt dann erst recht für Untertheorien von* A_n^2*, z.B. erster Stufe.)*

Beweis: Annahme: Es gibt eine Aussage dieser Art, sei η eine solche. Sei $\mathfrak{A}$ der n-dimensionale kartesische Raum $\mathfrak{L}_n(\mathbb{R})$ und $\mathfrak{B}$ der n-dimensionale (hyperbolische) Kleinsche Raum $\mathfrak{K}_n(\mathbb{R})$ (2.4). Beide Strukturen sind Modelle von A_n^2. Nach Annahme ist η also, ebenso wie A10, gültig in $\mathfrak{A}$, aber nicht $\mathfrak{B}$. Dasselbe trifft zu für die Redukte (1.20) $\mathfrak{A}^-$ bzw. $\mathfrak{B}^-$ von $\mathfrak{A}$ bzw. $\mathfrak{B}$ auf die Sprache L(B) (1.18). Nun ist aber $\mathfrak{B}^-$ eine Unterstruktur von $\mathfrak{A}^-$ (vgl. 6.17, 18(i)). Nach 6.12(i) müßte η also doch in $\mathfrak{B}^-$ gültig sein. Damit ist die Annahme zum Widerspruch geführt. □

6.43 Das untere Dimensionsaxiom in L(B). Nach 6.21 ist das (in der Sprache L(B) formulierte) untere Dimensionsaxiom Dim_n^- eine $\exists\forall$-Aussage. Wir kommen nun zu dem schon in 6.22(ii) und I.1.7.5 (Bem. zu A8 und A9) angekündigten Resultat (s. 6.48, 47, 50), daß sich dieses Axiom für $n \ge 3$ nicht äquivalent durch eine $\forall\exists$-Aussage von L(B) ersetzen läßt.

6.44 Satz (Kordos 1969). *Sei* $3 \le n \in \mathbb{N}$, *sei* $\mathfrak{B}$ *ein n-dimensionaler affiner Raum (mit Anordnung) über einem angeordneten Körper* $\mathfrak{f}$ *und* $\mathfrak{A}$ *ein* $(n-1)$*-dimensionaler Unterraum von* $\mathfrak{B}$. *Weiter sei* η *eine* $\forall\exists$*-Aussage der Sprache* L(B), *die in* $\mathfrak{B}$ *gültig ist. Dann ist* η *auch in* $\mathfrak{A}$ *gültig.*

Zum Beweis wird folgendes Lemma verwendet.

6.45 Lemma. *Seien n,* $\mathfrak{B}$, $\mathfrak{A}$ *wie oben, und sei N eine endliche Menge von Punkten von* $\mathfrak{B}$. *Dann gibt es eine Abbildung* φ *von N auf eine (natürlich ebenfalls endliche) Teilmenge M von* $\mathfrak{A}$, *so daß für jedes* $a,b,c \in N$ *gilt:*

(1) $a=b$ *gdw* $\varphi(a)=\varphi(b)$,

(2) $B_{\mathfrak{B}}abc$ *gdw* $B_{\mathfrak{A}}\varphi(a)\varphi(b)\varphi(c)$,

(3) *wenn* $a \in \mathfrak{A}$, *so* $\varphi(a)=a$;

d.h., φ ist ein Isomorphismus von N auf M (genauer: ein Isomorphismus zwischen den Unterstrukturen von $\mathfrak{B}$, die durch Einschränkung auf diese Punktmengen entstehen), der Punkte von $\mathfrak{A}$ fest läßt.

Beweis: In $\mathfrak{B}$ sei ein affines Koordinatensystem so gewählt, daß $A=U_{\mathfrak{A}}$ die von den ersten n-1 Achsen aufgespannte Hyperebene ist, sei dann L die n-te Achse. Für jedes $a,b,c \in N$ bedeute $C(abc)$ den von den Punkten a, b, c aufgespannten Unterraum, das ist also jeweils eine einpunktige Menge, eine Gerade oder eine Ebene. Der Einfachheit halber sei $\bar{C}(abc)$ stets eine Ebene, in der $C(abc)$ enthalten ist. Sei $\mathfrak{E}$ die endliche Menge dieser Ebenen. Sei $\mathfrak{E}_1$ die Menge derjenigen Ebenen aus $\mathfrak{E}$, die eine zu L parallele Gerade enthalten. Dann schneidet jede Ebene aus $\mathfrak{E}_1$ die Hyperebene A in einer gewissen Geraden; sei $\mathfrak{G}_1$ die endliche Menge dieser Geraden. Sei r ein Punkt von A, der auf keiner dieser Geraden liegt (die Existenz eines solchen Punktes läßt sich streng zeigen wie unten die von s [s. (6)] durch Übertragung der hier für die Ebenenmenge $\mathfrak{E}$ ausgeführten Überlegung auf die Geradenmenge $\mathfrak{G}_1$). Sei K die Parallele zu L durch r. Dann enthält K keinen Punkt einer Ebene aus $\mathfrak{E}_1$ und hat mit jeder Ebene aus $\mathfrak{E}_2 := \mathfrak{E} - \mathfrak{E}_1$ höchstens einen Punkt (Schnittpunkt) gemein. Sei S die Menge dieser Schnittpunkte, und sei s auf K so gewählt, daß gilt:

(4) s ist verschieden von den (endlich vielen) Punkten von S,

(5) die n-te Koordinate von s ist positiv und größer als die n-ten Koordinaten sämtlicher Punkte aus N.

Wegen (4) gilt dann

(6) s liegt in keiner der Ebenen aus $\mathfrak{E}$.

Sei nun φ die Zentralprojektion mit dem Zentrum s von N in A, d.h. die Abbildung derart, daß $\varphi(a)$ stets der Schnittpunkt von $L(sa)$ mit A ist (der wegen (5) existiert), und sei M die Menge der Bildpunkte. Dann gelten (3) und die Richtungen (→) der Behauptungen (1) und (2) (wegen (5)) offenbar. Zum Beweis der Umkehrungen (indirekt) verwenden wir, daß s wegen (6) in keinem der Räume $C(abc)$ $(a,b,c \in N)$ liegt. Daraus ergibt sich, daß verschiedene Punkte a, b von N durch φ wieder in verschiedene Punkte und nicht-kollineare Punkte wieder in nicht-kollineare Punkte übergeführt werden. Sind schließlich a, b, c kollineare Punkte von N, aber nicht $B_{\mathfrak{B}}abc$, so erhält man die Zwischenbeziehung und gewisse Verschiedenheitsfeststellungen für eine geeignete Permutation dieser Punkte, und nach dem schon Bewiesenen gilt das Entsprechende für die Bilder, damit ist nicht $B_{\mathfrak{A}}\varphi(a)\varphi(b)\varphi(c)$. Somit leisten φ und M das Verlangte. □

Beweis von 6.44: Sei η von der Form $\forall x_1 \ldots \forall x_k \exists y_1 \ldots \exists y_l \alpha$ (α quantorenfrei). Zum Beweis von $\models_{\mathfrak{A}} \eta$ seien $a_1, \ldots, a_k$ beliebige Punkte von $\mathfrak{A}$. Wegen der Gültigkeit von η in $\mathfrak{B}$ ist dann $\models_{\mathfrak{B}} \alpha[a_1, \ldots, a_k, b_1, \ldots, b_l]$ für geeignete Punkte $b_1, \ldots, b_l$ von $\mathfrak{B}$. Sei nun $N=\{a_1, \ldots, a_k, b_1, \ldots, b_l\}$, und seien φ, M gemäß 6.45 gewählt. Da φ ein Isomorphismus ist, ist auch $\models_{\mathfrak{A}} \alpha[\varphi(a_1), \ldots, \varphi(b_l)]$ (nach 6.45(1), (2) bleiben die Wahrheitswerte der in α enthaltenen Primformeln dieselben). Nach 6.45(3) ist $\varphi(a_\kappa)=a_\kappa$ ($\kappa=1, \ldots, k$). Damit haben die Punkte $b'_\lambda=\varphi(b_\lambda) \in \mathfrak{A}$ ($\lambda=1, \ldots, l$) die gewünschte Eigenschaft $\models_{\mathfrak{A}} \alpha[a_1, \ldots, a_k, b'_1, \ldots, b'_l]$. □

6.46 Definition. Sei C eine Klasse von angeordneten Körpern. Dann seien die *affinen Geometrien* $\Gamma'_n(C)$ (n-dimensional), $\Gamma'(C)$ (dimensionsfrei), $\Gamma'_{\leq n}(C)$ bzw. $\Gamma'_{\geq n}(C)$ (mit entsprechenden Dimensionsbeschränkungen) analog zu 3.52 erklärt als Theorie der Klasse der affinen Räume über Körpern aus C von Dimension n, von beliebiger endlicher Dimension ≥ 2, von Dimension $\leq n$ bzw. von Dimension $\geq n$. (Statt der kartesischen Räume $\mathfrak{L}_n(\mathfrak{f})$ in 3.52 werden die entsprechenden affinen Räume $\mathfrak{A}_n^B(\mathfrak{f})$ verwendet.)

6.47 Beispiele. Spezialfälle sind die früher betrachteten affinen Geometrien (mit Anordnung) $A\mathfrak{g}o = \Gamma'(OF)$, $A\mathfrak{g}o^* = \Gamma'(RF)$ und entsprechende Geometrien mit Dimensionsforderungen, die durch die angegebenen Indizes ausgedrückt werden.

6.48 Satz (Kordos 1969). *Sei C eine nicht-leere Klasse von angeordneten Körpern, und sei $3 \leq n \in \mathbb{N}$. In den affinen Geometrien $\Gamma'(C)$ und $\Gamma'_{\leq m}(C)$ mit $n \leq m$ ist das untere Dimensionsaxiom Dim_n^- nicht äquivalent zu einer $\forall\exists$-Aussage.*

Beweis: Sei T eine der betrachteten Geometrien, sei $\mathfrak{f} \in C$ gewählt, und sei $\mathfrak{B} = \mathfrak{A}_n^B(\mathfrak{f})$ und $\mathfrak{A}$ ein $(n-1)$-dimensionaler Unterraum von $\mathfrak{B}$. Dann sind $\mathfrak{A}$ und $\mathfrak{B}$ Modelle von T, und Dim_n^- gilt in $\mathfrak{B}$, aber nicht in $\mathfrak{A}$. Somit kann Dim_n^- nicht in (allen Modellen von) T äquivalent zu einer $\forall\exists$-Aussage sein, deren Gültigkeit sich ja nach 6.44 von $\mathfrak{B}$ auf $\mathfrak{A}$ übertragen würde. □

6.49 Anmerkungen. (i) Statt der Znetralprojektionen bei Kordos 1969 kann man im Beweis von 6.45 auch Parallelprojektionen verwenden. Dazu hat man zu zeigen, daß es zu endlich vielen Ebenen in einem affinen Raum von Dimension ≥ 3 stets eine Projektionsrichtung gibt, die in keiner der Ebenen enthalten ist, d.h. eine Gerade L derart, daß keine der Ebenen eine zu L parallele Gerade enthält. Ein Beweis dafür wurde dem Verfasser von

H. N. Gupta mitgeteilt.

(ii) In Kordos 1969 wird noch festgestellt: (a) Die Resultate 6.44 und 45 übertragen sich auf unendlichdimensionale Räume (da eine endliche Punktmenge sich stets in einen endlichdimensionalen Unterraum einbetten läßt). (b) Diese Resultate übertragen sich auch auf affine Räume ohne Anordnung als Strukturen für die Sprache L(Col) (mit der Kollinearität als einzigem Grundbegriff). Dazu braucht man offenbar nur vorauszusetzen, daß der Körper $\mathfrak{F}$ unendlich viele Elemente besitzt, damit die Gerade K im Beweis von 6.45 unendlich viele Punkte enthält; an Stelle von 6.45(5) genügt die Bedingung, daß die n-te Koordinate von s verschieden von 0 und von den n-ten Koordinaten sämtlicher Punkte aus N ist. Die hier formulierten unteren (und oberen) Dimensionsaxiome lassen sich auch als Aussagen der Sprache L(Col) auffassen; 6.48 überträgt sich dann auf entsprechende affine Geometrien über Klassen von Körpern unter Ausschluß der Charakteristik 2 (vgl. 2.8, 5.3(i)), die mindestens einen unendlichen Körper enthalten.

(iii) In 6.45 und 44 kann man statt der affinen Räume $\mathfrak{B}$ und $\mathfrak{A}$ auch kartesische Räume, d.h. Strukturen für die Sprache L(D,B), verwenden; die Behauptungen (die sich nach wie vor auf die eingeschränkte Sprache L(B) beziehen) ergeben sich offenbar genauso. Daraus erhalten wir analog zu 6.48:

<u>6.50 Satz</u>. *Seien C und n wie in* 6.48. *In den euklidischen Geometrien* $\Gamma(C)$ *und* $\Gamma_{\leq m}(C)$ *mit* $n \leq m$ *(z.B. in* P, P*, $P^*_{\leq n}$*) und damit auch in allen Untertheorien (z.B.* A, $A_{<n}$, Pm*) ist das untere Dimensionsaxiom* Dim_n^- *nicht äquivalent zu einer* ∀∃*-Aussage der Sprache* L(B).

<u>6.51 Vereinigungen von affinen Räumen</u>. Im Unterschied zu früheren ähnlichen Resultaten (6.17, 36, 37) erfolgte der Beweis von 6.48 ohne Konstruktion einer ω-Kette, d.h. ohne Benutzung des Satzes von Łoś, Suszko u. Chang. Wir können nun erstmals die nicht-triviale Richtung (vgl. 6.15) dieses Satzes, und zwar in der Version 6.11, anwenden und gelangen damit zu einem Resultat, das wohl auf den ersten Blick verblüffend erscheint.

<u>6.52 Satz</u> (nach einer Idee von Tarski, s. Schwabhäuser u. Szczerba 1974). *Sei C eine nicht-leere "axiomatische Klasse" von angeordneten Körpern, d.h. eine Klasse, die durch ein Axiomensystem in der ersten Stufe (hier in der Sprache* $L_{OF}:=L(+,\cdot,-,0,1;<)$*) festgelegt ist (z.B. die Klasse* OF *aller angeordneten oder die Klasse* RF *der reell-abgeschlossenen Körper).*

Sei $3\leq n\in\mathbb{N}$. Dann gibt es eine ω-Kette K von n-dimensionalen affinen Räumen (mit Anordnung) über Körpern aus C derart, daß die Vereinigung $\bigcup K$ wieder ein affiner Raum über einem Körper aus C ist, aber eine kleinere Dimension m (mit $2\leq m<n$) besitzt.

Beweis: Sei $T = \Gamma_n'(C)$, $\Gamma = \Gamma_{\leq n}'(C)$ und $\alpha = \mathrm{Dim}_n^-$. Dann gilt die Voraussetzung von 6.11, aber die Bedingung 6.11(i) wird nach 6.48 verletzt. Damit wird auch die Bedingung 6.11(ii) verletzt, d.h., es gibt eine ω-Kette von Modellen von T, deren Vereinigung ein Modell von Γ, aber nicht Modell von $\{\alpha\}$ ist; sei K eine solche Kette. Da C eine axiomatische Klasse ist, lassen sich die Axiome für C (die zusätzlich zu den Axiomen für OF gebraucht werden) durch entsprechende geometrische Reduzierten in der Sprache L(B) ausdrücken (vgl. 3.37(iii), 51). Damit (vgl. 2.14(ii)) ist jedes Modell von T bzw. Γ (bis auf Isomorphie) ein affiner Raum (mit Anordnung) mit Dimension n bzw. $\leq n$ über einem Körper aus C. Insbesondere ist die Vereinigung $\bigcup K$ ein Modell von Γ, in dem Dim_n^- falsch wird, d.h. ein affiner Raum der gewünschten Art. Somit leistet K das Verlangte. □

6.53 Anmerkungen. (i) In diesem Beweis wurde nicht benötigt, daß Γ ein $\forall\exists$-Axiomensystem besitzt (das als System der "übrigen Axiome" von T verwendet werden kann). Tatsächlich ist das gar nicht der Fall, vgl. 6.62, 64(iii).

(ii) Dieser modelltheoretische Beweis ist verhältnismäßig abstrakt. Mehr Information liefert die anschließend ausgeführte direkte Konstruktion solcher Ketten, die außerdem zu der folgenden Verschärfung von 6.52 führt. Insbesondere ergibt sich, daß man die Dimension m mit $2\leq m<n$ (und natürlich auch $m=n$) beliebig vorschreiben kann.

6.54 Satz (Schwabhäuser u. Szczerba 1974). *Vor.:* (i) $2\leq m\leq n\in\mathbb{N}$.

(ii) $\langle \mathfrak{F}_\nu \mid \nu\in\mathbb{N}\rangle$ *ist eine (ω-)Kette von angeordneten Körpern, so daß der Körpergrad* $(\mathfrak{F}_{\nu+1}:\mathfrak{F}_\nu)$ *(der auch unendlich sein kann, s. 4.86(ii)) mindestens* $(n-m+1)^2-1$ *ist.*

Beh.: Es gibt eine Folge $\langle \mathfrak{A}_\nu \mid \nu\in\mathbb{N}\rangle$*, so daß stets gilt:*

(a) $\mathfrak{A}_\nu$ *ist Unterstruktur von* $\mathfrak{A}_{\nu+1}$ *(Ketteneigenschaft),*

(b) $\mathfrak{A}_\nu$ *ist ein n-dimensionaler affiner Raum (mit Anordnung) über* $\mathfrak{F}_\nu$,

(c) $\bigcup_{\nu\in\mathbb{N}} \mathfrak{A}_\nu$ *ist ein m-dimensionaler affiner Raum über* $\bigcup_{\nu\in\mathbb{N}} \mathfrak{F}_\nu$.

Beweis: Gelte (i). Über die Kette $\langle \mathfrak{F}_\nu \mid \nu\in\mathbb{N}\rangle$ von angeordneten Körpern setzen wir zunächst voraus, daß stets $(\mathfrak{F}_{\nu+1}:\mathfrak{F}_\nu)\geq g$ ist (g wird später festgelegt). Sei $\mathfrak{f} = \bigcup_{\nu \in \mathbb{N}} \mathfrak{f}_\nu$ und $\mathfrak{V}$ ein m-dimensionaler Vektorraum über $\mathfrak{f}$ mit der Basis $<b_1,\dots,b_m>$. Sei $\mathfrak{V}'_\nu$ der m-dimensionale Vektorraum über $\mathfrak{f}_{\nu+1}$, der aus den Vektoren von $\mathfrak{V}$ von der Form $\sum_{\mu=1}^{m} x_\mu b_\mu$ mit Koeffizienten x_μ aus $\mathfrak{f}_{\nu+1}$ besteht. Entstehe $\mathfrak{V}''_\nu$ aus $\mathfrak{V}'_\nu$ durch Einschränkung der Multiplikation mit Skalaren auf (die Skalare von) $\mathfrak{f}_\nu$. Nach 4.86 besitzt $\mathfrak{V}''_\nu$ dieselben Vektoren wie $\mathfrak{V}'_\nu$, ist jedoch ein Vektorraum über $\mathfrak{f}_\nu$ mit der Dimension $(\mathfrak{V}''_\nu : \mathfrak{f}_\nu) = m\cdot(\mathfrak{f}_{\nu+1} : \mathfrak{f}_\nu) \geq m\cdot g$. Wir setzen jetzt $m\cdot g \geq n$ voraus. Die Vektoren $b_1,\dots, b_m$ liegen auch in $\mathfrak{V}''_\nu$ und sind dort linear unabhängig (da das sogar in $\mathfrak{V}$ der Fall ist).

Sei nun $\mathfrak{V}_\nu$ ein n-dimensionaler Unterraum von $\mathfrak{V}''_\nu$, über den wir zunächst nur voraussetzen, daß er eine Basis der Form $<b_1,\dots,b_m,c_1^{(\nu)},\dots,c_k^{(\nu)}>$ besitzt, wobei $k=n-m$. (Er wird später durch eine geeignete Wahl der zusätzlichen Basisvektoren $c_\kappa^{(\nu)}$ festgelegt.) Die oberen Indizes (ν) lassen wir im folgenden weg, da die $c_\kappa^{(\nu)}$ in einer für alle ν einheitlichen Art beschrieben werden. Als Vektor von $\mathfrak{V}'_\nu$ hat c_κ eine Basisdarstellung der Form

(1) $c_\kappa = \sum_{\mu=1}^{m} c_{\kappa\mu} b_\mu$ mit $c_{\kappa\mu} \in \mathfrak{f}_{\nu+1}$ $(\kappa=1,\dots,k)$.

Eine geeignete Wahl der Vektoren c_κ läßt sich somit erreichen durch eine geeignete Wahl der Koeffizienten $c_{\kappa\mu}$.

Sei schließlich $\mathfrak{a}$ der durch $\mathfrak{V}$ bestimmte affine Raum (mit Vektoren als Punkten wie in 2.8(ii)) und $\mathfrak{a}_\nu$ der durch $\mathfrak{V}_\nu$ bestimmte affine Raum $(\nu \in \mathbb{N})$. Dann ist offensichtlich $\mathfrak{a}_\nu$ ein n-dimensionaler affiner Raum über $\mathfrak{f}_\nu$, $\mathfrak{a}$ ein m-dimensionaler affiner Raum über $\mathfrak{f}$ und $\mathfrak{a} = \bigcup_{\nu \in \mathbb{N}} \mathfrak{a}_\nu$.

Es bleibt zu zeigen, daß stets $\mathfrak{a}_\nu$ eine Unterstruktur von $\mathfrak{a}_{\nu+1}$ ist. Da für die Punktmengen dieser Räume die gewünschte Inklusion gilt, bedeutet das, daß Punkte a, b, c von $\mathfrak{a}_\nu$ in der Zwischenbeziehung von $\mathfrak{a}_\nu$ stehen genau dann, wenn sie in der Zwischenbeziehung von $\mathfrak{a}_{\nu+1}$ stehen. Hiervon ist die Richtung $(\rightarrow)$ trivial. Zum Beweis der Umkehrung können wir wie im Beweis von 6.45 benutzen, daß drei kollineare Punkte in irgendeiner Reihenfolge in der Zwischenbeziehung stehen müssen. Damit genügt es zu zeigen, daß die entsprechende Bedingung für die Kollinearität gilt, oder, gleichbedeutend:

(*) Für beliebige x, y von $\mathfrak{V}_\nu$ gilt: Wenn x, y in $\mathfrak{V}_{\nu+1}$ linear abhängig sind, so sind sie auch linear abhängig in $\mathfrak{V}_\nu$.

Tatsächlich ergibt sich die Bedingung (*) noch nicht aus den bisherigen Voraussetzungen, wenn $m<n$ ist. Um sie zu erfüllen, werden gerade die c_κ

in geeigneter Weise gewählt. Zu diesem Zweck betrachten wir beliebige Vektoren x, y von $\mathfrak{V}_\nu$ mit Basisdarstellungen

$$(2)\qquad x = \sum_{\mu=1}^{m} x_\mu b_\mu + \sum_{\kappa=1}^{k} u_\kappa c_\kappa \ , \quad y = \sum_{\mu=1}^{m} y_\mu b_\mu + \sum_{\kappa=1}^{k} v_\kappa c_\kappa \ .$$

Nach (1) ergeben sich dann in $\mathfrak{V}_{\nu+1}$ die Basisdarstellungen

$$(3)\qquad x = \sum_{\mu=1}^{m} \bar{x}_\mu b_\mu \ , \quad y = \sum_{\mu=1}^{m} \bar{y}_\mu b_\mu \ ,$$

wobei

$$(4)\qquad \bar{x}_\mu = x_\mu + \sum_{\kappa=1}^{k} u_\kappa c_{\kappa\mu} \ , \quad \bar{y}_\mu = y_\mu + \sum_{\kappa=1}^{k} v_\kappa c_{\kappa\mu} \ .$$

Die lineare Abhängigkeit von x und y in $\mathfrak{V}_{\nu+1}$ bedeutet nun, daß die Koordinatenmatrix

$$(5)\qquad \begin{pmatrix} \bar{x}_1 \ldots \bar{x}_m \\ \bar{y}_1 \ldots \bar{y}_m \end{pmatrix}$$

nicht den Rang 2 hat, d.h., daß alle ihre zweireihigen Unterdeterminanten verschwinden. Diese Unterdeterminaten ergeben sich nach (4) zu

$$(6)\qquad \bar{D}_{\lambda\mu} = \begin{vmatrix} \bar{x}_\lambda & \bar{x}_\mu \\ \bar{y}_\lambda & \bar{y}_\mu \end{vmatrix} = \begin{vmatrix} x_\lambda & x_\mu \\ y_\lambda & y_\mu \end{vmatrix} + \sum_{\kappa=1}^{k} c_{\kappa\mu} \cdot \begin{vmatrix} x_\lambda & u_\kappa \\ y_\lambda & v_\kappa \end{vmatrix} + \sum_{\kappa=1}^{k} c_{\kappa\lambda} \cdot \begin{vmatrix} u_\kappa & x_\mu \\ v_\kappa & y_\mu \end{vmatrix}$$

$$+ \sum_{\substack{\iota,\kappa=1 \\ \iota\neq\kappa}}^{k} c_{\iota\lambda} \cdot c_{\kappa\mu} \cdot \begin{vmatrix} u_\iota & u_\kappa \\ v_\iota & v_\kappa \end{vmatrix} \qquad (1 \le \lambda < \mu \le m) .$$

Die lineare Abhängigkeit von x und y in $\mathfrak{V}_\nu$ ist gleichwertig mit der entsprechenden Bedingung für die zugehörige Koordinatenmatrix

$$(7)\qquad \begin{pmatrix} x_1 \ldots x_m & u_1 \ldots u_k \\ y_1 \ldots y_m & v_1 \ldots v_k \end{pmatrix} .$$

Wir wählen nun die $c_{\kappa\mu}$, falls $m<n$, folgendermaßen.

Sei $g=k^2+2k=(n-m+1)^2-1$. (Man erhält leicht, daß dann auch die frühere Voraussetzung $m\cdot g \ge n$ erfüllt ist.) Sei a ein Element von $\mathfrak{k}_{\nu+1}$, welches über $\mathfrak{k}_\nu$ (transzendent oder) mindestens vom Grad g ist (so daß die Potenzen $1, a, a^2, \ldots, a^{g-1}$ über $\mathfrak{k}_\nu$ linear unabhängig sind). Ein solches a existiert. Ist nämlich der Körper $\mathfrak{k}_{\nu+1}$ über $\mathfrak{k}_\nu$ algebraisch, so muß er eine endliche algebraische Erweiterung $\mathfrak{k}'_\nu$ von $\mathfrak{k}_\nu$ enthalten, die auch schon von einem Grad $\ge g$ über $\mathfrak{k}_\nu$ ist und außerdem separabel (da angeordnete Körper die Charakteristik Null haben). Damit erhalten wir ein solches a nach dem Satz vom primitiven Element (s. etwa van der Waerden 1971, §46). Wir setzen

$$c_{\kappa\mu} = \begin{cases} a^{\kappa(k+1)} & , \text{ falls } \mu=1, \\ a^{\kappa} & , \text{ falls } \mu=2, \\ 0 & , \text{ falls } 2<\mu\leq m. \end{cases}$$

Zum Beweis von (*) sei $\bar{D}_{\lambda\mu} = 0$ für alle betrachteten λ, μ.

Im Spezialfall $\lambda=1$, $\mu=2$ erhalten wir aus (6)

$$(8) \quad \begin{vmatrix} x_1 & x_2 \\ y_1 & y_2 \end{vmatrix} + \sum_{\kappa=1}^{k} a^{\kappa} \cdot \begin{vmatrix} x_1 & u_\kappa \\ y_1 & v_\kappa \end{vmatrix} + \sum_{\kappa=1}^{k} a^{\kappa(k+1)} \cdot \begin{vmatrix} u_\kappa & x_2 \\ v_\kappa & y_2 \end{vmatrix}$$

$$+ \sum_{\substack{\iota,\kappa=1 \\ \iota\neq\kappa}}^{k} a^{\iota(k+1)+\kappa} \cdot \begin{vmatrix} u_\iota & u_\kappa \\ v_\iota & v_\kappa \end{vmatrix} = 0.$$

Die hierbei als Faktoren vor den Determinanten stehenden k^2+k+1 verschiedenen Elemente 1, a^{κ}, $a^{\kappa(k+1)}$, $a^{\iota(k+1)+\kappa}$ $(\iota,\kappa=1,\ldots,k;\ \iota\neq\kappa)$ kommen aber unter den $(k+1)^2-1$ verschiedenen Elementen $a^{\iota(k+1)+\kappa}$ $(\iota,\kappa=0,\ldots,k;$ $\langle\iota,\kappa\rangle\neq\langle k,k\rangle)$ vor; sie sind also nach Wahl von a linear unabhängig über $\mathfrak{f}_\nu$. Da die in (8) auftretenden Determinanten in $\mathfrak{f}_\nu$ liegen, müssen sie also verschwinden (d.h. gleich Null sein).

Im Spezialfall $\lambda=1$, $\mu>2$ erhalten wir aus (6)

$$(9) \quad \begin{vmatrix} x_1 & x_\mu \\ y_1 & y_\mu \end{vmatrix} + \sum_{\kappa=1}^{k} a^{\kappa(k+1)} \cdot \begin{vmatrix} u_\kappa & x_\mu \\ v_\kappa & y_\mu \end{vmatrix} = 0 .$$

Wie vorher ergibt sich auch hier, daß die auftretenden Determinanten verschwinden.

Betrachten wir nun (6) für beliebige λ, μ. Darin verschwindet nun jede hinter einem Summationszeichen auftretende Determinante, da sie in (8) oder (9) (bis auf einen Vorzeichenfaktor) vorkommt. Damit muß auch die erste Determinante auf der rechten Seite von (6) verschwinden. Somit verschwinden alle zweireihigen Unterdeterminanten von (7), und (*) ist bewiesen.

Es bleibt noch zu zeigen, daß die Vektoren $b_1,\ldots, b_m, c_1,\ldots, c_k$ - wobei die c_κ durch (1) gegeben werden - für unsere Wahl der $c_{\kappa\mu}$ tatsächlich linear unabhängig in $\mathfrak{V}''_\nu$ sind, wie es früher vorausgesetzt wurde. Sei also x in (2) der Nullvektor, d.h. $\bar{x}_\mu=0$ für $1\le\mu\le m$. Im Spezialfall $\mu=1$ erhalten wir aus (4) (ähnlich wie vorher aus (6)), daß $u_\kappa=0$ ist für $1\le\kappa\le k$. Betrachten wir (4) sodann für beliebiges μ, so erhalten wir, daß auch $x_\mu=0$ ist, wie zu zeigen. □

6.55 Zusatz. *In 6.54 läßt sich die Behauptung durch folgende Feststellung ergänzen:*
(d) *Der n-dimensionale Raum* $\mathfrak{A}_\nu$ *ist stets als Unterstruktur (natürlich nicht als Unterraum) enthalten in einem m-dimensionalen Unterraum* $\mathfrak{A}_{\nu+1}$.

Beweis: Der durch $\mathfrak{W}'_\nu$ im vorigen Beweis bestimmte affine Raum leistet offenbar das Verlangte. □

Verwenden wir zwei angeordnete Körper an Stelle der ganzen Kette, so erhalten wir analog

6.56 Korollar. *Sei* $2\le m\le n$ *und* $(\mathfrak{k}':\mathfrak{k})\ge(n-m+1)^2-1$. *Dann ist der n-dimensionale affine Raum* $\mathfrak{A}_n^B(\mathfrak{k})$ *(mit Anordnung) über* $\mathfrak{k}$ *isomorph einbettbar in* $\mathfrak{A}_m^B(\mathfrak{k}')$ *(d.h. isomorph zu einer Unterstruktur dieses m-dimensionalen Raumes).*

6.57 Anmerkungen. (i) Vermutlich läßt sich die untere Schranke $g=(n-m+1)^2-1$ in den letzten beiden Sätzen noch verkleinern (vielleicht auf k^2+k+1, vgl. den Beweis). Sie läßt sich aber jedenfalls nicht verkleinern im Spezialfall $m=2$, $n=3$ (in diesem Falle ist $k=1$, $g=3$). Betrachten wir nämlich die Faktoren 1, c_{12}, c_{11}, die dann in (6) auftreten (und die stets eingeführt werden können, wenn $\mathfrak{A}_\nu$, $\mathfrak{A}_{\nu+1}$ die Eigenschaften (a), (b), (d) haben). Wäre nun $(\mathfrak{k}_{\nu+1}:\mathfrak{k}_\nu)\le 2$, so müßten diese Faktoren über $\mathfrak{k}_\nu$ linear abhängig sein; dann könnte man aber eine Matrix der Form (7) konstruieren, deren zweireihige Unterdeterminaten gerade solche Werte (nicht alle Null) haben, die $\bar{D}_{12}$ in (6) zu Null machen, und hätte damit ein Gegenbeispiel zu (*).

(ii) Aus dem Beweis von 6.54 sieht man, daß dieser Satz sich auf affine Räume (ohne Anordnung) über Körpern mit der Kollinearität als einzigem Grundbegriff überträgt; statt der Gradbedingung kann man dabei voraussetzen, daß in $\mathfrak{k}_{\nu+1}$ stets ein Element a existiert, das über $\mathfrak{k}_\nu$ mindestens vom Grade $(n-m+1)^2-1$ ist. Diese Voraussetzung ist sicher erfüllt, wenn $\mathfrak{k}_{\nu+1}$ transzendent über $\mathfrak{k}_\nu$ ist.

(iii) Ketten der betrachteten Art existieren z.B. in axiomatischen Klassen von Körpern (s. 6.52) mit oder ohne Anordnung, die wenigstens einen unendlichen Körper als Element enthalten. Zu jedem unendlichen solchen $\mathfrak{K}$ in einer solchen Klasse K gibt es nämlich nach dem Satz von Löwenheim-Skolem-Tarski (1.54) einen Körper $\mathfrak{K}'$ in K mit größerer Mächtigkeit; dieser muß dann transzendent über $\mathfrak{K}$ sein. Damit kann man 6.52 auch als Folgerung aus 6.54 erhalten.

<u>6.58 Das Parallelenaxiom für affine Geometrien</u>. Auch in e b e n e n affinen Geometrien (in denen das untere Dimensionsaxiom vom Typ ∃ ist, s. 6.21(iii)) kommt man nicht mit ∀∃-Axiomensystemen in der Sprache L(B) aus. Das wurde von Szczerba gefunden (6.62) mit Hilfe der folgenden Konstruktion, die sich auf die Existenzforderung im euklidischen Parallelenaxiom bezieht.

In affinen Geometrien wird als *Parallelenaxiom* üblicherweise gefordert, daß es zu jeder Geraden A durch jeden Punkt außerhalb A g e n a u eine Parallele (im euklidischen Sinne, 12.2, 3, d.h. Nicht-Schneidende in derselben Ebene) gibt. Hiervon läßt sich die *Existenzforderung* in der Sprache L(B) (oder L(Col)) ausdrücken als die Aussage

(1) $E^{\geq} := \forall a \forall c \forall d[\neg \mathrm{Col}\, acd \to \exists b\; ab \overset{\vee}{\|} cd]$,

wobei

(2) $ab \overset{\vee}{\|} cd := \neg \exists x[\mathrm{Col}\, abx \wedge \mathrm{Col}\, cdx]$
$\wedge \exists y[(\mathrm{Col}\, ady \wedge \mathrm{Col}\, bcy) \vee (\mathrm{Col}\, acy \wedge \mathrm{Col}\, bdy)]$

(hierbei drückt das zweite Konjunktionsglied die Komplanarität der betrachteten Punkte aus und kann für ebene Geometrien weggelassen werden). Auf Grund der Quantorenverschiebung nach 1.60 erhält man, daß $E^{\geq}$ eine Aussage vom Typ ∀∃∀ ist (vgl. 6.64(iii)). Dagegen ist die *Eindeutigkeitsforderung*

(3) $E^{\leq} := \forall a \forall c \forall d[\neg \mathrm{Col}\, acd \to \forall b_1 \forall b_2 (ab_1 \overset{\vee}{\|} cd \wedge ab_2 \overset{\vee}{\|} cd \to \mathrm{Col}\, a_1 b_2)]$

noch vom Typ ∀∃.

In der absoluten Geometrie A läßt sich die Existenzforderung $E^{\geq}$ bekanntlich beweisen (s. I.12.10).

<u>6.59 Satz</u> (Szczerba, mündlich mitgeteilt, vgl. auch 1972a).

Vor.: (i) $\mathfrak{K}$, $\mathfrak{K}'$ *sind angeordnete Körper mit* $\mathfrak{K} \subseteq \mathfrak{K}'$, *und* u *ist ein Element von* $\mathfrak{K}'$, *das (in* $\mathfrak{K}'$*) größer ist als jedes Element von* $\mathfrak{K}$.

(ii) $\mathfrak{A}$ *ist isomorph zur affinen Ebene* $\mathfrak{A}_2^B(\mathfrak{K})$.

Beh.: Es gibt eine Struktur $\mathfrak{B}$*, so daß gilt:*

(a) $\mathfrak{A} \subseteq \mathfrak{B}$,

(b) $\mathfrak{B}$ *ist isomorph zur affinen Ebene* $\mathfrak{A}_2^B(\mathfrak{K}')$,

(c) *je zwei Geraden von* $\mathfrak{A}$ *haben einen gemeinsamen Punkt in* $\mathfrak{B}$.

<u>6.60 Vorbemerkung</u>. Zum Beweis verwenden wir, daß man - analog zu 2.9(iii) - auch bei Räumen mit Anordnung durch Hinzufügen bzw. Herausnehmen einer "uneigentlichen" Hyperebene von einem affinen Raum zu einem projektiven Raum und umgekehrt unter Erhaltung der Dimension und des Grundkörpers folgendermaßen übergehen kann.

Sei $2 \leq n \in \mathbb{N}$, $\mathfrak{K}$ ein angeordneter Körper und H eine beliebige Hyperebene im projektiven Raum $\mathfrak{P} = \mathfrak{P}_n^{|}(\mathfrak{K})$ (mit Anordnung, s. 2.9(vi)). Sei $\mathfrak{A}$ die folgendermaßen "durch Herausnehmen von H aus $\mathfrak{P}$ " entstehende Struktur für die Sprache L(B):

(1) $U_{\mathfrak{A}} = U_{\mathfrak{P}} - H$ (die Menge der nicht auf H liegenden Punkte von $\mathfrak{P}$),

(2) $B_{\mathfrak{A}}abc$ gdw $b=a$ oder $b=c$ oder es gibt ein $d \in H$ mit $ac \,|_{\mathfrak{P}}\, bd$.

Dann ist $\mathfrak{A}$ isomorph zum n-dimensionalen affinen Raum $\mathfrak{A}_n^B(\mathfrak{K})$ (mit Anordnung, s. 2.8(ii)) über $\mathfrak{K}$. Für die Kollinearitäten dieser beiden Räume (die mittels $|_{\mathfrak{P}}$ gemäß 2.9(3) bzw. mittels $B_{\mathfrak{A}}$ wie üblich gemäß I.4.10 definiert sind) gilt

(3) $Col_{\mathfrak{A}}$ ist die Einschränkung von $Col_{\mathfrak{P}}$ auf $U_{\mathfrak{A}}$.

Für die zugehörigen Räume ohne Anordnung (mit dem Grundbegriff Col) gilt somit das in 2.9(iii) Gesagte; insbesondere sind Geraden in $\mathfrak{A}$ zueinander echt parallel gdw sie sich in einem Punkt von H schneiden.

<u>Beweis von 6.59</u>: In der projektiven Ebene $\bar{\mathfrak{L}} := \mathfrak{P}_2^{|}(\mathfrak{K}')$ (mit Anordnung) sei ein homogenes Koordinatensystem eingeführt (2.9(vi), (iv)). Die projektive Ebene $\bar{\mathfrak{A}} = \mathfrak{P}_2^{|}(\mathfrak{K})$ über $\mathfrak{K}$ fassen wir auf als die Unterstruktur von $\bar{\mathfrak{L}}$, die aus den Punkten mit Koordinaten in $\mathfrak{K}$ besteht. Als Geraden von $\bar{\mathfrak{A}}$ (die nur bis auf Isomorphie festgelegt werden müssen) verwenden wir diejenigen Geraden von $\bar{\mathfrak{L}}$, die wenigstens zwei (und damit unendlich viele) Punkte von $\bar{\mathfrak{A}}$ enthalten (damit ist die Menge der Geraden von $\bar{\mathfrak{A}}$ eine Teilmenge der Geraden von $\bar{\mathfrak{L}}$ in Analogie zu den entsprechenden Punktmengen). Wir nehmen o.B.d.A. an (s. 6.61), daß die affine Ebene $\mathfrak{A}$ aus $\bar{\mathfrak{A}}$ durch Herausnehmen der "uneigentlichen Geraden" U mit der Gleichung $x_0=0$ entsteht (6.60); ebenso entstehe die affine Ebene $\mathfrak{L}$ über $\mathfrak{K}'$ aus $\bar{\mathfrak{L}}$. Die affinen Koordinaten x_1, x_2 eines Punktes von $\mathfrak{A}$ bzw. $\mathfrak{L}$ ergeben sich dann aus dem zugehörigen Tripel $\langle x_0, x_1, x_2 \rangle$ homogener Koordinaten mit $x_0=1$.

Sei nun K die Gerade von $\bar{\mathfrak{L}}$ mit der Gleichung

(1) $u \cdot x_0 - x_1 + u^{-1} \cdot x_2 = 0.$

Als Gerade von $\mathfrak{L}$ hat sie dann in affinen Koordinaten die Gleichung

(2) $u - x_1 + u^{-1} \cdot x_2 = 0$

(Man kann sie sich anschaulich vorstellen als eine Gerade, die die erste Achse in einem Punkt "rechts" von allen Punkten von $\mathfrak{A}$ [nämlich im Punkt mit dem Koordinatenpaar $\langle u,0 \rangle$] schneidet und die zur zweiten Achse "weniger geneigt" ist als alle nicht dazu parallelen Geraden von $\mathfrak{A}$.) Sei schließlich $\mathfrak{B}$ die affine Ebene, die aus $\bar{\mathfrak{L}}$ durch Herausnehmen der Geraden K entsteht; nach 6.60 gilt dann schon die Bedingung (b).

Weiter gilt:

(3) Für Punkte von $\mathfrak{A}$, d.h. für $x_1, x_2 \in \mathfrak{f}$, wird die linke Seite der Gleichung (2) stets positiv.

Für solche Punkte ist nämlich $|x_1 - u^{-1} \cdot x_2| \leq |x_1| + |u^{-1} \cdot x_2| \leq |x_1| + |x_2| \in \mathfrak{f}$, also $u > x_1 - u^{-1} \cdot x_2$. Wie aus der analytischen Geometrie bekannt, wird durch das Vorzeichen der linken Seite einer Geradengleichung die Seite (Halbebene) der zugehörigen Geraden festgelegt. Aus (3) ergibt sich also

(4) In der affinen Ebene $\mathfrak{L}$ liegen alle Punkte von $\mathfrak{A}$ auf derselben Seite der Geraden K.

Insbesondere gilt

(5) K enthält keinen Punkt von $\mathfrak{A}$.

Für den Schnittpunkt p von U und K in $\bar{\mathfrak{L}}$ erhält man aus den Geradengleichungen das Koordinatentripel $\langle 0,1,u \rangle$ (oder irgendein Vielfaches davon), somit liegt p nicht in der projektiven Ebene $\bar{\mathfrak{A}}$ über $\mathfrak{f}$. Da jeder Punkt von $\bar{\mathfrak{A}}$ in $\mathfrak{A}$ oder auf U liegt, gilt sogar

(6) K enthält keinen Punkt von $\bar{\mathfrak{A}}$.

Sind nun L_1, L_2 irgendwelche verschiedenen Geraden von $\mathfrak{A}$, so schneiden sie sich in $\bar{\mathfrak{A}}$; damit liegt der Schnittpunkt nicht auf K und somit in $\mathfrak{B}$, womit die Bedingung (c) gezeigt ist.

Es bleibt (a) zu zeigen. Die gewünschte Inklusion der Punktmengen ergibt sich sofort aus (5). Aus 6.60 ergibt sich:

(7) $B_{\mathfrak{A}}abc$ gdw $b=a$ oder $b=c$ oder es gibt ein $d \in U$ mit $ac \,|_{\bar{\mathfrak{A}}}\, bd$; entsprechendes gilt für $B_{\mathfrak{L}}$, U, $|_{\bar{\mathfrak{L}}}$ und für $B_{\mathfrak{B}}$, K, $|_{\bar{\mathfrak{L}}}$.

Auch im ersten Fall kann man $|_{\overline{\mathfrak{L}}}$ statt $|_{\overline{\mathfrak{A}}}$ verwenden (da $\overline{\mathfrak{A}}$ Unterstruktur von $\overline{\mathfrak{L}}$ ist).

Seien nun a, c Punkte von $\mathfrak{A}$ und o.B.d.A. $a \neq c$, und seien d bzw. e die Schnittpunkte der Geraden $L(ac)$ mit K bzw. mit U (in $\overline{\mathfrak{L}}$). Wegen (4) ist nicht $B_{\mathfrak{A}}adc$ und somit

(8) nicht $ac|_{\overline{\mathfrak{L}}}de$.

Nun gilt für fünf paarweise verschiedene Punkte a, b, c, d, e auf einer Geraden die Gleichung

(9) $(ac,bd)\cdot(ac,de) = (ac,be)$

(vgl. 2.9(v)). (8) bedeutet nach 2.9(vi), daß das Doppelverhältnis (ac,de) positiv ist. Damit haben die anderen beiden Doppelverhältnisse in (9) dasselbe Vorzeichen, d.h.

(10) $ac|_{\overline{\mathfrak{L}}}bd$ gdw $ac|_{\overline{\mathfrak{L}}}be$.

Für Punkte b von $\mathfrak{A}$ ergibt sich daraus mittels (7) die gewünschte Beziehung

(11) $B_{\mathfrak{B}}abc$ gdw $B_{\mathfrak{A}}abc$. □

<u>6.61 Anmerkung</u>. Ist die zu Anfang "ohne Beschränkung der Allgemeinheit" gemachte Annahme über $\mathfrak{A}$ noch nicht erfüllt, so liefert der Beweis jedenfalls eine isomorphe Einbettung von $\mathfrak{A}$ in die konstruierte Struktur $\mathfrak{B}$. Man kann dann den Isomorphismus auf eine geeignete Obermenge der Punktmenge von $\mathfrak{A}$ fortsetzen und erhält, daß eine zu dem konstruierten $\mathfrak{B}$ isomorphe Struktur das in der Behauptung Verlangte leistet.

<u>6.62 Satz</u> (Szczerba, s. 6.59). *Vor.: C ist eine nicht-leere Klasse von angeordneten Körpern mit:*

(*) *Zu jedem* $\mathfrak{f}$ *aus C gibt es ein* $\mathfrak{f}'$ *aus C und ein u, so daß* 6.59(i) *gilt.*

Beh.: Die ebene affine Geometrie $\Gamma_2'(C)$ *besitzt kein* $\forall\exists$*-Axiomensystem.*

<u>6.63 Zusatz</u>. (i) *Die Bedingung* (*) *aus der Voraussetzung von* 6.62 *ist insbesondere erfüllt, wenn C eine nicht-leere axiomatische Klasse von angeordneten Körpern (s.* 6.52*) ist.*

(ii) *Beispiele für die in der Behauptung genannten Geometrien sind insbesondere* $Afo_2 = \Gamma_2'(OF)$ *und* $Afo_2^* = \Gamma_2'(RF)$ *(vgl.* 6.47*).*

<u>Beweis von 6.62</u>: Sei $< \mathfrak{f}_i | i \in \mathbb{N} >$ eine ω-Kette von angeordneten Körpern

aus C, so daß in $\mathfrak{F}_{i+1}$ stets ein Element vorhanden ist, das größer ist als alle Elemente von $\mathfrak{F}_i$ (eine solche Kette existiert wegen (*)). Auf Grund von 6.59 erhalten wir daraus eine Folge $K = \langle \mathfrak{A}_i \mid i \in \mathbb{N} \rangle$, so daß stets gilt:

(a) $\mathfrak{A}_i \subseteq \mathfrak{A}_{i+1}$,

(b) $\mathfrak{A}_i$ ist isomorph zur affinen Ebene $\mathfrak{A}_2^B(\mathfrak{F}_i)$ über $\mathfrak{F}_i$,

(c) je zwei Geraden von $\mathfrak{A}_i$ haben einen gemeinsamen Punkt in $\mathfrak{A}_{i+1}$.

Damit ist K eine ω-Kette von Modellen von $\Gamma_2'(C)$. Sei $\mathfrak{A}$ die Vereinigung von K. Je zwei Geraden von $\mathfrak{A}$ gehören dann schon zu einem gewissen $\mathfrak{A}_i$, haben also einen gemeinsamen Punkt im zugehörigen $\mathfrak{A}_{i+1}$ und damit erst recht in $\mathfrak{A}$. Somit gilt in $\mathfrak{A}$ nicht die Existenzforderung des euklidischen Parallelenaxioms, also ist $\mathfrak{A}$ kein affiner Raum und kein Modell von $\Gamma_2'(C)$. Aus 6.10 ergibt sich dann die Behauptung. □

Beweis von 6.63(i): Sei $\mathfrak{F}$ aus C beliebig. Wir bilden das Diagramm $D(\mathfrak{F})$ (5.4) und nehmen zur Diagrammsprache $L_{\mathfrak{F}}$ (mit $L = L_{OF}$ eine zusätzliche Individuenkonstante u hinzu. In der so entstehenden Sprache bilden wir die Formelmenge

(1) $\Theta = \Sigma_C \cup D(\mathfrak{F}) \cup \{\bar{c} < u \mid c \in \mathfrak{F}\}$,

wobei Σ_C ein Axiomensystem für C ist. Dann besitzt jede endliche Teilmenge Θ' ein Modell; ein solches erhält man nämlich aus $\mathfrak{F}$ (genauer $\mathfrak{F}_{\mathfrak{F}}$), indem man die Konstante u interpretiert als ein Element von $\mathfrak{F}$, das größer ist als diejenigen (endlich vielen) Elemente c von $\mathfrak{F}$, für die die zugehörigen Konstanten $\bar{c}$ in Θ' vorkommen. Nach dem Endlichkeitssatz (1.52) besitzt auch Θ ein Modell $\bar{\mathfrak{F}}'$. Dieses ist ein angeordneter Körper aus C mit zusätzlichen festen Elementen, der (nach 5.6) einen zu $\mathfrak{F}$ isomorphen Unterkörper $\mathfrak{F}_0$ besitzt sowie ein Element (mit dem u interpretiert wird), das größer ist als alle Elemente von $\mathfrak{F}_0$. Daraus erhält man (wie in 6.61) einen isomorphen angeordneten Körper $\mathfrak{F}'$, der eine Erweiterung von $\mathfrak{F}$ ist und damit die gewünschten Eigenschaften hat. □

6.64 Anmerkungen. (i) In den Spezialfällen C=RF und (damit auch) C=OF von 6.63(ii) kann man die Bedingung 6.62(*) auch aus dem ersten Beispiel in 3.19 erhalten.

(ii) Satz 6.59 läßt sich durch eine entsprechend abgeänderte Konstruktion auch für Räume mit beliebigen endlichen Dimensionen $n \geq 2$ erhalten. Auf diesem Wege kann man auch eine entsprechende Übertragung von 6.62 beweisen, die sich aber für $n > 2$ schon aus 6.48 ergibt.

(iii) Aus Szczerba 1972 (wo allgemeinere Geometrien behandelt werden), kann man ein Axiomensystem für $\mathcal{A}fo_2$ der Form $\Gamma \cup \{E^{\geq}\}$ erhalten, wobei das System Γ der "übrigen Axiome" aus $\forall\exists$-Aussagen besteht. Damit gelten diese Axiome natürlich auch in der Vereinigung der im Beweis von 6.62 konstruierten Kette. Auf Grund der folgenden Betrachtung über die Stetigkeitsaxiome erhält man ein Axiomensystem derselben Art auch für $\mathcal{A}fo_2^*$ (s. 6.86(iv)). Mittels 6.11 ergibt sich daraus für beide Geometrien, daß das Axiom $E^{\geq}$ auf Grund der übrigen Axiome nicht äquivalent zu einer $\forall\exists$-Aussage ist. Es ist jedoch äquivalent zu einer $\exists\forall$-Aussage (s. Szczerba 1972, Theorem 6).

6.65 Abbau von Termen. Als Vorbereitung für die nächste Konstruktion betrachten wir ein Verfahren für den Abbau von Termen in beliebigen Sprachen mit Standardformalisierung.

6.66 Satz. *Vor.:* (i) $L=\langle F,R,r\rangle$ *ist eine Sprache in Standardformalisierung mit der Variablenmenge* V *(1.1). Sei* F_o *die Menge der Individuenkonstanten (nullstelligen Operationszeichen), und sei* $F'=F$ *und* $V'=V$ *oder* $F'=F-F_o$ *und* $V'=V\cup F_o$. *(Damit wird zugelassen, daß die Individuenkonstanten im folgenden wahlweise wie die übrigen Operationszeichen oder wie Variablen behandelt werden.)*

(ii) Ξ *ist ein Präfixtyp, dessen letztes Grundzeichen* $\forall$ *oder* $\exists$ *ist (das bedeutet, daß zumindest für den letzten Quantorenblock von pränexen Formeln dieses Typs eine beliebige Länge zugelassen ist, s.* 6.2(ii)*).*

Beh.: Zu jeder pränexen Formel α *von* L *vom Typ* Ξ *läßt sich eine dazu logisch äquivalente pränexe Formel* β *vom selben Typ mit denselben freien Variablen angeben, in der die Operationszeichen* f *aus* F' *nur in Primformeln der Form* $fx_1\ldots x_{r(f)} \doteq x$ *vorkommen, wobei* $x_1,\ldots,x_{r(f)},x \in V'$.

Zum Beweis verwenden wir

6.67 Definition. Gelte 6.66(i). Der *Grad* $gr(\tau)$ eines Terms τ von L (genauer, der F'-*Grad*) wird induktiv eingeführt durch

(1) $gr(x)=0$ für $x\in V'$,

(2) $gr(f\tau_1\ldots\tau_{r(f)}) = 1+\max(gr(\tau_1),\ldots,gr(\tau_{r(f)}))$ für $f\in F'$.

Beweis von 6.66: Sei α wie angegeben. Die Konstruktion von β geschieht (wie in 1.60) durch logisch äquivalente Umformungen in endlich vielen Schritten. Wir verwenden dazu (außer denen von 1.60) die folgenden Äquivalenzen, wobei $r=r(R)$ bzw. $r=r(f)$ gesetzt ist und als quantifizierte

Variablen (x_ρ bzw. x) jeweils "neue Variablen" benutzt werden (d.h. solche, die vor Anwendung der betreffenden Äquivalenz in der bis dahin konstruierten Formel noch nicht vorkommen).

(1) (a) $R\tau_1 \ldots \tau_r$ Äq $\exists x_1 \ldots \exists x_r [\bigwedge_{\rho=1}^{r} \tau_\rho \doteq x_\rho \wedge Rx_1 \ldots x_r]$,

(b) $R\tau_1 \ldots \tau_r$ Äq $\forall x_1 \ldots \forall x_r [\bigwedge_{\rho=1}^{r} \tau_\rho \doteq x_\rho \rightarrow Rx_1 \ldots x_r]$,

(c) $R\tau_1 \ldots \tau_r$ Äq $\forall x_1 \ldots \forall x_r [\bigvee_{\rho=1}^{r} \neg\tau_\rho \doteq x_\rho \vee Rx_1 \ldots x_r]$;

(2) (a) $\sigma \doteq \tau$ Äq $\exists x[\sigma \doteq x \wedge \tau \doteq x]$,

(b) $\sigma \doteq \tau$ Äq $\forall x[\sigma \doteq x \rightarrow \tau \doteq x]$,

(c) $\sigma \doteq \tau$ Äq $\forall x[\neg\sigma \doteq x \vee \tau \doteq x]$;

(3) (a) $f\tau_1 \ldots \tau_r \doteq x$ Äq $\exists x_1 \ldots \exists x_r [\bigwedge_{\rho=1}^{r} \tau_\rho \doteq x_\rho \wedge fx_1 \ldots x_r \doteq x]$,

(b) $f\tau_1 \ldots \tau_r \doteq x$ Äq $\forall x_1 \ldots \forall x_r [\bigwedge_{\rho=1}^{r} \tau_\rho \doteq x_\rho \rightarrow fx_1 \ldots x_r \doteq x]$,

(c) $f\tau_1 \ldots \tau_r \doteq x$ Äq $\forall x_1 \ldots \forall x_r [\bigvee_{\rho=1}^{r} \neg\tau_\rho \doteq x_\rho \vee fx_1 \ldots x_r \doteq x]$.

(Zu den jeweiligen Teilen (a) und (b) vgl. den Satz 3.38 über die Elimination des bestimmten Artikels, (c) entsteht aus (b) mittels 1.60(2) und (4).) Wenn in (1) oder (3) ein τ_ρ schon in V' liegt, so kann man natürlich auf der rechten Seite die Variable x_ρ mit dem zugehörigen Quantor und den Bestandteil $\tau_\rho \doteq x$ bzw. seine Negation weglassen.

Die Konstruktion besteht aus folgenden Teilen.

1. Herstellung eines Kerns, der nur mittels der Junktoren $\wedge$ und $\vee$ aufgebaut ist aus Primformeln und Negationen von Primformeln.

Das geschieht wie in 1.60 (mittels der Äquivalenzen 1.60(1) bis (5)).

2. Herstellung eines Kerns derselben Art, in dem Terme positiven Grades nur auf der linken Seite von Gleichungen vorkommen.

Sei ε eine im Kern auftretende Primformel, die noch nicht die gewünschte Bauart hat. Dann läßt sich darauf (1) oder (2) (oder eine Vereinfachung mit weniger Variablen) anwenden. Wenn ε nicht negiert auftritt, so wird Teil (a) bzw. (c) angewendet, falls das letzte Grundzeichen von Ξ das Zeichen $\exists$ bzw. $\forall$ ist; wenn ε negiert auftritt, ist es gerade umgekehrt. Die neuen quantifizierten Variablen werden anschließend wie in 1.60

(mittels 1.60(6) bis (15)) nach vorn hinter die Variablen des schon vorhandenen Präfixes gebracht. Dadurch wird der letzte Quantorenblock verlängert, aber an seiner Art ($\forall$-Block oder $\exists$-Block) wird nichts geändert, und es entsteht wieder eine pränexe Formel vom Typ Ξ. Falls ε negiert auftrat, wird außerdem das vor der eckigen Klammer von (a) bzw. (c) entstandene Negationszeichen vor die neuen Primformeln gebracht (mittels 1.60(3) bis (5)), damit der Kern wieder die in 1. angegebene Bauart erhält.

Dieser Schritt wird so oft ausgeführt (für jede Primformel höchstens einmal), bis die in 2. gewünschte Bauart erreicht ist.

3. Herstellung eines Kerns derselben Art, in dem nur noch Terme vom Grad ≤ 1 vorkommen.

Sei $\tau \doteq x$ eine im Kern auftretende Gleichung, in der τ den größten vorkommenden Grad t mit $t>1$ hat. Darauf läßt sich (3) (oder eine Vereinfachung) anwenden. Wie vorher wird Teil (a) oder (c) angewendet je nachdem, welche Art von Quantoren benötigt wird, und die neuen quantifizierten Variablen werden anschließend nach vorn und ggf. das Negationszeichen "nach innen" gebracht. Dadurch wird die Anzahl der Terme vom Grad t um Eins erniedrigt.

Nach endlich vielen Schritten fallen also zunächst alle Terme vom Grad t weg und schließlich alle Terme vom Grad >1. Damit ist die in 3. gewünschte Bauart erreicht, die entstandene Formel β leistet das Verlangte. □

6.68 Vorbemerkung. Im folgenden sollen noch die elementaren Stetigkeitsaxiome (A11' in I.1.2) betrachtet werden. In Tarski 1959 (S. 24) wurde schon festgestellt, daß sich diese in der euklidischen Geometrie P äquivalent durch $\forall\exists$-Aussagen ersetzen lassen. Von Szczerba wurde dasselbe gezeigt für eine große Klasse von Geometrien, und zwar mit Hilfe eines wesentlich allgemeineren Satzes (6.71, s.u.), der besagt, daß man eine Art von "geometrischen Reduzierten" (vgl. 3.36, 48ff.) unter Erhaltung des Präfixtyps konstruieren kann. Dieses Ergebnis wurde angekündigt in Kordos u. Szczerba 1969. Die Ausführung eines Beweises erfolgt hier nach einer brieflichen Mitteilung von Szczerba von 1969 mit seiner freundlichen Genehmigung.

6.69 Definition. (i) Eine Menge S von Punkten eines affinen Raumes mit Anordnung oder eines kartesischen Raumes $\mathfrak{A}$ heißt bekanntlich *konvex*, falls sie mit beliebigen Punkten a, c auch stets jeden Punkt ihrer Verbindungsstrecke (d.h. jeden Punkt zwischen a und c) als Element enthält; sie heiße *planar konvex*, falls sie außerdem drei nicht-kollineare Punkte enthält. Mit $\mathfrak{A}|S$ werde dann die *Einschränkung* von $\mathfrak{A}$ auf S bezeichnet, d.h. die Unterstruktur von $\mathfrak{A}$ mit der Punktmenge (Trägermenge) S.

(ii) Unter der *konvexen Geometrie* Cv verstehen wir die Theorie der Klasse aller Strukturen der Form $\mathfrak{A}_n^B(\mathfrak{f})|S$, wobei $2 \leq n \in \mathbb{N}$, $\mathfrak{f}$ ein angeordneter Körper und S eine planar konvexe Teilmenge des affinen Raumes $\mathfrak{A}_n^B(\mathfrak{f})$ ist. Eine Struktur dieser Form nennen wir dann einen *konvexen Raum* über $\mathfrak{f}$ (wir nennen diesen n-dimensional, falls S nicht in einem $(n-1)$-dimensionalen affinen Unterraum von $\mathfrak{A}_n^B(\mathfrak{f})$ enthalten ist). Jeder solche Raum ist damit eine Struktur für die Sprache L(B).

6.70 Anmerkung. Beispiele für konvexe Räume sind die affinen Räume $\mathfrak{A}$ von Dimension ≥ 2 selbst ($S = U_{\mathfrak{A}}$) sowie die hyperbolischen Kleinschen Räume als Strukturen für L(B) (2.4, 4.58). Aus dem Darstellungssatz von Pejas für die absoluten Geometrien A_n (s. 2.14(iv)) ergibt sich insbesondere, daß jedes Modell bis auf Isomorphie die Einschränkung eines kartesischen Raumes $\mathfrak{C}_n(\mathfrak{f})$ auf eine geeignete planar konvexe Menge ist. Aus $\mathfrak{C}_n(\mathfrak{f})$ entsteht aber durch Weglassen der Streckenkonkongruenz der affine Raum $\mathfrak{A}_n^B(\mathfrak{f})$. Somit ergibt sich (mittels 3.77), daß die konvexe Geometrie Cv eine Untertheorie von A ist. Außerdem ist sie Untertheorie von $A\mathfrak{f}o = \Gamma'(OF)$ (6.46, 47) und von $\Gamma(OF)$ (3.52) und damit von allen hier betrachteten Geometrien, in denen die Zwischenbeziehung als Grundbegriff vorkommt und die Existenz nicht-kollinearer Punkte (Dimension ≥ 2) vorausgesetzt wird.

6.71 Satz (Szczerba, s. 6.68). *Vor.: Ξ ist ein Präfixtyp, der nur aus den Grundzeichen $\forall$ und $\exists$ zusammengesetzt ist (das bedeutet, daß für alle Quantorenblöcke von pränexen Formeln dieses Typs eine beliebige Länge zugelassen ist, s. 6.2(ii)) und der mindestens ein solches Zeichen enthält.*

Beh.: Zu jeder pränexen Aussage α vom Typ Ξ der Sprache L_{OF} für angeordnete Körper (3.20) läßt sich eine pränexe Aussage γ von L(B) vom selben Typ angeben, so daß für jeden angeordneten Körper $\mathfrak{f}$ und jeden konvexen Raum $\mathfrak{A}$ über $\mathfrak{f}$ gilt:

(1) $\models_{\mathfrak{f}} \alpha$ *gdw* $\models_{\mathfrak{A}} \gamma$.

6.72 Vorbemerkung. Der Beweis wird in mehreren Schritten geführt, die zunächst in den folgenden Sätzen behandelt werden. Der wichtigste Schritt ist natürlich wieder die Einführung einer Punktrechnung (wie in I.14). In einem beliebigen konvexen Raum $\mathfrak{A}_n^B(\mathfrak{f})|S$ ist allerdings im allgemeinen nicht mehr eine ganze Zahlengerade des affinen Raumes $\mathfrak{A}_n^B(\mathfrak{f})$ enthalten, aber immer noch eine Teilstrecke einer Zahlengeraden G, und durch eine projektive Transformation läßt sich der "Halbkörper" der nicht-negativen Elemente von G in eine solche Teilstrecke einbetten. Daher wird 6.71 zurückgeführt auf die Betrachtung von (angeordneten) Halbkörpern, die anschließend durch eine geeignete Punktrechnung geometrisch beschrieben werden. Das gewünschte Resultat ergibt sich dann unmittelbar aus den Sätzen 6.76 und 6.81.

6.73 Definition. Sei $\mathfrak{f}$ ein angeordneter Körper. Unter dem durch $\mathfrak{f}$ bestimmten *Halbkörper* $\mathfrak{f}^+$ verstehen wir dann die folgendermaßen festgelegte Struktur $\mathfrak{A}$ für die Sprache $L_{OF}^+ := L(+,\cdot,0,1;<)$:

(1) $U_{\mathfrak{A}}$ ist die Menge der nicht-negativen Elemente von $\mathfrak{f}$,

(2) die Operationen und Relationen von $\mathfrak{A}$ entstehen aus den entsprechenden Operationen und Relationen von $\mathfrak{f}$ durch Einschränkung auf die Menge $U_{\mathfrak{A}}$.

($U_{\mathfrak{A}}$ ist in bezug auf die Operationen von $\mathfrak{f}$, außer der hier weggelassenen Inversenbildung $-_{\mathfrak{f}}$, abgeschlossen, also entstehen durch die Einschränkung tatsächlich entsprechende Operationen auf $U_{\mathfrak{A}}$, wie es für die Definition einer Struktur [1.13] gebraucht wird).

6.74 Anmerkung. Bekanntlich ist ein angeordneter Körper $\mathfrak{f}$ durch $\mathfrak{f}^+$ schon eindeutig bis auf Äquivalenz (vgl. 3.17(v)) festgelegt. (Das kann man auch aus den unten durch 6.75(1) bis (6) ausgedrückten Rechengesetzen erhalten.)

6.75 Lemma. (i) *Zu jedem Term* τ *von* L_{OF} *lassen sich Terme* τ', τ'' *von* L_{OF}^+ *(d.h. ohne Minuszeichen) mit höchstens denselben Variablen wie in* τ *angeben, so daß die Gleichung* $\tau \doteq \tau' - \tau''$ *ein Satz von* Th(OF) *ist (d.h. in jedem angeordneten Körper gilt).*

(ii) *Zu jeder quantorenfreien Formel* α *von* L_{OF} *läßt sich eine dazu (in* Th(OF)*) äquivalente quantorenfreie Formel* β *von* L_{OF}^+ *mit höchstens denselben Variablen angeben.*

Beweis: (i) ergibt sich durch Induktion über den Termaufbau auf Grund der folgenden Sätze von Th(OF):

(1) $\tau \doteq \tau - 0$ (angewendet für $\tau \in V \cup \{0,1\}$),

(2) $(\sigma'-\sigma'')+(\tau'-\tau'') \doteq (\sigma'+\tau')-(\sigma''+\tau'')$,

(3) $(\sigma'-\sigma'')\cdot(\tau'-\tau'') \doteq (\sigma'\cdot\tau'+\sigma''\cdot\tau'')-(\sigma'\cdot\tau''+\sigma''\cdot\tau')$,

(4) $-(\sigma'-\sigma'') \doteq \sigma''-\sigma'$.

(ii) ergibt sich dann durch Induktion über den Formelaufbau auf Grund der folgenden Äquivalenzen für die Primformeln (wobei T=Th(OF) gesetzt ist):

(5) $\sigma'-\sigma'' \doteq \tau'-\tau''$ $Äq_T$ $\sigma'+\tau'' \doteq \sigma''+\tau'$,

(6) $\sigma'-\sigma'' < \tau'-\tau''$ $Äq_T$ $\sigma'+\tau'' < \sigma''+\tau'$. □

6.76 Satz. *Vor.: Ξ ist ein Präfixtyp wie in* 6.71.

Beh.: Zu jeder pränexen Aussage α von L_{OF} vom Typ Ξ läßt sich eine pränexe Aussage β von L^+_{OF} vom selben Typ angeben, so daß für jeden angeordneten Körper $\mathfrak{F}$ gilt:

(1) $\models_{\mathfrak{F}} \alpha$ *gdw* $\models_{\mathfrak{F}^+} \beta$.

Zum Beweis: In L_{OF} verwenden wir die Abkürzungen

(2) $\forall_+ x\gamma := \forall x[x\geq 0 \rightarrow \gamma]$,

(3) $\exists_+ x\gamma := \exists x[x\geq 0 \wedge \gamma]$.

Die so gebildeten Formeln werden bisweilen als "beschränkte Quantifizierungen" (Generalisierung bzw. Partikularisierung) von γ bezeichnet (durch die "beschränkten Quantoren" $\forall_+$, $\exists_+$ wird ausgedrückt, daß der Variabilitätsbereich beschränkt wird - in diesem Beispiel auf die Menge der nicht-negativen Elemente des betrachteten angeordneten Körpers).

Sei α von der angegebenen Form. Jeder darin auftretenden Variablen x seien zwei Variablen x', x" zugeordnet, so daß die verschiedenen Variablen x, y zugeordneten Variablen x', x", y', y" stets paarweise verschieden sind (man kann auch eine entsprechende Zuordnung ξ von der Menge V aller Variablen in V^2 betrachten wie bei der Bildung der arithmetischen Reduzierten in 3.25 für n=2).

Sei ε der Kern von α. Sei ε^* diejenige quantorenfreie Formel von L_{OF}, die aus ε entsteht, indem für jede auftretende Variable x der Term x'-x" substituiert wird, und sei $\bar{\varepsilon}$ eine dazu äquivalente quantorenfreie Formel von L^+_{OF} gemäß 6.75(ii). Sei $\bar{\alpha}$ die Aussage, die aus α entsteht, indem jede Quantifizierung $\forall$x bzw. $\exists$x durch $\forall_+ x'\forall_+ x''$ bzw. $\exists_+ x'\exists_+ x''$ und außerdem ε durch $\bar{\varepsilon}$ ersetzt wird. Dann ist $\bar{\alpha}$ äquivalent zu α (das ergibt sich daraus, daß jedes Element eines angeordneten Körpers als

Differenz zweier nicht-negativer Elemente dargestellt werden kann).

Sei schließlich β die Aussage, die aus $\bar{\alpha}$ dadurch entsteht, daß die beschränkten Quantifizierungen $\forall_+ z$, $\exists_+ z$ wieder durch die entsprechenden einfachen ("unbeschränkten") Quantifizierungen $\forall z$ bzw. $\exists z$ ersetzt werden. Dann ist β eine Aussage von L^+_{OF} vom Typ Ξ (die Quantorenblöcke sind doppelt so lang wie die entsprechenden von α), die bei Interpretation in $\mathfrak{F}^+$ "dasselbe besagt" wie $\bar{\alpha}$ (und damit α) bei Interpretation in $\mathfrak{F}$. (Eine Präzisierung dieser anschaulichen Betrachtung läßt sich mit Hilfe von assoziierten Belegungen nach dem Muster von 3.26 durchführen und kann dem Leser überlassen bleiben; die Benutzung der beschränkten Quantifizierungen ist dann entbehrlich.) Somit leistet β das Verlangte. □

<u>6.77 Definition</u> (nach Szczerba 1972 und der in 6.68 genannten Mitteilung). (i) In der Sprache L(B) definieren wir folgende Formeln.

(1) $\mathrm{Ar}_{oeut}a_1 \ldots a_n := \neg\, \mathrm{Col}\, out \wedge \mathrm{B}oeu \wedge \neq(oeu) \wedge \bigwedge_{\nu=1}^{n} [\mathrm{B}oa_\nu u \wedge a_\nu \neq u]$ $(n \in \mathbb{N})$

($a_1, \ldots, a_n$ liegen *arithmetisch* in bezug auf o, e, u, t; für $n=0$ soll die allgemeine Konjunktion ganz wegfallen, in diesem Falle sagen wir auch, daß o, e, u, t ein *Bezugssystem* [für eine Punktrechnung] bilden).

(2) $\mathrm{Su}_{oeut}abc := \mathrm{Ar}_{oeut}abc \wedge \exists p \exists q \exists r \exists s \{\mathrm{B}upt \wedge \mathrm{B}uqt \wedge \neq(upq)$
$\wedge\, \mathrm{B}orp \wedge \mathrm{B}arq \wedge \mathrm{B}bsp \wedge \mathrm{B}csq$
$\wedge\, [\mathrm{B}urs \vee \mathrm{B}usr]\}$

(c ist eine *geometrische Summe* von a und b in bezug auf o, e, u, t, Abb. 157).

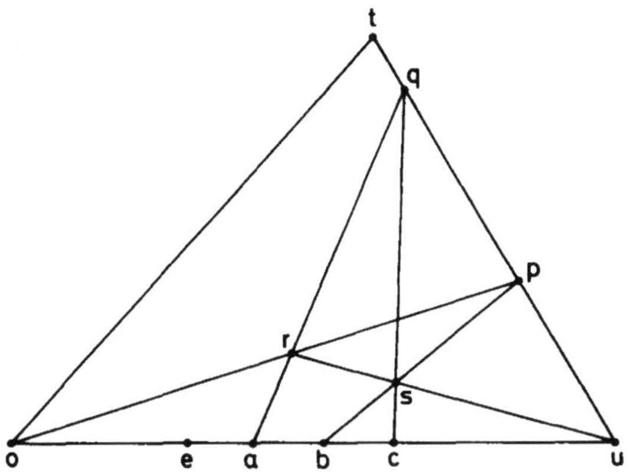

Abb. 157

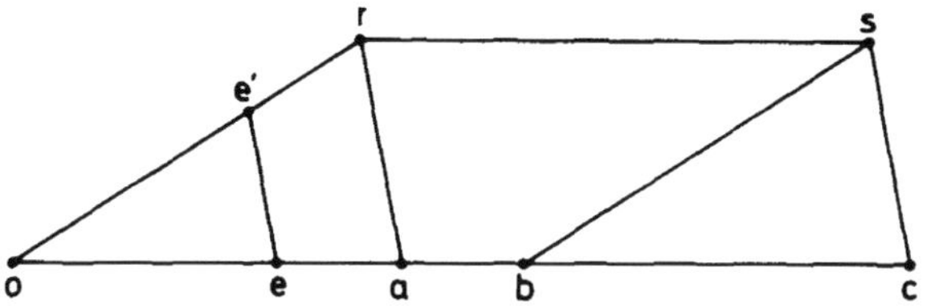

Abb. 158

(3) $Pr_{oeut}abc := Ar_{oeut}abc \wedge \exists p \exists q \exists r \exists s\{Bupt \wedge Buqt \wedge \neq(upq) \wedge Berp \wedge Barq \wedge Bbsp \wedge Bcsq \wedge [Bors \vee Bosr]\}$

(*c* ist ein *geometrisches Produkt* von *a* und *b* in bezug auf *o*, *e*, *u*, *t*, Abb. 159).

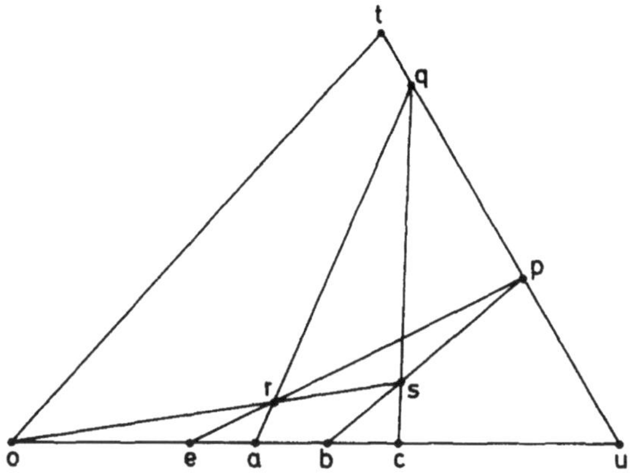

Abb. 159

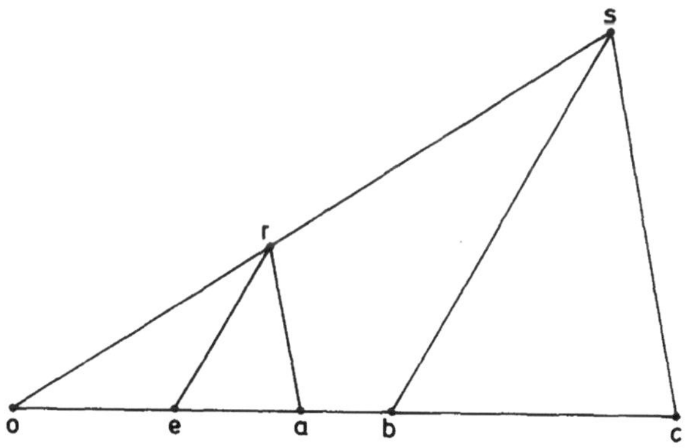

Abb. 160

(4) $a <_{oeut} b := Ar_{oeut}ab \wedge Boab \wedge a \neq b$

(*a* ist *geometrisch kleiner* als *b* in bezug auf *o*, *e*, *u*, *t*).

(ii) Die obigen Abkürzungen für Formeln verwenden wir dann auch in der Metasprache (mit Kursivbuchstaben für Punkte) zur Bezeichnung der Relationen, die in einem konvexen Raum durch diese Formeln definiert werden. Für ein festgehaltenes Bezugssystem sind das Relationen auf $G^+ = G^+(o,u) := \{x \mid \mathrm{Ar}_{oeut}x\}$.

<u>6.78 Anmerkung</u>. Sei $\mathfrak{A} = \mathfrak{A}_n^B(\mathfrak{F}) \mid S$ ein beliebiger konvexer Raum über $\mathfrak{F}$, und seien darin Punkte o, e, u, t eines Bezugssystems beliebig gewählt (solche existieren nach Def. 6.69).

Betrachten wir die gemäß 6.77(ii) analog zu 6.77(2) und (3) formulierten Definitionen für die Relationen Su_{oeut} und Pr_{oeut}. Aus den Forderungen in den ersten beiden Zeilen der jeweiligen Definitionen ergeben sich dann folgende Feststellungen:

(1) Alle genannten Punkte liegen in der konvexen Hülle des Dreiecks out (ihre Existenz in $\mathfrak{A}$ ist gleichbedeutend mit der in $\mathfrak{A}_n^B(\mathfrak{F})$).

(2) Durch die gegebenen Punkte a, b, c sowie p, q sind r und s eindeutig bestimmt (als Schnittpunkte von Geraden).

(3) Für die Summe : (a) Wenn $a=o$, so $r=o=a$.
(b) Wenn $a \neq o$, so ist ($\neg\mathrm{Col}\ oar$ und damit) $Bupq$.
Für das Produkt : (a) Wenn $a=e$, so $r=e=a$.
(b) Wenn $a \neq e$, so ist ($\neg\mathrm{Col}\ ear$ und damit)

$$a \underset{e}{\simeq} u \quad \text{gdw} \quad Bupq.$$

Die Ebene des Dreiecks out ist ein zweidimensionaler Unterraum von $\mathfrak{A}_n^B(\mathfrak{F})$; wir können sie gleich $\mathfrak{A}_2^B(\mathfrak{F})$ setzen und betrachten sie als eingebettet in die projektive Ebene $\mathfrak{P}_2^{|}(\mathfrak{F})$. Daraus entstehe durch Herausnehmen der Geraden $L(ut)$ gemäß 6.60 die affine Ebene $\mathfrak{B}$, die zu $\mathfrak{A}_2^B(\mathfrak{F})$ isomorph ist. Darin ist G^+ die Menge der nicht-negativen Punkte der Zahlengeraden G (I.14) mit dem Nullpunkt o und dem Einheitspunkt e. Die Definitionen 6.77(2) und (3) erweisen sich nun als eine Übertragung der früheren Definitionen aus I.14 für die geometrische Summe und das geometrische Produkt, eingeschränkt auf G^+ (mit Ausnahme des Falles $a=e$ für das Produkt, für den eine andere, aber offenbar gleichwertige Charakterisierung gegeben wird); denn die Existenz eines Schnittpunkts (p bzw. q) auf $L(ut)$ ist ja gleichbedeutend mit der Parallelität von Geraden in der affinen Ebene $\mathfrak{B}$ (vgl. Abb. 158, 160). Die Wahl der Punkte p und q bedeutet dabei die Festlegung einer Projektionsrichtung or bzw. er und ar.
Wir können nun die Resultate von I.14, die sich für beliebige affine Ebenen mit Anordnung ergeben, auf $\mathfrak{B}$ anwenden. Aus der Unabhängigkeit

der früheren Definitionen von der Projektionsrichtung (I.14.15 und 20) erhalten wir:

(4) Bis auf die jeweils in (3)(b) genannten Einschränkungen sind Punkte p und q mit den Eigenschaften in der ersten Zeile der Definition 6.77(2) bzw. (3) beliebig wählbar (so daß sie mit geeigneten Punkten r, s gemäß (2) das Verlangte leisten); man kann insbesondere als einen von beiden den Punkt t wählen.

Daraus ergibt sich (zu den eindeutig bestimmten Punkten r, s vgl. 3.38)

6.79 Satz. *Die* $\exists$*-Formeln* (2), (3) *in* 6.77 *sind in* Cv *äquivalent zu entsprechenden* $\forall$*-Formeln*

(2') $\mathrm{Su}^{\forall}_{\mathrm{oeut}}\mathrm{abc} := \mathrm{Ar}_{\mathrm{oeut}}\mathrm{abc} \wedge \forall p\forall q\forall r\forall s\{\mathrm{Bupt} \wedge \mathrm{Buqt} \wedge \nparallel(\mathrm{upq}) \wedge \mathrm{Borp} \wedge \mathrm{Barq} \wedge \mathrm{Bbsp} \wedge \mathrm{Bcsq} \to [\mathrm{Burs} \vee \mathrm{Busr}]\}$,

(3') $\mathrm{Pr}^{\forall}_{\mathrm{oeut}}\mathrm{abc} := \mathrm{Ar}_{\mathrm{oeut}}\mathrm{abc} \wedge \forall p\forall q\forall r\forall s\{\mathrm{Bupt} \wedge \mathrm{Buqt} \wedge \nparallel(\mathrm{upq}) \wedge \mathrm{Berp} \wedge \mathrm{Barq} \wedge \mathrm{Bbsp} \wedge \mathrm{Bcsq} \to [\mathrm{Bors} \vee \mathrm{Bosr}]\}$.

Aus I.14.41 ergibt sich außerdem

6.80 Satz. *Sei* $\mathfrak{A}$ *ein konvexer Raum über* $\mathfrak{f}$ *, und sei darin* Ar_{oeut}. *Dann bildet* $G^{+}(o,u)$ *mit den Relationen gemäß* 6.77(2) *bis* (4) *sowie dem Nullelement o und dem Einselement e einen zu* $\mathfrak{f}^{+}$ *isomorphen Halbkörper.*

6.81 Satz. *Vor.:* Ξ *ist ein Präfixtyp, dessen erstes und dessen letztes Grundzeichen eins der Zeichen* $\forall$, $\exists$ *ist (das bedeutet, daß zumindest für den ersten und den letzten Quantorenblock von pränexen Formeln dieses Typs beliebige Längen zugelassen sind, diese Blöcke können zusammenfallen, d.h.,* Ξ *darf auch aus nur einem Zeichen bestehen, s.* 6.2(ii)*).*

Beh.: Zu jeder pränexen Aussage β *von* L^{+}_{OF} *(6.73) vom Typ* Ξ *läßt sich eine pränexe Aussage* γ *von* L(B) *vom selben Typ angeben, so daß für jeden angeordneten Körper* $\mathfrak{f}$ *und jeden konvexen Raum* $\mathfrak{A}$ *über* $\mathfrak{f}$ *gilt:*

(1) $\models_{\mathfrak{f}^{+}} \beta$ *gdw* $\models_{\mathfrak{A}} \gamma$.

Beweis: Sei β wie angegeben, und sei dazu β' konstruiert wie in 6.66 (durch Abbau von Termen) mit $L=L^{+}_{OF}$, $F_{O}=\{0,1\}$, $F'=\{+,\cdot\}$, $V'=V\cup\{0,1\}$. Dann ist β' wieder vom Typ Ξ und enthält nur noch Primformeln der Formen

$$\left.\begin{array}{ll} (1) & x+y \doteq z, \\ (2) & x \cdot y \doteq z, \\ (3) & x \doteq y, \\ (4) & x<y, \end{array}\right\} \text{ wobei } x,y,z \in V'.$$

Sei nun V_P die Menge der Punktvariablen der Sprache L(B). In Analogie zur Bildung der geometrischen Reduzierten in 3.36 seien jetzt vier Punktvariablen o, e, u, t für das Folgende fest gewählt, und sei η eine ein-eindeutige Abbildung von V' in $V_P-\{u,t\}$ mit $\eta(0)=o$, $\eta(1)=e$; zur Abkürzung setzen wir $\eta(x)=\bar{x}$.

Sei π das Präfix und β^* der Kern von β'. Wir setzen o.B.d.A. voraus, daß β^* nur mittels $\wedge$ und $\vee$ aufgebaut ist aus Primformeln und Negationen von Primformeln der angegebenen Form (tatsächlich liefert der Beweis von 6.66 ein solches β').

Wir konstruieren zunächst eine Formel γ^* von L(B) aus β^* durch folgende Ersetzungen. Eine nicht-negierte Primformel der Form (1) wird ersetzt durch $Su_{oeut}\bar{x}\bar{y}\bar{z}$ bzw. $\overset{\forall}{Su}_{oeut}\bar{x}\bar{y}\bar{z}$, falls $\exists$ bzw. $\forall$ das letzte Grundzeichen von Ξ ist, für eine negierte Primformel der Form (1) wird gerade umgekehrt verfahren. Eine Primformel der Form (2) wird in analoger Weise ersetzt durch $Pr_{oeut}\bar{x}\bar{y}\bar{z}$ bzw. $\overset{\forall}{Pr}_{oeut}\bar{x}\bar{y}\bar{z}$. Für jede dieser Ersetzungen werden dabei neue Variablen als quantifizierte Variablen (p, q, r, s in 6.77(2), (3) bzw. 6.79) verwendet. Die Primformeln der Form (3) bzw. (4) werden ersetzt durch $\bar{x} \doteq \bar{y}$ bzw. $\bar{x} <_{oeut} \bar{y}$.

Sei weiter π' das Präfix, das aus π entsteht, indem jede Variable x durch das entsprechende $\bar{x}$ ersetzt wird, und sei γ' die pränexe Formel, die aus $\pi'\gamma^*$ entsteht, indem die neuen quantifizierten Variablen von γ^* mittels 1.60(6) bis (15) nach vorn (hinter π') gebracht werden. Wie im Beweis von 6.66 wird dadurch nur der letzte Quantorenblock von π' verlängert, d.h., γ' ist wieder vom Typ Ξ.

γ' enthält im allgemeinen noch die freien Variablen o, e, u, t. Aus der Konstruktion ergibt sich mittels 6.80 die folgende Feststellung:

(*) Sei $\mathfrak{f}$ ein angeordneter Körper, $\mathfrak{A}$ ein konvexer Raum über $\mathfrak{f}$, und sei darin Ar_{oeut}. Dann ist
$\models_{\mathfrak{f}^+} \beta'$ gdw $\models_{\mathfrak{A}} \gamma'[o,e,u,t]$.

(Ein strenger Beweis dafür läßt sich mit Hilfe von assoziierten Belegungen nach dem Muster von 3.49 und 3.51(ii) durchführen und kann dem

Leser überlassen bleiben.)

Sei nun γ die pränexe Aussage von L(B), die aus

(1) $\exists o \exists e \exists u \exists t[Ar_{oeut} \wedge \gamma']$ oder

(2) $\forall o \forall e \forall u \forall t[Ar_{oeut} \rightarrow \gamma']$

durch Quantorenverschiebung (gemäß 1.60(10), (8) bzw. (16), (17), angewandt auf die Quantoren in γ') entsteht, und zwar werde (1) bzw. (2) verwendet, falls $\exists$ bzw. $\forall$ das e r s t e Grundzeichen von Ξ ist. Damit wird der erste Block von π' um vier Variablen verlängert, aber in seiner Art nicht geändert, d.h., γ ist wieder vom Typ Ξ. Da (*) für beliebige Bezugssysteme gilt, erhält man daraus

(**) Seien $\mathfrak{f}$ und $\mathfrak{A}$ wie in (*). Dann ist

$\models_{\mathfrak{f}^+} \beta'$ gdw $\models_{\mathfrak{A}} \gamma$.

Wegen der Äquivalenz von β und β' leistet also γ das Verlangte. □

6.82 Definition. Sei T irgendeine Erweiterung der konvexen Geometrie *Cv* mit den Grundbegriffen B oder D und B. Eine Menge Θ von Aussagen der Sprache L(T) von T heiße ein *Schema von Körperstetigkeitsaxiomen* für T, falls für jedes Modell $\mathfrak{A}$ von T der Form $\mathfrak{A}_n^B(\mathfrak{f})|S$ bzw. $\mathfrak{L}_n(\mathfrak{f})|S$, wobei n, $\mathfrak{f}$, S wie in 6.69, gilt:

(1) $\mathfrak{A}$ ist Modell von Θ gdw $\mathfrak{f}$ reell-abgeschlossen ist.

6.83 Satz (Szczerba, s. 6.68). *Es läßt sich eine rekursive Menge* Σ_{fc} *von* $\forall\exists$*-Aussagen der Sprache* L(B) *angeben, die ein Schema von Körperstetigkeitsaxiomen für die konvexe Geometrie* Cv *(und damit für jede Erweiterung von* Cv*) ist.*

Beweis: Für angeordnete Körper $\mathfrak{f}$ wurden in 3.20 die Zusatzaxiome β' ("$\mathfrak{f}$ ist euklidisch") und γ_n für ungerade $n \geq 3$ ("in $\mathfrak{f}$ hat jedes Polynom n-ten Grades eine Nullstelle") betrachtet. Das waren $\forall\exists$-Aussagen der Sprache L_{OF}, die zusammen ausdrücken, daß $\mathfrak{f}$ reell-abgeschlossen ist. Die Menge Σ_{fc} der aus diesen Zusatzaxiomen gemäß 6.71 gebildeten pränexen Aussagen von L(B) leistet dann das Verlangte. □

6.84 Definition. Mit A11" bezeichnen wir die Menge derjenigen elementaren Stetigkeitsaxiome des Schemas A11' (I.1.2), die der Sprache L(B) angehören (d.h., die Formeln $\alpha(x)$, $\beta(y)$ zur Beschreibung von "Unterklasse" und "Oberklasse" eines Schnitts sind in L(B) gebildet, während in A11' Formeln von L(D,B) zugelassen waren).

6.85 Folgerung. *In der euklidischen Geometrie P und in der hyperbolischen Geometrie H ist das Schema* A11' *der elementaren Stetigkeitsaxiome äquivalent zu* Σ_{fc}, *d.h. zu einer Menge von Zusatzaxiomen vom Präfixtyp* ∀∃. *Dasselbe gilt für die affine Geometrie* $A\mathfrak{f}o$ *und* A11".

Beweis: Auf Grund der Darstellungssätze I.16.15 und II.2.5 ist A11' ein Schema von Körperstetigkeitsaxiomen für *P* und *H*. Damit ergibt sich die gewünschte Äquivalenz für alle endlichdimensionalen Modelle von *P* und *H*, was nach 3.77 genügt. Das Entsprechende ergibt sich auch für $A\mathfrak{f}o$ und A11". □

6.86 Anmerkungen. (i) Dieses Resultat läßt sich nicht auf beliebige Erweiterungen der konvexen Geometrie übertragen, es wird z.B. falsch für die absolute Geometrie *A* und die Erweiterungen Ps_n (semi-euklidische Geometrien). Aus dem Beweis von 6.36 ergibt sich nämlich, daß Ps_n Modelle der oben betrachteten Form über reell-abgeschlossenen Körpern besitzt; das sind also Modelle von Σ_{fc}. Andererseits ist das Grenzparallelenaxiom Gp in den Modellen von Ps_n stets falsch (6.31(iii)), Gp ergibt sich aber als Folgerung aus einem Axiom des Schemas A11".

(ii) Allgemein gilt in der konvexen Geometrie *Cv* aber noch eine Richtung der vorher betrachteten Äquivalenz, nämlich: Aus A11" folgen die Axiome von Σ_{fc}. Als Spezialfälle von A11" erhält man nämlich diejenigen elementaren Stetigkeitsaxiome für die "Zahlenhalbgeraden" $G^+(o,u)$ (6.77(ii)), in denen nur von der geometrischen Addition, Multiplikation und Kleiner-Beziehung die Rede ist. Für den damit beschriebenen Halbkörper $\mathfrak{f}^+$ (6.80) wird durch diese Spezialfälle ausgedrückt, daß $\mathfrak{f}$ reell-abgeschlossen ist.

(iii) Als Folgerung aus (ii) und 6.85 ergibt sich in *P* und in *H* die Äquivalenz von A11' mit A11"; zur Axiomatisierung von P^* und von H^* genügt also statt A11' das Schema A11" (zusammen mit Axiomen für *P* bzw. *H*).

(iv) Ein Axiomensystem der in 6.64(iii) erwähnten Art für $A\mathfrak{f}o_2^*$ erhält man nach 6.83 durch Hinzunahme von Σ_{fc}. Andererseits ist, wie in Szczerba u. Tarski 1965 festgestellt wird (s.a. Szczerba 1972a, S. 845), die Existenzforderung $E^{\geq}$ des euklidischen Parallelenaxioms eine Folgerung aus Γ (s. 6.64(iii)) und einem Axiom von A11" (nämlich aus $\Gamma\cup\{Gp\}$), und damit läßt sich auch $\Gamma\cup$A11" als Axiomensystem für $A\mathfrak{f}o_2^*$ verwenden. Aus 6.62, 63 ergibt sich dann: Auf Grund des Systems Γ der übrigen Axiomen ist A11" nicht äquivalent zu einer Menge von ∀∃-Aussagen.

<u>6.87 Ein Entscheidbarkeitsresultat</u>. Nach 3.58, 60, 86 sind die n-dimensionalen euklidischen Geometrien $\Gamma_n(OF)$, P_n und P'_n und die zugehörigen dimensionsfreien Geometrien $\Gamma(OF)$, P und P' unentscheidbar. Beschränkt man sich dagegen auf $\forall$-Aussagen, so verschwinden die Unterschiede zwischen diesen Geometrien und den (nach 3.31, 78) entscheidbaren Geometrien P_n^* bzw. P^*. Es gilt nämlich

<u>6.88 Satz</u> (nach Tarski 1959, Thm. 7). *Sei* $\mathrm{Aus}^\forall$ *die Menge der* $\forall$*-Aussagen der Sprache* L(D,B).

(i) *Ist* T *eine Theorie mit* $\Gamma_n(OF)\subseteq T\subseteq P_n^*$ *(z.B.* $\Gamma_n(OF)$, P_n *oder* P'_n*), so ist*

(1) $\mathrm{Aus}^\forall\cap T = \mathrm{Aus}^\forall\cap P_n^*$

(d.h., die Sätze vom Präfixtyp $\forall$ *sind in den beiden Geometrien* T *und* P_n^* *dieselben)* ($2\leq n\in\mathbb{N}$).

(ii) *Ist* T *eine Theorie mit* $\Gamma(OF)\subseteq T\subseteq P^*$ *(z.B.* $\Gamma(OF)$, P *oder* P'*), so ist*

(2) $\mathrm{Aus}^\forall\cap T = \mathrm{Aus}^\forall\cap P^*$.

<u>Beweis</u>: Zu (i): Nach 3.17(v) gibt es zu einem beliebigen angeordneten Körper $\mathfrak{F}$ stets einen reell-abgeschlossenen angeordneten Erweiterungskörper $\mathfrak{F}'$. Nun ist jedes Modell von $\Gamma_n(OF)$ bis auf Isomorphie ein kartesischer Raum $\mathfrak{L}_n(\mathfrak{F})$ und damit eine Unterstruktur eines geeigneten Modells $\mathfrak{L}_n(\mathfrak{F}')$ von P_n^*. Ist also eine $\forall$-Ausssge α ein Satz von P_n^* (d.h. in jedem Modell dieser Geometrie gültig), so ist sie schon ein Satz von $\Gamma_n(OF)$ (d.h. in jedem Modell davon gültig), da sich die Gültigkeit von $\forall$-Aussagen auf Unterstrukturen überträgt (6.12). Damit ergibt sich die Inklusion $\mathrm{Aus}^\forall\cap P_n^*\subseteq\mathrm{Aus}^\forall\cap\Gamma_n(OF)$, die zusammen mit der Voraussetzung über T das Gewünschte liefert.

Zu (ii): Aus der Charakterisierung von $\Gamma(OF)$ und P^* als Durchschnitte der zugehörigen n-dimensionalen Geometrien mit $2\leq n\in\mathbb{N}$ (3.52, 77) ergibt sich mittels (i) zunächst

(3) $\mathrm{Aus}^\forall\cap\Gamma(OF) = \mathrm{Aus}^\forall\cap P^*$.

Aus der Voraussetzung über T ergibt sich dann (2). □

<u>6.89 Korollar</u> (nach Tarski 1959, Thm. 8). *Für jede Geometrie* T *wie in* 6.88(i) *oder* (ii) *ist die Menge ihrer Sätze vom Präfixtyp* $\forall$ *entscheidbar.*

<u>6.90 Anmerkungen</u>. (i) Der Beweis von 6.88(i) beruhte lediglich darauf, daß sich jedes Modell der betrachteten Untertheorie als Unterstruktur in ein Modell der gegebenen Obertheorie einbetten läßt. Damit übertragen

sich die Resultate 6.88, 89 auch auf hyperbolische und elliptische Geometrien sowie auf affine und projektive Geometrien mit Anordnung (als Theorien von Kleinschen, affinen bzw. projektiven Räumen). Mittels 3.17(iii) ergibt sich dasselbe für affine und projektive Geometrien ohne Anordnung, in denen der Grundkörper als formal-reell vorausgesetzt wird.

(ii) Ist $2 \leq n < m$, so läßt sich jeder n-dimensionale kartesische Raum als Unterstruktur in einen m-dimensionalen kartesischen Raum einbetten; daraus ergibt sich wie in 6.88 noch die Inklusion

(1) $\mathrm{Aus}^{\forall} \cap P_m \subseteq \mathrm{Aus}^{\forall} \cap P_n$.

Die umgekehrte Inklusion gilt nicht, da das obere Dimensionsaxiom Dim_n^+ äquivalent zu einer $\forall$-Aussage ist (6.23(ii)). Übertragungen auf andere Geometrien liegen auf der Hand.

7. Allgemeine affine Geometrie

7.1 Überblick. (i) In diesem Abschnitt geht es nicht um eine bestimmte metamathematische Fragestellung oder Methode, sondern um die Anwendung verschiedener metamathematischer Methoden auf im wesentlichen eine bestimmte Geometrie, die für solche Anwendungen besonders geeignet ist, und zwar die (ebene) allgemeine affine Geometrie ("general affine geometry") GA_2, die von Szczerba u. Tarski 1965 eingeführt wurde. Viele Beweise können hier allerdings nur skizziert werden, andere werden ganz weggelassen - vor allem solche von rein geometrischer Art - , und es wird dann nur über die Ergebnisse berichtet.

(ii) Die Grundidee für die Einführung der allgemeinen affinen Geometrie ist, daß das euklidische Parallelenaxiom A10 bzw. $E^{\leq}$ (Eindeutigkeitsforderung, 6.58(3)) aus einem geeigneten Axiomensystem für die affine Geometrie $A\mathit{fo}_2^*$ weggelassen wird, so daß ein gemeinsamer Teil einer "euklidischen" und einer "hyperbolischen" affinen Geometrie entsteht in Analogie zur (metrischen) absoluten Geometrie A_2^*.

(iii) Im ersten Teil dieses Abschnitts (bis 7.30) werden Ergebnisse aus Szczerba u. Tarski 1965, 1979 behandelt (in der zweiten Arbeit von 1979 sind viele Beweise ausgeführt, über die in der ersten von 1965 und auch hier nur berichtet wird). Da ein Darstellungssatz nicht zur Verfügung steht (s. 7.8), wird eine Übersicht gegeben über das, was über beliebige Modelle bekannt ist (7.3 bis 10), und es werden spezielle Modelle und Methoden zu ihrer Beschreibung betrachtet (7.11ff., 17f., 20ff.). Ein wichtiges Resultat ist die erbliche Unentscheidbarkeit von GA_2 (7.16), die - im Gegensatz zu früheren Unentscheidbarkeitsresultaten - nicht durch Rückführung auf die (hier entscheidbare) Theorie der zugrundeliegenden Körper gezeigt wird, sondern "rein geometrisch" durch Konstruktion eines speziellen Modells (7.17). Andererseits ergibt sich die Entscheidbarkeit für zahlreiche Erweiterungen von GA_2, die durch definierbare Modellklassen festgelegt werden (7.20ff.), und - als Folgerung - für die Menge der Sätze von GA_2 vom Präfixtyp $\forall\exists$ (7.26f.). Für die Menge der Sätze vom Präfixtyp $\forall$ wird sogar gezeigt, daß sie in allen widerspruchsfreien Erweiterungen von GA_2 (und von GA_2^-, s.u.) übereinstimmen (7.28f.). Außerdem ergibt sich, daß GA_2 den Unvollständigkeitsgrad $2^{\aleph_0}$ besitzt (7.14) und nicht finitisierbar ist (7.19).

(iv) Die im zweiten Teil dieses Abschnitts (ab 7.31) behandelten Resultate wurden im wesentlichen erzielt von Szczerba, zum Teil gemeinsam mit A. Prestel, während eines Gastaufenthalts von Szczerba an der Universität Bonn im Sommer 1972, bei dem sich Szczerba, Prestel und der Verfasser zu regelmäßigen Diskussionen trafen. Einige Verbesserungen an den Beweisen wurden 1979 auf einer Tagung im Mathematischen Forschungsinstitut Oberwolfach diskutiert. Diesem Institut sowie dem Deutschen Akademischen Austauschdienst (DAAD), der den damaligen Gastaufenthalt in Bonn finanzierte, sei an dieser Stelle gedankt. Ein Teil dieser Resultate, insbesondere über GA_2^+ (s.u.) wird behandelt in Prestel u. Szczerba 1979. Einige Ergänzungen des Verfassers wurden innerhalb der Anmerkungen hinzugefügt.

Bei diesen Resultaten geht es vor allem um eine besonders interessierende Obertheorie und eine Untertheorie von GA_2, die hier mit GA_2^+ bzw. GA_2^- bezeichnet werden (Def. s. 7.33). Zunächst wird gezeigt, daß die betrachteten Geometrien und Obertheorien keine $\forall\exists$-Axiomensysteme besitzen (7.31, 34(v)). Außerdem wird GA_2^+ als nicht rekursiv axiomatisierbar nachgewiesen (7.35); damit wird insbesondere die bis dahin offene Frage, ob GA_2^+ überhaupt eine echte Erweiterung von GA_2 ist, positiv beantwortet. Durch Betrachtung spezieller Modelle werden dann Ergebnisse gewonnen über die Präfixtypen von Aussagen, durch die sich die genannten Theorien unterscheiden (7.38ff.), sowie (vgl. 7.54(ii)) von Aussagen, die als Zusatzaxiome verwendet werden können, um GA_2 aus GA_2^- zu erhalten (7.43f.). In diesem Zusammenhang ergeben sich auch einige zusätzliche Resultate über die schon im ersten Teil beim Unentscheidbarkeitsbeweis behandelten "Standardstrukturen" (7.45ff.).

(v) Einige seinerzeit (s. (iv)) noch offene Probleme wurden inzwischen von Prestel gelöst. Darauf wird zum Schluß hingewiesen (7.54ff.).

<u>7.2 Definition</u>. Die (ebene) *allgemeine affine Geometrie* GA_2 ist die Theorie erster Stufe mit der Sprache L(B), die auf den im folgenden angegebenen Axiomen B1 bis B8 und dem Schema B9 axiomatisch aufgebaut ist.

B1 (Identitätsaxiom für die Zwischenbeziehung)
$\forall a \forall b[Baba \rightarrow a \doteq b]$.

B2 (Transitivitätsaxiom)
$\forall a \forall b \forall c \forall d[Babc \wedge Bbcd \wedge b \neq c \rightarrow Babd]$.

B3 (Konnexitätsaxiom)
$\forall a \forall b \forall c \forall d[Babc \wedge Babd \wedge a \neq b \rightarrow Bbcd \vee Bbdc]$.

B4 (Verlängerungsaxiom)
$\forall b \forall c \exists a[Babc \wedge a \neq b]$.

B5 (Axiom von Pasch, "äußere Form", wie AP in I.1.2 [hinter A7])
$\forall a \forall b \forall c \forall d \forall e \exists f[Babc \wedge Bdce \rightarrow Bafd \wedge Bebf]$.

B6 (Desarguessches Axiom. Abb. 161)
$\forall o \forall a \forall b \forall c \forall a' \forall b' \forall c' \forall d \forall e \forall f[Boaa' \wedge Bobb' \wedge Bocc'$
$\wedge Babd \wedge Ba'b'd \wedge Bace \wedge Ba'c'e \wedge Bbcf \quad Bb'c'f$
$\wedge \neg Boab \wedge \neg Babo \wedge \neg Bboa \wedge \neg Bobc \wedge \neg Bbco \wedge \neg Bcob$
$\wedge \neg Boca \wedge \neg Bcao \wedge \neg Baoc \wedge a \neq a' \rightarrow Bdef]$.

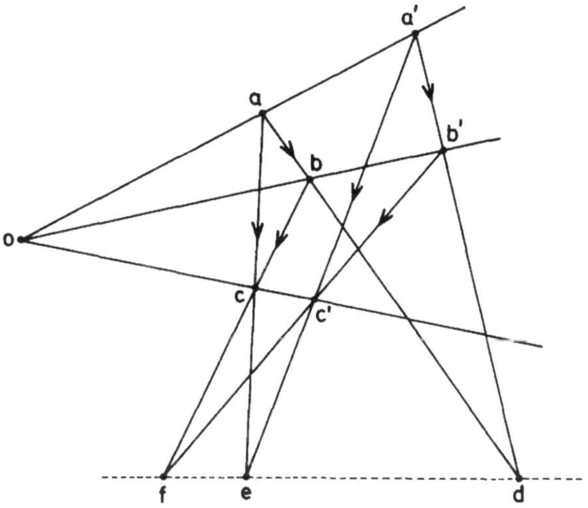

Abb. 161

B7 (Unteres Dimensionsaxiom)
$\exists a \exists b \exists c[\neg Babc \wedge \neg Bbca \wedge \neg Bcab]$.

B8 (Oberes Dimensionsaxiom, vgl. $A9^{(2)}$ in I.1.7.5 [Bem. zu A8 und A9])
$\forall a \forall b \forall c \forall x \exists y\{[Bbyc \wedge (Baxy \vee Bayx \vee Bxay)]$
$\vee [Bayc \wedge (Bbyx \vee Bxby)] \vee [Bayb \wedge (Bcyx \vee Bxcy)]\}$.

B9 (Schema der elementaren Stetigkeitsaxiome, wie A11" in 6.84) besteht aus allen Aussagen der Form

$$\forall a_1 \ldots \forall a_n \{\exists b \forall x \forall y[\alpha(x) \wedge \beta(y) \rightarrow Bbxy] \rightarrow \exists c \forall x \forall y[\alpha(x) \wedge \beta(y) \rightarrow Bxcy]\},$$

wobei $\alpha(x)$ eine Formel der Sprache L(B) ist, in der die Variablen $a_1, \ldots, a_n$, x, aber nicht b, c, y frei vorkommen können, und $\beta(y)$ analog mit Vertauschung von x und y ($n \in \mathbb{N}$).

7.3 Topologie. Zur Beschreibung von Modellen wird eine Topologie der affinen Ebenen $\mathfrak{A}_2^B(\mathfrak{F})$ verwendet, die sich in naheliegender Weise einführen läßt, z.B. mit Hilfe von offenen Dreiecken als Umgebungen.

7.4 Definition. (i) Unter einem *offenen Dreieck* T mit den nicht-kollinearen Ecken a, b, c verstehen wir die Menge seiner "inneren Punkte", d.h. der Punkte x mit $\exists u[Bbuc \wedge \neq(buc) \wedge Baxu \wedge \neq(axu)]$.

(ii) Eine Teilmenge S der Menge aller Punkte (in einem gegebenen Modell) heiße *offen*, falls es zu jedem Punkt x von S ein offenes Dreieck T gibt mit $x \in T \subseteq S$.

7.5 Anmerkung. In einer beliebigen affinen Ebene $\mathfrak{A}_2^B(\mathfrak{F})$ wird durch 7.4 eine Topologie eingeführt, und die offenen Dreiecke aus 7.4(i) sind offene Mengen im Sinne dieser Topologie (Übung!). (Dasselbe gilt für Modelle der Geometrien GA_2 und GA_2^- [s.7.33(ii)]).

7.6 Definition. Sei $\mathfrak{F}$ ein angeordneter Körper. Eine Struktur $\mathfrak{A}$ für die Sprache L(B) heiße eine *(affine) konvex-offene Ebene über* $\mathfrak{F}$ ("*restricted affine space over* $\mathfrak{F}$ " bei Szczerba u. Tarski 1965), falls sie die Einschränkung $\mathfrak{A}_2^B(\mathfrak{F})|S$ (vgl. 6.69) der affinen Ebene $\mathfrak{A}_2^B(\mathfrak{F})$ über $\mathfrak{F}$ auf eine nicht-leere, offene und konvexe Teilmenge S ist; Mengen S dieser Art nennen wir auch *konvexe (Teil-)Bereiche (von* $\mathfrak{A}_2^B(\mathfrak{F})$*)*.

7.7 Satz (Szczerba u. Tarski 1965/79). (i) *Jedes Modell von* GA_2 *ist isomorph zu einer konvex-offenen Ebene über einem reell-abgeschlossenen (angeordneten) Körper* $\mathfrak{F}$.

(ii) *Jede konvex-offene Ebene über dem Körper* $\mathbb{R}$ *der reellen Zahlen ist ein Modell von* GA_2.

(iii) *Ist* $\mathfrak{F}$ *ein reell-abgeschlossener Körper, der nicht zu* $\mathbb{R}$ *isomorph ist, so gibt es konvex-offene Ebenen über* $\mathfrak{F}$, *die nicht Modelle von* GA_2 *sind.*

Im Unterschied zu den früher betrachteten Geometrien ist jedoch ein Darstellungssatz für GA_2 noch nicht bekannt; die folgenden Probleme sind bisher ungelöst.

7.8 Probleme. (i) (Darstellungssatz) Man gebe eine "befriedigende" Charakterisierung der Modelle von GA_2 bis auf Isomorphie. Eine solche "befriedigende" Charaktersierung soll natürlich mit anderen Mitteln als nur der Definition des Modellbegriffs erfolgen, etwa mit Mitteln aus der Algebra, der Topologie und der Analysis. Nach 7.7(i) ist eine solche Charakterisierung gleichwertig mit einer Charakterisierung der konvexen Bereiche S (in Abhängigkeit vom Grundkörper $\mathfrak{f}$), die zu Modellen von GA_2 führen. Von Prestel wurde inzwischen ein Resultat über die "Schwierigkeit" dieses Problems erzielt, s. 7.57f. (Siehe auch 7.50.)

(ii) (Abschwächung von (i)) Man gebe eine befriedigende Charakterisierung der Modelle von GA_2 bis auf elementare Äquivalenz.

(iii) Man charakterisiere eine geeignete Klasse K von Strukturen derart, daß $GA_2 = \mathrm{Th}(K)$ (s. 1.35(ii)) ist. Das ist wieder eine Abschwächung von (ii); ist nämlich K eine nicht-leere Strukturklasse und $\mathfrak{B}$ ein beliebiges Modell von $\mathrm{Th}(K)$, so ist $\mathfrak{B}$ nicht notwendig elementar äquivalent zu einer Struktur aus K, aber jedenfalls elementar äquivalent zu einem Ultraprodukt von Strukturen aus K (s. etwa Schwabhäuser 1972, Satz 5.2.11). Vgl. 7.34(ii).

7.9 Anmerkungen (zum Beweis von 7.7). (i) Die wesentlichen Schritte im Beweis von 7.7(i) sind (wie auch in früheren Darstellungssätzen) die geometrische Konstruktion eines angeordneten Körpers und die Einführung eines Koordinatensystems in einem beliebigen Modell $\mathfrak{A}$ von GA_2.

Nehmen wir zunächst an, daß $\mathfrak{A}$ schon eine konvex-offene Ebene $\mathfrak{A}_2^B(\mathfrak{f})|S$ über einem angeordneten Körper ist, und betrachten wir die affine Ebene $\mathfrak{A}_2^B(\mathfrak{f})$ als eingebettet in die zugehörige projektive Ebene $\mathfrak{P}_2^{I}(\mathfrak{f})$ (6.60). Durch die Punktrechnung gemäß 6.77ff. erhält man dann einen angeordneten Halbkörper $\mathfrak{f}_{\mathfrak{A}}^{+}$ (der zu $\mathfrak{f}^{+}$ isomorph ist). Daraus erhält man einen angeordneten Körper $\mathfrak{f}_{\mathfrak{A}}$ (isomorph zu $\mathfrak{f}$, seine Elemente sind Äquivalenzklassen von "differenzengleichen" Paaren von Punkten einer Zahlenhalbgeraden $G^{+}(o,u)$).

Außerdem kann man ein projektives Koordinatensystem (vgl. 2.9(iv)) festlegen durch Wahl von drei Grundpunkten b_o, b_1, b_2 und einem Einheitspunkt c in $\mathfrak{A}$. Mit Hilfe der Punktrechnung kann man zunächst Koordinaten (in $\mathfrak{F}_{\mathfrak{A}}$) für Punkte auf den Seiten des "Basisdreiecks" $b_o b_1 b_2$ einführen; die Koordinaten eines beliebigen Punktes p kann man dann erhalten, indem man durch p zwei Geraden legt, die jeweils zwei Seiten(strecken) des Basisdreiecks schneiden. Die Abbildung, die jedem Punkt von $\mathfrak{A}$ ein entsprechendes Koordinatentripel x (oder die Äquivalenzklasse aller Vielfachen von x) zuordnet, liefert dann einen Isomorphismus von $\mathfrak{A}$ in $\mathfrak{P}_2^1(\mathfrak{F}_{\mathfrak{A}})$.

Ein beliebiges projektives Koordinatensystem in $\mathfrak{P}_2^1(\mathfrak{F}_{\mathfrak{A}})$ kann man erhalten, indem man den in $\mathfrak{A}$ gewählten Punkten b_o, b_1, b_2, c beliebige Koordinatentripel zuordnet, von denen je drei "nicht kollinear" sind (d.h. nicht-kollineare Punkte beschreiben). Daraus ergibt sich noch, daß der konvexe Bereich S - durch Koordinaten innerhalb $\mathfrak{P}_2^1(\mathfrak{F}_{\mathfrak{A}})$ ausgedrückt - durch die Wahl eines Koordinatensystems eindeutig festgelegt wird. Ohne diese Wahl ist er nur eindeutig bis auf projektive Transformationen festgelegt. Insbesondere kann man jede Gerade außerhalb S als uneigentliche Gerade gemäß 6.60 verwenden und zu entsprechenden affinen Koordinaten übergehen (vgl. Beweis von 6.59).

Aus den bisherigen Betrachtungen ergibt sich nur, daß sich durch endlich viele Aussagen von L(B) (Sätze über die Punktrechnung und über Koordinaten) ausdrücken läßt, daß jedes Modell $\mathfrak{A}$ der Menge dieser Aussagen isomorph zu einer konvex-offenen Ebene über einem angeordneten (Grund-)Körper ist. Es bleibt zu zeigen, daß sich diese Aussagen aus den angegebenen Axiomen folgern lassen. Ein Beweis ist dargestellt in Szczerba u. Tarski 1979. Aus den Stetigkeitsaxiomen ergibt sich natürlich (vgl. 6.86(ii)), daß der Grundkörper reell-abgeschlossen ist.

(ii) Mit denselben Methoden läßt sich ein allgemeinerer Darstellungssatz aus Szczerba 1972 für eine "schwache allgemeine affine Geometrie" (weak general affine geometry) WGA_2 beweisen. Bei Hinzunahme eines weiteren Axioms zu WGA_2 ergeben sich als Modelle gerade die konvex-offenen Ebenen über beliebigen angeordneten Körpern, sonst allgemeinere Modelle.

(iii) Die in (i) beschriebene Konstruktion liefert insbesondere das Resultat, daß man in einem konvexen Teilbereich S einer affinen Ebene die ganze umgebende projektive Ebene bis auf Isomorphie beschreiben kann. Dieses Resultat ist aus der projektiven Geometrie bekannt und wird bisweilen ohne Verwendung von Koordinaten unter Benutzung des Desarguesschen Satzes bewiesen (vgl. auch den Abschnitt "Konstruktionen in begrenzter Ebene" in Prüfer 1953, S. 44).

(iv) Die Behauptung 7.7(ii) ergibt sich durch einfaches Nachprüfen der

Axiome; statt des Schemas B9 erhält man in den konvex-offenen Ebenen über $\mathbb{R}$ sogar das entsprechende volle Stetigkeitsaxiom zweiter Stufe (A11 in I.1.2).

(v) Zum Beweis von 7.7(iii) sei $\mathfrak{f}$ reell-abgeschlossen, aber nicht isomorph zu $\mathbb{R}$. Dann gilt in $\mathfrak{f}$ nicht das volle Axiom (zweiter Stufe) vom Dedekindschen Schnitt, d.h., es gibt Mengen X, Y, die (als Unter- bzw. Oberklasse) in $\mathfrak{f}$ einen Dedekindschen Schnitt ohne Schnittelement bilden. Seien X, Y so gewählt. In der affinen Ebene $\mathfrak{A}_2^B(\mathfrak{f})$ sei ein affines Koordinatensystem gewählt, und sei S die Menge derjenigen Punkte, deren erste Koordinate in X liegt. Dann ist $\mathfrak{A} := \mathfrak{A}_2^B(\mathfrak{f})|S$ eine konvex-affine Ebene über $\mathfrak{f}$. Auf der ersten Koordinatenachse gibt es jedoch keinen ersten Punkt, der nicht in $\mathfrak{A}$ liegt. Auf Grund des folgenden Resultats 7.10(ii) kann somit $\mathfrak{A}$ kein Modell von GA_2 sein.

(vi) Aus (v) (und (ii)) ergibt sich, daß das Schema B9 auf Grund der übrigen Axiome echt stärker ist als das Schema Σ_{fc} der Körperstetigkeitsaxiome (6.83).

Ohne eine genaue Charakterisierung der Modelle von GA_2 zu haben, kann man jedenfalls folgendes darüber aussagen.

<u>7.10 Definitionen und Sätze</u>. (i) Das Grenzparallelenaxiom Gp (6.27(6)) ist ein Satz von GA_2 (d.h. in jedem Modell gültig).

(ii) Sei $\mathfrak{A} = \mathfrak{A}_2^B(\mathfrak{f})|S$ ein Modell von GA_2 (gemäß 7.7(i)). Seien p, q verschiedene Punkte von $\mathfrak{A}$, und sei K bzw. K^+ die dadurch in $\mathfrak{A}$ bzw. in $\mathfrak{A}_2^B(\mathfrak{f})$ bestimmte Halbgerade (von der Form $H(pq)$). Wenn dann K' nicht ganz in $\mathfrak{A}$ enthalten ist, so gibt es auf K^+ einen eindeutig bestimmten "ersten" Punkt, der nicht in $\mathfrak{A}$ liegt, d.h. einen Punkt r, so daß jeder echt zwischen p und r liegende Punkt in $\mathfrak{A}$ liegt, aber nicht r selbst.

Dieser Punkt heiße der durch K, durch K^+ oder durch p und q bestimmte *Randpunkt* von $\mathfrak{A}$ oder von S und werde mit r_K, r_{K^+} oder r_{pq} bezeichnet. Wir übertragen diese Begriffsbildung auf den Fall, daß K^+ ganz in $\mathfrak{A}$ enthalten (d.h. gleich K) ist; r_K sei dann der Schnittpunkt der Trägergeraden $\bar{K}$ von K mit der uneigentlichen Geraden. Tatsächlich ist r_K ein Randpunkt von S im Sinne der durch 7.4 in $\mathfrak{A}_2^B(\mathfrak{f})$ eingeführten bzw. der in natürlicher Weise auf $\mathfrak{P}_2^!(\mathfrak{f})$ fortgesetzten Topologie, und jeder Randpunkt von S läßt sich durch eine von p ausgehende Halbgerade bestimmen.

(iii) Seien $\mathfrak{A}$ und K wie in (ii). Dann gibt es durch den Randpunkt r_K wenigstens eine Gerade G von $\mathfrak{P}_2^{\mathrm{I}}(\mathfrak{K})$, auf der kein Punkt von S liegt. Wegen der Konvexität liegen dann alle Punkte von S auf derselben Seite von G, abgesehen von dem Fall, daß G die uneigentliche Gerade ist (in diesem Fall sind Seiten von G gar nicht erklärt, da die Zwischenbeziehung nur in $\mathfrak{A}_2^B(\mathfrak{K})$, aber nicht in $\mathfrak{P}_2^{\mathrm{I}}(\mathfrak{K})$ gegeben ist). Eine solche Gerade wird eine *Stützgerade* von S genannt; wir verwenden diese Bezeichnung auch für den Ausnahmefall der uneigentlichen Geraden.

Zum Beweis: (i) ergibt sich aus einem speziellen Stetigkeitsaxiom des Schemas B9. (ii) ergibt sich als Folgerung aus Gp. r_K läßt sich nämlich charakterisieren als der in $\mathfrak{P}_2^{\mathrm{I}}(\mathfrak{K})$ vorhandene Schnittpunkt von $\bar{K}$ mit einer echten Grenzparallelen zu K in $\mathfrak{A}$. (iii) ergibt sich wieder durch Anwendung eines geeigneten Stetigkeitsaxioms des Schemas B9, wobei die Beschreibung der umgebenden projektiven Ebene von S aus (vgl. 7.9(iii)) ausgenutzt wird. (Dagegen ergibt sich (iii) - ebenso wie (i) - noch nicht aus den Körperstetigkeitsaxiomen, vgl. 7.34(iii)).

7.11 Satz. *Sei $\mathfrak{K}$ ein beliebiger reell-abgeschlossener Körper. Für folgende Teilmengen S von $\mathfrak{A}_2^B(\mathfrak{K})$ ist dann $\mathfrak{A} = \mathfrak{A}_2^B(\mathfrak{K})|S$ jedenfalls ein Modell von GA_2:*

1. *die Menge aller Punkte (d.h. $\mathfrak{A} = \mathfrak{A}_2^B(\mathfrak{K})$),*
2. *jede Halbebene von $\mathfrak{A}_2^B(\mathfrak{K})$ im Sinne von I.9.14 (die von einer Geraden von $\mathfrak{A}_2^B(\mathfrak{K})$ begrenzt wird),*
3. *jeden nicht-leeren Durchschnitt von endlich vielen solchen Halbebenen, insbesondere also Parallelstreifen, Winkelräume und das Innere von konvexen Polygonen,*
4. *das Innere jedes Kreises.*

Zum Beweis: Das ergibt sich durch direkte Nachprüfung oder aus dem allgemeineren Resultat 7.24 (s.a. 7.20).

7.12 Geradlinige Randstücke und Ecken. Viele der folgenden Überlegungen beruhen darauf, daß man geradlinige Randstücke und Ecken eines Modells durch Formeln von GA_2 beschreiben kann

7.13 Satz. *Seien folgende Abkürzungen für Formeln eingeführt.*

(1) $\mathrm{Eukl(bpc)} := \neg\, \mathrm{Col\,bpc} \wedge \forall a \forall t\{\mathrm{Bbac} \wedge \mathrm{Bpat} \rightarrow \exists x \exists y[\mathrm{Bpbx} \wedge \mathrm{Bpcy} \wedge \mathrm{Bxty}]\}$ (der Winkel $\sphericalangle bpc$ verhält sich *euklidisch* [d.h., für ihn gilt die Forderung des Euklidischen Axioms A10], Abb. 162).

(2) $\mathrm{Hyp}(pa) := p \neq a \wedge \forall b \forall c\{\neg \mathrm{Col}\, bpc \wedge \mathrm{B}bac \wedge \neq(bac) \rightarrow \neg\, \mathrm{Eukl}(bpc)\}$
(die Halbgerade H(*pa*) verhält sich *hyperbolisch*, Abb. 163, 164).

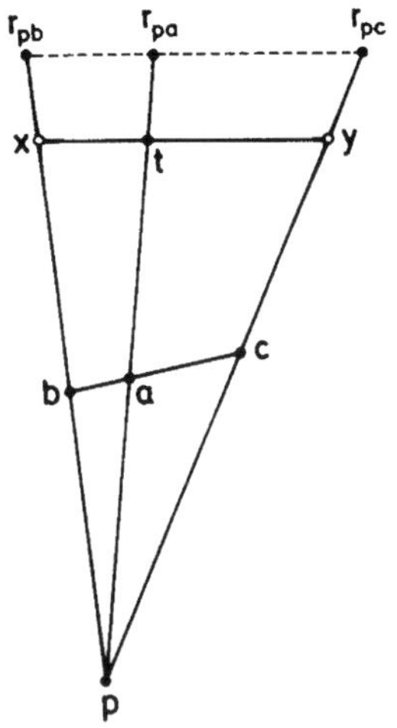

Abb. 162

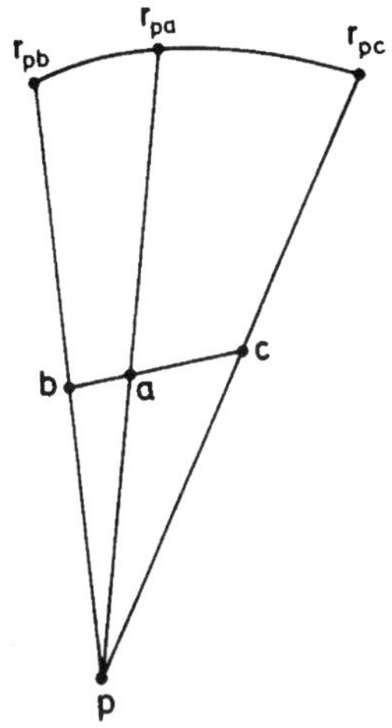

Abb. 163

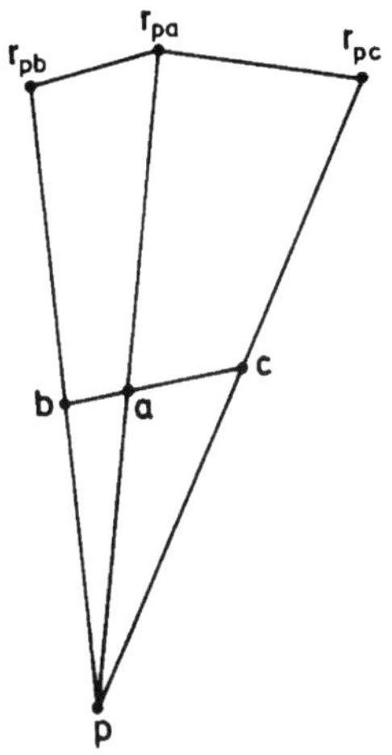

Abb. 164

(i) *Seien b, p, c nicht-kollineare Punkte (d.h. ∢bpc ein nicht-entarteter Winkel) in einem Modell* $\mathfrak{A} = \mathfrak{A}_2^B(\mathfrak{K})|S$ *von* GA_2. *Dann wird die Formel* (1) *durch diese Punkte in* $\mathfrak{A}$ *erfüllt* ($\models_{\mathfrak{A}}$ Eukl[*bpc*], *vgl.* 1.16) *genau dann, wenn das durch den Winkel ∢bpc bestimmte Randstück* $\{r_{pa} \mid \mathrm{B}_{\mathfrak{A}} bac\}$ *geradlinig ist, d.h. die so bestimmte Menge von Randpunkten* (7.10(ii)) *enthalten ist in einer Geraden von* $\mathfrak{P}_2^{\,|}(\mathfrak{K})$.

(ii) *Die Formel* (2) *drückt dann offenbar (in demselben Sinne) aus, daß die Halbgerade A*=H(*pa*) *nicht im Inneren eines ("noch so kleinen") euklidischen Winkels liegt. Ist insbesondere A der gemeinsame Schenkel von zwei aneinander liegenden euklidischen Winkeln, so drückt* (2) *aus, daß der durch A bestimmte Randpunkt* r_A *eine Ecke von S ist (Abb.* 164*).*

(iii) *Die in* (1) *für jeden Punkt a zwischen b und c gestellte Forderung läßt sich äquivalent ersetzen durch dieselbe Forderung für nur einen echt zwischen b und c liegenden Punkt; damit ergibt sich in* GA_2 *die Äquivalenz*

(3) $\mathrm{Hyp}(pa) \leftrightarrow p \neq a \wedge \forall b \forall c\{\neg \mathrm{Col}\, bpc \wedge \mathrm{B}bac \wedge \neq(bac)$
$\rightarrow \exists t[\mathrm{B}pat \wedge \neg \exists x \exists y(\mathrm{B}pbx \wedge \mathrm{B}pcy \wedge \mathrm{B}xty)]\}$.

Zum Beweis: Das ergibt sich ohne Schwierigkeiten aus den Definitionen (1), (2) sowie der Konvexität und der Offenheit von *S* und kann dem Leser überlassen bleiben.

<u>7.14 Satz</u> (Szczerba u. Tarski 1965/79). *Die allgemeine affine Geometrie GA_2 ist unvollständig, und zwar vom "Unvollständigkeitsgrad" $2^{\aleph_0}$, das bedeutet, sie hat genau $2^{\aleph_0}$ verschiedene widerspruchsfreie Vervollständigungen (d.h. Erweiterungen mit derselben Sprache, die widerspruchsfrei und vollständig sind, vgl. 3.82).*

<u>Zum Beweis</u> (Skizze): Für eine beliebige reelle Zahl $x>0$ bedeute $\mathfrak{A}_x$ die Struktur $\mathfrak{A}^B_2(\mathbb{R})|S_x$, wobei S_x das Innere des Fünfecks mit den Ecken $a=\langle 2,-1\rangle$, $b=\langle 0,-1\rangle$, $c=\langle 0,1\rangle$, $d=\langle 2,1\rangle$, $e=\langle 4+x,0\rangle$ ist (die Punkte von $\mathfrak{A}^B_2(\mathbb{R})$ seien dabei mit Koordinatenpaaren identifiziert, Abb. 165).

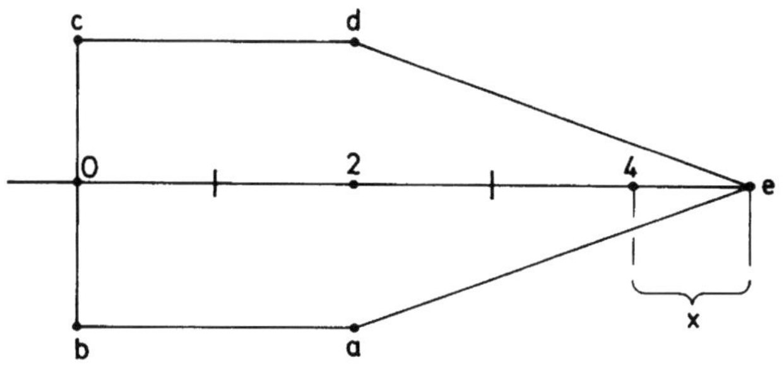

Abb. 165

Nach 7.7(ii) (oder 7.11.3.) ist $\mathfrak{A}_x$ ein Modell von GA_2, also $Th(\mathfrak{A}_x)$ eine Erweiterung von GA_2, die nach 1.39 (Beisp.) widerspruchsfrei und vollständig ist. Gemäß 7.9(i) kann man den Grundkörper geometrisch beschreiben und ein projektives Koordinatensystem einführen. Damit kann man Formeln α, β, γ der Sprache L(B) (mit Parametern für Körperelemente) konstruieren, deren Zutreffen in $\mathfrak{A}_x$ besagt, daß ein Punkt mit gegebenen Koordinaten ein Punkt von S_x, ein Randpunkt bzw. eine Ecke von S_x ist. Schließlich kann man mit diesen Mitteln auch die Ecke e von den übrigen Ecken unterscheiden (aus der Definition von S_x ergibt sich, daß jede projektive Transformation, die die Mengen der Ecken in sich überführt, die Ecke e fest läßt). Damit kann man das projektive Koordinatensystem auch so wählen, daß die Ecken die oben angegebenen Koordinaten haben.

Seien nun $\mathfrak{A}_x$, $\mathfrak{A}_y$ zwei beliebige Modelle der betrachteten Art mit $x<y$. Sei dann q eine rationale Zahl mit $x<q<y$. Da jede rationale (und allgemeiner jede reelle algebraische) Zahl in $Th(\mathbb{R})$ definierbar ist (vgl. 3.22(v)), kann man mit den genannten Mitteln schließlich eine Aussage von L(B) konstruieren, die besagt, daß der Punkt $\langle q,0\rangle$ zum Modell gehört, d.h. eine Aussage, die in $\mathfrak{A}_y$, aber nicht in $\mathfrak{A}_x$ gültig ist. Somit sind $Th(\mathfrak{A}_x)$ und $Th(\mathfrak{A}_y)$ verschiedene Theorien.

Mit den Theorien Th($\mathfrak{A}_x$) für alle positiven reellen x hat man somit schon $2^{\aleph_0}$ widerspruchsfreie und vollständige Erweiterungen von GA_2 in der Sprache L(B). Das sind natürlich noch nicht alle solchen Erweiterungen (Vervollständigungen). Es gibt aber auch nicht mehr als $2^{\aleph_0}$ solche Erweiterungen; denn jede solche ist eine Menge von Formeln, d.h. eine Teilmenge der Menge $\mathrm{Fml}_{L(B)}$. Diese Menge ist abzählbar und hat somit nur $2^{\aleph_0}$ Teilmengen. □

7.15 Anmerkung. Die Eckenzahl fünf ist die kleinste, für die sich die Konstruktion des vorigen Beweises durchführen läßt. Die zu konvexen (nicht-entarteten) Vierecken S gehörenden Modelle $\mathfrak{A} = \mathfrak{A}_2^B(\mathbb{R})\,|\,S$, sind nämlich noch alle untereinander isomorph (da sich die Vierecke durch projektive Transformationen ineinander überführen lassen), sie liefern also immer dieselbe Theorie Th($\mathfrak{A}$). Dasselbe gilt natürlich erst recht für nicht-entartete Dreiecke.

7.16 Satz (Szczerba u. Tarski 1965/79). *Die allgemeine affine Geometrie GA_2 ist "erblich unentscheidbar", d.h., sie selbst und alle ihre Untertheorien mit derselben Sprache* L(B) *sind unentscheidbar.*

Zum Beweis ("geometrisch", vgl. 7.1(iii)): Nach 3.54 genügt es zu zeigen, daß die Struktur der natürlichen Zahlen mit Addition und Multiplikation sich definieren läßt in einem geeigneten Modell von GA_2 (und damit von allen Untertheorien). Das wird in 7.18 für das Modell $\mathfrak{I}_{\mathbb{N}}^{\mathbb{R}}$ gezeigt.

7.17 Definition. (i) Sei $\mathfrak{F}$ ein reell-abgeschlossener angeordneter Körper, und seien die Punkte der affinen Ebene $\mathfrak{A}_2^B(\mathfrak{F})$ mit Koordinatenpaaren (in bezug auf ein affines Koordinatensystem) identifiziert. Sei dann $E_{\mathbb{N}}^{\mathfrak{F}}$ die Menge der Punkte ("Ecken", s.a. (ii))

$$a = \langle 2,0\rangle, \quad b = \left\langle \frac{1+\sqrt{7}}{2}, \frac{1-\sqrt{7}}{2} \right\rangle, \quad c = \langle 0,-2\rangle,$$

$$d = \langle -2,0\rangle, \quad f = \left\langle \frac{1-\sqrt{7}}{2}, \frac{1+\sqrt{7}}{2} \right\rangle, \quad g_n = \left\langle \frac{2n}{\sqrt{n^2+1}}, \frac{2}{\sqrt{n^2+1}} \right\rangle \quad (n \in \mathbb{N}).$$

Weiter sei $S_{\mathbb{N}}^{\mathfrak{F}}$ das Innere der konvexen Hülle von $E_{\mathbb{N}}^{\mathfrak{F}}$ (Abb. 166). Dann heiße die Struktur $\mathfrak{I}_{\mathbb{N}}^{\mathfrak{F}} = \mathfrak{A}_2^B(\mathfrak{F})\,|\,S_{\mathbb{N}}^{\mathfrak{F}}$ die *Standardstruktur* (zur Interpretation der natürlichen Zahlen) *über* $\mathfrak{F}$.

(ii) In $\mathfrak{A}_2^B(\mathfrak{F})$ seien noch folgende Punkte eingeführt.

$$o = \langle 0,1\rangle, \quad e = \left\langle \tfrac{1}{2}, \tfrac{1}{2} \right\rangle, \quad u = \langle 1,0\rangle, \quad t = \langle 0,0\rangle .$$

Außerdem werde die Streckenkongruenz $D_{\mathfrak{F}}^2$ nach I.1.4(2) eingeführt, wodurch aus $\mathfrak{A}_2^B(\mathfrak{F})$ die kartesische Ebene $\mathfrak{C}_2(\mathfrak{F})$ entsteht, so daß die vorher betrachteten affinen Koordinaten(paare) zu kartesischen Koordinaten(paaren) werden. Die obige Charakterisierung (i) von $E^{\mathfrak{f}}_{\mathbb{N}}$ (und damit von $\gamma^{\mathfrak{f}}_{\mathbb{N}}$) erweist sich dann (durch einfache Rechnung) als gleichwertig mit der folgenden.

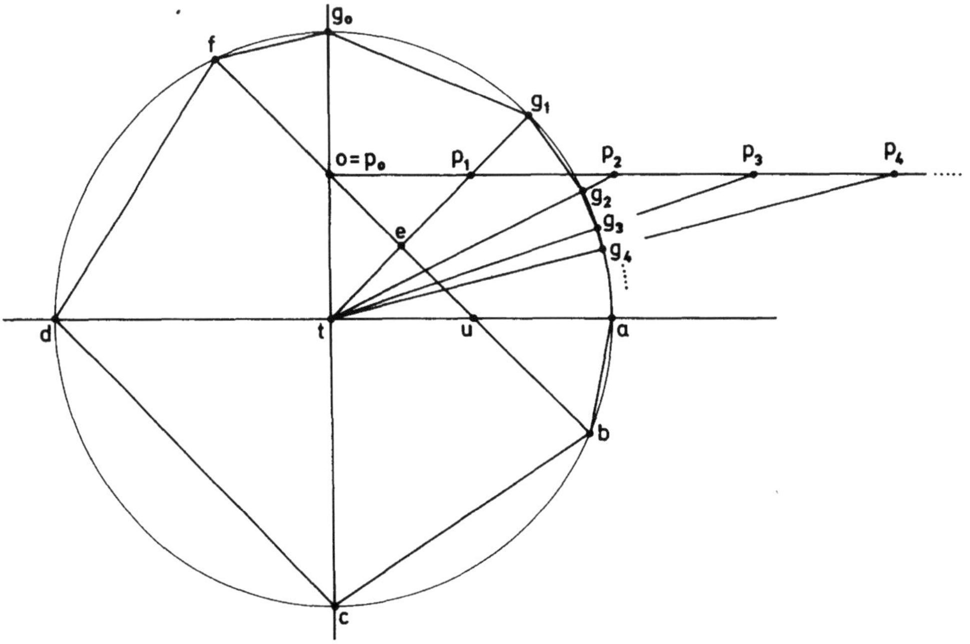

Abb. 166

Sei K der Kreis um den Koordinatenursprung t mit dem Radius 2. Dann sind a, c, d, g_0 die Schnittpunkte von K mit den Koordinatenachsen in der Reihenfolge, die durch die vorher angegebenen Koordinaten festgelegt ist. b und f sind die Schnittpunkte von K mit den Halbgeraden H(ou) bzw. H(uo). Sei p_n=<n,1>; dann ist g_n der Schnittpunkt von K mit der Halbgeraden H(tp_n) ($n \in \mathbb{N}$).

Insbesondere liegen also alle Punkte von $E^{\mathfrak{f}}_{\mathbb{N}}$ auf dem Kreis K, und e ist der Schnittpunkt von H(tp_1) mit L(ou).

<u>7.18 Satz</u>. (i) *In der Standardstruktur* $\gamma^{\mathfrak{f}}_{\mathbb{N}}$ *(wobei* $\mathfrak{f}$ *reell-abgeschlossen) ist die Struktur der natürlichen Zahlen mit Addition und Multiplikation definierbar.*

(ii) *Die Standardstruktur* $\gamma^{\mathbb{R}}_{\mathbb{N}}$ *(Spezialfall mit* $\mathfrak{f}=\mathbb{R}$*) ist ein Modell von* GA_2 *("Standardmodell").*

Zum Beweis: (ii) ergibt sich unmittelbar aus 7.7(ii). Zum Beweis von (i) sei $\mathfrak{f}$ beliebig (reell-abgeschlossen). Sei s zunächst ein beliebiger Punkt von $\gamma^{\mathfrak{f}}_{\mathbb{N}}$. Die Halbgeraden von s aus durch die in 7.17 genannten Ecken (Bezeichnung $\bar{H}(sa)$, $\bar{H}(sb)$ usw.) lassen sich dann charakterisieren als die (Verlängerungen in $\mathfrak{A}_2^B(\mathfrak{f})$ der) von s ausgehenden hyperbolischen Halbgeraden im Sinne von 7.13. (In dem weniger interessanten Fall, daß die Anordnung von $\mathfrak{f}$ nicht-archimedisch ist, gibt es Halbgeraden $H(sx)$, die zwischen $\bar{H}(sa)$ und jedem $\bar{H}(sg_n)$ liegen; es läßt sich zeigen, daß diese nicht hyperbolisch sind und keine Randpunkte von $\gamma^{\mathfrak{f}}_{\mathbb{N}}$ bestimmen [7.10(ii)], so daß $\gamma^{\mathfrak{f}}_{\mathbb{N}}$ kein Modell von GA_2 ist. Die übrigen Halbgeraden $H(sx)$ [d.h. alle bei archimedischer Anordnung] bestimmen jedenfalls Randpunkte, und die Überlegungen von 7.13 lassen sich dafür direkt anwenden, auch wenn $\gamma^{\mathfrak{f}}_{\mathbb{N}}$ kein Modell von GA_2 [vgl. 7.48] sein sollte.) Die Halbgerade $\bar{H}(sa)$ läßt sich damit insbesondere charakterisieren als die einzige von s ausgehende hyperbolische Halbgerade, für die es auf einer Seite keine nächstgelegene von s ausgehende hyperbolische Halbgerade gibt. Man kann nun einen Umlaufssinn um s festlegen, bei dem mit der anderen Seite von $\bar{H}(sa)$ begonnen wird. Indem man hyperbolische Halbgeraden in diesem Sinne zählt, kann man sogar für jede Ecke q ($q=a,b,c,\ldots$, es genügt bis g_1) die zugehörige Halbgerade $\bar{H}(sq)$ charakterisieren, d.h. eine Formel $\alpha_q(s,x)$ der Sprache L(B) konstruieren mit

(1) $\models_{\gamma^{\mathfrak{f}}_{\mathbb{N}}} \alpha_q[s,x]$ gdw $H(sx)\subseteq\bar{H}(sq)$.

Damit lassen sich auch die Punkte t, o, u von 7.17(ii) eindeutig charakterisieren als Schnittpunkte von Verbindungsgeraden geeigneter Ecken sowie e als Schnittpunkt von $\bar{H}(tg_1)$ mit $L(ou)$.

Wir verwenden nun o, e, u, t als Bezugssystem für eine Punktrechnung im Sinne von 6.77. Außerdem betrachten wir die in der affinen Ebene $\mathfrak{A}_2^B(\mathfrak{f})$ durch o und p_1 bestimmte Zahlengerade L im Sinne von I.1.14; sei L^+ die zugehörige Zahlenhalbgerade. Die Zentralprojektion mit dem Zentrum t ist dann ein Isomorphismus von L^+ auf $G^+(o,u)$, da sie o, p_1 und den uneigentlichen Punkt von L überführt in o, e, u. Die natürlichen Elemente von $G^+(o,u)$ sind also gerade die Bilder der natürlichen Elemente p_n von L^+, sie lassen sich somit charakterisieren als diejenigen Punkte x mit $\mathrm{Ar}_{oeut}x$, für die $H(tx)$ eine hyperbolische Halbgerade ist. Die Addition und die Multiplikation für diese natürlichen Elemente entstehen durch Einschränkung der Operationen, die gemäß 6.77 für $G^+(o,u)$ definiert sind. Damit läßt sich die Struktur der natürlichen Zahlen mit Addition und Multiplikation (bis auf Isomorphie) in $\gamma^{\mathfrak{f}}_{\mathbb{N}}$ durch geeignete Formeln von L(B) definieren. □

7.19 Satz (Szczerba u. Tarski 1965/79). GA_2 *ist nicht endlich axiomatisierbar.*

Beweis: Wäre GA_2 endlich axiomatisierbar, so wäre es auch die affine Geometrie $\mathcal{A}\!f\!o_2^*$, die aus GA_2 durch Hinzunahme des euklidischen Parallelenaxioms $E^{\leq}$ entsteht (7.1(ii)). Es läßt sich aber zeigen - ebenso wie für P_n^* in 3.33 -, daß das nicht der Fall ist. □

7.20 Definition (Definierbare Modelle und Modellklassen, nach Szczerba u. Tarski 1965). Sei $\mathfrak{F}$ ein beliebiger angeordneter Körper, und seien die Punkte der affinen Ebene $\mathfrak{A}_2^B(\mathfrak{F})$ wieder mit Koordinatenpaaren $\langle x,y\rangle \in U_{\mathfrak{F}}^2$ identifiziert. Sei $k\in\mathbb{N}$, und seien $\alpha(x,y,t_1,\ldots,t_k)$ und $\beta(t_1,\ldots,t_k)$ Formeln der Sprache L_{OF} (für angeordnete Körper), in denen höchstens die angebenen Variablen frei vorkommen. Unter der über $\mathfrak{F}$ durch α mit Hilfe der Parameter $t_1,\ldots,t_k \in \mathfrak{F}$ *definierten Punktmenge* verstehen wir die Menge

(1) $S^{\mathfrak{F}}_{\alpha,t_1,\ldots,t_k} := \{\langle x,y\rangle \mid \models_{\mathfrak{F}} \alpha[x,y,t_1,\ldots,t_k]\}$.

Zum Beispiel läßt sich das Innere eines konvexen n-Ecks durch eine geeignete Formel α definieren mit Hilfe der $2n$ Koordinaten der n Ecken als Parameter ($k=2n$).

Unter der über $\mathfrak{F}$ durch α und β *(uniform) definierten Klasse* von Punktmengen verstehen wir die Klasse

(2) $K^{\mathfrak{F}}_{\alpha,\beta} := \{S^{\mathfrak{F}}_{\alpha,t_1,\ldots,t_k} \mid \models_{\mathfrak{F}} \beta[t_1,\ldots,t_k]\}$.

Die Formel β drückt also - anschaulich gesagt - eine Bedingung aus, der die Parameter genügen müssen, damit die Punktmenge (1) in die Klasse (2) aufgenommen wird.

Die erste Konstruktion ist für das Folgende natürlich nur von Interesse (7.6, 7), wenn gilt:

(*) die Punktmenge (1) ist ein konvexer Bereich.

Für die dadurch bestimmte konvex-offene Ebene und für eine analog zu (2) definierte Klasse von solchen Ebenen verwenden wir dann die Bezeichnungen

(3) $\mathfrak{A}^{\mathfrak{F}}_{\alpha,t_1,\ldots,t_k} = \mathfrak{A}_2^B(\mathfrak{F}) \mid S^{\mathfrak{F}}_{\alpha,t_1,\ldots,t_k}$,

(4) $C_{\alpha,\beta} = \{\mathfrak{A}^{\mathfrak{F}}_{\alpha,t_1,\ldots,t_k} \mid \models_{\mathfrak{F}} \beta[t_1,\ldots,t_k]\}$.

7.21 Anmerkungen. (i) Die Bedingung 7.20(*) läßt sich offenbar ausdrücken durch die folgende mittels α gebildete Formel $\gamma_\alpha(t_1,\ldots,t_k)$ von L_{OF}:

(1) $\exists\mathfrak{x}\,\alpha(\mathfrak{x},\mathfrak{t})$
$\wedge\forall\mathfrak{x}\forall\mathfrak{y}\{\alpha(\mathfrak{y},\mathfrak{t})\rightarrow\exists\mathfrak{z}[Ba(\mathfrak{x},\mathfrak{y},\mathfrak{z})\wedge\mathfrak{z}\neq\mathfrak{y}\wedge\alpha(\mathfrak{z},\mathfrak{t})]\}$
$\wedge\forall\mathfrak{x}\forall\mathfrak{y}\forall\mathfrak{z}\,\{\alpha(\mathfrak{x},\mathfrak{t})\wedge\alpha(\mathfrak{z},\mathfrak{t})\wedge Ba(\mathfrak{x},\mathfrak{y},\mathfrak{z})\rightarrow\alpha(\mathfrak{y},\mathfrak{t})\}$;

dabei sind $\mathfrak{x}$, $\mathfrak{y}$, $\mathfrak{z}$, $\mathfrak{t}$ in naheliegender Weise als Abkürzungen für die Paare $<x_1,x_2>$, $<y_1,y_2>$, $<z_1,z_2>$ bzw. das k-tupel $<t_1,\ldots,t_k>$ von Variablen verwendet, und $Ba(\mathfrak{x},\mathfrak{y},\mathfrak{z})$ bzw. $\mathfrak{z}\neq\mathfrak{y}$ bedeuten Formeln, die die Zwischenbeziehung bzw. die Verschiedenheit von Punkten mit Hilfe ihrer Koordinaten ausdrücken (vgl. 3.25(1)).

(ii) Damit ist 7.20(4) eine nicht-leere Klasse von konvex-offenen Ebenen genau dann, wenn gilt:

(2) Die Aussagen $\exists t_1\ldots\exists t_k\beta$ und $\forall t_1\ldots\forall t_k(\beta\rightarrow\gamma_\alpha)$ sind Sätze von $Th(\mathfrak{F})$.

Ist außerdem $\mathfrak{F}$ reell-abgeschlossen, so läßt sich $Th(\mathfrak{F})$ in (2) durch die Theorie Th(RF) aller reell-abgeschlossenen Körper ersetzen (wegen der Vollständigkeit derselben ist ja $Th(\mathfrak{F})=Th(RF)$).

7.22 Satz (Szczerba u. Tarski 1965/79). *Sei* $\mathfrak{F}$ *ein reell-abgeschlossener (angeordneter) Körper und* $C^{\mathfrak{F}}_{\alpha,\beta}$ *eine gemäß* 7.20 *definierte nicht-leere Klasse von konvex-offenen Ebenen. Für die Theorie* $T=Th(C^{\mathfrak{F}}_{\alpha,\beta})$ *dieser Klasse gelten dann folgende Behauptungen.*

(i) T *ist eine widerspruchsfreie endliche Erweiterung von* GA_2.

(ii) T *ist nicht endlich axiomatisierbar.*

(iii) T *ist entscheidbar.*

(iv) *Wenn je zwei Strukturen* $\mathfrak{A},\mathfrak{B}\in C^{\mathfrak{F}}_{\alpha,\beta}$ *zueinander isomorph (oder wenigstens elementar äquivalent) sind, so ist* T *vollständig.*

(v) *Wenn* $\mathfrak{F}'$ *irgendein anderer reell-abgeschlossener Körper ist, so ist auch* $T=Th(C^{\mathfrak{F}'}_{\alpha,\beta})$.

Zum Beweis verwenden wir

7.23 Lemma. *Seien* α, β *wie vorher. Dann läßt sich eine Konstruktion von "arithmetischen Reduzierten"* $Rda(\gamma)$ *angeben, so daß für jeden angeordneten Körper* $\mathfrak{F}$ *und jede Aussage* γ *von* L(B) *gilt:*

(1) $\gamma\in Th(C^{\mathfrak{F}}_{\alpha,\beta})$ *gdw* $\models_{\mathfrak{F}} Rda(\gamma)$.

Zum Beweis (Skizze): Die Idee der Konstruktion ist folgende. Durch Rda(γ) wird ausgedrückt, daß für jedes $t_1, \dots, t_k$ mit der Eigenschaft β gilt:

(2) γ ist gültig in der konvex-offenen Ebene $\mathfrak{A}_2^B(\mathfrak{F})|S^{\mathfrak{F}}_{\alpha,t_1,\dots,t_k}$.

Die Bedingung (2) läßt sich in der Sprache L_{OF} formalisieren, indem man Punktvariablen a_ν wie üblich durch Zahlenvariablen x_ν, y_ν (für die Koordinaten) ersetzt und entsprechende Quantifizierungen ($\forall x_\nu \forall y_\nu$ bzw. $\exists x_\nu \exists y_\nu$) der Beschränkung $\alpha(x_\nu, y_\nu, t_1, \dots t_k)$ unterwirft. Eine genaue Ausführung kann ähnlich zu früheren Reduziertenkonstruktionen erfolgen. □

Zum Beweis von 7.22: Mit Hilfe des Lemmas ergeben sich (v) und (iii) aus der Vollständigkeit und der Entscheidbarkeit der Theorie Th(RF) der reell-abgeschlossenen Körper (3.22(i)). (iv) ergibt sich unmittelbar aus der Definition der elementaren Äquivalenz (T läßt sich in diesem Falle als die Theorie einer einzigen Struktur darstellen, vgl. 1.37, 39).

Zu (i): Nach 7.21(ii) ist auch $C^{\mathbb{R}}_{\alpha,\beta}$ eine nicht-leere Klasse von konvex-offenen Ebenen, die nach 7.7(ii) Modelle von GA_2 sind. Damit ist $Th(C^{\mathbb{R}}_{\alpha,\beta})$ und wegen (v) auch T jedenfalls eine Erweiterung von GA_2. Um zu zeigen, daß diese Erweiterung endlich ist (1.36), genügt es, ein geeignetes Zusatzaxiom η zu finden. Dafür genügt eine Aussage, die folgendes ausdrückt.

(*) Es gibt vier Bezugspunkte für ein projektives bzw. affines Koordinatensystem (s. 7.9(i)) und Elemente $t_1, \dots, t_k$ des zugehörigen geometrisch eingeführten angeordneten Körpers $\mathfrak{F}_{oeut}$ mit der Eigenschaft $\beta[t_1, \dots, t_k]$, so daß ein Punkt mit den affinen Koordinaten $x, y \in \mathfrak{F}_{oeut}$ stets genau dann existiert, wenn $\alpha[x, y, t_1, \dots, t_k]$.

Mit Hilfe der Konstruktion von geometrischen Reduzierten gemäß 6.68, 71 kann man tatsächlich eine solche Aussage η von L(B) erhalten.

Zum Beweis von (ii) nehmen wir an, daß T endlich axiomatisierbar und γ die Konjunktion der endlich vielen Axiome ist. Nach 7.23 ist dann Rda(γ) in $\mathfrak{F}$ gültig und somit ein Satz von Th(RF). Da diese Theorie nicht endlich axiomatisierbar ist (3.33), sind Rda(γ) und die beiden Aussagen von 7.21(2) auch gültig in einem gewissen angeordneten Körper $\mathfrak{F}''$, der nicht reell-abgeschlossen ist. Nach 7.23 ist dann $\gamma \in Th(C^{\mathfrak{F}''}_{\alpha,\beta})$. Damit hat γ ein Modell in der nicht-leeren Klasse $C^{\mathfrak{F}''}_{\alpha,\beta}$, das nach 7.7(i) kein Modell von GA_2 ist, im Widerspruch zur Annahme, daß durch γ eine Erweiterung von GA_2 axiomatisiert wird. □

7.24 Korollar. *Vor.: $\mathfrak{A}$ ist eine konvex-offene Ebene, die über einem reell-abgeschlossenen Körper $\mathfrak{K}$ durch eine Formel α mit Hilfe gewisser Parameter (gemäß 7.20(3)) definiert wird.*

Beh.: $\mathfrak{A}$ ist ein Modell von GA_2 und sogar von $Th(C^{\mathbb{R}}_{\alpha,\beta})$ für ein geeignetes β (und damit auch der Theorie GA_2^+, s. 7.33(i)).

Beweis: Sei β die Formel γ_α von 7.21(i). Dann gilt 7.21(2). Nach Definition ist $\mathfrak{A}$ Modell von $Th(C^{\mathfrak{K}}_{\alpha,\beta})$ und nach 7.22(v) auch von $Th(C^{\mathbb{R}}_{\alpha,\beta})$. □

7.25 Anmerkungen. (i) In 7.24 braucht $\mathfrak{A}$ jedoch nicht elementar äquivalent zu einer Struktur aus $C^{\mathbb{R}}_{\alpha,\beta}$ zu sein (vgl. 7.8(iii)) und nicht einmal elementar äquivalent zu irgendeiner konvex-offenen Ebene über $\mathbb{R}$. Das zeigt folgendes Beispiel (Szczerba, mündliche Mitteilung).

Sei $\mathfrak{K}$ ein reell-abgeschlossener nicht-archimedisch angeordneter Körper und sei x darin ein positives "unendlich kleines" Element, d.h. $0<x\ll 1$ (vgl. 6.36). Sei die "Fünfecksstruktur" $\mathfrak{A}_x = \mathfrak{A}_2^B(\mathfrak{K})|S_x$ über $\mathfrak{K}$ festgelegt wie im Beweis von 7.14 durch die dort angegebenen Ecken. Wie dort sei auch ein Grundkörper $\mathfrak{K}_{oeut}$ und ein Koordinatensystem geometrisch eingeführt, und sei o.B.d.A. $\mathfrak{K} = \mathfrak{K}_{oeut}$. Dann läßt sich das Körperelement x in $\mathfrak{A}_x$ eindeutig charakterisieren durch eine Formel $\gamma(y)$, die ausdrückt, daß der Punkt mit den Koordinaten $4+y>0$ und 0 ein Randpunkt ist. Somit gelten in $\mathfrak{A}_x$ gewisse Aussagen η_n, die ausdrücken, daß das durch γ definierte Element positiv und kleiner als $\frac{1}{n}$ ist ($n\in\mathbb{N}$). In einer beliebigen konvex-offenen Ebene $\mathfrak{B}$ über $\mathbb{R}$ können diese Aussagen nicht alle gelten; denn sie besagen zusammen, daß es ein unendlich kleines Element des Grundkörpers gibt. Somit ist $\mathfrak{A}_x$ nicht elementar äquivalent zu einem solchen $\mathfrak{B}$.

(ii) Für die Strukturklassen der Form $C^{\mathfrak{K}}_{\alpha,\beta}$, die man in 7.22 verwenden kann, werden in Szczerba u. Tarski 1965/79 zahlreiche Beispiele behandelt (darunter auch affine Ebenen, Halbebenen, offene Kreisscheiben, vgl. 7.11). Ein Zusatzaxiom für die zugehörige Theorie läßt sich dabei jeweils direkt angeben und ist wesentlich einfacher als das aus dem Beweis von 7.22(i) gewonnene. Insbesondere das folgende Beispiel benutzen wir anschließend in Anwendungen.

Sei $Kp^{\leq n}(\mathfrak{K})$ bzw. $Kp^n(\mathfrak{K})$ die Klasse der offenen konvexen Polygone mit höchstens n bzw. genau n Ecken über dem reell-abgeschlossenen Körper $\mathfrak{K}$ (d.h. in $\mathfrak{A}_2^B(\mathfrak{K})$) ($3\leq n\in\mathbb{N}$); dabei werde unter einem offenen konvexen Polygon das Innere eines (als Streckenzug oder als konvexe Hülle definierten) konvexen Polygons verstanden, wobei dieses Innere nicht leer ist. Jede solche Klasse läßt sich in der Form 7.20(2) darstellen. Sei dann $Po^{\leq n}$ bzw. Po^{n} die Theorie der dadurch bestimmten Klasse 7.20(4) von Strukturen (die Angabe von $\mathfrak{f}$ ist hierbei nach 7.22(v) entbehrlich). Aus 7.13 erhält man ohne Schwierigkeiten, daß $Po^{\leq n}$ aus GA_2 durch Hinzunahme des folgenden Zusatzaxioms χ_n entsteht:

(1) $\exists t \exists u_1 \ldots \exists u_n \forall x \forall y \exists z \{ \bigwedge_{\nu=1}^{n} u_\nu \neq t \wedge [Bxyz \vee Byzx \vee Bzxy] \wedge \bigvee_{\nu=1}^{n} Bu_\nu tz \}$

(es gibt n von t ausgehende Halbgeraden, so daß jede Gerade eine dieser Halbgeraden oder den Punkt t trifft).

Für Po^{n} läßt sich entsprechend das Zusatzaxiom $\chi_n \wedge \neg \chi_{n-1}$ verwenden.

<u>7.26 Satz</u> (Szczerba u. Tarski 1965/79). *Sei* $3 \leq n \in \mathbb{N}$ *und* η *eine Aussage von* L(B) *vom Präfixtyp* $\forall^n \exists$ *(6.2) d.h. von der Form* $\forall x_1 \ldots \forall x_k \exists y_1 \ldots \exists y_l \alpha(x_1, \ldots, y_l)$, *wobei* $k \leq n$, $l \in \mathbb{N}$ *beliebig und* α *quantorenfrei. Dann ist* η *ein Satz von* GA_2 *genau dann, wenn es ein Satz von* $Po^{\leq n}$ *ist. (Siehe auch* 7.34(v).*)*

<u>Beweis</u>: Die Richtung ($\rightarrow$) ergibt sich unmittelbar daraus, daß $Po^{\leq n}$ eine Erweiterung von GA_2 ist. Zum Beweis der Umkehrung ($\leftarrow$) sei $\mathfrak{A} = \mathfrak{A}_2^B(\mathfrak{f})|S$ ein (bis auf Isomorphie, 7.7(i)) beliebiges Modell von GA_2, und seien darin $x_1, \ldots, x_k$ beliebige Punkte. Da S konvex und offen ist, gibt es ein offenes konvexes Polygon mit höchstens n Ecken, in dem $x_1, \ldots, x_k$ (als innere Punkte!) liegen und das noch ganz in S enthalten ist; sei P ein solches und $\mathfrak{B} = \mathfrak{A}_2^B(\mathfrak{f})|P$. Nun ist $\mathfrak{B}$ ein Modell von $Po^{\leq n}$. Damit ist η in $\mathfrak{B}$ gültig, d.h., es ist $\models_{\mathfrak{B}} \alpha[x_1, \ldots, x_k, y_1, \ldots, y_l]$ für gewisse Punkte y_λ von $\mathfrak{B}$. Wegen $\mathfrak{B} \subseteq \mathfrak{A}$ ist dann auch $\models_{\mathfrak{A}} \alpha[x_1, \ldots, x_k, y_1, \ldots, y_l]$. Die Existenz solcher Punkte y_λ war gerade zu zeigen. □

Im Unterschied zu 7.16 ergibt sich daraus

<u>7.27 Satz</u> (Szczerba u. Tarski 1965/79). *Die Menge der Sätze von* GA_2 *vom Präfixtyp* $\forall\exists$ *ist entscheidbar.*

<u>Beweis</u>: Nach 7.22(iii) ist jede der Geometrien $Po^{\leq n}$ entscheidbar; der Beweis liefert darüber hinaus ein einheitliches Verfahren, mit dem man für jedes $n \geq 3$ und jede Aussage η feststellen kann, ob η ein Satz von $Po^{\leq n}$ ist. Damit ergibt sich das Gewünschte aus 7.26 (vgl. den Beweis von 3.78). □

Für $\forall$-Aussagen hat man statt 7.26 sogar

7.28 Satz (Szczerba u. Tarski 1965). *In je zwei widerspruchsfreien Erweiterungen von GA_2 stimmen die Sätze vom Präfixtyp ∀ überein. (Siehe auch 7.34(v).)*

Zum Beweis genügt

7.29 Satz (nach Szczerba u. Tarski 1979). *In beliebigen Strukturen der Form $\mathfrak{A}_2^B(\mathfrak{k})|S$, wobei $\mathfrak{k}$ reell-abgeschlossen und S eine nicht-leere, offene Teilmenge der affinen Ebene $\mathfrak{A}_2^B(\mathfrak{k})$, gelten dieselben ∀-Aussagen.*

Das ergibt sich aus der elementaren Äquivalenz der affinen Ebenen über (verschiedenen) reell-abgeschlossenen Körpern (3.65(i)) und dem folgenden Lemma.

7.30 Lemma. *Sei S eine nicht-leere, offene Teilmenge der affinen Ebene $\mathfrak{A}_2^B(\mathfrak{k})$ über dem angeordneten Körper $\mathfrak{k}$. Eine beliebige ∀-Aussage $\eta = \forall x_1 \ldots \forall x_k \alpha$ ist dann in $\mathfrak{A} = \mathfrak{A}_2^B(\mathfrak{k})|S$ gültig genau dann, wenn sie in $\mathfrak{A}_2^B(\mathfrak{k})$ gültig ist.*

Beweis: Die Richtung (←) ergibt sich unmittelbar aus der Unterstrukturbeziehung $\mathfrak{A} \subseteq \mathfrak{A}_2^B(\mathfrak{k})$ (6.12). Zum Beweis der Umkehrung (→) genügt die Feststellung, daß sich jede endliche Teilmenge $\{x_1,\ldots,x_k\}$ (der Menge der Punkte) von $\mathfrak{A}_2^B(\mathfrak{k})$ durch eine reguläre affine Abbildung (also einen Isomorphismus bezüglich B) überführen läßt in eine Teilmenge einer Umgebung eines beliebigen Punktes und damit in eine Teilmenge von S. □

7.31 Satz (Szczerba, s. 7.1(iv)). *Keine widerspruchsfreie Erweiterung von GA_2 (also auch nicht GA_2 selbst) besitzt ein ∀∃-Axiomensystem. (Siehe auch 7.32, 34(v).)*

Beweis: Sei T eine solche Erweiterung.

Fall 1: $T = Ago_2^*$ (ebene affine Geometrie über reell-abgeschlossenen Körpern). Dieser Fall wurde schon behandelt in 6.62, 63.

Fall 2: $T \neq Ago_2^*$. Als Erweiterung von GA_2 besitzt dann T ein Modell, das eine konvex-offene Ebene über einem reell-abgeschlossenen Körper $\mathfrak{k}$ (7.7(i)), aber nicht die ganze affine Ebene über $\mathfrak{k}$ ist. Nach dem Satz von Löwenheim-Skolem-Tarski (1.54) besitzt T auch ein Modell derselben Form (z.B. ein abzählbares), bei dem $\mathfrak{k}$ nicht isomorph zu $\mathbb{R}$ ist. Sei das Modell $\mathfrak{A} = \mathfrak{A}_2^B(\mathfrak{k})|S$ so gewählt. In $\mathfrak{A}$ sei $H(pa)$ eine Halbgerade, die einen eigentlichen Randpunkt r bestimmt (7.10(ii)). In $\mathfrak{A}_2^B(\mathfrak{F})$ sei ein affines Koordinatensystem eingeführt und seien Punkte mit Koordinatenpaaren identifiziert, so daß $p=\langle 0,0\rangle$ und $r=\langle 1,0\rangle$ ist. Seien nun X und Y die Unter- bzw. Oberklasse eines Dedekindschen Schnitts ohne Schnittelement in $\mathfrak{F}$ (vgl. 7.9(v)), und seien o.B.d.A. $0,1 \in X$. Für beliebiges t aus der Menge X^+ der positiven Elemente von X sei dann φ_t die affine Abbildung mit $\varphi_t(x)=t\cdot x$, weiter sei S_t das Bild von S bei der Abbildung φ_t und $\mathfrak{A}_t=\mathfrak{A}_2^B(\mathfrak{F})|S_t$. Somit ist jede solche Struktur $\mathfrak{A}_t$ isomorph zu $\mathfrak{A}$ und damit ebenfalls ein Modell von T. Außerdem ist $\mathfrak{A}_{t_1}$ Unterstruktur von $\mathfrak{A}_{t_2}$ für $0<t_1\leq t_2 \in X^+$. Somit ist $K=\{\mathfrak{A}_t \mid t\in X^+\}$ eine Kette von Modellen von T. Sei $\mathfrak{L}$ die Vereinigung dieser Kette. In $\mathfrak{L}$ betrachten wir nun wieder die von p ausgehende Halbgerade durch a. Diese besteht nach Konstruktion genau aus den Punkten der Form $\langle x,0\rangle$ mit $x\in X^+$, sie bestimmt also keinen Randpunkt von $\mathfrak{L}$. Nach 7.10(ii) ist also $\mathfrak{L}$ nicht Modell von GA_2 und damit auch nicht von T. Das war nach 6.9 gerade zu zeigen. □

7.32 Anmerkung. Der Spezialfall $T=GA_2$ wird in Szczerba u. Tarski 1979 behandelt. Der Beweis ist dann wesentlich einfacher: Für die Anwendung von 6.9 wird eine geeignete Kette von Halbebenen direkt angegeben; diese leistet auch für $T=GA_2^-$ (s. 7.34(v)) das Verlangte.

7.33 Definition. (i) Sei $C^{\mathbb{R}}$ die Klasse aller konvex-offenen Ebenen über $\mathbb{R}$. Mit GA_2^+ bezeichnen wir die Theorie dieser Klasse (1.35(ii)), d.h.

(1) $GA_2^+ = Th(C^{\mathbb{R}})$.

(ii) Mit GA_2^- werde die Theorie bezeichnet, die axiomatisch aufgebaut ist auf den Axiomen B1,..., B8 (7.2), dem Grenzparallelenaxiom Gp (6.27(6)) und dem Schema Σ_{fc} der Körperstetigkeitsaxiome (6.83), d.h.

(2) $GA_2^- = Cn(\{B1,\dots,B8,Gp\}\cup\Sigma_{fc})$.

7.34 Anmerkungen. (i) GA_2^+ ist jedenfalls eine Obertheorie von GA_2 (nach 7.7(ii)) und GA_2^- eine Untertheorie von GA_2 (nach 7.10(i), 6.86(ii)).

(ii) Die Theorie GA_2^+ wurde schon in Szczerba u. Tarski 1965 diskutiert; die Frage, ob sie gleich GA_2 ist, war damals noch offen. Der folgende Satz 7.35 zeigt, daß das nicht der Fall ist. Damit hat jedenfalls die Klasse $C^{\mathbb{R}}$ nicht die in Problem 7.8(iii) gewünschte Eigenschaft.

(iii) Aus dem Beweis von 7.7(i) (vgl. 7.9(i)) ergibt sich, daß auch jedes Modell von GA_2^- eine konvex-offene Ebene über einem reell-abgeschlossenen Körper ist. Aus dem Grenzparallelenaxiom Gp ergibt sich, daß in diesen Modellen auch die Randpunkte gemäß 7.10(ii) existieren.

(iv) Trotzdem ist GA_2^- eine echte Untertheorie von GA_2 (Prestel u. Szczerba, s. 7.1(iv)). Das ergibt sich hier aus Satz 7.43. Es läßt sich auch zeigen durch Konstruktion eines Modells von GA_2^-, das in einem Randpunkt r keine Stützgerade besitzt (vgl. 7.10(iii)). Dazu kann man (wie in 7.9(v)) einen Dedekindschen Schnitt ohne Schnittelement verwenden und die Steigungen der Stützgeraden auf einer Seite von r in der Unterklasse und auf der anderen Seite in der Oberklasse dieses Schnitts geeignet wählen (Übung!).

(v) Die Sätze 7.26, 28, 31 gelten sogar für GA_2^- an Stelle von GA_2. Die Beweise verlaufen (mit Benutzung von (iii)) völlig analog.

(vi) Unter den Axiomen von GA_2^- ist Gp das einzige, das nicht vom Präfixtyp $\forall\exists$ ist. Es läßt sich auch nicht durch ein Axiom dieses Typs ersetzen (wegen 7.31, s. (v)).

7.35 Satz (Prestel u. Szczerba 1979, s. 7.1(iv)). *Die Theorie* GA_2^+ *ist nicht rekursiv axiomatisierbar* (3.9(ii)). *Damit ist sie insbesondere eine echte Obertheorie der Theorie* GA_2, *die ja nach Definition ein rekursives Axiomensystem besitzt.*

Zum Beweis: Zunächst ist die Theorie Th($\mathcal{T}_{\mathbb{N}}^{\mathbb{R}}$) der Standardstruktur über $\mathbb{R}$ unentscheidbar (das ergibt sich wie im Beweis von 7.16 aus 7.18). Außerdem ist diese Theorie vollständig (s. 1.39, Beisp.) und somit nach 3.13 nicht rekursiv axiomatisierbar. Da aus einer rekursiven Menge (hier einem Axiomensystem) durch Hinzunahme von endlich vielen Elementen stets wieder eine rekursive Menge entsteht, genügt der Beweis des folgenden Satzes.

7.36 Satz (Prestel u. Szczerba s.o.). *Die Theorie* Th($\mathcal{T}_{\mathbb{N}}^{\mathbb{R}}$) *der Standardstruktur über* $\mathbb{R}$ *ist eine endliche Erweiterung* (1.36) *von* GA_2^+.

Zum Beweis: Mit den in 7.9(i), (iii) und 7.13 genannten Methoden (geometrische Einführung eines Körpers und eines Koordinatensystems sowie Beschreibung der umgebenden Ebene und des Randes) läßt sich eine Aussage η von L(B) konstruieren, deren Gültigkeit in einem beliebigen Modell $\mathfrak{A}$ von GA_2 gleichwertig ist mit folgenden Bedingungen für ein geeignet gewähltes Koordinatensystem:

(1) Die Punkte a, b, c, d, f, g_o mit den in 7.17 angegebenen Koordina-

ten sind Ecken von $\mathfrak{A}$, die Randstücke dazwischen (in dieser Reihenfolge) sind geradlinig;

(2) die Koordinaten aller weiteren Ecken genügen (ebenfalls) der (Kreis-)Gleichung $x^2+y^2=4$;

(3) für die Punkte o, e, u, t mit den in 7.17 angegebenen Koordinaten und beliebige Punkte x, x' gilt: Ist $\mathrm{Su}_{oeut}xex'$ (d.h. $x'=x+1$ im Sinne der Punktrechnung in bezug auf o, e, u, t) und bestimmt die Halbgerade $\mathrm{H}(tx)$ eine Ecke p von $\mathfrak{A}$, so bestimmt $\mathrm{H}(tx')$ wieder eine Ecke p', und zwar die "nächste", d.h., das Randstück zwischen p und p' ist geradlinig.

Ist nun $\mathfrak{k}$ ein reell-abgeschlossener archimedisch angeordneter Körper und $\mathfrak{A}$ eine konvex-offene Ebene über $\mathfrak{k}$, so ist η in $\mathfrak{A}$ gültig genau dann, wenn $\mathfrak{A}$ isomorph zur Standardstruktur $\gamma_{\mathbb{N}}^{\mathfrak{k}}$ ist. Insbesondere wird die Standardstruktur $\gamma_{\mathbb{N}}^{\mathbb{R}}$ innerhalb der Klasse $C^{\mathbb{R}}$ durch η bis auf Isomorphie charakterisiert. Somit ist $\mathrm{Th}(\gamma_{\mathbb{N}}^{\mathbb{R}})$ die endliche Erweiterung, die aus GA_2^+ durch Hinzunahme des Axioms η entsteht, d.h. $\mathrm{Cn}(GA_2^+\cup\{\eta\}) = \mathrm{Th}(\gamma_{\mathbb{N}}^{\mathbb{R}})$; denn für eine beliebige Formel α von $\mathrm{L(B)}$ sind die folgenden Bedingungen untereinander gleichwertig:

(4) $GA_2^+\cup\{\eta\} \models \alpha$;

(5) $\eta\to\alpha$ ist ein Satz von GA_2^+;

(6) $\eta\to\alpha$ ist gültig in jeder Struktur der Klasse $C^{\mathbb{R}}$;

(7) α ist gültig in $\gamma_{\mathbb{N}}^{\mathbb{R}}$. □

7.37 Anmerkung und Definition. Auf Grund der in 7.35 festgestellten Verschiedenheit der Theorien GA_2 und GA_2^+ muß es Aussagen η geben, *durch die sich diese Theorien unterscheiden lassen*, d.h., daß η ein Satz von GA_2^+, aber nicht von GA_2 ist ($\eta\in GA_2^+-GA_2$). Es erhebt sich die Frage, wie kompliziert solche Aussagen sein müssen. Der Beweis von 7.35 gibt darüber keine Auskunft; er wurde ja gar nicht durch Angabe einer unterscheidenden Aussage geführt. Von Prestel und Szczerba wurde inzwischen eine unterscheidende Aussage vom Präfixtyp $\forall\exists\forall$ gefunden, in einer erweiterten Sprache mit Individuenkonstanten (für die Punkte eines Bezugssystems) auch eine vom Typ $\exists\forall\exists$ (s. Prestel u. Szczerba 1979, dabei wurde statt $\gamma_{\mathbb{N}}^{\mathfrak{k}}$ ein anderes Modell verwendet mit einem Rand aus Stücken von Tangenten, die den Kreis K gerade in den Punkten von $E_{\mathbb{N}}^{\mathfrak{k}}$ [7.17] berühren). Der folgende Satz zeigt, daß diese Ergebnisse in einem gewissen Sinne optimal sind, nämlich, daß die Präfixtypen $\forall\exists$ und $\exists\forall$ sowie einige speziellere Typen noch nicht zur Unterscheidung geeignet sind.

7.38 Satz (Prestel u. Szczerba 1979 für $\forall^3\exists$ und $\forall^2\exists\forall$; Szczerba, s. 7.1(iv), für $\exists^1\forall^2\exists$). *Die Theorien* GA_2 *und* GA_2^+ *lassen sich nicht unterscheiden durch Aussagen der Präfixtypen* $\forall\exists$, $\exists^1\forall^2\exists$ *und* $\forall^2\exists\forall$.

Beweis: (i) Für den Typ $\forall\exists$: Für $3\leq n\in\mathbb{N}$ läßt sich die Theorie $Po^{\leq n}$ der offenen konvexen Polygone mit höchstens n Ecken darstellen als die Theorie dieser Polygone über $\mathbb{R}$ in der Form $Th(C^{\mathbb{R}}_{\alpha,\beta})$ (7.25(ii)) und ist somit (wegen $C^{\mathbb{R}}_{\alpha,\beta}\subseteq C^{\mathbb{R}}$) eine Obertheorie von GA_2^+. Damit ergibt sich die Behauptung unmittelbar aus 7.26.

Für den nächsten Typ benutzen wir folgendes Lemma.

7.39 Lemma. *Sei* $\mathfrak{f}$ *ein angeordneter Körper. Für* $i=1,2$ *sei* S_i *ein offenes konvexes Dreieck (d.h.* $S_i\in Kp^3(\mathfrak{f})$, 7.25(ii)) *in der affinen Ebene* $\mathfrak{A}_2^B(\mathfrak{f})$, $a_i\in S_i$ *und* $\mathfrak{A}_i=\mathfrak{A}_2^B(\mathfrak{f})|S_i$. *Dann gibt es einen Isomorphismus* φ *von* $\mathfrak{A}_1$ *auf* $\mathfrak{A}_2$, *der* a_1 *in* a_2 *überführt.*

Beweis: Zunächst gibt es eine projektive Transformation, die die Ecken von S_1 in die Ecken von S_2 (sogar in einer vorgegebenen Reihenfolge) und a_1 in a_2 überführt, sei ψ eine solche. Mit Hilfe der Charakterisierung 6.60(2) der Zwischenbeziehung (durch die Trennbeziehung und die uneigentliche Gerade) erhält man leicht, daß die uneigentliche Gerade wieder in eine ganz außerhalb von S_2 liegende Gerade übergeführt wird und daß die Einschränkung φ von ψ auf S_1 ein Isomorphismus (bezüglich B) der gewünschten Art ist. □

7.40 Korollar. *Ist* $\mathfrak{A}$ *eine Dreiecksstruktur wie in* 7.39 *und ist darin eine beliebige Aussage der Form* $\exists x\gamma$ *gültig, so ist auch* $\forall x\gamma$ *in* $\mathfrak{A}$ *gültig (hierbei darf die Formel* $\gamma=\gamma(x)$ *weitere Quantoren enthalten).*

Beweis: Sei x ein (nach Vor. existierender) Punkt mit $\models_{\mathfrak{A}}\gamma[x]$ und sei $y\in\mathfrak{A}$ beliebig. Sei dann φ gemäß 7.39 ein Isomorphismus von $\mathfrak{A}$ auf sich, der x in y überführt. Dann ist auch $\models_{\mathfrak{A}}\gamma[y]$. □

Beweis von 7.38 (Fortsetzung): (ii) Für den Typ $\exists^1\forall^2\exists$: Sei $\eta=\exists x\gamma$ eine beliebige Aussage dieses Typs, die Satz von GA_2^+ ist. Dann ist η insbesondere gültig in jeder Dreiecksstruktur über $\mathbb{R}$. Nach 7.40 ist sogar $\eta'=\forall x\gamma$ in jeder solchen Struktur gültig, d.h. ein Satz von $Po^{\leq 3}$. Nach Voraussetzung ist η' vom Präfixtyp $\forall^3\exists$, also nach 7.26 schon ein Satz von GA_2. Damit ist erst recht η ein Satz von GA_2, d.h., GA_2 und GA_2^+ lassen sich nicht durch η unterscheiden.

(iii) Für den Typ $\forall^2\exists\forall$: Annahme: η ist eine Aussage dieses Typs und $\eta \in GA_2^+ - GA_2$. Wegen $\eta \notin GA_2$ ist dann $\models_{\mathfrak{A}} \neg\eta$ für ein gewisses Modell $\mathfrak{A} = \mathfrak{A}_2^B(\mathfrak{f})|S$ von GA_2 (7.7(i)). Die Aussage $\neg\eta$ ist (nach 1.60(6), (7)) logisch äquivalent zu einer pränexen Aussage η' der Form $\exists x \exists y \gamma(x,y)$, wobei $\gamma(x,y)$ eine Formel vom Präfixtyp $\forall\exists$ ist. Seien a, b entsprechende Punkte von $\mathfrak{A}$, d.h. $\models_{\mathfrak{A}} \gamma[a,b]$. Sei G eine Gerade von $\mathfrak{A}$ durch die Punkte a, b (eindeutig bestimmt, wenn $a \neq b$). Wir zerlegen G in zwei Halbgeraden und betrachten die von diesen Halbgeraden bestimmten Randpunkte r und r' von S (7.10(ii)). Nach 7.9(i) läßt sich eine Stützgerade von S durch r' (7.10(iii)) als die uneigentliche Gerade U verwenden; dementsprechend sei o.B.d.A.

(1) $r' \in U$.

In der affinen Ebene $\mathfrak{A}_2^B(\mathfrak{f})$ sei nun ein affines Koordinatensystem eingeführt und seien Punkte x mit den zugehörigen Koordinatenpaaren $\langle x_1, x_2 \rangle$ identifiziert, so daß folgende Bedingungen gelten:

(2) G ist enthalten in der ersten Koordinatenachse L_1.

(3) Wenn r nicht auf U liegt, so ist die zweite Koordinatenachse L_2 eine Stützgerade von S durch r (7.10(iii)) und G die "positive Halbachse" $\{x \mid x_1 > 0,\ x_2 = 0\}$ (Abb. 167).

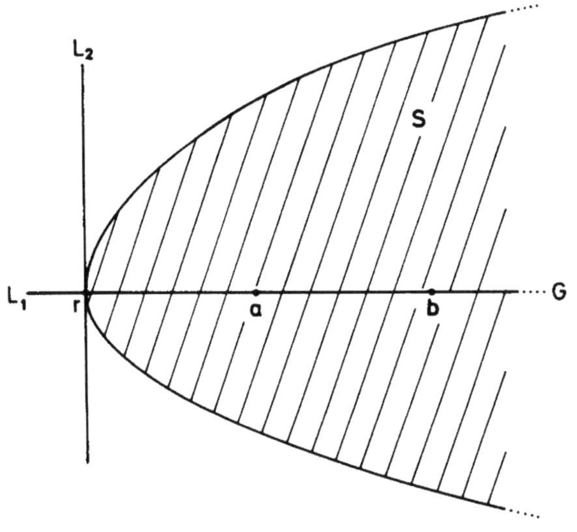

Abb. 167

Wenn r (ebenso wie r') auf U liegt (d.h. $r = r'$), muß G somit die ganze erste Koordinatenachse L_1 sein.

Sei $\bar{L}$=L(B,a,b) eine Sprache mit zwei zusätzlichen Individuenkonstanten a, b. Für beliebige positive Elemente t von $\mathfrak{f}$ sei φ_t die affine Abbildung mit $\varphi_t(x)=\langle x_1, t\cdot x_2\rangle$, weiter sei S_t das Bild von S bei der Abbildung φ_t, $\mathfrak{A}_t = \mathfrak{A}_2^B(\mathfrak{f})|S_t$ und $\bar{\mathfrak{A}}_t$ die Struktur für $\bar{L}$, die aus $\mathfrak{A}_t$ entsteht, indem die zusätzlichen Konstanten a, b als die Punkte a bzw. b interpretiert werden. Da φ_t die erste Koordinatenachse und damit insbesondere die Punkte a, b fest läßt, ist jede solche Struktur $\bar{\mathfrak{A}}_t$ isomorph zu $\bar{\mathfrak{A}}_1$. Außerdem ist $\bar{\mathfrak{A}}_{t_1}$ Unterstruktur von $\bar{\mathfrak{A}}_{t_2}$ für $t_1 \le t_2$. Somit ist

$K=\{\bar{\mathfrak{A}}_t \mid t>0,\ t\in\mathfrak{f}\}$ eine Kette von Strukturen für $\bar{L}$. Nach Konstruktion ist die $\forall\exists$-Aussage $\gamma(a,b)$ von $\bar{L}$ gültig in $\bar{\mathfrak{A}}_1$ (wegen $\mathfrak{A}_1 = \mathfrak{A}$), auf Grund der festgestellten Isomorphie also auch in jedem $\bar{\mathfrak{A}}_t$ und nach 6.9 somit auch in der Vereinigung $\bar{\mathfrak{L}}$ der Kette K. Damit ist die Aussage $\eta' = \exists x \exists y \gamma(x,y)$ und folglich auch $\neg\eta$ gültig im Redukt (1.20) $\mathfrak{L}$ von $\bar{\mathfrak{L}}$ auf die Sprache L(B), das ist die Struktur $\mathfrak{A}_2^B(\mathfrak{f})|U_{\mathfrak{L}}$ mit der Trägermenge $U_{\mathfrak{L}} = \cup\{S_t \mid t>0,\ t\in\mathfrak{f}\}$. Wegen der Offenheit von S gibt es zu jedem Punkt x auf G auch Punkte x', x'' von S mit derselben ersten Koordinate auf beiden Seiten von G. Somit ist $U_{\mathfrak{L}}$ die ganze Ebene (im Falle $r\in U$) oder die Halbebene $\{\langle x_1,x_2\rangle \mid x_1>0\}$ (im Falle $r\notin U$). In jedem Falle ist $\mathfrak{L}$ ein (besonders einfach) definierbares Modell im Sinne von 7.20, nach 7.24 also Modell einer Theorie der Form $\mathrm{Th}(C^{\mathbb{R}}_{\alpha,\beta})$ und erst recht Modell der Untertheorie $GA_2^+ = \mathrm{Th}(C^{\mathbb{R}})$ (vgl. (i)). Nach Voraussetzung ist also η gültig in $\mathfrak{L}$, was zusammen mit der Gültigkeit von $\neg\eta$ den gewünschten Widerspruch ergibt. □

<u>7.41 Anmerkungen</u>. (i) Ebenso wie die benutzten Sätze (vgl. 7.34(v)) lassen sich auch die Beweise (i), (ii) von 7.38 für die Präfixtypen $\forall\exists$ und $\exists^1\forall^2\exists$ von GA_2 auf GA_2^- übertragen. Dagegen läßt sich der Beweis (iii) für den Präfixtyp $\forall^2\exists\forall$ nicht übernehmen; denn darin wurde die Existenz von Stützgeraden verwendet, die für beliebige Modelle von GA_2^- nicht mehr zur Verfügung steht (vgl. 7.34(iv)). Man erhält aber zumindest einen entsprechenden Beweis für den Präfixtyp $\forall^1\exists\forall$, wenn man den Punkt a (statt a und b in (iii)) als Koordinatenursprung wählt und zur Konstruktion der Kette K die Abbildungen φ_t mit $\varphi_t(x)=t\cdot x$ verwendet. Damit ergibt sich

(ii) Die Sätze von den Präfixtypen $\forall\exists$, $\exists^1\forall^2\exists$ und $\forall^1\exists\forall$ sind in allen drei Theorien GA_2^-, GA_2 und GA_2^+ dieselben.

<u>7.42 Problem</u>. Man konstruiere eine Aussage η von einem möglichst einfachen Präfixtyp, durch die sich GA_2^- und GA_2 unterscheiden lassen.

7.43 Satz (Szczerba, s. 7.1(iv)). (i) *Für jede Menge* Θ *von* $\forall\exists$*-Aussagen ist*

(1) $Cn(GA_2^- \cup \Theta) \neq GA_2$.

(ii) *Für jede endliche Menge* Θ *von* $\exists\forall$*-Aussagen gilt ebenfalls* (1).

7.44 Anmerkung. Für eine beliebige Menge Θ von Aussagen ist $Cn(GA_2^- \cup \Theta)$ die Theorie, die aus GA_2^- durch Hinzunahme der Aussagen aus Θ als "Zusatzaxiome" entsteht. 7.43 besagt also, daß man die Theorie GA_2 aus GA_2^- nicht durch Hinzunahme von Zusatzaxiomen eines genügend einfachen Präfixtyps erhalten kann (vgl. auch 7.54(ii)). Insbesondere ergibt sich aus jeder der beiden Behauptungen (i), (ii) die Verschiedenheit der Theorien GA_2^- und GA_2. Hat man diese Verschiedenheit schon auf irgendeinem Wege gezeigt (vgl. 7.34(iv)), so kann man mit Hilfe von 7.41(ii) sogar die folgende Verschärfung von 7.43 erhalten:

(*) Sei Θ eine (endliche oder unendliche) Menge von Aussagen der Sprache L(B), in der jedes Element von einem der Präfixtypen $\forall\exists$, $\exists^1\forall^2\exists$, $\forall^1\exists\forall$ ist. Dann ist

$$Cn(GA_2^- \cup \Theta) \neq GA_2.$$

Insbesondere ist in 7.43(ii) die Voraussetzung entbehrlich, daß Θ endlich ist ($\exists\forall$-Aussagen sind ja auch vom komplizierteren Präfixtyp $\forall^1\exists\forall$).

Der folgende direkte Beweis für den älteren Satz 7.43 erscheint trotzdem interessant wegen der verwendeten Methode und der dabei erzielten zusätzlichen Resultate. Für diesen Beweis (s. hinter 7.49 und 7.53) beweisen wir zunächst zwei Lemmata.

7.45 Lemma (Szczerba). *Vor.:* (i) $\mathfrak{k}$ *ist ein reell-abgeschlossener* *archimedisch angeordneter Körper (d.h. isomorph zu einem Unterkörper von* $\mathbb{R}$*).*

(ii) Θ *ist eine Menge von* $\forall\exists$*-Aussagen mit* $\Theta \subseteq GA_2$ *(d.h. eine Menge von Sätzen von* GA_2*).*

Beh.: Die Standardstruktur $\gamma_{\mathbb{N}}^{\mathfrak{k}}$ *(7.17) ist ein Modell von* $GA_2^- \cup \Theta$.

7.46 Anmerkung. Die Voraussetzung, daß die Anordnung archimedisch ist, ist hierbei nicht entbehrlich (vgl. die Bemerkung über nicht-archimedische Anordnungen im Beweis von 7.18).

Beweis von 7.45: In der affinen Ebene $\mathfrak{A}_2^B(\mathfrak{k})$, in der Punkte mit Koordinatenpaaren identifiziert sind, sei S_n das offene konvexe Polygon mit den $n+6$ Ecken $a, b, c, d, f, g_0, \ldots, g_n$ von 7.17 und $\mathfrak{A}_n = \mathfrak{A}_2^B(\mathfrak{F})|S_n$ $(n \in \mathbb{N})$. Nach 7.25(ii) ist jede der Strukturen $\mathfrak{A}_n$ ein Modell von GA_2. Damit ist $K=\{\mathfrak{A}_n | n \in \mathbb{N}\}$ eine Kette von solchen Modellen mit der Vereinigung $\mathfrak{P}_{\mathbb{N}}^{\mathfrak{F}}$. Die Elemente von Θ und die Axiome von GA_2^- sind als Sätze von GA_2 in jedem $\mathfrak{A}_n$ gültig. Aus 6.9 ergibt sich die Gültigkeit dieser Aussagen in der Vereinigung $\mathfrak{P}_{\mathbb{N}}^{\mathfrak{F}}$, soweit sie vom Präfixtyp $\forall\exists$ sind. Das einzige Axiom, das nicht von diesem Typ ist, ist das Axiom Gp über die Existenz von Grenzparallelen (vgl. 7.34(iv)). Es bleibt also zu zeigen, daß auch Gp in $\mathfrak{P}_{\mathbb{N}}^{\mathfrak{F}}$ gültig ist, das bedeutet (vgl. 7.10), daß jede Halbgerade in $\mathfrak{P}_{\mathbb{N}}^{\mathfrak{F}}$ einen Randpunkt bestimmt. Die Verlängerung einer solchen Halbgeraden (in $\mathfrak{A}_2^B(\mathfrak{F})$) schneidet aber ein geradliniges Randstück zwischen gewissen benachbarten Ecken von $\mathfrak{P}_{\mathbb{N}}^{\mathfrak{F}}$ (die für alle hinreichend großen n schon benachbarten Ecken von $\mathfrak{A}_n$ sind), d.h., die Halbgerade bestimmt tatsächlich einen Randpunkt. □

7.47 Definition. Eine reelle Zahl x sei mit Hilfe einer Basis b ($b \in \mathbb{N}$, $b \geq 2$, z.B. $b=10$ oder $b=2$) in der üblichen Weise als b-adischer Bruch dargestellt, d.h. in der Form

$$(1) \quad x = \pm \sum_{\nu=0}^{\infty} \frac{z_\nu}{b^\nu}, \quad \text{m.a.W.} \quad x = \pm \lim_{n\to\infty} x_n \quad \text{mit} \quad x_n = \sum_{\nu=0}^{n} \frac{z_\nu}{b^\nu},$$

wobei $z_\nu \in \mathbb{N}$, $z_\nu \leq b-1$ für alle $\nu \geq 1$, aber nicht $z_\nu = b-1$ für alle ν oberhalb eines gewissen n.

Dann heißt x *berechenbar* oder *rekursiv*, falls die (durch x und b eindeutig festgelegte) zahlentheoretische Funktion f mit $f(\nu)=z_\nu$ $(\nu \in \mathbb{N})$ rekursiv (3.3) ist.

Diese Definition erweist sich als unabhängig von der Wahl der Basis b; in der Literatur sind auch andere, gleichwertige Definitionen üblich (s. etwa Hermes 1978, §36).

Beispiele für berechenbare reelle Zahlen sind alle reellen algebraischen Zahlen, außerdem die transzendenten Zahlen π und e sowie e^x, falls x schon berechenbar ist (das läßt sich zeigen mit Hilfe der aus der Analysis bekannten Berechnungsmethoden).

7.48 Lemma (Szczerba). *Vor.: $\mathfrak{F}$ ist ein reell-abgeschlossener Unterkörper von $\mathbb{R}$ und $\mathfrak{P}_{\mathbb{N}}^{\mathfrak{F}}$ ist ein Modell von GA_2.*

Beh.: (i) *In $\mathfrak{F}$ liegen alle berechenbaren reellen Zahlen.*

(ii) (Folgerung) *$\mathfrak{F}$ hat einen unendlichen Transzendenzgrad über $\mathbb{Q}$.*

<u>Beweis</u> (Skizze): Zu (i): Seien x und f gegeben wie in 7.47 und o.B.d.A. $x>0$. Aus Sätzen über rekursive Funktionen (s. etwa Hermes 1978, §29, Beh. (3)) erhält man zunächst eine Formel $\delta(n,z)$, durch die die Funktion f in der Struktur $\langle\mathbb{N},+,\cdot\rangle$ der natürlichen Zahlen definiert wird. Diese Struktur ist nach 7.18 im Modell $\mathfrak{r}^{\mathfrak{f}}_{\mathbb{N}}$ definierbar, und zwar (bis auf Isomorphie) als Unterstruktur des dort geometrisch konstruierten Halbkörpers $\mathfrak{g}^{+}$, der (nach 6.80) zu $\mathfrak{f}^{+}$ isomorph ist. Mit diesen Mitteln läßt sich δ in eine entsprechende Formel δ' der Sprache L(B) übersetzen. Damit kann man dann in $\mathfrak{g}^{+}$ einen Schnitt definieren, der die Zahl x (bzw. den beim betrachteten Isomorphismus entsprechenden Punkt von $\mathfrak{g}^{+}$) festlegt; zur Definition der Unterklasse kann man nämlich eine Formel $\alpha(u)$ konstruieren (vgl. 7.49), die ausdrückt, daß u kleiner-gleich einem Näherungsbruch x_n (s. 7.47(1)) für hinreichend großes n ist (alle diese Näherungsbrüche sind rational und liegen damit auf jeden Fall in $\mathfrak{f}$). Das zugehörige Stetigkeitsaxiom des Schemas B9 liefert dann, daß auch x in $\mathfrak{f}$ liegt.

Zu (ii): Nach einem Satz von Lindemann (s. etwa Le Veque 1956, Corollary 3, S. 186), sind die Zahlen $e^{c_1},\ldots, e^{c_n}$ stets algebraisch unabhängig über $\mathbb{Q}$ (und über dem Körper der algebraischen Zahlen), wenn nur $c_1,\ldots, c_n$ algebraische Zahlen und linear unabhängig über $\mathbb{Q}$ sind. Wählt man also eine abzählbare und über $\mathbb{Q}$ linear unabhängige Menge M von reellen algebraischen Zahlen, z.B.

(1) $M=\{c_\nu \mid \nu\in\mathbb{N}\}$ mit $c_\nu = \sqrt[2^\nu]{2}$,

so ist $N=\{e^{c_\nu} \mid \nu\in\mathbb{N}\}$ eine abzählbare und algebraisch unabhängige Menge, die nach den Beispielen zu 7.47 aus berechenbaren reellen Zahlen besteht und damit wegen (i) in $\mathfrak{f}$ enthalten ist. □

<u>7.49 Anmerkung</u>. Als Verschärfung von 7.48 wurde inzwischen die Gleichwertigkeit der folgenden beiden Bedingungen gefunden (Schwabhäuser 1979b, s.a. 7.59):

(i) $\mathfrak{r}^{\mathfrak{f}}_{\mathbb{N}}$ ist ein Modell von GA_2.

(ii) $\mathfrak{f}$ ist ein Modell der Theorie R_2 in der schwachen zweiten Stufe, die aufgebaut ist auf den Axiomen für angeordnete Körper und dem Schema der Stetigkeitsaxiome (mit Unter- und Oberklassen, die sich durch Formeln im Rahmen der schwachen zweiten Stufe definieren lassen).

Die Theorie R_2 wurde untersucht in Schwabhäuser 1979. Dort wurde vor allem die Nicht-Finitisierbarkeit von R_2 (und von R_N, s. (iii)) gezeigt (s.a. 8.22). Als Nebenresultat wurden die Gleichwertigkeit von (ii) mit den folgenden Bedingungen (iii) und (iv) und die anschließend angegebenen Folgerungen behandelt. Dabei wurden auch Techniken bereitgestellt (s. dort, 4.23, S. 98), die zur Konstruktion der Formel $\alpha(u)$ aus dem Beweis von 7.48(i) führen

Die folgenden beiden Bedingungen sind ebenfalls gleichwertig mit (i) und (ii):

(iii) $\mathfrak{F}$ ist ein Modell der Theorie R_N, die aus R_2 dadurch entsteht, daß nur Formeln erster Stufe verwendet werden, aber dafür ein Prädikat N für die Eigenschaft, eine natürliche Zahl zu sein (mit Standardinterpretation).

(iv) Eine geeignete "Übersetzung" $s(\mathfrak{F}) \subseteq \mathfrak{P}(\omega)$ von $\mathfrak{F}$ ist ein Modell der Arithmetik A_ω zweiter Stufe im Sinne von Grzegorczyk, Mostowski u. Ryll-Nardzewski 1958.

Als Folgerungen aus jeder der Bedingungen (i) bis (iv) ergeben sich:

(v) $\mathfrak{F}$ ist archimedisch angeordnet, also isomorph zu einem Unterkörper von $\mathbb{R}$, und reell-abgeschlossen.

(vi) (Verschärfung von 7.48(i)) In $\mathfrak{F}$ liegen alle sog. hyperarithmetischen reellen Zahlen.

(vii) (ohne Voraussetzung der Kontinuumhypothese) Zu jeder Mächtigkeit κ mit $\aleph_0 \leq \kappa \leq 2^{\aleph_0}$ gibt es wenigstens eine Struktur $\mathfrak{F}$ der Mächtigkeit κ mit den Eigenschaften (i) bis (iv); insbesondere gibt es abzählbare Strukturen dieser Art (vgl. 1.57(i)).

(viii) Der Durchschnitt aller Strukturen $\mathfrak{F}$ mit den Eigenschaften (i) bis (iv) - aufgefaßt als Unterstrukturen von $\mathbb{R}$ - ist der Körper $\mathfrak{F}_H$ der hyperarithmetischen reellen Zahlen. $\mathfrak{F}_H$ selbst hat jedoch nicht die Eigenschaften (i) bis (iv).

Eine genauere Charakterisierung der (von $\mathbb{R}$ verschiedenen) Strukturen mit den Eigenschaften (i) bis (iv) und ebenso der Modelle von A_ω ist von großen Interesse; bisher ist darüber wenig bekannt.

<u>Beweis von 7.43(i)</u>: Sei Θ eine Menge von $\forall\exists$-Aussagen und o.B.d.A. $\Theta \subseteq GA_2$ (sonst gilt 7.43(1) ohnehin). Sei $\mathfrak{F}$ ein reell-abgeschlossener Unterkörper von $\mathbb{R}$, der nicht alle berechenbaren reellen Zahlen enthält, z.B. der Körper der reellen algebraischen Zahlen. Nach 7.45 ist dann $\gamma^{\mathfrak{F}}_{\mathbb{N}}$ ein Modell von $GA_2^- \cup \Theta$ und damit auch der darauf aufgebauten Theorie $Cn(GA_2^- \cup \Theta)$.

Nach 7.48 ist $\gamma^{\mathfrak{f}}_{\mathbb{N}}$ dagegen kein Modell von GA_2. Damit ist 7.43(1) gezeigt. □

7.50 Anmerkung. Die soeben konstruierte Struktur $\gamma^{\mathfrak{f}}_{\mathbb{N}}$ besitzt selbstverständlich durch jeden Randpunkt (wenigstens) eine Stützgerade und hat damit die in 7.10(i) bis (iii) genannten Eigenschaften. Daraus ergibt sich, daß diese Eigenschaften jedenfalls nicht ausreichen zur Charakterisierung der Modelle von GA_2 (vgl. 7.8(i)).

Zum Beweis von 7.43(ii) benutzen wir noch die folgenden beiden Sätze aus der mathematischen Logik.

7.51 Lemma. *Eine Konjunktion ϑ endlich vieler Formeln vom Präfixtyp $\exists\forall$ ist stets logisch äquivalent zu einer einzigen Formel vom selben Typ, in der höchstens dieselben freien Variablen wie in ϑ vorkommen.*

Beweis: Sei $\vartheta = \bigwedge_{\nu=1}^{n} \vartheta_\nu$, wobei $\vartheta_\nu = \exists\mathfrak{y}_\nu \forall\mathfrak{x}_\nu \varepsilon_\nu$ mit quantorenfreien ε_ν (s. 6.2), und seien o.B.d.A. die Variablen in den Blöcken $\mathfrak{y}_1, \mathfrak{x}_1, \ldots, \mathfrak{y}_n, \mathfrak{x}_n$ paarweise verschieden und verschieden von den freien Variablen von ϑ (was sich durch gebundene Umbenennungen [vgl. 1.9] erreichen läßt). Mittels 1.60(10), (14) (Verschiebung von $\exists$) sowie (8), (12) (Verschiebung von $\forall$) erhält man dann die Äquivalenzen

(1) $\vartheta \;\text{Äq}\; \exists\mathfrak{y}_1 \ldots \exists\mathfrak{y}_n \bigwedge_{\nu=1}^{n} \forall\mathfrak{x}_\nu \varepsilon_\nu$

$\quad\;\text{Äq}\; \exists\mathfrak{y}_1 \ldots \exists\mathfrak{y}_n \;\forall\mathfrak{x}_1 \ldots \forall\mathfrak{x}_n \bigwedge_{\nu=1}^{n} \varepsilon_\nu$

und damit das Gewünschte.

Sind hierbei die Blöcke $\forall\mathfrak{x}_\nu$ alle von einer Länge $\leq i$, so läßt sich auch (durch gebundene Umbenennungen und ggf. Hinzunahme nicht benötigter Variablen gemäß 1.61(1)) erreichen, daß diese Blöcke untereinander gleich, etwa gleich $\forall\mathfrak{x}$, sind. Statt 1.60(8), (12) kann man dann die Äquivalenz

(2) $\forall x\gamma \wedge \forall x\delta \;\text{Äq}\; \forall x(\gamma \wedge \delta)$

verwenden (vgl. 6.21(3)) und erhält die Äquivalenz

(3) $\vartheta \;\text{Äq}\; \exists\mathfrak{y}_1 \ldots \exists\mathfrak{y}_n \;\forall\mathfrak{x} \bigwedge_{\nu=1}^{n} \varepsilon_\nu$

von ϑ mit einer einzigen Formel vom Typ $\exists\forall^i$ (kürzer als in (1)). □

Analog kann man offenbar für beliebige Präfixtypen verfahren, die nur aus Grundzeichen der Form $\exists$, $\forall$, $\forall^i$ zusammengesetzt sind. Damit ergibt sich

<u>7.52 Korollar</u>. *Lemma 7.51 überträgt sich vom Typ ∃∀ auf jeden Präfixtyp, in dem die Grundzeichen* $\exists^{i}$ *(*$i\in\mathbb{N}$*) nicht vorkommen (das bedeutet, daß für ∃-Blöcke eine beliebige Länge zugelassen ist, s.* 6.2(ii)*).*

<u>7.53 Satz</u> (Satz von Löwenheim-Skolem-Tarski für elementare Unterstrukturen, Tarski u. Vaught 1957, Verschärfung von 1.54).

(i) (abwärts) *Vor.:* $\mathfrak{B}$ *ist eine (unendliche) Struktur für die Sprache* L *(erster Stufe).* κ *ist eine Kardinalzahl mit* $|L|\leq\kappa\leq|\mathfrak{B}|$ *(Mächtigkeiten s.* 1.53*).* M *ist eine Teilmenge von* $U_{\mathfrak{B}}$ *mit* $|M|\leq\kappa$.

Beh.: Es gibt eine elementare Unterstruktur $\mathfrak{A}$ *von* $\mathfrak{B}$ *(*$\mathfrak{A}\prec\mathfrak{B}$*, s.* 3.72*) mit der Mächtigkeit* $|\mathfrak{A}|=\kappa$ *mit* $M\subset U_{\mathfrak{A}}$.

(ii) (aufwärts) *Vor.:* $\mathfrak{A}$ *ist eine unendliche Struktur für die Sprache* L *(erster Stufe).* κ *ist eine Kardinalzahl mit* $|\mathfrak{A}|\leq\kappa$ *und* $|L|\leq\kappa$.

Beh.: Es gibt eine echte elementare Erweiterung $\mathfrak{B}$ *von* $\mathfrak{A}$ *mit der Mächtigkeit* $|\mathfrak{B}|=\kappa$.

Zum Beweis s. etwa Schwabhäuser 1972, S. 62ff.

<u>Beweis von 7.43(ii)</u>: Sei $\mathfrak{A}$ gemäß 7.53(i) eine abzählbare elementare Unterstruktur der Standardstruktur $\mathfrak{r}_{\mathbb{N}}^{\mathbb{R}}$, d.h. $\mathfrak{A}\prec\mathfrak{r}_{\mathbb{N}}^{\mathbb{R}}$ und $|\mathfrak{A}|=\aleph_0$. Da $\mathfrak{A}$ Unterstruktur von $\mathfrak{r}_{\mathbb{N}}^{\mathbb{R}}$ ist, ist der in $\mathfrak{A}$ geometrisch konstruierte Körper $\mathfrak{F}$ (bis auf Isomorphie) ein Unterkörper von $\mathbb{R}$ und damit archimedisch angeordnet. Wegen $\mathfrak{A}\prec\mathfrak{r}_{\mathbb{N}}^{\mathbb{R}}$ ist $\mathfrak{A}$ erst recht elementar äquivalent zu $\mathfrak{r}_{\mathbb{N}}^{\mathbb{R}}$ und damit ein Modell von GA_2. Insbesondere überträgt sich die Gültigkeit der Aussage η aus dem Beweis von 7.36 von $\mathfrak{r}_{\mathbb{N}}^{\mathbb{R}}$ auf $\mathfrak{A}$, somit (s. dort) ist $\mathfrak{A}$ isomorph zur Standardstruktur $\mathfrak{r}_{\mathbb{N}}^{\mathfrak{F}}$, und wir können o.B.d.A. $\mathfrak{A}=\mathfrak{r}_{\mathbb{N}}^{\mathfrak{F}}$ mit $\mathfrak{F}\subseteq\mathbb{R}$ annehmen (es läßt sich zeigen, daß das tatsächlich der Fall ist). Wegen $|\mathfrak{A}|=\aleph_0$ ist auch $\mathfrak{F}$ abzählbar und hat nach 7.48 einen unendlichen Transzendenzgrad über $\mathbb{Q}$, also den Transzendenzgrad $\aleph_0$. Somit läßt sich $\mathfrak{F}$ aus $\mathbb{Q}$ erhalten durch Adjunktion von abzählbar vielen transzendenten Elementen a_i ($i\in\mathbb{N}$) und Bilden der reell-abgeschlossenen Hülle (3.17(v)) innerhalb $\mathbb{R}$. Wir setzen nun

(1) $\mathfrak{F}_i=[\mathbb{Q}(a_0,\ldots,a_i)]^{RC}$ und $\mathfrak{A}_i=\mathfrak{r}_{\mathbb{N}}^{\mathfrak{F}_i}$ $(i\in\mathbb{N})$,

wobei $[\mathfrak{K}]^{RC}$ die reell-abgeschlossene Hülle von $\mathfrak{K}$ in $\mathbb{R}$ bedeute. Dann sind $\{\mathfrak{F}_i\,|\,i\in\mathbb{N}\}$ und $\{\mathfrak{A}_i\,|\,i\in\mathbb{N}\}$ ω-Ketten mit

(2) $\bigcup_{i\in\mathbb{N}}\mathfrak{F}_i=\mathfrak{F}$, $\bigcup_{i\in\mathbb{N}}\mathfrak{A}_i=\mathfrak{A}$.

Nach Konstruktion hat $\mathfrak{F}_i$ über $\mathbb{Q}$ einen endlichen Transzendenzgrad, nach

7.48(ii) ist also $\mathfrak{A}_i$ (im Gegensatz zu $\mathfrak{A}$) kein Modell von GA_2, aber nach 7.45 (mit $\Theta=\emptyset$) ein Modell von GA_2^-.

Sei nun Θ gemäß Voraussetzung eine endliche Menge von $\exists\forall$-Aussagen, sei ϑ' die Konjunktion der Elemente von Θ, und sei ϑ gemäß 7.51 eine dazu logisch äquivalente $\exists\forall$-Aussage. Nehmen wir an, daß statt 7.43(1) Gleichheit eintritt, d.h.

(3) $Cn(GA_2^- \cup \{\vartheta\}) = GA_2$.

Dann sind die Strukturen $\mathfrak{A}_i$ Modelle von GA_2^-, aber nicht von $GA_2^- \cup \{\vartheta\}$; also muß die Aussage ϑ in diesen Strukturen falsch werden, d.h. in den $\mathfrak{A}_i$ ist $\neg\vartheta$ gültig. $\neg\vartheta$ ist aber eine Aussage vom Präfixtyp $\forall\exists$ und damit nach 6.9 auch gültig in der Vereinigung $\mathfrak{A}$. Da $\mathfrak{A}$ Modell von GA_2 ist, müßte wegen (3) aber auch ϑ in $\mathfrak{A}$ gültig sein. Damit ist die Annahme zum Widerspruch geführt. □

7.54 Anmerkungen. (i) Satz 7.43 und die Verschärfung 7.44(*) besagen, daß die dort genannten Mengen Θ noch nicht "kompliziert" genug sind, um in Frage zu kommen als Systeme von Zusatzaxiomen, mit denen man GA_2 aus GA_2^- erhält, d.h.

(1) $Cn(GA_2^- \cup \Theta) = GA_2$.

Diese Resultate ließen sich jedoch - nach dem Nachweis der Verschiedenheit der betrachteten Theorien - schon aus dem Ununterscheidbarkeitsresultat 7.41(ii) erhalten (s. 7.44).

(ii) Allgemein läßt sich aber für verschiedene Theorien nur sagen, daß die Feststellung ihrer Ununterscheidbarkeit durch Aussagen eines Präfixtyps stärker ist als eine entsprechende Behauptung über die Kompliziertheit von Zusatzaxiomen. Man muß also bei den Mengen Θ mit der Eigenschaft (1) mit komplizierteren Präfixtypen rechnen als bei den unterscheidenden Aussagen in Problem 7.42.

(iii) Während der in 7.1(iv) erwähnten Diskussionen in Bonn wurden insbesondere folgende Fragen aufgeworfen.

(2) (Szczerba 1972) Muß eine Menge Θ mit der Eigenschaft (1) unendlich sein? (D.h., daß GA_2 keine endliche Erweiterung von GA_2^- ist.)

(3) (Prestel 1972) Muß eine Menge Θ mit der Eigenschaft (1) Aussagen beliebig großer Quantorentiefe enthalten?

Dabei heißt η von der *Quantorentiefe m*, falls η in A_m oder in E_m liegt (6.2(ii)), d.h., falls die zugehörige pränexe Normalform m Quantorenblöcke enthält, wobei ∀- und ∃-Blöcke abwechseln.

Eine teilweise Lösung wurde von Prestel mit dem folgenden Resultat 7.55 (s.a. 7.56(i)) gefunden. Damit wird insbesondere (2) positiv beantwortet, und eine positive Antwort auf die Frage (3) ergibt sich unter zusätzlichen mengentheoretischen Annahmen wie z.B. V=L ("jede Menge ist konstruktibel"). (Hierbei bedeutet L die Klasse der konstruktiblen Mengen, siehe z.B. Handbook 1977, S. 455.)

7.55 Satz (Prestel, mündliche Mitteilung 1980). *Jede konstruktible Menge* Θ *von Aussagen von* L(B) *mit der Eigenschaft* 7.54(1) *enthält Elemente von beliebig großer Quantorentiefe (d.h. zu jedem* $m \in \mathbb{N}$ *ein* η *mit* $\eta \notin A_m$*).*

Zum Beweis werden Resultate aus Apt 1972 verwendet.

7.56 Anmerkungen. (i) Stellt man die Frage 7.54(3) nur für rekursive Mengen Θ von Zusatzaxiomen (vgl. 3.10), so wird sie durch 7.55 ebenfalls positiv beantwortet (ohne zusätzliche Annahmen), da jede rekursive Menge von Aussagen von L(B) auch konstruktibel ist.

(ii) Verwendet man GA_2^+ als Obertheorie in 7.54(1) (statt GA_2) und GA_2 oder (wie bisher) GA_2^- als Untertheorie, so ergibt sich aus 7.35 sofort, daß es überhaupt kein rekursives System Θ von Zusatzaxiomen gibt.

Über die "Schwierigkeit" des Problems, die Modelle von GA_2 zu charakterisieren (vgl. 7.8(i)) wurden von Prestel die folgenden in 7.58 zusammengefaßten Resultate bewiesen.

7.57 Definition. Es bedeute L_S die Sprache (erster Stufe), die aus der Sprache L_{OF} für angeordnete Körper (3.20) durch Hinzunahme des zweistelligen Relationszeichens S entsteht. Sei $\mathfrak{f}$ ein angeordneter Körper und S eine beliebige zweistellige Relation auf der Trägermenge $U_{\mathfrak{f}}$, d.h. eine Menge von Punkten $\langle x,y\rangle$ der affinen Ebene $\mathfrak{A}_2^B(\mathfrak{f})$ (vgl. 7.6). Mit $(\mathfrak{f},S)$ bezeichnen wir dann die Struktur $\mathfrak{A}$ für die Sprache L_S, die aus $\mathfrak{f}$ durch Hinzunahme der Relation S zur Interpretation von S entsteht (d.h. $U_{\mathfrak{A}}=U_{\mathfrak{f}}$, $S_{\mathfrak{A}}=S$, $\zeta_{\mathfrak{A}}=\zeta_{\mathfrak{f}}$ für die übrigen Operations- und Relationszeichen).

Bedeute DS" das Schema der "*elementaren Axiome vom Dedekindschen Schnitt*" für die Sprache L_S, das aus dem entsprechenden Schema DS' für die Sprache L_{OF} (s. 3.24(i), (ii)) dadurch entsteht, daß Formeln der Erweiterungssprache L_S zugelassen werden als (die einen Schnitt definierenden Formeln) $\alpha(u)$ und $\beta(v)$. Die Redeweise "*in* $\mathfrak{k}$ *ist jeder S-definierbare Schnitt realisiert*" bedeute dann, daß $(\mathfrak{k}, S)$ ein Modell von DS" ist (damit ist insbesondere $\mathfrak{k}$ Modell von DS', also reell-abgeschlossen).

7.58 Satz (Prestel 1981). (i) (Charakterisierungslemma) *Sei* $\mathfrak{k}$ *ein reell-abgeschlossener Körper und S ein konvexer Teilbereich von* $\mathfrak{A}_2^B(\mathfrak{k})$. *Dann bestimmt S ein Modell von* GA_2 *(gemäß 7.7(i), d.h.* $\mathfrak{A}_2^B(\mathfrak{k})|S$ *ist ein Modell) gdw in* $\mathfrak{k}$ *jeder S-definierbare Schnitt realisiert ist.*

(ii) *Es gibt keine Aussage* ρ *der Sprache* L_S, *die genau die (gemäß 7.7(i) beschriebenen) Modelle von* GA_2 *charakterisiert, d.h., die folgender Bedingung genügt:*

(1) *Für jeden reell-abgeschlossenen Körper* $\mathfrak{k}$ *und jeden konvexen Teilbereich S von* $\mathfrak{A}_2^B(\mathfrak{k})$ *gilt:*

$\mathfrak{A}_2^B(\mathfrak{k})|S$ *ist ein Modell von* GA_2 *gdw* ρ *in* $(\mathfrak{k}, S)$ *gültig ist.*

(iii) (Verschärfung) *Es gibt auch im Rahmen der schwachen zweiten Stufe über* L_S *(1.44) keine entsprechende Aussage* ρ *mit der Eigenschaft* (1).

7.59 Anmerkung (zum Beweis). Zum Beweis von (ii) werden Mengen *N* betrachtet, die bei nicht-archimedisch angeordneten Körpern die Menge der natürlichen Zahlen in einem gewissen Sinne ins Transfinite fortsetzen. Es werden Resultate von Montague 1961 über die Theorie der reell-abgeschlossenen Körper mit einem entsprechenden zusätzlichen Prädikat *N* verwendet, und es werden Modelle betrachtet, die aus den Standardstrukturen $\mathfrak{T}_{\mathbb{N}}^{\mathfrak{k}}$ von 7.17 entstehen durch Verwendung solcher Mengen *N* statt $\mathbb{N}$. Aus diesen Überlegungen ergibt sich übrigens auch das Ergebnis 7.49 als Spezialfall. Zum Beweis von (iii) werden außerdem wieder Resultate von Apt 1972 verwendet.

7.60 Definition und Sätze (Szczerba u. Tarski 1965/79). Unter den zahlreichen Beispielen für Erweiterungen von GA_2 werden auch die folgenden behandelt (Szczerba u. Tarski 1979, S. 187f.). Sei $T_{(3)}$ die Theorie, die aus GA_2 durch Hinzunahme des folgenden Zusatzaxioms entsteht:

(1) $\forall p \forall a \forall d \exists b[\neg \mathrm{Col}\, pad \rightarrow \mathrm{Babd} \wedge \mathrm{Eukl}(apb)]$
(jede Halbgerade $H(pa)$ läßt sich nach jeder der beiden Seiten von $L(pa)$ zu einem ["hinreichend kleinen"] Winkel $\sphericalangle apb$ ergänzen, der sich euklidisch verhält [s. 7.13]).

Sei $T_{(4)}$ die Theorie, die aus $T_{(3)}$ durch Hinzunahme aller Axiome $\neg\chi_n$ gemäß 7.25(1) (die Menge der Punkte ist kein offenes konvexes Polygon mit höchstens n Ecken) mit $3 \leq n \in \mathbb{N}$ entsteht.

Dann ist $T_{(3)}$ eine Untertheorie aller Theorien $Po^{\leq n}$ von 7.25(ii) (der Polygone mit höchstens n Ecken) mit $3 \leq n \in \mathbb{N}$. Die Modelle von $T_{(3)}$ der Form $\mathfrak{A}_2^B(\mathfrak{f})|S$ sind für $\mathfrak{f} \cong \mathbb{R}$ genau die offenen konvexen Polygone, für $\mathfrak{f} \not\cong \mathbb{R}$ gibt es dagegen auch andere Modelle. Die Theorie $T_{(4)}$ besitzt Modelle der angegebenen Form, aber keine mit $\mathfrak{f} \cong \mathbb{R}$. $T_{(3)}$ hat den Unvollständigkeitsgrad $2^{\aleph_0}$ (vgl. 7.14). $T_{(4)}$ ist keine endliche Erweiterung von GA_2. $T_{(3)}$ und $T_{(4)}$ sind nicht endlich axiomatisierbar.

Das bei Szczerba u. Tarski 1979 noch als offen bezeichnete Problem der Entscheidbarkeit von $T_{(3)}$ und $T_{(4)}$ wird gelöst durch

7.61 Satz (Prestel, briefliche Mitteilung 1980). *Die Theorien* $T_{(3)}$ *und* $T_{(4)}$ *sind unentscheidbar.*

Zum Beweis wird mit einer geeigneten Ultraprodukt-Konstruktion gezeigt, daß $T_{(4)}$ (und damit auch $T_{(3)}$) verträglich ist mit der Theorie R aus Tarski, Mostowski u. Robinson 1953. Daraus ergeben sich nach einem Satz von Cobham (s. Vaught 1962) die gewünschten Unentscheidbarkeiten.

Immer noch offen ist meines Wissens das folgende Problem.

7.62 Problem (Szczerba u. Tarski 1965, S. 177). Gibt es eine (echte) widerspruchsfreie Erweiterung von GA_2, die endlich axiomatisierbar ist?

7.63 n-dimensionale allgemeine affine Geometrien. Die meisten der für GA_2 erzielten Resultate lassen sich übertragen auf entsprechende n-dimensionale allgemeine affine Geometrien GA_n. Darauf wird am Schluß von Szczerba u. Tarski 1979 eingegangen.

8. Hinweise auf weitere Ergebnisse

8.1 Überblick. In diesem Abschnitt werden Hinweise gegeben auf Pasch-freie Geometrien (ab 8.2), auf die euklidische Geometrie in der schwachen zweiten Stufe (ab 8.19) und auf konstruktive Geometrien (8.24). Zum Schluß (8.25) folgen Literaturhinweise auf Arbeiten, die in diesem Buch noch nicht an anderer Stelle erwähnt wurden.

8.2 Pasch-freie Geometrien. Die elementare (ebene, euklidische) *Pasch-freie Geometrie* - hier mit $P\mathfrak{f}$ bezeichnet - wurde eingeführt in Szczerba u. Szmielew 1970 (s.a. Szmielew 1974, 1974a, 1980). Sie ist aufgebaut auf einem Axiomensystem, das durch Weglassen des Axioms von Pasch (in einer äußeren Form, vgl. I.1.2 [hinter A7], bezeichnet mit (P)) entsteht aus einem gewissen Axiomensystem für die ebene euklidische Geometrie P_2 (s. 2.1; das System wurde gegenüber früher verwendeten Systemen wie z.B. $\{A_1,\ldots,A10\}$ in geeigneter Weise abgeändert). Dabei wurde ein besonders schönes Ergebnise über die Rolle des Axioms von Pasch bei der Charakterisierung der Modelle gefunden, das in dem folgenden Darstellungssatz zum Ausdruck kommt.

8.3 Satz. (i) *Eine Struktur* $\mathfrak{A}$ *für die Sprache* L(D,B) *ist ein Modell der Pasch-freien Geometrie* $P\mathfrak{f}$ *genau dann, wenn sie isomorph ist zu einer kartesischen Ebene* $\mathfrak{L}_2(\mathfrak{f})$ *(s.u.,* 8.6(ii)*) über einem formal-reellen pythagoreischen normierten semi-geordneten (s.* 8.4*) Körper* $\mathfrak{f}$.

(ii) *In einem Modell der genannten Art gilt darüber hinaus das Axiom* (P) *von Pasch gdw* $\mathfrak{f}$ *sogar ein angeordneter Körper ist.*

(iii) (Folgerung) *Durch Hinzunahme des Axioms von Pasch erhält man aus* $P\mathfrak{f}$ *die ebene euklidische Geometrie* P_2.

Dabei werden (normierte) semi-geordnete Körper folgendermaßen eingeführt.

8.4 Definition. Eine Struktur $\mathfrak{F}$ für die Sprache $L_{OF}=L(+,\cdot,-,0,1;<)$ der angeordneten Körper (3.20) heißt ein *semi-geordneter Körper* (und < die *Semi-Ordnung* von $\mathfrak{F}$), falls folgende Axiome in $\mathfrak{F}$ gelten:

(1) die Axiome für Körper,

(2) die Axiome für (lineare, irreflexive) Anordnungen (bzgl. <),

(3) $x<y \rightarrow x+z < y+z$
(Monotoniegesetz für die Addition).

Bekanntlich ist $\mathfrak{F}$ ein angeordneter Körper gdw darüber hinaus noch gilt:

(4) $x<y \wedge 0<z \rightarrow x\cdot z < y\cdot z$
(Monotoniegesetz für die Multiplikation).

Die Axiome (2) und (3) besagen zusammen, daß die additive Gruppe des Körpers (1) durch < zu einer angeordneten Gruppe gemacht wird. Für semigeordnete Körper wird also keinerlei Zusammenhang (Verträglichkeit) von < mit der Multiplikation gefordert.

An Stelle der Kleinerbeziehung < kann man als Grundbegriff natürlich auch die Kleiner-Gleich-Beziehung ≤ oder einen der einstelligen Begriffe Ps (Positivität) oder Pn (für die Eigenschaft, nicht-negativ ["positiv oder Null"] zu sein) verwenden. Jeder dieser vier Begriffe ist mit Hilfe jedes anderen und der Operationen +, -, 0 definierbar (für < mittels Ps vgl. I.14.38). Dementsprechend sind die Axiome abzuändern. In Szczerba u. Szmielew 1970 und einigen anderen Arbeiten wird Pn verwendet und $x\in P$ statt Pn x geschrieben; die entsprechende Menge P wird dann der *Semi-Positiv-Bereich* von $\mathfrak{F}$ genannt. An die Stelle von (2) und (3) treten dann die Axiome

(2') $x\in P \vee -x\in P$,

(3') $x\in P \wedge -x\in P \rightarrow x\doteq 0$,

(4') $x\in P \wedge y\in P \rightarrow x+y\in P$.

An die Stelle von (4) tritt entsprechend

(5') $x\in P \wedge y\in P \rightarrow x\cdot y\in P$.

Ein semi-geordneter Körper $\mathfrak{F}$ heißt *normiert* (s.a. 8.7(i)), falls sein Einselement im Semi-Positiv-Bereich liegt (d.h. $0<1$ bzw. $1\in P$; bekanntlich gilt das für angeordnete Körper).

8.5 Anmerkungen. (i) Damit kommt die vorher erwähnte Rolle des Axioms von Pasch zum Ausdruck: Es entspricht bei der gewählten Axiomatisierung genau dem Monotoniegesetz für die Multiplikation (4) bzw. dem entsprechenden Axiom (5') für angeordnete Körper.

(ii) Schon bei Hilbert (s. Hilbert 1977) werden die Axiome für die Zwischenbeziehung ("Anordnungsaxiome") eingeteilt in "lineare Anordnungsaxiome" (d.h. solche, die sich nur auf die Anordnung von Punkten auf einer Geraden beziehen) und das Axiom von Pasch als einziges "ebenes Anordnungsaxiom". Diese zunächst heuristische Einteilung bleibt sinnvoll für die hier betrachteten Axiomensysteme und findet ihr Gegenstück in der Charakterisierung der Modelle: Das Axiom von Pasch ist das einzige, durch das ein Zusammenhang zwischen der Anordnung des Grundkörpers und der Multiplikation hergestellt wird; in allen anderen Axiomen (linearen Anordnungsaxiomen und Axiomen in den Grundbegriffen B und D) wird die Anordnung nur für die additive Gruppe des Grundkörpers gebraucht.

(iii) Um diese Ergebnisse zu erhalten, wird für kartesische Ebenen (und Räume) die folgende Definition zugrunde gelegt, die sich von der Definition I.1.4 dadurch unterscheidet, daß zur Festlegung der Zwischenbeziehung "nicht zuviel über die Multiplikation" des Grundkörpers verwendet wird. Für pythagoreische a n g e o r d n e t e Körper liefern beide Definitionen dasselbe, so daß die jetzige Definition auch für den Darstellungssatz I.16.15 geeignet ist.

<u>8.6 Definition</u>. Sei $\mathfrak{F}$ ein pythagoreischer semi-geordneter Körper und $n \in \mathbb{N}$ beliebig. Wir setzen $F = U_{\mathfrak{F}}$ und verwenden für n-tupel $a \in F^n$ die üblichen Bezeichnungen aus der Vektorrechnung (vgl. I.1.3).

(i) Unter der *Norm* $\|a\|$ eines Vektors $a \in F^n$ verstehen wir das Körperelement $x \in F$ mit den Eigenschaften

(1) $x^2 = \sum_{\nu=1}^{n} a_\nu^2$ und

(2) $0 \leq x$.

(Da $\mathfrak{F}$ pythagoreisch ist, gibt es stets ein solches Element x, und es wird durch (2) eindeutig festgelegt wegen 8.4(2'), (3').)

(ii) Unter dem *n-dimensionalen kartesischen Raum* $\mathfrak{L}_n(\mathfrak{F})$ über $\mathfrak{F}$ verstehen wir die folgendermaßen festgelegte Struktur $\mathfrak{A}$ für die Sprache L(D,B):

(3) $U_{\mathfrak{A}} = F^n$,

(4) $D_{\mathfrak{A}}abcd$ gdw $\|a-b\| = \|c-d\|$,

(5) $B_{\mathfrak{A}}abc$ gdw $\|a-b\| + \|b-c\| = \|a-c\|$.

(Bedingung (4) ist natürlich gleichwertig mit der früheren Festlegung I.1.4(2) der Streckenkongruenz.)

Unter der *kartesischen Ebene* über $\mathfrak{f}$ verstehen wir insbesondere den zweidimensionalen Raum $\mathfrak{L}_2(\mathfrak{f})$.

8.7 Anmerkungen. (i) Ersetzt man in einem semi-geordneten Körper $\mathfrak{f}$ den Semi-Positiv-Bereich P durch $-P := \{-x \mid x \in P\}$, so entsteht offenbar wieder ein semi-geordneter Körper, der mit $-\mathfrak{f}$ bezeichnet werden möge. Genau einer von beiden ist normiert, und beide liefern dieselben kartesischen Räume (d.h. $\mathfrak{L}_n(\mathfrak{f}) = \mathfrak{L}_n(-\mathfrak{f})$). Daher kann die Forderung der Normiertheit in der Charakterisierung 8.3(i) der Modelle auch weggelassen werden (aber nicht beim Übergang zu 8.3(ii)).

(ii) Die bisherigen Sätze sagen noch nichts darüber, ob das Axiom (P) von Pasch unabhängig von den übrigen Axiomen ist, d.h., ob es überhaupt Modelle von $P\mathfrak{s}$ gibt, in denen (P) falsch wird. Tatsächlich wurde kurz vorher von Szczerba folgendes Ergebnis erzielt, nach dem (P) sogar unabhängig ist von der "*vollen Pasch-freien Geometrie (zweiter Stufe)*", die aus $P\mathfrak{s}$ durch Hinzunahme des Stetigkeitsaxioms A11 (I.1.2) entsteht.

8.8 Satz (Szczerba 1970). *Sei f eine unstetige Lösung der Funktionalgleichung $f(x+y) = f(x)+f(y)$ auf $\mathbb{R}$ (solche f existieren) und $\mathfrak{f}$ der semi-geordnete Körper, der von $\mathbb{R}$ mit dem Semi-Positiv-Bereich $f(\mathbb{R}^+) = \{f(x) \mid x \geq 0\}$ gebildet wird. Dann ist $\mathfrak{L}_2(\mathfrak{f})$ ein Modell von $P\mathfrak{s} \cup \{A11\}$, in dem* (P) *falsch wird ("Unabhängigkeitsmodell").*

8.9 Anmerkung. In Adler 1973 wird gezeigt, daß sogar jedes solche Modell von der vorher genannten Art sein muß. (Siehe auch 8.18.)

Eine andere Beschreibung für die zweite Stufe wird in dem folgenden Satz gegeben.

8.10 Definition. Ein semi-geordneter Körper $\mathfrak{f}$ heiße *vollständig semi-geordnet* (engl. *continuously semi-ordered*), falls für seine Semi-Ordnung das Axiom vom Dedekindschen Schnitt gilt; die Semi-Ordnung selbst heißt dann auch *(schnitt-)vollständig*.

8.11 Satz (nach Szczerba 1971c). (i) *$\mathfrak{L}_2(\mathfrak{f})$ ist ein Modell von $P\mathfrak{s} \cup \{A11\}$ gdw $\mathfrak{f}$ ein formal-reeller pythagoreischer vollständig semi-geordneter Körper ist.*

(ii) *Jeder vollständig semi-geordnete Körper $\mathfrak{f}$ (insbesondere auch jeder formal-reelle und pythagoreische) läßt sich bis auf Isomorphie aus dem angeordneten Körper $\mathbb{R}$ durch eine Abänderung der Multiplikation und, falls $\mathfrak{f}$ nicht normiert ist, des Einselements erhalten; er ist damit von der Mächtigkeit des Kontinuums.*

(iii) *Jeder Körper der Charakteristik* 0 *von der Mächtigkeit des Kontinuums (insbesondere auch jeder formal-reelle und pythagoreische) läßt sich durch Hinzunahme einer Relation* $<$ *zu einem (normierten) vollständig semi-geordneten Körper machen.*

In Szczerba 1971a, 1971, 1971b sind außerdem die folgenden Resultate enthalten, die auf gewissen Abänderungen der Konstruktion aus Szczerba 1970 beruhen.

8.12 Satz (Szczerba s.o.). *Die Zwischenbeziehung* B *ist nicht definierbar mit Hilfe der Streckenkongruenz* D *in der vollen Pasch-freien Geometrie (zweiter Stufe)* $Cn^2(P\mathfrak{f} \cup \{A11\})$.

(ii) $P\mathfrak{f}$ *hat den Unvollständigkeitsgrad* $2^{\aleph_0}$ *(vgl. 7.14).*

(iii) $P\mathfrak{f}$ *ist unentscheidbar.*

8.13 Anmerkung. Die Unentscheidbarkeit von $P\mathfrak{f}$ läßt sich natürlich auch erhalten aus der Unentscheidbarkeit der endlichen Erweiterung P_2 (vgl. 3.60, 56). Ein solcher Beweis benutzt jedoch als besonders starkes Hilfsmittel die Unentscheidbarkeit der Theorie der pythagoreischen angeordneten Körper. Der Beweis in Szczerba 1971b ist dagegen unabhängig von diesem Hilfsmittel (das damals noch nicht zur Verfügung stand); er beruht darauf, daß für den Körper $\mathbb{R}$ ein Semi-Positiv-Bereich P konstruiert wird, mit dessen Hilfe die Menge der rationalen Zahlen und damit (nach 3.57(i)) die Menge der natürlichen Zahlen definierbar ist.

8.14 Zusatzeigenschaften. Für die Modelle der elementaren Pasch-freien Geometrie $P\mathfrak{f}$ bzw. die zugehörigen semi-geordneten Körper wurde eine Reihe von interessanten Zusatzeigenschaften untersucht. Dabei ist bemerkenswert, daß einige Äquivalenzen zwischen solchen Eigenschaften, die man für angeordnete Körper bzw. die Geometrie P_2 kennt, jetzt nicht mehr gelten. Beispiele dafür sind die folgenden beiden Sätze.

8.15 Satz (Szmielew 1970, 1970a). (i) *In $P\mathfrak{f}$ ergibt sich aus dem Kreisaxiom* (CA) *(s. I.1.2) bereits das Axiom von Pasch.*

(ii) *Die (für* P_2 *durch* (CA) *ausgedrückte) Eigenschaft, daß der semi-*

geordnete Grundkörper euklidisch ist, läßt sich für $P\mathfrak{f}$ durch das folgende "Zwei-Kreis-Axiom" ausdrücken:

(C_2) *Wenn eine Gerade durch den Mittelpunkt einer Seite eines beliebigen Dreiecks geht, so schneidet sie wenigstens einen der beiden Kreise, die über den anderen beiden Dreiecksseiten als Durchmesser errichtet sind.*

(iii) (Folgerung) *In $P\mathfrak{f}$ gilt die Äquivalenz*

(1) $(CA) \leftrightarrow (P) \wedge (C_2)$.

(iv) (C_2) *läßt sich in $P\mathfrak{f}$ äquivalent ersetzen durch den Spezialfall, daß die betrachtete Gerade auf der ersten Dreiecksseite senkrecht steht und daß die Dreiecksecken kollinear sind.*

<u>8.16 Satz</u> (Gupta u. Prestel 1972, 1972a, s.a. Prestel 1972). *Sei $\mathfrak{F}$ ein beliebiger formal-reeller pythagoreischer normierter semi-geordneter Körper (wie in* 8.3(i)*). Wie in* 8.6 *sei die Norm für Vektoren aus F^n eingeführt und außerdem ein inneres Produkt $a \circ b$ durch*

(1) $$a \circ b = \sum_{\nu=1}^{n} a_\nu \cdot b_\nu \ .$$

$\mathfrak{R}_n(\mathfrak{F})$ bedeute den damit gebildeten n-dimensionalen Raum; für das Folgende werde eine feste endliche Dimension $n \geq 2$ vorausgesetzt. Der Betrag $|x|$ eines Körperelements x sei erklärt wie die Norm in 8.6 *für $n=1$* .

(i) *Die folgenden Bedingungen (für $\mathfrak{F}$ selbst, $\mathfrak{R}_n(\mathfrak{F})$ bzw. den kartesischen Raum $\mathfrak{L}_n(\mathfrak{F})$) sind untereinander äquivalent.*

(2) *Wenn $y \geq 0$, so ist auch stets $x^2 \cdot y \geq 0$* .
Dafür sagt man nach Prestel 1972, daß die Semi-Ordnung *quadratisch* ist.

(3) $\|a+b\| \leq \|a\| + \|b\|$ (Dreiecksungleichung).

(4) $|a \circ b| \leq |\, \|a\| \cdot \|b\| \,|$ (Schwarzsche Ungleichung)
(die Betragsstriche auf der rechten Seite sind jetzt nicht entbehrlich!).

(5) *Der Höhenfußpunkt im rechtwinkligen Dreieck liegt zwischen den Endpunkten der Hypotenuse.*

(6) *Im rechtwinkligen Dreieck ist jede Kathete kleiner-gleich der Hypotenuse.*

(ii) *Ebenso sind die folgenden Bedingungen untereinander äquivalent.*

(7) *Es ist stets* $x^2 \geq 0$.

(8) $(a \circ b)^2 \leq a^2 \cdot b^2$ ("schwache Schwarzsche Ungleichung")
(dabei ist $a^2 := a \circ a$ *gesetzt).*

(iii) *Die Bedingungen unter* (i) *sind echte Abschwächungen von* 8.4(4) *("$\mathfrak{f}$ ist angeordneter Körper"); die Bedingungen unter* (ii) *sind echte Abschwächungen von denen unter* (i).

8.17 Weitere Modelle. Zum Beweis für die "Echtheit" der Abschwächungen in 8.16(iii) wurden natürlich entsprechende spezielle Modelle konstruiert. Es entstand nun das Problem, eine Übersicht über möglichst viele Modelle von $P\mathfrak{f}$ oder - was nach 8.3(i) gleichwertig ist - über die zugehörigen semi-geordneten Körper zu bekommen. Eine umfangreiche Untersuchung darüber ist enthalten in der Habilitationsschrift Prestel 1972, s.a. 1973, 1974. Sie enthält mehrere Konstruktionsverfahren, die zu zahlreichen "echten" semi-geordneten Körpern (d.h. solchen, die nicht angeordnet sind) führen. Die Ergebnisse aus 8.16 sind als Spezialfälle darin enthalten. Außerdem wurden dort weitere Bedingungen gefunden, durch die die Bedingungen aus 8.16(i) in anderer Richtung echt abgeschwächt werden. Es kann hier nicht auf alle Ergebnisse dieser Arbeit eingegangen werden; es seien nur noch zwei Konstruktionsverfahren hervorgehoben: 1. Ein Verfahren, das Hilfsmittel aus der Bewertungstheorie benutzt. Damit lassen sich insbesondere alle quadratischen Semi-Ordnungen (8.16(i)) erhalten. 2. Ein Verfahren, das auf der Methode des Forcing beruht. Damit werden insbesondere zahlreiche *elementar vollständige* Semi-Ordnungen konstruiert, d.h. solche, in denen das Axiom vom Dedekindschen Schnitt für in der ersten Stufe definierbare Unter- und Oberklassen gilt. Solche Semi-Ordnungen sind nicht quadratisch, sofern sie keine Anordnungen sind.

8.18 Determiniertheitsaxiom. In der Mengenlehre, die hier für Modellkonstruktionen verwendet wird, wird üblicherweise das Auswahlaxiom (AC) zugrunde gelegt, dessen Anwendung im allgemeinen (wie auch in diesem Buch) nicht ausdrücklich erwähnt wird. Manchmal wird aber auch das sog. Determiniertheitsaxiom (AD) ("axiom of determinacy") zugrunde gelegt, das dem Auswahlaxiom widerspricht (s. etwa Handbook 1977, S. 369). Als interessante Folgerung aus der Charakterisierung in 8.9 ergibt sich in Adler 1973, daß unter Voraussetzung von (AD) das Axiom von Pasch in $P\mathfrak{f} \cup \{A11\}$ beweisbar wird. (Tatsächlich wird zum Nachweis der Existenz der unstetigen Lösungen f aus 8.8 das Auswahlaxiom verwendet.)

8.19 Euklidische Geometrie in der schwachen zweiten Stufe. Diese (ebene) Geometrie P_2^{S2} wurde eingeführt von Tarski 1959 (S. 24, ursprünglich bezeichnet mit E_2') als eine axiomatisch aufgebaute Theorie schwacher zweiter Stufe. Ihre Sprache $L^{S2}(D,B)$ entsteht aus der für die elementare Geometrie P_2^* durch Hinzunahme von Variablen für endliche Punktmengen (genauer s. 1.44). Ihre Axiome lassen sich formal genauso beschreiben wie die für P_2^*, d.h., es sind A1 bis A10 und ein zu A11' (s. I.1.2) analoges Schema $A11^{(S2)}$ der Stetigkeitsaxiome; der Unterschied ist, daß in diesem Schema jetzt Formeln $\alpha(x)$ und $\beta(y)$ der erweiterten Sprache zugelassen sind. $A11^{(S2)}$ läuft also hinaus auf ein Axiom vom Dedekindschen Schnitt auf beliebigen Geraden für diejenigen Unter- und Oberklassen, die sich in der schwachen zweiten Stufe definieren lassen. Das erweist sich als echte Verschärfung des elementaren Schemas A11'. Mit Hilfe von (jeweils einem Axiom von) $A11^{(S2)}$ erhält man nämlich für die Modelle von P_2^{S2}:

(1) Der in der üblichen Weise geometrisch eingeführte Grundkörper $\mathfrak{F}_{oe}$ (s. I.14.40) ist archimedisch angeordnet.

(2) Die Menge der "natürlichen Zahlen" (natürlichen Vielfachen des Einselements) von $\mathfrak{F}_{oe}$ ist definierbar.

Damit ergibt sich (vgl. 3.54)

8.20 Satz (Tarski 1959). *P_2^{S2} ist unentscheidbar.*

Zwei Fragen, die in Tarski 1959 als offene Probleme formuliert wurden, werden mit den folgenden beiden Sätzen beantwortet.

8.21 Satz (Mostowski 1961). *P_2^{S2} ist unvollständig.*

8.22 Satz (Schwabhäuser 1979). (i) *P_2^{S2} ist nicht finitisierbar (d.h. nicht endlich axiomatisierbar,* 3.9(i)*).*

(ii) *Die Modelle von P_2^{S2} sind bis auf Isomorphie die kartesischen Ebenen über den angeordneten Körpern $\mathfrak{F}$ aus* 7.49(ii) *bis* (iv).

8.23 Anmerkungen zum Beweis. (i) In Mostowski 1961 wird u.a. die Theorie $Th^{S2}(\mathfrak{L}_2(\mathbb{R}))$ des speziellen Modells $\mathfrak{L}_2(\mathbb{R})$ betrachtet, und es wird gezeigt, daß diese kein sog. analytisches Axiomensystem und damit - im Gegensatz zu P_2^{S2} - erst recht kein rekursives Axiomensystem besitzt. Somit ist diese Theorie eine echte Erweiterung von P_2^{S2}, woraus sich 8.21 ergibt.

(ii) Die früher für elementare Theorien verwendete Methode zum Nachweis der Nicht-Finitisierbarkeit (vgl. 3.33) beruhte auf dem Lemma 3.34 und damit auf dem Endlichkeitssatz. Diese Methode ist hier nicht mehr anwendbar, da der Endlichkeitssatz und 3.34 für die schwache zweite Stufe (und höhere Stufen) nicht gelten.

Der Beweis für 8.22(i) in Schwabhäuser 1979 geschieht durch eine schrittweise Rückführung auf die entsprechenden Resultate für die in 7.49 genannten Theorien R_2, R_N, A_ω; für A_ω ("Arithmetik zweiter Stufe") ist das Resultat enthalten in Zbierski 1978 (vgl. auch Apt 1972). Um die jeweilige Rückführung (ausgeführt von R_2 ab) zu erhalten, werden endlich axiomatisierbare Untertheorien der genannten Theorien betrachtet und gewisse Formelübersetzungen (Reduziertenbildungen, vgl. 3.25, 36) zwischen diesen konstruiert, mit deren Hilfe sich aus der Finitisierbarkeit von R_2 auch die von R_N und von A_ω ergeben würde. In diesem Zusammenhang ergibt sich auch 8.22(ii).

8.24 Konstruktive Geometrien. Elementare konstruktive Geometrien im Sinne von Moler u. Suppes 1968 sind Geometrien, in denen Existenzbehauptungen nur formuliert werden durch Angabe der betreffenden Objekte als Werte von (Konstruktions-)Operationen, die als Grundbegriffe verwendet werden, wie z.B. die Operation der Streckenabtragung und die Operation der Bildung des Schnittpunkts von Geraden. An die Stelle von Existenzformeln $\exists x\alpha(x)$ treten in der konstruktiven Geometrie Formeln der Form $\alpha(\tau)$, wobei τ ein (i.a. mehrfach zusammengesetzter) Term ist. In die Sprache der Theorie werden nur quantorenfreie Formeln aufgenommen, die wie ihre Generalisierten (1.24) behandelt werden. Von Engeler 1973 (s.a. 1975) wurden Erweiterungen dieser Sprachen zu sog. *algorithmischen Sprachen* eingeführt; diese entstehen dadurch, daß Disjunktionen mit unendlich vielen Gliedern zugelassen werden, die in einem gewissen Sinne das Halten eines Algorithmus (in Abhängigkeit von der Eingabebelegung) beschreiben. Die betrachteten (abstrakten) Algorithmen arbeiten dabei auf beliebigen Modellen der Theorie, benutzen die dort gegebenen Operationen und Relationen und werden durch Flußdiagramme beschrieben. Die ganze algorithmische Sprache ist somit nicht mehr ein Teil einer Sprache erster Stufe, aber ein Teil einer "infinitären" Sprache der Form $L_{\omega_1\omega}$ (in einer solchen werden beliebige Konjunktionen und Disjunktionen mit abzählbar vielen Gliedern zugelassen, s. etwa Keisler 1971). Genauere Ausführungen über entsprechende "algorithmische Theorien" und die Behandlung mehrerer konstruktiver Geometrien sind enthalten in Seeland 1978.

8.25 Hinweise auf weitere Literatur. Zur Ergänzung sei auf die folgenden Arbeiten hingewiesen, die noch nicht an früherer Stelle erwähnt wurden:

Bernays 1928, 1948, 1953, 1959, Gupta 1967, 1969, Piesyk 1977, Rautenberg 1963, 1965, 1966, 1968, Schwabhäuser 1969, Szczerba 1977a, Szmielew 1977.

Literaturverzeichnis

Adler, A.
1973 Determinateness and the Pasch axiom. Canad. Math. Bull. 16, 159-160.

Apt, K. R.
1972 Non-finite axiomatizability of the second-order arithmetic. Bull. Acad. Polon. Sci. Sér. Sci. Math. Astronom. Phys. 20, 347-348.

Bachmann, F.
1959 Aufbau der Geometrie aus dem Spiegelungsbegriff. Berlin, Göttingen, Heidelberg: Springer-Verlag.
1964 Zur Parallelenfrage. Abh. Math. Sem. Univ. Hamburg 27, 173-192.
1964a Modelle der ebenen absoluten Geometrie. Jber. Deutsch. Math.-Verein. 66, 152-170.

Baldus, R.
1964 Nichteuklidische Geometrie. 4. Aufl. Berlin: W. de Gruyter.

Barwise, J. u.a.: s. Handbook

Baumann, H. u. Schwabhäuser, W.
1970 Eine Vereinfachung des hyperbolischen Parallelenaxioms. Arch. Math. (Basel) 21, 634-640.

Bell, J. L. u. Slomson, A. B.
1969 Models and ultraproducts. Amsterdam, London: North-Holland Publ. Co.

Bernays, P.
1928 Besprechung zu "Die Grundbegriffe der reinen Geometrie in ihrem Verhältnis zur Anschauung" von R. Strohal. Die Naturwissenschaften, Heft 12, 197-203.
1948 Bemerkungen zu den Grundlagen der Geometrie. Studies and Essays presented to R. Courant on his 60th Birthday, New York: Interscience Publishers, S. 29-44.
1953 Über die Verwendung der Polygoninhalte an Stelle eines Spiegelungsaxioms in der Axiomatik der Planimetrie. Elem. Math. 8, 102-107.
1959 Die Mannigfaltigkeit der Direktiven für die Gestaltung geometrischer Axiomensysteme. The Axiomatic Method, Proc. of the 1957/58 International Symposium at Berkeley, Amsterdam: North-Holland Publ. Co., S. 1-15.

Bernays, P. u.a.: s. Hilbert.

Beth, E. W.
1953 On Padoa's method in the theory of definition. Indag. Math. 15, 330-339.

Beth, E. W. u. Tarski, A.
1956 Equilaterality as the only primitive notion of Euclidean geometry. Indag. Math. 18, 462-467.

Bolyai, J.
1832 Appendix scientiam spatii absolute veram exhibens: a veritate aut falsitate Axiomatis XI Euclidei (a priori haud unquam decidenda) independentem; adjecta ad casum falsitatis, quadratura circuli geometrica. Als Anhang in: Wolfgang Bolyai, Tentamen juventutem studiosam in elementa matheseos ... introducendi, Bd. 1, Maros Vâsârhely 1832. Deutsche Übersetzung in Engel u. Stäckel 1913.

Bonola, R. u. Liebmann, H.
1921 Die nichteuklidische Geometrie. 3. Aufl. Leipzig u. Berlin: B. G. Teubner.

Borsuk, K. u. Szmielew, W.
1960 Foundations of geometry. Amsterdam: North-Holland Publ. Co.

Chang, C. C. u. Keisler, H. J.
1973 Model theory. Amsterdam, London: North-Holland Publ. Co., New York: American Elsevier Publ. Co.

Collins, G. E.
1975 Quantifier elimination for real closed fields by cylindrical algebraic decomposition. Automata Theory and Formal Languages (Lecture Notes in Computer Science 33), Berlin, Heidelberg, New York: Springer-Verlag, S. 134-183.

Day, M. M.
1973 Normed linear spaces. 3. Aufl. Berlin, Heidelberg, New York: Springer-Verlag.

Dedekind, R.
1872 Stetigkeit und irrationale Zahlen. Braunschweig: Vieweg.

Dehn, M.
1900 Die Legendreschen Sätze über die Winkelsumme im Dreieck. Math. Ann. 53, 404-439.

Doraczyńska, E.
1974 A complete theory of the midpoint operation. Bull. Acad. Polon. Sci. Sér. Sci. Math. Astronom. Phys. 22, 1195-1200.
1977 On constructing a field in hyperbolic geometry. Ibid. 25, 1109-1114.

Engel, F. u. Stäckel, P.
1895 Die Theorie der Parallellinien von Euklid bis auf Gauß. Leipzig: B. G. Teubner.
1898/1913 Urkunden zur Geschichte der nichteuklidischen Geometrie I/II. Leipzig: B. G. Teubner.

Engeler, E.
1973 On the structure of algorithmic problems. Lecture Notes in Computer Science No. 2, Berlin, Heidelberg, New York: Springer-Verlag, S. 2-15.
1975 Algorithmic logic. Mathematical Centre Tracts 63, ETH Zürich, S. 57-85.
1983 Metamathematik der Elementarmathematik. Berlin, Heidelberg, New York: Springer-Verlag.

Enriques, F.
1911 Fragen der Elementargeometrie. Leipzig u. Berlin: B. G. Teubner.

Felscher, W.
1978/78a/79 Naive Mengen und abstrakte Zahlen I/II/III. Mannheim, Wien, Zürich: Bibliographisches Institut.

Fladt, K.
1920 Kleine Mitteilungen: Der Saccheri-Legendresche Satz. Zeits. f. Math. u. Naturwiss. Unterricht 51, 124.

Gerretsen, J. C. H.
1942 Die Begründung der Trigonometrie in der hyperbolischen Ebene. Nederl. Akad. Wetensch. Proc. Ser. A 45, 360-366, 479-483, 559-566.

Gödel, K.
1940 The consistency of the axiom of choice and of the generalized continuum hypothesis with the axioms of set theory (Ann. Math. Studies 3). Princeton: Princeton University Press (4. Aufl. 1958).

Grätzer, G.
1968 Universal algebra. Princeton, Toronto, London, Melbourne: D. van Nostrand Co.

Grzegorczyk, A., Mostowski, A. u. Ryll-Nardzewski, C.
1958 The classical and the ω-complete arithmetic. J. Symbolic Logic 23, 188-206.

Gupta, H. N.
1965 Contributions to the axiomatic foundations of geometry. Doctoral Dissertation, University of California, Berkeley.

1965a An axiomatization of finite-dimensional Cartesian spaces over arbitrary ordered fields. Bull. Acad. Polon. Sci. Sér. Sci. Math. Astronom. Phys. 13, 551-552.
1967 On an interesting property of vector spaces over formally real, non-Pythagorean fields. Ibid. 15, 119-121.
1969 On some axioms in foundations of Cartesian spaces. Canad. Math. Bull. 12, 831-836.

Gupta, H. N. u. Piesyk, Z.
1965 An axiomatization of two-dimensional Cartesian spaces over arbitrary ordered fields. Bull. Acad. Polon. Sci. Sér. Sci. Math. Astronom. Phys. 13, 549-550.

Gupta, H. N. u. Prestel, A.
1972 On a class of Pasch-free Euclidean planes. Bull. Acad. Polon. Sci. Sér. Sci. Math. Astronom. Phys. 20, 17-23.
1972a Triangle and Schwarz inequality in Pasch-free Euclidean geometry. Ibid. 20, 999-1003.

Handbook
1977 Handbook of mathematical logic, Hrsg. Jon Barwise. Amsterdam, New York, Oxford: North-Holland Publ. Co.

Hauschild, K.
1974 Rekursive Unentscheidbarkeit der Theorie der pythagoräischen Körper. Fund. Math. 82, 191-197.
1977 Addendum, betreffend die rekursive Unentscheidbarkeit der Theorie der pythagoräischen Körper. Als Manuskript gedruckt, Akademie der Wissenschaften der DDR, Zentralinstitut für Mathematik und Mechanik, Berlin, Dezember 1977.

Hecht, S.
1973 Dimensionsfreie euklidische Geometrie ohne Streckenabtragung. Staatsexamensarbeit, Bonn.

Heffter, L.
1950 Grundlagen und analytischer Aufbau der Projektiven, Euklidischen, Nichteuklidischen Geometrie. 2. Aufl. Leipzig: B. G. Teubner.

Henkin, L.
1962 Symmetric Euclidean relations. Indag. Math. 24, 549-553.

Hermes, H.
1969/76 Einführung in die mathematische Logik. 2./4. Aufl. Stuttgart: B. G. Teubner.
1978 Aufzählbarkeit, Entscheidbarkeit, Berechenbarkeit. 3. Aufl. Berlin, Heidelberg, New York: Springer-Verlag.

Hessenberg, G.
1905 Beweis des Desarguesschen Satzes aus dem Pascalschen. Math. Ann. 61, 161-172.

Hilbert, D.
1899/1977 Grundlagen der Geometrie. 1. Aufl. Leipzig/12. Aufl. Stuttgart: B. G. Teubner.

Hilbert, D. u. Bernays, P.
1968/70 Grundlagen der Mathematik I/II. 2. Aufl. Berlin, Heidelberg, New York: Springer-Verlag.

Hjelmslev, J.
1907 Neue Begründung der ebenen Geometrie. Math. Ann. 64, 449-474.

Jaśkowski, S.
1948 Une modification des définitions fondamentales de la géométrie des corps de M. A. Tarski. Ann. Soc. Polon. Math. 21, 298-301.
1949 Geometria brył. Czasopismo matematyka Nr. 1(3), 1-6.
1949a Quelques problèmes actuels concernant les fondements des mathématiques. Časopis Pěst. Mat. Fys. 74, 74-78.

Keisler, H. J.
1971 Model theory for infinitary logic. Amsterdam, London: North-Holland Publ. Co.

Keisler, H. J. u.a.: s. Chang.

Kordos, M.
1969 On the syntactic form of dimension axiom for affine geometry. Bull. Acad. Polon. Sci. Sér. Sci. Math. Astronom. Phys. 17, 833-837.
1973 Elliptic geometry as a theory of one binary relation. Ibid. 21, 609-614.

Kordos, M. u. Szczerba, L. W.
1969 On the $\Pi\Sigma$-axiom systems of hyperbolic and some related geometries. Bull. Acad. Polon. Sci. Sér. Sci. Math. Astronom. Phys. 17, 175-180.

Legendre, A. M.
1833 Réflexions sur différentes manières de démontrer la théorie des parallèles ou le théorème sur la somme des trois angles du triangle. Mémoires de l'Académie des Sciences, Bd. 12, Paris, S. 367-410.

Liebmann, H.
1923 Nichteuklidische Geometrie. 3. Aufl. Berlin, Leipzig: W. de Gruyter.

Liebmann, H. u.a.: s. Bonola.

Lindenbaum, A. u. Tarski, A.
1936 Über die Beschränktheit der Ausdrucksmittel deduktiver Theorien. Ergebnisse eines Mathematischen Kolloquiums 7, 15-22.

Lindenbaum, A. u.a.: s. Tarski.

Lingenberg, R.
1969 Grundlagen der Geometrie I. Mannheim, Wien, Zürich: Bibliographisches Institut.

Lorenz, J. F.
1791 Grundriß der reinen und angewandten Mathematik, Bd. I. Helmstädt: Carl Gottfried Fleckeisen.

Makowiecka, H.
1965 On a primitive notion of one-dimensional geometries over some fields. Bull. Acad. Polon. Sci. Sér. Sci. Math. Astronom. Phys. 13, 43-47.
1975a The theory of bi-proportionality as a geometry. Ibid. 23, 657-664.
1975b An elementary geometry in connection with decomposition of a plane. Ibid. 23, 665-674.
1975c On primitive notions in n-dimensional elementary geometries. Ibid. 23, 675-682.
1975d Quaternary relations in weak Euclidean geometries. Ibid. 23, 683-692.
1975e The norm relation in weak Euclidean geometry with the Axiom on Two Circles. Ibid. 23, 1181-1187.
1976 A general property of ternary relations in elementary geometry. Ibid. 24, 163-169.
1977 On minimal systems of primitives in elementary Euclidean geometry. Ibid. 25, 269-277.

Menger, K.
1938 A new foundation of non-Euclidean, affine, real projective, and Euclidean geometry. Proc. Nat. Acad. Sci. U.S.A. 24, 486-490.

Moler, N. u. Suppes, P.
1968 Quantifier-free axioms for constructive plane geometry. Compositio Math. 20, 143-152.

Mollerup, J.
1904 Die Beweise der ebenen Geometrie ohne Benutzung der Gleichheit und Ungleichheit der Winkel. Math. Ann. 58, 479-496.

Montague, R.
1961 Semantical closure and non-finite axiomatizability I. Infinitistic Methods, Oxford, London, New York, Paris: Pergamon Press, Warszawa: Pańcstwowe Wydawnictwo Naukowe, S. 45-69.

Mostowski, A.
1961 Concerning the problem of axiomatizability of the field of real numbers in the weak second-order logic. Essays on the Foundations of Mathematics, dedicated to A. A. Fraenkel on his 70th Birthday, Jerusalem: The Magnes Press, The Hebrew University, S. 269-286.

Mostowski, A. u.a.: s. Grzegorczyk, Tarski.

Oberschelp, A.
1968 Aufbau des Zahlensystems. Göttingen: Vandenhoeck & Ruprecht.

Pejas, W.
1961 Die Modelle des Hilbertschen Axiomensystems der absoluten Geometrie. Math. Ann. 143, 212-235.

Pieri, M.
1908 La geometria elementare istituita sulle nozioni di 'punto' e 'sfera'. Memorie di Matematica e di Fisica della Società Italiana delle Scienze, ser. 3, 15, 345-450.

Piesyk, Z.
1961 The existential and universal statements on parallels. Bull. Acad. Polon. Sci. Sér. Sci. Math. Astronom. Phys. 9, 761-764.
1977 On the betweenness relation in a set of cardinality $\neq 4$. Ibid. 25, 667-670.

Piesyk, Z. u.a.: s. Gupta.

Pinter, C.
1978 Properties preserved under definitional equivalence and interpretations. Z. Math. Logik Grundlagen Math. 24, 481-488.

Podehl, E. u. Reidemeister, K.
1934 Eine Begründung der ebenen elliptischen Geometrie. Abh. Math. Sem. Univ. Hamburg 10, 231-255.

Prestel, A.
1972 Untersuchungen über Pasch-freie Geometrie und semi-geordnete Körper. Habilitationsschrift, Skripten des Seminars für Logik und Grundlagenforschung, Universität Bonn.
1973 Quadratische Semi-Ordnungen und quadratische Formen. Math. Z. 133, 319-342.
1974 Euklidische Geometrie ohne das Axiom von Pasch. Abh. Math. Sem. Univ. Hamburg 41, 82-109.
1981 Zur Axiomatisierung gewisser affiner Geometrien. Enseignement Math. (2) 27, 125-136. *) siehe Seite 467.

Prestel, A. u. Szczerba, L. W.
1979 Non-axiomatizability of real general affine geometry. Fund. Math. 104, 193-202.

Prestel, A. u.a.: s. Gupta.

Prüfer, H.
1953 Projektive Geometrie. 2. Aufl. Leipzig: Akad. Verlagsgesellschaft Geest & Portig.

Rautenberg, W.
1961 Unentscheidbarkeit der euklidischen Inzidenzgeometrie. Z. Math. Logik Grundlagen Math. 7, 12-15.
1962 Über metatheoretische Eigenschaften einiger geometrischer Theorien. Ibid. 8, 5-41.
1963 Konstruktionen in der hyperbolischen Geometrie. Math. Nachr. 25, 151-158.
1965 Ein Beweis des Satzes von Pappus-Pascal in der affinen Geometrie. Math.-Phys. Semesterber. 12, 197-210.
1966 Über Hilberts Schnittpunktsätze. Z. Math. Logik Grundlagen Math. 12, 57-59.
1968 The elementary Archimedean scheme in geometry and some of its applications. Bull. Acad. Polon. Sci. Sér. Sci. Math. Astronom. Phys. 16, 837-843.

Reidemeister, K. u.a.: s. Podehl.

Robinson, A.
1956 Complete theories. Amsterdam: North-Holland Publ. Co.
1963 Introduction to model theory and to the metamathematics of algebra. Amsterdam: North-Holland Publ. Co.

Robinson, J.
1949 Definability and decision problems in arithmetic. J. Symbolic Logic 14, 98-114.

Robinson, R. M.
1959 Binary relations as primitive notions in elementary geometry. The Axiomatic Method, Proc. of the 1957/58 International Symposium at Berkeley, Amsterdam: North-Holland Publ. Co., S. 68-85.

Robinson, R. M. u.a.: s. Tarski.

Royden, H. L.
1959 Remarks on primitive notions for elementary Euclidean and non-Euclidean plane geometry. The Axiomatic Method, Proc. of the 1957/58 International Symposium at Berkeley, Amsterdam: North-Holland Publ. Co., S. 86-96.

Ryll-Nardzewski, C. u.a.: s. Grzegorczyk.

Saccheri, G.
1733 Euclides ab omni naevo vindicatus: sive conatus geometricus quo stabiliuntur prima ipsa universae geometricae principia. Milano: Paolo Antonio Montano.

Schaal, H.
1976/76a Lineare Algebra und Analytische Geometrie I/II. Braunschweig: Vieweg.

Schmidt, A.
1938 Über deduktive Theorien mit mehreren Sorten von Grunddingen. Math. Ann. 115, 485-506.

Schröter, K.
1956 Theorie des bestimmten Artikels. Z. Math. Logik Grundlagen Math. 2, 37-56.

Schur, F.
1909 Grundlagen der Geometrie. Leipzig u. Berlin: B. G. Teubner.

Schütte, K. u. van der Waerden, B. L.
1953 Das Problem der dreizehn Kugeln. Math. Ann. 125, 325-334.

Schwabhäuser, W.
1956 Über die Vollständigkeit der elementaren euklidischen Geometrie. Z. Math. Logik Grundlagen Math. 2, 137-165.
1959 Entscheidbarkeit und Vollständigkeit der elementaren hyperbolischen Geometrie. Ibid. 5, 132-205.
1962 Über die Nichtdefinierbarkeit einiger Begriffe der hyperbolischen Geometrie mit elementaren Mitteln. Ibid. 8, 43-55.
1965 On models of elementary elliptic geometry. The Theory of Models, Proc. of the 1963 International Symposium at Berkeley, Amsterdam: North-Holland Publ. Co., S. 312-328.
1965a Metamathematical methods in foundations of geometry. Logic, Methodology and Philosophy of Science, Proc. of the 1964 International Congress at Jerusalem, Amsterdam: North-Holland Publ. Co., S. 152-165.
1967 On the problem of definability of collinearity in terms of the midpoint relation. Indag. Math. 29, 183-187.
1969 The connection between two geometrical axioms of H. N. Gupta. Proc. Amer. Math. Soc. 22, 233-234.
1970 Axiomatisierbarkeit der dimensionsfreien euklidischen Geometrie über geeigneten Körpern. J. reine angew. Math. 242, 134-147.
1971/72 Modelltheorie I/II. Mannheim, Wien, Zürich: Bibliographisches Institut.
1972a Mathematische Logik. Grundlagen der modernen Mathematik, Darmstadt: Wissenschaftliche Buchgesellschaft, S. 79-109.
1979 Non-finitizability of a weak second-order theory. Fund. Math. 103, 83-102.
1979a Modellvollständigkeit der Mittelpunktsgeometrie und der Theorie der Vektorgruppen. Abh. Math. Sem. Univ. Hamburg 48, 213-224.
1979b A class of undecidable models of general affine geometry. Abstract, 1979 International Congress of Logic, Methodology and Philosophy of Science, Hannover.

Schwabhäuser, W. u. Szczerba, L. W.
1974 An affine space as union of spaces of higher dimension. Proceedings of the Tarski Symposium, held 1971 at Berkeley, Providence, Rhode Island: American Mathematical Society, S. 133-138.
1975 Relations on lines as primitive notions for Euclidean geometry. Fund. Math. 82, 347-355; Korrektur ibid. 87 (1975), 283.

Schwabhäuser, W. u.a.: s. Baumann.

Scott, D.
1956 A symmetric primitive notion for Euclidean geometry. Indag. Math. 18, 456-461.
1959 Dimension in elementary Euclidean geometry. The Axiomatic Method, Proc. of the 1957/58 International Symposium at Berkeley, Amsterdam: North-Holland Publ. Co., S. 53-67.

Seeland, H.
1978 Algorithmische Theorien und konstruktive Geometrie (Dissertation). (Hochschul-Sammlung Naturwissenschaft: Informatik Band 2) Stuttgart: Hochschul-Verlag.

Sette, A. M. u. Szczerba, L. W.
1978 Algebraic characterization of interpretability. Notas e Communicações de Matemática Nº 94, Universidade Federal de Pernambuco, Recife, Brasil.

Sierpiński, W.
1958/65 Cardinal and ordinal numbers. 1. Aufl. Warszawa: Państwowe Wydawnictwo Naukowe/ 2. Aufl. ibid. u. New York: Hafner.

Slomson, A. B. u.a.: s. Bell.

Stäckel, P. u.a.: s. Engel.

Strommer, J.
1960 Zur Vereinfachung des Parallelenaxioms. Ann. Univ. Sci. Budapest. Eötvös Sect. Math. 3-4, 315-318.
1962 Vereinfachung des hyperbolischen Parallelenaxioms. Ann. Mat. Pura Appl. (4) 57, 179-186.

Suppes, P. u.a.: s. Moler.

Szász, P.
1958 Unmittelbare Einführung Weierstrassscher homogener Koordinaten in der hyperbolischen Ebene auf Grund der Hilbertschen Endenrechnung. Acta Math. Acad. Sci. Hungar. 9, 1-28.
1959 Direct introduction of Weierstrass homogeneous coordinates in the hyperbolic plane, on the base of the endcalculus of Hilbert. The Axiomatic Method, Proc. of the 1957/58 International Symposium at Berkeley, Amsterdam: North-Holland Publ. Co., S. 97-113.
1962 Einfache Herstellung der hyperbolischen Trigonometrie in der Ebene auf Grund der Hilbertschen Endenrechnung. Ann. Univ. Sci. Budapest. Eötvös Sect. Math. 5, 79-85.

Szczerba, L. W.
1970 Independence of Pasch's axiom. Bull. Acad. Polon. Sci. Sér. Sci. Math. Astronom. Phys. 18, 491-498.
1971 Incompleteness degree of elementary Pasch-free geometry. Ibid. 19, 213-215.
1971a Undefinability of order in Pasch-free geometry. Ibid. 19, 315-317.
1971b Undecidability of elementary Pasch-free geometry. Ibid. 19, 469-474.
1971c The Pasch-free geometry with the Continuity Axiom. Ibid. 19, 613-616.
1972 Weak general affine geometry. Ibid. 20, 753-761.
1972a A paradoxial model of Euclidean affine geometry. Ibid. 20, 845-851.
1977 Interpretability of elementary theories. Logic, Foundations of Mathematics, and Computability Theory (Proceedings of the 1975 International Congress of Logic, Methodology and Philosophy of Science at London, Ont., Canada, part 1), Dordrecht, Boston: D. Reidel Publ. Co., S. 129-145.
1977a Notions in geometry. Math. Chronicle 6, 1-5.

Szczerba, L. W. u. Szmielew, W.
1970 On the Euclidean geometry without the Pasch Axiom. Bull. Acad. Polon. Sci. Sér. Sci. Math. Astronom. Phys. 18, 659-666.

Szczerba, L. W. u. Tarski, A.
1965 Metamathematical properties of some affine geometries. Logic, Methodology and Philosophy of Science, Proc. of the 1964 International Congress at Jerusalem, Amsterdam: North-Holland Publ. Co., S. 166-178.
1979 Metamathematical discussion of some affine geometries. Fund. Math. 104, 155-192.

Szczerba, L. W. u.a.: s. Kordos, Prestel, Schwabhäuser, Sette.

Szmielew, W.
1959 Absolute calculus of segments and its metamathematical implications. Bull. Acad. Polon. Sci. Sér. Sci. Math. Astronom. Phys. 7, 213-220.
1959a Some metamathematical problems concerning elementary hyperbolic geometry. The Axiomatic Method, Proc. of the 1957/58 International Symposium at Berkeley, Amsterdam: North-Holland Publ. Co., S. 30-52.
1961 A new analytic approach to hyperbolic geometry. Fund. Math. 50, 129-158.
1962 New foundations of absolute geometry. Logic, Methodology and Philosophy of Science, Proc. of the 1960 International Congress at Stanford, Stanford: Stanford University Press, S. 168-175.
1970 The Pasch Axiom as a consequence of the Circle Axiom. Bull. Acad. Polon. Sci. Sér. Sci. Math. Astronom. Phys. 18, 751-758.
1970a A statement on two circles as the geometric analog of Euclid's field property. Ibid. 18, 759-764.
1974 The role of the Pasch Axiom in the foundations of Euclidean geometry. Proceedings of the Tarski Symposium, held 1971 at Berkeley, Providence, Rhode Island: American Mathematical Society, S. 123-132.
1974a The order and the semi-order of n-dimensional Euclidean space in the axiomatic and model-theoretic aspects. Grundlagen der Geometrie und algebraische Methoden (internat. Koll. Potsdam 1973). Potsdamer Forschungen - Reihe B, Heft 3, Wiss. Schriftenreihe der Pädagogischen Hochschule "Karl Liebknecht" Potsdam 1974, S. 69-79.
1977 Oriented and nonoriented linear orders. Bull. Acad. Polon. Sci. Sér. Sci. Math. Astronom. Phys. 25, 659-665.
1980 Concerning the order and the semi-order of n-dimensional Euclidean space (ed. by M. Moszyńska). Fund. Math. 107, 47-56.

Szmielew, W. u.a.: s. Borsuk, Szczerba.

Tarski, A.
1929 Les fondements de la géométrie des corps. Mémoirs du Premiér Congrès Polonais de Mathématique en 1927, Supplément aux Annales de la Soc. Polonaise de Mathématique, Kraków, S. 29-33.
1940 The completeness of elementary algebra and geometry (Nachdruck nach Korrekturabzügen für ein Buch, das 1940 erscheinen sollte). Paris: 1967: Institut Blaise Pascal.
1948/51 A decision method for elementary algebra and geometry. 1. Aufl. Santa Monica: RAND Corporation/2. Aufl. Berkeley, Los Angeles: University of California Press.
1949 Undecidability of the theories of lattices and projective geometries. J. Symbolic Logic 14, 77-78.
1956 A general theorem concerning primitve notions of Euclidean geometry. Indag. Math. 18, 468-474.
1958 Some model-theoretical results concerning weak second-order logic. Notices Amer. Math. Soc. 5, 673.
1959 What is elementary geometry? The Axiomatic Method, Proc. of the 1957/58 International Symposium at Berkeley, Amsterdam: North-Holland Publ. Co., S. 16-29.
1974 Französische Übersetzung von Tarski 1940 von G. Kalinowski, enthalten (als Artikel XIX, pp. 203-242) in: Logique, sémantique, métamathématique, 1923-1944, Vol. 2. Philosophies pour l'Age de la Science, Paris: Librairie Armand Colin.

Tarski, A. u. Lindenbaum, A.
1927 Sur l'indépendance des notions primitives dans les systèmes mathématiques. Ann. Soc. Polon. Math. 5, 111-113.

Tarski, A., Mostowski, A. u. Robinson, R. M.
1953 Undecidable theories. Amsterdam: North-Holland Publ. Co.

Tarski, A. u. Vaught, R. L.
1957 Arithmetical extensions of relational systems. Compositio Math. 13, 81-102.

Tarski, A. u.a.: s. Beth, Lindenbaum, Szczerba.

Vaught, R. L.
1962 On a theorem of Cobham concerning undecidable theories. Logic, Methodology and Philosophy of Science, Proc. of the 1960 International Congress at Stanford, Stanford: Stanford University Press, S. 14-25.

Vaught, R. L. u.a.: s. Tarski.

Veblen, O.
1904 A system of axioms for geometry. Trans. Amer. Math. Soc. 5, 343-384.
1914 The foundations of geometry. Monographs on Topics of Modern Mathematics, ed. by J. W. A. Young, New York: Longsman, Green, and Co. Nachdruck New York 1955: Dover Publications.

Le Veque, W. J.
1956 Topics in number theory II. Reading, Mass.: Addison-Wesley Publ. Co.

van der Waerden, B. L.
1971 Algebra I. 8. Aufl. Berlin, Heidelberg, New York: Springer-Verlag.

van der Waerden, B. L. u.a.: s. Schütte.

Zbierski, P.
1978 Axiomatizability of second-order arithmetic with ω-rule. Fund. Math. 100, 51-57.

Ziegler, M.
1982 Einige unentscheidbare Körpertheorien. Enseignement Math. (2) 28, 269-280. *) s.u.

*) Die Arbeiten Prestel 1981 und Ziegler 1982 sind auch erschienen (auf S. 341-352 bzw. 381-392) in
1982 Logic and algorithmic. - An International Symposium held in Honour of Ernst Specker, Zurich, February 5-11, 1980. - Monographie de l'Enseignement Mathématique N° 30. Université de Genève.

Symbolverzeichnis

Einteilung:
Bezeichnungen allgemeiner Art
logische Zeichen
Sprachen, Formeln, Formelmengen
Mengen, Strukturen, Algebra
Spezielle Mengen, Strukturen und Strukturklassen
Räume
Geometrien
verwendete Axiome und Axiomensysteme
geometrische Begriffe

Bezeichnungen allgemeiner Art

logische Zeichen

Sprachen, Formeln, Formelmengen

Mengen, Strukturen, Algebra

Spezielle Mengen, Strukturen und Strukturklassen

Räume

Namensverzeichnis

Sachverzeichnis

Hier sind die Seitenzahlen angegeben (für Verweise innerhalb des Textes werden dagegen Nummern von Unterabschnitten verwendet, vgl. S. vi).

(Forts. nächste Seite)

Math Books by Ishi Press

How To Solve It: A New Aspect of Mathematical Method ISBN 4-87187-830-9

Mathematical Discovery on Understanding, Learning, and Teaching Problem Solving, Volume I ISBN 4-87187-831-7

Mathematical Discovery on Understanding, Learning, and Teaching Problem Solving, Volume II ISBN 4-87187-832-5

Mathematics and Plausible Reasoning: Vol. I: Induction and Analogy in Mathematics ISBN 4-87187-833-3

Mathematics and Plausible Reasoning: Vol. II: Patterns of Plausible Inference ISBN 4-87187-834-1

Differential and Integral Calculus, Vol. One ISBN 4-87187-838-4

Differential and Integral Calculus, Vol. 2 ISBN 4-87187-835-X

Geometries, Groups and Algebras in the Nineteenth Century – A History ISBN 4-87187-836-8

Foundations of Projective Geometry ISBN 4-87187-837-6

Metamathematische Methoden in der Geometrie ISBN 4-87187-707-8

Math Theory of Chess S-Version, Parts 1,2 ISBN 4-87187-708-6

The Trachtenberg Speed System of Basic Mathematics ISBN 4-87187-709-4

Algebra Can Be Fun by Perelman ISBN 4-87187-710-8

The Consistency of the Continuum Hypothesis by Kurt Gödel ISBN 0-923891-53-6

Calculus with Analytic Geometry by Angus E. Taylor Vol. 1 ISBN 0-923891-24-2

Calculus with Analytic Geometry by Angus E. Taylor Vol. 2 ISBN 0-923891-25-0

General Topology by John L. Kelley ISBN 0-923891-55-2

Introduction to Metamathematics by Kleene ISBN 0-923891-57-9

36282707R00286

Printed in Poland
by Amazon Fulfillment
Poland Sp. z o.o., Wrocław